Ocean
geology

 Jurassic Cretaceous Lower
Tertiary Upper
Tertiary

⌒ Convergent
boundary

╪ Divergent
boundary

╲ Fracture
zone

○ Hotspot

● Seamount

Dedication

I am deeply indebted to W. Kenneth Hamblin, who was a mentor, a colleague, and a friend. Without his passion for understanding Earth, careful illustration, and photography, this book would have been impossible. This first edition of *Dynamic Earth: An Introduction to Physical Geology* published by Jones & Bartlett Learning owes much to its predecessor, *Earth's Dynamic Systems*, a groundbreaking book conceived by Ken in 1974.

Brief Contents

Contents

Preface

Planet Earth was first photographed from space more than 40 years ago by the astronauts of the Apollo mission to the Moon. Since then a variety of satellites have given us supernatural eyes in space. With sophisticated sensors we can "see" through clouds, observe features on the ocean floor, and selectively image temperature, water vapor, ocean currents, and even patterns of ancient river systems now buried beneath the sand of the Sahara Desert. New technology has permitted us to penetrate deep into the interior of our planet and effectively X-ray its internal structure. We can now "see" hot material in the interior moving in huge convective systems, creating ocean basins, volcanic islands, and mountain systems.

With this era of enlightenment comes an increased awareness of how our planet is continually changing and a fresh awareness of how fragile it is. Earth is a finite sphere with limited resources, so it is impossible for the population to grow indefinitely. We may find more oil, gas, and coal by improved detection methods, yet nature requires more than a million years to concentrate the oil we consumed in one year. We have created nuclear waste, yet are unsure how to dispose of it safely. What can we do about the fact that rivers today transport more agricultural and industrial waste than natural sediment? Is greenhouse heating real? The answers to these questions can be found only if we understand Earth's dynamic geologic systems with their many interdependent and interconnected components.

There are two major pathways for the flow of energy and matter on the planet: (1) the hydrologic system—the circulation of water over Earth's surface and through its atmosphere powered by energy from the Sun, and (2) the tectonic system—the movement of material powered by heat from Earth's interior. Everything discussed in this book is related to these unifying themes.

Dynamic Earth: An Introduction to Physical Geology introduces these systems and will help students to understand and participate in the solutions to some of the problems our society faces. It is written for students taking their first college course in physical geology at both two- and four-year schools.

The book is divided in four parts. In Part I, we present Earth's materials and how they are created by geologic systems. In Part II, we discuss the hydrologic system by examining subsystems chapter by chapter. Plate tectonics is the theme of Part III, with separate chapters on divergent, transform, and convergent boundaries, as well as mantle plumes—the subsystems of the tectonic system. In Part IV, we look back and apply the principles learned to see how Earth's resources formed and just how different our home in space is when compared with other planets.

Special attention has been placed on the illustrations so that the student can more fully experience the excitement and satisfaction of visualizing and understanding geology. Many of the photographs were taken specifically for this book. Photographs not otherwise credited were taken by W. Kenneth Hamblin. Ken traveled the world driven by his desire to photographically capture Earth's geologic wonders in ways that illuminate geologic processes and that are not just pretty pictures. We strive to present a fuller perspective in visualizing geology by using panoramic photographs of the landscape. Panoramic photographs begin each chapter and provide the student with a visual summary of the theme of the chapter.

An important element in the book is the use of digital topographic maps. The flood of new data on seafloor and continental topography by defense and other governmental agencies has opened a new window through which we can see the continents and seafloor in a more accurate and detailed fashion than ever before. These colored, shaded relief maps are a visual and intellectual feast for those who carefully study them.

A major feature of this book is a series of short essays that illustrate the scientific method. We call these summary messages *GeoLogic*, as we attempt to show the logic behind using simple observations to come to important conclusions about the way Earth works. The illustrations in these sections are designed to take students from "seeing" to "understanding." In addition, we use *"The State of the Art"* essays to distill the important techniques used in modern geology into a few words and couple them with some of the most spectacular and informative images available today. In these short essays, we show how geologists come to understand the world around us, emphasizing not just "what we know" but "how we know" it as well.

The real test of any textbook is how well it helps the student learn. I welcome opinions from students and instructors who have used this book. Please address your comments, criticisms, and suggestions to:

Eric H. Christiansen (eric_christiansen@byu.edu)
Department of Geological Sciences
Brigham Young University
Provo, Utah 84602

Instructor Resources

Compatible with Windows® and Macintosh® platforms, the Instructor's Media CD provides instructors with the following:

- The PowerPoint Image Bank provides the illustrations, photographs, and tables (to which Jones & Bartlett Learning holds the copyright or has permission to reproduce digitally) inserted into PowerPoint slides. You can quickly and easily copy individual images or tables into your existing lecture presentations.
- The PowerPoint Lecture Outline presentation package provides lecture notes and images for each chapter of *Dynamic Earth*. Instructors with the Microsoft PowerPoint software can customize the outlines, art, and order of presentation.

The Test Bank, provided as text files (with LMS-compatible options available), is offered online as a secure download. Please contact your sales representative for more information.

Acknowledgments

Special thanks are expressed to the following colleagues for their many helpful comments and suggestions.

James L. Baer, Brigham Young University

David M. Best, Northern Arizona University

Austin Boyd, University of Arkansas, Fort Smith

Donald Burt, Arizona State University

Edith Chasen-Cerreta, St. John's University

Winton Cornell, University of Tulsa

Peter S. Dahl, Kent State University

George H. Davis, University of Arizona

Rene DeHon, Texas State University

Steven Dent, Northern Kentucky University

Justin Devery, Alvin College

Martha Eppes, University of North Carolina, Charlotte

C. Patrick Ervin, Northern Illinois University

James R. Firby, University of Nevada at Reno

Ronald A. Harris, Brigham Young University

Charles W. Hickcox, Emory University

Merton Hill, Saddleback College

Roger D. Hoggan, Ricks College

Roger Hooke, University of Minnesota

Warren D. Huff, University of Cincinnati

Lois Breur Krase, Clemson University

David London, University of Oklahoma

David B. Loope, University of Nebraska

Harmon Maher, University of Nebraska, Omaha

Thomas Morris, Brigham Young University

David A. Mustart, San Francisco State University

David Nellis, University of Massachusetts, Boston

David Nash, University of Cincinnati

Richard S. Naylor, Northeastern University

Hallan C. Noltimier, Ohio State University

William Osei, Algoma University

Alfred Pekarek, St. Cloud State University

Irina Popova-Goll, Texas A&M University

Arthur L. Reesman, Vanderbilt University

John R. Reid, University of North Dakota

Beth Rinard, Tarlteton State University

James Roche, Louisiana State University

Malcolm J. Rutherford, Brown University

Gary Steinhardt, Purdue University

Bryan Tapp, University of Tulsa

Gregory Tucker, University of Colorado

Kenneth Van Dellen, Macomb Community College

Dayanthe Weeraratne, California State University, Northridge

Ed Wehling, Anoka Ramsey Community College

Teresa Williams-Drame, Merritt College

Karen Yip, Houston Community College Southwest

The artists who contributed to *Earth's Dynamic Systems*, the predecessor of this book, deserve special recognition including: William L. Chesser, Robert Pack, Dale Claflin, Russell McMullin, Barney McKay Design Group, James Miller, Rick Showalter, Kirsten Thompson, and Keryn Ross.

Lastly, we would like to thank the editorial staff at Jones & Bartlett Learning for playing a critical role in the development of this book. We are especially grateful for the assistance of Erin O'Connor, Rachel Isaacs, Raven Heroux, Lauren Miller, and Carolyn Pershouse.

To the Student: How to Use This Book

One of the most difficult problems you face in beginning a course in a new subject is to identify fundamental facts and concepts and separate them from supportive material. This problem is often expressed by the question, "What do I need to know?" We have attempted to overcome this problem by presenting the material in each chapter in a manner that will help you recognize immediately the essential concepts.

Pedagogy

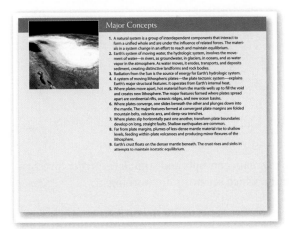

Outline of Major Concepts
To help you focus on the key points, we have identified them at the beginning of each chapter under the title of "Major Concepts."

Thesis Statements
A brief statement of the main idea of each section is in a colored box.

The Discovery of Time

Time is measured by change. Because rocks are themselves records of change, they mark the passage of geologic time. The interpretation of rocks as products and records of events in Earth's history is based on the principle of uniformitarianism, which states that the laws of nature do not change with time.

We are all aware of change in the physical and biological worlds. Were things unchanging and motionless, we would not be aware of time. Time is measured by change, and

Guiding Questions
Experience has shown that the most successful students are those who read with a specific purpose—those who read to answer a question. Consequently, we have developed guiding questions that are presented in the margins next to the appropriate text material. The questions are intended to guide you in your study, stimulate your curiosity, and help focus attention on important concepts.

For example, thin beds of volcanic ash are common in some richly fossiliferous shales and limestones of the western interior of the United States. Here, several different fossil species permit us to determine precisely the position of thin units in which the fossils occur within the geologic column. Comparisons with fossils found on other continents are also possible, so that their Late Cretaceous age is well established. The sequence of interbedded volcanic ash layers can be dated very precisely by radiometric means. Therefore, a numerical age for each successive fossil zone can be established. In this case, individual fossil zones have radiometric ages known within about 1% and provide tight constraints on the numerical ages of several important parts of the geologic time scale.

In other formations, sedimentary rocks lack fossils that are diagnostic of a certain period of geologic time. For example, the Morrison Formation of the western United States is famous for its abundant and distinctive dinosaur fauna, including stegosaurs, allosaurs, and apatosaurs, but it has few fossils that identify its exact position in the standard geologic column. Estimates of its age ranged from Jurassic to Cretaceous. However, volcanic ash beds are common in this unit, and careful radiometric dating has established that the Morrison Formation and its dinosaur remains range from 155 to 148 million years old. They are thus Late Jurassic in age (Figure 8.8).

How are igneous rocks used to calibrate the relative time scale?

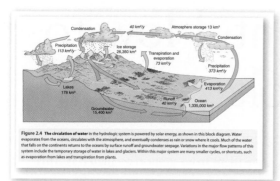

Figure 2.4 **The circulation of water** in the hydrologic system is powered by solar energy, as shown in this block diagram. Water evaporates from the oceans, circulates through the atmosphere, and eventually condenses as rain or snow where it cools. Much of the water that falls on the continents returns to the oceans by surface runoff and groundwater seepage. Variations in the major flow patterns of this system include the temporary storage of water in lakes and glaciers. Within this major system are many smaller cycles, or shortcuts, such as evaporation from lakes and transpiration from plants.

Illustrations

You will find that careful study of the figures and captions is one of the most useful methods of reviewing the content of the chapter.

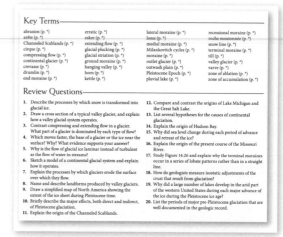

Key Terms

Important terms are printed in bold type. In the Key Terms section at the end of each chapter, the terms are listed alphabetically, with the number of the page on which each appears. These terms are also defined in the glossary at the end of the book.

Review Questions

These discussion questions are intended to reinforce the main concepts and stimulate further investigation by pointing out some of the intriguing questions on which scientists are working.

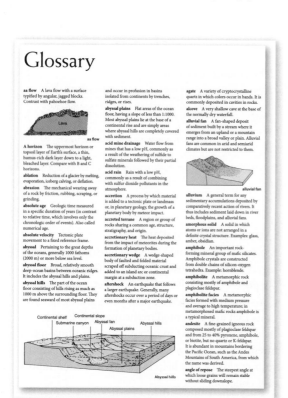

Illustrated Glossary

At the end of the book there is an illustrated glossary defining approximately 800 important geologic terms. Many of the terms are accompanied by an illustration that will help in visualizing the definition and meaning of the term. This glossary, if properly used, can be a convenient and significant aid in learning the basic vocabulary of geology.

Internet Tools

Jones & Bartlett Learning has prepared extensive electronic support for your studies: the *Dynamic Earth* website (go.jblearning.com/DynamicEarthCWS). It is designed to help you organize, prioritize, and apply what you learn from the text and in class. The website hosts a variety of review materials, including multiple choice questions and critical thinking questions linked to external sources of geologic data and images. The quizzes can be automatically graded to help you gauge your progress. To help focus your efforts further, refined feedback is supplied for right and wrong answers. It also provides you with chapter summaries, an interactive glossary for key term review, interactive flashcards, crossword puzzles, and a host of topic-specific Internet addresses and search terms to help you utilize this vast source of information. Access to this site is free with every new print copy of the text, and is also available for purchase separately.

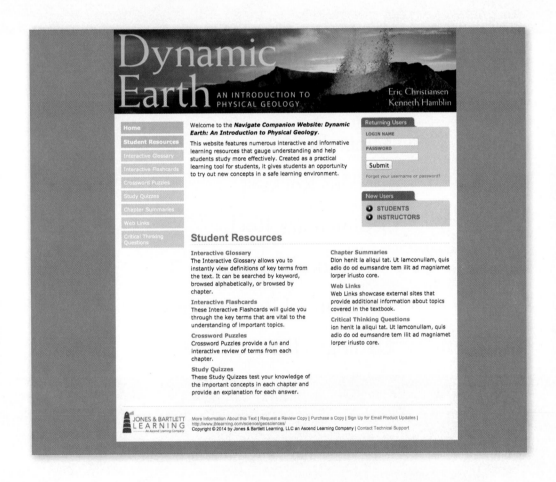

Part I
Earth Materials

We begin this study of planet Earth with an examination of Earth's composition and take you on a brief tour across its surface and into its deepest interior. Earth is a major system of energy and matter with many interconnected and interrelated subsystems. These systems are the mechanism by which energy and matter interact to create the infinite array of landforms and landscapes, structures and stones. In this part of the book, you will be introduced to each major system, to the energy that drives the flow of matter and makes Earth truly dynamic, and to rocks and minerals—the materials that this energy acts upon. But even after a rock body is formed, it is not static. Forces in the tectonic system can contort and deform it, and we will guide you through a basic understanding of the resulting structures.

Here you will be introduced to the great planetary systems–plate tectonics and the hydrologic system. Finally, you will see that geology is also a historical science. Our present is affected by our past; and the present will shape the future. This is what makes geology so fascinating—it is the study of changes that affect every aspect of our lives.

1 Planet Earth

From the beginning of our species, we have sought to understand our surroundings, whether that was the soil beneath our feet or the stars in the heavens. Above, the Milky Way shimmers in silent splendor above the Dolomites in northeastern Italy. The Milky Way is a gigantic galaxy consisting of billions of stars, some stars are forming today and others dying explosively. Many of these stars have their own systems of planets, moons, and comets orbiting them. Earth is but one planet of many. Our planet orbits a star that is on the fringes of the galaxy. Only since the beginning of space exploration in the 1960s have we seen our planet as it really is, a tiny blue sphere suspended in nearly empty space. But the planet remains dynamic and ever changing; its interior is still partially molten and volcanoes sporadically erupt across its surface.

Of course, Earth is not completely alone even in our Solar System. As you will see in this chapter, we have landed on the Moon, mapped the surfaces of Mercury, Venus, and Mars, and surveyed the diverse landscapes of several asteroids, and the moons of Jupiter, Saturn, Uranus, and Neptune. Every object in the solar system contains part of a record of planetary origin and evolution that helps us understand our own planet.

Back on Earth, we have also extended our explorations to the vast unknown of the ocean floor. We have mapped its landforms and structure, gaining insight into its origin and history. We now know that the rocks below the ocean floor are completely different from those below the surface of the continents. We also have peered into Earth's depths using indirect methods. We have traced

the paths of earthquake-generated seismic waves, measured the amount of heat that escapes from inside Earth, and recorded the pulse of the magnetic field. Consequently, we have discovered how Earth's interior churns slowly and how such movements affect processes at the very surface of the planet.

Inspired by this cosmic perspective of our planet, we can develop an all-encompassing view of how Earth operates as a constantly changing dynamic system. In this chapter, we start to do this by comparing and contrasting other planets with Earth. We also describe the major features of continents and ocean basins and view Earth's internal structure—all features that make Planet Earth unique in the solar system. You will see that our planet—our home—is a system of moving gas, liquid, and solids with numerous interconnected and interdependent components.

Major Concepts

1. A comparison of Earth with other inner planets reveals the distinguishing characteristics of our planet and shows what makes it unique.
2. Earth's atmosphere is a thin shell of gas surrounding the planet. It is a fluid in constant motion. Other planets have atmospheres, but Earth's is unique because it is 78% nitrogen and 21% oxygen.
3. The hydrosphere is another feature that makes Earth unique. Water moves in a great, endless cycle from the ocean to the atmosphere, over the land surface, and back to the sea again.
4. The biosphere exists because of water. Although it is small compared with other layers of Earth, it is a major geologic force operating at the surface.
5. Continents and ocean basins are the largest-scale surface features of Earth.
6. The continents have three major components: (a) ancient shields, (b) stable platforms, and (c) belts of folded mountains. Each reveals the mobility of Earth's crust.
7. The major structural features of the ocean floor are: (a) the oceanic ridges, (b) the vast abyssal floor, (c) long, narrow, and incredibly deep trenches, (d) seamounts, and (e) continental margins.
8. Earth is a differentiated planet, with its materials segregated into layers according to density. The internal layers classified by composition are (a) crust, (b) mantle, and (c) core. The major internal layers classified by physical properties are (a) lithosphere, (b) asthenosphere, (c) mesosphere, (d) outer core, and (e) inner core. Material within each of these units is in motion, making Earth a changing, dynamic planet.

Introduction to Geology

Geology is the science of Earth. It concerns all of Earth: its origin, its history, its materials, its processes, and the dynamics of how it changes.

Geology is an incredibly fascinating subject. It is concerned with such diverse phenomena as volcanoes and glaciers, rivers and beaches, earthquakes and landslides, and even the history of life. Geology is a study about what happened in the past and what is happening now—a study that increases our understanding of nature and our place in it.

Yet geology does much more than satisfy intellectual curiosity. We are at a point in human history when Earth scientists have a responsibility to help solve some of society's most pressing problems. These include finding sites for safe disposal of radioactive waste and toxic chemicals, determining responsible land use for an expanding population, and providing safe, plentiful water supplies. Geology is being called upon to guide civil engineers in planning buildings, highways, dams, harbors, and canals. Geology helps us recognize how devastation caused by natural hazards, such as landslides, earthquakes, floods, and beach erosion, can be avoided or mitigated. Another driving force in our attempt to understand Earth is the discovery of natural resources. All Earth materials, including water, soils, minerals, fossil fuels, and building materials, are "geologic" and are discovered, exploited, and managed with the aid of geologic science.

Geologists also offer key information about the entire global system, especially past climate change and likely causes and effects of future climate modification. Perhaps, in the end, more fully comprehending nature is as important as the discovery of oil fields and mineral deposits.

Let us begin by exploring why Earth is unique among the planetary bodies of the solar system. We will then examine some of its important characteristics: its size, composition, atmosphere, hydrosphere, and the structure of its interior.

Earth Compared with Other Planets

Among the inner planets (Mercury, Venus, Earth, the Moon, and Mars), Earth is unique because of its size and distance from the Sun. It is large enough to develop and retain an atmosphere and a hydrosphere. Temperature ranges are moderate, such that water can exist on its surface as liquid, solid, and gas.

The Solar System

A map of the solar system (**Figure 1.1**) shows the Sun and the major planetary bodies. This is Earth's cosmic home, the place of its origin and development. All of the planets in the solar system were created at the same time and from the same general material. The massive Sun, a star that generates heat by nuclear fusion, is the center of the system. Because of the Sun's vast gravitational influence, all of the planets orbit around it. As seen from above their north poles, the planets move counterclockwise about the Sun in slightly elliptical orbits. Moreover, all orbit in the same plane as the Sun's equator, except for the dwarf planets Pluto and Eris (note the different inclinations of their orbits).

The diagram of the solar system in Figure 1.1 is not, of course, to scale. The orbits are distorted, and the sizes of the planetary bodies are greatly exaggerated and shown in perspective. In reality, the orbits are extremely large compared with the planets' sizes. A simple analogy may help convey the size and structure of the solar system. If the Sun were the size of an orange, Earth would be roughly the size of a grain of sand orbiting 9 m (30 ft) away. Jupiter would be the size of a pea revolving 60 m (200 ft) away. Pluto would be like a grain of silt 10 city blocks away. The nearest star would be the size of another orange more than 1600 km (1000 mi) away.

Until recently, the planets and their moons were mute astronomical bodies, only small specks viewed in a telescope. But today, they are new worlds as real as our own, because we have landed on their surfaces and studied them with remotely controlled probes. One of the most fundamental facts revealed by our exploration of the solar system is that the sizes and compositions of the planets vary systematically with distance from the Sun (**Figure 1.2**). The **inner planets** (**Figure 1.3**) include Mercury and the planetlike Moon, with their cratered surfaces; Venus, with its extremely hot, thick atmosphere of carbon dioxide and numerous volcanoes; Earth, with cool blue seas, swirling clouds and multicolored lands; and Mars, with huge canyons, giant extinct volcanoes, frigid polar ice caps, and long, dry river beds. The large **outer planets**—Jupiter, Saturn, Uranus, and Neptune—are giant balls of gas, with majestic rings and dozens of small satellites composed mostly of ice. The most distant of the "traditional" planets, Pluto, is small and similar to these icy moons. In fact, the International Astronomical Union demoted Pluto to "dwarf planet" status in 2006 and grouped it with other icy bodies in the outer solar system. This decision was precipitated by the discovery of Eris, an icy planetary body even bigger than Pluto. These small icy objects constitute a distinctive class that dominates the outer solar system. Indeed, water ice is the most common "rock" in the outer solar system. We use the **density** of a planet or moon to examine these dramatic differences in composition. (Density is a measure of mass per unit volume: g/cm^3; Figure 1.3). For example, the densities of the rocky inner planets are quite high (over $3\ g/cm^3$) compared to the gas- and ice-rich outer planets which have densities less than about $1.5\ g/cm^3$.

Our best evidence tells us that Earth formed, along with the rest of the solar system, about 4.6 billion years ago. Nonetheless, only the inner planets are even vaguely like Earth. The compositions (dominated by dense solids with high melting points) of the inner planets make them radically different from the outer planets, made of low-temperature ices as well as gas. Although the inner planets are roughly of the same general size, mass, and composition, they vary widely in ways that are striking and important to us as living creatures. Why is Earth so different from its neighbors? Why does it alone have abundant liquid water, a dynamic crust, an oxygen-rich atmosphere, and perhaps most unique, that intricate web of life, the biosphere?

How do the inner planets differ from the outer planets?

Figure 1.1 **Our solar system** consists of one star, an orbiting family of eight planets, dozens of dwarf planets like Pluto and Eris, at least 180 moons, hundreds of thousands of asteroids, and billions of meteoroids and comets (not shown here). The inner planets (Mercury, Venus, Earth with its Moon, and Mars) are composed mostly of rocky materials. The outer planets (Jupiter, Saturn, Uranus, and Neptune) are much larger, are composed mostly of gas and liquid, and have no solid surfaces. Pluto, Eris, and the satellites in the outer solar system are composed mostly of water ice. Some are so cold (–230°C) that they have methane ice or nitrogen ice at their surfaces.

All planetary bodies in the solar system are important in the study of Earth because their chemical compositions, surface features, and other characteristics show how planets evolve. They provide important insight into the forces that shaped our planet's history.

Figure 1.2 **The planets of the solar system** vary in size and composition with distance from the Sun. The inner planets are small and rocky, whereas the outer planets are much larger and composed mostly of hydrogen and helium. The dwarf outer planters are composed mostly of ice.

Earth

From a planetary perspective, Earth is a small blue planet bathed in a film of white clouds and liquid water (Figure 1.3). In this remarkable view, we see Earth motionless, frozen in a moment of time, but there is much more action shown here than you might imagine. The blue water and swirling white clouds dominate the scene and underline the importance of moving water in the Earth system. Huge quantities of water are in constant motion, in the sea, in the air (as invisible vapor and condensed as clouds), and on land. You can see several complete cyclonic storms spiraling over thousands of square kilometers, pumping vast amounts of water into the atmosphere. When this water becomes precipitation on land, it flows back to the sea in great river systems that erode and sculpt the surface.

Large parts of North and South America are visible in this view. The major climactic zones of our planet are clearly delineated. For example, the great deserts are visible at the top of the scene, extending across the southern United States and Mexico. Much of the vast tropical rain forest of South America is seen beneath the discontinuous cloud cover. Also, large portions of the north polar ice cap are clearly visible.

Earth is just the right distance from the Sun for its temperature to let water exist as a liquid, a solid, and a gas. Water in any of those forms is part of the hydrosphere. If Earth were closer to the Sun, our oceans would evaporate; if it were farther from the Sun, our oceans would freeze solid. However, there is plenty of liquid water on Earth, and it is liquid water, as much as anything else, that makes Earth unique among the planets of the solar system. Heated by the Sun, water moves on Earth in great cycles. It evaporates from the huge oceans into the atmosphere, precipitates over the landscape, collects in river systems, and ultimately flows back to the oceans. As a result, Earth's surface stays "young," being constantly changed by water and eroded into intricate systems of river valleys. This dynamism is in remarkable contrast to other planetary bodies, the surfaces of which are dominated by the craters of ancient meteorite impacts (Figure 1.3).

The presence of water as a liquid on Earth's surface throughout its long history also enabled life to evolve. And life, strange as it may seem, has profoundly changed the composition of Earth's atmosphere. Here is the mechanism: photosynthesis by countless plants removes large quantities of carbon dioxide from the atmosphere. As part of this process, the plants "exhale" oxygen. In addition, many forms of marine life remove carbon dioxide from seawater to make their shells, which later fall to the seafloor and form limestone.

How has the exploration of space changed our view of Earth and its geologic systems?

Earth

Earth is a delicate blue ball wrapped in filmy white clouds. The water and swirling clouds that dominate Earth's surface underline the importance of water in Earth's systems. The cold polar regions are buried with ice, and the warm tropics are speckled with clouds and greenery. The rocks of the high continents are strongly deformed and older than the rocks of the ocean basins. Earth has active volcanoes, a dynamic interior, and no large impact craters are visible on its surface.

Diameter 12,800 km; Density 5.55 g/cm³

Venus

Venus is often considered Earth's twin because of its similar size and density, but the two planets are not identical. This image of Venus shows its cloudy atmosphere partially stripped away to reveal a radar map of the solid surface made by an orbiting satellite. Venus has high plateaus, folded mountain belts, many volcanoes, and relatively smooth volcanic plains, but it has no water and few meteorite impact craters.

Diameter 12,100 km; Density 5.25 g/cm³

Mars

Mars is much smaller than Earth and Venus but has many fascinating geologic features—evidence that its surface has been dynamic in the past. Three huge extinct volcanoes, one more than 28 km high, can be seen in the left part of this image. An enormous canyon extends across the entire hemisphere—a distance roughly equal to that from New York to San Francisco. These features reveal that today's windy, desert Martian surface has been dynamic in the past, but ancient meteorite impact craters (visible in the upper right part of the image) have not been completely obliterated by younger events.

Diameter 6800 km; Density 3.9 g/cm³

Mercury

Mercury is similar to the Moon, with a surface dominated by ancient impact craters and younger smooth plains presumably made from floods of lava. Like the Moon, Mercury lacks an atmosphere and hydrosphere.

Diameter 4900 km; Density 5.44 g/cm³

Moon

The Moon has two contrasting provinces: bright, densely cratered highlands and dark, smooth lava plains. We know from rock samples brought back by the *Apollo* astronauts that the dark smooth plains are ancient floods of lava that filled many large meteorite impact craters and spread out over the surrounding area. The volcanic activity thus occurred after the formation of the heavily cratered terrain, but was not sufficient to obliterate all of the impact craters. Today the Moon is a geologically quiet body with no atmosphere or liquid water.

Diameter 3500 km; Density 3.3 g/cm³

Figure 1.3 The surfaces of the inner planets, shown at the same scale, provide insight into planetary dynamics.

Another characteristic of Earth is that it is dynamic. Its interior and surface continually change as a result of its **internal heat**. In marked contrast, many other planetary bodies have changed little since they formed because they are no longer hot inside. Most of Earth's heat comes from natural **radioactivity**. The breakdown of three elements—potassium, uranium, and thorium—is the principal source of this heat. Once generated, this heat flows to the surface and is lost to space. Another source of heat has been inherited from the formation of the planets. Heat was deposited in each of the planets by the infall of countless meteorites to form a larger and larger planet. This **accretionary heat** may have melted the early planets, including Earth. Larger planets have more internal heat and retain it longer than smaller planets.

Earth's internal heat creates slow movements within the planet. Its rigid outer layer (the lithosphere) breaks into huge fragments, or plates, that move. Over billions of years, these moving plates have created ocean basins and continents. The heat-driven internal movement also has deformed Earth's solid outer layers, creating earthquakes, mountain belts, and volcanoes. Thus, Earth has always been a dynamic planet, continuously changing as a result of its internal heat and the circulation of its surface water.

Look at the view of Earth from space (on p. 52). Of particular interest in this view is the rift system of East Africa. The continent is slowly being ripped apart along this extensive fracture system. Where this great rift separates the Arabian Peninsula from Africa, it has filled with water, forming the Red Sea. The rift extends from there southward across most of the continent (it is mostly obscured by clouds in the equatorial region). Some animals that evolved in the East African rift valleys spread from there and learned to live in all of the varied landscapes of the planet. This was their first home, but they have since walked on the Moon.

What can surface features tell us about planetary dynamics?

The Other Inner Planets

In stark contrast to the dynamic Earth, some of the other inner planets are completely inactive and unchanging. For example, the Moon and Mercury (Figure 1.3) are pockmarked with thousands of craters that record the birth of the planets about 4.55 billion years ago. This was a period when planetary bodies swept up what remained of the cosmic debris that formed the Sun and its planets. As the debris struck each body, **impact craters** formed.

The Moon and Mercury are so small that they were unable to generate and retain enough internal heat to sustain prolonged geologic activity. They rapidly cooled and lost the ability to make volcanoes. Their smooth lava plains are ancient by comparison with those on Earth. Consequently, their surfaces have changed little in billions of years. They retain many meteorite impact craters formed during the birth of the solar system. Neither planet has a hydrosphere or an atmosphere to modify them. Thus, these small planets remain as "fossils" of the early stages in planetary development. The footprints left on the Moon by the *Apollo* astronauts will remain fresh and unaltered for millions of years.

Mars is larger and has more internal heat and a thin atmosphere (Figure 1.3). Its originally cratered surface has been modified by volcanic eruptions, huge rifts, and erosion by wind and, in its distant past, running water. Today, Mars is too cold and the atmospheric pressure too low for water to exist as a liquid. Large polar ice caps mark both poles. In many ways, Mars is a frozen wasteland with a nearly immobile crust. Consequently, its ancient impact craters were never completely obliterated.

Venus is larger still and has more internal energy, which moves the crust and continually reshapes its surfaces (Figure 1.3). Venus is only slightly smaller than Earth and closer to the Sun. A thick carbon dioxide-rich atmosphere holds in the solar energy that reaches the surface and makes the temperature rise high enough to melt lead (around 500°C or 900°F). The atmospheric pressure is 90 times that on Earth. Unlike the smaller planets, Venus has no heavily cratered areas. Its ancient impact craters must have been destroyed by deformation or by burial below lava flows. Its surface is apparently young. Because of its large size, it has cooled quite slowly, so that volcanoes may even be active today. On the other hand, no evidence of water has been found on Venus; it has no oceans, no rivers, no ice caps, and only a very little water vapor. Only Earth has large amounts of liquid water that have influenced its development throughout history.

This, then, is Planet Earth in its cosmic setting—only a pale blue dot in space, part of a family of planets and moons that revolve around the Sun. It is a minor planet bound to an ordinary star in the outskirts of one galaxy among billions. Yet, from a human perspective, it is a vast and complex system that has evolved over billions of years, a home we are just beginning to understand. Learning about Earth and the forces that change it—the intellectual journey upon which you are about to embark—is a journey we hope you will never forget. Our study of the diverse compositions and conditions of the planets should remind us of the delicate balance that allows us to exist at all. Are we intelligent enough to understand how our world functions as a planet and to live wisely within those limits?

Earth's Outermost Layers

The outermost layers of Earth are the atmosphere, hydrosphere, and biosphere. Their dynamics are especially spectacular when seen from space.

Views of Earth from space like the one in Figure 1.3 reveal many features that make Earth unique, and they provide insight into our planet's history of change. The atmosphere is the thin, gaseous envelope that surrounds Earth. The hydrosphere, the planet's discontinuous water layer, is seen in the vast oceans. Even parts of the biosphere—the organic realm, which includes all of Earth's living things—can be seen from space, such as the dark green tropical forest of equatorial Africa. The lithosphere—the outer, solid part of Earth—is visible in continents and islands.

One of the unique features of Earth is that each of the planet's major realms is in constant motion and continual change. The atmosphere and the hydrosphere move in dramatic and obvious ways. Movement, growth, and change in the biosphere can be readily appreciated—people are part of it. But Earth's seemingly immobile lithosphere is also in motion, and it has been so throughout most of the planet's history.

The Atmosphere

Perhaps Earth's most conspicuous features, as seen from space, are the **atmosphere** and its brilliant white swirling clouds (Figure 1.3). Although this envelope of gas forms an insignificantly small fraction of the planet's mass (less than 0.01%), it is particularly significant because it moves easily and is constantly interacting with the ocean and land. It plays a part in the evolution of most features of the landscape and is essential for life. On the scale of the illustration in Figure 1.3, most of the atmosphere would be concentrated in a layer as thin as the ink with which the photo is printed.

The atmosphere's circulation patterns are clearly seen in Figure 1.3 by the shape and orientation of the clouds. At first glance, the patterns may appear confusing, but upon close examination we find that they are well organized. If we ignore the details of local weather systems, the global atmospheric circulation becomes apparent. Solar heat, the driving force of atmospheric circulation, is greatest in the equatorial regions. The heat causes water in the oceans to evaporate, and the heat makes the moist air less dense, causing it to rise. The warm, humid air forms an equatorial belt of spotty clouds, bordered on the north and south by zones that are cloud-free, where air descends. To the north and south, cyclonic storm systems develop where warm air from low latitudes confronts cold air around the poles.

Our atmosphere is unique in the solar system. It is composed of 78% nitrogen, 21% oxygen, and minor amounts of other gases, such as carbon dioxide (only 0.035%) and water vapor. The earliest atmosphere was much different. It was essentially oxygen-free and consisted largely of carbon dioxide and water vapor. The present carbon dioxide-poor atmosphere developed as soon as limestone began to form in the oceans, tying up the carbon dioxide. Oxygen was added to the atmosphere later, when plants evolved. As a result of photosynthesis, plants extracted carbon dioxide from the primitive atmosphere and expelled oxygen into it. Thus, the oxygen in the atmosphere is and was produced by life.

The Hydrosphere

The **hydrosphere** is the total mass of water on the surface of our planet. Water covers about 71% of the surface. About 98% of this water is in the oceans. Only 2% is in streams, lakes, groundwater, and glaciers. Thus, it is for good reason that Earth has been called "the water planet." It has been estimated that if all the irregularities of Earth's surface were smoothed out to form a perfect sphere, a global ocean would cover Earth to a depth of 2.25 km.

Again, it is this great mass of water that makes Earth unique. Water permitted life to evolve and flourish; every inhabitant on Earth is directly or indirectly controlled by it. All of Earth's weather patterns, climate, rainfall, and even the amount of carbon dioxide in the atmosphere are influenced by the water in the oceans. The hydrosphere is in constant motion; water evaporates from the oceans and moves through the atmosphere, precipitating as rain and snow, and returning to the sea in rivers, glaciers, and groundwater. As water moves over Earth's surface, it erodes and transports weathered rock material and deposits it. These actions constantly modify Earth's landscape. Many of Earth's distinctive surface features are formed by action of the hydrosphere.

How are Earth's atmosphere and hydrosphere different from those on other planets?

The Biosphere

The **biosphere** is the part of Earth where life exists. It includes the forests, grasslands, and familiar animals of the land, together with the numerous creatures that inhabit the sea and atmosphere. Microorganisms such as bacteria are too small to be seen, but they are probably the most common form of life in the biosphere. As a terrestrial covering, the biosphere is discontinuous and irregular; it is an interwoven web of life existing within and reacting with the atmosphere, hydrosphere, and lithosphere. It consists of more than 1.5 million described species and perhaps as many as 3 million more not yet described. Each species lives within its own limited environmental setting (**Figure 1.4**).

What is the biosphere? How does it affect Earth dynamics?

Almost the entire biosphere exists in a narrow zone extending from the depth to which sunlight penetrates the oceans (about 200 m) to the snow line in the tropical and subtropical mountain ranges (about 6000 m above sea level). At the scale of the photograph in Figure 1.2, the biosphere—all of the known life in the solar system—would be in a thin layer no thicker than the paper on which the image is printed.

Certainly one of the most interesting questions about the biosphere concerns the number and variety of organisms that compose it. Surprisingly, the truth is that no one knows the answer. Despite more than 250 years of systematic research, estimates of the total number of plant and animal species vary from 3 million to more than 30 million. Of this number, only 1.5 million species have been recorded. The diversity is stranger than you may think. Insects account for more than one-half of all known species, whereas there are only 4000 species of mammals, or about 0.025% of all species. Observation shows that there are more species of small animals than of large ones. The smallest living creatures—those invisible to the unaided eye, such as protozoa, bacteria, and viruses—contribute greatly to the variety of species. The biosphere is a truly remarkable part of Earth's systems.

The main factors controlling the distribution of life on our planet are temperature, pressure, and chemistry of the local environment. However, the range of environmental conditions in which life is possible is truly amazing, especially the range of environments in which microorganisms can exist (Figure 1.4B).

Although the biosphere is small compared with Earth's other major layers (atmosphere, hydrosphere, and lithosphere), it has been a major geologic force. Essentially all of the present atmosphere has been produced by the chemical activity of the biosphere. The composition of the oceans is similarly affected by living things; many marine

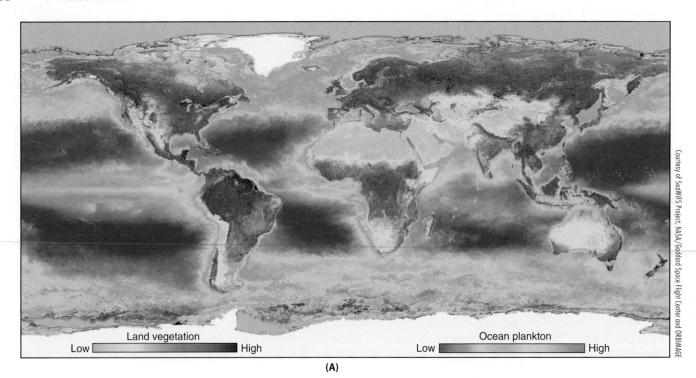

(A)

(A) This map of the biosphere was produced from data derived from satellite sensors. Land vegetation increases from tan to yellow to green to black. Escalating concentrations of ocean phytoplankton are shown by colors ranging from purple to red. Phytoplankton are microscopic plants that live in the surface layer of the ocean and form the foundation of the marine food chain. Note the particularly high concentrations of phytoplankton in polar waters (red and yellow) and the very low concentrations (purple and blue) in the mid-latitudes.

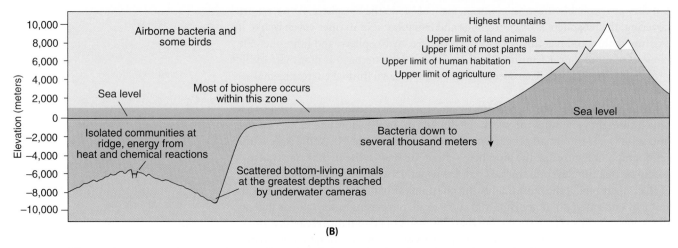

(B)

(B) Most of the biosphere exists within a very thin zone from 100 m below sea level to about 2000 m above sea level.

Figure 1.4   These global views of Earth's biosphere emphasize that life is widespread and has become a powerful geologic force.

organisms extract calcium carbonate from seawater to make their shells and hard parts. When the organisms die, their shells settle to the seafloor and accumulate as beds of limestone. In addition, the biosphere formed all of Earth's coal, oil, and natural gas. Thus, much of the rock in Earth's crust originated in some way from biological activity.

A historical record of the biosphere is preserved, sometimes in remarkable detail, by fossils that occur in rocks. Indeed, the number of living species today represents less than 10% of the number of species that have existed since life first developed on Earth.

The Geosphere: Earth's Internal Structure

The geosphere is made of Earth's solid materials—its rocks. They are separated into layers according to composition and mechanical properties. From outside in, the compositional layers are (1) crust, (2) mantle, and (3) core. Layers based on physical properties are (1) lithosphere, (2) asthenosphere, (3) mesosphere, (4) outer core, and (5) inner core.

Studies of earthquake waves, meteorites that fall to Earth, magnetic fields, and other physical properties show that Earth's interior consists of a series of shells of different compositions and mechanical properties. Earth is called a **differentiated planet** because of this separation into layers. How did Earth become differentiated? First, recall that the density of liquid water is 1 g/cm³. The density of most rocks at the surface is about three times as great, just under 3 g/cm³. But the overall density of Earth is about 5.5 g/cm³. Clearly, Earth consists of internal layers of increasing density toward the center. The internal layers were produced as different materials rose or sank so that the least-dense materials were at the surface and the most dense were in the center of the planet. Thus, gravity is the motive force behind Earth's differentiated structure.

In the discussion below we take you on a brief tour of the **geosphere**, which is defined as the solid materials that make up the planet. We will go to the very center of Earth, which lies at a depth of about 6400 km. Chemical properties define one set of layers, and mechanical behavior defines a different set. **Figure 1.5** shows the layers based on chemical properties on the left and those based on mechanical properties on the right. An understanding of both types of layers is vital.

Internal Structure Based on Chemical Composition

Geologists use the term **crust** for the outermost compositional layer (Figure 1.5, left). The base of the crust heralds a definite change in the proportions of the various elements that compose the rock but not a strong change in its mechanical behavior or physical properties.

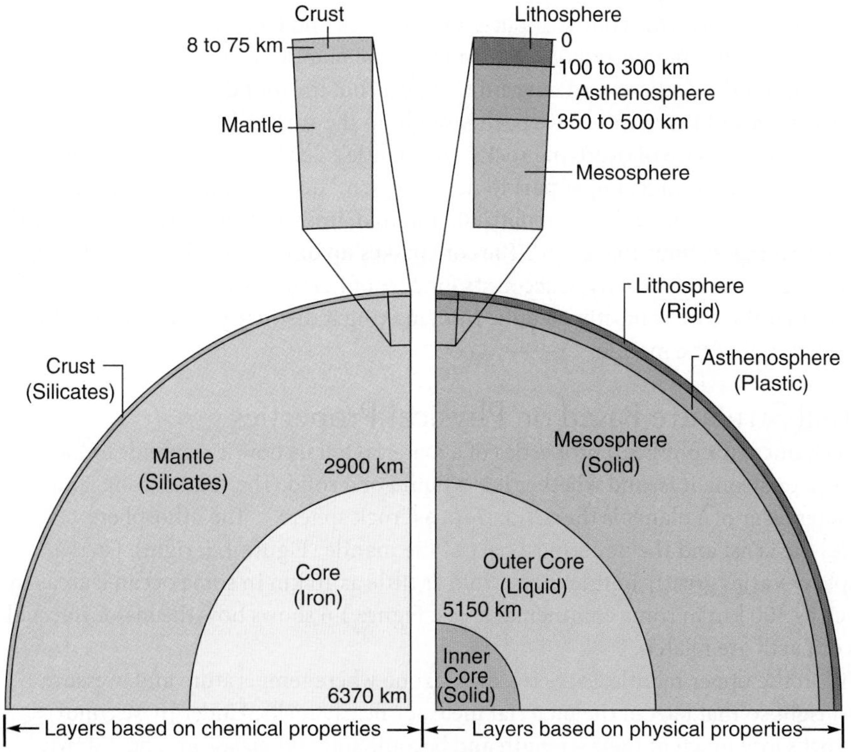

Why is it important to distinguish between the physical and chemical layers of Earth's interior?

Figure 1.5 The internal structure of Earth consists of layers of different composition and layers of different physical properties. The left side shows the layering based on chemical composition. These consist of a crust, mantle, and core. The right side shows the layering based on physical properties such as rigidity, plasticity, and whether it is solid or liquid. (Note that the two divisions, chemical and physical, do not coincide except at the core mantle boundary.)

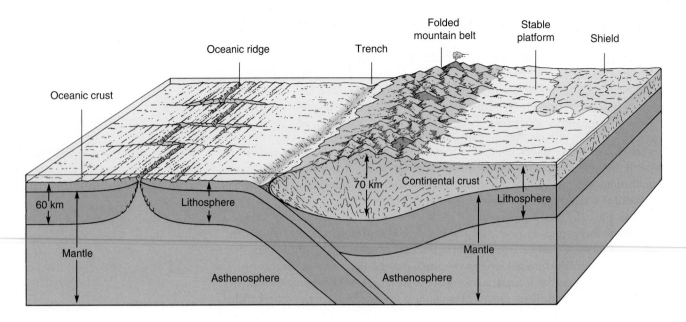

Figure 1.6 The outermost layers of the solid Earth, based on physical characteristics, are the asthenosphere and the lithosphere. The asthenosphere is hot, close to its melting point, and is capable of plastic flow. The lithosphere above it is cooler and rigid. It includes the uppermost part of the mantle and two types of crust: thin, dense oceanic crust and thick, buoyant continental crust.

Moreover, the crust of the **continents** is distinctly different from the crust beneath the **ocean basins** (**Figure 1.6**). Continental crust is much thicker (as much as 75 km), is composed of less-dense "granitic" rock (about 2.7 g/cm³), is strongly deformed, and includes the planet's oldest rocks (billions of years in age). By contrast, the oceanic crust is only about 8 km thick, is composed of denser volcanic rock called basalt (about 3.0 g/cm³), is comparatively undeformed by folding, and is geologically young (200 million years or less in age). These differences between the continental and oceanic crusts, as you shall see, are fundamental to understanding Earth.

The next major compositional layer of Earth, the **mantle**, surrounds or covers the core (Figure 1.5, left). This zone is about 2900 km thick and constitutes the great bulk of Earth (82% of its volume and 68% of its mass). The mantle is composed of silicate rocks (compounds of silicon and oxygen) that also contain abundant iron and magnesium. Fragments of the mantle have been brought to the surface by volcanic eruptions. Because of the pressure of overlying rocks, the mantle's density increases with depth from about 3.2 g/cm³ in its upper part to nearly 5 g/cm³ near its contact with the core.

Earth's **core** is a central mass about 7000 km in diameter. Its density increases with depth but averages about 10.8 g/cm³. The core makes up only 16% of Earth's volume, but, because of its high density, it accounts for 32% of Earth's mass. Indirect evidence indicates that the core is mostly metallic iron, making it distinctly different from the silicate material of the mantle.

Internal Structure Based on Physical Properties

What layers of Earth are most significant to planetary dynamics?

The mechanical (or physical) properties of a material tell us how it responds to force, how weak or strong it is, and whether it is a liquid or a solid. The solid, strong, and rigid outer layer of a planet is the **lithosphere** ("rock sphere"). The lithosphere includes the crust and the uppermost part of the mantle (Figure 1.5, right). Earth's lithosphere varies greatly in thickness, from as little as 10 km in some oceanic areas to as much as 300 km in some continental areas. Figure 1.6 shows how the major internal layers of Earth are related.

Within the upper mantle, there is a major zone where temperature and pressure are just right so that part of the material melts, or nearly melts. Under these conditions, rocks lose much of their strength and become soft and plastic and flow slowly. This zone of easily deformed mantle is known as the **asthenosphere** ("weak sphere").

The boundary between the asthenosphere and the overlying lithosphere is mechanically distinct but does not correspond to a fundamental change in chemical composition. The boundary is simply a major change in the rock's mechanical properties.

The rock below the asthenosphere is stronger and more rigid than in the asthenosphere. It is so because the high pressure at this depth offsets the effect of high temperature, forcing the rock to be stronger than the overlying asthenosphere. The region between the asthenosphere and the core is the **mesosphere** ("middle sphere").

Earth's core marks a change in both chemical composition and mechanical properties. On the basis of mechanical behavior alone, the core has two distinct parts: a solid **inner core** and a liquid **outer core**. The outer core has a thickness of about 2270 km compared with the much smaller inner core, with a radius of only about 1200 km. The core is extremely hot, and heat loss from the core and the rotation of Earth probably cause the liquid outer core to flow. This circulation generates Earth's magnetic field.

Major Features of the Continents

> Continents consist of three major structural components: (1) shields, (2) stable platforms, and (3) folded mountain belts. Continental crust is less dense, thicker, older, and more deformed than oceanic crust.

If Earth did not have an atmosphere and a hydrosphere, two principal regions would stand as its dominant features: ocean basins and continents. The ocean basins, which occupy about two-thirds of Earth's surface, have a remarkable topography, most of which originated from extensive volcanic activity and Earth movements that continue today. The continents rise above the ocean basins as large platforms. The ocean waters more than fill the ocean basins and rise high enough to flood a large part of the continents. The present shoreline, so important to us geographically and so carefully mapped, has no simple relation to the structural boundary between continents and ocean basins.

In our daily lives, the position of the ocean shoreline is very important. But from a geologic viewpoint, the elevation of the continents with respect to the ocean floor is much more significant than the position of the shore. The difference in elevation of continents and ocean basins reflects their fundamental difference in composition and density. Continental "granitic" rocks are less dense (about 2.7 g/cm³) than the basaltic rocks of the ocean basins (about 3.0 g/cm³). That is, a given volume of continental rock weighs less than the same volume of oceanic rock. This difference causes the **continental crust** to be more buoyant—to rise higher—than the denser **oceanic crust** in much the same way that ice cubes float in a glass of water because ice is less dense than water. Moreover, the rocks of the continental crust are older (some as old as 4.0 billion years old) than the rocks of the oceanic crust.

What are the fundamental structural features of continents?

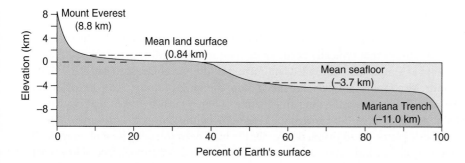

Figure 1.7 A graph of the elevation of the continents and ocean basins shows that the average height of the continents is only 0.8 km above sea level. Only a small percentage of Earth's surface rises above the average elevation of the continents or drops below the average elevation of the ocean floor.

The elevation and area of the continents and ocean basins now have been mapped. These data can be summarized in various forms. **Figure 1.7** shows that the average elevation of the continents is 0.8 km above sea level, and the average elevation of the seafloor (depth of the ocean) is about 3.7 km below sea level. Only a relatively small percentage of Earth's surface rises significantly above the average elevation of the continents or drops below the average depth of the ocean. If the continents did not rise quite so high above the ocean floor, the entire surface of Earth would be covered with water.

Shields

Where are Earth's oldest rocks found? Why?

The extensive flat region of a continent, in which complexly deformed ancient crystalline rocks are exposed, is known as a **shield** (**Figure 1.8**). All of the rocks in the shield formed long ago—most more than 1 billion years ago. Moreover, these regions have been relatively undisturbed for more than a half-billion years except for broad, gentle warping. The rocks of the shields are highly deformed igneous and metamorphic rock; they are also called the **basement complex**.

Without some firsthand knowledge of a shield, visualizing the nature and significance of this important part of the continental crust is difficult. **Figure 1.9** shows part of the Canadian shield of the North American continent as seen from space. It will help you to comprehend the extent, the complexity, and some of the typical features of shields. First, a shield is a regional surface of low relief that generally has an elevation within a few hundred meters of sea level. (**Relief** is the elevation difference between the low and the high spots.) Resistant rocks may rise 50 to 100 m above their surroundings.

A second characteristic of shields is their complex internal structure and complex arrangements of rock types. Many rock bodies in a shield once were molten, and others have been compressed and extensively deformed while still solid. Much of the rock in shields was formed several kilometers below the surface. They are now exposed only because the shields have been subjected to extensive uplift and erosion.

Stable Platforms

What are the most significant features of stable platforms?

When the basement complex is covered with a veneer of sedimentary rocks, a **stable platform** is created. The layered sedimentary rocks are nearly horizontal and commonly etched by dendritic (treelike) river patterns (**Figure 1.10**). These broad areas have been relatively stable throughout the last 500 million or 600 million years; that is, they have not been uplifted a great distance above sea level or submerged far below it—hence the term stable platform. In North America, the stable platform lies between the Appalachian Mountains and the Rocky Mountains and extends northward to the Lake Superior region and into western Canada. Throughout most of this area, the sedimentary rocks are nearly horizontal, but locally they have been warped into broad domes and basins (Figure 1.8). Sometimes it is useful to group the shield and stable platform together in what is called a **craton.**

Folded Mountains

Some of the most impressive features of the continents are the young **folded mountain belts** that typically occur along their margins. Most people think of a mountain as simply a high, rugged landform, standing in contrast to flat plains and lowlands. Mountains, however, are much more than high country. To a geologist, the term *mountain belt* means a long, linear zone in Earth's crust where the rocks have been intensely deformed by horizontal stress during the slow collision between two lithospheric plates. In addition, they generally have been intruded by molten rock. The topography can be high and rugged, or it can be worn down to a surface of low relief. To a sightseer, the topography of a mountain belt is everything, but to a geologist, it is not as important as the extent and style of its internal deformation. The great folds and fractures in mountain belts provide evidence that Earth's lithosphere is, and has been, in motion.

Figure 1.8 The major surface features of Earth reflect the structure of the lithosphere. The continental crust rises above the ocean basins and forms continents. They have as their major structural features shields, stable platforms, and folded mountain belts. The continents are formed mostly of granitic rock. The oceanic crust forms the ocean floor. Its major features include the oceanic ridge, the abyssal floor, seamounts, and trenches. It is composed primarily of basalt.

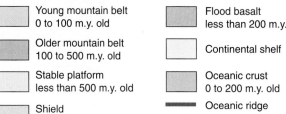

Young mountain belt
0 to 100 m.y. old

Older mountain belt
100 to 500 m.y. old

Stable platform
less than 500 m.y. old

Shield
greater than 500 m.y. old

Flood basalt
less than 200 m.y. old

Continental shelf

Oceanic crust
0 to 200 m.y. old

Oceanic ridge

Trench

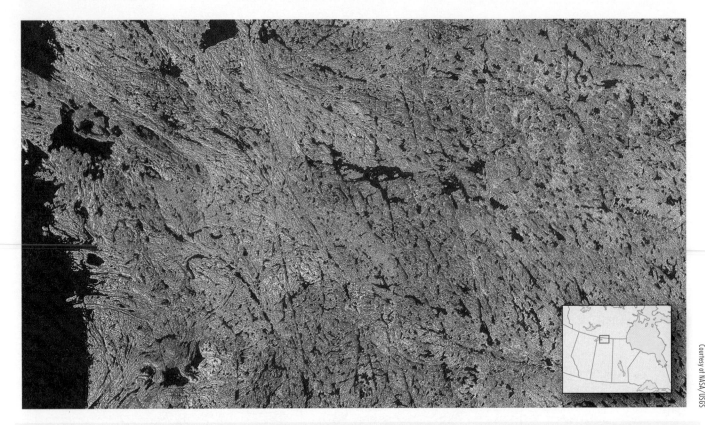

Courtesy of NASA/USGS

Figure 1.9 **The Canadian shield** is a fundamental structural component of North America. It is composed of complexly deformed crystalline rock bodies, eroded to an almost flat surface near sea level, as shown in this false-color satellite image. Throughout much of the Canadian shield, the topsoil has been removed by glaciers, and different rock bodies are etched in relief by erosion. The resulting depressions commonly are filled with water, forming lakes and bogs that emphasize the structure of the rock bodies. Dark tones show areas of metamorphic rock. Light pink tones show areas of granitic rock.

Can a folded mountain belt be a lowland?

Figure 1.10 illustrates some characteristics of folded mountains and the extent to which the margins of continents have been deformed. On this map of the Appalachian Mountains, the once horizontal layers of rock have been deformed by compression and are folded like wrinkles in a rug. Erosion has removed the upper parts of the folds, so the resistant layers form zigzag patterns similar to those that would be produced if the crest of the wrinkles in a rug were sheared off.

The crusts of the Moon, Mars, and Mercury lack this type of deformation. All of their impact craters, regardless of age, are circular—proof that the crusts of these planets have not been strongly deformed by compressive forces. Their crusts, unlike that of Earth, appear to have been fixed and immovable throughout their histories. However, Venus is like Earth in this respect and has long belts of folded mountains.

Summary of the Continents

The broad, flat continental masses that rise above the ocean basins have an almost endless variety of hills and valleys, plains and plateaus, and mountains. Yet from a regional perspective, the geologic differences between continents are mostly in size and shape and in the proportions of shields, stable platforms, and folded mountain belts.

Let us now briefly review the major structural components of the continents by examining North and South America (Figure 1.8). North America has a large shield, most of which is in Canada. Most of the Canadian shield is less than 300 m above sea level.

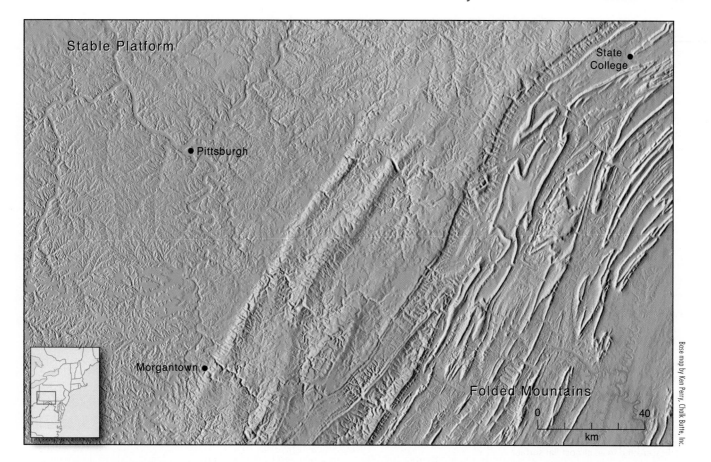

Figure 1.10 **The stable platform and folded mountain belt** in the eastern United States are clearly shown on this topographic map. The layered rocks of the stable platform are nearly horizontal, but on a regional scale, they are warped into a large structural basin that has been highly dissected by intricate networks of river valleys (upper left). The rocks of the Appalachian Mountains, in contrast, have been compressed and folded (center and lower right). The folded layers are expressed by long, narrow ridges of resistant sandstone that rise about 300 m above the surrounding area (lower right). Erosion has removed the upper parts of many folds, so their resistant limbs (which form the ridges) are exposed in elliptical or zigzag outcrop patterns.

The rocks in the Canadian shield formed between 1 and 4 billion years ago. The stable platform extends through the central United States and western Canada and is underlain by sedimentary rocks, slightly warped into broad domes and basins. The Appalachians are an old folded mountain belt that formed about 250 million years ago. The Rocky Mountains form part of another folded mountain belt (the Cordillera) that dominates western North America and extends into South America. The Rockies started forming about 60 million years ago, and parts of this belt are still active.

In many ways, the structure of South America (Figure 1.8) is similar to that of North America. The continent consists of a broad shield in Brazil and Venezuela, and stable platforms in the Amazon basin and along the eastern flanks of the Andes Mountains. The Andes Mountains are part of the Cordilleran folded mountain belt that extends from Alaska to the southern tip of South America. The continent has no mountain belt along the eastern margin like the Appalachian Mountains in North America. More than 90% of South America drains into the Atlantic Ocean by way of the Amazon River system.

Before going on, you should briefly review the major structural features of each of the other continents and examine how they are similar and how they are different (see the shaded relief map inside the back cover).

STATE OF THE ART   Mapping the Continents from Space

Topographic maps that show elevations have always been important to geologists studying the continents. Until the mid-1900s, such maps were painstakingly constructed by careful fieldwork using survey instruments. It might take weeks to cover an area of 100 km², assuming access was good. Later in the 1930s and 1940s, aerial photographs replaced or supplemented these field techniques, but some fieldwork was still necessary. However, many remote areas and less-developed countries remained largely unmapped.

In February 2000, astronauts on the Space Shuttle using imaging radar revolutionized mapmaking. In just 9 days, they collected the data for the most accurate topographic map ever made of much of the planet. Radar signals were bounced off Earth's surface and then received by two different antennas, one inside the spacecraft and the other on a 60-m-long boom extended from the shuttle. A computer then combined these separate images to prepare a three-dimensional topographic map, just as your brain combines two separate images, one from each eye, to construct a 3-D image of your surroundings.

A key advantage to radar is that it can "see" the surface through clouds and in darkness. Another major advantage is speed. Shuttle radar captured the topographic data for an area the size of Rhode Island in only two seconds and for an area 100,000 km² in a minute. In nine days, the Shuttle mapped nearly 80% of Earth's land surface. In many areas these will be the highest resolution maps available. Before the Shuttle mission, less than 5% of Earth's surface had been mapped at a comparable scale.

The National Aeronautics and Space Administration (NASA), U.S. Department of Defense, and the German and Italian space agencies supported the project. The most detailed maps, showing objects just 3 m across, may only be available to the U.S. military. Such data will be used, for example, to guide cruise missiles through complex terrains and assist in troop deployment. Lower resolution maps (10 to 30 m resolution)

Courtesy of NASA

will be used to study Earth on a global scale in a way never before possible, including topics such as tectonics, flooding, erosion rates, volcanic and landslide hazards, earthquakes, and climate change.

The Shuttle radar topographic map below shows the dramatic difference between the new data (30 m resolution on the right) and the previous map (on the left) for the tropical rain forests of central Brazil. This region is near the city of Manaus on the great Amazon River. With the new map, you can see the delicate branching patterns of a multitude of stream valleys. The dark, smooth areas are reservoirs behind large dams. Most of the small valleys are not visible on the earlier topographic map.

0   10   20
km

Courtesy of JPL/NIMA/NASA

Major Features of the Ocean Basins

Oceanic crust differs strikingly from continental crust in rock types, structure, landforms, age, and origin. The major features of the ocean floor are (1) the oceanic ridge, (2) the abyssal floor, (3) seamounts, (4) trenches, and (5) continental margins.

The ocean floor, not the continents, is the typical surface of the solid Earth. If we could drain the oceans completely, this fact would be obvious. The seafloor holds the key to the evolution of Earth's crust, but not until the 1960s did we recognize that fact and obtain enough seafloor data to see clearly its regional characteristics. This new knowledge caused a revolution in geologists' ideas about the nature and evolution of the crust, a revolution as profound as Darwin's theory of evolution.

Until about the 1940s, most geologists believed that the ocean floor was simply a submerged version of the continents, with huge areas of flat abyssal plains covered with sediment eroded from the land mass. Since then, great advances in technology and exploration have been used to map the ocean basins in remarkable detail, as clearly as if the water had been removed (see the inside cover of this book). These maps show that submarine topography is as varied as that of the continents and in some respects is more spectacular.

Along with this kind of mapping, we have collected samples of the oceanic crust with drill rigs, dredges, and submarines. We have learned that the oceanic crust is mostly basalt, a dense volcanic rock, and that its major topographic features are somehow related to volcanic activity. These features make the oceanic crust entirely different from the continental crust. Moreover, the rocks of the ocean floor are young, in geologic terms. Most are fewer than 150 million years old, whereas the ancient rocks of the continental shields are more than 600 million years old. We have discovered that the rocks of the ocean floor have not been deformed into folded mountain belts—in marked contrast to the complexly deformed rocks in the mountains and basement complex of the continents.

How do the landforms on the ocean floor differ from those on continents?

The Oceanic Ridge

The **oceanic ridge** is perhaps the most striking and important feature on the ocean floor. It extends continuously from the Arctic Basin, down the center of the Atlantic Ocean, into the Indian Ocean, and across the South Pacific. You can see it clearly in Figure 1.8 and on the map inside the back cover. The oceanic ridge is essentially a broad, fractured rise, generally more than 1400 km wide. Its higher peaks rise as much as 3000 m above their surroundings. A huge, cracklike **rift valley** runs along the axis of the ridge throughout much of its length, which totals about 70,000 km. In addition, great fracture systems, some as long as 4000 km, trend perpendicular to the ridge.

What are the highest and lowest parts of the ocean floor?

The Abyssal Floor

The oceanic ridge divides the Atlantic and Indian oceans roughly in half and traverses the southern and eastern parts of the Pacific. On both sides of the ridge are vast areas of broad, relatively smooth deep-ocean basins known as the **abyssal floor**. This surface extends from the flanks of the oceanic ridge to the continental margins and generally lies at depths of about 4000 m.

The abyssal floor can be subdivided into two sections: the abyssal hills and the abyssal plains. The **abyssal hills** are relatively small ridges or hills, rising as much as 900 m above the surrounding ocean floor. They cover from 80% to 85% of the seafloor, and thus, they are the most widespread landforms on Earth. Near the continental margins, land-derived sediment completely covers the abyssal hills, forming flat, smooth **abyssal plains**.

Trenches

The deep-sea **trenches** are the lowest areas on Earth's surface. The Mariana Trench, in the Pacific Ocean, is the deepest part of the world's oceans—11,000 m below sea level—and many other trenches are more than 8000 m deep. Trenches have attracted the attention of geologists for years, not only because of their depth, but also because they represent fundamental structural features of Earth's crust. As illustrated in **Figure 1.11**, the trenches are invariably adjacent to chains of volcanoes called **island arcs** or to coastal mountain ranges of the continents. Why? We will see in subsequent chapters how the trenches are involved in the planet's most intense volcanic and seismic (earthquake) activity, and how the movement of Earth's lithospheric plates causes it all.

Seamounts

Isolated peaks of submarine volcanoes are known as **seamounts**. Some seamounts rise above sea level and form islands, but most are submerged and are known only from oceanographic soundings. Although many may seem to occur at random, most, such as the Hawaiian Islands, form chains along well-defined lines. Islands and seamounts testify to the extensive volcanic activity that is ongoing throughout the ocean basins. They also provide important insight into the dynamics of the inner Earth.

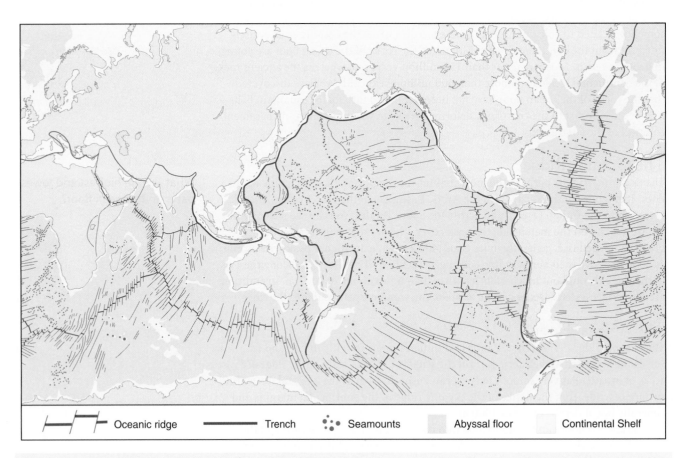

| ⊢—┼┼—⊣ Oceanic ridge | ───── Trench | ⦂∴⦂ Seamounts | Abyssal floor | Continental Shelf |

Figure 1.11 The major features of the ocean floor include the oceanic ridge, the deep-sea trenches, and the abyssal floor. Seamounts rise above the deep-ocean floor and are formed by volcanic eruptions.

Continental Margins

The zone of transition between a continent and an ocean basin is a **continental margin**. The submerged part of a continent is called a **continental shelf**, essentially a shallow sea that extends around a continent for many kilometers. You can clearly see the continental shelf around the continents in Figure 1.8 and on the map inside the back cover. Geologically, the continental shelf is part of the continent, not part of the ocean basin. At present, continental shelves form 11% of the continental surface, but at times in the geologic past, these shallow seas were much more extensive.

The seafloor descends in a long, continuous slope from the outer edge of the continental shelf to the deep-ocean basin. This **continental slope** marks the edge of the continental rock mass. Continental slopes are found around the margins of every continent and around smaller fragments of continental crust, such as Madagascar and New Zealand. Look at Figure 1.8 and study the continental slopes, especially those surrounding North America, South America, and Africa. You can see that they form one of Earth's major topographic features. On a regional scale, they are by far the longest and highest slopes on Earth. Within this zone, from 20 to 40 km wide, the average relief above the seafloor is 4000 m. In the trenches that run along the edges of some continents, relief on the continental slope is as great as 10,000 m. In contrast to the shorelines of the continents, the continental slopes are remarkably straight over distances of thousands of kilometers.

To ensure that you understand the basic features of the oceans, refer again to Figure 1.11 and the map inside the back cover and study the regional relationships of the oceanic ridges, abyssal plains, trenches, and seamounts of each of the major oceans. For example, the topography of the Atlantic Ocean floor shows remarkable symmetry in the distribution of the major features (Figure 1.11). It is dominated by the Mid-Atlantic Ridge, a broad rise in the center of the basin. Iceland is a part of the Mid-Atlantic Ridge that reaches above sea level. South of Iceland, the ridge separates the ocean floor into two long, parallel sub-basins that are cut by fracture zones stretching across the entire basin. Abyssal hills lie on either side the ridge, and abyssal plains occur along the margins of the continental platforms. In the South Atlantic, two symmetrical chains of seamounts extend from the continental margins to the oceanic ridge and come together to form an open V. Deep trenches flank volcanic island arcs off the north and south margins of South America. The symmetry of the Atlantic Basin even extends to the continental margins: the outlines of Africa and Europe fit those of South America and North America.

How are continental margins different from the rest of the seafloor?

STATE OF THE ART   Mapping the Ocean Floor from Space

Earth's uncharted frontiers lie at the floor of the oceans, and for most of human history they were as inaccessible as the stars. Thanks to several new techniques, geologists are now seeing the topography of the ocean floor.

The technique that is easiest to understand involves echo sounding. This is a special type of sonar—sound waves are timed while they travel to the ocean bottom and bounce back. Research ships tow long trails of "hydrophones" behind them to detect the signals. The result is a narrow strip map showing the elevation of the seafloor directly beneath the ship. A more recent innovation involves the mapping of a swath of the seafloor several kilometers wide. In either case, repeated traverses are necessary to accumulate enough data to compile a good topographic map. It would take about 125 years to map all of the ocean basins using this method.

A completely new way to make global maps of the seafloor involves sounding techniques carried out by an orbiting spacecraft instead of a ship. The satellites use radar—to carefully map the elevation of the sea surface. These maps show that the surface of the ocean bulges outward and inward, mimicking the topography of the underlying ocean floor. Although these bumps are too small to be seen with your eyes, they can be measured by a radar altimeter on the satellite. A radar satellite can map with a vertical resolution of only about 3 cm and can map the entire ocean in 1.5 years. The satellite altimeter emits a pulse of radar at the ocean surface, and the time for its reflection back to the satellite is measured. The width of the pulse is several kilometers wide and averages out local irregularities caused by ocean waves. To make accurate elevation

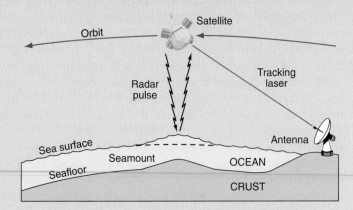

measurements, the satellite itself is tracked from ground stations using lasers.

The data are then processed with a computer to calculate the topography of the underlying seafloor. The maps provide the first view of the ocean-floor structures in many remote areas of the Earth. The map shown here and on the inside back cover was constructed in this way.

Why does the sea surface bulge? Earth's gravitational field is not constant everywhere. The gravitational acceleration at any spot on Earth's surface is proportional to the mass that lies directly beneath it. Thus, if a high seamount or ridge lies on the ocean floor, it has enough mass to pull water toward it, piling up the water immediately above it. Such a bulge may be several meters high. On the other hand, because water has a density less than that of rock, a point above a deep trough in the ocean floor has less mass directly below it and shows up as a shallow trough on the sea surface.

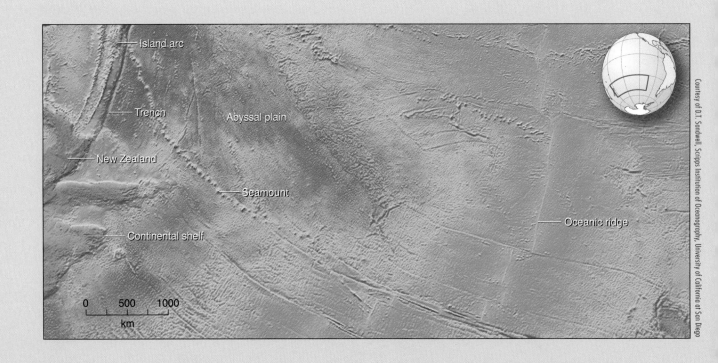

Courtesy of D.T. Sandwell, Scripps Institution of Oceanography, University of California at San Diego

The Ecosphere—A Model of Planet Earth

One way to simplify this vast array of details is to consider a model of Earth in its simplest form, in which the fundamental components—energy, rock, air, water, and life—are the only elements.

Courtesy of Stan Matcheon

Having surveyed the important components of Planet Earth, you should now look back and consider what these facts imply for life in general and humans in particular. To help you understand, let us contemplate a simple model of Earth—the ecosphere.

An **ecosphere** is a small glass globe, about the size of a large cantaloupe, containing five essential elements: energy, air, water, sand, and living things (algae, seaweed, shrimp, snails, and a variety of microorganisms). The globe is sealed, forming a closed system in which plants and animals are self-sustaining (**Figure 1.12**). Just like a planet, nothing enters or leaves the system except sunlight and heat. You cannot add oxygen. You can never clean the water or replace the seaweed or remove dead organisms. You can never add food or remove waste. The plants and animals are on their own small planet: an isolated world in miniature.

Experiments have shown that if even one of the five parts is missing, the shrimp will not survive and the entire system will fail. The biological cycle is shown in the diagram in Figure 1.12. The key to the system is energy in the form of light. Light energy powers photosynthesis, the chemical mechanism through which algae make their own food from carbon dioxide and water and release oxygen into the water. The shrimp breathe the oxygen in the water and feed on the algae and bacteria. The bacteria break down the animal waste into nutrients that the algae use in their growth. The shrimp, snails, and bacteria also give off carbon dioxide, which the algae use to produce oxygen. Thus, the cycle is repeated and constantly renews itself. The shrimp and snails are masters of this little world—as long as they do not overpopulate or contaminate their environment. In this closed system, plants and animals grow, reproduce, and die, but the self-renewing cycle continues. The ecosphere in Figure 1.12 operated well for more than three years, until it was moved to a spot near a window. There it received more sunlight. The algae grew too fast and upset the critical balance, causing everything to die. Some ecospheres have sustained themselves for more than 10 years.

As you might suspect, an ecosphere is much the same as Planet Earth—a closed, self-contained system with a few basic parts. The only real external input is energy from sunlight. Our ecosphere—the lithosphere, atmosphere, hydrosphere, and biosphere—provides the rest. When astronauts look back at our planet from space, they see an ecosphere made of continents and oceans, forests, and polar bodies of ice, all enclosed in a thin blue dome of gases, bathed in sunlight.

These five "spheres" interact to form a single dynamic system in which components are interconnected in fascinating ways with an amazing strength: change in one sphere can affect the others in unsuspected ways.

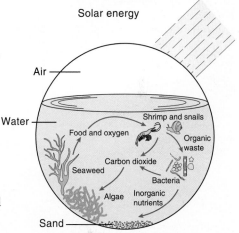

Figure 1.12 An ecosphere is a small model of Earth. It is a closed system of air, water, sand, and living organisms. Sunlight enters, providing the energy needed by the algae to make oxygen from carbon dioxide and water. Shrimp and snails breathe the oxygen and consume some algae and bacteria. Bacteria break down the animal waste into nutrients, which are used by the algae. The shrimp, snails, and bacteria also give off carbon dioxide, which the algae use to make more oxygen. This is an interlocking cycle of constant decay and renewal. If any component is missing, or is too far out of proportion, the system collapses and the entire biosphere of this tiny planet becomes extinct.

GeoLogic   Meteorites and Earth's Interior

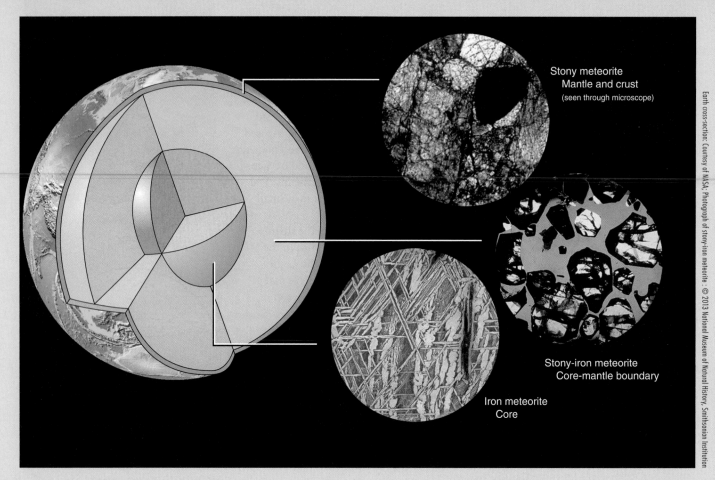

Stony meteorite
Mantle and crust
(seen through microscope)

Stony-iron meteorite
Core-mantle boundary

Iron meteorite
Core

Earth cross-section: Courtesy of NASA; Photograph of stony-iron meteorite : © 2013 National Museum of Natural History, Smithsonian Institution

Can you imagine any part of Earth more inaccessible than the core? Our best estimates place it at a depth of almost 3000 km. The deepest mines or even caves extend only a few kilometers deep—in spite of what you may have read in Jules Vernes' Journey to the Center of the Earth. Our deepest drill holes only penetrate to a depth of 15 or 20 km. And yet geologists believe they know a great deal about Earth's interior. For example they believe that Earth's core is made mostly of iron, that it supports a huge magnetic field, and that it has a temperature of 5000°C. While we cannot justify all of these propositions here, we can present a few of the facts that lead to the logical conclusion that Earth's core is made of molten metal.

Observations

1. There are three fundamentally different types of meteorites (rocks that fall from space). The most common are similar to rocks found at Earth's surface and are called stony meteorites. Other meteorites are mixtures of stony materials and shiny metal, and a third type is made solely of metal.

2. When cut open, polished and etched with acid, metallic meteorites have spectacular crystalline structures that reveal they formed at high temperatures from molten metal and then crystallized slowly in the solid state.

3. Detailed chemical studies show that these metallic meteorites are made principally of only two elements, iron and nickel.

4. Iron meteorites are also extremely dense, a cubic centimeter weighs nearly 8 g/cm³, compared to a typical surface rock that weighs less than 3 g/cm³.

Interpretations

Each of these bits of evidence points to a logical interpretation that has implications for the nature of Earth's interior. Apparently, iron meteorites formed by: (1) partial melting of a planet (implied by high temperature of formation); (2) gravitational sinking of the molten metal to near its center (high density and slow cooling rate caused by a thick insulating layer that allowed heat to escape slowly), and finally (3) cooling and crystallization (crystalline structure). Thus, we have concluded that when you hold an iron meteorite in your hand, you are actually holding a piece of the once molten core of another planet. The other types of meteorites appear to have come from the mantles and crusts of small planets that were like Earth in having differentiated interiors.

The average density of the Earth is 5.5 g/cm³. The average density of surface rocks is about 2.8 g/cm³. How does this support the interpretation above?

Key Terms

abyssal floor (p. 23)
abyssal hill (p. 23)
abyssal plain (p. 23)
accretionary heat (p. 11)
asthenosphere (p. 16)
atmosphere (p. 12)
basement complex (p. 18)
biosphere (p. 13)
continent (p. 16)
continental crust (p. 17)
continental margin (p. 25)

continental shelf (p. 25)
continental slope (p. 25)
core (p. 16)
craton (p. 18)
crust (p. 15)
density (p. 7)
differentiated planet (p. 15)
ecosphere (p. 27)
folded mountain belt (p. 18)
geology (p. 6)
geosphere (p. 15)

hydrosphere (p. 13)
impact crater (p. 11)
inner core (p. 17)
inner planets (p. 7)
internal heat (p. 11)
island arc (p. 24)
lithosphere (p. 16)
mantle (p. 16)
mesosphere (p. 17)
ocean basin (p. 16)
oceanic crust (p. 17)

oceanic ridge (p. 23)
outer core (p. 17)
outer planets (p. 7)
radioactivity (p. 11)
relief (p. 18)
rift valley (p. 23)
seamount (p. 24)
shield (p. 18)
stable platform (p. 18)
trench (p. 24)

Review Questions

1. Why are some planets geologically active today and others inactive?

2. Why are the atmosphere and the oceans considered as much a part of Earth as is solid rock?

3. Study the view of Earth in Figure 1.3. Sketch a map showing: (a) major patterns of atmospheric circulation, (b) low-latitude deserts, (c) the tropical belt, (d) the Red Sea rift, and (e) the Antarctic ice cap.

4. Draw two diagrams of Earth's internal structure. Draw one to show its internal structure based on chemical composition and draw another showing its structure based on mechanical (physical) properties.

5. Draw a cross section showing the lithosphere's major structural features: the continental crust, shields, stable platforms, and folded mountain belts, together with the oceanic ridge, the abyssal floor, and deep-sea trenches.

6. Make a table comparing the differences in the age, thickness, density, composition, and structure of oceanic and continental crust.

7. Briefly describe the distinguishing features of continental shields, stable platforms, and folded mountain belts.

8. Using the map in the back of the book, describe the locations of the shield, stable platforms, and folded mountain belts of Asia, Africa, Australia, and Europe.

9. Briefly describe the distinguishing features of the oceanic ridge, the abyssal floor, trenches, seamounts, and continental margins.

10. Describe the major elements of an ecosphere and how it functions. Relate these elements to their counterparts on the real Earth.

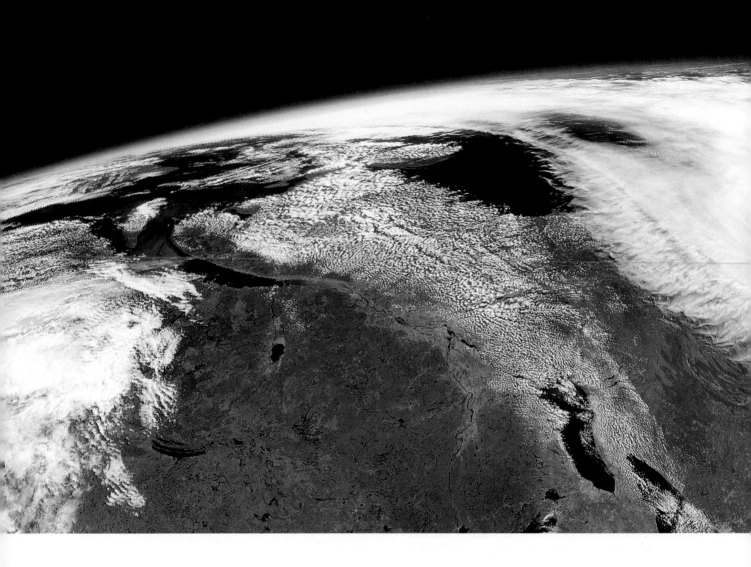

2 Geologic Systems

Earth is a dynamic planet because the materials of its various layers are in motion. The effects of both the hydrologic and the tectonic systems are dramatically expressed in this space photograph of eastern North America. The most obvious motion is that of the surface fluids: air and water. The complex cycle by which water moves from the oceans into the atmosphere, to the land, and back to the oceans again is the fundamental movement within the hydrologic system. The energy source that drives this system is the Sun. Its energy evaporates water from the oceans and causes the atmosphere to circulate, as shown above by the swirling clouds of hurricane Dennis. Water vapor is carried by the circulating atmosphere and eventually condenses to fall as rain or snow, which gravity pulls back to Earth's surface. Still acted on by the force of gravity, the water then flows back to the oceans in several subsystems (rivers, groundwater, and glaciers). In every case, gravity causes the water to flow from higher to lower levels.

Earth's lithosphere may appear to be permanent and stationary, but like the hydrosphere, it is in constant motion, albeit much, much slower. There is now overwhelming evidence that the entire lithosphere moves, and as it does, continents split and the fragments drift thousands of kilometers across Earth's surface perchance to collide with one another. The great Appalachian Mountain chain, visible here as a series of parallel ridges and valleys, formed when two continents collided hundreds of millions of years ago. The folded and crumpled rock layers formed a

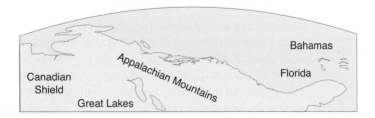

high mountain belt that was slowly eroded away by streams. Subsequently, the margin of North America formed as it rifted away from Africa. In fact, all of the structural features of our planet are the result of a simple system of moving lithospheric plates. Movement in this plate tectonic system is driven by the loss of internal heat energy.

The concept of a natural system, as developed for the study of geology, provides a framework for understanding how each part of Earth works and why it is constantly changing. Geologic systems are governed by natural laws that provide the keys to understanding Earth and all of its varied landscapes and processes.

In this chapter, we consider the fundamentals of natural systems. We also explore the basics of the hydrologic system and the tectonic system as the ultimate causes of geologic change.

Major Concepts

1. A natural system is a group of interdependent components that interact to form a unified whole and are under the influence of related forces. The materials in a system change in an effort to reach and maintain equilibrium.
2. Earth's system of moving water, the hydrologic system, involves the movement of water—in rivers, as groundwater, in glaciers, in oceans, and as water vapor in the atmosphere. As water moves, it erodes, transports, and deposits sediment, creating distinctive landforms and rock bodies.
3. Radiation from the Sun is the source of energy for Earth's hydrologic system.
4. A system of moving lithospheric plates—the plate tectonic system—explains Earth's major structural features. It operates from Earth's internal heat.
5. Where plates move apart, hot material from the mantle wells up to fill the void and creates new lithosphere. The major features formed where plates spread apart are continental rifts, oceanic ridges, and new ocean basins.
6. Where plates converge, one slides beneath the other and plunges down into the mantle. The major features formed at convergent plate margins are folded mountain belts, volcanic arcs, and deep-sea trenches.
7. Where plates slip horizontally past one another, transform plate boundaries develop on long, straight faults. Shallow earthquakes are common.
8. Far from plate margins, plumes of less-dense mantle material rise to shallow levels, feeding within-plate volcanoes and producing minor flexures of the lithosphere.
9. Earth's crust floats on the denser mantle beneath. The crust rises and sinks in attempts to maintain isostatic equilibrium.

Geologic Systems

A system is a group of interdependent materials that interact with energy to form a unified whole. Most geologic systems are open; that is, they can exchange matter and energy across their boundaries.

The world is a unified whole. Nothing in or on it exists as an isolated entity. Everything is interconnected. We may know myriad details about the many separate items found on or inside Earth, but most of us do not understand how the pieces are interrelated and fit together. Without some concept of how the world functions as a whole, we easily miss important relationships between seemingly isolated phenomena, such as the critical connections among rainfall, temperature, and landslides. To understand Earth and how it functions, we need a model or framework, a plan or map that shows how things are interrelated and how things operate. Such a framework is provided by the concept of the **system**.

What is a geologic system?

There are many different kinds of systems. You are undoubtedly familiar with many natural and artificial systems. An engineer may think of a system as a group of interacting devices that work together to accomplish a specific task. In your home, there is an electrical system, a plumbing system, and a heating system. Each functions as an independent unit in some way. Each transfers material or energy from one place to another, and each has a driving force that makes the system operate. In another example, a biologist conceives of a similar kind of system, but it is one composed of separate organs made of living tissues that work together. The circulatory system is composed of the heart, blood vessels, and other organs that together move blood through the body.

In the physical sciences, we speak of systems in very general terms; a system is that part of space in which we are interested. The space may contain various materials acted on by energy in different ways. By defining a system, we identify the extent of

(A) A closed system, such as a cooling lava flow, exchanges only radiant heat. Here, heat from the lava is lost to the atmosphere.

(B) Open systems, such as a river, exchange energy and matter. In a river, water and sediment are collected from the drainage area and flow through the system to the sea. Most geologic systems are open systems.

Figure 2.1 **Natural systems** may be closed or open.

the material being considered and the energy involved so that we can more clearly understand any changes.

In each of these cases, a system is composed of individual items or components that work together to make a unified whole. In accomplishing specific tasks, material and energy move about and change from one form to another. Such a system is **dynamic**, in motion, rather than static or unchanging.

A *natural system* is a bit more complicated than a typical engineering system. For example, a geologic system may have real boundaries, such as the top and bottom of a flowing stream of water or the walls of a body of molten rock (**Figure 2.1**). Or it may have arbitrary boundaries defined for the specific purpose of study. Everything outside of the system's boundaries is the surroundings or environment and is not considered part of the system.

Earth's systems obviously are so broad in scope, and cover so many phenomena in the natural world, that we should be careful about how we use the term. Two types of systems are important in geology: (1) a **closed system** exchanges only heat (no matter); and (2) an **open system** exchanges both heat and matter with its surroundings. In a closed system, such as a cooling lava flow, heat is lost, but new material is neither added nor lost (Figure 2.1A). However, most geologic systems are open systems, in which matter and energy freely flow across the system's boundaries. A river system, for example (Figure 2.1B), gains water from springs, snowmelt, and rainfall as it flows toward the ocean.

Earth itself is a system. It is a sphere of matter with distinct boundaries. Earth has been an essentially closed system since the end of the heavy meteorite bombardment some 4 billion years ago. Since then, no significant new material has entered the system (except meteorites and space dust), and, just as important, significant quantities have not left the system. Since the planet formed, however, its materials have experienced tremendous change. Solar energy enters this nearly closed system and causes matter (air and water) to move and flow in distinctive patterns. Heat energy from within Earth also causes motion resulting in earthquakes, volcanism, and shifting continents. Thus, a space photograph of our planetary home seen as a whole is a powerful image of a natural system—an image that imparts a sense of the oneness in a natural system.

On a much smaller scale, a river and all its branching tributaries is a natural system. The floor of the stream and the upper surface of the flowing water form some of its boundaries. Matter enters this system from the atmosphere as rain, snow, or groundwater and then flows through the river channel and leaves the system

What is a dynamic system? Can you give an example?

Why is Planet Earth considered to be a natural system?

as it enters the sea. As long as rain falls, the system will be supplied with matter, gravitational potential energy, and kinetic energy. The ultimate energy source for a river system is energy from the Sun. Its energy heats water in the ocean, evaporating it and lifting it into the atmosphere, and transporting it to the continents. The force of gravity causes the water to flow downslope to the sea.

Most other geologic systems are complex open systems like river systems. One type of complexity results from subsystems. For example, a river system is only part of the much larger hydrologic system that includes all possible paths of worldwide water movement. Atmospheric circulation of water vapor is another important subsystem of the hydrologic system. Oceanic currents are another; glaciers and groundwater are others. Each is a subset of the overall circulation of water and energy at Earth's surface.

Direction of Change in Geologic Systems

In all natural systems change occurs in the direction necessary to establish and maintain equilibrium—a condition of the lowest possible energy.

The very essence of Earth's geologic systems is the flow of energy and the movement of matter. As a result, materials on and in Earth are changed or rearranged. Yet, this change does not occur at random. It occurs in a definite, predictable way. By carefully examining a system, we can see how one component is connected to another in an invisible web. The individual threads of this network are so tightly interdependent that a change in any component, even a small change, causes change in the rest of the system. Predicting and understanding these changes is an important reason for using the system approach.

What determines the direction of change in a dynamic geologic system? For example, does water flow downhill or uphill? Does hot air rise or sink? Although the answers to these questions seem self-evident, they seem simple only because of your experience with natural systems driven by gravity. You have thousands of experiences each day that reveal many of the principles of gravity. These experiences allow you to predict what will happen in many different situations.

However, because you lack experience with other natural systems, there are many questions regarding direction of change that are more difficult for you to answer. For example, under what conditions of temperature or pressure does one mineral convert to a different mineral? At what temperature does rock melt? Or water freeze? Why does heat flow from one rock to another or from one region to another? When will solid rock break to cause an earthquake? In short, how can we predict the direction of change in any natural system?

What is equilibrium in a natural system?

Most of these questions can be answered, or at least better understood, because of one very simple principle. Changes in natural systems have a universal tendency to move toward a state of **equilibrium**—a condition of the lowest possible energy. This pattern holds for the landscape, earthquakes, volcanoes, flowing water, and many other geologic phenomena. This governing principle has been clearly established through painstaking experimentation by thousands of scientists working over several centuries. Thus, if we can deduce which of several possible conditions is lowest in energy, we can predict the direction of change in a natural system.

Another way to think about equilibrium is to consider it a condition in which the net result of the forces acting on a system is zero. It is a state of no permanent change in any characteristic of the system. Systems not in equilibrium tend to change in a direction to reach equilibrium. To better understand this idea, think of two boulders on a hillside. One sits high on the side of the hill and has much gravitational potential energy. Another sits on the valley floor and has very little gravitational potential energy. Which boulder is more likely to change its position? Obviously, only a small perturbation could send the first boulder rolling down the hillside. But any force exerted on the second boulder would cause only a modest and temporary change in position, and it would then roll back to its original position. The second boulder has low gravitational potential energy and is at an equilibrium position.

Now, imagine a third boulder that sits in a slight depression on the hillside. It is in a **metastable** position. A very small force would be insufficient to change its position permanently, but a larger force would push it over the brink and allow it to crash down the hill to a stable position.

A hot lava flow cools for similar reasons. It loses heat energy to its surroundings in order to reach equilibrium with its environment. If a change upsets this equilibrium, the system will naturally change in such a direction as to reestablish equilibrium under the new conditions.

In all such transformations, some energy is lost to the environment, generally as heat. Often, the lost heat energy is no longer available to cause change. A fundamental natural law holds that any system tends to "run down," meaning that it gradually loses energy of the sort that can cause change.

If you look carefully at a geologic system, you should be able to identify its equilibrium state. For example, what would be the equilibrium landscape formed by a river system? The state that would provide the very least gravitational potential agency would be one of flatness. Thus, a perfectly flat landscape with no hills, ridges, or valleys would be the equilibrium landscape. Of course, that state may never be perfectly achieved, because of the inability of erosion to keep up with other changes imposed on the river system.

In summary, the total energy of a system must decrease for a spontaneous change to occur. The change will proceed until equilibrium is attained and the energy is at

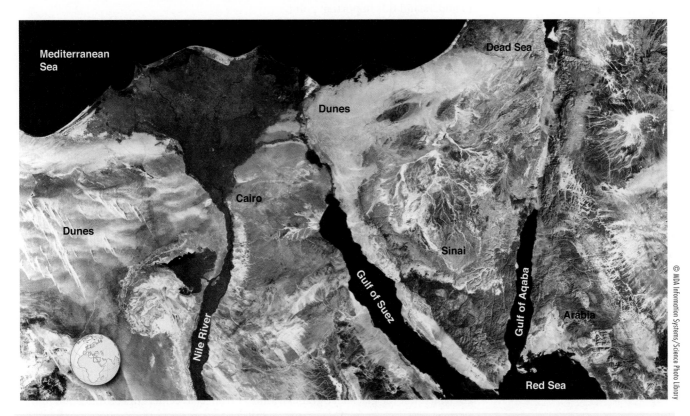

Figure 2.2 Earth's geologic systems are evident in this space photograph. The effects of the hydrologic system are revealed by river systems, even in this desert region of the Middle East. The Sinai Peninsula is etched by delicate networks of stream valleys, which disappear into the sandy desert along the shores of the Mediterranean Sea. Elsewhere, stream erosion has etched out the fractures in the rocks, but the drainage patterns are masked by wind-blown sand. The Nile River is flanked by farmlands (red). The Nile carries a tremendous volume of sediment to the sea, where it is deposited in a huge delta (dark red because of vegetation). The Nile River and its delta are a dramatic expression of the hydrologic system as running water erodes the highlands of central Africa and transports the sediment to the sea. An additional expression of the hydrologic system is the wave action along the delta front that reworks the sediment brought to the sea by the Nile and redeposits it as beaches and barrier bars. In this arid region, linear windblown sand dunes have developed on either side of the Nile Delta. The tectonic system is expressed by the rift of the Red Sea and the fracture system extending northward up the Gulf of Aqaba and into the Dead Sea–Jordan River valley. The Arabian Peninsula is moving to the northeast and, as it splits and moves away from Africa, a new ocean basin (the Red Sea) is born. The movement of tectonic plates is a clear expression of the fundamental dynamics of Earth's interior.

a minimum. The most stable state is always the one with the lowest energy. In other words, all materials attempt to achieve a balance with the chemical and physical forces exerted upon them, and they will change to arrive eventually at equilibrium. This effort results in progressive changes in any planetary material that is exposed to an environment different from that in which it formed. Although this equilibrium state is the preferred state of all systems, there are many intermediate or metastable states, adding to the complex problem of understanding Earth's dynamic systems.

Systems, Equilibrium, and Geology

The dual concepts of systems and equilibrium, as developed for the study of geology, provide a framework for understanding how each part of Earth works and why it is constantly changing. Order can be seen in all scales of time and space. Nothing is random. Everything, from a grain of sand on a beach to a lake, mountain range, or canyon, is there because it was formed in a systematic way by an organized interaction of matter and energy. Dynamic geologic systems are governed by natural laws, which provide the keys to understanding Earth and all its landscapes and processes.

The major geologic systems are the hydrologic system and the tectonic system. Perhaps no other place on Earth illustrates the operation of these two grand systems as well as the Middle East (**Figure 2.2**). This satellite photograph may at first appear to be a chaotic jumble of colors and textures, but careful examination shows that every feature is a product of these two dynamic systems. Carefully read the figure caption to understand this important point.

The Hydrologic System

> The hydrologic system is the complex cycle through which water moves from the oceans, to the atmosphere, over the land, and back to the oceans again. Water in the hydrologic system—moving as surface runoff, groundwater, glaciers, waves, and currents—erodes, transports, and deposits surface rock material.

The complex motion of Earth's surface water—the **hydrologic system**—operates on a global scale. It unites all possible paths of water into a single, grand system of motion. The term hydrologic is rooted in the Greek term *hydor* for "water." The basic elements of the system can be seen from space (**Figure 2.3**) and are diagrammed in **Figure 2.4**. The system operates as energy from the Sun heats water in the oceans, the principal reservoir for Earth's water. As it is heated, the water evaporates. Most of the water vapor condenses and returns directly to the oceans as rain. Atmospheric circulation carries the rest over the continents, where it is precipitated as rain, sleet, hail, or snow.

Water that falls on the land can take a variety of paths. The greatest quantity returns to the atmosphere by evaporation, but the most visible return is to the oceans by surface runoff in river systems, which funnel water back to the oceans. Some water also seeps into the ground and moves slowly through pore spaces in the soil and rocks, where it is available to plants. Part of the water is used by the plants, which "exhale" it into the atmosphere, but much of it slowly seeps into streams and lakes. In polar regions or in high mountains, water can be temporarily trapped on a continent as glacial ice, but the glacial ice gradually moves from cold centers of accumulation into warmer areas, where melting occurs and the water returns to the oceans as surface runoff.

In short, water in the hydrologic system is constantly moving as vapor, rain, snow, surface runoff, groundwater, and glaciers, or even in ocean waves and currents. As it moves across the surface, it erodes and transports rock material and then deposits it as deltas, beaches, and other accumulations of sediment. Consequently, the surface materials, as well as the water, are in motion—motion that results in a continuously changing landscape.

One of the best ways to gain an accurate conception of the magnitude of the hydrologic system is to study space photography. These photographs provide a view of the system in operation on a global scale. A traveler arriving from space would see

Figure 2.3 The hydrosphere, the thin film of water that makes Planet Earth unique, is essential for life. Earth is just the right distance from the Sun for water to exist as a liquid, solid, and gas. If it were closer to the Sun, our oceans would evaporate; if it were farther from the Sun, the oceans would freeze solid. Drawing energy from the Sun, it moves in great cycles from the oceans to the atmosphere and over the landscape in river systems, ultimately returning to the oceans.

that the surface of Earth is predominantly water (Figure 2.3). The movement of water from the oceans to the atmosphere is expressed in the flow patterns of the clouds. The atmosphere and moving cloud cover are among the most distinctive features of Earth as it is viewed from space. The great river systems stand out markedly when compared with the surfaces of the Moon and Mercury, where impact craters dominate stark landscapes. Without hydrologic systems, the surfaces of these planets have remained largely unmodified for billions of years.

In addition, the swirling, moisture-laden clouds carry an enormous amount of energy. For instance, the kinetic energy produced by a hurricane amounts to roughly 100 billion kilowatt-hours per day. That is much more than the energy used by all of the people of the world in one day.

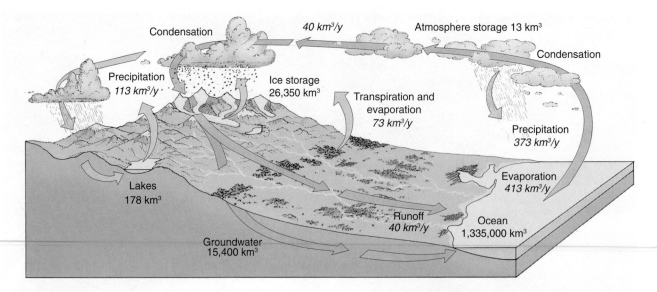

Figure 2.4 **The circulation of water** in the hydrologic system is powered by solar energy, as shown in this block diagram. Water evaporates from the oceans, circulates with the atmosphere, and eventually condenses as rain or snow where it cools. Much of the water that falls on the continents returns to the oceans by surface runoff and groundwater seepage. Variations in the major flow patterns of this system include the temporary storage of water in lakes and glaciers. Within this major system are many smaller cycles, or shortcuts, such as evaporation from lakes and transpiration from plants.

How much water is in motion in the hydrologic system?

Another way to grasp the magnitude of the hydrologic system is to consider the volume of water involved. From measurements of rainfall and stream discharge, together with measurements of heat and energy transfer in bodies of water, scientists have calculated that if the hydrologic system were interrupted and water did not return to the oceans by precipitation and surface runoff, sea level would drop 1 m per year. All of the ocean basins would be completely dry within 4000 years. The "recent" ice ages demonstrate this point clearly: The hydrologic system was partly interrupted, as much of the water that fell on the Northern Hemisphere froze and accumulated to form huge continental glaciers, preventing the water from flowing immediately back to the oceans as surface runoff. Consequently, sea level dropped more than 100 m during the most recent ice age.

As a final observation, consider what the hydrologic system has produced. It eroded millions of tons of rock from the Grand Canyon, carved the fantastic peaks of the Himalaya mountains, and deposited the vast Mississippi delta. The hydrologic system also created the small streams and valleys around your hometown. There is little, if any, surface on Earth that is not affected in some way by the work of the hydrologic system. When you go for a walk in the country, it is very likely you will be walking over a surface formed by running water, a young surface (geologically speaking) that is still developing.

Major Subsystems of the Hydrologic System

The enormous energy of the hydrologic system is apparent in each subsystem by which water moves—rivers, glaciers, groundwater, oceans, and wind. All erode, transport, and deposit material and create new landforms in the process. We will explore the details of each major geologic system in subsequent chapters, but for now let us examine some results of the water movement over Earth.

Atmosphere–Ocean System Earth's oceans are vast reservoirs of liquid water that together with the gases in the atmosphere create the **climate system**. Circulation in these envelopes of fluid is driven by heat from the Sun. The uneven heating of Earth's surface causes the atmosphere to convect, winds to blow, causes evaporation of huge quantities of water vapor into the atmosphere, and drives ocean currents. In addition, variations in this convection system set up a regular pattern for the distribution of precipitation and temperature around the entire globe. Thus, the climate is controlled by the materials and energy in this system. In turn, the climate controls how the hydrologic system operates in a local area.

River Systems   Most water precipitated onto the land returns directly to the oceans through surface drainage systems—**river systems**. The amount of water in Earth's rivers appears vast, but in fact it is startlingly small; it is only about 0.0001% of the total water on Earth, or 0.005% of the water not in the oceans. Water flows through rivers very rapidly, at an average rate of 3 m per second. At this rate, water can travel through the entire length of a long river in a few weeks. This means that, although the volume of the water in rivers at any given time is small, the total volume passing through river systems in a given period can be enormous. As a result, most of the landscape is dominated by features formed by running water.

From viewpoints on the ground, we cannot appreciate the prevalence of stream channels on the surface of Earth. From space, however, we readily see that stream valleys are the most abundant landforms on the continents. In arid regions, where vegetation and soil cover do not obscure our view, the intricate network of stream valleys is most impressive **(Figure 2.5)**. Most of the surface of every continent is somehow related to the slope of a stream valley, which collects and funnels surface runoff toward the ocean.

Another important aspect of a river system is that it provides the fluid medium that transports huge amounts of sand, silt, and mud to the oceans. These sediments form the great deltas of the world, which are records of the amount of material washed off the continents by rivers. The Nile Delta is a classic example (Figure 2.2). The Nile

What are the major components of the hydrologic system and how do they operate?

Courtesy of NASA

Figure 2.5 **River systems** are clear records of how the hydrologic system sculpts the land. They testify to the magnitude of this vast interconnected system of moving water, for few areas on land are untouched by stream erosion. In this photograph of a desert region, details of the delicate network of tributaries are clearly shown. On the Moon, Mercury, and Mars, craters dominate the landscape, but on the continents of Earth, stream valleys are the most abundant landforms.

Figure 2.6 Valley glaciers, such as these in Alaska, occur where more snow accumulates each year than is melted in the summer. Over many years, this cycle allows accumulating snow to build the glaciers. Valley glaciers originate in the snowfields of high mountain ranges and slowly flow as large tongues of ice down preexisting stream valleys. The moving ice is a powerful agent of erosion and modifies the valleys in which it flows. The dark lines on the glaciers are rock debris derived from the valley walls.

Courtesy of U.S. Department of Agriculture

River is confined to a single channel far upstream from Cairo. It then splits into a series of branching channels, from which the sediment carried by the river is eventually deposited as new land in the Mediterranean Sea. The main channels slowly shift their courses back and forth across the delta, and the older extensions of the delta are eroded back by ocean waves and currents.

Why are glaciers considered to be part of the hydrologic system?

Glacial Systems   In cold climates, precipitation falls as snow, most of which remains frozen and does not return immediately to the ocean as surface runoff. If more snow falls each year than melts during the summer months, huge bodies of ice build up to form glaciers (**Figure 2.6**). Large valley glaciers originate from snowfall in high mountains and slowly flow down valleys as rivers of ice. **Glacier systems** greatly modify the normal hydrologic system because the water that falls upon the land does not return immediately to the ocean as surface runoff. It is not until the glaciers melt at their lower end that water flows back to the sea, seeps into the ground, or evaporates.

At present, the continent of Antarctica is almost entirely covered with a continental glacier, a sheet of ice from 2.0 to 2.5 km thick. It covers an area of 13 million km^2—an area larger than the United States and Mexico combined. An ice sheet similar to that now on Antarctica covered a large part of North America and Europe during the last ice age, and it retreated only within the last 18,000 years. As the ice moved, it modified the landscape by creating many lakes and other landforms in Canada and the northern United States, including the Great Lakes.

Water in the form of ice constitutes about 80% of the water not in the oceans, or about 2% of Earth's total water—far more than is in our streams and rivers. Water in glaciers moves very slowly and may remain in a glacier for thousands or even millions of years. Present estimates suggest that water resides in a glacier for about 10,000 years on average.

Groundwater Systems   Another segment of the hydrologic system is the **groundwater system**—the water that seeps into the ground and moves slowly through the pore spaces in soil and rocks. Surprisingly, about 20% of the water not in the oceans occurs as groundwater. As it slowly moves, groundwater dissolves soluble rocks (such as limestone) and creates caverns and caves that can enlarge and collapse to form surface depressions called sinkholes. This type of dissolution-generated landform is common

Figure 2.7 **Groundwater** is a largely invisible part of the hydrologic system because it occupies small pore spaces in the soil and rocks beneath the surface. It can, however, dissolve soluble rocks, such as limestone, to form complex networks of caves and subterranean passageways. As the caverns enlarge, their roofs may collapse, forming circular depressions called sinkholes. The hundreds of lakes shown in this false-color photograph of the area west of Cape Canaveral, Florida, occupy sinkholes and testify to the effectiveness of groundwater as a geologic agent.

in Kentucky, Florida, Indiana, and western Texas and is easily recognized from the air **(Figure 2.7)**. Sinkholes commonly create a pockmarked surface somewhat resembling the cratered surface of the Moon. They may also become filled with water and form a series of circular lakes.

Shoreline Systems   The hydrologic system also operates in **shoreline systems** along the shores of all continents, islands, and inland lakes through the unceasing work of waves. The oceans and lakes are bodies of mobile water subject to a variety of movements—waves, tides, and currents. All of these movements erode the coast and transport vast quantities of sediment (for example, the Nile Delta in Figure 2.2). The effects of shoreline processes are seen in wave-cut cliffs, shoreline terraces, deltas, beaches, bars, and lagoons.

Eolian (Wind) Systems   The hydrologic system also operates in the arid regions of the world. In many deserts, river valleys are still the dominant landform. There is no completely dry spot on Earth. Even in the most arid regions some rain falls, and climate patterns change over the years. River valleys can be obliterated, however, by dunes of wind-blown sand that cover parts of the desert landscape (Figure 2.2).

The circulation of the atmosphere forms the **eolian system**. Wind can transport enormous quantities of loose sand and dust, leaving a distinctive record of the wind's activity. In the broadest sense, the wind itself is part of the hydrologic system, a moving fluid on the planet's surface.

The Tectonic System

The tectonic system involves the movement of the lithosphere, which is broken into a mosaic of separate plates. These plates move independently, separating, colliding, and sliding past one another. The margins of the plates are sites of considerable geologic activity, such as seafloor spreading, continental rifting, mountain building, volcanism, and earthquakes.

Geologists have long recognized that Earth has its own source of internal energy. It is repeatedly manifested by earthquakes, volcanic activity, and folded mountain belts. But it was not until the middle 1960s that a unifying theory developed to explain Earth's dynamics. This theory, known as **plate tectonics**, provides a master plan of Earth's internal dynamics. The term tectonics, like the related word architecture, comes from the Greek *tektonikos* and refers to building or construction. In geology, tectonics is the study of the formation and deformation of Earth's crust that results in large-scale features.

Evidence for this revolutionary theory of lithospheric movement comes from many sources. It includes data on the structure, topography, and magnetic patterns of the ocean floor; the locations of earthquakes; the patterns of heat flow in the crust; the locations of volcanic activity; the structure and geographic fit of the continents; and the nature and history of mountain belts.

The basic elements of the tectonic system are simple and can be easily understood by carefully studying **Figure 2.8**. The lithosphere, which includes Earth's crust and part of the upper mantle, is rigid, but the underlying asthenosphere slowly flows. A fundamental tenet of plate tectonics is that the segments, or **plates**, of the rigid

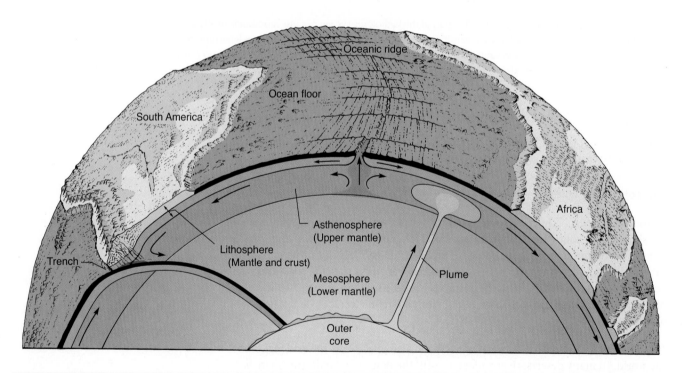

Figure 2.8 **The tectonic system** is powered by Earth's internal heat. The asthenosphere is more plastic than either the overlying lithosphere or the underlying lower mantle. Above the plastic asthenosphere, relatively cool and rigid lithospheric plates split and move apart as single mechanical units (along the ocean ridges). As this happens, molten rock from the asthenosphere wells up to fill the void between the lithospheric plates and thus creates new lithosphere. Very slow convection occurs in the mantle. Some plates contain blocks of thick, lower-density continental crust, which cannot sink into the denser mantle. As a result, where a plate carrying continental crust collides with another plate, the continental margins are deformed into mountain ranges. Plate margins are the most active areas on Earth—the sites of the most intense volcanism, seismic activity, and crustal deformation. Locally, convection in the deep mantle creates rising mantle plumes.

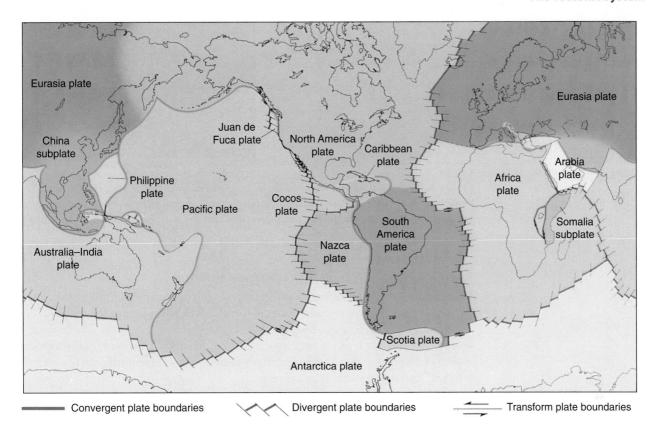

| Convergent plate boundaries | Divergent plate boundaries | Transform plate boundaries |

Figure 2.9 A mosaic of plates forms Earth's lithosphere, or outer shell. The plates are rigid, and each moves as a single unit. There are three types of plate boundaries: (1) the axis of the oceanic ridge, where the plates are diverging and new oceanic crust is generated (red lines); (2) transform faults, where the plates slide past each other (the short lines slicing across the divergent boundaries); and (3) subduction zones, where the plates are converging and one descends into the asthenosphere (blue lines).

lithosphere are in constant motion relative to one another and carry the lighter continents with them.

Plates of oceanic lithosphere form as hot mantle material rises along mid-oceanic ridges; they are consumed in **subduction** zones, where one converging plate plunges downward into the hotter mantle below (Figure 2.8). The descent of these plates is marked by deep-sea trenches that border island arcs and some continents. Where plates slide by one another, large fractures form. The movement and collision of plates accounts for most of Earth's earthquakes, volcanoes, and folded mountain belts, as well as for the drift of its continents.

From the standpoint of Earth's dynamics, the boundaries of plates are where the action is. As seen in **Figure 2.9**, plate boundaries do not necessarily coincide with continental boundaries, although some do. There are seven very large plates and a dozen or more small plates (not all of which are shown in Figure 2.9). Each plate is as much as a few hundred kilometers thick. Plates slide over the more mobile asthenosphere below, generally at rates between 1 and 10 cm per year. Because the plates are internally quite rigid, they become most deformed along their edges.

The basic source of energy for tectonic movement is believed to be Earth's internal heat, which is transferred by **convection**. In a simple model of Earth's convecting interior, hot mantle material rises to the lithosphere's base, where it then moves laterally, cools, and eventually descends to become reheated, continuing the cycle. A familiar example of convection can be seen while heating a pot of soup **(Figure 2.10)**. Heat applied to the base of the pot warms the soup at the bottom, which therefore expands and becomes less dense. This warm fluid rises to the top and is forced to move laterally while it cools. Consequently, it becomes denser and sinks, setting up a continuing cycle of convection.

What is the source of energy for the tectonic system?

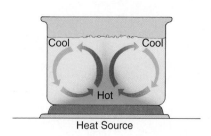 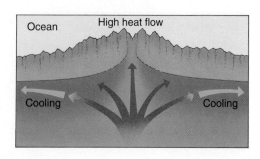

Figure 2.10 **Convection in the mantle** can be compared to convection in a pot of soup. Heat from below causes the material to expand and thus become less dense. The warm material rises by convection and spreads laterally. It then cools, and thus becomes denser, and sinks. It is reheated as it descends, and the cycle is repeated.

Major Subsystems of the Tectonic System

Many features of the ocean basins and continents can be nicely explained by the plate tectonic system. We will consider many of them in detail elsewhere in the text, but let us look briefly at the major features of the planet and how they fit into the tectonic system. The different types of plate boundaries are, in effect, subsystems of the tectonic system. Each creates specific geologic phenomena. We have illustrated each boundary type with an example from the continents.

Divergent Plate Boundaries   The plates move apart at **divergent plate boundaries**, which coincide with mid-oceanic ridges (Figure 2.9). Hot molten material from the deeper mantle wells up to fill the void. Some of this material erupts on the seafloor as lava. The molten rock solidifies and forms new lithosphere. The mid-oceanic ridges stand high because their material is hot and, therefore, less dense than the colder adjacent oceanic crust.

The most intense volcanism on Earth occurs at divergent plate boundaries, but it is largely concealed below sea level. When oceanic earthquake locations are plotted on a map, they outline with dramatic clarity the divergent plate boundaries (**Figure 2.11**). Most of these are shallow earthquakes, quite unlike those found where plates converge.

Most divergent boundaries occur on the seafloor (**Figure 2.12**), but continental rifts also develop where divergent boundaries form on the continents. Such a continental rift eventually creates a new ocean basin. The great rift of the Red Sea (Figure 2.2) displays many features of a continental rift. The Red Sea is an extension of the mid-oceanic ridge of the Indian Ocean, which splits the Sinai and Arabian peninsulas from Africa. Take the time to locate the area shown in this remarkable photograph (Figure 2.2) on the topographic map on the inside covers of this book. The structure of the area is dominated by the long, linear fault valley that forms the north end of the Red Sea and Gulf of Suez. Note the sharp contrast where faults have juxtaposed young, light-colored sediments against the ancient shields as this region is slowly ripping asunder. New seafloor is forming on the floor of the Red Sea. This rift expresses dramatically the tensional stresses in the lithosphere and the way these stresses affect Earth's surface.

Transform Plate Boundaries   The oceanic ridges are commonly broken and offset along lines perpendicular to the ridges. These offsets are large faults expressed by their own high ridges and deep valleys. **Transform plate boundaries** occur where plates horizontally slide past one another (**Figure 2.13**). Shallow earthquakes are common along all transform boundaries (Figure 2.11), but volcanic eruptions are uncommon.

Most transform plate boundaries are on the seafloor, but the best-known example of this type of fault on a continent is the great San Andreas Fault system in California (Figure 2.13). The fault zone is marked by sharp linear landforms, such as straight and narrow valleys, straight and narrow ridges, and offset stream valleys. The San Andreas Fault system is an active boundary between the Pacific plate to the west and the North America plate to the east. The Pacific plate is moving at about 6 cm per

What are the major subsystems of the tectonic system and how do they operate?

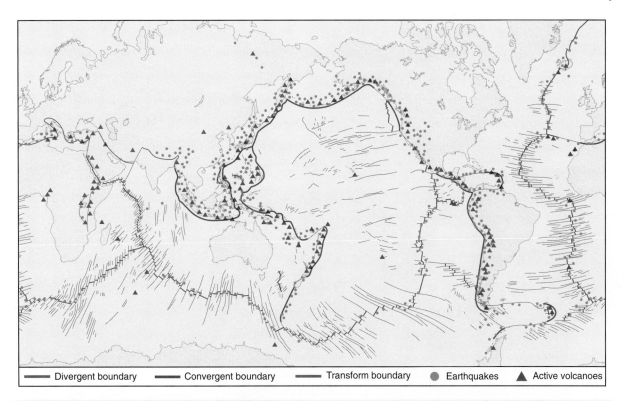

			●	▲
Divergent boundary	Convergent boundary	Transform boundary	Earthquakes	Active volcanoes

Figure 2.11 Earthquakes and active volcanoes outline plate margins with remarkable fidelity. At divergent plate boundaries, shallow earthquakes, submarine volcanic eruptions, and tensional fractures occur. Transform boundaries have shallow earthquakes but generally lack active volcanoes. Along convergent margins, there are deep earthquakes, volcanic eruptions, trenches on the seafloor, and folded mountain belts. Isolated areas of volcanism and earthquakes reveal the locations of active mantle plumes.

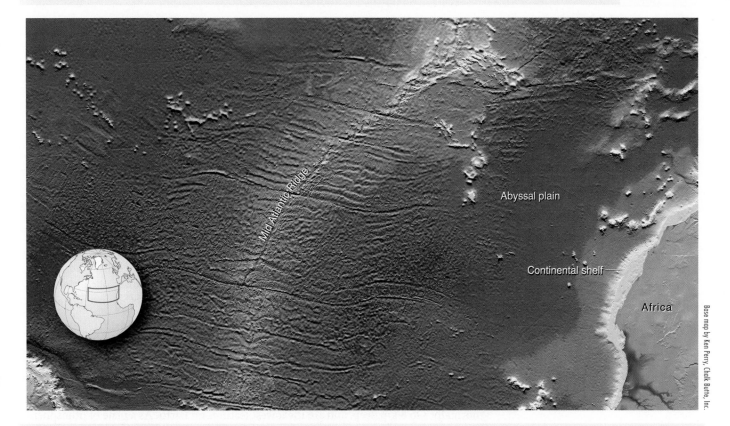

Base map by Ken Perry, Chalk Butte, Inc.

Figure 2.12 The Mid-Atlantic Ridge is a divergent plate boundary and marks the spot where new lithosphere is forming and where two plates are separating. The North America plate is slowly moving west and Africa on the Eurasia plate is moving east. Earthquakes and volcanoes are concentrated along the crest of the ridge. Transform faults cut the ridge and offset it.

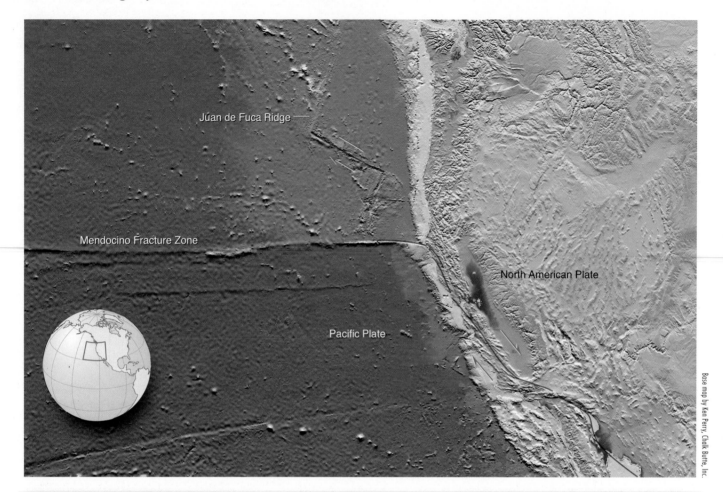

Juan de Fuca Ridge

Mendocino Fracture Zone

North American Plate

Pacific Plate

Base map by Ken Perry, Chalk Butte, Inc.

Figure 2.13 The San Andreas Fault system in California is part of a long transform plate boundary that separates the North America plate from the Pacific plate. It connects a divergent boundary in the Gulf of California with the Mendocino transform fault and the Juan de Fuca ridge. At least a dozen major fault systems can be seen as linear mountain trends. Movement along the San Andreas Fault is horizontal; that is, one block of Earth's slides laterally past another.

year, relative to the North America plate. As stress builds between the plates, the rock bodies deform until they break. This sudden release along the fault causes earthquakes like those so common in California. Another transform boundary cuts the Asian continent from the Gulf of Aqaba to the Dead Sea and creates a valley obvious from space (Figure 2.2).

Convergent Plate Boundaries   Plates move toward one another along **convergent plate boundaries**. Along such plate margins, geologic activity is far more varied and complicated than at transform plate boundaries (**Figure 2.14**). Intense compression ultimately rumples the lithosphere and builds high folded mountain belts. Preexisting rocks become altered by heat and pressure. The net result is the growth of continents. Where two plates converge, one tips down and slides beneath the other in a process known as *subduction*.

It is clear that earthquakes and volcanoes dramatically outline convergent plate margins (Figure 2.11). The simplest form of convergence involves two plates with oceanic crust. Such subduction zones in the western and northern Pacific region lie along the volcanic islands of Tonga, the Marianas, and the Aleutians. Trenches form where the downgoing plate plunges into the mantle. These are long, narrow troughs, normally 5 to 8 km deep, and are the lowest features on Earth. As a plate of lithosphere slips into the mantle, it becomes heated and dehydrated. Some rock material melts, becomes less dense, and rises, and some erupts to form a string of volcanic islands called an island arc.

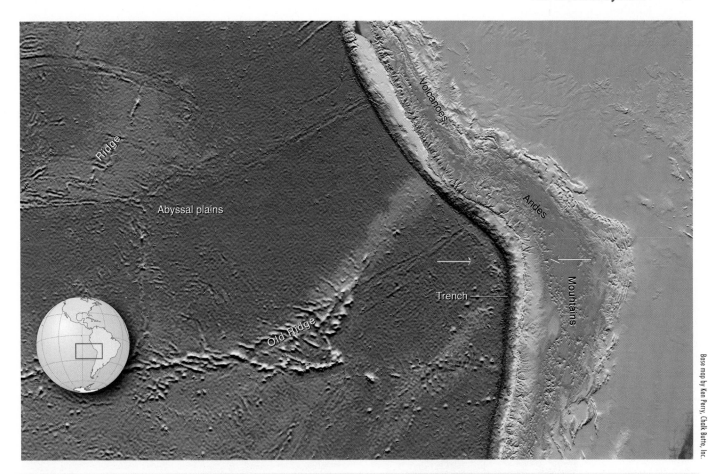

Base map by Ken Perry, Chalk Butte, Inc.

Figure 2.14 The Andes Mountains were formed by the subduction of the Nazca plate beneath South America at a convergent plate margin. Layers of sedimentary rock, which were originally horizontal, have been elevated and compressed into folds that were subsequently eroded. The resistant layers appear as ridges in the eastern Andes. Folded mountain belts such as the Andes are one of the most significant results of converging plates, but if you look carefully you can also see the relatively smooth volcanic plains and isolated volcanic cones that show the role played by volcanism at convergent plate margins.

If the oceanic plate dives beneath a continent, the molten rock may form a chain of volcanoes on the continental margin; the Cascades of California-Oregon-Washington are an example. The remarkable series of deep-sea trenches and associated volcanic arcs make the "ring of fire" that almost surrounds the Pacific Ocean (Figure 2.11).

As each subducting plate grinds its way downward, earthquakes are produced. The deepest of all earthquakes, almost 700 km deep, occur at convergent plate boundaries. Plate tectonics can thus readily explain why the Andes mountains of South America are tormented by repeated volcanic eruptions and earthquakes (Figure 2.14). They are forming where two tectonic plates converge. The same is true for the western coasts of Central America. It is equally clear that the earthquakes and volcanic eruptions in the Mediterranean area occur at a convergent plate margin.

Where moving plates converge, the rocks in the crust may also become deformed. The crust in continents and in island arcs is buoyant (it is less dense than oceanic crust) and resists subduction back into the dense mantle. Consequently, this kind of crust becomes intensely compressed and folded at some convergent plate margins. The structures of the Andes Mountains (Figure 2.14) of South America vividly express this type of deformation. The complex system of ridges and valleys in the eastern Andes is produced by folded sedimentary rock layers deformed by the collision of two plates. The folded layers now appear like wrinkles in a rug. The Appalachians were formed in a similar manner.

A younger mountain belt that extends from Alaska through the Rockies and Central America and into the Andes of South America was produced by the encounter of the American plates with the Pacific, Cocos, and Nazca plates. This is a geologically young mountain system, with many parts still being deformed as the plates continue to move.

Within-Plate Tectonics and Mantle Plumes   Within the moving plates, the continental and oceanic crust experience little tectonic or volcanic activity as they move away from mid-oceanic ridges. However, **plumes** of hot rock rising from the deep mantle (Figure 2.8) may create isolated volcanoes and gently warp the interior of a plate. An excellent example is the Hawaiian Island chain in the Pacific Ocean (**Figure 2.15**). The huge volcanoes and geysers of Yellowstone National Park in western North America may also overlie a mantle plume. Earthquakes related to the volcanoes in these areas are also common (Figure 2.11), but large deep earthquakes are rarely felt in such within-plate regions.

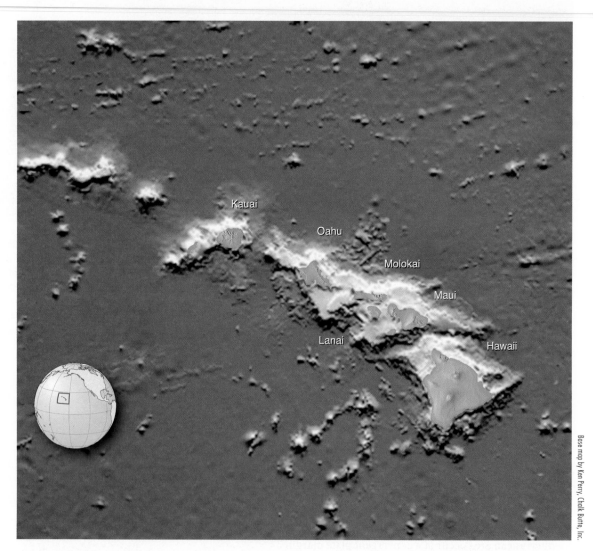

Base map by Ken Perry, Chalk Butte, Inc.

Figure 2.15 The Hawaiian Islands formed far from any plate boundary and are thought to lie above a plume of hot material rising through the mantle. As the lithosphere slowly moves northeast, it carries the older volcanoes away from the hot spot. Volcanoes are still active on the large southern island of Hawaii, but not on the more eroded Maui, and other islands to the northeast where the action of the hydrologic system is dominant.

STATE OF THE ART   Heat Flow Measurements and Plate Tectonics

Earth is a giant heat engine. Not only does it create its own heat, but its tectonic system is driven by the flow of this heat. The outer core is molten and convects vigorously. Portions of the mantle melt. Volcanoes erupt hot lava. Hot springs bubble and boil. Metallic ore deposits form from hot fluids. These facts underscore the importance of understanding Earth's internal heat. To do this, geologists evaluate *heat flow* by measuring the amount of thermal energy released through a given area (in milliwatts/m²). Heat flow is usually measured by lowering a sensitive thermometer down a deep drill hole and recording the temperature—sometimes called the *geothermal gradient.*

The map summarizes decades of careful measurements and shows fundamental relationships between heat flow and global tectonic setting (compare with Figure 2.11). It is obvious that heat flow is not distributed uniformly across the planet. But why do these patterns emerge? Where is Earth's internal heat coming from anyway?

Let us answer the last question first. Today, most of the heat released from Earth is generated by radioactive decay of three elements found in small quantities in almost all rocks—potassium, uranium, and thorium. The heat is created when small quantities of matter are converted to energy. Even though the mantle contains very low concentrations of these radioactive elements, it is so thick and massive that the mantle is the dominant source of Earth's thermal energy. Therefore, heat flow (and the temperature gradient) is high where the hot mantle is near the surface.

This principle helps us understand why heat flow is so high below oceanic regions where the cold lithosphere is thin compared to that beneath the continents. Another way of thinking about this oceanic thermal anomaly is to remember that oceanic lithosphere is hot when it forms by igneous processes at ocean ridges. It is also much younger than most continental lithosphere and has not yet lost all of its heat. Moreover, the highest heat flow regions are the mid-ocean ridges—especially the East

Courtesy of H.N. Pollock

Pacific rise and the ridge in the Indian Ocean. At the ridge, the lithosphere is young, thin, and the site of active volcanism.

The zones of lowest heat flow correspond to ancient central parts of the continents—the shields (compare with the map on the inside front cover). Apparently, the lithosphere is cool and thick under the ancient shields.

Another interesting pattern is visible upon careful examination of the heat flow over trenches (or subduction zones). Note that heat flow is low over the subduction zones near Indonesia and the west coast of South America. Why is heat flow low near these zones of active volcanoes? Probably because the subduction of cold oceanic lithosphere refrigerates the mantle and reduces the heat flow.

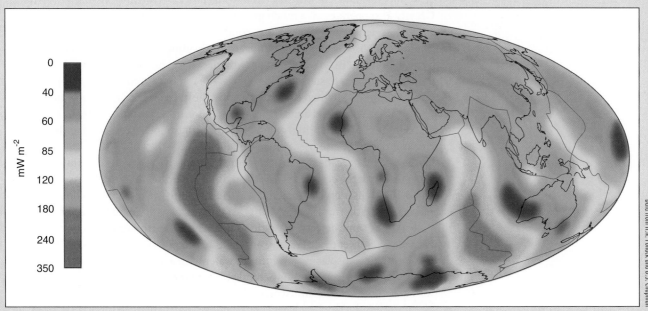

mW m⁻²

| 0 |
| 40 |
| 60 |
| 85 |
| 120 |
| 180 |
| 240 |
| 350 |

Data from H.N. Pollock and D.S. Chapman

Plates and Plate Motion

What are the major plates and how are they moving?

Take a moment to study Figure 2.9 again. You will want to become very familiar with this map because it shows a new geography—the geography of tectonic plates. As you have seen, most of Earth's major features can be understood from the interactions of these plates in the tectonic system.

In Figure 2.9, you can see that seven major lithospheric plates are recognized—the North America, South America, Pacific, Australia, Africa, Eurasia, and Antarctica plates—together with several smaller ones. Let us take a brief tour, so you will know what to look for.

1. The divergent plate boundaries are marked by oceanic ridges, which extend from the Arctic south through the central Atlantic and into the Indian and Pacific oceans. Movement of the plates is away from the crest of the oceanic ridge.
2. The North America and South America plates are moving westward and interacting with the Pacific, Juan de Fuca, Cocos, and Nazca plates along the west coast of the Americas.
3. The Pacific plate is moving northwestward away from the oceanic ridge toward a system of deep trenches in the western Pacific basin.
4. The Australia plate includes Australia, India, and the northeastern Indian Ocean. It is moving northward, causing India to collide with the rest of Asia to produce the high Himalaya Mountain ranges and the volcanic arc of Indonesia.
5. The Africa plate includes the continent of Africa, plus the southeastern Atlantic and western Indian oceans. It is moving northward and colliding with the Eurasia plate.
6. The Eurasia plate, which consists of Europe and most of Asia, moves eastward.
7. The Antarctica plate includes the continent of Antarctica, plus the floor of the Antarctic Ocean. It is unique in that it is nearly surrounded by oceanic ridges.

Gravity and Isostasy

> Gravity plays a fundamental role in Earth's dynamics. It is intimately involved with differentiation of the planet's interior, isostatic adjustments of the crust's elevation, plate tectonics, and downward flow of water in the hydrologic system.

How is isostasy a reflection of an equilibrium state?

Gravity is one of the great fundamental forces in the universe. It played a vital role in the formation of the solar system, the origin of the planets, and the impact of meteorites that dominated their early history. Since then, gravity has been a constant force in every phase of planetary dynamics, and it is a dominant factor in all geologic processes operating on and within Earth—glaciers, rivers, wind, and even volcanoes.

Gravity also operates on a much grander scale within Earth's crust. It causes "lighter" (less dense) portions, such as continents, to stand higher than the rocks of the "heavier," denser ocean floor. Similarly, the loading of Earth's crust at one place with thick sediment in a river delta, or with glacial ice, or with water in a deep lake will cause that region to subside. Conversely, the removal of rock from a mountain range by erosion will lighten the load, causing the deep crust to move upward to take its place. This gravitational adjustment of Earth's crust is **isostasy** (Greek *isos*, "equal"; *stasis*, "standing"). Earth's lithosphere therefore continuously responds to the force of gravity as it tries to maintain a gravitational balance.

Isostasy occurs because the crust is more buoyant than the denser mantle beneath it. Each portion of the crust displaces the mantle according to its thickness and density (**Figure 2.16**). Denser crustal material sinks deeper into the mantle than does less-dense crustal material. Alternatively, thicker crustal material will extend to greater depth than thin crust of the same density. Isostatic adjustments in Earth's crust can be compared to adjustments in a sheet of ice floating on a lake as you skate on it. The layer of ice bends down beneath you, displacing a volume of water with a weight equal to your weight. As you move ahead, the ice rebounds behind you, and the displaced water flows back.

As a result of isostatic adjustment, high mountain belts and plateaus are commonly underlain by thicker crust that extends deeper into the mantle than do

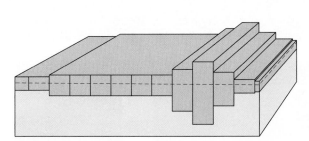

(A) Low-density blocks float on a denser liquid. If the blocks have equal densities, the thicker blocks rise higher and sink deeper than the thinner blocks.

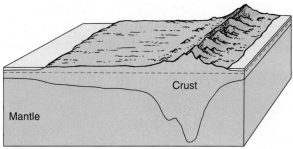

(B) High mountains in low-density crust are balanced by a deep root that extends into the mantle.

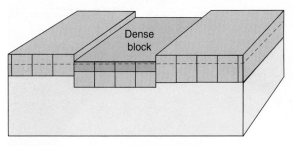

(C) Floating blocks of unequal density. The block with denser (green) portions sinks and the surface is lower than the adjacent blocks, even though the thickness is the same.

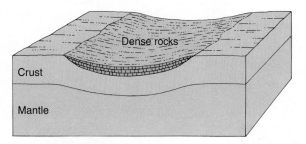

(D) A deep basin may form if the rocks beneath it are denser (green) than surrounding rocks (brown).

Figure 2.16 Isostasy is the universal tendency of segments of Earth's crust to establish a condition of gravitational balance. Differences in both thickness and density can cause isostatic adjustments in Earth's crust.

areas of low elevation. Any thickness change in an area of the crust—such as the removal of material by erosion or the addition of material by sedimentation, volcanic eruption, or accumulations of large continental glaciers—causes an isostatic adjustment.

The construction of Hoover Dam, on the Colorado River, is a well-documented illustration of isostatic adjustment. The added weight of water and sediment in the reservoir was sufficient to cause measurable subsidence. From the time of the dam's construction in 1935, 24 billion metric tons of water, plus an unknown amount of sediment, accumulated in Lake Mead, behind the dam. In a matter of years, this added weight caused the crust to subside in a roughly circular area around the lake. Continental glaciers are another clear example of isostatic adjustment of the crust. The weight of an ice sheet several thousand meters thick disrupts the crustal balance and depresses the crust beneath. In both Antarctica and Greenland, the weight of the ice has depressed the central part of the land masses below sea level. A similar isostatic adjustment occurred in Europe and North America during the last ice age, when continental glaciers existed there. Parts of both continents, such as Hudson Bay and the Baltic Sea, are still below sea level. Now that the ice is gone, however, the crust is rebounding at a rate of 5 to 10 m/1000 yr.

Tilted shorelines of ancient lakes provide another means of documenting isostatic rebound. Lake Bonneville, for example, was a large lake in Utah and Nevada during the ice age but has since dried up, leaving such small remnants as Utah Lake and Great Salt Lake. Shorelines of Lake Bonneville were level when they were formed but have been tilted in response to unloading as the water was removed.

The concept of isostasy, therefore, is fundamental to studies of the crust's major features—continents, ocean basins, and mountain ranges. It also is fundamental to understanding the response of the crust to erosion, sedimentation, glaciation, and the tectonic system.

How do we know that isostatic adjustments occur?

GeoLogic   Earth's Systems from Space

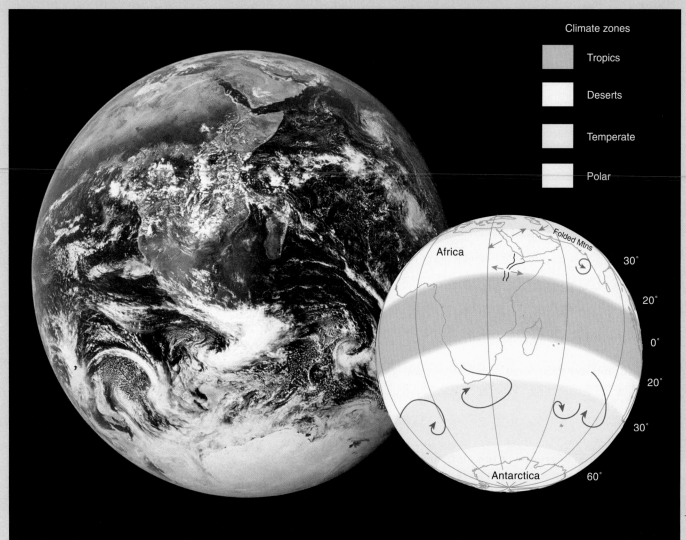

Climate zones

Tropics

Deserts

Temperate

Polar

Africa

Folded Mtns

30°

20°

0°

20°

30°

Antarctica   60°

Courtesy of NASA

A view of planet Earth from space gives us a truly global view of the geologic systems that shape the planet.

Observations

1. The continents and ocean basins are Earth's most prominent features.
2. The planet's water is seen in the vast blue oceans and the white polar ice caps. Water cycles through the atmosphere as shown by the bright swirling clouds.
3. Climate zones are expressed as regular patterns in the distribution of green vegetation on land, of the amount and shapes of clouds in the atmosphere, and ice at the poles.

Interpretations

Two major geologic systems shape the Earth—the hydrologic system and the tectonic system. The tectonic system created the lithosphere with its huge ocean basins and high continental platforms, which are underlain by rocks of different compositions, structures, and ages. Locally, new ocean basins are forming where continents are rifting apart as shown by the separation of Africa from Arabia. Elsewhere, plates are colliding to form new continents and mountain belts like the one barely visible in southern Iran.

The hydrologic system endlessly modifies the surface features of the lithosphere with rain, wind, waves, flowing water, and ice. The role played by the climate in the operation of the hydrologic system is clear. In the cloud-spotted tropics, air heated by the Sun rises in vertical convection cells making abundant rainfall and fueling the growth of vegetation in the biosphere. Cloud-free deserts lie north and south of the tropics. At the South Pole, the cold climate has created the vast Antarctic glacier. Cyclonic storms, which appear as clouds resembling huge commas, show the prevailing wind patterns generated by solar radiation and by Earth's rotation.

Key Terms

climate system (p. 38)
closed system (p. 33)
convection (p. 43)
convergent plate boundary (p. 46)
divergent plate boundary (p. 44)
dynamic system (p. 33)
eolian system (p. 41)
equilibrium (p. 34)

glacier system (p. 40)
groundwater system (p. 40)
hydrologic system (p. 36)
isostasy (p. 50)
metastable (p. 35)
open system (p. 33)
plate (p. 42)
plate tectonics (p. 42)

plume (p. 48)
river system (p. 39)
shoreline system (p. 41)
subduction (p. 43)
system (p. 32)
transform plate boundary (p. 44)

Review Questions

1. Consider the gravitational interactions among Earth, the Sun, and the Moon. Does this constitute a system? If so, what are its boundaries? Is it open or closed? What forms of energy are involved?
2. Diagram the paths by which water circulates in the hydrologic system.
3. What energy drives the hydrologic system?
4. Approximately how much water evaporates from the ocean each year?
5. Describe the major landforms resulting from (a) rivers, (b) groundwater, (c) glaciers, and (d) wind.
6. Draw a diagram (cross section) showing (a) converging plates and (b) diverging plates.
7. On a map such as the one in Figure 2.9, identify the three fundamental kinds of plate boundaries.

8. What surface features distinguish each kind of plate boundary?
9. Explain how the Alps, mid-ocean ridges, deep-sea trenches, island arcs, and volcanoes are related to plate tectonics.
10. Describe the geologic processes that occur above a mantle plume.
11. Why do the materials inside Earth convect?
12. Make a list of the many roles played by gravity in geologic systems.
13. Explain isostasy, and give two examples of isostatic adjustment of Earth's crust in recent geologic time.

3 Minerals

We live in a world of minerals—they are everywhere around us. Gems and jewelry are minerals. Gravel and sand are minerals. Mud is a mixture of microscopic minerals. Ice is a mineral, and even dust in the air we breathe is made up of tiny mineral grains. Minerals sustain our lives and provide continuously for society. The houses in which we live, the automobiles we drive, as well as the roads and other structures of our society, and almost everything we touch are made of minerals or material derived from minerals. Indeed, on average, every person on Earth uses, directly or indirectly, 10 metric tons of minerals each year.

But the importance of minerals extends far beyond their value as economic deposits. Minerals are also the substance of Earth's natural systems. They are the building blocks of rocks. Taken together, a group of mineral grains forms a rock. A photograph of the rock we call basalt taken through a microscope spans the top of these pages. It illustrates this fundamental idea. Each of the variously colored grains represents a different mineral. Although the minerals interlock to form a tight, coherent mass, each has distinguishing properties. The large gray and white grains are feldspar; the bright pink and green grains are olivine; the small black grains are

iron-titanium oxide. Each grain of feldspar has much in common with all of the other grains of its mineral species. For example, all grains of feldspar have the same internal arrangement of atoms and have the same chemical and physical properties even though each grain varies greatly in size and shape. The arrangement of atoms inside feldspar helps to define its characteristic shape, its hardness and density, color, cleavage, and luster—properties you will learn more about in this chapter.

All of Earth's dynamic processes involve the growth and destruction of minerals as matter changes from one state to another. As Earth's surface weathers and erodes, some minerals are destroyed and others grow in their place. Olivine and plagioclase in this photo grew from a molten silicate liquid in a lava flow. As sediments accumulate in the oceans, minerals also grow from solution. Other minerals grow from watery solutions flowing through solid rock. Deep below Earth's surface, high pressure and temperature remove atoms from the crystal structures of some minerals and cause them to recombine as new minerals. As tectonic plates move and continents drift, minerals are created and destroyed by a variety of processes. Some knowledge of Earth's major minerals, therefore, is essential to understanding Earth's dynamics.

In this chapter, we survey the general characteristics of minerals and the physical properties that identify them. We then explore the major rock-forming minerals in preparation for a study of the major rock types.

Major Concepts

1. An atom is the smallest unit of an element that possesses the properties of the element. It consists of a nucleus of protons and neutrons and a surrounding cloud of electrons.
2. An atom of a given element is distinguished by the number of protons in its nucleus. Isotopes are varieties of an element, distinguished by the different numbers of neutrons in their nuclei.
3. Ions are electrically charged atoms, produced by a gain or loss of electrons.
4. Matter exists in three states: (a) solid, (b) liquid, and (c) gas. The differences among the three are related to the degree of ordering of the atoms.
5. A mineral is a natural solid possessing a specific internal atomic structure and a chemical composition that varies only within certain limits. Each type of mineral is stable only under specific conditions of temperature and pressure.
6. Minerals grow when atoms are added to the crystal structure as matter changes from the gaseous or the liquid state to the solid state. Minerals dissolve or melt when atoms are removed from the crystal structure.
7. All specimens of a mineral have well-defined physical and chemical properties (such as crystal structure, cleavage or fracture, hardness, and density).
8. Silicate minerals are the most important minerals and form more than 95% of Earth's crust. The most important silicates are feldspars, micas, olivine, pyroxenes, amphiboles, quartz, and clay minerals. Important nonsilicate minerals are calcite, dolomite, gypsum, and halite.
9. Minerals grow and are broken down under specific conditions of temperature, pressure, and chemical composition. Consequently, minerals are a record of the changes that have occurred in Earth throughout its history.

Matter

> An atom is the smallest unit of an element that possesses the properties of the element. It consists of a nucleus of protons and neutrons and a surrounding cloud of electrons. There are three states of matter: gas, liquid, and solid. Each state is distinguished by unique physical properties. Processes in Earth's dynamics mostly involve the changing of matter from one state to another.

To understand the dynamics of Earth and how rocks and minerals are formed and changed through time, you must have some knowledge of the fundamental structure of matter and how it behaves under various conditions. The solid materials that make up Earth's outer layers are called rocks. Most rock bodies are mixtures, or aggregates, of minerals. A mineral is a naturally occurring compound with a definite chemical formula and a specific internal structure. Because minerals, in turn, are composed of atoms, to understand minerals we must understand something about atoms and the ways in which they combine.

Atoms

What is the structure of an atom?

An **atom** is the smallest fraction of an element that can exist and still show the characteristics of that element. Atoms are best described by abstract models constructed from mathematical formulas involving probabilities. They are much too small to be seen with optical microscopes; recently, however, images of atoms have been made. An example is shown in **Figure 3.1**. In its simplest form, an atom is characterized by a relatively small **nucleus** of tightly packed protons and neutrons, with a surrounding cloud of electrons. These are the principal building blocks of atoms, but many other subatomic particles have been identified in recent years.

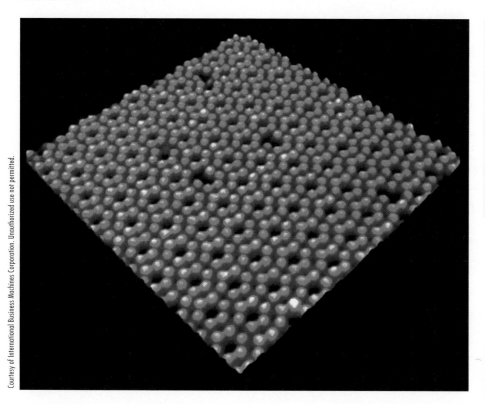

Figure 3.1 Image of atoms of silicon produced by a scanning tunneling microscope at the IBM Research Center, Yorktown Heights, New York. The blue spots are individual silicon atoms, which are arranged in a regular pattern that repeats itself across the surface. You can also see the hexagonal arrangement of groups of the atoms. Locally, flaws in the structure are also visible. Images such as this are helpful in understanding the structure of different minerals.

The distinguishing feature of an atom of a given element is the number of protons in the nucleus. The number of electrons and neutrons in an atom of a given element can vary, but the number of protons is always the same. Each **proton** carries a positive electrical charge, and the mass of a proton is taken as the unit of atomic mass, approximately $1.66 \propto 10^{-24}$ g. The **neutron**, as its name indicates, is electrically neutral and has approximately the same mass as the proton. The **electron** is a much smaller particle, with a mass approximately 1/1850 the mass of the proton. It carries a negative electrical charge equal in intensity to the positive charge of the proton. Because the electron is so small, for practical purposes, the entire mass of the atom is considered to be concentrated in the protons and neutrons of the nucleus. The **atomic mass** is simply the sum of the number of neutrons and protons.

Hydrogen is the simplest of all elements. It consists of one proton in the nucleus and one orbiting electron (**Figure 3.2**). The next heaviest atom is helium, with two protons, two neutrons, and two electrons. Each subsequently heavier element contains more protons, neutrons, and electrons. **Figure 3.3** is a simplified chart of all naturally occurring elements. The elements are arranged in rows, with increasingly heavier elements to the right and bottom. This chart is commonly called the periodic table. The distinguishing feature of an element is the number of protons in the nucleus of each of its atoms, often called the **atomic number**. The number of electrons and neutrons in the atoms of a given element can vary, but the number of protons is constant.

Atoms normally have the same number of electrons as protons and thus do not carry an electrical charge. As the number of protons increases in progressively heavier atoms, the number of electrons also increases. The electrons fill a series of energy-level shells around the nucleus, each shell having a maximum capacity. The progressive filling of these shells is reflected in the rows of the periodic chart (Figure 3.3). The electrons in the outer shells control the chemical behavior of the element.

Isotopes

Although the number of protons in each atom of a given element is constant, the number of neutrons in the nucleus can vary. This means that atoms of a given element are not all exactly alike. Iron atoms, for example, have 26 protons but individual atoms may have 28, 30, 31, or 32 neutrons. These varieties of iron are examples of **isotopes**; they all

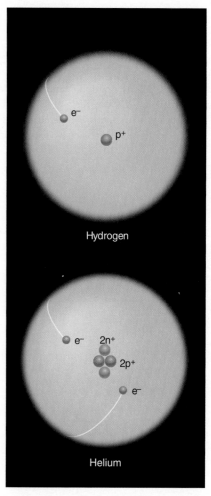

Figure 3.2 The atomic structures of hydrogen and helium illustrate the major particles of an atom. Hydrogen has one proton (p) in a central nucleus and one orbiting electron (e). Helium has two protons (p), two neutrons (n) in the nucleus, and two orbiting electrons (e).

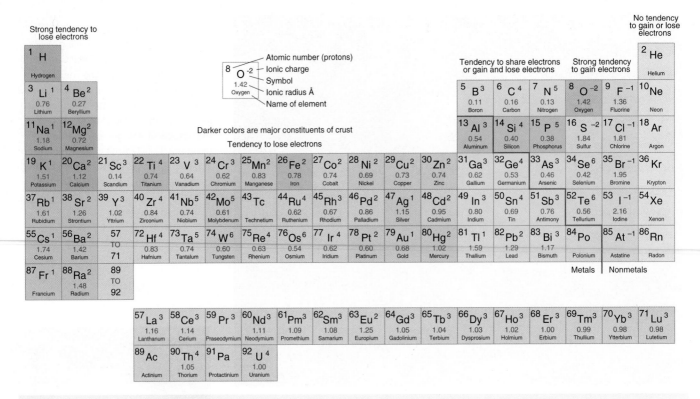

Figure 3.3 The periodic table of the elements shows the name and symbol of all of the naturally occurring elements. The lightest and simplest elements are in the upper left; across and toward the bottom, each element is progressively more complex, with increasing numbers of nuclear particles and electrons. The elements are separated into rows according to the outermost electron shell. Also shown is the charge of the common ion and the radius for that ion. These properties of an element control how it combines with other elements to form minerals.

have the properties of iron but differ from one another in mass. Most common elements exist in nature as mixtures of isotopes. Some isotopes are unstable, emitting particles and energy as they experience radioactive decay to form new, more stable isotopes.

Ions

What are the distinguishing characteristics of an isotope? Of an ion?

Atoms that have as many electrons as protons are electrically neutral, but atoms of most elements can gain or lose electrons in their outermost shells. If electrons are gained or lost, an atom loses its electrical neutrality and becomes charged. These electrically charged atoms are **ions**. The loss of an electron makes a positively charged ion because the number of protons then exceeds the number of negatively charged electrons. If an electron is gained, the ion has a negative charge. The electrical charges of ions are important because the attraction between positive ions and negative ions is the bonding force that sometimes holds matter together. Like atoms, ions have distinctive sizes that reflect the number of particles in the nucleus and the number of electrons in the surrounding cloud. Ionic size and ionic charge control how elements fit together to make solid minerals (Figure 3.3).

Bonding

An atom is most stable if its outermost shell is filled to capacity with electrons. The inner shell can hold no more than 2 electrons. The next shell can hold 8 electrons and is full in neon (atomic number 10). In heavier elements, the next shell can have 18 electrons, and the shell after that one can have 32 electrons. Neon, for example, has 10 protons in the nucleus and 10 electrons, of which 2 are in the first shell and 8 are in the second shell. A neon atom does not have an electrical charge. Its two electron shells are complete because the second shell has a limit of 8 electrons. As a result, neon does not interact chemically with other atoms. Argon and the other noble gases (the right column on the periodic chart) also have 8 electrons in their outermost shell, and they normally do not

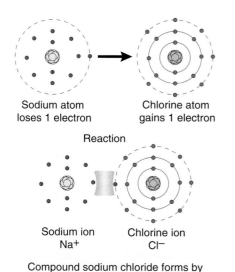

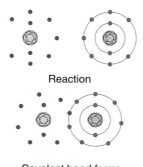

Sodium atom loses 1 electron

Chlorine atom gains 1 electron

Reaction

Sodium ion Na+

Chlorine ion Cl⁻

Compound sodium chloride forms by electrical attraction between Na+ and Cl⁻

Reaction

Covalent bond forms by sharing electrons

(A) The formation of an ionic bond in sodium and chloride ions by transfer of an electron from the outermost shell of a sodium atom to the outermost shell of a chlorine atom results in a stable outer shell for each ion.

(B) Covalent bonds form when two atoms share electrons. The bond between silicon and oxygen, so common in minerals, is largely of this type.

combine with other elements. Most elements, however, have an incomplete outermost shell. Their atoms readily lose or gain electrons to achieve a structure like that of argon, neon, and the other inert gases, with 8 electrons in the outermost shell.

For example, an atom of sodium has only 1 electron in its outermost shell but 8 in the shell beneath (**Figure 3.4**). If it could lose the lone outer electron, the sodium atom would have a stable configuration like that of the inert gas neon. The chlorine atom, in contrast, has 7 electrons in its outermost shell, and if it could gain an electron, it too would attain a stable electron configuration. Whenever possible, therefore, sodium gives up an electron and chlorine gains one. The sodium atom thus becomes a positively charged sodium ion, and the chlorine atom becomes a negatively charged chloride ion. With opposite electrical charges, the sodium ions and chloride ions attract each other and bond together to form the compound sodium chloride (common salt, also known as the mineral halite). (A **compound** has more than one element in its structure.) This type of bond, between ions of opposite electrical charge, is known as an **ionic bond**. Such bonds commonly develop between elements that lie far from one another on the periodic table.

Atoms can also attain the electron arrangement of a noble gas, and thus attain stability, by sharing electrons. No electrons are lost or gained, and no ions are formed. Instead, an electron cloud surrounds both nuclei. This type of bond is a **covalent bond** and typically develops between elements that are near one another on the periodic table. Bonds between two atoms of the element may be of this type; the bonds in an oxygen molecule (O_2) are a good example. The bonds between carbon and hydrogen in organic materials are also of this type. Many bonds found in natural substances are intermediate between covalent and ionic bonds. Electrons are "pulled" closer to the nucleus of one ion than to the other. As a consequence, one part of the molecule may have a slight charge. The Si–O bond that is so common in minerals is like this.

A third type of bond is the **metallic bond**. In a metal, each atom contributes one or more outer electrons that move relatively freely throughout the entire aggregate of ions. A given electron is not attached to a specific ion pair but moves about. This sea of negatively charged electrons holds the positive metallic ions together in a crystalline structure and is responsible for the special characteristics of metals, including their high electrical conductivity and ductile behavior. Except for a few native elements (such as gold), few minerals have metallic bonds.

States of Matter

The principal differences between solids, liquids, and gases involve the degree of ordering of the constituent atoms. In the typical **solid**, atoms are arranged in a rigid framework. The arrangement in crystalline solids is quite different. The atomic structure of a crystal consists of a regular, repeating, three-dimensional pattern known as a **crystal structure**. However, there are some **amorphous solids** in which the atomic arrangement is random. Glass is an example of an amorphous solid that lacks a clearly defined crystalline structure. In such solids, each atom occupies a more or less fixed position but has a vibrating motion. Changes in crystalline solids occur as the temperature or pressure changes. For example, as temperature rises, the vibration of atoms in the structure increases, and atoms move farther and farther apart. Eventually the bonds between two atoms may break and they become free and able to glide past one another. Melting ensues, and the crystalline solid passes into the liquid state.

In a **liquid**, the basic particles are in random motion, but they are packed closely together. They slip and glide past one another or collide and rebound, but they are held together by forces of attraction greater than those in gases. This force of attraction explains why density generally increases and compressibility decreases as matter changes from gas to liquid to solid. If a liquid is heated, the motion of the particles increases, and individual atoms or molecules become separated as they move about at high speeds.

In a **gas**, the particles are in rapid motion and travel in straight lines until their direction is changed by collision. Because the individual atoms or molecules are separated by empty spaces and are comparatively far apart, gases can be markedly compressed and can exert pressure. Gases have the ability to expand indefinitely, and the continuous rapid motion of the particles results in rapid diffusion.

Water undoubtedly provides the most familiar example of matter changing through the three basic states. At pressures prevailing on Earth's surface, water changes from a solid, to a liquid, to a gas in a temperature range of only 100°C. Most people are familiar with the effects of temperature changes on the state of matter because of their experience with water as it freezes, melts, and boils. Fewer people are familiar with the effects of pressure. Under great pressure, water will remain liquid at temperatures as high as 371°C.

The combined effects of temperature and pressure on water are shown in the phase diagram in **Figure 3.5**. An interesting and very important feature of water is the fact that as it freezes, the solid is actually less dense than the liquid. As a result, water ice floats rather than sinks. The expansion of water during freezing is important for weathering and in the moderation of Earth's climate. Because polar ice floats on the

Why can gaseous, liquid, and solid forms of a substance have such different physical properties and still have the same composition?

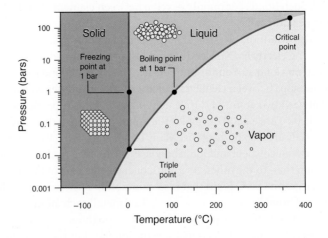

Figure 3.5 Temperature and pressure determine the state in which matter exists. In this diagram, the ranges of temperature and pressure for the various phases of water are shown. The triple point is the point at which all three phases are in equilibrium. Beyond the critical point the liquid and gas phases cannot be distinguished. Similar phase diagrams can be constructed for other minerals.

sea, it creates an insulating layer that slows the cooling of the rest of the sea. If ice did not float, Earth's oceans may have frozen solid during the ice ages.

Other forms of matter in the solid Earth are capable of similar changes, but usually their transitions from solid, to liquid, to gas occur at comparatively high temperatures. At normal room temperature and pressure, 93 of the 106 elements are solids, 2 are liquids, and 11 are gases. Diagrams similar to Figure 3.5, constructed from laboratory work on other minerals, provide important insight into the processes operating at the high temperatures and pressures below Earth's surface.

The Nature of Minerals

A mineral is a natural inorganic solid with a specific internal structure and a chemical composition that varies only within specific limits. All specimens of a given mineral—regardless of where, when, or how they were formed—have the same physical properties (including cleavage, crystal form, hardness, density, color, luster, and streak). Minerals also have restricted stability ranges.

Minerals are the solid constituents of Earth. Many people think of minerals only as exotic crystals in museums or as valuable gems and metals; but grains of sand, snowflakes, and salt particles are also minerals, and they have much in common with gold and diamonds. A precise definition is difficult to formulate, but for a substance to be considered a **mineral**, it must meet the conditions listed above and described in greater detail below. The differences among minerals arise from the kinds of atoms they contain and the ways those atoms are arranged in a crystalline structure.

Natural Inorganic Solids

By definition, only naturally occurring inorganic solids are minerals—that is, natural elements or inorganic compounds in a solid state. Synthetic products, such as artificial diamonds, are therefore not minerals in the strict sense. Organic compounds, such as coal and petroleum, which lack a crystal structure, are also not considered to be minerals. This criterion is not as important as most of the others. After all, there is little difference between a synthetic and a natural gem, other than where they formed. All of its structural, physical, and chemical properties are shared with its natural counterparts. Likewise, there are organic solids that have all of the characteristics of minerals.

The Structure of Minerals

The key words in the definition of mineral are *internal structure*. Minerals can consist of a single element, such as gold, silver, copper, diamond, or sulfur. However, most are compounds of two or more elements. The component atoms of a mineral have a specific arrangement in a definite geometric pattern. All specimens of a given mineral have the same internal structure, regardless of when, where, and how they were formed. This property of minerals was suspected long ago by mineralogists who observed the many expressions of order in **crystals**. Nicolaus Steno (1638–1687), a Danish monk, was among the first to note this property. He found from numerous measurements that each of the different kinds of minerals has a characteristic crystal form. Although the size or shape of a mineral's crystalline form may vary, similar pairs of crystal faces always meet at the same angle. This is known as the *law of constancy of interfacial angles*.

Later, René Haüy (1743–1822), a French mineralogist, accidentally dropped a large crystal of calcite and observed that it broke along three sets of planes only, so all the fragments had a similar shape (see Figure 3.9). He then proceeded to break other calcite crystals in his own collection, plus many in the collections of his friends, and found that all of the specimens broke in exactly the same manner. All of the fragments, however small, had the shape of a rhombohedron. To explain his observations, he assumed that calcite is built of innumerable infinitely small rhombohedra packed together in an orderly manner; he concluded that the cleavage of calcite is related to the ease of parting of such units from adjacent layers. His discovery was a remarkable

Why is the structure of a mineral so important?

advance in understanding crystals. Today we know that cleavage planes are planes of weakness in the crystal structure and that they are not necessarily parallel to the crystal faces. Cleavage planes do, however, constitute a striking expression of the orderly internal structure of crystals.

To understand the importance of structure in a mineral, consider the characteristics of diamond and graphite (**Figure 3.6**). These two minerals are identical in chemical composition. Both consist of a single element, carbon (C). Their crystal structures and physical properties, however, are very different. In diamond, which forms only under high pressure, the carbon atoms are packed closely, and the covalent bonds between the atoms are very strong. Their structure explains why diamonds are extremely hard—the hardest natural substance known. In graphite, the carbon atoms form layers that are loosely bound. Because of weak bonds, the layers separate easily, so graphite is slippery and flaky. Because of its softness and slipperiness, graphite is used as a lubricant and is also the main constituent of common "lead" pencils. The important point to note is that different structural arrangements of exactly the same elements produce different minerals with different properties. This ability of a specific chemical substance to crystallize in more than one type of structure is known as **polymorphism**.

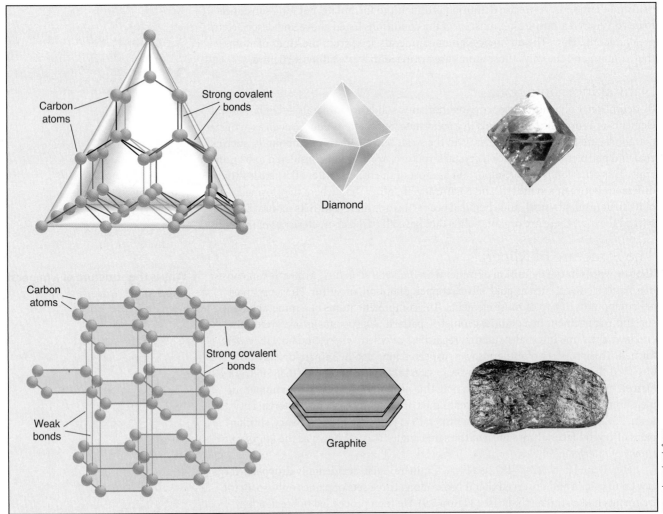

Figure 3.6 The internal structure of a mineral controls its physical properties. Diamond and graphite have exactly the same chemical composition, but the carbon atoms are arranged differently and held together by different types of bonds. Graphite is made of sheets of carbon stacked on top of one another. It is soft and black. Diamond, the hardest mineral known, is made of carbon atoms bound together in a tight tetrahedral framework. Most grains of diamond are transparent.

STATE OF THE ART X-Ray Diffraction and the Structure of Minerals

With modern methods of **X-ray diffraction**, we can determine precisely a mineral's internal structure and learn much about the arrangement of its atoms. Diffraction involves the bending of X rays as they pass through a crystalline substance.

The technique is illustrated in the figure below. When a narrow beam of X rays is passed through a mineral grain, the X rays are diffracted by the framework of atoms. The individual ions are spaced very closely in the rigid network, close enough to bend X rays—like a diffraction grating bends light rays. The diffracted rays cause constructive and destructive interference—in effect, concentrating the energy of the X rays in some areas and dispersing it in others. After they leave the crystal, the X rays expose a photographic plate or are detected with a scanning device and plotted as shown. From the pattern made by the spots or from measurements of the position and height of the peaks, the systematic orientation of planes of atoms within the crystal can be deduced. Such measurements are so precise that the distances between atoms can be measured and the size and shape of the electron cloud calculated. Detailed models of crystal structures showing the position of each different atom can thus be constructed.

The X-ray diffraction instrument is now the most basic device for determining the internal structure of minerals, and geologists use it extensively for precise mineral identification and analysis.

Two typical examples of X-ray spectra are shown on the chart. The lower curve is the X-ray diffraction pattern for the mineral quartz. The peaks are created by constructive interference of the X rays and correspond to specific atomic spacings that are the result of the nearly covalent silicon–oxygen bond. The peak positions do not directly reveal the kinds of atoms,

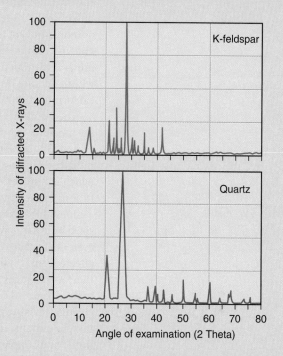

only their distances and arrangements. The more complicated X-ray diffraction pattern was formed from a specimen of feldspar. Because feldspars have a much greater variety of elements, bond types, and structural elements, the diffraction pattern is also more complicated.

X-ray diffraction analysis is the definitive technique that shows us that each mineral species has its own distinctive structure that is repeated many times in every grain of the mineral. It reveals the great symmetry and order found in the mineral kingdom.

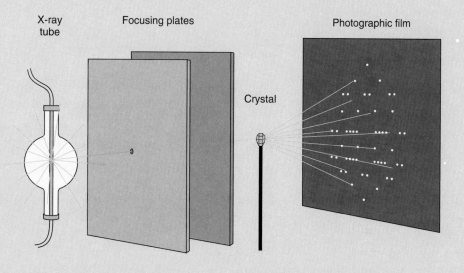

The Composition of Minerals

A mineral has a definite chemical composition, in which specific elements occur in definite proportions. Thus, a precise chemical formula can be written to express the chemical composition—for example, SiO_2, $CaCO_3$, and so on. The chemical composition of some minerals can vary, but only within specific limits. In these minerals, two or more kinds of ions can substitute for each other in the mineral structure, a process known as **ionic substitution**. Ionic substitution results in a chemical change in the mineral without a change in the crystal structure, so substitution can occur only within definite limits. The composition of such a mineral can be expressed by a chemical formula that specifies ionic substitution and how the composition can change.

The suitability of one ion to substitute for another is determined by several factors, the most important being the size and the electrical charge of the ions in question (Figure 3.3 and **Figure 3.7**). Ions can readily substitute for one another if their ionic radii differ by less than 15%. If a substituting ion differs in charge from the ion for which it is substituted, the charge difference must be compensated for by other substitutions in the same structure in order to maintain electrical neutrality.

Ionic substitution is somewhat analogous to substituting different types of equal-sized bricks in a wall. The substitute brick may be composed of glass, plastic, or whatever, but because it is the same size as the original brick, the structure of the wall is not affected. An important change in composition has, however, occurred, and as a result there are changes in physical properties. In minerals, ionic substitution causes changes in hardness and color, for example, without changing the internal structure.

Ionic substitution is common in rock-forming minerals and is responsible for mineral groups, the members of which have the same structure but varying composition. For example, in the olivine group, with the formula $(Mg, Fe)_2SiO_4$, ions of iron (Fe^{+2}) and magnesium (Mg^{+2}) can substitute freely for one another because they have similar charges and sizes (Figure 3.7). The total number of Fe^{+2} and Mg^{+2} ions is constant in relation to the number of silicon (Si^{+4}) and oxygen (O^{-2}) ions in the olivine, but the ratio of iron to magnesium may vary in different samples. The common minerals feldspar, pyroxene, amphibole, and mica each constitute a group of related minerals in which ionic substitution produces a range of chemical composition.

The Physical Properties of Minerals

What determines the physical properties of a mineral?

Because a mineral has a definite chemical composition and internal crystalline structure, all specimens of a given mineral, regardless of when or where they were formed, have the same physical and chemical properties. If ionic substitution occurs, variation in physical properties also occurs, but because ionic substitution can occur

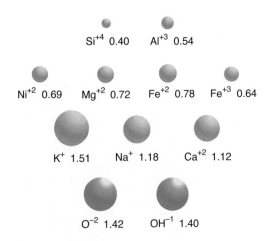

Si^{+4} 0.40 Al^{+3} 0.54

Ni^{+2} 0.69 Mg^{+2} 0.72 Fe^{+2} 0.78 Fe^{+3} 0.64

K^+ 1.51 Na^+ 1.18 Ca^{+2} 1.12

O^{-2} 1.42 OH^{-1} 1.40

Figure 3.7 The relative size and electrical charge of ions are important factors governing the suitability of one ion to substitute for another in a crystal structure. Silicon can be replaced by aluminum, iron by magnesium or nickel, and sodium by calcium.

only within specific limits, the range in physical properties also can occur only within specific limits. This means that one piece of quartz, for example, is as hard as any other piece, that it has the same density, and that it breaks in the same manner, regardless of when, where, or how it was formed.

The more significant and readily observable physical properties of minerals are crystal form, cleavage, hardness, density, color, luster, and streak.

If a crystal is allowed to grow in an unrestricted environment, it develops natural **crystal faces** and assumes a specific geometric **crystal form**. The shape of a crystal is a reflection of the internal structure and is an identifying characteristic for many mineral specimens (**Figure 3.8**). If the atoms are arranged in a long chain, the crystal may be shaped like a needle. If the atoms are arranged in a boxlike network, the crystal will likely be in the form of a cube. If the space for growth is restricted, however, smooth crystal faces cannot develop.

Cleavage is the tendency of a crystalline substance to split or break along smooth planes parallel to zones of weak bonding in the crystal structure (**Figure 3.9**). If the bonds are especially weak in a given plane, as in graphite, mica, or halite, perfect cleavage occurs with ease. Breaking the mineral in any direction other than along a cleavage plane is difficult (Figure 3.9). In other minerals, the differences in bond strength are not great, so cleavage is poor or imperfect. Cleavage can occur in more than one direction, but the number and direction of cleavage planes in a given mineral species are always the same. Some minerals have no weak planes in their crystalline structure, so they do not have cleavage and break along various types of fracture surfaces. Quartz, for example, characteristically breaks by **conchoidal fracture**—that is, along curved surfaces, like the curved surfaces of chipped glass. Cleavage planes and crystal faces should not be confused with the facets found on gems. Facets are produced by grinding and polishing the surface of a mineral grain and do not necessarily correspond to cleavage directions. For example, diamond lacks cleavage altogether but can be polished so that a single crystal will have many shiny faces.

Hardness is a measure of a mineral's resistance to abrasion. It is in effect a measure of the strength of the atomic bonds in a crystal. This property is easily determined and is used widely for field identification of minerals. More than a century ago, Friedrich Mohs (1773–1839), a German mineralogist, assigned relative numbers to 10 common minerals in order of their hardness. He assigned the number 10 to diamond, the hardest mineral known. Softer minerals were ranked in descending order, with talc, the softest mineral, assigned the number 1. The Mohs hardness scale (**Table 3.1**) provides a standard for testing minerals for preliminary identification. Gypsum, for example, has a hardness of 2 and can be scratched by a fingernail (**Figure 3.10**). More exacting measures of hardness show that diamond is by far the hardest mineral.

Density is the ratio of the weight of a substance to its volume. For example, at room temperature, 1 cm³ of water weighs 1 g; the density is thus 1 g/cm³. On the other hand, 1 cm³ of solid lead weighs a little over 11 g, and thus its density is 11 g/cm³.

Density is one of the more precisely defined properties of a mineral. It depends on the kinds of atoms making up the mineral and how closely they are packed in the crystal structure. Clearly, the more numerous and compact the atoms, the higher the density. Most common rock-forming minerals have densities that range from 2.65 g/cm³ (for quartz) to about 3.37 g/cm³ (for magnesium olivine). Iron-rich olivine is even denser (4.4 g/cm³) because iron has a higher atomic weight than magnesium. Some metallic minerals have much higher densities. For example, native gold has a density of about 20 g/cm³ and native iron has a density of almost 8 g/cm³. At high pressures, the densities of most minerals increase because the atoms are forced to be closer together. At high temperatures, their densities decrease as the atoms move farther apart.

Color is one of the more obvious properties of a mineral. Unfortunately, it is not diagnostic. Most minerals are found in various hues, depending on such factors as subtle variations in composition and the presence of inclusions and impurities. Quartz, for example, ranges through the spectrum from clear, colorless crystals to purple, red, white, yellow, gray, and black.

(A) Prismatic tourmaline [Na(Li,Al)$_3$Al$_6$(BO$_3$)$_3$Si$_6$O$_{18}$(OH)$_4$].

(B) Tetrahedrons of sphalerite (ZnS).

(C) Needles of the rare mineral crocoite (PbCrO$_4$).

(D) Radiating clusters of long slender needles of the zeolite mineral mordenite (Ca,Na$_2$,K$_2$)(Al$_2$Si$_{10}$)(O$_{24}$·7H$_2$O).

(E) Cubes of pyrite (FeS$_2$), commonly known as fool's gold.

Photographs by Jeffrey A. Scovil

Figure 3.8 Crystal form is an important physical property showing the arrangement of atoms in a mineral.

(A) One plane of cleavage in mica produces thin plates or sheets.

(B) Two planes of cleavage at right angles in feldspar produce blocky fragments.

(C) Three planes of cleavage at right angles in halite produce cubic fragments.

(D) Cleavage of calcite occurs in three planes that do not intersect at right angles, forming rhombohedrons.

Figure 3.9 **Cleavage** reflects planes of weakness within a crystal structure.

Table 3.1 Mohs Hardness Scale

Hardness	Mineral	Test
1	Talc	
2	Gypsum	Fingernail
3	Calcite	Copper coin
4	Fluorite	
5	Apatite	Knife blade or glass plate
6	K-feldspar	
7	Quartz	Steel file
8	Topaz	
9	Corundum	
10	Diamond	

Figure 3.10 Hardness reflects the strength of the atomic bonds inside the mineral. Gypsum has a hardness of 2 on Mohs hardness scale. It is a very soft mineral and can easily be scratched with a fingernail.

Luster describes the appearance of light reflected from a mineral's surface. Luster is described only in subjective, imprecise terms. There are two basic kinds of luster: metallic and nonmetallic. Minerals with a metallic luster shine like metals. Nonmetallic luster ranges widely, including vitreous (glassy), porcelainous, resinous, and earthy (dull). The luster of a mineral is controlled by the kinds of atoms and by the kinds of bonds that link the atoms together. Many minerals with covalent bonds have a brilliantly shiny luster, called adamantine luster, as in diamond. Ionic bonds create more vitreous luster, as in quartz. Metallic bonding in native metals, such as gold, also has its characteristic luster.

Streak refers to the color of a mineral in powder form and is usually more diagnostic than the color of a large specimen. For example, the mineral pyrite (fool's gold) has a gold color but a black streak, whereas real gold has a gold streak—the same color as that of larger grains. Streak is tested by rubbing a mineral vigorously against the surface of an unglazed piece of white porcelain. Minerals softer than the porcelain leave a streak, or line, of fine powder. For minerals harder than porcelain, a fine powder can be made by crushing a mineral fragment. The powder is then examined against a white background.

Magnetism is a natural characteristic of only a few minerals, like the common iron oxide magnetite. Although only a few minerals can be identified using this property, magnetism is an important physical property of rocks that is used in many investigations of how Earth works.

Stability Ranges

Another important feature of each mineral is that it is stable only over a fixed range of conditions. We call a mineral **stable** if it exists in equilibrium with its environment. In such a case, there is little tendency for further change. The environment that exists when a mineral crystallizes determines which of the many thousands of minerals will form. The environmental conditions that determine whether a particular mineral is stable are mainly pressure, temperature, and composition.

We have already examined the **stability ranges** for the various states of water (see Figure 3.5), and we can use similar phase diagrams to represent the range of conditions over which a specific mineral is stable. **Figure 3.11** shows the names and stability fields for various minerals with the chemical formula SiO_2. Quartz is the most common of these minerals because it is stable over the range of temperatures and pressures found near Earth's surface. However, if the temperature is increased to 1300°C at a pressure of 1000 bars (a depth in Earth of about 3 km), the arrangement of the atoms in quartz will change to form a different mineral called tridymite, which has its own structure and distinctive physical properties. For example, quartz has a density of 2.65 g/cm³ and tridymite has a density of about 2.26 g/cm³. If the temperature is increased to 1600°C, still another change occurs as tridymite converts to cristobalite with a density of 2.33 g/cm³. In the absence of water, pure SiO_2 melts only at a temperature higher than 1700°C. Changes in pressure can also induce minerals to break down and form new species that are stable under the new conditions. Metamorphic processes are driven by the tendency for minerals to react and change as their environment changes.

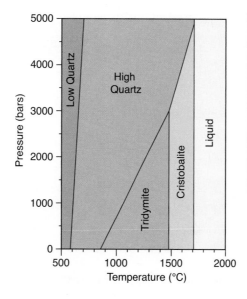

Figure 3.11 The stable form of SiO₂ depends on pressure and temperature. The colored areas show the range of temperature and pressure over which each of five different minerals are stable. Quartz, for example, is stable at intermediate temperatures over a wide range of pressures. Other minerals of the same composition (SiO_2), but with different atomic arrangements, are stable at other pressures and temperatures.

Although minerals have distinctive stability ranges, they may remain in existence far from those conditions. A mineral existing outside of its stability range is called **metastable**. Metastability occurs if the reactions to form new minerals from preexisting minerals are very slow. Such barriers are common at Earth's surface, where low temperatures make atomic movements and reactions very sluggish in solids. Thus, tridymite has been found at temperatures far below the range of temperatures shown in Figure 3.11. Moreover, feldspars are common at Earth's surface, even though clay minerals are more stable in the presence of water. Despite these reaction barriers at low temperature, it is useful to keep in mind the approximate range of temperatures and pressures over which a given mineral is stable.

The Growth and Destruction of Minerals

Minerals grow as matter changes from a gaseous or liquid state to a solid state or when one solid recrystallizes to form another. They break down as the solid changes back to a liquid or a gas. All minerals came into being because of specific physical and chemical conditions, and all are subject to change as these conditions change. Minerals, therefore, are an important means of interpreting the changes that have occurred in Earth throughout its history.

Crystal Growth

Even though minerals are inorganic, they can grow. Growth is accomplished by **crystallization,** which occurs by the addition of ions to a crystal face. As noted above, an environment suitable for crystal growth includes (1) proper concentration of the kinds of atoms or ions required for a particular mineral and (2) proper temperature and pressure.

The time-lapse photographs in **Figure 3.12** show how crystals grow from a liquid in an unrestricted environment. Although the size of each crystal increases, its form and internal structure remain the same. New atoms are added to the faces of the crystal, parallel to the plane of atoms in the basic structure. Some crystal faces, however, grow faster than others. As a result of these different growth rates, the crystal may become elongated in one direction. Thus, the ideal crystal shape reflects not only the arrangement of atoms inside the crystal, but it also controlled by which faces grow faster or slower. You can see that all of the crystals in Figure 3.12 have the same idealized shape, because they are all the same mineral.

How can a mineral, which is inorganic, grow?

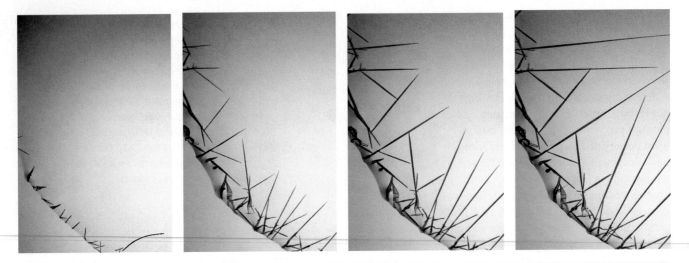

Figure 3.12 Crystal growth can be recorded by time-lapse photography. Each crystal grows as atoms in the surrounding liquid lock onto the outer faces of the crystal structure.

In contrast, where space is restricted, a crystal may not grow to form its ideal crystal shape. Where a growing crystal encounters a barrier (such as another crystal), it simply stops growing. This process is illustrated in Figure 3.12. Note how the vertical crystal grew between 10 and 30 seconds. At 30 seconds, it has impinged on a nearly horizontal crystal and stopped growing. However, the more horizontal crystal grows throughout the sequence because there were no restrictions to its growth.

Figure 3.13 shows how crystal growth occurs in a restricted environment. A crystal growing from a liquid in a restricted space assumes the shape of the confining area, and well-developed crystal faces do not form. The external form of the crystal can thus take on practically any shape, but its internal structure is in no way modified. The mineral's internal structure remains the same; its composition is unaffected, and no changes in its physical and chemical properties occur. The only modification is a change in the shape of the crystal.

Crystal growth in restricted spaces is common for rock-forming minerals. In a still molten lava flow or in an aqueous solution, many crystals grow at the same time and must compete for space. As a result, in the later stages of growth, crystals in rocks commonly lack well-defined crystal faces and typically interlock with adjacent crystals to form a strong, coherent mass (**Figure 3.14**). This interlocking texture is

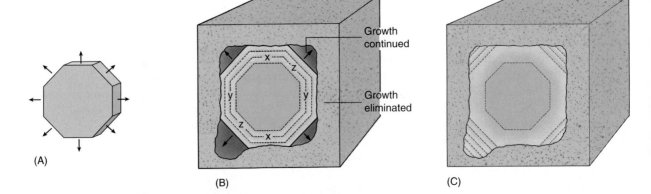

Figure 3.13 Crystals growing in a restricted environment do not develop perfect crystal faces. **(A)** Where growth is unrestricted, all crystal faces grow with equal facility. **(B)** In a restricted environment, growth on certain crystal faces, such as x and y, is terminated but growth on the faces labeled z continues. **(C)** The final shape of the crystal is determined by the geometry of the available space in which it grows.

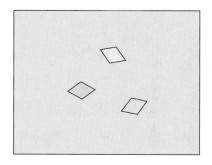  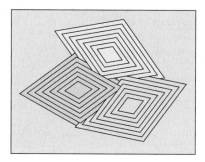

Figure 3.14 **Interlocking texture** develops if crystals grow in a restricted environment. Crystals grow into one another when they are forced to compete for space. Such textures are common in igneous rocks which form from molten magma.

especially common in igneous rocks, which form by crystallization from molten rock material.

Most crystals are rather small, measuring from a few tenths of a millimeter to several centimeters in diameter. Some are so small they can be seen only when enlarged thousands of times with a high-powered electron microscope (**Figure 3.15**). Where crystallization occurs from a mobile fluid in an unrestricted environment, however, crystals can grow to enormous sizes (**Figure 3.16**). Such a texture is illustrated in the chapter opening photomicrograph. The gray and white grains are feldspar and the bright pink, blue, and yellow grains are olivine and pyroxene. The early grown olivine and feldspar crystals acquired their ideal crystal shapes, but the later grown blue and brown pyroxenes lack simple crystal shapes because they filled in between the earlier grains.

Destruction of Crystals

Mineral grains can be destroyed in many different ways. Minerals **melt** by removal of outer atoms from the crystal structure as they enter a less organized liquid state. The heat that causes a crystal to melt increases atomic vibrations enough to break the

Courtesy of George W. Bolger, PetroTech Associates

(A) Sand grains magnified 50 times. Small crystals of clay form between the grains.

(B) Hexagonal crystals of clay in the pore space between sand grains, magnified 1000 times.

Figure 3.15 **Submicroscopic crystals** of hexagonal plates of clay growing in the pore spaces between sand grains can be seen with an electron microscope. Each crystal contains all of the physical and chemical properties of the mineral, even though each one is extremely small.

Figure 3.16 Large crystals can form where there is ample space for growth, as in caves. These crystals of gypsum are more than 1 m long.

bonds holding an atom to the crystal structure. Similarly, atoms can be "pried" loose and carried away by a solvent, usually (in geologic processes) water. Crystals begin to break down or dissolve at the surface and the reaction moves inward.

Mineral grains can also be destroyed as their constituent atoms become rearranged in the solid state. Such **recrystallization** processes are especially common deep inside the crust and mantle, where heat and pressure cause some crystal structures to collapse and new minerals, with a denser, more compact atomic structure (**Figure 3.17**) to form in their place. In this case, the atoms do not move far, but new bonds form and new internal structures are created. The new mineral grains have different physical properties, like cleavage, luster, hardness, and density.

Silicate Minerals

More than 95% of Earth's crust is composed of silicate minerals, a group of minerals containing silicon and oxygen linked in tetrahedral units, with four oxygen atoms to one silicon atom. Several fundamental configurations of tetrahedral groupings are single chains, double chains, two-dimensional sheets, and three-dimensional frameworks.

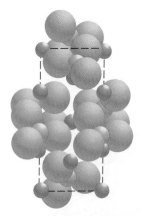

(A) Open structure at low pressure.

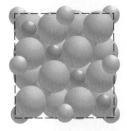

(B) Densely packed structure at high pressure.

Figure 3.17 Under high pressure, the atomic structure of a mineral can collapse into a denser form, in which the atoms are more closely packed. Although the physical properties change, the chemical composition may remain the same.

Although more than 4000 minerals have been identified, 95% of the volume of Earth's crust is composed of a group of minerals called the silicates. This should not be surprising because silicon and oxygen constitute nearly three-fourths of the mass of Earth's crust (**Table 3.2**) and therefore must predominate in most rock-forming minerals. Silicate minerals are complex in both chemistry and crystal structure, but all contain a basic building block called the silicon–oxygen tetrahedron. Nearly covalent Si–O bonds form a complex ion [$(SiO_4)^{4-}$] in which four large oxygen ions (O^{2-}) are arranged to form a four-sided pyramid with a smaller silicon ion (Si^{4+}) bonded between them (**Figure 3.18**). This geometric shape is known as a tetrahedron. The major groups of silicate minerals differ mainly in the arrangement of such silicate tetrahedrons in their crystal structures.

Perhaps the best way to understand the unifying characteristics of the **silicates**, as well as the reasons for the differences, is to study the models shown in **Figure 3.19**. These were constructed on the basis of X-ray studies of silicate crystals. **Silicon–oxygen tetrahedrons** combine to form minerals in two ways. In the simplest combination, the oxygen ions of the tetrahedrons form bonds with other elements, such as iron or magnesium. Olivine is an example. Most silicate minerals, however, are formed

Table 3.2 Concentrations of the Most Abundant Elements in Earth's Crust (by weight)

Element	Percentage
O	46.60
Si	27.72
Al	8.13
Fe	5.00
Ca	3.63
Na	2.83
K	2.59
Mg	2.09
Ti	0.44
H	0.14
P	0.12
Mn	0.10
S	0.05
C	0.03

Source: Data from Mason and Moore, *Principles of Geochemistry*, 4th ed. New York: Wiley, 1982.

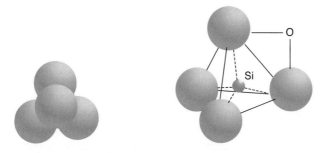

Figure 3.18 The silicon–oxygen tetrahedron is the basic building block of the silicate minerals. In this figure, the diagram on the right is expanded to show the position of the small silicon atom. Four large oxygen ions are arranged in the form of a pyramid (tetrahedron), with a small silicon ion covalently bonded into the central space between them. This is the most important building block in geology because it is the basic unit for 95% of the minerals in Earth's crust.

by the sharing of an oxygen ion between two adjacent tetrahedrons. In this way, the tetrahedrons form a larger ionic unit, just as beads are joined to form a necklace. The sharing of oxygen ions by the silicon ions results in several fundamental configurations of tetrahedral groups. These structures define the major silicate mineral groups:

1. Isolated tetrahedrons (example: olivine)
2. Single chains (example: pyroxene)
3. Double chains (example: amphibole)
4. Two-dimensional sheets (examples: micas, chlorite, and clays)
5. Three-dimensional frameworks (examples: feldspars and quartz)

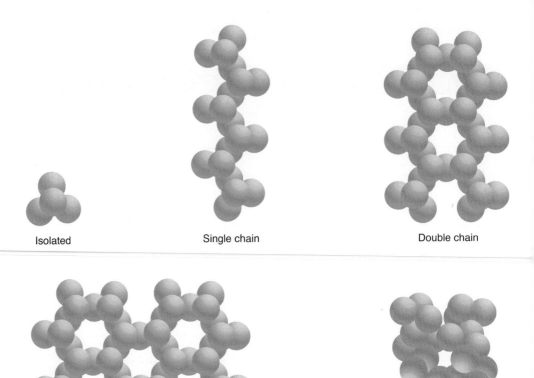

Isolated   Single chain   Double chain

Two-dimensional sheet   Three-dimensional framework

Figure 3.19 Silicon–oxygen tetrahedral groups can form various structures by the sharing of oxygen ions among silicon ions. A small silicon ion lies at the center of each tetrahedral unit. In general, various types of metal ions complete the mineral structure; they are not shown here.

The unmatched electrons of the silicate tetrahedron are balanced by various metal ions, such as ions of calcium, sodium, potassium, magnesium, and iron. The silicate minerals thus contain silicon–oxygen tetrahedrons linked in various patterns by metal ions. Considerable ionic substitution can occur in the crystal structure. For example, sodium can substitute for calcium, or iron can substitute for magnesium. Minerals of a major silicate group can thus differ chemically from one another but have a common silicate structure.

Rock-Forming Minerals

Fewer than 20 kinds of minerals account for the great bulk of Earth's crust and upper mantle. The most common silicate minerals are feldspars, quartz, micas, olivine, pyroxenes, amphiboles, and clay minerals. Important nonsilicate minerals are calcite, dolomite, halite, and gypsum.

Most of Earth's crust and upper mantle are composed of silicate minerals in which the common elements—such as iron, magnesium, sodium, calcium, potassium, and aluminum—combine with silicon and oxygen. The identification of these minerals presents some special problems. Rock-forming minerals rarely have well-developed *crystal faces* because (1) they grow by crystallization from melts (e.g., magmas) or from aqueous solutions (e.g., seawater) and vigorously compete for space; (2) they are abraded as they are transported as sediment; or (3) they are deformed under high temperature and pressure. In addition, most rock-forming mineral grains are small, generally less than the size of your little fingernail, so their physical properties may be difficult to see without a hand lens or microscope. Further complications arise because most rock-forming mineral groups have variable compositions attributable to ionic substitution in the crystal structure. As a result, color, hardness, and other physical properties may be variable.

It is important for you to become familiar with the general characteristics of each of the major rock-forming mineral groups (feldspars, quartz, micas, olivine, pyroxenes, amphiboles, clays, calcite, dolomite, halite, and gypsum) and to know something about their physical properties, their mode of origin, the environment in which they form, and their genetic significance. Some of the characteristics of these, as well as other important but less common minerals, are listed in **Table 3.3**. You will find the following summary of each mineral group to be much more meaningful if you examine a specimen of a rock containing the mineral while you study the written description.

A careful examination of the minerals that make up granite is a good beginning. The polished surface of granite (**Figure 3.20**) shows that the rock is composed of myriad mineral grains of different sizes, shapes, and colors. Although the minerals interlock to form a tight, coherent mass, each has distinguishing properties.

Felsic Silicate Minerals

One large group of silicate minerals includes the major constituents of continental crust: feldspars and quartz. These are commonly known as **felsic minerals**. (They are sometimes called sialic because they are rich in *silicon* and *aluminum*.) In addition to being the major constituents of continental crust, the felsic minerals also have low densities and crystallize at low temperatures in magmas.

Feldspars are the most abundant minerals in granite, a common crustal rock. The granite in Figure 3.20 consists largely of a pink, porcelainous mineral that has a rectangular form and a milky-white mineral that is somewhat smaller but similarly shaped. These are feldspars (German, "field crystals"), the most abundant minerals in Earth's crust, comprising about 50%. The feldspars have good cleavage in two directions, a porcelainous luster, and a hardness of about 6 on the Mohs hardness scale. The crystal structure involves a complex three-dimensional framework of silicate tetrahedrons (Figure 3.19). Considerable ionic substitution gives rise to two major types of feldspars: potassium feldspar (K-feldspar) and plagioclase feldspar. Potassium feldspar ($KAlSi_3O_8$) is commonly pink in granitic rocks. **Plagioclase** feldspar (shown in gray in the sketch) permits complete substitution of sodium (Na) for calcium (Ca) in the crystal structure, giving rise to a compositional range from $NaAlSi_3O_8$ to $CaAl_2Si_2O_8$. Moreover, most grains of plagioclase have distinctive, closely spaced striations on their cleavage planes. Plagioclase in granite is rich in sodium. Feldspars (with a density of 2.7 g/cm^3) are common in most igneous rocks, in many metamorphic rocks, and in some sedimentary rocks. Consequently, the continental crust has a characteristically low density (ranging from 2.6 to 2.7 g/cm^3), controlled by the shear abundance of feldspar and quartz.

Quartz forms the glassy, irregularly shaped grains in Figure 3.20. It usually grows in the spaces between the other minerals. As a result, quartz in granite typically lacks well-developed crystal faces. When quartz crystals are able to grow freely, their form is elongated, has six sides, and terminates in a point, but well-formed crystals are rarely found in rocks. In sandstone, quartz is abraded into rounded sand grains.

Quartz is abundant in all three major rock types. It has the simple composition SiO_2 and is distinguished by its hardness (7), its conchoidal fracture, and its glassy luster. Pure quartz crystals are colorless, but slight impurities produce a variety of colors. Quartz is made of silicate tetrahedrons linked together in a tight framework. All of the bonds are between Si and O; it includes no other elements. As a result, quartz is very hard, and, because all of the bonds have the same strength, it lacks cleavage. Quartz is stable both mechanically (it is very hard and lacks cleavage) and chemically (it does not react with elements at or near Earth's surface). It is therefore a difficult mineral to alter or break down once it has formed.

Table 3.3 Earth's Common Minerals

Name	Composition	Cleavage/Fracture	Color	Hardness	Density (g/cm³)	Comments
Amphibole	$Ca_2(Mg,Fe)_5Si_8O_{22}(OH)_2$	Two at 60° and 120°	Black to green	5–6	3.2	
Bauxite	$AlO(OH)$	One perfect	White	6.5	3.4	Aluminum ore, mineral diaspore
Beryl	$Be_3Al_2Si_6O_{18}$	One poor	Green, blue, red	8	2.7	Emerald is gem variety Hexagonal prisms
Biotite	$K(Mg,Fe)_3AlSi_3O_{10}(OH)_2$	One perfect	Black to dark brown	2.5–3	3	Splits into thin sheets
Calcite	$CaCO_3$	Three perfect Rhombohedral	Colorless, white	3	2.7	Bubbles in dilute acid
Chalcopyrite	$CuFeS_2$	Fracture	Brassy, golden yellow Metallic luster	4	4.3	Copper ore
Chlorite	$(Mg,Fe)_5Al_2Si_3O_{10}(OH)_8$	One perfect	Green	2	2.5	Foliated masses
Clay	$Al_2Si_2O_5(OH)_4$	One perfect	White to brown	2	2.0–2.5	Common in soils
Corundum	Al_2O_3	Fracture	Brown or blue	9	4	Rubies and sapphires
Diamond	C	Fractures	Transparent Adamantine luster	10	3.5	Hardest mineral known
Dolomite	$CaMg(CO_3)_2$	Three perfect	Transparent to white	3.5–4	2.8	Bubbles in acid when powdered
Fluorite	CaF_2	Perfect	Transparent, green, purple, yellow	4	3.2	Fluorine ore
Galena	PbS	Three perfect Cubic	Black to silver Metallic luster	2.5	7.6	Lead ore
Garnet	$Ca_3Al_2Si_3O_{12}$	Conchoidal fracture	Red to brown Vitreous luster	6.5–7	3.6	
Graphite	C	One perfect	Black	1–2	2.1	Compare with diamond
Gypsum	$CaSO_4 \cdot 2H_2O$	One perfect Two good	Transparent to white	2	2.3	Used in plasterboard
Halite	$NaCl$	Three perfect Cubic	Transparent to white	2.5	2.2	Table salt

Table 3.3 Earth's Common Minerals (*continued*)

Name	Composition	Cleavage/ Fracture	Color	Hardness	Density (g/cm³)	Comments
Hematite	Fe_2O_3	None	Red to silvery gray Metallic or earthy	6	5.3	Iron ore
K-feldspar	$KAlSi_3O_8$	Two at right angles	White to gray or pink	6	2.6	
Kyanite	Al_2SiO_5	One perfect One poor	White to light blue	5–7	3.6	Long-bladed aggregates
Magnetite	Fe_3O_4	Conchoidal irregular	Black Metallic luster	6	5.2	Magnetic
Muscovite	$KAl_3Si_3O_{10}(OH)_2$	One perfect	Colorless to light brown	2–2.5	2.8	Splits into translucent sheets
Olivine	$(Mg,Fe)_2SiO_4$	Conchoidal	Green to brown	6.5	3.4	Gem peridot
Plagioclase	$NaAlSi_3O_8$ $CaAl_2Si_2O_8$	Two at right angles	White to gray	6	2.7	Striations on cleavage planes Most common mineral at surface
Pyrite	FeS_2	Uneven fractures	Brassy to golden yellow	6.5	5	Fool's gold; well-formed cubes common
Pyroxene	$(Mg,Fe)SiO_3$	Two at about 90°	Green to dark brown or black	6	3.3	
Quartz	SiO_2	Conchoidal fracture	Colorless, also gray, purple, other	7	2.7	Six-sided elongate crystals
Serpentine	$Mg_6Si_4O_{10}(OH)_8$	Splintery fracture Asbestos fibrous	Green to brown Silky or waxy luster	2.5	2.5	
Sillimanite	Al_2SiO_5	One perfect	Colorless to white	6–7	3.2	Long, slender crystals
Staurolite	$Fe_2Al_9Si_4O_{22}(OH)_2$	One poor	Brown to red	7	3.8	
Talc	$Mg_3Si_4O_{10}(OH)_2$	One perfect	White to light green	1	2.8	Soft, soapy masses
Zeolite	Complex hydrous silicates	One perfect	Colorless to light green	4–5	2.2	Earthy, but may form radiating crystals in cavities

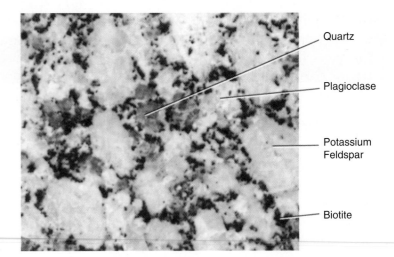

Quartz

Plagioclase

Potassium
Feldspar

Biotite

(A) A polished surface of granite, shown at actual size, displays mineral grains of different sizes, shapes, and colors.

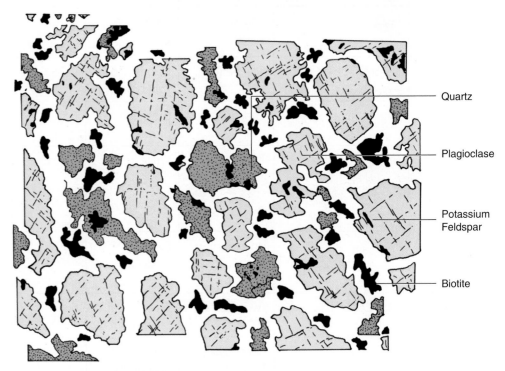

Quartz

Plagioclase

Potassium
Feldspar

Biotite

(B) An exploded diagram of (A) shows the relative size and the shape of individual mineral grains.

Figure 3.20 Mineral grains in granite, a common rock in continental crust, form a tight, interlocking texture because each mineral is forced to compete for space as it grows. The most common minerals in granite are the felsic minerals: quartz, plagioclase feldspar, and potassium feldspar.

Micas are the tiny black, shiny grains in Figure 3.20. These distinctive minerals are potassium aluminum silicates. Micas are readily recognized by their perfect one-directional cleavage, which permits breakage into thin, elastic flakes. Mica is a complex silicate with a sheet structure, which is responsible for its perfect cleavage. Two common varieties occur in rocks: **muscovite** [$KAl_3Si_3O_{10}(OH)_2$], which is white or colorless and is found along with felsic minerals, and **biotite** [$K(Mg,Fe)_3AlSi_3O_{10}(OH)_2$], a black mica, rich in iron and magnesium that belongs to the category of mafic minerals discussed below. Both types of mica contain water in the form of hydroxyl ions (OH^-). The densities of these minerals are also distinctive, with biotite (about 3 g/cm^3) denser than muscovite (about 2.8 g/cm^3). Mica is abundant in granites and in many metamorphic rocks and is also a significant constituent of many sedimentary rocks.

Mafic Silicate Minerals

Another category of silicate minerals is the **mafic minerals**, so named because they contain much *magnesium* and *iron*. These minerals contrast with felsic minerals and generally range from dark green to black and have high densities. Biotite is classified in this general group, together with the olivine, pyroxene, and amphibole. In granite, biotite is common, but the other mafic minerals are rare or absent. The mafic minerals are common, however, in Earth's mantle and in oceanic crust. They generally crystallize at higher temperatures and have higher densities than felsic minerals. Let us examine basalt, a common mafic volcanic rock, to see what these minerals are like (**Figure 3.21**).

Olivine is the only mineral clearly visible in the hand specimen in Figure 3.21; it is a green, glassy mineral. Olivine is a silicate in which iron and magnesium substitute freely in the crystal structure. The composition is expressed as $(Mg,Fe)_2SiO_4$. Olivine is composed of isolated Si–O tetrahedrons linked together by magnesium or iron ions

(A) In a hand specimen, only a few large grains of green olivine can be seen. The dark spots are gas bubbles frozen into the once molten rock.

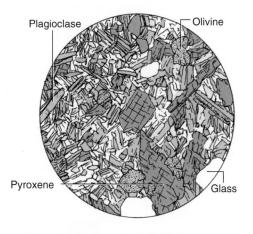

(B) Viewed through a microscope, the mineral grains form an interlocking texture. Plagioclase feldspar crystals typically form small lathlike grains between the mafic minerals.

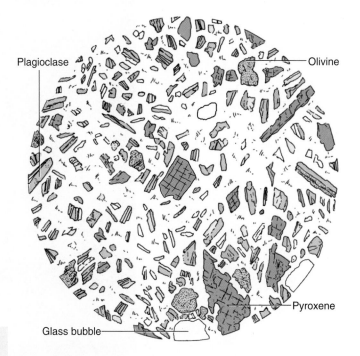

Figure 3.21 Mineral grains in basalt are microscopic and are dominated by mafic minerals. Basalt is a mafic volcanic rock common in the oceanic crust.

(C) An exploded diagram of (B) shows the size and shape of individual mineral grains.

(Figure 3.19). This hard mineral is characterized by an olive-green color (if magnesium is abundant) and a glassy luster. In rocks, it rarely forms crystals larger than a millimeter in diameter. Like most mafic minerals, olivine has a relatively high density (about 3.3 g/cm³) and typically forms at high temperatures. It is probably a major constituent of the upper mantle. At depths of about 400 km in the mantle, olivine is no longer stable and recrystallizes to form an even denser mineral with the same elemental composition.

Pyroxenes are high-temperature minerals also found in many mafic rocks in the crust and mantle. In Figure 3.21, pyroxene occurs as microscopic crystals, but some basalts contain larger grains of this mineral, which typically range from dark green to black. Their internal structure consists of single chains of linked Si–O tetrahedrons (Figure 3.19). Pyroxene crystals commonly have two directions of cleavage that intersect at right angles.

Amphiboles (**Figure 3.22**) have much in common with the pyroxenes. Their chemical compositions are similar, except that amphiboles contain hydroxyl ions (OH⁻) and pyroxenes do not. The minerals also differ in structure. The internal structure consists of double chains of silicon–oxygen tetrahedrons (Figure 3.19). The amphiboles produce elongate crystals that cleave perfectly in two planes, which are not at right angles. Amphibole ranges from green to black. This mineral is common in many igneous and metamorphic rocks. Hornblende [NaCa(Mg,Fe)$_5$AlSi$_7$O$_{22}$(OH)$_2$] is the most common variety of amphibole. The density of a typical amphibole is about 3.2 g/cm³.

A dangerous form of amphibole is asbestos, once used widely to make fireproof fabrics, tiles, and as insulation in buildings. Miners working in old dusty mines became sick as small cleavage fragments of a specific type of this mineral became lodged in their lungs, especially in conjunction with cigarette smoking. The incidence of this noncancerous lung disease in modern mines with dust controls is much lower. Fortunately, most asbestos used in construction consists of an entirely different mineral, and the hazard to people is much less than commonly supposed.

Figure 3.22 Amphibole crystals were among the first to crystallize in this "granitic" rock and therefore have well-developed crystal faces. The largest grain is about 3 cm long.

Clay Minerals

The **clay minerals** form another important group of silicate minerals. They are a major part of the soil and are thus encountered more frequently in everyday experience than many other minerals. Clay minerals form at Earth's surface, where air and water react with various silicate minerals, breaking them down to form clay and other products. Like the micas, the clay minerals are sheet silicates (Figure 3.19), but their crystals are usually microscopic and are most easily detected with an electron microscope (Figure 3.15). More than a dozen clay minerals can be distinguished on the basis of their crystal structures and variations in composition. A common clay mineral, kaolinite, has the formula $Al_4Si_4O_{10}(OH)_8$ and a low density of about 2.6 g/cm^3.

Nonsilicate Minerals

Some important rock-forming minerals are not silicate minerals. Most of these minerals are carbonates or sulfates and typically form at low temperatures and pressures near Earth's surface.

Calcite is composed of calcium carbonate ($CaCO_3$), the principal mineral in limestone. It can precipitate directly from seawater or is removed from seawater by organisms as they use it to make their shells. Calcite is dissolved by groundwater and reprecipitated as new crystals in caves and fractures in rock. It is usually transparent or white, but the aggregates of calcite crystals that form limestone contain various impurities that give them gray or brown hues. Calcite is common at Earth's surface and is easy to identify. It is soft enough (hardness of 3) to scratch with a knife, and it effervesces in dilute hydrochloric acid. It has perfect cleavage in three planes, which are not at right angles, so that cleaved fragments form rhombohedra (see Figure 3.9). Besides being the major constituent of limestone, calcite is the major mineral in the metamorphic rock marble. Calcite has a density of about 2.7 g/cm^3.

Dolomite is a carbonate of calcium and magnesium [$CaMg(CO_3)_2$]. Large crystals form rhombohedra, but most dolomite occurs as granular masses of small crystals. Dolomite is widespread in sedimentary rocks, forming when calcite reacts with solutions of magnesium carbonate in seawater or groundwater. Dolomite can be distinguished from calcite because it effervesces in dilute hydrochloric acid only if it is in powdered form. Dolomite has a density of nearly 2.9 g/cm^3, denser than many silicate minerals.

Halite and **gypsum** are the two most common minerals formed by evaporation of seawater or saline lake water. Halite, common salt ($NaCl$), is easily identified by its taste. It also has one of the simplest of all crystal structures; the sodium and chloride ions form a cubical array. Most physical properties of halite are related to this structure. Halite crystals cleave in three planes, at right angles, to form cubic or rectangular fragments (Figure 3.9). Salt, of course, is very soluble and readily dissolves in water.

Gypsum is composed of calcium sulfate and water ($CaSO_4 \cdot 2H_2O$). It forms crystals that are generally colorless, with a glassy or silky luster. It is a very soft mineral and can be scratched easily with a fingernail. It cleaves perfectly in one plane to form thin, non-elastic plates (Figure 3.10). See the GeoLogic discussion in this chapter for more information about the internal structure of gypsum. Gypsum occurs as single crystals, as aggregates of crystals in compact masses (alabaster), and as a fibrous form (satin spar).

Oxide minerals lack silicon as well and include several economically important iron oxides, such as magnetite and hematite (Table 3.3). Magnetite is particularly interesting because it is one of only a very few minerals that are naturally magnetic.

A wide variety of other minerals have been identified, including silicates, carbonates, oxides, sulfides, and sulfates. There are literally thousands of naturally formed minerals; some seem rare and exotic because of their color, crystal form, and hardness, and others seem more mundane because they occur as minor constituents in common rocks. Some we consider precious, such as gold, silver, diamonds, and rubies; others are important in high technology. In addition to providing documents of Earth's history, minerals are at the foundation of all human societies—from pre-Paleolithic times, in which minerals were used for tools, to modern technological societies that require vast amounts of metals and construction materials.

GeoLogic   Internal Structure of Minerals

Gypsum ($CaSO_4 \cdot 2H_2O$) (1 cm across). Cleavage planes look like topographic steps.

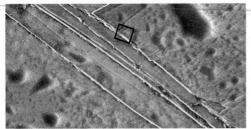

Gypsum seen with scanning electron microscope (150 microns across). Cleavage planes still visible as plateaus separating different cleavage sheets.

Gypsum seen with atomic force microscope (10 microns across). Each cleavage sheet consists of even thinner layers.

Gypsum seen with scanning tunneling microscope (15 nanometers across). Individual sulfate ions seen as hills in precise geometric arrangement. Lighter area is an atom high "plateau," magnified from the sheets seen in the cleavage planes.

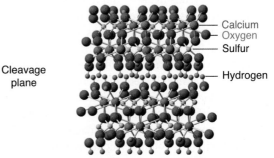

Calcium
Oxygen
Sulfur
Hydrogen
Cleavage plane

Atomic model of the internal structure of gypsum. Calcium ions (blue) are linked to sulfate groups (sulfur yellow and oxygen red). Sulfur–oxygen bonds are strong and covalent. Cleavage planes are created by weak bonds between these tightly bonded sheets. The layers are bound together by weak hydrogen bonds. Hydrogens are small and pink.

Seeing is believing—a phrase we often use to discount the unseeable. But how can we understand the internal structure of minerals at the atomic scale where distances are measured in nanometers (10^{-9} m)? These images take you on a tour through inner space, from the surface of a mineral into its deep interior.

Observations

1. At the lowest magnification with an optical microscope, you can see the nature of a cleavage plane in gypsum—a smooth lustrous break.
2. If we zoom in closer with a scanning electron microscope, you can see that the cleavage plane is not quite as smooth as it first looked, but you can still see broad flat plateaus.
3. Zooming in closer with an atomic force microscope, we can see that the planar structure is preserved at the micron scale (10^{-6} m). You can see that the mineral grew as a series of layers controlled by its internal structure. Each "layer" has a relatively smooth surface but is only a molecule thick.

4. At even higher resolution, the atomic force microscope shows the "smooth surface" is a series of humps and swales—individual groups of atoms—packed into a precise geometric network. The "step" shown in brighter colors is one atomic layer thick and lies on top of other similar layers below it.

Interpretations

1. The last step of the journey is not a real image, but rather an interpretive model that has been constructed from the information gleaned by studying the other images and from X-ray diffractometry. Each group of atoms is bound together by a strong electrical charge emanating from a cloud of electrons. This electron cloud gives the atomic groups their shapes in the images. Finally, the images and measurements show us that the physical properties of a mineral are controlled by its internal atomic structure. For example, strong bonds form hard minerals and where weak bonds are aligned like those between the hydrogen ions shown in the model, the mineral cleaves easily in that direction.

Key Terms

amorphous solid (p. 60)
amphibole (p. 80)
atom (p. 56)
atomic mass (p. 57)
atomic number (p. 57)
biotite (p. 78)
calcite (p. 81)
clay mineral (p. 81)
cleavage (p. 65)
color (p. 65)
compound (p. 59)
conchoidal fracture (p. 65)
covalent bond (p. 59)
crystal (p. 61)
crystal faces (p. 65)

crystal form (p. 65)
crystallization (p. 69)
crystal structure (p. 60)
density (p. 65)
dolomite (p. 81)
electron (p. 57)
feldspar (p. 75)
felsic minerals (p. 75)
gas (p. 60)
gypsum (p. 81)
halite (p. 81)
hardness (p. 65)
ion (p. 58)
ionic bond (p. 59)
ionic substitution (p. 64)

isotope (p. 57)
liquid (p. 60)
luster (p. 68)
mafic mineral (p. 79)
magnetism (p. 68)
melt (p. 71)
metallic bond (p. 59)
metastable (p. 69)
mica (p. 78)
mineral (p. 61)
muscovite (p. 78)
neutron (p. 57)
nucleus (p. 56)
olivine (p. 79)
oxide mineral (p. 81)

plagioclase (p. 75)
polymorphism (p. 62)
proton (p. 57)
pyroxene (p. 80)
quartz (p. 75)
recrystallization (p. 72)
silicates (p. 72)
silicon–oxygen tetrahedron
   (p. 72)
solid (p. 60)
stability range (p. 68)
stable (p. 68)
streak (p. 68)
X-ray diffraction (p. 63)

Review Questions

1. Contrast atoms, ions, and isotopes.
2. Give a brief but adequate definition of a mineral.
3. Explain the meaning of "the internal structure of a mineral."
4. Why does a mineral have a definite chemical composition?
5. What common element might substitute for Ca in plagioclase feldspar? Why?
6. How do geologists identify minerals too small to be seen in a hand specimen?
7. Briefly explain how minerals grow and are destroyed.
8. Explain the origin of cleavage in minerals.
9. Describe the silicon–oxygen tetrahedron. Why is it important in the study of minerals?
10. Discuss the implications of a mineral's limited stability range for the kinds of minerals found at progressively greater depths in Earth's mantle.

11. Why is color of little use in identifying minerals? What are some better diagnostic properties?
12. What are silicate minerals? List the silicate minerals that are most abundant in rocks.
13. Why are feldspars so abundant in Earth's crust?
14. Construct a table listing the distinguishing characteristics of quartz, feldspar, biotite, amphibole, pyroxene, mica, and clay.
15. Study Figure 3.20, and explain why most of the mineral grains in granite have irregular shapes even though they still have an orderly atomic structure.
16. What is the difference between a mineral and a rock?

4 Igneous Rocks

Igneous rocks are records of the thermal history of Earth. Their origin is closely associated with the tectonic system, and they play an important role in the spreading of seafloor, the origin of mountains, and the evolution of continents. The best-known examples of igneous activity are volcanic eruptions, in which liquid rock material works its way to the surface and erupts from volcanic vents such as those shown above.

Less obvious, although just as important, are the enormous volumes of liquid rock that never reach the surface but remain trapped in the crust, where they cool and solidify. Granite is the most common variety of this type of igneous rock and is typically exposed in eroded mountain belts and in the roots of ancient mountain systems now preserved in the shields.

Two volcanoes, Juriques and Licancabur, dominate the skyline of this panorama of the high Andes Mountains in South America. These steep-sided volcanoes are only two of the hundreds of volcanoes along the borders between Bolivia, Chile, and Argentina. The volcanoes here are part of the famous ring of fire that encircles much of the Pacific Ocean. Even more importantly, they lie above a wet slab of subducting oceanic lithosphere that is being thrust beneath the continent. Subduction has created a folded mountain belt, thick crust, and a long chain of

active explosive volcanoes. Young volcanoes like Licancabur are dramatic proof that Earth's interior is still warm and active. In this spectacular image, you can see how igneous processes created the landscape. Molten flows of oozing lava and highly explosive pyroclastic flows created the composite volcanoes over the last 12,000 years. The magma formed deep in the mantle when solid rock melted at depths of more than 100 km; the temperature must have exceeded 1200°C. The molten magma then rose toward the surface through a series of fractures and pipes because its density was lower than that of the surrounding solid mantle. Alternating eruptions of viscous lava and flowing ash formed when the magma reached the surface, gradually building up the high conical mountains. Lava flows extend more than 10 km away from the summit and some rapidly moving, turbulent flows of ash dashed down slopes to reach at least 15 km away. The volcanoes dammed river drainages to create small lakes like Laguna Verde in the foreground. In this high and dry region, lakes may completely evaporate away only to be formed again later. Evaporation of the salty water leaves white deposits of salt and other minerals on the fluctuating shorelines. Because of the clear sky and dry climate, an array of 66 radio telescopes scans the skies from a site near Licancabur.

In this chapter, we study the major types of igneous rocks and what they reveal about the thermal activity of Earth. We pay particular attention to the compositions and distinctive textures of igneous rocks and how we can read from the interlocking network of grains the history of how the hot liquid became part of the solid crust.

Major Concepts

1. Magma is molten rock that originates from the partial melting of the lower crust and the upper mantle, usually at depths between 10 and 200 km below the surface.
2. The texture of a rock provides important insight into the cooling history of the magma. The major textures of igneous rocks are (a) glassy, (b) aphanitic, (c) phaneritic, (d) porphyritic, and (e) pyroclastic.
3. Most magmas are part of a continuum that ranges from mafic magma to silicic magma.
4. Silicic magmas produce rocks of the granite-rhyolite family, which are composed of quartz, K-feldspar, Na-plagioclase, and minor amounts of biotite or amphibole.
5. Basaltic magmas produce rocks of the gabbro-basalt family, which are composed of Ca-plagioclase and pyroxene with lesser amounts of olivine and little or no quartz.
6. Magmas with composition intermediate between mafic and silicic compositions produce rocks of the diorite-andesite family.
7. Basalt, the most abundant type of extrusive rock, typically either erupts from fissures to produce relatively thin lava flows that cover broad areas or erupts from central vents to produce shield volcanoes and cinder cones. Volcanic features developed by intermediate to silicic magmas include viscous lava flows, ash-flow tuff, composite volcanoes, and collapse calderas. The abundance of water in silicic magma is critical to its development and eruption.
8. Masses of igneous rock formed by the cooling of magma beneath the surface are called intrusions or plutons. The most important types of intrusions are batholiths, stocks, dikes, sills, and laccoliths.
9. The wide variety of magma compositions is caused by variations in (a) the composition of the source rocks, (b) partial melting, (c) fractional crystallization, (d) mixing, and (e) assimilation of solid rock into the molten magma.
10. Most basaltic magma is generated by partial melting of the mantle at divergent plate boundaries and in rising mantle plumes. Most intermediate to silicic magma is produced at convergent plate boundaries. Partial melting of continental crust at rifts and above plumes can also produce silicic magma.

The Nature of Igneous Rocks

Igneous rocks form from magma—molten rock material consisting of liquid, gas, and crystals. A wide variety of magma types exists, but important end members are (1) basaltic magma, which is typically very hot (from 900°C to 1200°C) and highly fluid, and (2) silicic magma, which is cooler (less than 850°C) and highly viscous.

The term *magma* comes from the Greek word that means "kneaded mixture," like a dough or paste. In its geologic application, it refers to hot, partially molten rock material (**Figure 4.1**). Most **magmas** are not entirely liquid but are a combination of liquid, solid, and gas. Crystals may make up a large portion of the mass, so magma could be thought of more accurately as slush, a liquid melt mixed with a mass of mineral crystals. Such a mixture has a consistency similar to that of freshly mixed concrete, slushy snow, or thick oatmeal. The movement of most magma is slow and sluggish.

What are the physical and chemical characteristics of molten rock?

Like most fluids, magma is less dense than the solid from which it forms, and because of buoyancy, it tends to migrate upward through the mantle and crust. Magma can intrude into the overlying rock by injection into fractures, it can dome the overlying rock, or it can melt and assimilate the rock it invades. Magma eventually cools and crystallizes to form **igneous rocks**. The rise of the magma may be halted

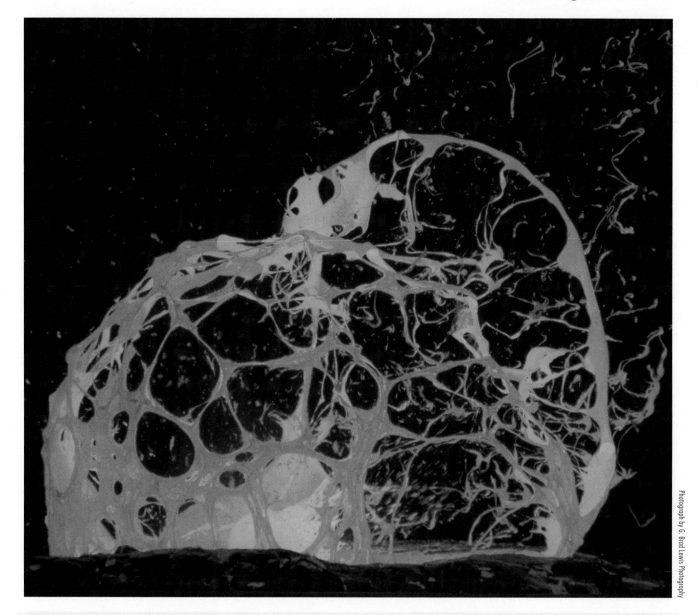

Figure 4.1 **Magma** is molten rock material that commonly contains gas and solids. Much of it cools deep in Earth's crust, but some of it works its way to the surface and is extruded as lava. Eruptions such as this permit us to study the nature of magma, its composition, and its physical characteristics and give us an insight into the origin of igneous rocks.

where it comes to density equilibrium with the surrounding rocks or where the roof rocks are too strong to allow the magma to penetrate farther. Magma that solidifies below the surface forms **intrusive rock**. When magma reaches the surface without completely cooling and flows out over the landscape as **lava**, it forms **extrusive rock**.

Chemical analyses of igneous rocks have revealed a wide variety of magma types, but most terrestrial magmas consist largely of molten silicates. The principal elements in such magmas are oxygen (O), silicon (Si), aluminum (Al), calcium (Ca), sodium (Na), potassium (K), iron (Fe), and magnesium (Mg). Two constituents—silica (SiO_2) and water (H_2O)—largely control the physical properties of magma, such as its density, **viscosity** (the tendency for a material to resist flow), and the manner in which it is extruded.

Although there is great variety in magma composition, we can illustrate much about silicate magma by examining only two extreme types. *Mafic magmas* contain about 50% SiO_2 and have temperatures ranging from about 1000°C to 1200°C. Mafic minerals, such as olivine and pyroxene, crystallize from such magmas. *Silicic magmas* contain between 65% and 77% SiO_2 and generally have temperatures lower than 850°C. Felsic minerals, such as feldspars and quartz, are the dominant minerals that crystallize from these magmas. Basaltic magmas are characteristically fluid, whereas silicic magmas are viscous. This is because silicic magmas have lower temperatures and greater amounts of SiO_2. The viscosity of magma is influenced by its SiO_2 content because silica tetrahedra bond or link together even before crystallization occurs, and the linkages offer resistance to flow. Temperature affects viscosity because as the temperature of magma drops, more and more linkages—prototypes of the minerals that crystallize—are made. Therefore, the higher the silica content or the lower the temperature, the greater the magma's viscosity.

Water vapor and carbon dioxide are the principal gases dissolved in magma. More than 90% of the gas emitted from hot magma is water (H_2O) and carbon dioxide (CO_2). Together, these **volatiles** (materials that are readily vaporized to form gases at Earth surface conditions) usually constitute from 0.1% to 5% by weight but may reach concentrations as high as 15% in some silicate magmas. These volatiles are important because they strongly influence the viscosity and melting point of a magma and the types of volcanic activity that can be produced. Dissolved water tends to decrease the viscosity of magma by breaking the Si–O bonds, which may otherwise form long, complex chains. Magmas rich in volatiles also tend to erupt more violently than volatile-poor magmas because of the explosive expansion of gas bubbles.

Where are most igneous rocks formed?

Igneous rocks are found in many parts of the globe, but they are actually formed in a few relatively restricted settings. On the continents, for example, most igneous rocks form at convergent plate margins where intrusions of magma feed overlying volcanoes. In North America (**Figure 4.2**), you can see ancient and recent examples. The intrusive rocks of western North America largely formed above an ancient subduction zone that no longer exists. Even older intrusive igneous rocks are exposed in the Canadian shield; they probably intruded the roots of mountain belts formed at ancient convergent plate margins. On the other hand, the young volcanic belt in the northwestern United States that extends into Alaska and the volcanic rocks of southern Mexico and Central America all erupted above still-active subduction zones. Igneous rocks are not common in the stable platform, but they may form in association with a mantle plume. For example, many geologists think that the lava flows in the Columbia River Plateau and Snake River Plain may have formed above a mantle plume that lies beneath the continent. A few other

Figure 4.2 Igneous rocks in North America are concentrated along recent and former convergent plate margins like those of the western (active) and eastern (inactive) mountain belts. Older, mostly intrusive, igneous rocks are found in the shields, especially that of eastern Canada. However, these igneous rocks probably formed at what were convergent plate boundaries. Igneous rocks are rare in the stable platform, but they can form above mantle plumes, like that thought to lie beneath Yellowstone Park.

Volcanic rocks   Plutonic rocks

volcanic rocks formed at rifts, such as the one that is forming the Gulf of California. In addition, the oceanic crust is almost entirely igneous rock formed at an oceanic rift.

Textures of Igneous Rocks

> The texture of a rock refers to the size, shape, and arrangement of its constituent mineral grains. The major textures in igneous rocks are glassy, aphanitic, phaneritic, porphyritic, and pyroclastic.

The texture of a rock can be compared to the texture of a piece of cloth. In this analogy, the mineral grains in a rock are likened to the yarn or threads that compose the cloth. The cloth's texture is determined by the weave (it may be open or tight; it may be knitted, woven, or felted), by the coarseness of the yarn, or by the mix of various yarns (coarse, fine, or a mixture of thicknesses). Careful examination of a piece of cloth will reveal how the cloth was constructed and the way in which it was formed. The color and the composition of the thread (whether it is silk, wool, or cotton), however, are separate properties of the cloth. Similarly, the **texture** of a rock is the size, shape, and arrangement of its constituent minerals. It is a characteristic separate and distinct from composition. Texture is important because the mineral grains bear a record of the energy changes involved in the rock-forming process and the conditions existing when the rock originated.

The genetic imprint left on the texture of a rock is commonly clear and easy to read. For example, rocks formed from a cooling liquid have a texture characterized by interlocking grains.

To illustrate the importance of texture, we will consider six examples of igneous rocks that have essentially the same chemical and mineralogical composition but different textures (**Figure 4.3**). In each rock, a chemical analysis would disclose about 48% O, 30% Si, 7% Al, and between 1% and 4% each of Na, K, Ca, and Fe. On the basis of chemical composition alone, these rocks would be considered the same; they differ in texture *only*. It is their texture that provides the most information about how each specimen was formed.

Glassy Texture

The nature of volcanic **glass** is illustrated in Figure 4.3A. The hand specimen displays a conchoidal fracture, with the sharp edges typical of broken glass. No distinct grains are visible, but viewed under a microscope, distinct flow layers are apparent. These result from the uneven concentration of innumerable, minute, "embryonic" crystals.

In the laboratory, melted rock or synthetic lava hardens to glass if it is quenched (or quickly cooled) from a temperature above that at which crystals would normally form. We can conclude that a **glassy texture** is produced by very rapid cooling. The randomness of the ions in a high-temperature melt is "frozen in" because the ions do not have time to migrate and organize themselves in an orderly, crystalline structure. Field observations of glassy rocks in volcanic regions support the hypothesis that rapid cooling produces glass. Small pieces of magma blown from a volcanic vent into the much cooler atmosphere harden to form glassy ash. A glassy crust forms on the surface of many lava flows, and glassy fragments form if a flow enters a body of water.

Aphanitic Texture

If crystal growth from a melt requires time for the ions to collect and organize themselves, then a crystalline rock indicates a slower rate of cooling than that of a glassy rock. The texture illustrated in Figure 4.3B is crystalline but extremely fine-grained—a texture referred to as **aphanitic** (Greek *a*, "not"; *phaneros*, "visible"). In hand specimens, few, if any, crystals can be detected in aphanitic textures. Viewed under a microscope, however, many crystals of feldspar and quartz are recognizable.

An aphanitic texture indicates relatively rapid cooling, but not nearly as rapid as the quenching that produces glass. Aphanitic textures are typical of the interiors of lava flows, in contrast to the glassy texture that forms on the surface or crust.

What does the texture of a rock indicate about the cooling history of a magma body?

Why do volcanic rocks have aphanitic textures?

(A) A glassy texture develops when molten rock cools so rapidly that the migration of ions to form crystal grains is hampered. Glassy texture typically forms on the crust of lava flows and in viscous magma. The sample shown here is obsidian.

(B) An aphanitic texture consists of mineral grains too small to be seen without a microscope. The sample shown here is rhyolite. Only a few grains are large enough to be seen. Most are microscopic. Aphanitic texture results from rapid cooling.

(C) A phaneritic texture consists of grains large enough to be seen with the unaided eye. All grains are roughly the same size, and they interlock to form a tight mass. The large crystals suggest a relatively slow rate of cooling.

(D) A pyroclastic texture forms when crystals, fragments of rock, and glass are blown out of a volcano as hot ash. The material may accumulate as an ash fall or as an ash flow. The black lenses of glass were pumice fragments that were squashed during welding of the hot ash.

Figure 4.3 Textures of igneous rocks provide important information concerning rock genesis. All of the silicic rocks presented here have roughly the same chemical composition but extremely different textures. The photographs show the actual size of the specimens.

Many aphanitic and glassy rocks have numerous small spherical or ellipsoidal cavities, **vesicles**. These are produced by gas bubbles trapped in the solidifying rock. As hot magma rises toward Earth's surface, the confining pressure diminishes, and dissolved gas (mainly H_2O steam) separates and collects in bubbles. The process is similar to the effervescence of champagne and soda pop when the bottles are opened. Vesicular textures typically develop in the upper part of a lava flow, just below the solid crust, where the upward-migrating gas bubbles are trapped. Even though vesicles change the outward appearance of the rock and indicate the presence of gas in a rapidly cooling lava, they do not change the basic aphanitic texture.

STATE OF THE ART   Microscopes Reveal Hidden Worlds of Geology

Geologists examine Earth and its materials on a vast range of scales—from that of the entire planet down to the smallest constituent of a grain of dust. To do this, they use several sophisticated instruments. One tool used extensively to examine small features in rocks and minerals is the microscope. The last few decades have seen the development of a wide variety of microscopes, but we mention only two here: the standard optical microscope and the electron microscope.

Optical microscopes are used to examine rocks with visible wavelengths of light and can be used at magnifications as large as about 500 times. To examine a rock with a microscope, the rock is usually sliced into a thin section, glued to a glass slide, and then ground and polished until it is only 30 microns thick. Under these conditions, many minerals are transparent or translucent, although a few minerals, such as the iron oxides and pyrite, remain opaque. A series of lenses bend rays of light transmitted through the thin section to create an enlarged image. Polarizing lenses add to the discriminating power. An optical microscope can give clear pictures only of specimens that are larger than the wavelength of light used.

The photograph below shows the wealth of information that can be seen. Not only are the identities of the individual grains revealed, but the texture is also obvious, which tells us about the history of the rock. By identifying the minerals, a geologist can classify the rock as igneous, sedimentary, or metamorphic and thereby decipher some of the rock's history. The network of intergrown crystals reveals that this specimen is an igneous rock that was at one time molten. The fine-grain size shows that the molten magma cooled quickly and the rock is probably volcanic rather than plutonic.

Electron microscopes create greatly magnified images of rock surfaces or mineral grains and are capable of magnifications to several tens of thousands of times. An electron microscope accomplishes this without using optical light at all. Instead, it uses electrons, which have wave characteristics equivalent to extremely short wavelengths. A rock specimen is placed in a small vacuum chamber and bombarded by a beam of electrons stripped from a heated filament at the top of the instrument. A series of magnetic lenses bend and focus the electron beam on the specimen, which then emits a shower of secondary electrons. These secondary electrons control the intensity of another beam of electrons inside a television picture tube to construct an image that we can see with our eyes. Tiny mineral grains are thus visible for study and the details of mineral structures can be seen. Moreover, some electrons in the target atoms may be kicked into higher energy levels and when they fall back to lower energy levels X rays are given off that can be used to identify the element and its concentration in the rock. A map of the microdistribution of the elements in the mineral can be made. The colorful electron microprobe image shows the concentration of magnesium in a single crystal of garnet. The colors show that the environment changed as the garnet crystallized so that the rim is poor in magnesium (blue) and the core is high (yellow). This information can be used to infer the temperature and pressure of crystallization.

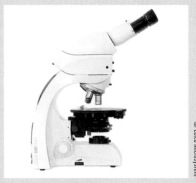

Optical microscope

© Leica Microsystem

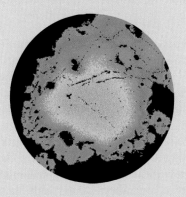

Courtesy of Cameca Instruments, Inc.

Electron microprobe

Phaneritic Texture

The specimen shown in Figure 4.3C is composed of grains large enough to be recognized without a microscope, a texture known as **phaneritic** (Greek *phaneros*, "visible"). The grains are approximately equal in size and form an interlocking mosaic. The equigranular texture suggests a uniform rate of cooling, and the large size of the crystals shows that the rate of cooling was very slow.

For cooling to take place at such a slow rate, the magma must have cooled far below the surface. Field evidence supports this conclusion, because magma that crystallizes after volcanic eruption produces only aphanitic and glassy textures. Rocks with phaneritic textures are exposed only after erosion has removed thousands of meters of covering rock.

Some intrusive igneous rocks have especially coarse grains—as much as a meter long. These **pegmatites** probably crystallized from water-rich magmas and are typically found in association with granite.

Porphyritic Texture

Some igneous rocks have grains of two distinct sizes. The larger, well-formed crystals are referred to as **phenocrysts**; the smaller crystals constitute the **matrix**, or the **groundmass**. This texture is known as porphyritic. It occurs in either aphanitic or phaneritic rocks.

A **porphyritic texture** usually indicates two stages of cooling. An initial stage of slow cooling, during which the large grains developed, is followed by a period of more rapid cooling, during which the smaller grains formed (see Figure 4.5F). The aphanitic matrix indicates that the cooling melt had sufficient time for all of its material to crystallize. The initial stage of relatively slow cooling produced the larger grains; the later stage of rapid cooling, when the magma was extruded, produced the smaller grains. Similarly, a phaneritic matrix with phenocrysts indicates two stages of cooling (see Figure 4.5A). An initial stage of very slow cooling was followed by a second stage, when cooling was more rapid but not rapid enough to form an aphanitic matrix.

Pyroclastic Texture

The texture shown in Figure 4.3D may appear at first to be that of a porphyritic rock with phenocrysts of quartz and feldspar. Under a microscope, however, the grains are seen to be broken fragments rather than interlocking crystals. Some fragments of glass are bent and flattened. This is a **pyroclastic texture** (Greek *pyro*, "fire"; *klastos*, "broken"), produced when explosive eruptions blow crystals and bits of still molten magma into the air as a mixture of hot fragments called ash. If the fragments are still hot when they are deposited, they will be welded (fused) together by the weight of the overlying rock.

Types of Igneous Rocks

Igneous rocks are classified on the basis of texture and composition. The major kinds of igneous rocks are granite, diorite, gabbro, rhyolite, andesite, and basalt.

A simple chart of the major types of igneous rocks is shown in **Figure 4.4**. The basis for this scheme of classification is texture and composition. Variations in composition are arranged horizontally, and variations in texture are arranged vertically. Rocks that cool below the surface are called *intrusive*, and those that cool at the surface are called *extrusive*. The rock names are printed in bold type, the size of which is roughly proportional to the relative abundance of the rock at the surface. Rocks in the same column have the same composition but different textures. Rocks in the same horizontal row have the same texture but different compositions. The chart shows that granite, for example, has a phaneritic texture and is composed predominantly of quartz, plagioclase, and K-feldspar. The type size indicates that it is the most abundant intrusive

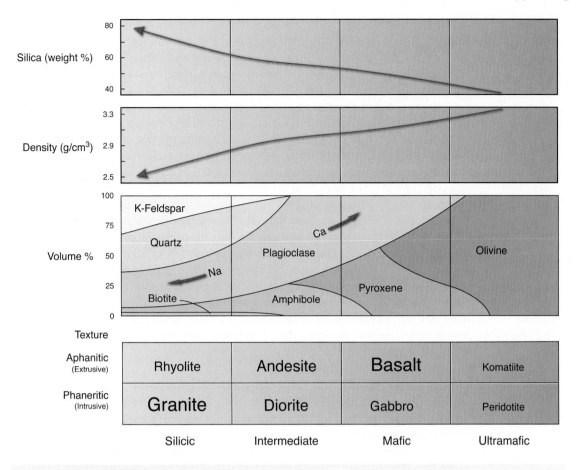

Figure 4.4 **The classification of common igneous rocks** is based on texture (shown vertically on the chart) and composition (shown horizontally). The size of type in which the names of the rocks are printed is roughly proportional to their abundance at Earth's surface.

igneous rock. Rhyolite has the same composition as granite but is aphanitic. Basalt has an aphanitic texture and is composed predominantly of Ca-plagioclase and pyroxene. It has the same composition as gabbro but is much more abundant at Earth's surface.

This classification attempts to show the natural, or genetic, relationships between the various rock types. As we saw in the preceding section, texture provides important information on the cooling history of the magma. Rocks that crystallize slowly are able to grow large crystals; those that cool rapidly have a fine-grained or glassy texture. The composition of a rock provides information about the nature and origin of the magma. Mafic magmas high in iron and magnesium and poor in silica generally originate from partial melting of the mantle; they are erupted in continental rift systems and along mid-oceanic ridges. Rocks richer in silica, such as andesite and rhyolite, or their intrusive equivalents, diorite and granite, typically form at convergent plate margins and in other settings, such as rifts or above hot spots, where continental crust can be partially melted by hot basalt.

What are the properties upon which igneous rocks are classified?

Rocks with Phaneritic Textures (Intrusive Rocks)

Granite is a coarse-grained igneous rock composed predominantly of feldspar and quartz (**Figure 4.5A**). K-feldspar is the most abundant mineral, and usually it is easily recognized by its pink color. Plagioclase is present in moderate amounts, usually distinguished by its white color and its porcelain-like appearance. Mica is conspicuous as black or bronze-colored flakes, usually distributed evenly throughout the rock. A very important property of granite is its relatively low density, about 2.7 g/cm^3, in contrast to basalt that has a density of 3.0 g/cm^3. This fact is important in considering the nature of continents and the contrast between continental crust and oceanic crust. Granite and related rocks make up the great bulk of the continental crust and basalt dominates the oceanic crust.

Diorite is similar to granite in texture (Figure 4.5C), but it differs in composition. Plagioclase feldspar is the dominant mineral, and quartz and K-feldspar are minor constituents. Amphibole is an important constituent, and some pyroxene may be present. In composition, diorite is intermediate between granite and gabbro. Its extrusive equivalent is andesite.

Gabbro is not commonly exposed at Earth's surface, but this mafic rock is a major constituent of the lower part of the oceanic crust and is present in some intrusions on the continents. It has a coarse-grained texture similar to that of granite, but it is composed almost entirely of pyroxene, calcium-rich plagioclase, and olivine. Gabbro is dark green, dark gray, or almost black because of the predominance of dark-colored minerals (Figure 4.5E). Because of the abundance of mafic minerals, the density of gabbro is quite high—about 3.0 g/cm³.

Peridotite is composed almost entirely of two minerals, olivine and pyroxene (Figure 4.5G). It is not common at Earth's surface or within the continental crust, but it is a major constituent of the mantle. Its high density (about 3.3 g/cm³), together with other physical properties, suggests that the great bulk of Earth's interior is composed of peridotite and closely related rock types. The Alps and St. Paul's Rocks (islands in the Atlantic Ocean) are two areas where small masses of peridotite from the mantle have been pushed through the crust to Earth's surface. Small pieces of peridotite are also found in basaltic lava flows. These fragments were ripped out of the mantle by rapidly rising magma.

Rocks with Aphanitic Groundmasses (Extrusive)

Rhyolite is an extrusive rock with an aphanitic groundmass that has the same silicic composition as granite (Figure 4.5B). It commonly contains a few phenocrysts of feldspar, quartz, and biotite and is therefore porphyritic. Because of their high silica contents and low temperatures, rhyolite lava flows are viscous. Instead of spreading in a long thin flow, rhyolite typically piles up in large, bulbous domes (see Figure 4.13). Obsidian with the composition of rhyolite is quite common in these lava flows. Rhyolite ash-flow tuffs are also common. Rhyolite is not common along the ocean ridges or oceanic islands but is more common on the continents.

Andesite is an aphanitic rock typically composed of plagioclase, pyroxene, and amphibole (Figure 4.5D). Like other fine grained igneous rocks, andesite is typically volcanic. Andesite has an intermediate chemical composition and usually contains little or no quartz and has the same composition as diorite (Figure 4.4). The texture of andesite is generally porphyritic, with phenocrysts of plagioclase feldspar and mafic minerals. It takes its name from the Andes Mountains, where volcanic eruptions have produced lavas with this composition in great abundance. Andesite is the next most abundant lava type after basalt and occurs most frequently along the convergent plate margins in island arcs and along continental margins. It is not found along oceanic ridges, and it is rare in oceanic islands or other intraplate settings related to mantle plumes.

Basalt is the most common aphanitic rock (Figure 4.5F). It is a very fine-grained, usually dark-colored rock that crystallizes from relatively fluid lava flows. The mineral grains in the aphanitic matrix are so small they can rarely be seen without a microscope. If a **thin section** (a thin, transparent slice of rock) is viewed through a

What is the difference between rhyolite and basalt?

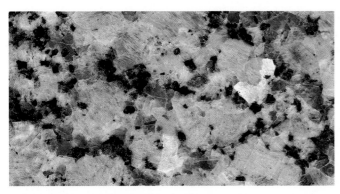

(A) Granite: K-feldspar, quartz, plagioclase, and biotite.

(B) Rhyolite: K-feldspar, plagioclase, quartz, biotite, and light-colored fine-grained groundmass.

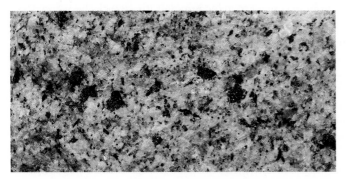

(C) Diorite: plagioclase, amphibole, quartz, and biotite.

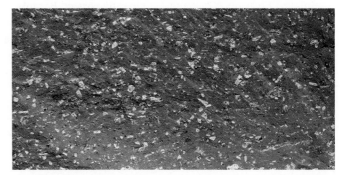

(D) Porphyritic andesite: plagioclase, pyroxene, and amphibole along with fine-grained, gray groundmass.

(E) Gabbro: pyroxene, plagioclase, and olivine.

(F) Porphyritic basalt: pyroxene, plagioclase, and olivine along with black vesicles and gray groundmass.

(G) Peridotite: olivine and pyroxene.

(H) Komatiite: olivine and pyroxene.

Figure 4.5 The major types of igneous rocks and their mineral constituents. Photographs are actual size.

microscope, the individual minerals can then be seen and studied. Many basalt flows are strongly vesicular—especially in flow tops and bottoms.

Basalt is a mafic rock composed predominantly of calcium-rich plagioclase and pyroxene, with smaller amounts of olivine. The plagioclase occurs as a mesh of elongate, lathlike crystals surrounding the more equidimensional pyroxene and olivine grains. In some cases, large crystals of olivine or pyroxene form phenocrysts, resulting in a porphyritic texture. Many basalt lavas have some glass, especially near the tops of flows. Basalt is the most common volcanic rock on Earth, because it is so abundant on the seafloor.

Komatiite is a rare extrusive rock (Figure 4.5H) found mostly in very ancient volcanic sequences exposed in the continental shields. It is composed mainly of the mineral olivine. The olivine phenocrysts may grow to exceptional lengths of as much as 50 cm, giving most komatiites a decidedly porphyritic character. The rest of the rock is made of fine crystals of pyroxene and rare plagioclase, giving it an appearance like basalt. Because of their high content of mafic minerals, these rocks have even less silica than basalt. Komatiites are important because they provide evidence for the high temperatures of volcanism early in Earth's history.

Rocks with Pyroclastic Textures

In what way are the products of explosive eruptions unique?

Explosive volcanic eruptions of rhyolitic and andesitic magmas commonly produce large volumes of fragmental material that are ejected high into the atmosphere. The fragments range from dust-sized pieces, or **ash**, to large blocks more than a meter in diameter. **Pumice** is a vesicular frothy glass common among the larger fragments. Some pumice fragments have densities low enough that they can float on water. **Tephra** deposits are composed of shards of volcanic glass, pumice, broken phenocrysts, and foreign rock fragments. The rock resulting from the accumulation of pyroclastic fragments is also known as **tuff**. Although of volcanic origin, tuff has many of the characteristics of sedimentary rocks because the fragments composing tuff may settle out from suspension in the air and commonly are stratified like sedimentary rocks. **Pyroclastic-fall tuff** is composed of volcanic ash that fell more or less vertically out of the atmosphere. These layers mantle the hills and valleys. In contrast, an **ash-flow tuff**, or **ignimbrite**, forms from particles that move laterally across the surface in a gas-charged flow in which movement resembles that of a lava flow but is much more rapid. In some ash-flow tuffs, the ash may be fused or welded together in a tight, coherent mass, and the glass fragments may be flattened and bent out of shape. This unique texture indicates that at the time of deposition, the ash fragments were hot enough to deform and fuse from the weight of the overlying ash (Figure 4.3D).

Extrusive Rock Bodies

Extrusive igneous rocks are those that form from magma extruded onto Earth's surface by volcanic eruptions. The rocks include lava flows and volcanic ash. Basaltic magmas are low in silica and are relatively fluid. The lava is typically extruded quietly from fissures and fractures. Silicic magmas are viscous, and their eruptions are typically explosive. The magma extrudes as thick lava flows, bulbous domes, or ash flows.

One of the most spectacular of all geologic processes is the extrusion of lava onto Earth's surface by volcanic eruptions (**Figure 4.6**). Throughout recorded history, more than 700 volcanoes have been known to be active, but this is only an instant in geologic time and ignores the region of the most intense volcanic activity on Earth—the region hidden beneath the oceans, where most eruptions go unnoticed. The importance of volcanic activity is that it testifies to the continuing dynamics of Earth, provides an important window on the planet's interior, and sheds light on the processes operating below the surface, in the lower crust and upper mantle.

Products of Basaltic Eruptions

Basaltic eruptions are probably the most common type of volcanic activity on Earth. The lava is generally extruded from fractures or fissures in the crust. Upon extrusion, it tends to flow freely downslope and spreads out to fill valleys and topographic depressions (Figure 4.6). This type of eruption occurs along fissures at the oceanic ridge, forming new oceanic crust where the tectonic plates move apart, and it is the major type of eruption in the volcanic plains of the continents related to hot spots and rifts. Basaltic lavas erupt with temperatures ranging between 1000°C and 1200°C. Lava can flow at speeds as high as 40 km/hr down steep slopes, but rates of 20 km/hr are considered unusually rapid. For example, the flow front of the 1998 basaltic lava in the Galapagos Islands moved an average of 170 m/hr. The fronts of Hawaiian basalt flows commonly move only a few meters per hour. Rapid flows are usually found in confined flows or inside lava tubes. As a flow moves downslope, it loses gas, cools, and becomes more viscous. Movement then becomes sluggish, and the flow soon comes to rest.

There are two common types of basaltic flows, referred to by the Hawaiian terms *aa* (pronounced ah'ah') and pahoehoe (pronounced pa ho'e ho'e) (Figure 4.6A, B). An **aa flow** moves slowly and is typically 3 to 10 m thick. The surface of the flow cools and forms a crust while the interior remains molten. As the flow continues to move, the hardened crust is broken into a jumbled mass of angular blocks and clinkers (Figure 4.6A). Gas within the fluid interior of the flow migrates toward the top, but it may remain trapped beneath the crust. These "fossil gas bubbles," called vesicles, make the rock light and porous. **Pahoehoe flows** are more fluid than aa flows. Many are less than 1 m thick, but they can be much thicker. As a pahoehoe flow moves, it develops a thin, glassy crust, which is wrinkled into billowy folds or surfaces that can resemble coils of rope. A variety of flow features (such as those shown in Figure 4.6B, G) can develop on the surface of the flow. Commonly, the crust of the flow buckles to form a **pressure ridge**, with a central fracture through which gas and lava can escape (Figure 4.6E). Both aa and pahoehoe flows can be erupted from the same vent. Many pahoehoe flows convert to aa when the surface cools and small crystals form or where flow rates increase when the flow drops over a steep slope.

The interior of a basaltic flow may be massive and nonvesicular. As a flow cools, it contracts and may develop a system of polygonal cracks, known as **columnar joints** that are similar in many ways to mud cracks (Figure 4.6C). In some flows, the sides and top freeze solid while the interior remains fluid. The fluid interior can break through the crust and flow out, leaving a long **lava tube** (Figure 4.6D). Instead of issuing from a central vent, basaltic lava is commonly extruded from a series of fractures in the crust known as **fissures** (Figure 4.6J). The fluid lava usually spreads out over a large area rather than building an isolated cone. In some places along the fissure, the rising lava may be concentrated and erupt like a lava fountain. The splashing of lava around the fountain can build up small conical mounds called **spatter cones** (Figure 4.6F). **Flood basalts** are some of Earth's most impressive volcanic deposits; single flows can be traced for hundreds of kilometers (**Figure 4.7**). Individual flows can be as much as 30 m thick, and a sequence of flows may stack up to be hundreds of meters thick, as in the Columbia River Plateau of Washington. The low viscosity of basalt and high eruption rates create these plains. Studies of the Moon and planets show that this type of volcanism is the most common not only on Earth but also on all the inner planets.

Droplets and globs of lava blown out from a volcanic vent may cool by the time they fall back to the ground. This material forms **volcanic ash** and dust, collectively known as tephra (Figure 4.6H). **Volcanic bombs** are the larger fragments (Figure 4.6I). As the tephra travels through the air, it is sorted according to size. The larger particles accumulate close to the vent and form a **cinder cone** (**Figures 4.8** and **4.9**), and the finer, dust-sized particles are transported afar by the wind. Cinder cones, which are generally less than 200 m high and 2 km in diameter, are relatively small features compared with large shield volcanoes and stratovolcanoes.

What kinds of lava flows and volcanoes form from basaltic lava flows?

(A) The surface of an aa flow consists of a jumbled mass of angular blocks that form when the congealed crust is broken as the flow slowly moves. Aa flows are viscous and much thicker than pahoehoe flows. This photograph shows a recent aa flow in Hawaii.

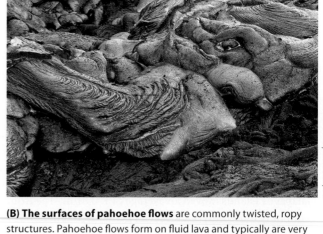

(B) The surfaces of pahoehoe flows are commonly twisted, ropy structures. Pahoehoe flows form on fluid lava and typically are very thin. The firm, hot plastic crust is wrinkled and folded by continued movement of the fluid interior.

(C) Hexagonal columnar joints commonly form by contraction when lava cools. The long axis of the column is approximately perpendicular to the cooling surface. These columns form the Giant's Causeway, Ireland.

(D) Lava tubes develop where the margin of the flow cools and solidifies and the interior, molten material is drained away.

(E) Pressure ridges develop in lava flows when the outer crust buckles as the flow surface folds. They commonly crack and release lava and gas from the interior of the flow.

(F) Spatter cones, or ramparts, form in local areas along fissures where globs of lava accumulate near a major vent.

Figure 4.6 A variety of features develop on basaltic lava flows and reflect the manner of flow, rates of cooling, amount of dissolved gases, and viscosity.

(H) Tephra is a general term referring to all pyroclastic material ejected from a volcano. It includes ash, dust, bombs, and rock fragments. It is commonly stratified.

(G) Pahoehoe lava flows on the island of Hawaii. The flow is moving away from the observer. The main flow forms a solid crust along its margins and upper surface. Hot liquid lava breaks through the crust, gradually cools, and forms a new crust. The process is then repeated downslope.

(I) Volcanic bombs are fragments of lava ejected in a liquid or plastic state. As they move through the air they twist and turn and form spindle-shaped masses.

Courtesy of USGS

(J) Fissure eruptions are the most common type of volcanic eruption on Earth. Lava is simply extruded through cracks or fissures in the crust. This type of eruption is typical of fluid basaltic magma and is the dominant eruption style along the oceanic ridge. This photograph shows a recent fissure eruption on the island of Hawaii.

Figure 4.7 Flood basalts cover large areas of the Columbia Plateau. Two thick flows and several thinner ones are exposed in this valley wall. Columnar joints form fractures through much of each flow.

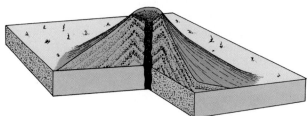

Figure 4.8 A cinder cone is a small volcano composed almost exclusively of pyroclasts blown out from a central vent, such as this one in southern Utah. The internal structure consists of layers of ash inclined away from the summit crater. The vent is commonly filled with solidified lava and fragmental debris.

If the extrusion of fluid basaltic lava dominates, a broad cone, or **shield volcano**, may form around a central vent or series of fissures (**Figure 4.10**). With each eruption, the fluid basaltic lava flows freely for some distance, spreading into a thin sheet, or tongue, before congealing. Shield volcanoes, therefore, have wide bases and gentle slopes (generally less than 10°). Their internal structure consists of innumerable thin basalt flows with comparatively little ash. The Hawaiian Islands are excellent examples of large shield volcanoes. They are enormous mounds of basaltic lava, rising as high as 10,000 m above the seafloor (Figure 4.9). The younger volcanoes typically have summit craters, or **calderas**, as much as 5 km wide and several hundred meters deep that result from subsidence following the eruption of magma from below (**Figure 4.11**).

The extrusion of basaltic lava into water produces a flow composed of a multitude of ellipsoidal masses referred to as **pillow lava** (**Figure 4.12**). The formation of pillow

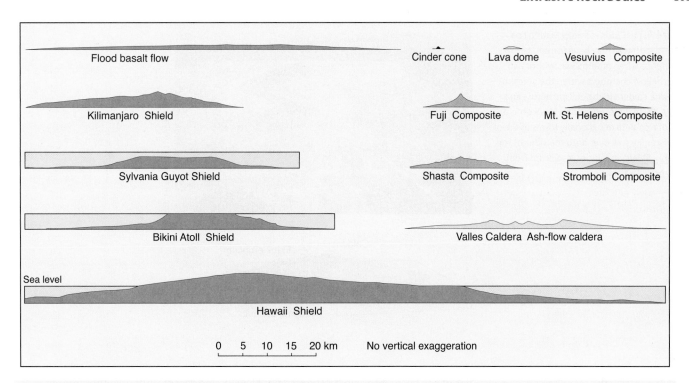

Figure 4.9 **Volcanoes** come in a wide variety of sizes and shapes. The smallest shown here are cinder cones and the largest are Hawaiian-type shield volcanoes. In between these extremes in size are small shield volcanoes, rhyolite lava domes, steep-sided composite volcanoes, ash-flow calderas, and vast plains covered by flood basalts.

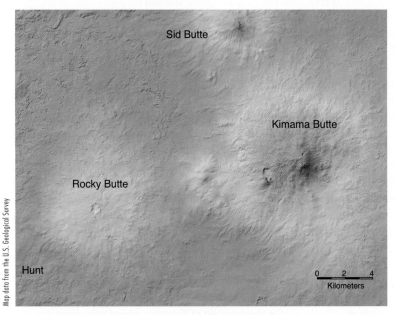

Map data from the U.S. Geological Survey

(A) The three small shields shown on this shaded relief map of the Snake River Palin of southern Idaho shows the typical size and shape.

(B) Kimama Butte is one of the tallest and steepest of the volcanoes here but it is only about 200 m high and about 8 km across. Some of the largest volcanoes, such as the Hawaiian Islands, are also shields.

Figure 4.10 **Shield volcanoes** are constructed of basaltic lava flows that erupt from a central vent or short fissures.

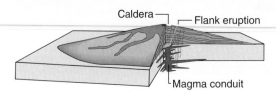

Figure 4.11 Calderas atop Mauna Loa volcano in Hawaii may have formed by subsidence when magma was withdrawn from shallow chambers within the volcano. The large caldera is about 3 km across and nearly 200 m deep. Eruption of lava onto the floor of a caldera occasionally forms a lake of lava that may spill over onto the sides of the volcano. Shield volcanoes also erupt from vents on their flanks.

basalt has been observed off the coast of Hawaii, and recent undersea photographs show that it is widespread on the seafloor, anywhere volcanic activity has occurred.

Eruptions can also occur beneath glaciers. The hot lava typically melts the ice which can accumulate to make a subglacial lake. Continued eruption can create piles of pillow lava, fragmented lava, and steam explosions. The ash thrown into the air by such explosions during the eruption of **a subglacial volcano** in Iceland disrupted air traffic across much of Europe in 2010. Catastrophic floods occur if the melt water breaks through or passes beneath the ice.

Products of Intermediate to Silicic Eruptions

The silica-rich magmas that produce andesite and rhyolite are relatively viscous, cool (600°C to 900°C), and water-rich. Consequently, their eruption and flow are quite different from basaltic lavas. Some silicic magmas are so viscous that small volumes hardly flow at all but instead form small bulbous **lava domes** over the volcanic vent (**Figure 4.13**).

Magmatic explosions are not driven by chemical reactions like manufactured explosives; instead, they are caused by the rapid expansion of gas bubbles at low pressure. The high viscosity of silicic magmas inhibits the escape of dissolved gas, so tremendous pressure builds up. Consequently, when eruptions occur, they are highly explosive and violent and commonly produce large quantities of tephra. Alternating layers of tephra and thick, viscous lava flows or domes typically produce a **composite volcano**, or **stratovolcano**, a high, steep-sided cone centered on the vent (**Figure 4.14**). This is probably the most familiar form of continental volcano, with such famous examples as Shasta, Fuji, Vesuvius, Etna, and Stromboli. A depression at the summit, the **crater**, usually marks the position of the vent. These volcanoes are long-lived and their eruptions infrequent. Although they are large structures, they are much smaller than the huge shields formed on the ocean floor (Figure 4.9).

Explosive eruptions of silicic volcanoes can blow out large volumes of ash and magma in a very short period. While the chamber empties, the roof becomes weak and unstable. As a result, the summit may collapse, forming a large caldera. Crater Lake, Oregon (**Figure 4.15**), for example, formed when a volcano's summit collapsed during its last major eruption and the resulting caldera filled with water. Wizard Island, a small cinder cone in the caldera, formed during subsequent minor eruptions. Other collapse calderas associated with rhyolite eruptions are as much as 50 km across. Fortunately, no eruption of the size necessary to create such a caldera has occurred during recorded history.

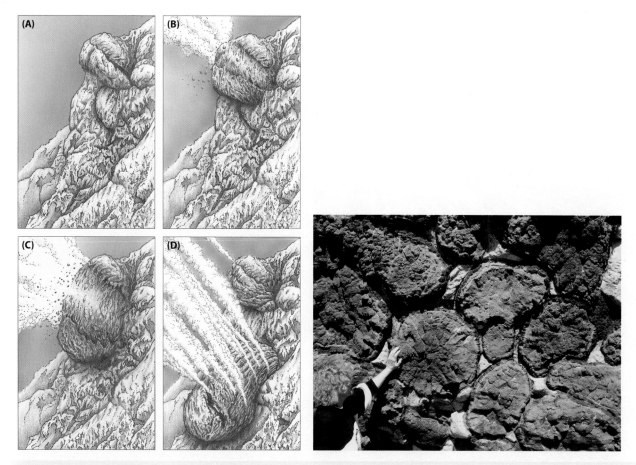

Figure 4.12 **Pillow basalt** is formed when lava extruded under water cools quickly, forming a series of ellipsoidal masses as shown in this sequence of sketches. **(A)** A new pillow buds from the front of a lava flow with a still molten interior. **(B)** The solid crust of the pillow splits and hot lava comes in contact with the cold water, sending a stream of water vapor bubbles toward the surface. **(C)** The gap widens as gravity pulls the pillow downward. **(D)** Eventually, the pillow breaks off and falls. Ultimately, a pile of pillows forms at the flow front and is then overridden by the rest of the flow. The photograph shows pillow lava that was originally formed on the ocean floor and is now exposed in New Zealand. Note the radial fractures and the quenched black glassy rim around each pillow.

(A) This small rhyolite dome, and the thick rhyolite lava flow behind it, erupted in northern California, near Mono Lake.

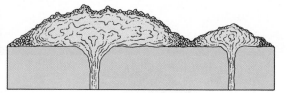

(B) Cross section showing the internal structure of a rhyolite dome. These domes inflate as magma rises from below, so the crust of the dome is continually stretched and fractured. The development is similar to that of inflating a balloon.

(C) Close-up photograph of glassy rhyolite flow showing contorted flow structures formed in viscous lava.

Figure 4.13 **Domes of silicic lava** form because silica-rich lava is viscous and resists flow. It therefore tends to pile up over the vent to form small bulbous domes, usually less than 1 km across.

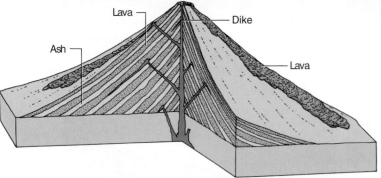

Figure 4.14 Composite volcanoes are built up of alternating layers of ash and lava flows, and intruded by lava domes. They are high, steep-sided cones such as Washington's Mt. St. Helens before its catastrophic eruption in 1980. A typical composite volcano is about 20 km across and may be 3 km high.

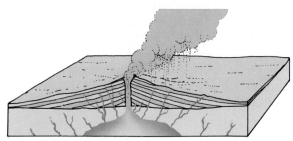

(A) Early explosive eruptions from the prehistoric volcano Mount Mazama created a high eruption column and ash fell out to form thin beds of ash.

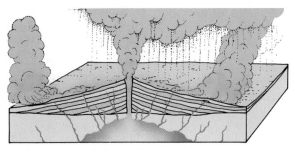

(B) Great eruptions of ash flows emptied more of the magma chamber, causing the top of the volcano to collapse.

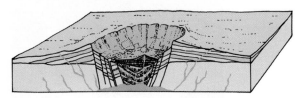

(C) The collapse of the summit into the partly drained magma chamber formed the caldera.

(D) A lake formed in the caldera, and subsequent minor eruptions produced small volcanic islands in the lake.

After H. Williams, F. J. Turner, and C. M. Gilbert

Figure 4.15   The evolution of the caldera at Crater Lake, Oregon, involved a series of great eruptions followed by the collapse of the summit into the magma chamber.

A spectacular type of eruption associated with silicic magma is the lateral flow of large masses of pumice and ash across the ground. This phenomenon is not a liquid lava flow or an ash fall (in which particles settle independently from the atmosphere) but a flow consisting of fragments of hot mineral grains, ash, and pieces of rock all suspended in hot gas. It moves rapidly close to the surface like a dense dust cloud. This type of eruption is therefore known as an **ash flow** (**Figure 4.16**). As magma works its way to the surface, confining pressure is released and bubbles of gas coming out of solution rapidly expand. Near the surface, the magma violently explodes, ejecting pieces of lava, bits of solid rock, crystals, and gas. This material is very hot, sometimes incandescent. Initially, explosions throw this material high into the atmosphere, but being denser than the air, it eventually falls and flows across the ground surface as a thick, dense cloud of hot ash. Ash flows can reach velocities greater than 250 km/hr. There is no outrunning an ash flow. When an ash flow comes to rest, the particles of hot crystal fragments, glass, and ash may fuse to form **welded tuff** (**Figures 4.17** and 4.3D). As it cools, the contracting mass can develop columnar jointing. Ash-flow tuffs can be very large. Some have carried ash as much as 100 km from their vents. Some flows form layers more than 100 m thick and cover thousands of square kilometers. A few ash-flow tuffs have volumes of more than 1000 km³. This is the equivalent of a cube 10 km on a side and probably erupted over the course of only a few weeks.

Ash-flow calderas are the largest silicic volcanoes on Earth. Constructed of far traveled sheets of tuff, these volcanoes form very low, very broad shields dominated by the central collapse structure (Figures 4.17 and 4.9)

Ash-flow eruptions are catastrophic events. A few fortunate geologists have had the opportunity to witness them from afar and to make direct observations of this type of extrusion. For example, Mount Lamington, in New Guinea, was considered extinct until it erupted in 1951. It had never been examined by geologists and was not even considered to be a volcano by the local inhabitants. When it did erupt, volcanic activity began with preliminary emissions of gas and ash, accompanied by earthquakes and landslides near the crater. Sensitive seismographs were soon installed near the crater to monitor Earth's movements, and aerial photographic records were made daily. Then,

Why don't basaltic eruptions produce ash flows?

on Sunday, January 21, 1951, a catastrophic explosion burst from the crater and produced an ash flow that completely devastated an area of about 200 km². Almost 3000 people died. The main eruption was observed and photographed at close quarters from passing aircraft, and a qualified volcanologist was on the spot within 24 hours. The ash flow descended radially from the summit crater, its direction of movement controlled to some degree by the topography. As the ash flow rushed downslope, it scoured and eroded the surface. Estimated velocities of 470 km/hr were calculated from the force

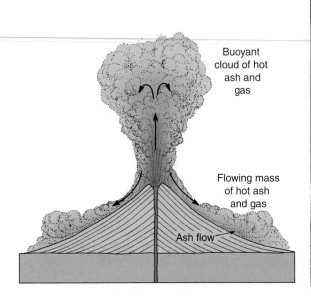

Figure 4.16 An ash flow is a hot mixture of highly mobile gas and ash that moves rapidly over the surface of the ground away from the vent. The ash rises into the air from the explosive force of the eruption but, being much denser than air, it moves en masse back to the surface and rushes down the slopes of the volcano as an ash flow. Less-dense gas and ash continue to move upward as a cloud into the atmosphere. This ash ultimately falls back to the surface as an ash fall. This photograph shows the eruption of a composite volcano on the north island of New Zealand.

Figure 4.17 Ash-flow tuffs can form sheets tens of meters thick that extend for many kilometers away from their source calderas. These rhyolite ash-flow sheets erupted several million years ago from Valles Caldera near Los Alamos in northern New Mexico. The block diagram shows the internal structure of a caldera. The ring faults allowed the roof of the magma chamber to drop like a piston during eruption of the tuff (gray). The tuff partially filled the caldera and a thinner sheet accumulated outside. Rhyolite domes (salmon) erupted along the ring fracture. A subsequent intrusion domed the floor of the caldera.

Figure 4.17 *(continued)*

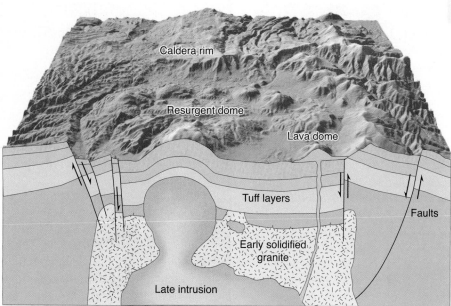

required to overturn certain objects. Entire buildings were ripped from their foundations, and automobiles were picked up and deposited in the tops of trees.

Intrusive Rock Bodies

> Igneous intrusions are masses of rock formed when magma cools beneath the surface. They are classified according to their sizes, shapes, and relationships to the older rocks that surround them. Important intrusive rock bodies are plutons, batholiths, stocks, dikes, sills, and laccoliths.

Magma is mobile, at times amazingly so. It rises because it is less dense than the surrounding rock. It can push aside surrounding rocks, force its way into cracks, and flow on the surface over distances of more than 100 km. It can move upward in the crust by melting away surrounding rocks or wedging and prying loose large blocks of rock, which it then replaces. When magma within the crust loses its mobility, it slowly cools and solidifies, forming a mass of igneous rock called an **intrusion** (**Figure 4.18**). Intrusions occur in a variety of sizes and shapes and are exposed at the surface only after the overlying rock has been removed by erosion.

What are the characteristics of the major types of intrusive rock bodies?

Plutons and Batholiths

Plutons are masses of intrusive igneous rock of any size. Ideally, each pluton represents one magma body crystallized to solid rock, but it is often hard to prove that only one batch of magma was involved. The true three-dimensional form of plutons is difficult to determine because of uncertainty about their extension deep below the surface. The bases of plutons are only rarely exposed at the surface, but evidence from gravity measurements and seismic studies, show that plutons exposed at the surface do not extend down into the mantle. They must therefore be less than 30 km thick and most are probably only a few kilometers thick.

A **stock** is a small pluton with an outcrop area of less than 100 km². Some stocks are known to be small protrusions rising from larger underlying plutons, but the downward extent of most intrusions is unknown. Large exposures (greater than 100 km²) of intrusive rock are called **batholiths**, but careful mapping shows that most batholiths are composite intrusions and consist of many individual plutons of different ages intruding one another. Many batholiths cover several thousand square kilometers. The Idaho batholith, for example, is a huge body of granite, exposed over an area of nearly 41,000 km² (**Figure 4.19**). The Coastal batholith of British Columbia

(A) Laccolith partially exposed by erosion, southern Utah. Layers of sedimentary rock were domed upward and shouldered aside by the intrusion.

(B) Volcanic neck exposed by erosion, central France. Necks form below volcanoes when magma solidifies in a pipe.

(C) Dike of dark mafic rock, Bar Harbor, Maine, forms a nearly vertical sheet.

(D) Sills of mafic rock with prominent columnar joints intruded into sedimentary rocks, Capitol Reef National Park, Utah, and form nearly horizontal sheets.

Figure 4.18 Magmatic intrusions may assume a variety of forms. Batholiths are large masses of coarsely crystalline rock that cools in the major magma chamber. Stocks are smaller masses and may be protrusions from a batholith. Dikes are discordant, tabular bodies formed as magma enters fractures and cools. Many dikes are related to conduits leading to volcanoes. Some radiate out from the volcanic neck; others form a circular pattern above a stock and are called ring dikes. Sills are layers of igneous rock squeezed in between layering. Laccoliths, dome-shaped bodies with flat floors, are formed where magma is able to arch up the overlying strata. Inclusions of the surrounding rock in the magma are called xenoliths. A pipe is a cylindrical conduit through which magma migrates upward.

(E) Sierra Nevada batholith in California. Parts of the wall and roof (dark-colored metamorphic rock) remain above the granite intrusion (tan rock low on the hillside).

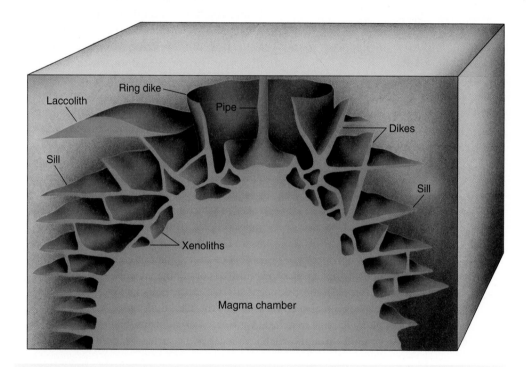

Figure 4.18 *(continued)*

is more than 2000 km long and 300 km wide and probably consists of several hundred separate intrusions.

Batholiths appear to be huge, slablike bodies, with horizontal extents much greater than their thicknesses. The map in Figure 4.19 gives a rough idea of the geometric form of some of the younger batholiths of western North America. The surface exposure of a granitic pluton can be elongate and elliptical or circular (**Figure 4.20**). It generally cuts across layering in the surrounding rock and is discordant. In some areas, however, the walls or roof of the batholith can be parallel to layering in the surrounding rocks and is said to be concordant.

Batholiths typically form in the deeper zones of mountain belts and are exposed only after considerable uplift and erosion. Some of the highest peaks of mountain ranges, such as the Sierra Nevada and the Coast Ranges of western Canada, are carved into granite batholiths (Figure 4.18E). These rocks originally cooled thousands of meters below the surface. The trend of a batholith usually parallels the axis of the

Figure 4.19 Batholiths are large intrusive bodies; most are composed of many small plutons. For example, the Sierra Nevada batholith of western North America contains hundreds of mapped intrusions. It was formed by multiple intrusive events over a span of about 50 million years. The Peninsular Range batholith, the Idaho batholith, and the Coast Range batholith are similar composite bodies.

mountain range, although the intrusion can cut locally across folds within the range. Extensive batholiths also are found in the shields of the continents (Figure 4.20). These exposures are considered to be the roots of ancient mountain ranges that have long since been eroded to lowlands.

Dikes

One of the most familiar signs of ancient igneous activity is a narrow, tabular body of igneous rock known as a **dike** (Figure 4.18C). All dikes are discordant; that is, they cut across preexisting structures such as layers in metamorphic or sedimentary rocks. A dike forms when magma enters a fracture and cools. The width of a dike can range from a fraction of a centimeter to hundreds of meters. The length is always much greater than the width. The largest known example is the Great Dike of Zimbabwe, which is 600 km long and has an average width of 10 km.

The emplacement of dikes is controlled by fracture systems within the surrounding rock. They commonly radiate from ancient volcanic necks and thus reflect the stresses associated with volcanic activity. Sometimes, upward pressure from a magma chamber produces circular or elliptical fracture systems, in which injected magma forms ring dikes (Figure 4.18). Large ring dikes can be as much as 25 km in diameter and thousands of meters deep. Dikes often occur in swarms related to continental rifting that may be hundreds of kilometers across and include numerous separate dikes. After erosion, the surface expression of a dike is usually a long narrow ridge. Dikes can also erode as fast as the surrounding rock, or even faster, and they can form long narrow trenches. A **volcanic neck** forms when magma solidifies in a pipe-like conduit through which lava reaches the surface (Figure 4.18B).

Sills

Rising magma follows the path of least resistance. If this path includes a bedding plane, which separates layers of sedimentary rock, magma may be injected between those layers to form a **sill**—a tabular intrusive body parallel to, or concordant with, the layering (Figure 4.18D). Sills range from a few centimeters to hundreds of meters thick and can extend laterally for several kilometers. A sill can resemble a buried lava flow lying within a sequence of sedimentary rock. It is an intrusion, however, squeezed between layers of older rock. The overlying rock is lifted by the intrusion. Many features evident at the contact with adjacent strata can be used to distinguish between a sill and a buried lava flow. For instance, rocks above and below a sill are commonly altered and recrystallized, and a sill shows no signs of weathering on its upper surface. Sills also commonly contain **inclusions**, blocks and pieces of the surrounding rocks. A buried lava flow, in contrast, has an eroded upper surface marked by vesicles; the younger, overlying rock commonly contains fragments of the eroded flow. Sills can form as local offshoots from dikes, or they can be connected directly to a stock or a batholith.

How would you tell a dike from a sill?

Courtesy of U.S. Geological Survey and EROS Data Center

Figure 4.20 **An isolated pluton exposed in Namibia** is shown in this satellite image. The intrusion is a large (about 20 km across), elliptical mass of highly fractured rock that intruded into older sedimentary rocks, which were then deeply eroded. The layered rocks are domed upward around the edge intrusion.

Laccoliths

When viscous magma is injected between layers of sedimentary rock, it may arch up the overlying strata. The resulting intrusive body, a **laccolith**, is lens-shaped, with a flat floor and an arched roof (Figure 4.18A). Laccoliths usually occur in blisterlike groups in areas of flat-lying sedimentary rocks. They can be several kilometers in diameter and thousands of meters thick. Typically, they are porphyritic.

The Origin and Differentiation of Magma

Magma is formed by melting preexisting solid rock. The wide range of magma compositions is the result of variations in the composition of the source rocks, partial melting, fractional crystallization, assimilation, and magma mixing. Differentiation of mafic magma generally forms silicic magma.

Origin of Magma

Magma can be produced by several processes, all of which involve an attempt to reach equilibrium between solid rock and its environment. Magma is often generated by one of these processes: (1) lowering the pressure; (2) raising the temperature; or (3) by changing the composition of the rock.

Although there are many ways by which these three changes can occur, we outline three of the most common here. As the mantle convects, some portions rise from deeper zones toward the surface. As this happens, the pressure becomes lower and lower and ultimately the mantle may become partially molten. The temperature can increase if hot mafic magma is intruded into the continental crust. This may make the crust hot enough to cause it to partially melt. In the third case, magma can be generated by adding a flux, such as water, to hot but solid mantle. For example, experiments show that the addition of only 0.1% water to dry mantle peridotite lowers its temperature of first melting by more than 100°C.

Differentiation of Magma

By now it is probably obvious to you that there are many different kinds of magmas—magmas that have different mineral constituents, magmas that have different element compositions, and magmas that have different temperatures. Moreover, many plutons, lava flows, and ash-flow tuffs reveal that the composition of magma in almost all magma chambers changes as time passes. We call the processes that cause these differences **magmatic differentiation**.

Why is there such great variability in the composition of magma?

One of the major causes of variability in magmas is the variability in composition of the **source rocks** from which the magma formed (**Figure 4.21**). Obviously, a

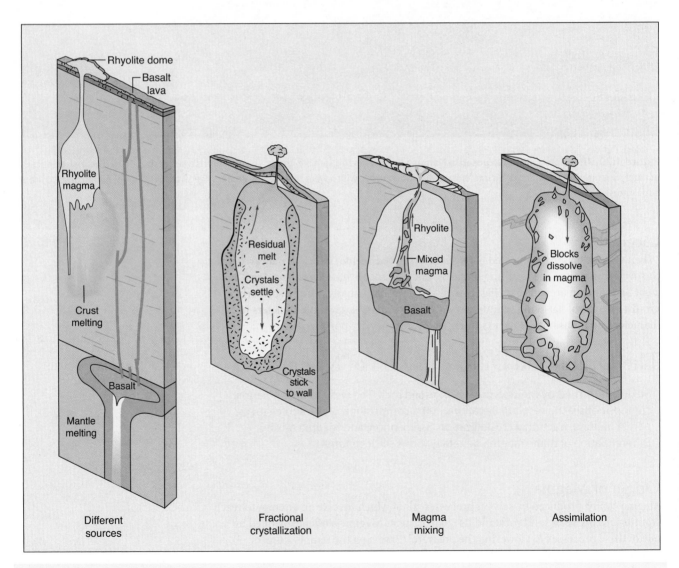

Figure 4.21 Magmatic differentiation is caused by several processes, including variations in the composition of the source rocks, fractional crystallization, magma mixing, and assimilation of wall rocks. Each of these is in turn controlled by the tectonic setting.

magma derived from the mantle will be very different (mafic) from a magma formed by melting of the continental crust (silicic). The molten fraction must be in equilibrium with the solid part; consequently, the melt is a reflection of the source from which it is derived. For example, granites derived from melting of sedimentary rocks are distinctive from those produced from other sources.

A second important cause of variation is **partial melting** of magma source rocks (Figure 4.21). Bear in mind that most rocks are composed of more than one mineral (Figure 4.4). Unlike water ice, which melts completely when heated above 0°C, a typical rock does not have a single melting temperature at which it becomes completely molten. Most natural rocks melt over a span of several hundred degrees, with the proportion of melting increasing with temperature. The liquid is different in composition than the solid source. The partial melt is enriched in components of the minerals that melt at low temperatures and depleted in elements that remain in the still-solid minerals. Because of this simple process of partial melting, the liquid is, in nearly all cases, richer in SiO_2 and less dense than the original solid rock. Thus, although the magma is a reflection of its source, it is not a perfect reflection that preserves the exact composition of the source. Consider an extreme but familiar example. A snowball mixed with sand consists of two fundamentally different kinds of minerals with different melting points—water ice and quartz sand. On a cold day, the ice and the quartz coexist as solids. If you place the ice ball in a sieve and raise the temperature slightly, the ice melts and pure liquid water flows away from the silicate minerals. As a result of the *partial* melting, the molten water does not have the same composition as the original solid mixture of snow and sand.

When magma cools and equilibrates with its environment, different minerals begin to crystallize at different temperatures and in a sequence that depends on the pressure and composition of the melt. Just as a rock does not melt at a single temperature, a magma does not crystallize completely at one temperature. The general order of crystallization of minerals from common magmas is summarized in **Figure 4.22**. When partial crystallization occurs, the crystal fraction can be separated from the remaining liquid, leaving a residual melt quite different from the parent magma. This process, **fractional crystallization**, usually makes daughter melts that are richer in SiO_2 than the parent melt. By this process, andesite and rhyolite can be sequentially derived from some basaltic magmas. Crystals can be removed from magma by simple gravitational settling or by adhering to the walls of the magma chamber. In either case, the liquid is generally less dense and can flow away from the solids and accumulate in the upper part of the magma chamber (Figure 4.21). Thus, even in a closed magmatic system the composition of the residual melt can become quite different from the original. The elements in some important gems and ores are concentrated by this progressive enrichment process.

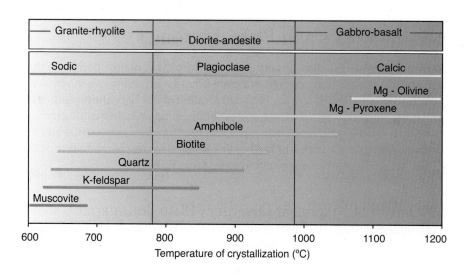

Figure 4.22 The temperature of crystallization of common minerals provides a key to the history of a magma. Olivine, pyroxene, and calcium-rich plagioclase crystallize at high temperatures in mafic magmas. Amphibole, plagioclase, and biotite crystallize at intermediate temperatures. Quartz, K-feldspar, sodium-rich plagioclase, and muscovite crystallize at low temperatures in silicic magmas. This sequence of crystallization is sometimes called Bowen's reaction series after the geologist whose experiments showed the temperature sequence. Of course, not every magma crystallizes all of these minerals.

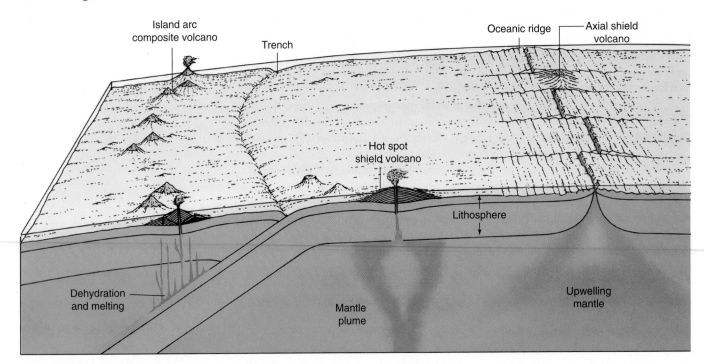

Figure 4.23 Magmas form in distinctive tectonic settings, generally related to plate boundaries. Basaltic magma originates by partial melting of the upper mantle at diverging plate margins. As mantle material moves upward in a convection cell, the peridotite (major rock in the mantle) begins to melt because of a decrease in pressure. The material that melts first produces magma of basaltic composition. Granitic magma

How would a magma derived from the mantle differ from one that formed by partial melting of the continental crust?

The magmatic differentiation processes described above can occur in closed systems, but you can probably imagine ways in which an open magmatic system can become differentiated as well. One obvious mechanism involves the mixing of two different magmas. **Magma mixing** can involve magmas as diverse as basalt and rhyolite (Figure 4.21). Sometimes, the mixing is complete enough to form a homogeneous intermediate, such as andesite. But just as often, the mixing process is evident in the form of blebs of mafic rock incompletely mixed with silicic rock to make a magmatic marble cake. Mixing of magmas is probably a universal process that occurs at all levels of the crust, from initial melting to final eruption.

Another open-system process that differentiates magmas is **assimilation** of the wall rock through which the magma passes (Figure 4.21). Chunks of rock surrounding a magma body may fall into the chamber and become dissolved in the magma. The composition of the magma is changed by the incorporation of these foreign materials.

Igneous Rocks and Plate Tectonics

At divergent plate margins, basaltic magma is formed as mantle peridotite rises and partially melts. Along convergent plate boundaries, distinctive magmas are generated in the mantle as a result of dehydration of the slab. These magmas may differentiate to form andesite and other silicic magmas. In mantle plumes, basaltic magma is produced when hot solids rise from the deep mantle and the pressure drops. The addition of hot basaltic magma to the continental crust can create rhyolitic magma in many different tectonic settings.

Generation of Magma at Divergent Plate Boundaries

Why are different types of magma produced at different types of plate boundaries?

As lithospheric plates move apart at divergent plate boundaries, solid mantle peridotite wells upward to fill the void. As it does, the pressure becomes lower and lower,

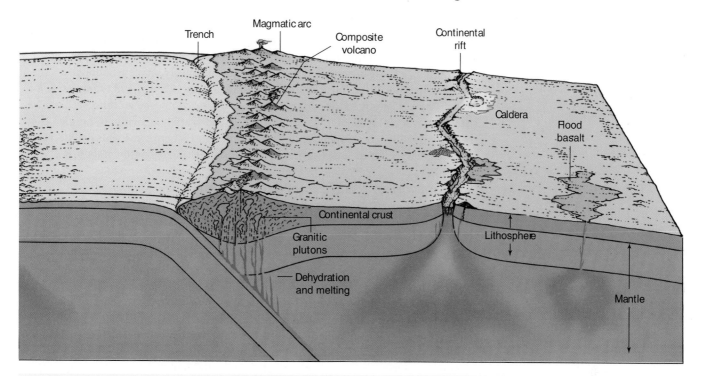

is generated at subduction zones by partial melting and fractional crystallization. As the oceanic crust (containing basalt, oceanic sediments, and water) descends into the mantle, it is heated and dehydrates. The resulting fluid is lighter than the surrounding rock and rises to cause melting in the overlying mantle.

and conditions appropriate for partial melting are reached at depths of 10 to 30 km (**Figure 4.23**). The first minerals to melt yield basaltic magma. Laboratory experiments on melting peridotite at these pressures show that basaltic magma is produced by 10% to 30% melting. The basaltic magma, being less dense than the still-solid portion of the peridotite, rises beneath the oceanic ridge and creates new oceanic crust by filling a small magma chamber just below the rift floor. Fractional crystallization to form silicic magma is inhibited by repeated injection of new batches of basaltic magma from below. Some magma is extruded to form small shield volcanoes, fissure-fed flows, and pillow basalts. The rest of the magma cools to form intrusive gabbro.

If rifting of a continental plate occurs, hot, dense basaltic magma from the mantle may lodge in or at the base of the crust and cause a second episode of partial melting. Experiments show that such crustal melts are rhyolitic. The low-density rhyolite may rise to the surface and erupt to form small lava domes, or it may accumulate in huge, shallow magma reservoirs that catastrophically erupt to form calderas. Thus, basalt and rhyolite are common at continental rifts, but magma intermediate between basalt and rhyolite is rare in this setting.

Generation of Magma at Convergent Plate Boundaries

At convergent plate boundaries, oceanic crust, composed of cold, wet basalt, and a veneer of marine sediment, descends deep into the mantle and is heated (Figure 4.23). At a depth of about 100–120 km, the temperature becomes high enough so that either the oceanic crust partially melts or the water in its minerals is driven out by dehydration. This low-density fluid rises into the overlying wedge of mantle peridotite and causes it to partially melt. The water acts as a flux to lower the mantle's melting point. Moreover, the resulting magma is rich in water and oxygen derived from the oceanic crust. Fractional crystallization of this mafic magma leads to more silicic andesite and

What kind of magma is most characteristic of a convergent margin?

rhyolite. In addition, if the magma also interacts with the continental crust, it may become even more silicic as SiO_2-rich crust is assimilated into the magma. Much of this magma stalls in the crust to form plutons that may coalesce into a huge composite batholith, like those unroofed by erosion at ancient convergent plate boundaries. The granites and diorites of California's Sierra Nevada batholith are good examples. Some magma makes its way to the surface to form andesitic composite volcanoes, characteristic of convergent plate boundaries; however, a wide variety of magmas, ranging from basalt to rhyolite, are erupted along with the andesite. Where continents collide, silica-rich magmas like granite develop from the partial melting of continental crust.

Generation of Magma in Mantle Plumes

The origin of magma in ocean islands found in the middle of the plates far from plate boundaries has long been a mystery. A distinctive type of basalt is found in oceanic islands and seamounts. The details of the composition of these basaltic lavas suggest to many geologists that they are generated by partial melting of rising plumes of solid mantle (Figure 4.23). Presumably, they rise because they are more buoyant and warmer than the rest of the mantle. As the material in the plume nears the surface, the plume partially melts to produce basalt that rises to the surface. The basaltic magma may erupt to form chains of shield volcanoes, like the Hawaiian Islands, or to form large provinces covered by flood basalts. Thus, the magma-generating process is very similar to that at oceanic ridges, but the composition and the shape of the zone of magma generation and eruption are significantly different.

If hot basaltic magma from a mantle plume impinges on the continental crust, rhyolite may be produced by the melting of continental materials. For example, the basaltic lavas of the Snake River Plain and the huge rhyolitic calderas that created the spectacular scenery of Yellowstone National Park may have been powered by heat derived from a mantle plume. These rhyolites or intrusive granites are similar to those found in rifts at divergent boundaries.

Do all igneous rocks form at some kind of plate boundary?

GeoLogic   Laccoliths of the Henry Mountains

Several large dome-shaped hills rise above the horizontal sedimentary strata of Utah's Colorado Plateau. They look like huge blisters. How did these distinctive mountains form? What did these mountains look like before erosion? What is the structure of the rocks below the surface? The surface exposures of these rocks provide a clear picture of what lies below each mountain and tell us how they formed.

Observations

1. Each mountain has a core of the coarse-grained igneous rock called diorite.
2. The sedimentary rocks form a nearly circular ring around the diorite.
3. The sedimentary beds are horizontal in the adjacent areas but are upturned to form a dome around the diorite core.
4. Locally, horizontal sedimentary rocks also form a floor below the diorite.

Interpretations

The diorite was once molten magma. This interpretation is based on the rock's composition, its high temperature minerals, and its texture. Coarse grains indicate the magma cooled slowly below the surface. The magma intruded into and between the layers of sedimentary rock, pushing them upward to form a dome. Originally, the sedimentary rocks completely covered the igneous rocks but erosion has removed much of the sedimentary cover and exposed the core of igneous rocks.

The geologic interpretation of these facts is summarized in the diagram below. Do you see the domal structure in the photograph? Can you infer the extent of the sedimentary rocks before erosion? Is the interpretation logical? Are there other logical interpretations?

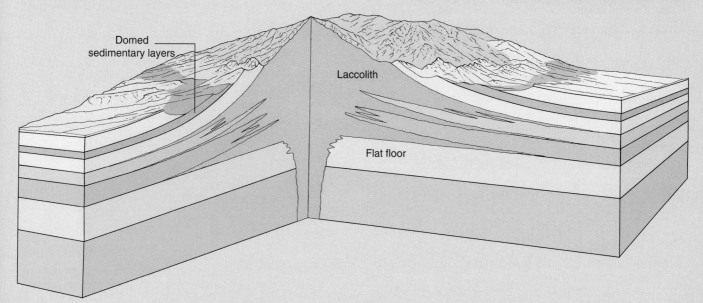

Domed sedimentary layers

Laccolith

Flat floor

Key Terms

aa flow (p. 97)
andesite (p. 94)
aphanitic texture (p. 89)
ash (p. 96)
ash flow (p. 105)
ash-flow caldera (p. 105)
ash-flow tuff (p. 94)
assimilation (p. 114)
basalt (p. 94)
batholith (p. 107)
caldera (p. 100)
cinder cone (p. 97)
columnar joint (p. 97)
composite volcano (p. 102)
crater (p. 102)
dike (p. 110)
diorite (p. 94)
extrusive rock (p. 87)
fissure (p. 97)

flood basalts (p. 97)
fractional crystallization (p. 113)
gabbro (p. 94)
glass (p. 89)
glassy texture (p. 89)
granite (p. 93)
groundmass (p. 92)
igneous rock (p. 86)
ignimbrite (p. 96)
inclusion (p. 110)
intrusion (p. 107)
intrusive rock (p. 87)
komatiite (p. 96)
laccolith (p. 111)
lava (p. 87)
lava dome (p. 102)
lava tube (p. 97)
magma (p. 86)

magma mixing (p. 114)
magmatic differentiation (p. 112)
matrix (p. 92)
pahoehoe flow (p. 97)
partial melting (p. 113)
pegmatite (p. 92)
peridotite (p. 94)
phaneritic texture (p. 92)
phenocryst (p. 92)
pillow lava (p. 100)
pluton (p. 107)
porphyritic texture (p. 92)
pressure ridge (p. 97)
pumice (p. 96)
pyroclastic-fall tuff (p. 96)
pyroclastic texture (p. 92)
rhyolite (p. 94)
shield volcano (p. 100)

sill (p. 110)
source rock (p. 112)
spatter cone (p. 97)
stock (p. 107)
stratovolcano (p. 102)
subglacial volcano (p. 102)
tephra (p. 96)
texture (p. 89)
thin section (p. 94)
tuff (p. 96)
vesicle (p. 90)
viscosity (p. 87)
volatile (p. 88)
volcanic ash (p. 97)
volcanic bomb (p. 97)
volcanic neck (p. 110)
welded tuff (p. 105)

Review Questions

1. Define the term magma.
2. Name the principal gases (volatiles) in magma.
3. Name two principal types of magma.
4. List the major types of igneous rock textures. Why is texture important in the study of rocks?
5. List the major types of igneous rocks, and briefly describe their textures, mineral proportions, and chemical compositions.
6. Describe some common surface features of basaltic flows.
7. Why does magma tend to rise upward toward Earth's surface?
8. Draw a series of diagrams showing the form and internal structure of (a) a cinder cone, (b) a composite volcano, and (c) a shield volcano.
9. Describe the events that are typically involved in the formation of a caldera.
10. Describe the extrusion of an ash flow.
11. Describe and illustrate the major types of igneous intrusions. What is the textural difference between intrusive rocks and extrusive rocks?
12. Explain how magma does not have the same composition as the rock from which it melts.
13. What is fractional crystallization?
14. Draw a simple diagram and explain how basaltic magma originates from the partial melting of the mantle at divergent plate boundaries or at a mantle plume.
15. Draw a simple cross section, and explain how granitic and andesitic magmas originate in a subduction zone.
16. Why can't basalt magma be produced by partial melting of granite?

5   Sedimentary Rocks

The geologic processes operating on Earth's surface produce only subtle changes in the landscape during a human lifetime, but over a period of tens of thousands or millions of years, the effect of these processes is considerable. Given enough time, the erosive power of the hydrologic system can reduce an entire mountain range to a featureless lowland. In the process, the eroded debris is transported by rivers and deposited as new layers of sedimentary rock.

A series of sedimentary rock layers may be thousands of meters thick. When exposed at the surface, each rock layer provides information about past events in Earth's history. Such is the case in the Moenkopi Formation of southern Utah shown in the panorama above. The various shades of red and white occur in the thin beds of siltstone and mudstone deposited on an ancient tidal flat about 220 million years ago. Thin layers of siltstone and shale each containing ripple marks, mud cracks, and rain imprints combine to tell the history recorded in the rock now exposed in this colorful cliff.

Apart from their scientific significance, the sedimentary rocks have been a controlling factor in the development of industry, society, and culture. Humans have used materials from

sedimentary rocks since the Neolithic Age; flint and chert played an important role in the development of tools, arrowheads, and axes. The great cathedrals of Europe are made from sedimentary rock, and the statues made by the artists of ancient Greece and Rome and during the Renaissance would have been impossible without limestone. Fully 85% to 90% of mineral products used by our society come from sedimentary rocks. Virtually our entire store of petroleum, natural gas, coal, and fertilizer come from sedimentary rocks. Sand, gravel, and limestone are the raw materials for cement. Sedimentary rocks are also important reservoirs for groundwater, and host important deposits of copper, uranium, lead, zinc, as well as gold and diamonds.

In this chapter, we study the remarkable record of Earth's history preserved in sedimentary rocks. Each bedding plane is a remnant of what was once the surface of Earth. Each rock layer is the product of a previous period of erosion and deposition. In addition, details of texture, composition, and fossils are important records of global change, showing how Earth evolved in the past and how it may change in the future. To interpret the sedimentary record correctly, we must first understand something about modern sedimentary systems, the sources of sediment, transportation pathways, and places where sediment is accumulating today, such as deltas, beaches, and rivers. The study of how modern sediment originates and is deposited provides insight into how ancient sedimentary rocks formed. Fossils preserved in sedimentary rocks not only reveal the environment of deposition but also the pace and course of evolution through Earth's long life.

Major Concepts

1. Sedimentary rocks form at Earth's surface by the hydrologic system. Their origin involves the weathering of preexisting rock, transportation of the material away from the original site, deposition of the eroded material in the sea or in some other sedimentary environment, followed by compaction and cementation.
2. Two main types of sedimentary rocks are recognized: (a) clastic rocks and (b) chemically precipitated rocks, including biochemical rocks.
3. Stratification is the most significant sedimentary structure. Other important structures include cross-bedding, graded bedding, ripple marks, and mud cracks.
4. The major sedimentary systems are (a) fluvial, (b) alluvial fan, (c) eolian, (d) glacier, (e) delta, (f) shoreline, (g) organic reef, (h) shallow marine, (i) submarine fan, and (j) deep marine.
5. Sedimentary rock layers can be grouped into formations, and formations can be grouped into sequences that are bound by erosion surfaces. These formations and sequences form an important interpretive element in the rock record.
6. Plate tectonics controls sedimentary systems by creating uplifted source areas, shaping depositional basins, and moving continents into different climate zones.

The Nature of Sedimentary Rocks

Sedimentary rocks form from fragments derived from other rocks and by precipitation from water. They typically occur in layers, or strata, separated one from the other by bedding planes and differences in composition.

Sedimentary rocks are probably more familiar than the other major rock types. Few people, however, are aware of the true nature and extent of sedimentary rock bodies.

The constituents of sedimentary rocks are derived from the mechanical breakdown and chemical decay of preexisting rocks. This **sediment** is compacted and cemented to form solid rock bodies. The original sediment can be composed of various substances:

1. Fragments of other rocks and minerals, such as gravel in a river channel, sand on a beach, or mud in the ocean
2. Chemical precipitates, such as salt in a saline lake or gypsum in a shallow sea
3. Organic materials formed by biochemical processes, such as vegetation in a swamp, coral reefs, and calcium carbonate precipitated in the ocean

Sedimentary rocks are important because they preserve a record of ancient landscapes, climates, and mountain ranges, as well as the history of the erosion of Earth. In addition, fossils are found in abundance in sedimentary rocks younger than 600 million years and provide evidence of the evolution of life through time. Earth's geologic time scale was worked out using this record of sedimentary rocks and fossils.

An excellent place to study the nature of sedimentary rocks is Arizona's Grand Canyon (**Figure 5.1**), where many distinguishing features are well exposed. Their most obvious characteristic is that they occur in distinct layers, or **strata** (singular **stratum**), many of which are more than 100 m thick. Rock types that are resistant to weathering and erosion form cliffs, and nonresistant rocks erode into gentle slopes.

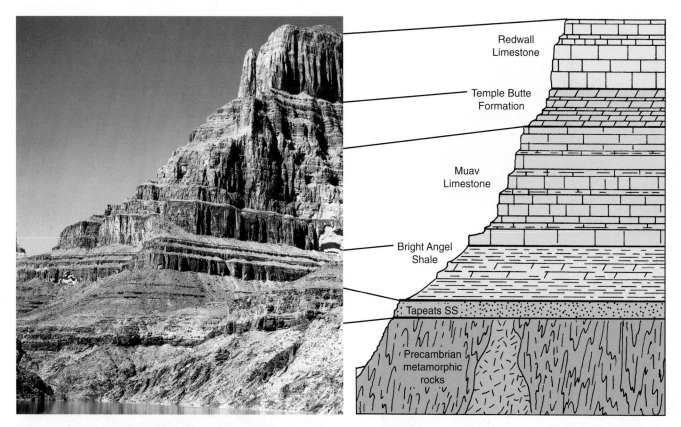

(A) Layers that are resistant to weathering and erosion (such as sandstone and limestone) erode into vertical cliffs. Rocks that weather easily (such as shale) form slopes or terraces.

(B) A cross section of the Grand Canyon graphically illustrates the major sedimentary layers or formations. The sedimentary strata are essentially horizontal and were deposited on older igneous and metamorphic rocks.

Figure 5.1   The layered series of sedimentary rocks exposed in the Grand Canyon, Arizona, is almost 2000 m thick and was deposited over a period of 300 million years. A similar sequence of sedimentary rocks is found on the stable platforms of most continents.

From Figure 5.1, you should be able to recognize the major layers or **formations** in the geologic cross section shown in the diagram. The formations exposed in the Grand Canyon can be traced across much of northern Arizona and parts of adjacent states. In fact, they cover an area of more than 250,000 km^2. A close view of sedimentary rocks in the canyon reveals that each formation has a distinctive texture, composition, and internal structure.

The major layers of the sandstone, limestone, and shale actually consist of smaller units separated by **bedding planes** that are marked by some change in composition, grain size, or color, or by other physical features. Animal and plant **fossils** are common in most of the rock units and can be preserved in great detail (**Figure 5.2**). The term *fossil* is generally used to refer to any evidence of former life (plant or animal). It may be direct evidence, such as shells, bones, or teeth, or it may be indirect, such as tracks and burrows produced by organic activity. The texture of most sedimentary rocks consists of mineral grains or rock fragments that show evidence of abrasion (Figure 5.2B) or consist of interlocking grains of the minerals calcite or dolomite. In addition, many layers show ripple marks (Figure 5.2C), mud cracks (Figure 5.2D), and other evidence of water deposition preserved in the bedding planes. All of these features show that sedimentary rocks form at Earth's surface in environments similar to those of present-day deltas, streams, beaches, tidal flats, lagoons, and shallow seas.

(A) Fossils found in sedimentary rocks include representatives of most types of marine animals.

(B) A microscopic view of sand grains in sediment shows the effects of transportation by running water. The grains are rounded and sorted to approximately the same size.

(C) Ripple marks preserved in sandstone suggest that the sediment was deposited by the current action of wind or water.

(D) Mud cracks form where sediment dries while it is temporarily exposed to the air. This structure is common on tidal flats, in shallow lake beds, and on stream banks.

Figure 5.2   A variety of features in sedimentary rocks show their origin at Earth's surface as a result of the hydrologic system. These include stratification, cross-bedding, ripple marks, mud cracks, fossils, and other features formed at the time the sediment was deposited.

Sedimentary rocks are widespread on the continents, covering about 75% of the surface of the continents; they therefore form most of the landscape. Nearly 100% of the ocean floor is blanketed with at least a thin layer of sediment. The map in **Figure 5.3** illustrates the distribution of sedimentary rocks in North America. You can see that the stable platform of the Great Plains and adjacent areas is completely covered by a relatively thin layer of sedimentary rocks. Most of these layers of rock are nearly horizontal. In addition, folded layers of sedimentary rocks are exposed in mountain belts. Sedimentary rocks are rare in the Canadian shield.

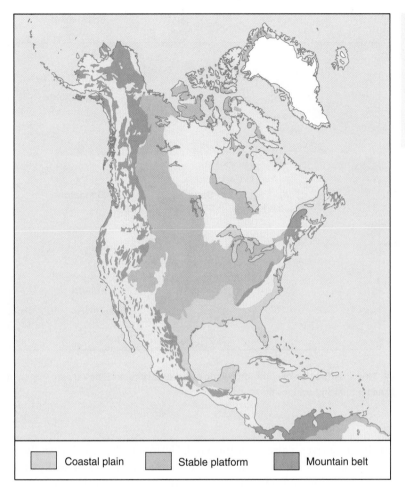

Figure 5.3 Sedimentary rocks in North America are widespread in the stable platforms. They are also found as deformed layers in the folded mountain belts. In the stable platform, the sedimentary sequence is generally less than 2 km thick. In folded mountain belts and on continental margins, they may be much thicker.

| Coastal plain | Stable platform | Mountain belt |

Types of Sedimentary Rocks

Sedimentary rocks are classified on the basis of the texture and composition of their constituent particles. Two main groups are recognized: (1) clastic rocks, formed from fragments of other rocks, and (2) chemical rocks and biochemical rocks.

Clastic Sedimentary Rocks

One important category of sedimentary rock consists of particles of gravel, sand, or mud. Rocks made up of such fragmental material are called clastic rocks. The term **clastic** comes from the Greek word *klastos*, meaning "broken," and describes the broken and worn particles of rock and minerals that were carried to the sites of deposition by streams, wind, glaciers, and marine currents.

In general, clastic rocks are subdivided according to grain size (**Figure 5.4**). From the largest grain size to the smallest, the types of clastic rocks are conglomerate, sandstone, and mudrock. The grain size of clastic sedimentary rocks is not controlled by progressive growth of grains as in igneous rocks, but is instead controlled by the size of clasts present in the source and by the carrying capacity of the transporting

How are the different types of sedimentary rocks distinguished and classified?

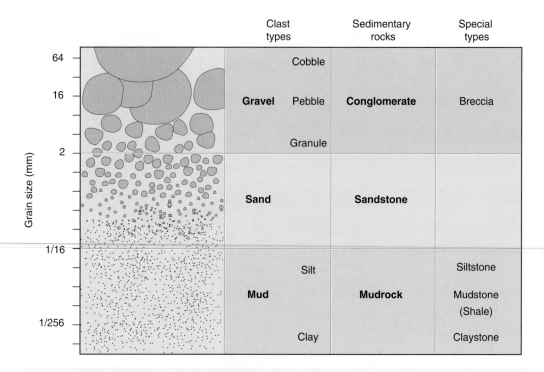

Figure 5.4 **The classification of common clastic sedimentary rocks** is based primarily on grain size and secondarily on textural and compositional variations.

In what way are clastic sedimentary rocks different from chemical precipitates?

medium—a river, a glacier, or the wind. Grains are deposited when the transporting medium loses its carrying capacity, commonly when its velocity decreases. Thus, sediment deposited from a fast-moving stream is coarser than that deposited in a quiet lagoon.

Conglomerate consists of consolidated deposits of gravel (fragments larger than 2 mm in diameter) with various amounts of sand and mud in the spaces between the larger grains (**Figure 5.5A**). The gravel is usually smooth and well rounded, suggesting the grains were rounded during transport. Most conglomerates are only crudely stratified and include beds and lenses of sandstone. High energy is required to transport large clasts like gravel, so conglomerate tends to be deposited in high-energy environments where water is flowing rapidly. Conglomerate accumulates today at the bases of many mountain ranges, in stream channels, and on some beaches.

Sandstone is probably the most familiar, though not the most abundant, sedimentary rock because it is well exposed, easily recognized, and generally resistant to weathering (Figure 5.5B). The sand grains range from 1/16 to 2 mm in diameter and can be composed of almost any material, so sandstones can be almost any color. Quartz grains (Figure 5.2B), however, are usually most abundant because quartz is a common constituent in many other rock types and because it is resistant to abrasion or chemical breakdown as the sediment particles are transported. The particles of sand in most sandstones are cemented by calcite, quartz, or iron oxide. Other grains may be feldspar (in a rock called arkose), pieces of chert, or small rock fragments.

The composition of a sandstone provides an important clue to its history. During prolonged transportation, small rock fragments and minerals that readily decompose, such as olivine, feldspar, and mica, break down into finer particles and are winnowed out, leaving only the stable quartz. Clean, well-sorted sandstone composed of well-rounded quartz grains indicates prolonged transportation, or even several cycles of erosion and deposition.

Mudrocks are fine-grained clastic rocks with grains less than 1/16 mm (0.0625 mm) across (Figure 5.4). Mudrocks are the most abundant sedimentary rocks. They are usually soft and weather rapidly to form slopes, so relatively few fresh,

(A) Conglomerate is a coarse-grained clastic sedimentary rock.

(B) Sandstone is a clastic sedimentary rock composed of sand-sized particles.

(C) Shale is a clastic sedimentary rock composed of very fine grains of clay or mud.

(D) Limestone is the most common nonclastic sedimentary rock. It is composed mostly of calcium carbonate.

(E) Limestone may also be made of abundant shell fragments.

(F) Gypsum precipitates as shallow bodies of water evaporate.

Figure 5.5 Major types of sedimentary rocks. Clastic rocks are shown on the left and biochemical chemical precipitates on the right. All of the rocks are shown at actual size.

unweathered exposures are found. They are frequently deposited in river floodplains and deltas and other **shallow-marine** settings. Many mudrocks also show evidence of burrowing by organisms.

What rock types commonly form in shallow-marine environments?

There are several important varieties of mudrocks. **Siltstone** is a fine-grained clastic rock coarser than **claystone**. Clasts in mudrocks tend to be more angular than those in sandstone. Many of the small clasts, especially in claystones, are flaky minerals like mica and clay. Tiny grains of quartz are another major constituent. A mudrock that contains very thin layers (**laminae**) is called **shale** (Figure 5.5C). Shales split easily along these layers to form small paper-thin sheets or flakes. The particles in claystones are generally too small to be clearly seen and identified even under a microscope.

Many types of shale are black and rich in organic material that accumulated in a variety of quiet-water, low-oxygen environments, such as lagoons and seas with poor circulation where oxygen-poor water accumulated. Red shales are colored with iron oxide and suggest oxidizing conditions in the environments in which they accumulate, such as river floodplains, tidal flats, lakes, and well-mixed oceans.

Biochemical and Chemical Sedimentary Rocks

The other major category of sedimentary rocks forms when chemical processes remove ions dissolved in water to make solid particles. Some are **biochemical rocks** with sediment formed during the growth of organisms such as algae, coral, or swamp vegetation. Others are inorganic **chemical precipitates** from lakes or shallow seas.

Limestone is by far the most abundant chemically precipitated rock (Figure 5.5D). It is composed principally of calcium carbonate ($CaCO_3$—dominantly calcite) and originates by both inorganic chemical and biochemical processes. Indeed, the distinction between biochemical and chemical rocks is rarely clear cut. Limestone has a great variety of textures, and many different types have been classified. Three important examples are described here: skeletal limestone, oolitic limestone, and microcrystalline limestone.

How do organisms make sedimentary rocks?

Some marine invertebrate animals construct their shells or hard parts by extracting calcium and carbonate ions from seawater. Corals, clams, algae, snails, and many other marine organisms construct their skeletons of calcium carbonate. After the organisms die, the shells accumulate on the seafloor. Over a long period of time, they build up a deposit of limestone with a texture consisting of shells and shell fragments (Figure 5.5E). These particles may then be cemented together as more calcite precipitates between the grains. This type of limestone, composed mostly of skeletal debris, can be several hundred meters thick and can extend over thousands of square kilometers. **Chalk**, for example, is a skeletal limestone in which the skeletal fragments are remains of microscopic plants and animals.

Other limestones are composed of small semispherical grains of calcium carbonate known as **oolites**. Oolites form where small fragments of shells or other tiny grains become coated with successive thin layers of $CaCO_3$ as they are rolled along the seafloor by waves and currents.

A third important type of limestone forms in quiet waters where calcium carbonate is precipitated by algae as tiny, needlelike crystals that accumulate on the seafloor as limy mud (**Figure 5.6A**). Soon after deposition, the grains commonly are modified by compaction and recrystallization. This modification produces microcrystalline limestone (or micrite), a rock with a dense, very fine-grained texture (Figure 5.5D). Its individual crystals can be seen only under high magnification. Other types of carbonate grains may be cemented together by microcrystalline limestone.

Inorganic limestone also is precipitated from springs and from the dripping water in caves to form beautifully layered rocks called travertine.

Most limestone forms on the shallow continental shelves where waters are warm and organic production is high (**Figure 5.7**). In contrast, carbonate sediment is rare in deep water and does not accumulate on the abyssal plains. In fact, calcite is not stable at the low temperatures and high pressures found on the deep seafloor. Calcite shells formed in surface waters fall toward the seafloor when the organism dies; but in deep water, calcite shells dissolve before they reach the bottom. Near the equator,

(A) Some kinds of algae produce calcium carbonate particles that accumulate to form limestone. These are found near the Kuril Islands of the north Pacific. Each leaflike structure is about 5 cm across.

(B) Diatoms are the shells of tiny single-celled algae that are made of silica. Some deep-marine sediments are dominated by diatoms like these seen through a microscope. Accumulations convert to chert.

Figure 5.6 **Algae** are important biochemical factories for the generation of sediment.

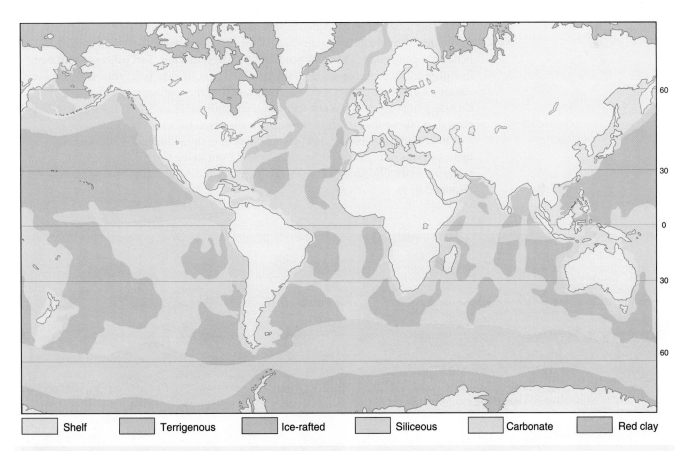

| | Shelf | | Terrigenous | | Ice-rafted | | Siliceous | | Carbonate | | Red clay |

Figure 5.7 **Marine sediments** form largely by biochemical precipitation. Carbonate sediments dominate at shallow depths (e.g., ocean ridges) and in warm near-shore waters. Elsewhere, siliceous sediment, which eventually forms chert, is typical in deeper water. These sediments form below areas of high biologic productivity. Most of the abyssal red clay is transported from the continents as wind-blown dust. Terrigenous sediment, derived from rivers on the continents, is coarser-grained clastic sediment.

calcite is not stable at depths below about 4500 m. Where the seafloor is shallower than this, as on oceanic ridges, carbonate sediment will accumulate (Figure 5.7).

Dolostone is a carbonate rock composed of the mineral dolomite, a calcium-magnesium carbonate $CaMg(CO_3)_2$. It is similar to limestone in general appearance, but reacts with acid only when powdered. Dolostone is commonly dull brownish yellow or light gray. It can develop by direct precipitation from seawater, but such environments are extremely rare. Instead, dolostone may form by the reaction of magnesium-bearing groundwater with calcium carbonate in limestone. The recrystallization generally destroys the original texture of the rock. In a fashion, dolostones are chemical precipitates formed from biochemical rocks.

Chert is a common rock composed of microcrystalline quartz. In a hand specimen, it is hard, dense, and typically breaks like glass, but under a high-power microscope, it has a fibrous or granular texture. Chert is usually white or shades of gray, tan, green, or red. Several varieties are recognized on the basis of color, including flint (black) and jasper (red). Because it fractures to make sharp edges, it has been shaped by many ancient people to make arrowheads, spear points, and tools. Chert commonly occurs as irregular nodules in limestone or as distinct thin layers in marine sedimentary rocks. Some nodular chert precipitates from pore fluids, particularly in carbonate rocks. However, most chert probably forms biochemically.

A distinctive type of deep-marine chert develops from deposits of siliceous shells of microscopic organisms, such as radiolaria and diatoms (Figure 5.6B). In the modern ocean, this kind of thinly bedded sediment dominates deep-marine environments where these tiny shells rain onto the seafloor. Siliceous marine sediment is thickest beneath regions of high biologic productivity. Deep-marine chert (siliceous sediment in Figure 5.7) is common along continental margins, in near-polar seas, and along the equator. Carbonate minerals do not accumulate on the seafloor where the ocean is very deep because calcite is not stable at great depths (Figure 5.7). If the water is deep enough, a falling shell made of carbonate is dissolved back into the seawater.

Other important biochemical components of many sedimentary rocks are hydrocarbons or organic compounds derived from living things. The decay of these materials in deeply buried sedimentary rocks produces oil, natural gas, and coal (**Figure 5.8**).

What kind of sedimentary rock typically forms on the deep-ocean floor?

Figure 5.8 Coal is an important biochemical precipitate. It forms by the decomposition of organic material buried within sedimentary rocks. Lush vegetation may form in an ancient swamp and then be converted by burial into coal. The coal beds in the Wyoming mine shown on the left are interlayered with sandstone and each is over 10 m thick.

Only a few important rock types form strictly by inorganic processes. **Rock salt** is made of the mineral halite (NaCl). It crystallizes when evaporation concentrates sodium and chlorine ions to the point that salt is stable in the residual brine. Strong evaporation creates saline lakes in closed desert basins (for example, the Great Salt Lake and the Dead Sea). Enhanced evaporation also occurs in restricted bays along the shore of the ocean. **Gypsum**, $CaSO_4 \cdot 2H_2O$, also originates from evaporation. It collects in layers as calcium sulfate is precipitated (Figure 5.5F) from water. Because **evaporites** (rocks formed by evaporation) accumulate only in restricted basins subjected to prolonged evaporation, they are important indicators of ancient climatic and geographic conditions.

Sedimentary Structures

Sedimentary rocks commonly show layering and other structures that form as sediment is transported. The most important sedimentary structures are stratification, cross-bedding, graded bedding, ripple marks, and mud cracks. Primary **sedimentary structures** provide key information about the conditions under which the sediment accumulated.

Stratification

One of the most obvious characteristics of sedimentary rocks is that they occur in distinct layers expressed by changes in color, texture, and the way they weather and erode. These layers are termed strata, or simply **beds**. The planes separating the layers are planes of stratification, or bedding planes. **Stratification** occurs on many scales and reflects the changes that occur during the formation of a sedimentary rock. Large-scale stratification is expressed by major changes in rock types (formations; Figure 5.1). For example, cliffs of limestone or sandstone can alternate with slopes of weaker shale.

The origin of stratification is quite simple. Different layers form because of some change that occurs during the process of deposition. But there are many types of changes that occur and operate on many different scales, so the construction of a detailed history of sedimentary rocks presents a real challenge to geologists. Changes in weather, changes in the seasons, and changes in climate all can produce stratification in a sedimentary basin. Tectonic changes such as uplift and subsidence of the continental platform, mountain building, and volcanism all produce changes in material transported to the sea, and all can produce different layers of sedimentary rock.

Why are sedimentary structures important in the study of sedimentary rocks?

Cross-Bedding

Cross-bedding is a type of stratification in which the layers within a bed are inclined at an angle to the upper and lower surfaces of the bed. The formation of cross-bedding is shown in **Figure 5.9**. As sand grains are moved by wind or water, they form small ripples or large **dunes**. These **sand waves** range in scale from small ripples less than a centimeter high to giant sand dunes several hundred meters high. Typically, they are asymmetrical, with the gentle slope facing the moving current. As the particles migrate up and over the sand wave, they accumulate on the steep down current face and form inclined layers. The direction of flow of the ancient currents that formed a given set of cross-strata can be determined by measuring the direction in which the strata are inclined. We can determine the patterns of ancient current systems by mapping the direction of cross-bedding in sedimentary rocks. Moreover, the style of cross-bedding changes with the sediment supply and with the flow conditions at the depositional site. Thus, the details of an ancient environment can be interpreted from careful study of the type of cross-bedding.

Graded Bedding

Another distinctive type of stratification, called **graded bedding**, displays a progressive decrease in grain size upward through a bed (**Figure 5.10**). This type of

What sedimentary structure forms from a turbidity current?

Figure 5.9 **Cross-bedding** is formed by the migration of sand waves (ripple marks or dunes). Particles of sediment, carried by currents, travel up and over the sand wave and are deposited on the steep down current face to form inclined layers.

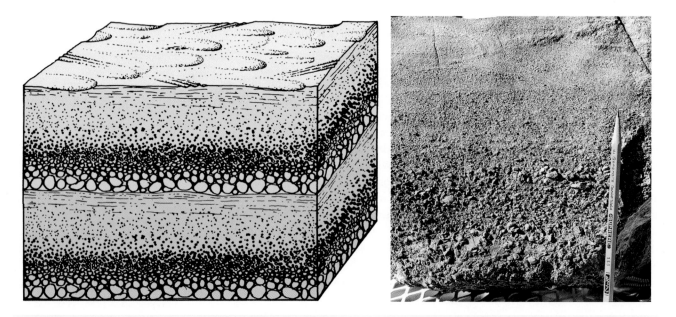

Figure 5.10 **Graded bedding** is produced by turbidity currents. It occurs in widespread layers, each layer generally less than a meter thick. Slumps off the deep continental slopes commonly produce great thicknesses of graded layers, which can easily be distinguished from sediment deposited in most other environments.

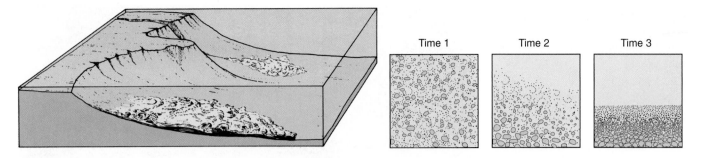

Figure 5.11 The movement of turbidity currents down the slope of the continental shelf can be initiated by a landslide or an earthquake. Sediment is moved largely in suspension. As the current slows, the coarse grains are deposited first, followed by the deposition of successively finer-grained sediment. Fine mud slowly settles out from suspension after the turbidity current stops. A layer of graded bedding is thus produced from a single turbidity current.

stratification commonly is produced on the deep-ocean floor by **turbidity currents**, which transport sediment from the continental slope to adjacent deep ocean forming bodies of rock called **turbidites**. A turbidity current is generated by turbid (muddy) water, which, being denser than the surrounding clear water, sinks beneath it and moves rapidly down the continental slope (**Figure 5.11**). The denser, muddy water moves out along the bottom of the basin and can flow for a considerable distance, even along the flat surface of an abyssal floor. As a turbidity current moves across the flat floor of a basin, its velocity at any given point gradually decreases. The coarsest sediment in the turbidity current is deposited first, followed by successively smaller particles. After the turbid water ceases to move, the sediment remaining suspended in the water gradually settles out. One turbidity current, therefore, deposits a single layer of sediment, which exhibits a continuous gradation from coarse material at the base to fine material at the top. Subsequent turbidity currents can deposit more layers of graded sediment, with sharp contacts between layers. The result is a succession of widespread, nearly horizontal turbidites, each a graded unit deposited by a single turbidity current. Turbidity currents also form where streams discharge muddy water into a clear lake or reservoir (**Figure 5.12**).

Small turbidity currents can be created by pouring muddy water down the side of a tank filled with clear water. The mass of muddy water moves down the slope of the tank and across the bottom at a relatively high speed, without mixing with the clear water.

Turbidity currents are commonly generated by earthquakes or submarine landslides, during which mud, sand, and even gravel are transported downslope. In 1929 one of the best-documented large-scale turbidity currents was triggered by an earthquake near the Grand Banks, off Newfoundland. Slumping of a large mass of soft sediment (estimated to be 100 km³) moved as a turbidity current down the continental slope and onto the abyssal plain, eventually covering an area of 100,000 km². As the turbidity current moved downslope, it broke a series of transatlantic cables at different times. The speed of the current, determined from the intervals between the times when the cables broke, was from 80 to 95 km/hr. This mass of muddy water formed a graded layer of sediment over a large area of the Atlantic floor.

Ripple Marks, Mud Cracks, and Other Surface Impressions

Ripple marks are commonly seen in modern streambeds, in tidal flats, and along the shores of lakes and the sea. Many are preserved in rocks and provide information concerning the environment of deposition, such as depth of water, ancient current directions, and trends of ancient shorelines (Figure 5.2C). **Mud cracks** are also commonly preserved in sedimentary rock and show that the sedimentary environment was occasionally exposed to the air during deposition. Mud cracks in rocks suggest that the original sediment was deposited in shallow lakes, on tidal flats, or on exposed stream banks (Figure 5.2D). Rain prints are even preserved in some mudrocks.

Figure 5.12 **Turbidity currents** into Lake Powell, Utah, result where the muddy water of the Colorado River enters the clear water of the lake.

Fossils and Trace Fossils

Another key feature of many sedimentary rocks are fossils of once living organisms (Figure 5.2A). Fossils often reveal much about the past environment, giving us hints about whether a deposit is marine or continental, what the water depth was when the sediment was deposited, and about temperature and salinity of the water. Beyond that, however, fossils in sedimentary rocks reveal the history of the evolution of life. Although the record is far from complete, everything we know about past life comes from reconstructions based on ancient fossils.

Tracks, trails, and borings of animals are typically associated with ripple marks and mud cracks and can provide additional important clues about the environment in which the sediment accumulated.

As can be seen, primary sedimentary structures and fossils are the clues or the tools used by geologists to interpret the conditions and environment at the site where the sediment is deposited.

Sedimentary Systems

Weathering of preexisting rocks, transportation, and deposition, followed by compaction and cementation are the major steps in the formation of sedimentary rocks. The major sedimentary systems are (1) fluvial, (2) alluvial fan, (3) eolian, (4) glacial, (5) delta, (6) shoreline, (7) organic reef, (8) shallow marine, and (9) deep marine. Each of these systems has a specific set of physical, chemical, and biological conditions and therefore develops distinctive rock types and fossil assemblages.

Sedimentary systems operate at Earth's surface through interactions of the hydrologic system and the crust. As a result of the transfer of energy between the various parts of a sedimentary system, new landforms and new bodies of sedimentary rock are created. Most of the energy that drives these systems ultimately comes from the Sun; gravitational and chemical potential energy are also transferred in various parts of the sedimentary system. It is useful to visualize a hypothetical sedimentary system as consisting of a source of sediment (weathering), a transport path for the sediment, a site of deposition, and the processes that compact and cement the sediment together to form a solid rock. Fortunately, many of these sedimentary processes operate today, and geologists actively study rivers, deltas, and oceans and other sedimentary systems in an effort to understand the characteristics of rocks formed in these environments.

Weathering

Weathering is the interaction between the elements in the atmosphere and the rocks exposed at Earth's surface. The atmosphere can mechanically break down the rock through processes such as ice wedging, and it can chemically decompose the rock by a variety of reactions. Note that weathering is the first step in the genesis of sedimentary rock. The atmosphere breaks down and decomposes preexisting solid rock and forms a layer of loose, decayed rock debris, or soil. This unconsolidated material can then be transported easily by water, wind, and glacial ice.

What are the major steps in the formation of sedimentary rock?

Transportation

Running water is the most effective form of sediment transport. All rivers carry large quantities of sediment toward the sea. This fact is readily appreciated if you consider the great deltas of the world, each formed from sediment transported by a river (**Figure 5.13**). Indeed, sediment is so abundant in most rivers that a river might best be thought of as a system of water and sediment rather than simply a channel of flowing water.

As clastic sediment is transported by a river, it is sorted and separated according to grain size and composition. Large particles accumulate in high-energy environments as gravel, medium-sized grains are concentrated as sand, and finer material settles out as mud. The grain size of the sediment correlates with the energy of the transporting medium. Thus, large particles are carried by rapidly moving streams with high amounts of kinetic energy; only small particles are transported by slowly moving streams. Wind, glaciers, and shoreline currents also transport sediment, but their activity is somewhat restricted to special climate zones. Components from dissolved minerals are carried in solution and are ultimately precipitated to form limestone or salt, for example.

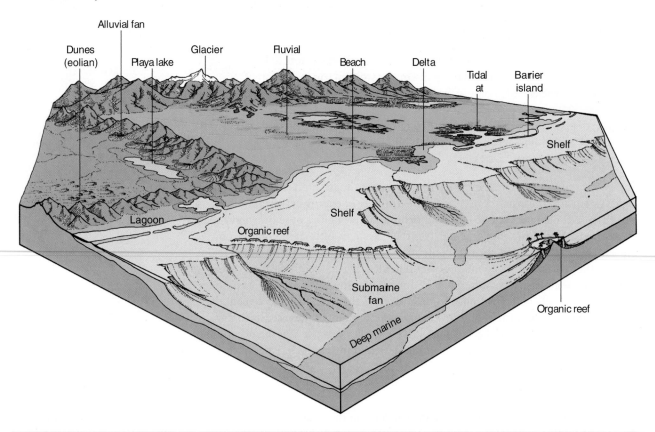

Figure 5.13 **The major** sedimentary systems are represented in this idealized diagram. Most sediment moves downslope from continental highlands toward the oceans, so the most important sedimentary systems are found along the shores and in the shallow seas beyond. Sedimentary systems can be categorized in three groups: continental, shoreline (transitional), and marine. Their important characteristics and the types of sediments that accumulate in each are outlined in Figures 5.14 to 5.25.

Deposition

Probably the most significant factor in the genesis of sedimentary rocks is the place where the sediment is deposited. The idealized diagram in Figure 5.13 shows the major depositional systems. The most important continental systems are river (**fluvial**) systems, **alluvial fans**, desert dunes, and margins of **glaciers**. Marine systems include the shallow marine, which cover parts of the continental platform, **reefs**, **submarine fans**, and the floors of the **deep-ocean basins**. Between continental areas and marine areas are the transitional, or mixed, environments that occur along the coasts and are influenced by both marine and nonmarine processes. These include **deltas**, **beaches**, **barrier islands**, **tidal flats**, and **lagoons** (Figure 5.13).

Each depositional system imprints specific characteristics on the rocks formed within it. A small depositional area within a system creates a **facies**—a body of rock with distinct chemical, physical, and biological characteristics created by the environment. Thus, a delta system produces many different facies, for example, sediment deposited in channels, between channels, and at the mouths of channels. Distinctive textures, compositions, sedimentary structures, and fossil assemblages develop in each facies. Illustrations of modern sedimentary systems, together with examples of the rocks they produce, are shown in **Figures 5.14** through **5.25**. Carefully study each of these photographs and the discussions in the captions.

(A) Point-bar deposits in a modern river.

(B) Ancient stream channel marked by sandstone lens interlayered with fine-grained clay and siltstone.

Figure 5.14 Fluvial systems. The great rivers of the world are the major channels by which erosional debris is transported from the continents to the oceans. Before reaching the ocean, most rivers meander across flat alluvial plains and deposit a considerable amount of sediment. Within this environment, sediment is deposited in stream channels, on bars, and on floodplains. Perhaps the most significant type of sedimentation occurs on bars on the insides of meander bends. Stream deposits have channels of relatively coarse sand or gravel cut into horizontal layers of fine silt and mud that were deposited on the flood plain.

(A) Modern alluvial fans in Death Valley, California.

(B) Ancient alluvial-fan deposits in central Utah.

Figure 5.15 Alluvial-fan systems. In many arid regions of the world, thick deposits of sedimentary rock accumulate in alluvial fans at the bases of mountain ranges. Deposition occurs here because stream channels widen and the slope decreases, causing the water to slow down and drop its sediment. Flash floods and debris flows are an important factor in this environment. Torrents from cloudbursts pick up the loose debris on the slopes of the mountain ranges and deposit it on the basin floor. The sediment in an alluvial fan characteristically is coarse-grained, and conglomerate is the most abundant rock type. In the central part of the basin, fine silt and mud can accumulate in temporary lakes and commonly are associated with the coarser fan deposits.

(A) Modern sand dunes of the Little Sahara, Utah.

(B) Ancient dune deposits in Zion National Park, Utah.

Figure 5.16 Eolian (wind) systems. Wind is a very effective sorting agent. Small silt and dust grains are lifted high in the air and may be transported thousands of kilometers before being deposited where wind velocity drops. Sand is transported close to the surface and eventually accumulates in dunes. Gravel cannot be moved effectively by wind. In arid regions, a major process is the migration of sand dunes. Sand is blown up and over the dunes and accumulates on the steep dune faces. Large-scale cross-strata that dip in a downwind direction are thus formed. Ancient dune deposits have large-scale cross-strata consisting of well-sorted, well-rounded sand grains. The most significant ancient wind deposits are sandstones that accumulated in large dune fields comparable to the present Sahara and Arabian deserts and the great deserts of Australia. These sandstones are vast deposits of clean sand that preserve the large-scale cross-beds developed by migrating dunes.

(A) The margins of a valley glacier in eastern Canada.

(B) Ancient glacial sediments in central Utah.

Figure 5.17 Glacial systems. A glacier transports large boulders, gravel, sand, and silt suspended together in the ice. This material is eventually deposited near the margins of the glacier as the ice melts. The resulting sediment is unsorted and unstratified, with angular individual particles that rest on the polished and striated floor of the underlying rock. Fine-grained particles dominate in many glacial deposits, but angular boulders and pebbles are invariably present. Streams from the melt waters of glaciers rework the unsorted glacial debris and redeposit it beyond the glaciers as stratified, sorted stream deposits. The unsorted glacial deposits are thus directly associated with well-sorted stream deposits from the melt waters.

(A) The delta of the Nile River, Egypt, forms where the river empties into the Mediterranean Sea.

(B) Ancient deltaic deposits in Cenozoic rocks of the Colorado Plateau.

Figure 5.18 Delta systems. One of the most significant depositional systems occurs where major rivers enter the oceans and deposit most of their sediment in marine deltas. A delta can be very large, covering areas of more than 36,000 km². Commonly, deltas are very complex and involve various distinct subenvironments, such as beaches, bars, lagoons, swamps, stream channels, and lakes. Because deltas are large features and include both marine and nonmarine subenvironments, a great variety of sediment types accumulate in them. Sand, silt, and mud dominate. A deltaic deposit can be recognized only after considerable study of the sizes and shapes of the various rock bodies and their relationships to each other. Both marine and nonmarine fossils can be preserved in a delta.

(A) A modern beach on Cape Hatteras along the Atlantic Coast of the United States.

(B) Ancient beach deposits in central Utah form resistant sandstone beds alternating with shale (slopes).

Figure 5.19 Shoreline systems. Much sediment accumulates in the zone where the land meets the ocean. Within this zone, a variety of subenvironments occurs, including beaches, bars, spits, lagoons, and tidal flats. Each has its own characteristic sediment. Where wave action is strong, mud is winnowed out, and only sand or gravel accumulates as beaches or bars. Beach gravels accumulate along shorelines, where high wave energy is expended. The gravels are well sorted and well rounded and commonly are stratified in low, dipping cross-strata. Ancient gravel beaches are relatively thin. They are widespread and commonly are associated with clean, well-sorted sand deposited offshore.

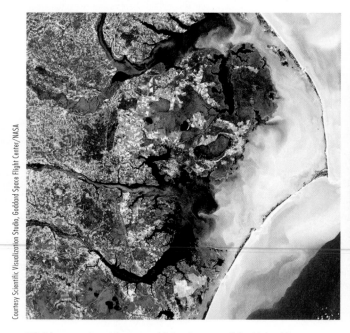

(A) A lagoon along the central Atlantic coast of the United States.

(B) Ancient lagoonal deposits with thick beds of coal in sandstone in eastern Wyoming.

Figure 5.20 **Lagoon systems**. Offshore bars and reefs commonly seal off part of the coast, forming lagoons. A lagoon is protected from the high energy of waves, so the water is relatively calm and quiet. Fine-grained sediment, rich in organic matter, accumulates as black mud. Eventually, the lagoon may fill with sediment and evolve into a swamp. Where the bottom vegetation provides enough organic matter, a coal deposit may form. The rise and fall of sea level shift the position of the barrier bar, and thus the organic-rich mud or coal formed in the lagoon or swamp is interbedded with sand deposited on the barrier island.

(A) A modern tidal flat in the Maritime Provinces of Canada.

(B) Ancient tidal flat deposits in southern Utah.

Figure 5.21 **Tidal-flat systems.** The tidal-flat environment is unique in being alternately covered with a sheet of shallow water and exposed to the air. Tidal currents are not strong. They generally transport only fine silt and sand and typically develop ripple marks over a broad area of the tidal flat. Mud cracks commonly form during low tide and are subsequently covered and preserved. Ancient tidal-flat deposits are thus characterized by accumulations of silt and mud in horizontal layers with an abundance of ripple marks and mud cracks. In restricted settings, evaporites can form on tidal flats.

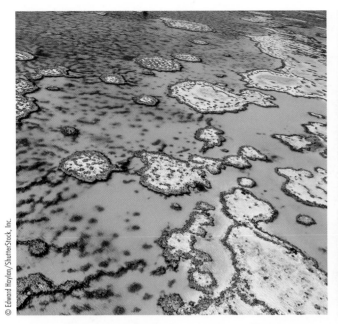

(A) Australia's Great Barrier Reef is a coral reef on the eastern shore.

(B) An ancient reef from the Paleozoic Era in the Guadalupe Mountains of west Texas.

Figure 5.22 Organic-reef systems. An organic reef is a solid structure of calcium carbonate constructed of shells and secretions of marine organisms. The framework of most reefs consists of a mass of colonial corals and forms a wall that slopes steeply seaward. Wave action continually breaks up part of the seaward face, and blocks and fragments of the reef accumulate as debris on the seaward slope. A lagoon forms behind the reef, toward the shore or toward the interior of the atoll (organic reef), and lime, mud, and evaporite salts may be deposited there. Gradual subsidence of the seafloor permits continuous upward growth of reef material to a thickness of as much as 1000 m. Because of their limited ecological tolerance (corals require warm, shallow water), fossil reefs are excellent indicators of ancient environments.

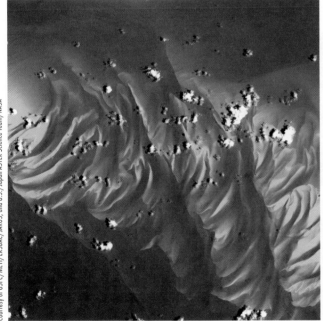

(A) A modern shallow-marine environment in the Bahamas as seen from space. The light ridges are drifts of carbonate sediment.

(B) Ancient shallow-marine sediments in eastern Nevada are made of many layers of limestone.

Figure 5.23 Shallow-marine systems. Shallow seas border most of the world's land area and can extend to the interior of a continent, as do Hudson Bay, the Baltic Sea, and the Gulf of Carpentaria (in northern Australia). The characteristics of the sediment deposited in the shallow-sea environment depend on the supply of sediment from the land and the local conditions of climate, wave energy, water circulation, and temperature. If there is a large supply of land-derived sediment, sand and mud accumulate. If sediment from the land is not abundant, limestone is deposited. Ancient shallow-marine deposits have thin, widespread, interbedded layers of sandstone, shale, and limestone.

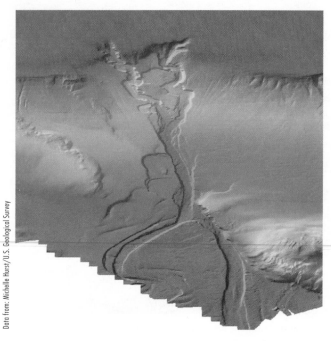

Data from: Michelle Hurst/U.S. Geological Survey

(A) Submarine fan with its distributary channels sits at the base of the continental slope offshore California.

(B) Ancient series of folded turbidites in southern France.

FIGURE 5.24 Continental slope-systems. Sediment on the continental slope is dominated by turbidites. The deposits are typically a series of graded beds, with each layer extending over a large area. Such beds are easily distinguishable from sediment deposited in most other environments. The turbidites accumulate to form a cone of debris at the foot of the continental slope called a submarine fan; most are at the mouths of submarine canyons. Channels filled with porous sandstone weave complicated patterns through the fan, as shown on this shaded relief map of a submarine fan off the coast of southern California. These sandstones that fill the channels are important hosts for oil in deep-marine environments.

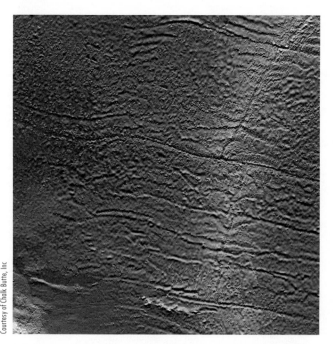

Courtesy of Chalk Butte, Inc

(A) Thin layers of deep-marine sediment form on abyssal plains on the flanks of the mid-ocean ridge (right).

(B) Thin beds of ancient deep-marine chert now exposed in coastal California. Such chert beds are interbedded with fine mudstones.

Figure 5.25 Deep-marine systems. Deep-marine sediment accumulates on the floor of the open ocean, far from continents. This material consists of shells of microscopic organisms and fine particles of mud that are carried in suspension and gradually sink. Biochemical chert in thin beds is a common deep-marine sediment. However, the most abundant sediment is fine-grained brown or red clay. Silt blown in by the wind is another important sediment in the deep-ocean basins. Calcium carbonate does not accumulate in the deepest oceans, but may be deposited on mid-ocean ridges where the water is shallower. In this environment, sediment accumulation rates are very low and the beds are typically thin.

Compaction and Cementation

The final stage in the formation of sedimentary rocks is the transformation of loose, unconsolidated sediment into solid rock. **Compaction** occurs when the weight of overlying material, which continually accumulates in a sedimentary environment, compresses the sediment buried beneath into a tight, coherent mass. Wet mud consists of 60% to 80% water, most of which is driven out during compaction. **Cementation** occurs when dissolved ions, carried by water seeping through pores, are precipitated. Common cementing minerals are calcite, quartz, and iron oxides. This post depositional crystallization of cement holds the grains of sediment together and is a fundamental process in transforming sediment into solid rock.

Stratigraphic Sequences

> Layers of sedimentary rocks can be grouped into formations, and formations can be grouped into sequences that are bounded by erosion surfaces. These formations and sequences form an important interpretive element in the rock record.

There are more than 17,000 formally recognized and named formations in the United States, each covering an area of up to 300,000 km². Each formation is a group of beds of a distinctive rock type that formed at a specific time and place. They are, thus, the fundamental rock units studied by geologists. On a regional scale, these formations are like a deck of cards scattered across a table with most cards overlapping another. In many areas, the total succession of sedimentary formations is thousands of meters thick. Stratification occurs on many scales, so there are typically many separate beds within a formation and innumerable thin laminae within the layers of each bed.

On a larger scale, formations and their stratification are produced by changes in the depositional system; each change causes a different type of sediment to form. One of the simpler and more common patterns in a vertical succession is the cycle of sandstone-shale-limestone-shale-sandstone (**Figure 5.26**). This pattern is produced by the advance (**transgression**) and retreat (**regression**) of a shallow sea across a continental margin. The base of the sedimentary layers is a preexisting surface produced by erosion—an **unconformity**. In Figure 5.26A, sand is accumulating on the floodplain of a river system and along the shore, fine-grained mud is carried farther and is accumulating just offshore, and calcium carbonate precipitates from solution beyond the mud zone. All three types of sediment are deposited simultaneously, each in a different environment.

As relative sea level rises, each environment shifts landward (Figure 5.26B, C). Beach sands are deposited over stream sediments, offshore mud is deposited over the previous beaches, and carbonate is deposited over the mud. As transgression continues, the layers of sand, mud, and carbonate are deposited farther and farther inland.

If relative sea level drops (Figure 5.26D), mud is deposited over limestone and nearshore sand over mud. The net result is a long wedge, or layer, of limestone encased in a wedge of shale, which in turn is encased in a wedge of sandstone. This package of sediment is bounded below and above by an unconformity. Subsequent uplift and erosion of the area reveal a definite **sequence** of rock (Figure 5.26E). Beginning at the basal unconformity, sandstone is overlain by shale and limestone, which in turn are overlain by shale and sandstone.

Sequence Stratigraphy

Traditionally, rock formations have been identified and classified on the basis of rock type—such as the limestone and sandstone layers that form the prominent cliffs in the photograph in Figure 5.1—but they can also be grouped into larger sequences of strata separated by major unconformities. The study of such sequences of rock is known as *sequence stratigraphy* and is an attempt to detect worldwide changes in sea level and to document tectonic movements that affect sea level in smaller regions. Sequences of rock bounded by unconformities reflect important events in Earth's history that have

Why do geologists study sedimentary sequences and not just the individual layers?

Figure 5.26 A sequence of sediments deposited by transgression and regression of a shallow sea is represented in these schematic diagrams. Sand accumulates along the beach, mud is deposited offshore, and calcite is precipitated farther offshore, beyond the mud. As the sea expands over the continent, these shallow-marine environments move inland, producing a vertical sequence of sand, mud, and carbonate sediment. When the sea recedes off the continent, mud is deposited over the carbonate sediment, and sand is deposited over the mud. The net result is a vertical sequence of sedimentary layers: sandstone, shale, limestone, shale, and sandstone.

(A) The sea begins to expand over an erosional unconformity. The original shoreline is marked by sand deposits that grade seaward into mud and lime (carbonate sediment).

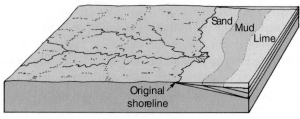

(B) The sea transgresses farther inland, depositing a sheet of sand overlain by mud (gray) and lime (tan).

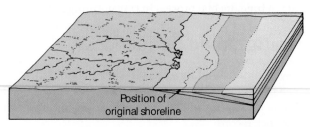

(C) With continued transgression of the sea, mud is deposited on top of the sand at the position of the original shoreline.

(D) A regression of the sea deposits shoreline sand over the offshore mud. Thus, the vertical succession of sediment at the position of the original shoreline is sand, mud, carbonate, mud, and sand. The package of sequence is bounded above and below by an unconformity.

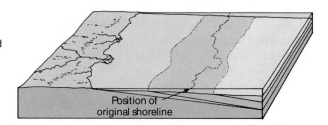

Sandstone

Shale

Limestone

(E) Much later the eroded exposure of a transgressive-regressive sequence shows the cycle from bottom to top: sandstone, shale, limestone, shale, and sandstone.

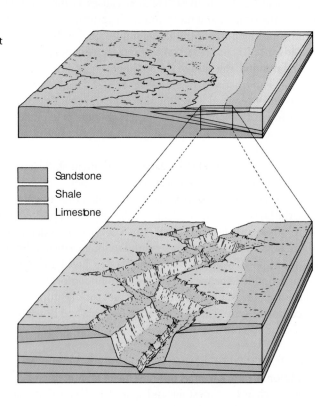

Figure 5.27 **Sequence stratigraphy** identifies the global changes of relative sea level. The major transgression-regression cycles are probably produced by large tectonic events that change the volume of the ocean basins. Shorter cycles of transgression and regression are superimposed on these major changes and probably are caused by glaciation and regional tectonic events.

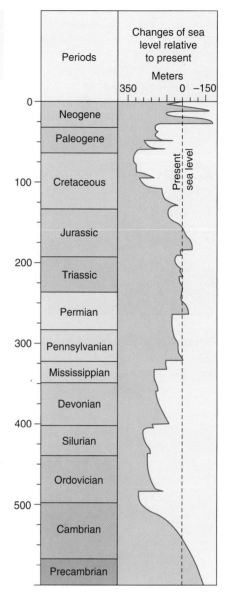

regional or even worldwide significance. Over much of the last 600 million years, the seas repeatedly transgressed and regressed across the continental shelves and over the platforms, leaving a record of shallow-marine deposits separated by erosional surfaces.

These sequences bounded by unconformities are created by major changes in the position of the shoreline. Such relative sea level changes can be caused by variations in the total amount of water in the oceans or by tectonic events. For example, the amount of water in the oceans drops during periods of glaciation when water is stored on continents. During periods of rapid seafloor spreading, ocean ridges become inflated; this reduces the volume of ocean basins, and forces water to spill over continental lowlands. Changes in ocean basin volume also occur when continents collide.

The concepts of sequence stratigraphy are extensively used by geophysicists interpreting seismic records of rocks in the subsurface. The principal erosional surfaces marking transgressions and regressions have been identified from seismic logs from all over the world. Consequently, a clear picture of changing sea level through time is emerging (**Figure 5.27**).

Sedimentary Systems and Plate Tectonics

Plate tectonics has a profound influence on the origin of sedimentary rocks in that it controls sediment sources, pathways, climate zones, and depositional systems. As a result, each major tectonic setting produces a distinctive sequence of sedimentary rocks.

Although the nature of sedimentary rocks depends on many factors (depth and velocity of water, ocean currents, biology, and sediment source area, for example), plate tectonics plays a major role in global aspects of sediment deposition and the sequences of rock that are produced. Tectonics control the extent of shallow seas upon the stable platform, the distribution of continental margins, the development of sedimentary basins, and the origin of mountain belts that are the sources of many sediments. Plate movement also controls rates of crustal uplift and subsidence, and therefore the rates of erosion and deposition. In addition, plate tectonics also has a prime control on the courses of rivers (which determine where the bulk of clastic sediment is deposited), and the topography and structure of continents in general. Let us consider some of the tectonic settings in which major bodies of sedimentary rocks form (**Figure 5.28**).

On the continents, sediment forms by weathering and is then transported across the stable platform from distant highlands and locally accumulates to form thin deposits that include stream sands and shallow-marine mudstones and limestones. Broad basins may form on the stable platform in which thicker sequences of sediment accumulate. However, most sediment carried all the way to the sea is deposited as shallow-marine sediment on the continental margin.

Continental rift valleys at incipient divergent plate boundaries receive a distinctive suite of conglomerate, sandstone, lake deposits, and evaporites (if the climate is arid). As the rift evolves into an open seaway, the continental margin subsides as it cools; shallow seas spread over the margin and across part of the stable platform to form a broad continental shelf. A thick wedge-shaped deposit of sediment forms as the margin continues to subside.

Farther from shore, on the continental slope, turbidity currents move sediment toward the abyssal plains to form deep-sea fans comprised of turbidites. Even farther from the continent, deep-ocean basins accumulate organic ooze (dominated by siliceous diatoms) and wind-blown dust. Carbonate sediments form in shallower parts of the oceans. These sediments slowly settle out of seawater to form thin layers of brown mud on the igneous part of the oceanic crust.

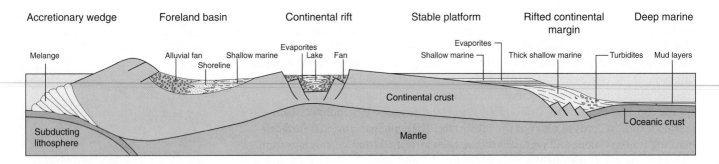

Figure 5.28 **Plate tectonics** exerts fundamental controls on sedimentary systems as shown in this cross section. The most important types of sedimentary basins form at convergent margins, divergent margins that evolve into passive continental margins, and shallow basins that form on the stable platform. Deep-ocean basins also have distinctive sediments.

STATE OF THE ART Ocean Drilling Programs: Probing the Deep Blue Sea

One of the most ambitious scientific endeavors in the study of Earth is the effort to drill into the floor of the deep-ocean basins. These programs employ specially designed research ships that carry huge drills and long strings of pipe. Imagine the difficulty in lowering a slender drilling stem 8000 m long to the floor of the ocean, then drilling another 2000 m or more through layers of siliceous organic ooze and then drilling into solid igneous rock. Once drilled, core segments about 10 m long and 10 cm in diameter are slowly pulled to the deck of the ship, where they are cut in two in preparation for detailed laboratory studies.

Ocean cruises of this sort require many crew members and scientific staff and extend over several months. The research is expensive and only a few nations have mounted this type of research. Since 1968 and continuing to this day, international deep-sea drilling programs have involved scientists from the United States, the United Kingdom, Japan, Germany, Australia, New Zealand, and other nations. Hundreds of thousands of meters of core have been recovered from thousands of holes drilled in many different locations.

One specially designed ship, the *JOIDES Resolution*, has drilled in the Atlantic, Pacific, Indian, and Arctic oceans. In addition to the huge drill, scientists on board have at their disposal a complete array of analytical instruments in the ship's twelve laboratories. They examine samples of sediment under microscopes to see what fossils they contain, X ray cores of sedimentary and igneous rocks to determine their composition, and probe the cores with magnetometers, thermometers, and other instruments. After the cruise, these rocks and sediment are stored in a core library and become valuable resources for scientists around the world.

The empty drill holes are often turned into "laboratories" themselves. Instruments are lowered into the holes to record the temperature, measure heat flow, take water samples, and generally record the physical and chemical properties of the oceanic crust.

Why spend months at sea and go to such great expense? The scientific payoff for such deep-sea drilling programs has been tremendous. Much of what we know about past climate change has come from painstaking studies of the sediment samples recovered by deep-sea drilling, including the timing of past glacial epochs and unraveling Milankovitch cycles, discussed elsewhere in the text. Ocean drilling has revealed evidence for the meteorite impact that marked the end of the Mesozoic Era. Other drilling opportunities have focused on understanding the origin of massive oceanic plateaus that some geologists think are produced by rapid submarine eruptions of flood lavas. Some projects drilled through the entire sedimentary cover to reveal exactly what types of igneous rocks are below the oceanic sediment and to allow for heat flow measurements that tell us the thickness of the lithosphere. These cores help us unravel Earth's past and present and conjecture about its future. In essence, these seafloor cores are cylindrical slices of Earth's history.

Photographs courtesy of the Ocean Drilling Program, Texas A&M University

Photographs courtesy of the Ocean Drilling Program, Texas A&M University

Photographs courtesy of the Ocean Drilling Program, Texas A&M University

How do plate tectonics influence the nature of sedimentary rocks?

At convergent plate boundaries, a folded mountain belt typically has a subsiding basin on its landward side. Erosional debris from the mountains accumulates rapidly in alluvial fans containing conglomerates and associated sandstones and shales. Subsidence in this basin may be so great that the region may be inundated by a shallow sea. On the ocean side of the mountain belt, the distance to the sea is short and there is no broad continental shelf. Large deltas seldom form on subducting continental margins because drainage basins are small. Instead, sediment transported from the mountain toward the boundary accumulates in the adjacent deep-sea trench where turbidites are common. Some of the sediment is consumed with the descending plate. However, much continental sediment is scraped off, and together with deep-marine mud and basalt on the descending plate, is plastered against the continent. These mixed deposits are known as *mélange*.

Thus, we see that each of the various tectonic settings—continental rift valleys, rifted and subsiding continental margins, deep-ocean floor, and convergent plate boundaries—produce distinctive basins that fill with unique kinds of sedimentary rock.

An indirect, but also profound, control tectonics has on sedimentation is the drifting of continents into different climate zones. This affects weathering, including the rate and nature of the sediment produced. For example, in the tropics, lush vegetation may form coal swamps. If the continent moves into the low-latitude deserts, wind-blown sand is deposited on the coal. If the continent continues northward, glacial sediments might eventually be deposited upon the desert sand. In short, plate tectonics influences not only the different types of sedimentary basins where sediment accumulates, but also the basic stratigraphic patterns in a sequence of sedimentary rock.

From a scientific point of view, sedimentary rocks are important for the geologic history they preserve. A record of the past is written in the layers of sedimentary rock. From this record, we can interpret such things as the origin and destruction of old mountain systems, the erosion of continents, climate changes, and even the evolution of life.

GeoLogic   Sedimentary Rocks at Dead Horse Point

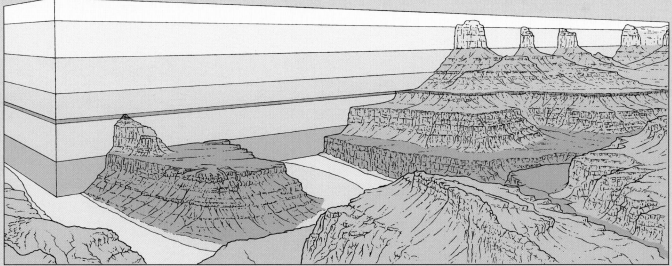

The spectacular scenery of Utah's Dead Horse Point is due to the colorful succession of horizontal sedimentary formations that were carved by stream erosion into vertical cliffs and slopes. To most people the visual impact of this scenery is enough and they are satisfied. But there is much more to this scene than you might think. The rocks and landforms are documents of history which, if understood, is more fascinating than the scenery.

Observations

1. The landscape was formed by stream erosion of these nearly horizontal sedimentary rock layers.
2. The rock units were once much more extensive than what we see here. Some formations once covered an area of more than 250,000 km².

3. Each major rock layer, or formation, was formed by a specific sedimentary process. Some formed by deposition by river systems, others by deposition on tidal flats, and others by deposition in deserts.

Interpretations

The geologic interpretation of this scene begins by trying to visualize the original extent of the rock sequences as they existed before erosion. The layers of rock also extended far beyond the area shown in the photograph, one superimposed upon another. Details of the sedimentary environments in which each rock unit was deposited come from studying the composition, texture, and structure of each major layer. Subsequent to deposition, compaction, and cementation, the rock layers were uplifted and the cliffs were formed by erosion of the hard resistant sandstone, whereas the slopes were carved on soft nonresistant shale.

Key Terms

alluvial fan (p. 136)
barrier island (p. 136)
beach (p. 136)
bed (p. 131)
bedding plane (p. 123)
biochemical rocks (p. 128)
cementation (p. 143)
chalk (p. 128)
chemical precipitate (p. 128)
chert (p. 130)
clastic (p. 125)
claystone (p. 128)
compaction (p. 143)
conglomerate (p. 126)
cross-bedding (p. 131)

deep-ocean basin (p. 136)
delta (p. 136)
dolostone (p. 130)
dune (p. 131)
eolian (p. 135)
evaporite (p. 131)
facies (p. 136)
fluvial (p. 136)
formation (p. 123)
fossil (p. 123)
glacier (p. 136)
graded bedding (p. 131)
gypsum (p. 131)
lagoon (p. 136)
lamina (p. 128)

limestone (p. 128)
mudrock (p. 126)
mud cracks (p. 133)
oolite (p. 128)
reef (p. 136)
regression (p. 143)
ripple marks (p. 133)
rock salt (p. 131)
sandstone (p. 126)
sand wave (p. 131)
sediment (p. 122)
sedimentary rock (p. 122)
sedimentary structure (p. 131)
sedimentary system (p. 135)
sequence (p. 143)

shale (p. 128)
shallow marine (p. 128)
siltstone (p. 128)
stratification (p. 131)
stratum (p. 122)
submarine fan (p. 136)
tidal flat (p. 136)
transgression (p. 143)
turbidite (p. 133)
turbidity current (p. 133)
unconformity (p. 143)
weathering (p. 135)

Review Questions

1. List the characteristics that distinguish sedimentary rocks from igneous and metamorphic rocks.
2. What is the principal mineral in sandstone? Why does this mineral dominate?
3. What is the major difference between the various kinds of clastic sedimentary rocks?
4. What is the principal cause of grain-size variations in clastic sediment?
5. How do limestones differ from clastic rocks?
6. What is the mineral composition of limestone?
7. How do evaporites form? How are they different from clastic sedimentary rocks?
8. Show, by a series of sketches, the characteristics of stratification, cross-bedding, and graded bedding.
9. How could you recognize an ancient turbidity current deposit now exposed in a mountain?
10. What rock types form in the following sedimentary systems: (a) delta, (b) lagoon, (c) alluvial fan, (d) eolian, (e) organic reef, and (f) deep marine?
11. How does plate tectonics control the character of sedimentary rocks deposited on a continental margin that is far from a modern plate boundary?
12. How does a transgression followed by a regression of the sea produce the vertical sequence sandstone, shale, limestone, shale, sandstone?

6 Metamorphic Rocks

Most of the rocks exposed in the continental shields and in the cores of mountain belts show evidence that their original igneous or sedimentary textures and compositions have changed. At the same time, many were ductilely deformed, as shown by contorted parallel bands of minerals resembling the swirled colors in marble cake. Other rocks recrystallized and developed large mineral grains, and the constituent minerals of many have strong fabrics with planar orientations called foliation. These are the hallmarks of recrystallization in the solid state, a process we call metamorphism. The result is a new rock type with a distinctive texture and fabric and new mineral compositions.

In the photograph above, metamorphic rocks are exposed along the valley carved by the Potomac River near Washington, D.C. The minerals in the metamorphic rocks did not crystallize from magma, but they formed at the high temperatures and pressures found deep in the crust. Note the strong vertical fabric of the gray rocks on the right. This planar foliation is characteristic of many metamorphic rocks. Complex contortions in the rock units show the degree to which

these rocks were deformed at high temperature. These are the roots of mountains built long ago when Africa collided with North America to make the Appalachian Mountains and close the Atlantic Ocean. The parents were deposited as horizontal beds of sediment about 600 million years ago. Later, the collision of the two tectonic plates pushed the sedimentary rocks to great depths, and there they recrystallized without melting at high temperature and under immense pressure. The rocks were folded and deformed; the bedding was destroyed; even the microscopic grain-to-grain textures changed. The change was as complete and striking as the metamorphosis of a caterpillar to a butterfly. Meanwhile, a folded mountain belt formed above the metamorphic zone. Subsequently, the mountains were slowly eroded away to eventually uncover the metamorphic rocks of the deep mountain roots. All of this history can be read by a simple realization of the metamorphic character of the rock.

As you study this chapter, you will see that events such as these formed the very foundation of each of the continents. The rocks of the shields and those in the deep parts of the stable platforms are mostly metamorphic rocks. Every aspect of metamorphic rock, from the small grain to the regional fabric of a shield, points toward the same theme: metamorphic rocks dramatically show the mobility of a dynamic crust.

Major Concepts

1. Metamorphic rocks can be formed from igneous, sedimentary, or previously metamorphosed rocks by recrystallization in the solid state. The driving forces for metamorphism are changes in temperature, pressure, and composition of pore fluids.
2. These changes produce new minerals, new textures, and new structures within the rock body. Careful study of metamorphic rocks reveals the thermal and deformation history of Earth's crust.
3. During metamorphism, new platy mineral grains grow in the direction of least stress, producing a planar texture called foliation. Rocks with only one mineral (such as limestone) or those that recrystallize in the absence of deforming stresses do not develop strong foliation but instead develop a granular texture. Mylonite develops where shearing along a fracture forms small grains by ductile destruction of larger grains.
4. The major types of foliated metamorphic rocks include slate, schist, gneiss, and mylonite; important nonfoliated (or granular) rocks include quartzite, marble, hornfels, greenstone, and granulite. They are distinguished by their textures and secondarily by their compositions.
5. Contact metamorphism is a local phenomenon associated with thermal and chemical changes near the contacts of igneous intrusions. Regional metamorphism is best developed in the roots of mountain belts along convergent plate boundaries.
6. Mineral zones are produced where temperature, pressure, or fluid compositions varied systematically across metamorphic belts or around igneous intrusions.
7. Distinctive sequences of metamorphic rocks are produced in each of the major plate tectonic settings.

The Nature of Metamorphic Rocks

> Metamorphic rocks form by recrystallization in the solid state because of changes in temperature, pressure, or the composition of pore fluids. New minerals form that are in equilibrium with the new environment, and a new rock texture develops in response to the growth of new minerals.

Many igneous and sedimentary rocks have recrystallized in the solid state—without melting—to such an extent that the diagnostic features of the original rock have been greatly modified or obliterated. Recrystallization occurs because of changes in temperature, pressure, and the chemical composition of the fluids that flow through them. We call these solid-state processes **metamorphism** (Greek, "changed form"). These solid-state reactions are akin to those that a potter uses to convert soft clay into hard ceramic. When a soft clay pot is placed in a kiln at a temperature near 1200°C, the clay minerals change into other minerals that are stable under those conditions. In other words the clay is metamorphosed. The recrystallization occurs without melting, but is sufficient to create a new material radically different than its precursor.

During metamorphism of rocks, most structural and textural features in the original rock—such as stratification, graded bedding, vesicles, and porphyritic textures—are destroyed. New minerals replaced those originally in the rock to create a new rock texture. These are **metamorphic rocks**, a major group of rocks that results largely from the constant motion of tectonic plates (**Figure 6.1**). Metamorphic rocks can be formed from igneous, sedimentary, or even previously metamorphosed rocks.

(A) Satellite image of metamorphic rocks in the Canadian Shield. Note the complex folds and fractures resulting from extensive crustal deformation while the rocks were at high temperature and pressure.

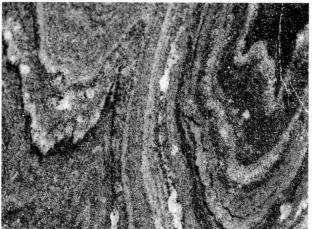

(B) Outcrop of metamorphic rocks at 5500-m level of Mount Everest in Tibet. The foliation in this rock formed by shear during the collision of India and Asia.

(C) Hand sample of a highly metamorphosed rock. Note that recrystallization in the solid state has concentrated light and dark minerals into layers which were then deformed and folded.

Figure 6.1 The characteristics of metamorphic rocks are shown on three different scales. Each shows features resulting from strong deformation and solid-state recrystallization caused by changes in temperature, pressure, or fluid composition.

Many people know something about various igneous and sedimentary rocks but only vaguely understand the nature of metamorphic rocks. All of us have seen many environments where new sedimentary rocks are forming; most have also seen a few igneous rocks form—when volcanoes erupt, for example. But the formation of metamorphic rocks takes place so deep within the crust that we are not familiar with these processes. Perhaps the best way to become acquainted with this group of rocks, and to appreciate their significance, is to study carefully Figure 6.1. The satellite image of

Figure 6.2 A stretched pebble formed during metamorphism of a conglomerate. The pebble was once nearly spherical and about the same size as the specimen shown to the side, but it was deformed at high confining pressure and temperature and stretched to six times its original length.

part of the Canadian Shield (Figure 6.1A) shows that the rocks have been distorted and compressed. Originally, these were sedimentary, and volcanic layers deposited horizontally. They have been deformed so intensely, however, that it is difficult to determine the original bottom or top of the rock sequence.

Figure 6.1B shows a more detailed view of metamorphic rocks. The alteration and deformation of the rock are evident in the alternating layers of light and dark minerals. These rocks were intensely sheared along almost horizontal planes while it was in a plastic or semiplastic state. The degree of **plastic deformation** possible during metamorphism is best seen by comparing the shapes of pebbles in a conglomerate with the shapes of pebbles in metamorphosed rock. In a metamorphosed rock, the original spherical pebbles in the conglomerate have been stretched into long, ellipsoidal blades (the long axis is as much as 30 times the original diameter, **Figure 6.2**). A definite preferred orientation of the grains shows that they recrystallized either under unequal stress (force applied to an area) or by flowing as a plastic.

The typical texture of metamorphic rocks does not show a sequence of formation of the individual minerals like that evident in igneous rocks. All grains in metamorphic rocks apparently recrystallize at roughly the same time, and they have to compete for space in an already solid rock body. As a result, the new minerals grow in the direction of lowest stress. Most metamorphic rocks thus have a layered, or planar, structure, resulting from recrystallization.

Metamorphic rocks make up a large part of the continental crust. Extensive exposures (**Figure 6.3**) are found in the vast shield areas of the continents. Deep drilling in the stable platform shows that the bulk of the continental crust is also made up of metamorphic rocks. In addition to those beneath the stable platforms of the continents and exposed in the shields, metamorphic rocks are also found in the cores of eroded mountain ranges, such as the Appalachian and Rocky Mountain chains. The widespread distribution of metamorphic rocks in the continental crust, especially among the older rocks, is evidence that Earth's crust has been deformed repeatedly. Large parts of the oceanic crust are also metamorphosed. Even the mantle is made mostly of a type of metamorphic rock.

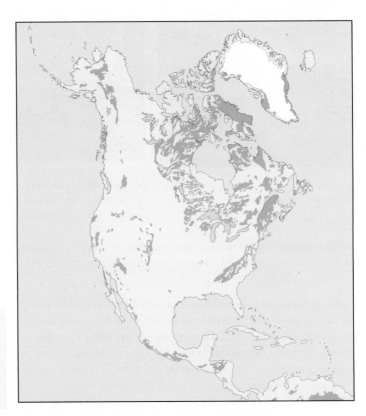

Figure 6.3 Metamorphic rocks are widely distributed in the Canadian Shield and in the cores of folded mountain belts such as the Appalachians of eastern North America. A blanket of sedimentary rocks covers the metamorphic rocks in the stable platform.

Origin of Metamorphic Rocks

> The driving forces for metamorphism are changes in temperature, pressure, and composition of the environment or strong deformation. These changes cause recrystallization in the solid state as the rock changes toward equilibrium with the new environment.

Metamorphism causes a series of changes in the texture and composition of a rock. The changes occur to restore equilibrium to rocks subjected to an environment different from the one in which they originally formed (**Figure 6.4**). Several agents of change act in combination and create distinctive metamorphic environments depending upon which factors are most important.

Temperature Changes

Heat is one of the most important factors in metamorphism. For example, as a rock's temperature increases, its minerals may become unstable and react with other minerals to form new mineral assemblages that are stable under the new conditions (Figure 6.4A). Below 200°C, reaction rates are low, and most minerals will remain unchanged for millions of years. As the temperature rises, however, chemical reactions become more vigorous. Crystal lattices are broken down and re-created using different combinations of ions and different atomic structures. As a result, new minerals appear. For example, if pressure is held constant at 2 kb and temperature increases, the mineral andalusite recrystallizes to sillimanite at about 600°C (**Figure 6.5**). When the sillimanite crystallizes, the bonding of atoms in the mineral is rearranged and new crystal forms result. If temperature continues to increase, the rock becomes partially molten at about 700°C, and layers of solid material mixed with layers of magma might form. The critical idea here is that different minerals are in equilibrium at different temperatures. The minerals in a rock, therefore, provide a key to the temperatures at which the rock was metamorphosed. This powerful interpretive tool is not without its problems, however. For example, with

A. Temperature change

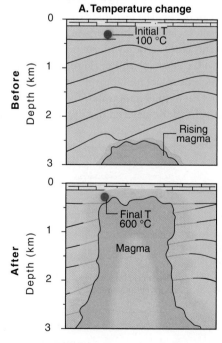

B. Pressure change

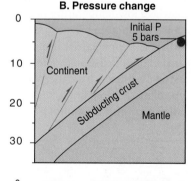

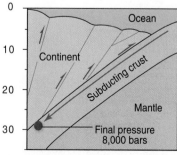

C. Composition change

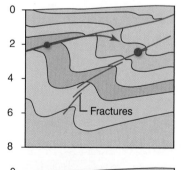

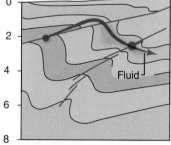

(A) Temperature changes when a magmatic body intrudes the shallow crust and causes recrystallization around the intrusion (region in light orange).

(B) Pressure changes can be caused by the collision of two plates, where minerals at low pressure (blue dot) are dragged to high pressure (red dot) in a subducting plate.

(C) Fluids carrying dissolved ions may flow from one spot (blue dot) to another (red dot), causing minerals along the flow path to recrystallize as they equilibrate with the fluid.

Figure 6.4 Metamorphic changes can occur as the result of changes in temperature, pressure, and in the composition of pore fluids, as the rocks attempt to reach equilibrium with the new conditions. These cross sections illustrate some of the changes.

Figure 6.5 **The stable polymorph of Al₂SiO₅** varies at different temperatures and pressures. Andalusite is stable at low temperatures and changes to sillimanite during metamorphism at higher temperatures. Higher pressure produces kyanite. At even higher temperatures, a metasedimentary rock partially melts to make migmatite. The arrows show possible pressure-temperature paths during metamorphism.

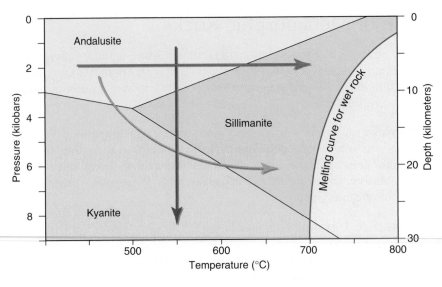

a decrease in temperature, the sillimanite becomes unstable; but, because reaction rates are lower at these lower temperatures, the sillimanite may persist for a long time without converting back to andalusite. In such cases, the mineral is said to be **metastable**.

How is heat added to cause metamorphism? The two most important ways are intrusion of hot magma and deep burial (Figure 6.4). Recall that magmas have temperatures that range from about 700°C to 1200°C depending on their compositions. The temperature of the country rocks around an intrusion increases as heat diffuses from the intrusion. Zones of different mineral assemblages in metamorphic rocks show that strong thermal gradients once existed around igneous intrusions. This kind of metamorphism is called **contact metamorphism** (**Figure 6.6A**).

Deep burial can also increase a rock's temperature. Temperature increases about 15°C to 30°C for each kilometer of depth in the crust. Even gradual burial in a sedimentary basin may take rocks formed at the surface to depths as great as several kilometers, where low-temperature metamorphism can occur. The tectonic processes that make folded mountain belts can bury rocks to even greater depths—tens of kilometers—where the temperature is much higher. In this case, metamorphism occurs over large areas, so geologists sometimes call it **regional metamorphism** (Figure 6.6B). It contrasts with the much smaller volumes involved in contact metamorphism. Because this type of metamorphism is key to the construction of folded mountain belts, it is also called **orogenic metamorphism** (Greek *oro*, "high or elevated").

What is the difference between regional and contact metamorphism?

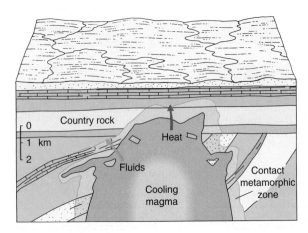

(A) Contact metamorphism occurs around hot igneous intrusions. Changes in temperature and composition of pore fluids cause preexisting minerals to change and reach equilibrium in the new environment. Narrow zones of altered rock extending from a few meters to a few hundred meters from the contact are produced.

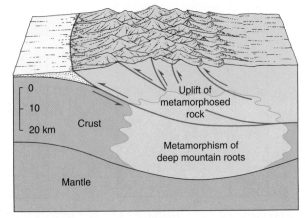

(B) Regional metamorphism develops deep in the crust, usually as the result of subduction or continental collision. Wide areas are deformed, subjected to higher pressures, and intruded by igneous rocks. Hot fluids may also cause metamorphic recrystallization.

Figure 6.6 **Metamorphic environments** are many and varied. Two major examples are shown here.

STATE OF THE ART   Rock Metamorphism in the Laboratory

In this chapter, phase diagrams are used as graphical summaries of the stability fields of minerals. Phase diagrams tell us much about the origin of metamorphic rocks, which have all recrystallized because of changes in their physical or chemical environment. But how do we know that kyanite is not stable at pressures higher than about 4 kilobars (Figure 6.5) or that garnet is stable in many rocks at temperatures of about 500°C (see Figure 6.15)? The answer is: we conduct laboratory experiments.

An important branch of geology involves the experimental determination of the stability ranges of minerals. One type of experimental apparatus is shown here. Small samples of pulverized rock are placed in a tiny metal capsule about the size of a vitamin pill. The capsule is usually made of gold or some other noble metal that remains stable at high temperatures. This small capsule is then placed inside a "bottle" with strong metal walls and a screw top. A fluid is pumped inside the bottle to increase the confining pressure. Heating filaments are used to control the temperature. Once the capsule is safely inside the "bomb," the pressure and temperature are brought up to the point of experimental interest, say 1 kilobar and 400°C, and maintained at that point for many hours. Some experiments last for weeks so that equilibrium can be achieved between the various solids and fluids in the capsule. At the end of the experiment, the capsule is rapidly cooled and the pressure is dropped back to normal conditions. If the temperature drop is rapid enough, the phases formed at high pressure and

temperature will persist as metastable minerals. The capsule is carefully opened to see what minerals were stable under the experimental conditions. The results are plotted on a pressure-temperature grid like the one shown here. Each point represents one experiment.

The major problem with such experiments is ensuring that equilibrium between the mineral phases and their environment actually occurred. To test this, several experiments are usually done with different starting minerals. Other tests involve starting the experiment from a high temperature or from a lower temperature. If equilibrium is achieved, every experiment at a given pressure and temperature will produce the same minerals.

You can see that many time-consuming experiments are needed to establish the stability field of a mineral. The experiments clearly show that many minerals indicate the specific temperature and pressure at which they formed and can be used to determine the history of changes a certain natural rock has experienced. For example, if sillimanite is present in a metamorphic rock (with the same composition as the experiment), then we can conclude that the rock recrystallized at a temperature above about 600°C. On the other hand, if andalusite is present and sillimanite is absent, the rock must have recrystallized at a lower temperature and a pressure between 0 and 4 kb. Such interpretations give us a better understanding of how mountain belts form and then erode away, uplifting the metamorphic rocks to the surface.

Courtesy of Malcolm Rutherford

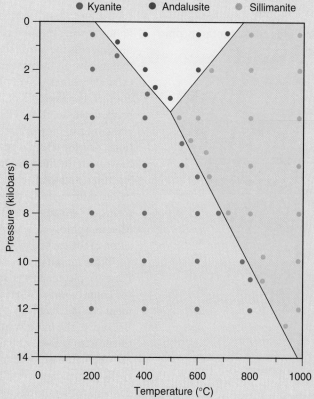

Pressure Changes

High pressure, deep within Earth, also causes significant changes in the properties of rocks that originally formed at the surface (Figure 6.4B). An increase in pressure can drive chemical reactions to produce new minerals with closer atomic packing and higher densities. The vertical blue arrow in Figure 6.5 shows a pressure increase at a constant temperature of 550°C. If a rock containing andalusite followed this pressure-temperature path, it would recrystallize to form sillimanite at 3 kb; kyanite would crystallize at about 5 kb (almost 20 km deep).

Pressure increases when rocks are buried deep beneath Earth's surface. Burial may be caused by prolonged sedimentation in a basin. Metamorphic rocks are also caused by increasing pressure during the stacking of thrust sheets at convergent plate boundaries or where oceanic crust is thrust deep into the mantle. The **confining pressure** is equal to the weight of the overlying rocks and causes these kinds of mineral changes.

If a rock experienced progressively lower pressure during uplift, theoretically it would undergo metamorphic changes to bring it to equilibrium at the lower pressure (Figure 6.5). However, these changes may be so slow that the high-pressure minerals remain metastable at the new lower pressure. An extreme example is that of diamond, which is stable only at pressures that exceed 30 kb, reached at depths of more than 100 km. Soft graphite is the stable form of carbon at 1 bar (atmospheric pressure), but the change from diamond to graphite is infinitesimally slow.

Temperature and confining pressure increase together in most environments where metamorphic rocks form. Such a path is shown with the sloping orange arrow in Figure 6.5. Along this pressure-temperature path, andalusite recrystallizes to form kyanite at about 450°C and 3.5 kb. Further increases in temperature and pressure make kyanite recrystallize to form sillimanite at about 600°C and 6 kb. If the rock continues to follow the sloping path of the curve in Figure 6.5, partial melting could occur to form small bodies of magma.

Obviously, metamorphism occurs under many different conditions that vary in intensity. We describe the relative intensity of metamorphic change as the **metamorphic grade**. Metamorphism that takes place at low pressure and temperature (~250°C to 400°C) low-grade metamorphism; high pressure and high temperature (above ~600°C) produce high-grade metamorphism.

Movement of Fluid

How can fluids cause metamorphic reactions?

Fluids flowing through rocks are also important agents of metamorphism. Especially important is the movement of water and carbon dioxide. In metamorphic processes that involve an increase in temperature, many minerals that contain H_2O or CO_2 eventually break down, providing a separate fluid that migrates from one place to another. For example, at high temperatures, calcite ($CaCO_3$) and clay [$Al_2Si_2O_5(OH)_4$] break down to release CO_2 and H_2O fluids and other ions (Figure 6.4C). Original crystals break down, and new crystal structures, which are stable under the new conditions, develop. If an ion becomes detached from a mineral's crystal structure, it may move with the fluid to some other place. The fluids move through tiny pore spaces, fractures, and along the margins of grains. The small amount of pore fluid transports material through the rock and causes minerals that find themselves in new environments to rearrange into new mineral structures to reach equilibrium with the fluid. Metamorphic recrystallization, accompanied by some change in the chemical composition of the rock, that is, by a loss or gain of certain elements is called **metasomatism**.

Other metamorphic reactions occur by the addition of volatile fluid components. For example, magmatic intrusions may release hot fluids that flow into the surrounding country rock. Consequently, minerals that are stable in the new chemical environment crystallize. Many types of metallic ore deposits are created by metasomatism. Because of the importance of hot water in the formation of such metasomatic rocks, the process is also known as **hydrothermal alteration**. Veins of white milky quartz are a common expression of the mobility of water and dissolved ions in metamorphic rocks. The quartz crystallized from a fluid flowing through a fracture. Gold or other valuable minerals may also crystallize with the quartz.

The circulation of hot seawater through cold oceanic crust probably produces more metasomatic rocks than all other processes combined. **Ocean ridge metamorphism** converts olivine and pyroxene into hydrated silicate minerals, including serpentine, chlorite, and talc (see Figure 6.18). This is the most characteristic kind of metamorphism in the oceanic crust. As much as one-fourth of the oceanic crust is metamorphosed in this way. This example shows that several different factors, in this case an increase in temperature and a change in fluids, may be involved in a single metamorphic environment (**Figure 6.7**).

Deformation

You have seen that changes in temperature, confining pressure, and fluid proportions can cause new minerals to crystallize while a rock is still in the solid state. In addition, deformation of rock can also cause metamorphism. The result is preserved in the grain-to-grain relationships—the texture. In many tectonic settings, there is directed or **differential stress** that acts to shorten and compress the rock, or, alternatively, to lengthen and extend the rock. In other words, the forces on the rock are not equal in all directions. Differential stress is usually the result of horizontal compression at zones of plate convergence or collision. At high temperature or confining pressure, a rock becomes **ductile** and may be deformed slowly if such a differential stress is applied. Mineral grains may move, rotate, or flatten, but more commonly new grains actually grow in new orientations. At low pressure or rapid rates of deformation, mineral grains may be strongly sheared. Deformation reorients mineral grains and forms a new rock texture.

Differential Stress Perhaps the most obvious sign of differential pressure is the distinct orientation of grains of platy minerals such as mica and chlorite. An important result of metamorphic deformation is the alignment and elongation of minerals in the direction of least stress (**Figure 6.8**). Because many metamorphic rocks form during deformation where stresses are not uniformly oriented, they develop textures in which the mineral grains have strongly preferred orientations (**Figure 6.9**). This orientation may impart a distinctly planar element to the rock, known as **foliation** (Latin *folium,* "leaf," hence "splitting into leaflike layers"). The planar structure can result from the alignment of platy minerals, such as mica and chlorite, or from alternating layers having different minerals (**gneissic foliation**).

Everything else being equal, the grain sizes in foliated rocks increase with the intensity of metamorphism; that is, they depend on the temperature and confining pressure. Grains range from microscopic to very coarse.

Foliation is a good record of rock deformation. It usually forms during recrystallization associated with regional horizontal compression. In most foliated metamorphic rocks, the mineral alignment is nearly perpendicular to the direction of compressional stress. The orientation of foliation, therefore, is closely related to the large folds and structural patterns of rocks. This relationship commonly extends from the largest folds down to microscopic structures. For instance, the foliation in slate is generally oriented parallel to the hinge planes of the folds, which can be many kilometers apart. A slice of the rock viewed under a microscope shows small wrinkles and folds having the same orientation as the larger structures mapped in the field.

Foliation is actually caused by several different mechanisms. For example, during solid state recrystallization platy minerals grow to become elongate perpendicular to the directed stress—growth is enhanced where the pressure is lowest. Some grains are also rotated during deformation to become aligned, like logs floating in a stream. Some ductile grains are also flattened by compression.

Shear stress is a distinctive type of differential stress which causes one part of a material to move laterally past another part: you can shear a deck of cards on a table by moving your hand parallel to the table. Intense shearing forms a group of relatively rare metamorphic rocks with textures formed by the destruction of grains rather than their growth. This type of rock may form in a tectonic shear zone where two walls of a fracture grind past one another at very high confining pressure. The progressive

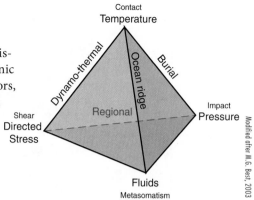

Figure 6.7 Metamorphism is caused by changes in temperature, pressure, fluid composition, or strong deformation. Different metamorphic environments involve one or more of these factors. Orogenic metamorphism lies within the tetrahedron because all four factors are important.

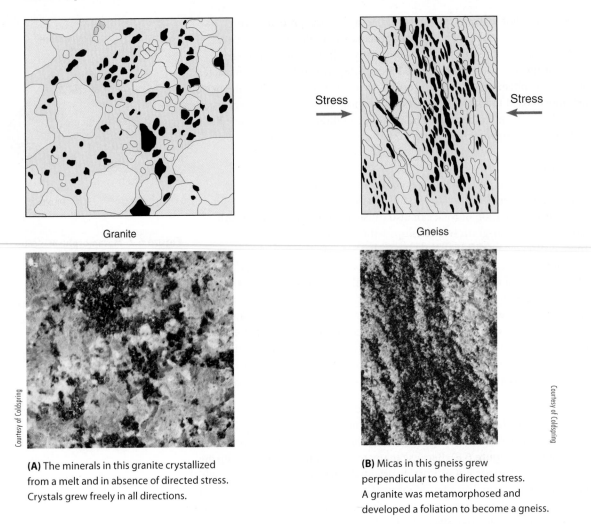

(A) The minerals in this granite crystallized from a melt and in absence of directed stress. Crystals grew freely in all directions.

(B) Micas in this gneiss grew perpendicular to the directed stress. A granite was metamorphosed and developed a foliation to become a gneiss.

Figure 6.8 **Foliation develops in metamorphic rocks** when platy minerals grow. Minerals such as mica grow perpendicular to the applied stress. For example, during compression, the foliation will be perpendicular to the directed stress.

destruction of grain shapes and reduction of grain sizes is characteristic of this type of deformation. The shearing dismembers and destroys preexisting mineral grains to make a very fine-grained rock called **mylonite** (Greek *mylon*, "to mill"). A microscope may be required to see the intensely strained individual grains (Figure 6.9B). Most mylonites form by pervasive ductile flow of solid rock. During deformation, zones of slippage develop within individual grains to allow them to flow. At high temperature, the rock deforms much like soft taffy. At lower temperatures, at which the deformation is dominated by brittle breakage, mylonitic rocks grade into tectonic breccias that have fragments with angular margins.

Uniform Stress Not all metamorphic rocks are foliated. Some metamorphic rocks form where the stress is fairly uniform in all directions and so no planar texture develops. The resulting texture is best described as granular, or, simply, **nonfoliated**. If the rocks have micas or other platy minerals, they are randomly oriented. For example, during contact metamorphism, there are no strong differential stresses and the metamorphic rocks are not strongly deformed.

The texture of nonfoliated metamorphic rocks reveals some of the results of crystallization in the solid state. Typical grains are polygonal, reflecting the mutual growth and competition for space. Grain boundaries are relatively straight, and triple junctions are common. The growth of quartz during the metamorphism of sandstone shows this kind of texture (Figure 6.9C). A familiar example of this process is the

 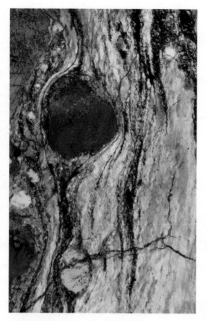 

(A) Strongly foliated schist with aligned grains of chlorite that grew in a differential stress field during contraction.

(B) Mylonites have grains that reflect destruction by shearing. The fine grains formed by crushing and shearing of larger grains, such as the large quartz grain.

(C) Nonfoliated texture results from growth without deformation. Solid-state growth produces polygonal grains with abundant triple junctions.

Figure 6.9 **Metamorphic textures** range widely, but all indicate crystallization in the solid state, as illustrated by these thin sections. Each view is 3 mm across.

growth of bread rolls as they bake in an oven. The outlines of the rolls become polygonal as they expand against one another, and they have straight boundaries; triple junctions occur where three rolls meet.

It should be apparent that real metamorphic environments on Earth are very dynamic and typically involve changes not in just one of these factors but contemporaneous changes in temperature, pressure, and fluid composition all while the rock sequence is being deformed (Figure 6.7).

Types of Metamorphic Rocks

The two major groups of metamorphic rocks—foliated and nonfoliated—are further subdivided based on the basis of mineral composition. The major types of foliated rocks are slate, schist, gneiss, and mylonite. Important nonfoliated rocks are quartzite, marble, hornfels, greenstone, and granulite.

Because of the great variety of original rock types and the variation in the kinds and degrees of metamorphism, many types of metamorphic rocks have been recognized. A simple classification of metamorphic rocks, largely based on texture, is usually sufficient for beginning students. The major rock names can then be qualified by prefixes listing the important minerals.

Foliated Rocks

Slate is a very fine-grained metamorphic rock, generally produced by the low-grade metamorphism of shale. It is characterized by excellent foliation, known as **slaty cleavage**, in which the planar element of the rock is a series of surfaces along which the rock can be easily split (**Figure 6.10A**). Slaty cleavage is produced by the parallel alignment of minute flakes of platy minerals, such as mica, chlorite, and talc. Zeolites also form in these low-grade rocks. The mineral grains are too small to be obvious

How does foliation differ from stratification?

(A) Slate is a fine-grained foliated rock. The foliation usually cuts across sedimentary bedding.

(B) Schist is a strongly foliated metamorphic rock with abundant platy minerals, usually muscovite or chlorite.

(C) Gneiss has a foliation defined by alternating layers of light (mostly feldspar and quartz) and dark (mafic silicates) layers. The layers do not conform to preexisting sedimentary beds.

(D) Quartzite is a nonfoliated metamorphic rock derived from quartz-rich sandstone. This quartzite also has a few relict pebbles.

(E) Metaconglomerate often displays highly elongated clasts.

(F) Marble is limestone that recrystallized during metamorphism. It consists of mostly calcite.

Figure 6.10 The major metamorphic rocks include foliated (A–C) and nonfoliated (D–F) varieties shown in their actual sizes.

without a microscope, but the parallel arrangement of small grains develops innumerable parallel planes of weakness, so the rock can be split into smooth slabs. Because of this property, slate has been used for blackboards and as roof and floor tiles.

Slaty cleavage should not be confused with the bedding planes of the parent rock. It is completely independent of the original (relict) bedding and commonly cuts across the original planes of sedimentary stratification. Relict bedding can be rather obscure in slates, but it is often expressed by textural changes resulting from interbedded, thin layers of sand or silt. Excellent foliation can develop in the shale part of the sedimentary sequence, in which clay minerals are abundant and are easily altered to mica. In thick layers of quartz sandstone, however, the slaty cleavage plane is generally poorly developed. Metamorphism of some volcanic sequences also produces slate.

A **phyllite** is a metamorphic rock with essentially the same composition as a slate, but the micaceous minerals are larger and impart a definite luster to the rock's plane of foliation. The large mineral grains result from enhanced growth at higher temperature and pressure than for slate. Like slates, most phyllites form from rocks that were originally shales.

Schist is a strongly foliated rock ranging in texture from medium-grained to coarse-grained. Foliation results from the parallel arrangement of relatively large grains of platy minerals, such as mica, chlorite, talc, and hematite, and is called **schistosity**. The mineral grains are large enough to be identified with the unaided eye and produce an obvious planar structure because of their overlapping subparallel arrangement (Figure 6.10B). The foliation of schist differs from that of slate mainly in the size of the crystals. The term schistosity comes from the Greek *schistos*, meaning "divided" or "divisible." As the name implies, rocks with this type of foliation break readily along the cleavage planes of the parallel platy minerals.

The mineral composition provides a basis for subdividing schists into many varieties, such as chlorite schist, mica schist, and amphibole schist. In addition to the platy minerals, significant quantities of quartz, feldspar, garnet, amphibole, sillimanite, graphite, and other minerals occur in schist. The mineral proportions are largely controlled by the original composition of the rock. Parent rock types include basalt, granite, shale, and tuff.

Schists result from a higher grade of regional metamorphism than the type that produces slates. Schists are one of the most abundant metamorphic rock types.

Gneiss is a coarse-grained, granular metamorphic rock in which foliation results from alternating layers of light and dark minerals, or gneissic layering (Figure 6.10C). The composition of most gneisses is similar to that of granite. The major minerals are quartz, feldspar, and mafic minerals such as biotite and amphibole. Feldspar commonly is abundant and, together with quartz, forms light-colored (white or pink) layers of polygonal grains. Mica, amphibole, and other mafic minerals form dark layers. Gneissic layering can be highly contorted because of deformation during recrystallization. When struck with a rock hammer, gneiss generally fractures across the layers, or planes of foliation, but where micas are abundant, it can break along the foliation.

Gneiss forms during high-grade metamorphism, and in some areas it grades into partially molten rock if the temperature of initial melting is reached (**Figure 6.11**). Such a rock that is partly igneous and partly metamorphic is known as **migmatite** (Greek *migma*, "mixed"). Migmatites are commonly deformed and crisscrossed by thin dikes or sills of magmatic rock. The migmatites may even grade into completely igneous rocks, such as granite.

The mineral composition of gneisses is varied because the possible parent rocks are so different from one another. Gneiss can form as the highest grade of metamorphism of shales, but more commonly the parents were plutonic and volcanic igneous rocks such as granite and basalt. For example, biotite gneiss is commonly derived from granite. Metasedimentary gneisses typically have garnet and other aluminum-rich silicates. Metamorphism of basalt or gabbro produces **amphibolite** gneisses, coarse-grained mafic rocks composed chiefly of amphibole and plagioclase. Because of the abundance of basalt, amphibolite is a fairly common metamorphic rock. Most amphibolites have a distinctive lineation caused by the alignment of elongate grains of amphibole. Some

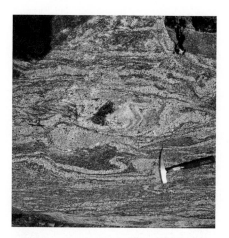

Figure 6.11 Migmatite is a mixed metamorphic and igneous rock. The light-colored pods and layers crystallized from granitic magma, and the darker zones consist of metamorphic rock rich in mafic minerals. Migmatite may form if the temperature and pressure are high enough to cause partial melting.

amphibolites develop a true foliation if mica or other platy minerals are abundant, but many are more or less massive with little foliation.

Mylonite is the hard, fine-grained metamorphic rock with a streaked or weakly foliated texture formed by intense shearing. Less-deformed, larger grains may survive as relicts embedded in a sheared groundmass. Very fine-grained mylonite forms sheet-like bodies that appear to be as structureless as chert, but the streaked and lineated appearance hints at its true origin. Mylonites form in shear zones in folded mountain belts and along transform fault plate boundaries.

Nonfoliated Rocks

Nonfoliated metamorphic rocks can form in two different ways. Some form by recrystallization in a uniform stress field. Others, probably most, lack a foliation because they are made of minerals that are equant in shape and not platy like micas and chlorite. For example, **quartzite** is a metamorphosed, quartz-rich sandstone (Figure 6.10D). It is not foliated because quartz grains, the principal constituents, do not form platy crystals. The individual grains commonly form a tight mass, so the rock breaks across the grains as easily as it breaks around them. Nonetheless, some sedimentary structures survive metamorphism, including cross-bedding and grain size variations. Pure quartzite is white or light-colored, but iron oxide and other minerals often impart various tones of red, brown, green, and other colors.

Metaconglomerate is not an abundant metamorphic rock. It is important in some areas, however, and illustrates the degree to which a rock can be deformed in the solid state. Under differential stress, individual pebbles are stretched into a mass that shows distinctive linear fabric (Figure 6.10E).

Marble is metamorphosed limestone or dolostone. Calcite, the major constituent of the parent rocks, is equidimensional, so marble is usually not foliated (Figure 6.10F). The grains are commonly large and compactly interlocked, forming a dense rock. The purest marbles are snow white, but many marbles contain a small percentage of minerals other than calcite that were present in the original sedimentary rock. These impurities result in streaks or bands and, when abundant, may impart a variety of colors to the marble. Thus, marbles may exhibit a range of colors including white, green, red, brown, and black. Because of its coloration and softness, marble is a popular building and monument stone. Most marble is found in areas of regional metamorphism where metamorphosed sedimentary rocks include schists and phyllites. Impure marbles contain a wide variety of other minerals.

Hornfels is a fine-grained, nonfoliated metamorphic rock that is very hard and dense. A lack of differential stress is the main reason these rocks are not foliated. Platy minerals, such as mica, can be present but they have random orientations. Commonly, grains of high-temperature minerals are present. Hornfelses are usually fine-grained and dark-colored and may resemble basalt, dark chert (flint), or even dark, fine-grained limestone. They result from thermal metamorphism of the wall rocks around igneous intrusions. The parent rock is usually shale.

Low-grade metamorphism converts the minerals in mafic igneous rocks (plagioclase, pyroxene, and olivine) to new minerals such as chlorite, epidote, and serpentine that are stable at low temperatures (about 200°C to 450°C) and in the presence of water. Because these abundant minerals are characteristically green, metamorphosed mafic rocks such as basalt have come to be called **greenstones**. These fine-grained rocks commonly lack pronounced foliation because of the low grade of metamorphism. Moreover, most greenstones form where differential stresses are absent. For example, much of the oceanic crust is metamorphosed by the interaction of hot water circulating passively through basaltic lava flows at an ocean ridge. Ancient greenstone belts in the continental shields record the low-grade metamorphism of basaltic lavas or incorporation of slivers of oceanic crust into a deformed mountain belt.

On the opposite end of the metamorphic spectrum, high-grade metamorphism produces a distinctly granular rock called **granulite**. Minerals that lack water, such as pyroxene and garnet, are characteristic of granulites; other common minerals include

Why are some strongly metamorphosed rocks not foliated?

How are the different types of metamorphic rocks distinguished and classified?

feldspars and quartz. Their parent rocks range from sedimentary to many kinds of igneous rocks. The most important implication of granulites is the extremely intense metamorphism that is required in their formation. Such high temperatures and confining pressures are achieved only in the lowermost parts of the continental crust. They cause micas to break down; the replacement of platy micas with equigranular pyroxene, garnet, and feldspar creates the unique texture of granulite. Granulite may form at temperatures as high as 700°C to 800°C.

Parent Material for Metamorphic Rocks

As you read through these descriptions you probably saw that pressure and temperature are not the only things that control what minerals occur in a metamorphic rock. The original composition of the rock is also important. An extreme example is apparent by comparing marble and quartzite. A limestone when metamorphosed does not convert into quartzite. And quartz-rich sandstones do not become calcite dominated marbles. Likewise, metamorphosed basaltic lavas are not rich in micas because they lack enough potassium to stabilize biotite or muscovite. The role of the parent rock is emphasized in **Figure 6.12**, which compares the minerals formed in different grades of metamorphism of basalt and shale—two very important parent rocks.

Despite these generalizations, interpreting the parents of some metamorphic rocks is complicated and presents some challenging problems. For example, a single type of parent rock can be changed into a variety of metamorphic rocks, depending on the grade of metamorphism and the type of deformation. For example, shale can be changed to slate, schist, gneiss (**Figure 6.13**), or even migmatite, if it gets hot enough. Contact metamorphism may also convert shale into hornfels. Alternatively, shale may

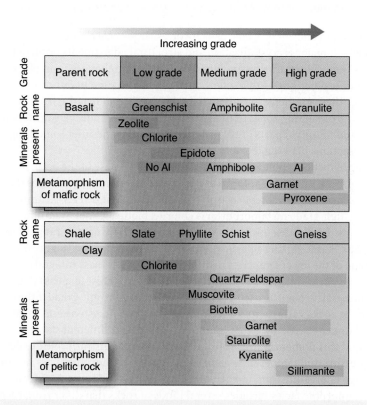

Figure 6.12 **The composition** of the parent rock is an important control on the minerals that form as metamorphic grade changes. The minerals that form as basalt is metamorphosed are different than those that form from shale. Minerals formed from basalt are poor in potassium and aluminum compared to those formed from metamorphism of shale. Metamorphic index minerals show the grade of metamorphism and are related to temperature and pressure. The sequence of index minerals for metamorphosed shale is commonly chlorite, biotite, garnet, staurolite, kyanite, and sillimanite with increasing grade.

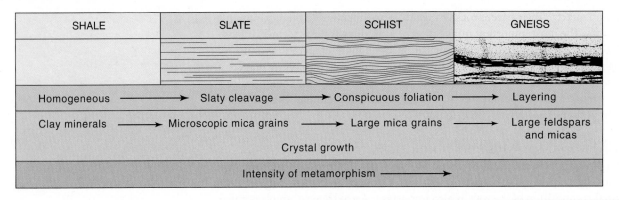

SHALE	SLATE	SCHIST	GNEISS

Homogeneous ⟶ Slaty cleavage ⟶ Conspicuous foliation ⟶ Layering

Clay minerals ⟶ Microscopic mica grains ⟶ Large mica grains ⟶ Large feldspars and micas

Crystal growth

Intensity of metamorphism ⟶

Figure 6.13 The metamorphism of shale involves a series of textural changes that depend on the intensity of temperature and pressure (the grade of metamorphism). Shale can change to slate, schist, or even gneiss.

be deformed to make mylonite if it is strongly sheared. Gneiss can form from many different kinds of rocks, such as shale, granite, or rhyolite. The chart in **Figure 6.14**, which relates parent rocks and metamorphic conditions to metamorphic rock types, gives a generalized picture of the origin of common metamorphic rocks.

Regional Metamorphic Zones

Regional or orogenic metamorphism involves large-scale recrystallization. The metamorphosed rocks commonly show mineralogic zones that reflect the differences in metamorphic grade (temperature and pressure) across the region.

What features of a rock indicate zones of different degrees of metamorphism?

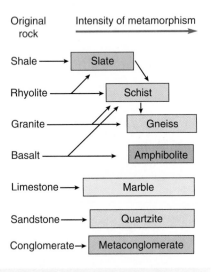

Figure 6.14 The source rocks for common metamorphic rocks are varied. In some cases, such as quartzite, marble, and metaconglomerate, the nature of the original rocks is easily determined. In other cases, such as schist and gneiss, it is difficult and sometimes impossible to determine the type of source rock. This simplified flowchart shows the origin of some of the common metamorphic rocks.

Metamorphism in ancient orogenic belts involves large-scale changes in thick masses of rock in which major recrystallization and structural adjustments occurred. Regional metamorphic rocks commonly show systematic changes from place to place—metamorphic zones—that reflect large gradients in temperature and confining pressure. These gradients are correlated with depth and distance from ancient heat sources. By mapping zones of differing metamorphic grade, geologists can locate the central and marginal parts of ancient mountain belts and infer something about ancient interactions between tectonic plates.

One type of metamorphic zonation can be defined because of the occurrence of an **index mineral**—a mineral that forms at a specific metamorphic grade. For the metamorphism of shale, a typical sequence of index minerals that reveals the transition from low-grade to high-grade metamorphism is chlorite, biotite, garnet, staurolite, kyanite, and sillimanite. Each index mineral is stable over a relatively narrow range of temperature, thus characterizing a particular grade of metamorphism (Figure 6.12). Along the way some of the lower temperature minerals disappear. These mineral changes are accompanied by textural changes from phyllite to slate to schist to gneiss (Figure 6.13).

Index minerals are not as useful for indicating metamorphic grade if the composition of the rocks varies across a region. For example, limestone and shale metamorphosed under exactly the same conditions would have different stable minerals. To get around this nomenclature problem, we use **metamorphic facies** to describe different pressure and temperature environments. A metamorphic facies represents a specific range of pressures and temperatures, as outlined in **Figure 6.15**. Consequently, a given metamorphic facies includes several kinds of rocks that differ from each other in chemical composition and mineral content, but all of the rocks in a given facies formed at roughly the same temperature and pressure. In this way, the metamorphosed limestone and the shale could be assigned to the same metamorphic facies if minerals in the rock showed that they were produced under similar conditions of temperature and pressure. It may be confusing, because the names of the facies are taken from metamorphosed mafic rocks, but we can apply the *facies name* to a metamorphic rock of any kind. Thus, it is not contradictory to say that a marble (from limestone),

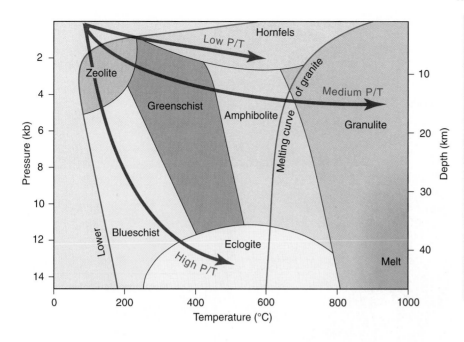

Figure 6.15 Metamorphic facies are defined by a set of minerals stable at a certain temperature and pressure (depth) and independent of rock composition. The arrows show three possible paths of metamorphism. If temperature increased moderately with pressure, the sequence of facies would be zeolite, greenschist, amphibolite, and granulite (the middle arrow typical of orogenic belts). If the increase in temperature with depth was slight, changes in metamorphic facies would follow the path indicated by the lower arrow, with the formation of blueschist and then eclogite (typical of subducted oceanic crust). Contact metamorphism is limited to zones of low pressure around shallow igneous intrusions (the upper arrow).

a garnet-sillimanite schist (from shale), and an amphibolite (from basalt) formed in the amphibolite facies. An interbedded sequence of limestone, shale, and basaltic lava flows metamorphosed at 550°C and 4 kb (about 15 km deep) would give exactly that set of rocks. **Figure 6.16** shows how the different metamorphic facies are exposed in New England—a deeply eroded mountain belt that formed during the Paleozoic Era. Granulite facies rocks are surrounded by amphibolite and then low-grade greenschist facies rocks.

Figure 6.15 shows the major metamorphic facies in relation to variations in confining pressure and temperature. The boundaries between the facies are gradational because of the complex nature of mineral reactions. The implications of each facies can be understood by tracing the metamorphic paths shown by the arrows. For example, contact metamorphism around shallow intrusions follows the upper, low-pressure path to very high temperatures (low P/T).

Most metamorphic rocks formed in folded mountain belts, however, recrystallized along the middle path (Medium P/T). The **zeolite facies** represents metamorphism at low temperature and pressure and is transitional from the changes in sediment resulting from compaction and cementation. The low temperature and pressure produce zeolite minerals. With a further increase in temperature and pressure, these minerals are soon altered as water is driven out of the mineral structure, an important trend during increasing metamorphism. The minerals characteristic of the **greenschist facies** then form at moderate pressure and still fairly low temperature. In metamorphosed basalt, this low-grade facies is typified by the minerals chlorite, talc, serpentine, muscovite, sodic plagioclase, and quartz (**Figure 6.17A**). The rocks are characteristically green because they have abundant green minerals—chlorite, talc, and serpentine. If temperature increases further along the middle curve in Figure 6.15, the minerals of the **amphibolite facies** form: in mafic rocks hornblende (a type of amphibole) forms (Figure 6.17B). With further increase in temperature (above 650°C), the minerals of the **granulite facies** form (Figure 6.17C). Pyroxene is an important mineral in this facies, along with sillimanite and garnet—depending on the original composition of the rock. The granulite facies represents the highest grade of metamorphism wherein most hydrous minerals like micas and amphiboles are not stable and water is released and flows away as a fluid. Under these conditions, melting may occur and magma may be produced.

The pressure–temperature path traced by the high P/T arrow produces a different sequence of metamorphic facies. In this case, temperature rises slowly with depth (pressure) and rocks of the **blueschist facies** form, so called because of the

Figure 6.16 Regional metamorphic gradients are displayed across large areas, as shown in this map of New England. Distinctive groups of minerals (facies) with different stability ranges show the pressures and temperatures of peak metamorphism. Compare the zonation with the phase diagram shown in Figure 6.15. This region once formed the roots of an ancient mountain belt before uplift and erosion exposed it to the surface.

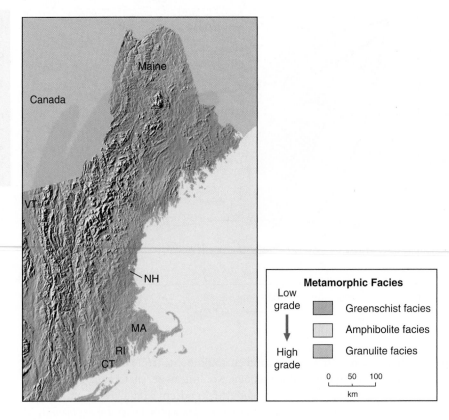

characteristic blue amphiboles that form in basaltic rocks metamorphosed under these conditions (Figure 6.17D). Distinctive blue-green pyroxenes also form. With further increase in temperature and pressure, the blueschist facies grades into the **eclogite facies**, consisting of dense feldspar-free rocks with pyroxene and garnet and granular textures (Figure 6.17E). This high pressure-low temperature path is followed by cold oceanic crust as it is subducted deep within the mantle.

Metamorphic Rocks and Plate Tectonics

Most metamorphic rocks develop because of plate collision deep in the roots of folded mountain belts. Subduction zone metamorphism occurs at high pressure but relatively low temperature. Ocean ridges, transform faults, and continental rift zones also develop distinctive types of metamorphic rocks.

We can never observe metamorphic processes in action because they occur deep within the crust. In the laboratory, however, we can study how minerals react to changes in temperature and pressure that simulate the conditions under which metamorphism occurs. These laboratory studies, together with field observations and studies of texture and composition, provide the rationale for interpreting metamorphic rocks in the framework of plate tectonics. **Figure 6.18** summarizes some of the major ideas concerning the relationships of metamorphic rocks to plate tectonics. According to the theory of plate tectonics, high confining pressures can be produced by tectonic burial at convergent plate boundaries. Temperatures are high near zones of magma intrusion or at great depth. Deformation and shearing occur where plates collide or where they slide past each other along fault zones and in deep subduction zones.

Regional metamorphism is best developed in the deep roots of folded mountain belts, which form at convergent plate boundaries. Recrystallization tends to produce nearly vertical foliations in a long belt parallel to the margins of the converging plates and perpendicular to the applied stress. Different kinds of metamorphic rocks are generated from different parent materials: sand, shale, and limestone along continental margins are converted into quartzite, schist or gneiss, and marble; volcanic sediments

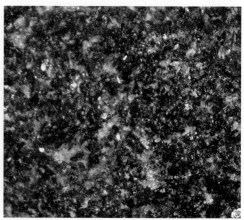

(A) Greenschist facies rocks are characteristic of low-grade metamorphism. The green color indicates an abundance of green minerals—chlorite, talc, serpentine, and epidote. Greenschist facies conditions are typical of ocean ridge metamorphism.

(B) Amphibolites are common in medium P/T environments and are dominated by black masses of amphibole, sometimes accompanied by garnet as shown here.

(C) Granulites are high-grade metamorphic rocks in which most of the hydrous minerals like amphibole and mica are not stable and pyroxene is stable. Plagioclase is the common feldspar. Granulite is not commonly foliated and has a massive granular texture.

(D) Blueschist facies rocks are characteristic of metamorphism in subduction zones. The distinctive blue mineral is a type of amphibole that is stable at high pressure but relatively low temperature.

(E) Eclogites are some of the most visually striking metamorphic rocks with their red garnets and green pyroxenes, which are only stable at high temperatures and relatively low temperatures. The lack of feldspar and presence of garnet makes eclogite fairly dense (> 3.2 g/cm³).

Courtesy of Woudloper

Figure 6.17 **Metamorphosed mafic rocks** have given their names to the different metamorphic facies.

and lava flows in island arcs change into greenstones, gneisses and amphibolites; and mixtures of deep-marine sediments and oceanic basalt from the oceanic crust in the subduction zone are converted into schists, amphibolites, and gneisses.

After the stresses from the converging plates are spent, erosion of the mountain belt occurs, and the mountain roots rise because of isostasy. Ultimately, the deep roots and their complex metamorphic rocks are exposed at the surface, forming a new segment of continental crust. Although the return of the root to the surface involves changes in confining pressure and temperature, metamorphic reaction rates are low because the changes are toward lower temperatures. Therefore, many high-grade metamorphic rocks reach the surface as metastable relicts, little changed from the peak in metamorphic temperatures and pressures. The entire process takes several hundred million years. Repetition of this process causes the continents to grow larger with each mountain-building event. The belts of metamorphic rocks in the shields are thus considered to be the record of ancient continental collisions (see Figure 6.3).

Is there only one kind of metamorphism at convergent plate margins?

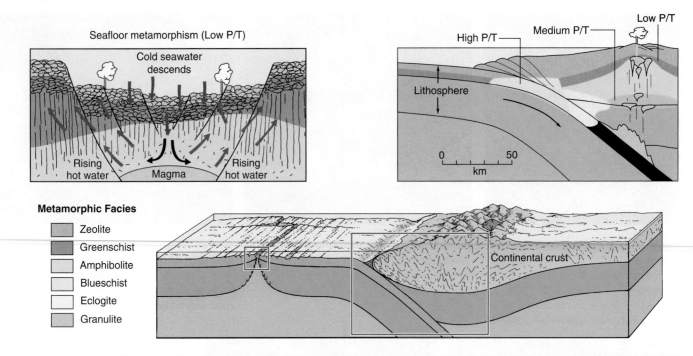

Figure 6.18 The origin of metamorphic rocks is strongly linked to plate tectonics. Oceanic crust is dragged deep into the mantle along a subduction zone to form blueschist. In the deep mountain roots, high temperatures and high pressures occur and develop schists and gneisses. Contact metamorphism develops around the margins of igneous intrusions. Ocean ridge metamorphism is caused by the circulation of seawater through hot basaltic rocks of the ocean floor.

Close to a subduction zone, sediments that have accumulated on the seafloor, together with fragments of basaltic oceanic crust, may be scraped off the descending plate. Locally, these rocks are crushed in a chaotic mass of deep-sea sediment, oceanic basalt, and other rock types. This jumbled association of rocks is called **mélange** (French for "mixture"). Slices of this material are apparently dragged to great depth by the relatively cold subducting slab, where they recrystallize along the high pressure/temperature path in Figure 6.15. The basalt in deeply subducted oceanic crust may convert to garnet-bearing eclogite. In fact, this dense eclogite may help drive plate subduction. In some cases, the metamorphic rocks then return rapidly to the surface as a mixed, broken up mass that includes blueschist and eclogite facies metamorphic rocks in a mélange. Farther inland from the subduction zone, in the mountain root, moderate-pressure and high-temperature metamorphism occurs, forming rocks of the greenschist, amphibolite, and granulite facies (Figure 6.18).

Mylonites can be produced by shearing along fracture zones developed at convergent plate margins. Shear zones are common in the ancient shields of the continents as well as along the transform faults that cut spreading ocean ridges.

What type of metamorphism dominates at divergent plate boundaries on the seafloor?

Another metamorphic environment that has a distinctive plate tectonic setting is found at and near mid-oceanic ridges (Figure 6.18). Here, ocean ridge metamorphism produces low grade metamorphic rocks at low pressure, mostly of the zeolite and greenschist facies. Hot fluids form when cold seawater flows through the hot igneous rocks near the ridge crest. The basaltic lavas and other rocks of the crust reequilibrate to form new minerals stable in the hot fluid, and much of the oceanic crust becomes metamorphosed.

Much smaller volumes of metamorphic rock are probably formed in the lower part of the crust at continental rift zones and above mantle plumes (Figure 6.18). High temperatures may be produced by the intrusion of mantle-derived magmas into the crust and by the rise of hot mantle below the rift zone. In this way, a small fraction of the lowermost continental crust may become metamorphosed in divergent rather than convergent environments.

GeoLogic   The Black Canyon of the Gunnison

The Black Canyon of the Gunnison

A remarkable sequence of metamorphic rocks is exposed in the steep walls of the Black Canyon of the Gunnison River in Colorado. Here, the characteristics of metamorphic rocks are there for all to see.

Observations

1. The canyon walls are made of high-grade metamorphic rock such as schist and gneiss.
2. The foliation of the metamorphic rocks results from aligned grains of muscovite and biotite in schist and by bands of different composition in gneiss.
3. Locally, beds of quartzite or layers with different grain size and texture reveal that these high grade metamorphic rocks were once beds of sedimentary rocks.
4. Careful mapping of the walls also shows that there are a multitude of folds and shear zones.
5. Radiometric dating shows that the metamorphic minerals crystallized about 1.7 billion years ago.
6. The metamorphic rocks are cut by thin light-colored granitic dikes.

Interpretations

Even a beginning geologist can use these facts to make a logical interpretation of the ancient history. More than 1.7 billion years ago, sedimentary rocks were deposited in an ancient ocean basin. These rocks were gradually buried to a depth of perhaps 15 km where the temperature was about 600°C to 700°C (indicated by the presence of garnet and staurolite, Figure 6.15). This dramatic change in pressure and temperature caused the minerals in the sedimentary rock to be unstable; they recrystallized to form new minerals and new foliated textures. Compression folded and deformed the hot rocks. Platy minerals grew perpendicular to the applied stress. Locally, the temperature was so high that the rocks melted and formed magma that rose and was injected into fractures and then cooled to become dikes.

What tectonic environment could produce such profound change? As we look into the Black Canyon are we seeing the roots of an ancient folded mountain belt formed billions of years ago at a convergent plate margin? Although the converging plates have long since disappeared, the evidence in the rocks remains to be seen today.

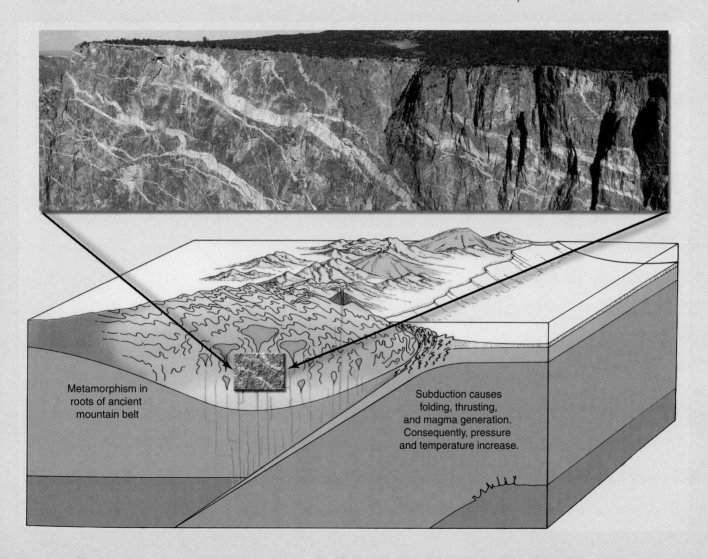

Metamorphism in roots of ancient mountain belt

Subduction causes folding, thrusting, and magma generation. Consequently, pressure and temperature increase.

Key Terms

amphibolite (p. 165)

amphibolite facies (p. 169)

blueschist facies (p. 169)

confining pressure (p. 160)

contact metamorphism
    (p. 158)

differential stress (p. 161)

ductile (p. 161)

eclogite facies (p. 170)

foliation (p. 161)

gneiss (p. 165)

gneissic foliation (p. 161)

granulite (p. 166)

granulite facies (p. 169)

greenschist facies (p. 169)

greenstone (p. 166)

hornfels (p. 166)

hydrothermal alteration
    (p. 160)

index mineral (p. 168)

marble (p. 166)

mélange (p. 172)

metaconglomerate (p. 166)

metamorphic facies (p. 168)

metamorphic grade (p. 160)

metamorphic rock (p. 154)

metamorphism (p. 154)

metasomatism (p. 160)

metastable (p. 158)

migmatite (p. 165)

mylonite (p. 162)

nonfoliated (p. 162)

ocean ridge metamorphism
    (p. 161)

orogenic metamorphism
    (p. 158)

phyllite (p. 165)

plastic deformation (p. 156)

quartzite (p. 166)

regional metamorphism
    (p. 158)

schist (p. 165)

schistosity (p. 165)

slate (p. 163)

slaty cleavage (p. 163)

zeolite facies (p. 169)

Review Questions

1. What causes metamorphic reactions?
2. Compare and contrast the characteristics of metamorphic rocks with those of igneous and sedimentary rocks.
3. What important variables cause changes associated with regional metamorphism? With contact metamorphism? With ocean ridge metamorphism?
4. Make a series of sketches showing the changes in texture that occur with regional metamorphism of (a) slate, (b) sandstone, (c) conglomerate, and (d) marble.
5. Contrast the texture of a schist and a mylonite. What accounts for the textural differences?
6. Define foliation and explain the characteristics of (a) slaty cleavage, (b) schistosity, (c) gneissic layering, and (d) mylonitic texture.
7. Describe the major types of metamorphic rocks.
8. Make a generalized flowchart showing the origin of the common metamorphic rocks.
9. Draw an idealized diagram of converging plates to illustrate the origin of regional metamorphic rocks.
10. What type of metamorphic rock would result if zeolite facies rocks were subjected to temperatures of about 800°C at a depth of 15 km as a result of tectonic processes?
11. You find the mineral sillimanite in gneiss in an old highly eroded orogenic belt. In what metamorphic facies did it form?
12. How does ocean ridge metamorphism change the composition of oceanic crust? What does this imply about the composition of subducted oceanic crust?
13. What evidence do you see that metamorphic crystallization takes place in the solid state without melting?

7 Deformed Rocks

Although many people think that Earth's crust is permanent and fixed, evidence of crustal movement comes in many forms and is there for all to see. In the Mediterranean area, some ancient harbors, such as Ephesus in Asia Minor, are now high and dry, several kilometers inland from the sea. Other ancient shorelines have been submerged well below low tide. Earthquakes are perhaps the most convincing evidence that the crust is moving. During earthquakes, the crust not only vibrates, but segments of it are fractured and displaced along faults. One impressive example was the movement along the San Andreas Fault during the 1906 San Francisco earthquake, which offset fences and roads by as much as 7 m. Another is the 2004 Sumatra earthquake when the seafloor broke and moved about 15 m. The raised beach terraces along the coast of southern California also testify to crustal movement in prehistoric times. There, ancient wave-cut cliffs and terraces—containing remnants of beaches with barnacles, shells, and sand—rise in a series of steps more than 500 m above the present shore.

Since the beginning of geologic studies more than 200 years ago, geologists have shown that rock layers in certain parts of the continents are folded, fractured, and deformed on a gigantic scale. The large-scale shape, geometry, and deformation of rock bodies are part of what we call their structure.

Deformation of the crust is most intense in the great mountain belts of the world, where sedimentary rocks, which were originally horizontal and below sea level, are now folded, contorted, fractured, and, in some places, completely overturned. In some mountains, large bodies of rock have been thrust several tens of kilometers over younger strata. The folded rocks in the world's major mountain belts (the Appalachians, the Rockies, the Andes, the Himalayas, the Urals, and the Alps) all exemplify this type of deformation. These folded and warped rock layers testify to the continuing motion of the lithosphere and the deformation it produces.

Spectacular examples of structural deformation are the large folds found in orogenic belts. In the photo above of the Alpstein Massif of Switzerland, the Mesozoic sedimentary rock layers were originally deposited on the seafloor, but were uplifted more than a kilometer above sea level and deformed into anticlines and synclines by horizontal compression at a convergent plate margin. Erosion has cut through the flexures to reveal their internal structure.

The extensive deformation of rock bodies in the shields and mountain belts shows that Earth's tectonic system has operated throughout geologic time, with shifting plates constantly deforming at their margins. In this chapter we examine how and why rocks deform, how we describe deformed rocks, and then give specific examples of deformed rock bodies.

Major Concepts

1. Deformation of Earth's crust is well documented in historical times by earthquakes along faults, by raised beach terraces, and by deformed rock bodies.
2. Rocks deform when applied stress exceeds their strength. They may deform by ductile flow or brittle fracture. Extensional stress causes rocks to stretch and thin. Contractional stress causes rocks to shorten and thicken.
3. Joints are fractures in rocks along which there is no horizontal or vertical displacement.
4. Faults are fractures, along which slippage or displacement has occurred. The three basic types are (a) normal faults, (b) thrust faults, and (c) strike-slip faults.
5. Folds in rock strata range in size from microscopic wrinkles to large structures hundreds of kilometers long. The major types of folds are (a) domes and basins, (b) plunging anticlines and synclines, and (c) complex folds.

Principles of Rock Deformation

> Rocks deform in response to differential stress. The resulting structure depends on the stress orientation. At high temperatures, ductile flow of rocks occurs. At low temperatures, brittle fractures form.

How are solid rocks deformed?

The folds and faults exposed in canyon walls and mountain ranges show that crustal rocks can be deformed on large scales and in dramatic ways (**Figure 7.1**). But why are some rocks warped into great folds and others only fractured or faulted? Why are some only gently folded, whereas others are complexly folded and faulted? In short, what factors control the type of deformation that rocks experience?

To understand this, you need to understand the forces on rocks. Force applied to an area is **stress**. Stress is the same thing as pressure and is a measure of the intensity of the force or of how concentrated the force is. Everyday experience tells us that solids will bend or break if too much stress is placed on them; that is, they deform if the stress exceeds their **strength** (their natural resistance to deformation). Rocks behave in the same way and deform in response to the forces applied to them.

All of Earth's rocks are under some type of stress, but in many situations the stress is equal in all directions and the rocks are not deformed. In many tectonic settings, however, the magnitude of stress is not the same in all directions and rocks experience **differential stress**. As a result, the rocks yield to the unequal stress and deform by changing shape or position. Geologists call the change in shape **strain**. In other words, differential stress causes strain.

Although strain proceeds by several complex phenomena, two end-member styles of deformation are recognized, with all gradations between them being possible. Under some conditions, rock bodies change shape by breaking to form continuous **fractures** and they lose cohesion; this is **brittle deformation** (**Figure 7.2**). We commonly see such behavior of solids in our daily experience: Chairs, baseball bats, pencils, and wooden beams break if too much force is applied to them. On the other hand, **ductile deformation** occurs when a rock body deforms permanently without fracturing or losing cohesion. The most obvious type of ductile deformation is the viscous flow of fluids, such as molten magma, but solids can also deform ductilely. At first it may seem strange that solids can flow. But bending of metal provides a familiar example. Consider sheet metal in a car fender. A minor collision commonly makes a dent, not a fracture in the fender. Likewise, rocks can flow in a solid state, under the right conditions. This type of solid state flow is usually called *plastic flow* and is accomplished by slow internal creep, gliding on imperfections in crystals, and recrystallization.

Figure 7.1 **A sequence of inclined beds** striking toward the background and dipping 40° to the east (left) forms the flank of the San Rafael Swell, in Utah. The diagram shows the form of the flexure and the upper beds, which have been partially removed by erosion.

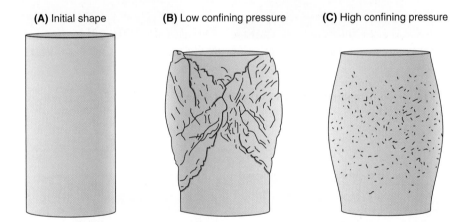

Figure 7.2 **Brittle versus ductile behavior** of rocks is controlled by external conditions. These marble cylinders were compressed in a laboratory under different confining pressures, but at the same temperature and with the same deforming stress. At low confining pressure, the cylinder deformed in a brittle fashion and fractured and faulted. At high confining pressure, the cylinder deformed in a ductile fashion as mineral grains flowed and recrystallized.

Depending on the temperature or pressure of the surroundings and the rate at which stress is applied, most types of rocks can deform by brittle fracture or ductile flow. Low pressures, low temperatures, and rapid deformation rates favor brittle deformation (Figure 7.2B). As a result, brittle structures are most common in the shallow crust. We use the term **shear** to describe slippage of one block past another on a fracture. High confining pressures, high temperatures, and low rates of deformation all favor ductile behavior (Figure 7.2C). Ductile deformation is more common in

What effect does confining pressure have on the style of rock deformation?

the mantle and deeper parts of the crust. Another example of this difference can be seen in the behavior of glass. When a glass rod is cold, it is strong and brittle fractures form when enough stress is applied (**Figure 7.3**). When the same glass rod is hot but not molten, it is weak and easy to bend. To visualize the role of the rate of deformation, consider taffy (or Silly Putty™) as an example. If warm taffy is pulled slowly and steadily, it is ductile and stretches continuously, without breaking, to form long, thin strands. On the other hand, if it is stretched rapidly, it may break and form brittle fractures.

It should be clear which types of rock structures form by ductile deformation and which form by brittle behavior. The flow of rocks in a solid state to form folds in metamorphic rocks is a good example of ductile behavior. The formation of fractures, joints, and faults are common expressions of deformation in the brittle regime.

We can also better understand rock structures if we consider the orientations of the stresses acting on a body (**Figure 7.4**). For example, **tension** occurs where the stresses point away from one another and tend to pull the rock body apart. In contrast, **compression** tends to press a body of rocks together. Three distinct types of

(A) A cold glass rod deforms by brittle fracture.

(B) A hot glass rod bends and shows ductile behavior without breaking.

Photographs by Stan Macbean

Figure 7.3 Temperature affects how solids deform.

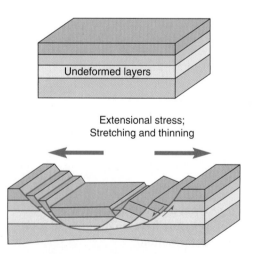

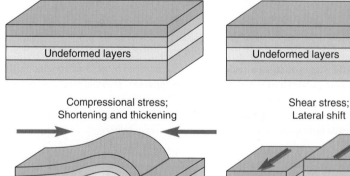

Undeformed layers

Extensional stress;
Stretching and thinning

Undeformed layers

Compressional stress;
Shortening and thickening

Undeformed layers

Shear stress;
Lateral shift

(A) Extension results in stretching rock bodies and produces brittle fractures in the upper crust that pass downward into ductile zones.

(B) Contraction (or horizontal compression) causes shortening and thickening and is manifested in faults and folds.

(C) Lateral-slip creates faults as blocks of crust slide horizontally past one another.

Figure 7.4 Extension, contraction, and lateral-slip produce fundamentally different types of structures in rocks. Moreover, each is caused by different stress orientations and dominates at different plate tectonic settings.

deformation occur because of differential stresses caused by tectonic processes (Figure 7.4). To visualize this, imagine how two adjacent blocks can interact. They can move away from one another (**extension**), move toward one another (**contraction**), or slip horizontally past one another (**lateral-slip**). Obviously, the orientation of the stresses acting on the blocks determines which of the three cases is dominant. In the simplest case, the stress orientations are directly related to plate tectonic settings. Extension is caused when the differential stresses point away from one another (Figure 7.4A). This type of deformation results in lengthening and is common at divergent boundaries. In brittle rocks it is expressed by fracturing and faulting, and in ductile rocks by stretching and thinning. Contraction is caused by horizontal compression when the differential stresses are directed toward one another (Figure 7.4B). Contraction is common at convergent boundaries and causes shortening and thickening of rock bodies, expressed as faults in brittle rocks and folds in ductile rocks. Lateral-slip is the kind of shear that occurs when rocks slide horizontally past one another along nearly vertical fractures and dominates at transform plate boundaries (Figure 7.4C).

Geometry of Rock Structures

The orientation of planar features in rocks, such as bedding planes, faults, and joints, can be defined by measurements of dip (the downward inclination of the plane) and strike (the direction or trend of the plane).

Many structural features of the crust are too large to be seen from one point on the ground. They are recognized only after the geometry of the rock bodies is determined from geologic mapping. At an outcrop, two fundamental observations—dip and strike—describe the orientations of bedding planes, fault planes, joints, and other planar features in the rock. The **dip** of a plane is the angle and direction of its inclination from the horizontal. The **strike** is the compass bearing of a horizontal line on the plane, such as a bedding plane or a fault. These two measurements together define the orientation of the planar surface in space.

The concept of dip and strike can be easily understood by referring to **Figure 7.5**, which shows an outcrop of tilted beds along a coast. The water provides a necessary

How do we measure the orientation of rock bodies, faults, and joints?

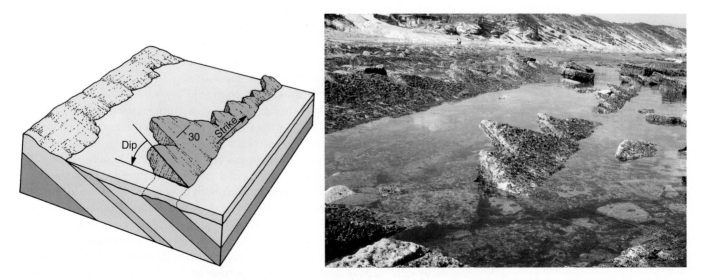

Figure 7.5 The concept of dip and strike can be understood by studying rock layers such as the ones shown in this photograph. The strike of a bed is the compass bearing of a horizontal line drawn on the bedding plane. It can readily be established by reference to the horizontal waterline in this example. The dip is the angle and direction of inclination on the bed, measured at right angles to the strike and is represented with a T-shaped symbol and accompanying dip in degrees.

reference to a horizontal plane. The trend of the waterline along the bedding plane is the direction of strike. The angle between the bedding plane and the water surface is the angle of dip. Figure 7.1 shows a sequence of beds striking south (to the top of the picture) and dipping 40° to the east (to the left of the picture). Another way to visualize dip and strike is to think of the roof of a building. The dip is the direction and amount of inclination of the roof, and the strike is the trend of the ridge.

In the field, dip and strike are measured with a geologic compass, which is designed to measure both direction and angle of inclination. A long crossbar shows the strike, a short line perpendicular to it shows the direction of dip, and the number represents the angle of dip (Figure 7.5). The symbols provide a way to view a map in three dimensions and simplify the construction of vertical cross sections through the crust.

A useful tool to interpret the structure of dipping layers of rock is called the **rule of Vs**. Look carefully at the sketch and photo in Figure 7.5. Erosion has cut V-shaped notches into the resistant layer of rock. Each V points in the direction the bed dips.

Joints

> Joints are tension fractures in brittle rocks along which no shear has occurred. They form at low pressure and are found in almost every exposure.

The simplest and most common structural features of rocks at Earth's surface are cracks or fractures, known as **joints**, along which little displacement (or slip) has occurred. Their most important feature is the absence of shear; no movement occurs parallel to the fracture surface. Joints form by the brittle failure of rocks at low pressure as stress accumulates and exceeds the rock's strength. They do not occur at random but are usually perpendicular to the direction of tension.

What is the difference between a joint and a fault?

Multiple sets of joints that intersect at angles ranging from 45° to 90° are very common. They divide rock bodies into large, roughly rectangular blocks. These joint systems can form remarkably persistent patterns extending over hundreds of square kilometers. Each set probably formed at a different time and under a different stress orientation.

The best areas to study joints are where brittle rocks, such as thick sandstones, have been fractured and their joint planes accentuated by erosion. The massive sandstones of the Colorado Plateau are excellent examples. Joints are expressed by deep, parallel cracks that have been enlarged by erosion; they are most impressive when seen from the air (**Figure 7.6**). In places, joints control the development of stream courses, especially secondary tributaries and areas of solution activity.

Joints result from broad regional upwarps, or from contraction or extension associated with faults and folds; some tensional joints result from erosional unloading and expansion of the rock. Columnar joints in volcanic rocks are produced by tensional stresses that are set up as the lava cools and contracts.

At first it may seem that joints are insignificant and uninteresting, but they have great economic importance. They can be paths for groundwater migration and for the movement and accumulation of petroleum. Analysis of joint patterns has been important in exploration and the development of these resources. Joints also control the deposition of copper, lead, zinc, mercury, silver, gold, and tungsten ores. Hot aqueous solutions associated with igneous intrusions migrate along joint systems and minerals crystallize along the joint walls, forming mineral **veins**. Modern prospecting techniques therefore include detailed analysis of fractures. Major construction projects are especially affected by joint systems within rocks, and allowances must be made for them in project planning. For example, dams must be designed so that the stresses caused by water storage tend to close any fractures in the bedrock foundation. Joint systems can be either an asset or an obstacle to quarrying operations. Closely spaced joints severely limit the sizes of blocks that can be removed. If a quarry follows the orientation of intersecting joints, however, the expense of removing building blocks is greatly reduced, and waste is held to a minimum.

Figure 7.6 Joint systems in resistant sandstones in Arches National Park, Utah, have been enlarged by weathering, forming long, narrow crevasses. The system of intersecting joints reflects the orientation of stress that deformed the rock body.

Faults

Faults are fractures in Earth's crust along which displacement has occurred. Three basic types of faults are recognized: (1) normal faults, (2) reverse faults, and (3) strike-slip faults. Normal faults are usually the result of extension, thrust faults the result of horizontal compression, and strike-slip faults the result of lateral slip.

Slippage (or shear) along brittle fractures in Earth's crust creates **faults** (**Figure 7.7**). Like other deformation features, they form by the application of differential stress. In a road cut, or in the walls of a canyon, a fault plane may be obvious, and the displaced, or offset, beds can easily be seen. Elsewhere, the surface expression of a fault can be very subtle, and detailed geologic mapping may be necessary before the precise location of such a fault can be established. Displacement along faults ranges from a few centimeters to hundreds of kilometers.

Faults grow by a series of small movements, which occur as stress built up in the crust is suddenly released in earthquakes. Displacement can also occur by an almost imperceptibly slow movement called tectonic creep.

Why are there three different types of faults?

Normal (Extensional) Faults

Along **normal faults**, the rocks above the fault plane (the **hanging wall**) move downward in relation to those beneath the fault plane (the **footwall**; Figure 7.7A and **7.8A**). Most normal faults are steeply inclined, usually between 65° and 90°. Their predominantly vertical movement commonly produces a cliff, or **scarp**, at the surface. Some normal faults dip at low angles and are called **detachment faults** (Figure 7.7B).

Normal faults are rarely isolated features. A group of parallel normal faults may develop a series of fault-bounded blocks. A narrow block dropped down between two normal faults is a **graben** (German, "trough" or "ditch"), and an upraised block is a **horst** (**Figure 7.8B**). A graben typically forms a conspicuous fault valley or basin marked by relatively straight, parallel walls. Horsts form plateaus bounded by faults. Normal faults usually juxtapose younger rocks over older rocks. In the field, this may

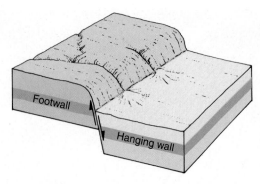

(A) In normal faults, the hanging wall moves downward in relation to the footwall.

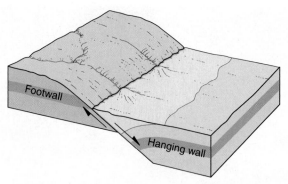

(B) Some normal faults have low angles and are called detachment faults.

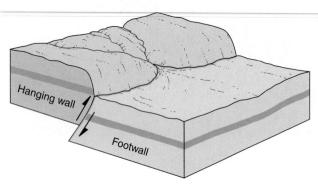

(C) In reverse faults, the hanging wall moves upward in relation to the footwall.

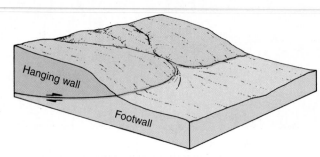

(D) Thrust faults are low-angle reverse faults.

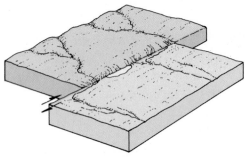

(E) In strike-slip faults, the displacement is horizontal as you can see on this right-lateral fault.

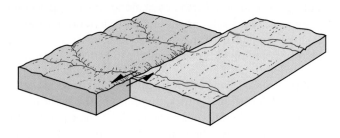

(F) Left-lateral strike slip faults are similar but displacement is to the left as you look across the fault.

Figure 7.7 The three major types of faults are distinguished by the direction of relative displacement.

What tectonic settling commonly produces normal faults?

be mistaken for normal layer-upon-layer deposition, especially where the normal fault dips at a low angle (Figure 7.4). However, careful fieldwork will also reveal that some rock layers are apparently missing along the fault. This is called *omission of strata*.

Large-scale normal faulting is the result of horizontal extensional stress, which stretches, thins, and pulls apart the lithosphere. Normal faults are common because rocks are weaker during extension than during compression. This type of extensional stress occurs on a global scale along divergent plate margins. Consequently, normal faults are the dominant structures along the oceanic ridge, in continental rift systems, and along rifted continental margins.

In the Basin and Range Province of western North America, normal faulting produced a series of grabens and horsts trending north from central Mexico to Oregon and Idaho. The horsts commonly form mountain ranges from 2000 to 4000 m high,

(A) Normal faults commonly form cliffs on the footwall or upthrown block. This is the Hurricane fault in southern Utah where the red Mesozoic strata are displaced downward about 1.5 km. The trace of the fault is along the base of the gray cliff.

(B) Grabens and horsts are commonly produced where the crust is extended. This area shows alternating horsts (ridges) and grabens (valleys).

(C) This high-angle reverse fault shows that the hanging wall has moved up relative to the footwall.

(D) Thrust faults form where the crust is under compression. Here, a series of thrust faults (shown in white) juxtapose Precambrian rocks on Paleozoic sandstones.

(E) Strike-slip faults are commonly expressed by a series of straight linear ridges and troughs that can be traced for long distances. Here, the San Andreas Fault in southern California offsets a drainage system.

Figure 7.8 The surface expression of faults is highly variable and influenced by the character of the rock units that have been displaced, the type and degree of erosion, and the nature of the exposed surface.

which are considerably dissected by erosion (**Figure 7.9**). Grabens form topographic basins, which are partly filled with erosional debris from the adjacent ranges. Between the Wasatch Range, in central Utah, and the Sierra Nevada, on the Nevada-California border, faulting has extended Earth's crust some 100 km during the last 15 million years. Modern earthquakes and displacement of young alluvial fans and surface soils show that many of the faults are still active. The great rift valley of east Africa is another example of large-scale normal faulting produced by a zone of extension in Earth's crust.

Reverse (Contractional) Faults

Faults in which the hanging wall has moved up and over the footwall are **reverse faults** (Figure 7.7C and **7.8C**). **Thrust faults** are low-angle reverse faults and dip at angles less than 45° (Figure 7.7D and **7.8D**). Movement on thrust faults is predominantly horizontal, and displacements can be more than 50 km.

Reverse faults result from horizontal compression with the maximum stress perpendicular to the trend of the fault. This shortens and thickens the crust. In contrast to normal faults, thrust faults usually place older over younger strata and instead of omitting layers, units are repeated in a vertical section (**Figure 7.10**). Where resistant rocks are thrust over nonresistant strata, a scarp is eroded on the upper plate. The scarp is not straight or smooth, as are cliffs produced by normal faulting. Rather, the outcrop of the fault surface typically is irregular in map view due to the low angle of dip (Figure 7.10B).

Most contractional faults form at convergent plate margins. Thrust faults are typically associated with folds and are prominent in all of the world's major folded mountain belts. They commonly evolve from folds in the manner diagrammed in Figure 7.10.

Why are thrust faults and folds found in the same regions?

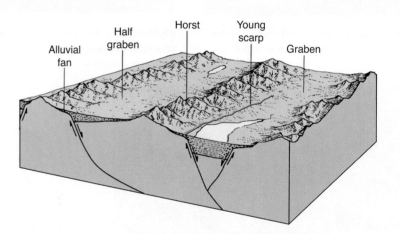

Figure 7.9   The Basin and Range province of the western United States has been shaped by a series of normal faults that block out horsts and grabens. Faults range from steep, high-angle faults to low-angle detachment faults.

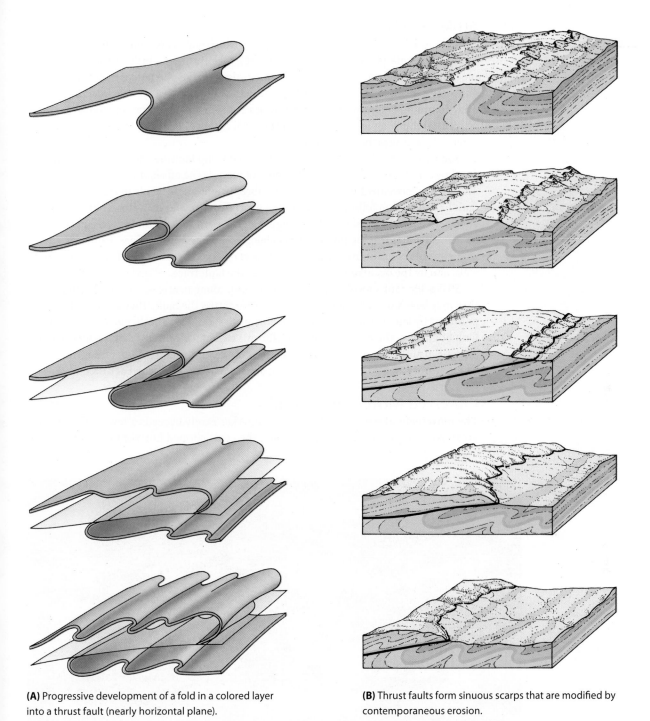

(A) Progressive development of a fold in a colored layer into a thrust fault (nearly horizontal plane).

(B) Thrust faults form sinuous scarps that are modified by contemporaneous erosion.

Figure 7.10 The evolution of thrust faults from folds is depicted in this sequence of diagrams. A fault forms when the strength of the folding rock layers is exceeded.

What kind of stress makes strike-slip faults?

Strike-Slip Faults

Strike-slip faults are high-angle faults along which slip is horizontal, parallel to the strike of the fault plane (Figure 7.7E and **7.8E**). Ideally, there is little or no vertical movement, so high cliffs do not usually form along strike-slip faults. Instead, these faults are expressed topographically by straight valleys or by a series of low ridges. They commonly mark sharp discontinuities in drainages and types of landscape. No crustal thinning or thickening is produced, except at bends in the fault where extension or contraction can occur.

Some topographic features produced by strike-slip faulting and subsequent erosion are shown in Figure 7.7E. One of the more obvious is the offset of the drainage pattern. The relative movement is often shown by abrupt right-angle bends in streams at the fault line. A stream follows the fault for a short distance and then turns abruptly and continues down the regional slope (Figure 7.8E). As the blocks move, some parts may be depressed to form sag ponds. Others buckle into low, linear ridges. Faults also disrupt patterns of groundwater movement, as is reflected by contrasts in vegetation and soils, and by the occurrence of springs along the fault trace.

Strike-slip faults result from horizontal shear along nearly vertical faults and they may be left or right lateral as viewed from across the fault. They commonly are produced by lateral-slip where one tectonic plate slides past another at a transform fault boundary. The most famous is California's seismically active San Andreas Fault. Strike-slip faults also join adjacent segments of the mid-ocean ridges. Others form where two segments of continental crust are stretched or shortened at different rates.

Observed Movement on Faults

The movement along faults during earthquakes rarely exceeds a few meters. In the great San Francisco earthquake of 1906, the crust slipped horizontally as much as

Courtesy of Glenn Embree

Figure 7.11 Recent displacement along a fault, at the base of the Lost River Range in southern Idaho, has produced the fresh cliff at the base of the mountain front. Cumulative movement on the fault during a vast period of time produced the mountain range.

7 m along the San Andreas Fault, so roads, fence lines, and orchards were off-set. Recent offsets along vertical faults in Nevada and Idaho have produced fresh scarps from 3 to 6 m high (**Figure 7.11**). The Good Friday earthquake in Alaska, in 1964, was accompanied by a 13-m uplift near Montague Island. The largest well-authenticated displacement during an earthquake appears to have occurred in 1899 near Yakutat Bay in Alaska, where beaches were raised as much as 15 m above sea level.

Movement along faults is not restricted to uplift during earthquakes, however. Precise surveys along the San Andreas Fault show slow shifting along the fault plane at an average rate of 4 cm/yr. Such slow movements, known as tectonic creep, break buildings constructed across the fault line and eventually result in considerable displacement.

The important point is that the total displacement on a fault that may amount to many kilometers does not occur in a single violent event. Rather, it is the result of numerous periods of displacement. On most faults, there are hundreds to thousands of years between major earthquakes. Slow tectonic creep may continue between these major events.

How long does it take for thousands of meters of displacement to occur on a normal fault?

Folds

> Folds are warps in rock strata during ductile deformation. They are three-dimensional structures ranging in size from microscopic crinkles to large domes and basins that are hundreds of kilometers across. Most folds develop by horizontal compression at convergent plate boundaries where the crust is shortened and thickened. Broad, open folds form in the stable interiors of continents, where the rocks are only mildly warped.

Almost every exposure of sedimentary rock shows some evidence that the strata have been deformed. In some areas, the rocks are slightly tilted; in others the strata are folded like wrinkles in a rug. Small flexures are abundant in sedimentary rocks and can be seen in mountainsides and road cuts and even in hand specimens. These warps in the strata are called **folds** and are a manifestation of ductile deformation in response to horizontal compression. This kind of deformation is also called *contraction*. Large folds cover thousands of square kilometers, and they can best be recognized from aerial or space photographs or from geologic mapping. Like faults, folds form slowly over millions of years, as rock layers gradually yield to differential stress and bend.

Folds are of great economic importance because they commonly form traps for oil and gas and may control localization of ore deposits. Consequently, it is of more than academic interest to understand folds.

Fold Nomenclature

Three general types of folds are illustrated in **Figure 7.12**. An **anticline**, in its simplest form, is uparched strata, with the two **limbs** (sides) of the fold dipping away from the crest. Rocks in an eroded anticline are progressively *older* toward the interior of the fold. **Synclines**, in their simplest form, are downfolds, or troughs, with the limbs dipping toward the center (Figure 7.12). Rocks in an eroded syncline are progressively *younger* toward the center of the fold. **Monoclines** are folds that have only one limb; horizontal or gently dipping beds are modified by simple steplike bends.

For purposes of description and analysis, it is useful to divide a simple fold into two more-or-less equal parts by an imaginary plane known as the **hinge plane**

STATE OF THE ART   Geologic Maps: Models of the Earth

The key tool used in almost all geologic investigations is a geologic map. A geologic map shows more than just locations and distances; it also shows more than elevation. Instead, a geologic map shows the distribution of rock types, faults, folds, and other geologic features, as well as the ages of the rocks. It is a scale model of the rock bodies, revealing much about their origin, internal structure, and deformation.

Geologic maps are constructed by careful fieldwork. A geologist working alone or as part of a small team collects critical data about rock types during traverses on foot across a field area. The locations and rock type of each outcrop are found and plotted on a map. In arid regions where vegetation is sparse, large areas can be mapped on an aerial photograph as shown here. By plotting the distribution of rock types along with strikes and dips of planar features, we reveal the three-dimensional structure of the rocks.

The goal of field geologists is to first recognize major rock bodies (formations) that can be mapped and then to determine the location and nature of the contacts between the rock units. They may do this by walking along the contacts in the field and marking their locations on a map or a photo. If the contacts are obscured by vegetation or by younger rocks, contacts may need to be inferred from the evidence gathered. Soils, selective growth of vegetation, stream patterns, and other geologic features can help. A geologist working on a map must also decide the relative ages of the rock units using the principles of superposition and crosscutting relationships.

Observations are recorded on maps and in field notebooks. A compass may be used to measure the strike and dip of deformed layers of sedimentary rock. A geologic hammer is used to collect samples and break off weathered surfaces. A stereoscope is used with a pair of aerial photographs to view the landscape in three dimensions. The data are then plotted on a topographic map. In some areas, where locations are

difficult to establish precisely because of vegetation cover, a GPS unit (Global Positioning System) and computer may be used to record locations or contacts. Regardless of how the data are collected and recorded, the end result is a scale model of Earth.

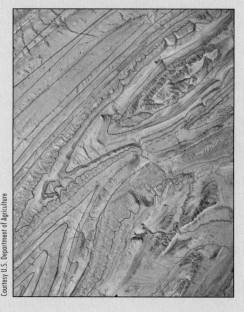

Courtesy U.S. Department of Agriculture

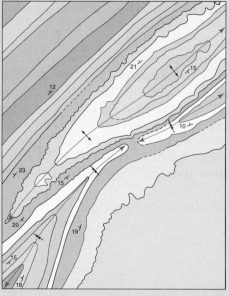

(**Figure 7.13**). The hinge marks the region of maximum curvature in the fold. The line formed by the intersection of the hinge plane and a bedding plane is the **hinge line**, and the downward inclination of the hinge line is called the **plunge**. A **plunging fold**, therefore, is a fold in which the hinge line is inclined.

In most folds, the hinge plane is not vertical but is inclined, and the fold is overturned and one limb is steeper than the other (Figure 7.12D). The direction the hinge plane is rotated from vertical indicates the direction the rocks were displaced. In other words, movement is toward the steep limb. In Figure 7.12D, you can see that the folded rocks were transported from left to right, just as they were in Figure 7.10.

The variety of folds produces dramatic landscapes and outcrop patterns in stratified rocks. Many of these are shown in the photographs in **Figure 7.14**.

Domes and Basins

In contrast to fold belts at convergent margins, the sedimentary rocks covering much of the continental interiors have been only gently warped into broad **domes** and **basins** many kilometers in diameter. One large basin covers practically all of the state of Michigan. Another underlies the state of Illinois. An elongate dome underlies central Tennessee, central Kentucky, and southwestern Ohio. Although these flexures in the sedimentary strata are extremely large, the configuration of the folds is known from

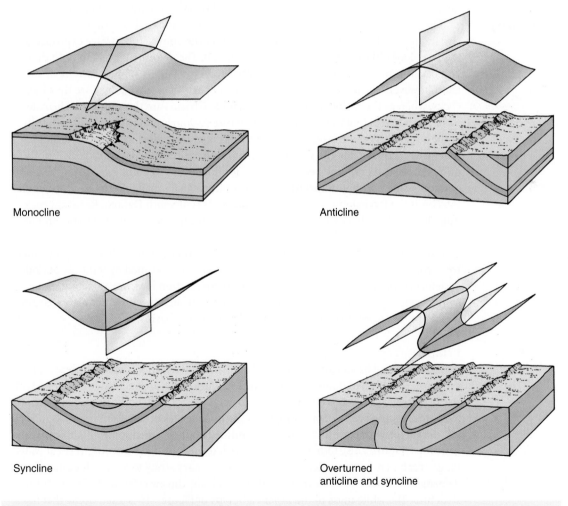

Monocline

Anticline

Syncline

Overturned
anticline and syncline

Figure 7.12 The nomenclature of folds is based on the three-dimensional geometry of the structure, although most exposures show only a cross section or map view.

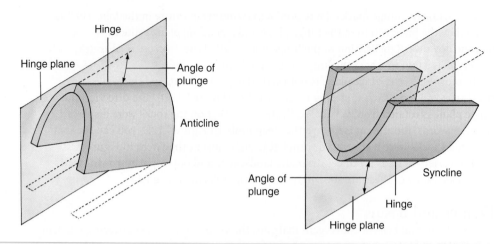

Figure 7.13 **The hinge plane of a fold** is an imaginary plane that divides the fold in equal parts. The line formed by the intersection of the hinge plane and a bedding plane is called the hinge. The downward inclination of the hinge is called the plunge.

How do we describe the geometry of folded rocks?

geologic mapping and from information gained through drilling. The nature of these flexures and their topographic expression are illustrated in **Figures 7.14** and **7.15**.

The form of a single bed warped into broad domes and basins is shown in perspective in Figure 7.15A. If erosion cuts off the tops of the domes, the surface exposure of the layer looks like the one shown in Figure 7.15B. The deformed layers in both domes and basins typically have circular or elliptical outcrop patterns. There is a major difference, however. The rocks exposed in the central parts of eroded domes are the oldest rocks, whereas the rocks exposed in the centers of basins are the youngest. The rule of Vs is also a useful tool to interpret the structure of domes and basins. Look carefully at the aerial photos of the dome in Figure 7.14B and the model in Figure 7.15. Note how stream erosion has cut V-shaped valleys into the resistant layers of rock. The "V" points in the direction the bed dips. Thus, for domes the Vs point away from the center and for basins they point inward. A classic example of a broad fold in the continental interior is the large dome that forms the Black Hills of South Dakota. Resistant rock units form ridges that can be traced completely around the core of the dome, and nonresistant formations make up the intervening valleys.

How these broad domes form is still enigmatic. Their circular shapes and distance from convergent margins are puzzling. Many could have formed by multiple periods of deformation. For example, gentle east–west contraction far from an active convergent margin, followed by north–south contraction related to a different plate margin, could produce a series of broad domes and intervening basins.

Diapirs

Some domes and basins form by vertical adjustments caused by density differences in the crust. Many small domes, like the one in Figure 7.14B, are associated with buoyant rise of material that is less dense than the overlying rock as it works its way upward along preexisting breaks. For example, in thick sequences of sedimentary rocks, beds of salt may deform and rise as **diapirs**, streamlined bodies shaped somewhat like inverted teardrops (**Figure 7.16**). Other salt bodies are like igneous dikes forming vertical walls. Plugs of salt may rise and pierce overlying sedimentary strata to form salt domes. The deformed sedimentary beds are faulted and typically dip away from the center of the structure. The white mass in the lower right part of Figure 7.19 is a salt dome that has reached the surface to flow like a glacier. The movement of salt has modified the seafloor south of the Mississippi delta on a grand scale, with subsidence basins and domes pockmarking the seafloor. Other small domes are formed by intrusions of magma.

(A) Anticlines and synclines are easily recognized when erosion cuts across the structures and exposes them in a vertical cross section, such as these folds in the Calico Hills of southern California.

(B) A small structural dome in west Texas has been truncated by erosion and is expressed as a series of circular ridges and valleys. The dome is only about 5 km across.

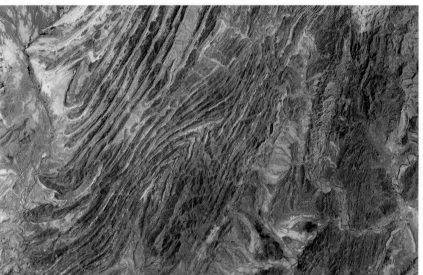

(C) Folds in the Sulaiman Ranges of western Pakistan are plunging anticlines and synclines formed when India collided with Asia. This area is about 30 km across.

Courtesy of U.S. Geological Survey

Courtesy of U.S. Geological Survey and EROS Data Center

(D) A plunging anticline forms a typical V-shape pattern pointing in the direction of plunge. This is an exceptionally clear example near St. George, Utah, where colorful beds form alternating ridges and valleys.

Figure 7.14 The surface expression of folds is extremely variable because of the great range in the size and shape of the structures and because of the variety of ways in which folds may be modified by erosion. Common types of folds are shown above.

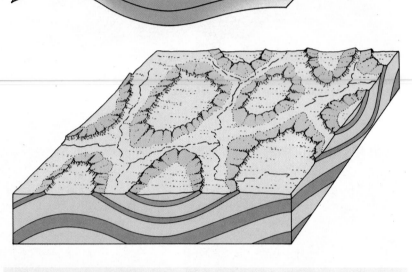

(A) A single folded bed is warped into broad domes and basins.

(B) As erosion proceeds, the tops of the domes are eroded first. The outcrop pattern of eroded domes and basins typically is circular or elliptical.

Figure 7.15 **The geometry and topographic expression of domes and basins** involve broad upwarps and downwarps of layered rocks. When eroded, the exposed rock forms circular or elliptical outcrop patterns.

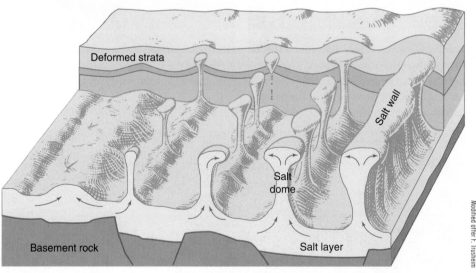

Figure 7.16 **Diapirs** form by the rise of low-density ductile materials such as salt and gypsum.
The overlying strata may be faulted and folded as they are pierced or domed upward. As salt domes form, the layer of salt is thinned.

Deformed strata

Salt wall

Salt dome

Basement rock

Salt layer

Salt layer

Modified after F. Trusheim

In the Persian Gulf countries, about 60% of the oil and gas is extracted from deformed rocks related to salt diapirs. Likewise, 65% of all recoverable oil worldwide is thought to be related to the movement of salt through the upper crust. Thus, understanding these structures is of great economic interest; they are not merely structural curiosities.

Fold Belts

Where contraction is intense (typically in **orogenic belts** at convergent plate boundaries), sedimentary rock layers are deformed into series of tight folds in long linear belts. The internal geometry of many fold belts is not exceedingly complex. In many ways, the folds resemble the wrinkles in a rug. However, complexity in the outcrop patterns of fold belts results from erosion, so folds may be difficult to recognize without some experience in geologic observation and interpretation. The diagrams in **Figure 7.17** illustrate a fold belt with plunging folds and its surface expression after the upper part has been removed by erosion. The outcrop of the eroded plunging anticlines and synclines forms a characteristic zigzag pattern. The nose of an anticline forms a V that points in the direction of plunge, and the oldest rocks are in the center of the fold (Figure 7.14D). The nose of a syncline forms a V that opens in the direction of plunge, and the youngest rocks are in the center of the fold. Together, the outcrop pattern, the strike and dip of the beds, and the relative ages of the rocks in the center of the fold make it possible to determine the structure's subsurface configuration. Thrust faults commonly form in association with these contractional folds.

Intense deformation in the cores of some mountain ranges produces complex folds like those in **Figure 7.18**. Some folds are refolded during the millions of years of deformation. Such structures commonly exceed 100 km across and can extend through a large part of a mountain belt. Details of such intensely deformed structures are extremely difficult to work out because of the complexity of the outcrop patterns. These complexities arise from multiple episodes of folding at elevated temperatures and pressures in lower crust. Under these conditions, rocks are much more likely to deform ductilely.

What is the outcrop pattern of a series of plunging anticlines and synclines?

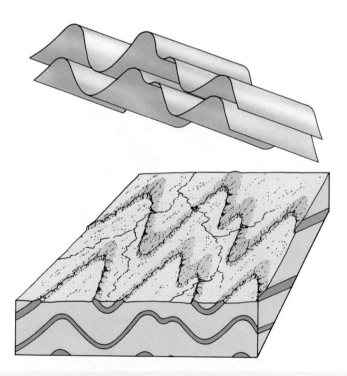

(A) The basic form of folded strata is similar to that of a wrinkled rug. In this diagram, the strata are compressed and plunge toward the background.

(B) If the tops of the folded strata are eroded away, a map of the individual layers shows a zigzag pattern at the surface. Rock units that are resistant to erosion form ridges, and nonresistant layers are eroded into linear valleys. In a plunging anticline, the surface map pattern of an anticline forms a V pointing in the direction of plunge.

Figure 7.17 **A series of plunging folds** forms a zigzag outcrop pattern.

Figure 7.18 illustrates the geometry and surface expression of a large complex fold. Figure 7.18A is a perspective drawing of a single bed in a complex fold. This **overturned fold** is a huge anticlinal structure with numerous minor anticlines and synclines forming indentations on the larger fold; Figure 7.18B shows the fold after it has been subjected to considerable erosion, which has removed most of the upper limb. Note the cross section of the structure on the mountain front and the outcrop pattern compared with that in Figure 7.18A. The topographic expression of complex folds is variable. They usually are expressed in a series of mountains (Figure 7.18C). Complex folds are common in the Swiss Alps (see the chapter opening photo), but they were recognized only after more than half a century of detailed geologic studies. They are also common in the roots of ancient mountain systems and thus are exposed in many areas of the shields.

An excellent example of an orogenic belt is the Zagros Mountains of southern Iran (**Figure 7.19**). The fold belt is only a small part of the Alpine Himalaya chain that extends from southern Europe and across southern Asia. The satellite photograph of the region shows that the sedimentary beds have been folded above a series of major thrust faults. Anticlines form the hills and are separated by intervening synclines. Other parts of this folded mountain belt include the Sulaiman Range of Pakistan (Figure 7.14C) and the Alps in France and Switzerland. The Valley and Ridge Province of the Appalachian Mountains in the eastern United States is another good example. The Appalachian fold belt extends from New England to central Alabama and consists of enormous parallel folds hundreds of kilometers long. Thrust faults slice through many of the folds. The region has been deeply eroded so that resistant formations form long, high ridges, and the weak formations form intervening valleys. Folds are the obvious expression of the ductile deformation of rocks in Earth's crust. All of these fold belts formed at convergent plate margins where an oceanic plate was thrust underneath a continental plate or where two continents collided.

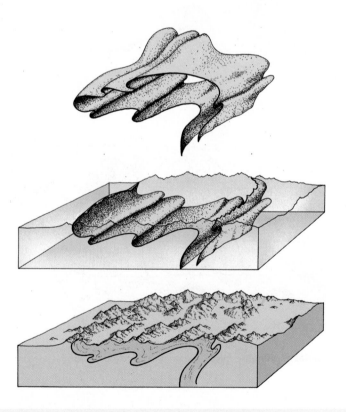

(A) Rocks that have been intensely deformed commonly consist of large overturned folds, with minor folds on the limbs. Some of these are refolded folds.

(B) The surface outcrop of the fold, after erosion has removed the upper surface, shows great complexity, so that its details and overall structure are difficult to recognize.

(C) The topographic expression of complex folds can be a series of linear mountain ridges.

Figure 7.18 Complexly deformed folds produce intricate outcrop patterns.

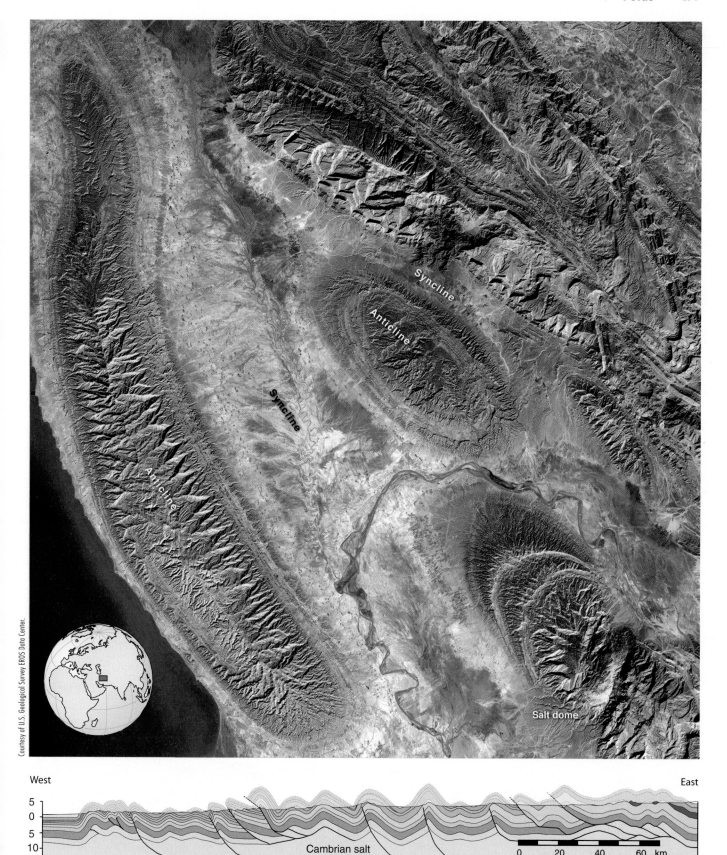

Courtesy of U.S. Geological Survey EROS Data Center.

Figure 7.19 A vast orogenic belt is being created by the collision of Arabia with southern Asia. This portion of the fold and thrust belt is part of the Zagros Mountains of Iran. The deformation accompanies the underthrusting of the Arabian subplate beneath Asia. Doubly plunging anticlines and synclines as well as elongate domes lie above major thrust faults. In the lower right, a light-colored salt dome has pierced through an anticline. The salt is not resistant and a valley has formed.

GeoLogic   The Keystone Thrust

The Keystone Thrust in Nevada is one of the most dramatic expressions of thrust faulting in the world. Here a sequence of dark bluish-gray limestones and dolomites are thrust over a sequence of reddish sandstone. The important questions are: (1) What was this area like before erosion, and (2) How did this structure form?

Observations

1. Older beds lie above younger beds! The fossils in the gray limestone beds show that they are about 350 million years *older* than the layers of red sandstone beneath them.
2. The contact between the two rock layers is not a sedimentary contact, but a zone of broken and crushed rock —a typical fault zone.
3. By carefully mapping the dip of the beds in the thrust sheet, a large fold is revealed.

4. The landscape has been eroded and huge volumes of rock have been removed from the area.

Interpretations

The geologic interpretation is shown in the lower diagram. The three major structural elements are (1) the older sequence of rocks, (2) the younger sequence of rocks, and (3) the thrust fault. If we project these major geologic features to their position prior to erosion, the magnitude of the structure becomes much more apparent. One hundred million years ago, a convergent plate margin existed along the western margin of North America. As a result of intense compression at the plate boundary, a thrust fault broke through these layers of rock. Gradually, over the course of the next few millions of years, the great slab of gray limestone was pushed up and over the reddish sandstone layers. The great folded mountain belt that once dominated this region is now dismembered by extensive erosion.

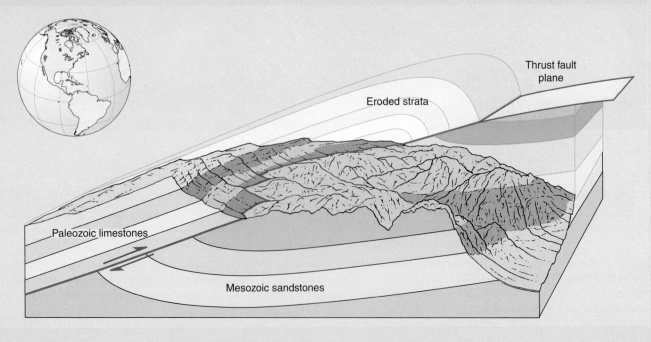

Thrust fault plane

Eroded strata

Paleozoic limestones

Mesozoic sandstones

Key Terms

anticline (p.191)
basin (p. 193)
brittle deformation (p. 180)
compression (p. 182)
contraction (p. 183)
detachment fault (p. 185)
diapir (p. 194)
differential stress (p. 180)
dip (p. 183)
dome (p. 193)
ductile deformation (p. 180)

extension (p. 183)
fault (p. 185)
fold (p. 191)
footwall (p. 185)
fracture (p. 180)
graben (p. 185)
hanging wall (p. 185)
hinge line (p. 193)
hinge plane (p. 191)
horst (p. 185)
joint (p. 184)

lateral-slip (p.183)
limb (p. 191)
monocline (p. 191)
normal fault (p. 185)
orogenic belt (p. 197)
overturned fold (p. 198)
plunge (p. 193)
plunging fold (p. 193)
reverse fault (p. 188)
rule of Vs (p. 184)
scarp (p. 185)

shear (p. 181)
strain (p. 180)
strength (p. 180)
stress (p. 180)
strike (p. 183)
strike-slip fault (p. 190)
syncline (p. 191)
tension (p. 182)
thrust fault (p. 188)
vein (p. 184)

Review Questions

1. List evidence that Earth's crust is in motion and has moved throughout geologic time.
2. What is the difference between a fault and a joint?
3. Compare the kinds of deformation produced by extension and those produced by contraction.
4. Contrast brittle and ductile behavior in rocks.
5. Do you think the fold shown in Figure 7.1 could have formed when the rocks were at the surface as they are now?
6. Explain the terms dip and strike.
7. Sketch a cross section of the structure of the rocks shown in Figure 7.14B.
8. List some of the surface features that commonly are produced by strike-slip faults.
9. Draw a simple block diagram of a normal fault, a thrust fault, and a strike-slip fault, and show the relative movement of the rock bodies along each. List the defining characteristics of each type of fault.
10. What are horsts and grabens? What global tectonic features are found where horsts and grabens most commonly are formed?
11. What global tectonic features are found where thrust faults commonly are formed?
12. What are the major features of an orogenic belt?
13. Make a perspective sketch of an anticline and the adjacent syncline, and label the following features: (a) hinge plane, (b) hinge, (c) angle of plunge, and (d) limbs.
14. Sketch the outcrop pattern of a plunging fold.
15. Describe the development of complex alpine-type folds.
16. What kinds of folds or other structures might form above a salt diapir?

8 Geologic Time

Some sciences deal with incredibly large numbers, others with great distances, still others with infinitesimally small particles. In every field of science, students must expand their conceptions of reality, a sometimes difficult, but very rewarding, adjustment to make. Geology students must expand their conceptions of the duration of time. Because life is short, we tend to think 20 years is a long time. A hundred years in most frames of reference is a very long time; yet, in studying Earth and the processes that operate on it, we must attempt to comprehend time spans of 1 million years, 100 million years, and even several billion years.

As a beginning student, you can see that rocks are records of time and begin to read this record. From the interrelationships of rocks, the events of Earth's history can be arranged in proper chronologic order. The part of Arizona's western Grand Canyon shown above is an excellent example. Several major geologic events are clearly expressed from the north rim of

the canyon. The oldest obvious event was the deposition of a sequence of horizontal marine sedimentary rocks into which the canyon is now cut. These rocks accumulated layer upon layer during much of the geologic time period known as the Paleozoic Era. The next event was the uplift of the rocks above sea level. Gradually, the great canyon and other details of the landscape formed by stream erosion as vast quantities of sediment were carried away and ultimately deposited in the ocean. With this brief introduction, you can begin to appreciate the fact that rocks and even landscapes are records of time. We will have much more to say about telling time in the Grand Canyon later in this chapter; remember that what is obvious here is only a fraction of the long history of the Grand Canyon region, which spans 1.8 billion (1,800,000,000) years.

How do scientists measure such long periods of time? Nature contains many types of time-measuring devices. Earth itself acts like a clock, rotating on its axis once every 24 hours. Fossils within a rock are a type of organic clock by which geologists can "tell time" and identify synchronous events in Earth's history. This is possible because life has evolved and changed over Earth's long history. Rocks also contain clocks that tick by systematic radioactive decay, which permit us to measure with remarkable accuracy the number of years that have passed since the minerals in a rock crystallized.

Major Concepts

1. The interpretation of past events in Earth's history is based on the principle that the laws of nature do not change with time.
2. Relative dating (determining the chronologic order of a sequence of events) is achieved by applying the principles of (a) superposition, (b) faunal succession, (c) crosscutting relations, (d) inclusions, and (e) succession in landscape development.
3. The standard geologic column was established from studies of rock sequences in Europe. It is now used worldwide. Rocks were originally correlated from different parts of the world largely based on the fossils they contain. Today, radiometric dating can be used to correlate major rock sequences.
4. Numerical time designates a specific duration of time in units of hours, days, or years. In geology, long periods of numeric time can be measured by radiometric dating.

The Discovery of Time

> Time is measured by change. Because rocks are themselves records of change, they mark the passage of geologic time. The interpretation of rocks as products and records of events in Earth's history is based on the principle of uniformitarianism, which states that the laws of nature do not change with time.

We are all aware of change in the physical and biological worlds. Were things unchanging and motionless, we would not be aware of time. Time is measured by change, and change occurs on many scales of time and space. In the natural world, countless clocks are ticking at different rhythms, measuring different time spans. We are acutely aware of the changes in our daily world: night to day, the seasons, and cycles of life from birth to death. However, there is a scale of deep time, spanning millions and even billions of years, measured by clocks that are less obvious to the human experience. It is the rhythm of Earth's dynamics: continents moving, mountains uplifting and eroding, volcanoes erupting, and seas expanding and contracting. It is also the clock of species evolving and disappearing into the oblivion of extinction. Our planet has a vast natural archive that reveals past episodes of many of these changes.

The great abyss of time—geologic time—was discovered in Edinburgh in the 1770s by a small group of scholars led by James Hutton (1726–1797). These men challenged the conventional thinking of their day, in which the age of Earth was accepted to be 6000 years, as established by Bishop James Ussher's (1581–1656) summation of biblical chronology. Hutton and his friends studied the rocks along the Scottish coast and observed that every formation, no matter how old, was the product of erosion from other rocks, older still. Their discovery showed that the roots of time were far deeper than anyone had supposed. It was perhaps the most significant discovery of the eighteenth century because it changed forever the way we look at Earth, the planets, the stars, and, consequently, the way we look at ourselves.

Uniformitarianism

The interpretation of rocks as products and records of events in Earth's history is based on a fundamental assumption of scientific inquiry: the principle of **uniformitarianism**, which states that *the laws of nature do not change with time*. We assume that the chemical and physical laws operating today have operated throughout all of time. The physical attraction between two bodies (gravity) acted in the past as it does today. Oxygen and hydrogen, which today combine under certain conditions to form water, did so in the past under those same conditions. Although scientific explanations have improved and changed over the centuries, natural laws and processes are constant and do not change. All chemical and physical actions and reactions occurring today are produced by the same causes that produced similar events 100 years or 5 million years ago.

Hutton's principle of uniformitarianism was radical for the time and slow to be accepted. In the late eighteenth century, before modern geology had developed, the Western world's prevailing view of Earth's origin and history was derived from the biblical account of creation. Earth was believed to have been created in 6 days and to be approximately 6000 years old. Creation in so short a time was thought to have involved forces of tremendous violence, surpassing anything experienced in nature. This type of creation theory is known as **catastrophism**. Foremost among its proponents was Baron Georges Cuvier (1769–1832), a noted French naturalist. Cuvier, an able student of fossils, concluded that each fossil species was unique to a given sequence of rocks. He cited this discovery in support of the theory that each fossil species resulted from a special creation and was subsequently destroyed by a catastrophic event.

This theory was generally supported by scholars until 1785, when Hutton challenged it. He saw evidence that Earth had evolved by gradual processes over an immense span of time, and he developed the concept that became known as the principle of uniformitarianism. According to Hutton, past geologic events could be explained by the natural processes operating today, such as erosion by running water, volcanism, and the gradual uplift of Earth's crust. Hutton assumed that these processes occurred in the distant past just as they occur now. He saw that, in the vast abyss of time, enormous work could be achieved by what seemed small and insignificant processes. Rivers could completely erase a mountain range. Volcanism and Earth movements could form new ones. Based on his observations of the rocks of Great Britain, he visualized "no vestige of a beginning—no prospect of an end." In a way, what Copernicus did for space, Hutton did for time. The universe does not revolve around Earth, and time is not measured by the life span of humans. Before Hutton, human history was all of history. Since Hutton, we know that we are but a tiny pinpoint on an extraordinarily long time line.

Sir Charles Lyell (1797–1875) based his *Principles of Geology* (1830–1833) on Hutton's uniformitarianism. Lyell's book established uniformitarianism as the method for interpreting the geologic and natural history of Earth. Charles Darwin (1809–1882) accepted Lyell's principles in formulating his theory of the origin of species and the descent of man. Modern science continues to make significant advances in understanding Earth, its long history, and how it was formed. As a result, the principle of uniformitarianism has been verified innumerable times.

The assumption of constancy of natural law is not unique to interpretations of geologic history; it is the logical essential in deciphering recorded history as well. We observe only the present and interpret past events on inferences based on present observations. We thus conclude that books or other records of history—such as fragments of pottery, cuneiform tablets, flint tools, temples, and pyramids—that were in existence before our arrival have all been the work of human beings, despite the fact that postulated past activities are outside the domain of any present-day observations. Having excluded supernaturalism, we draw these conclusions because humans are the only known agent capable of producing the effects observed. Similarly, in geology we conclude that ripple marks in sandstone in the Appalachian Mountains were in fact formed by streams, or that coral fossils found in limestone exposed in the high Rocky Mountains are indeed the skeletons of corals that lived in a now-nonexistent sea.

Many features of rocks serve as records, or documents, of past events in Earth's history; for those who listen, the rocks still echo the past. Igneous rocks are records

Why did scientists discard the idea of catastrophism as proposed by Cuvier?

of thermal events; the texture and composition of an igneous rock show whether volcanic eruptions occurred or if the magma cooled beneath the surface. Sedimentary rocks record changing environments on Earth's surface—the rise and fall of sea level, changes in climate, and changes in life forms. A layer of coal is a record of lush vegetation growth, commonly in a swamp. Limestone, composed of fossil shells, indicates deposition in a shallow sea. Salt is precipitated from seawater or from saline lakes only in an arid climate, so a layer of salt carries specific climatic connotations. The list of examples could go on and on. For more than two centuries, geologists have extracted from the rocks a remarkably consistent record of events in Earth's history: a record of time.

Unconformities

Geologic time is continuous; it has no gaps. In any sequence of rocks, however, many major discontinuities (unconformities) reveal significant interruptions in rock-forming processes.

James Hutton was a very perceptive observer who recognized the historical implications of the relationships between rock bodies. He not only recognized the vastness of time recorded in the rocks of Earth's crust, but he also recognized breaks, or gaps, in the record. In 1788, Hutton, together with Sir James Hall and John Playfair, visited Siccar Point in Berwickshire, Scotland, and saw for the first time the Old Red Sandstone resting upon the upturned edges of the older strata (**Figure 8.1**).

Figure 8.1 Angular unconformity at Siccar Point, southeastern Scotland. It was here that the historical significance of an unconformity was first realized by James Hutton in 1788. Note that the older "primary" rocks are nearly vertical and that the younger "secondary" strata were deposited on the eroded surface formed on the older rocks.

This exposure proved that the older rocks (primary strata) had been uplifted, deformed, and partly eroded away before the deposition of the "secondary strata." They soon discovered comparable relationships in other parts of Great Britain. This relationship between rock bodies became known as an **angular unconformity**.

To appreciate the significance of an angular unconformity, consider what the angular discordance, or difference, implies by studying the sequence of diagrams in **Figure 8.2**. At least four major events are involved in the development of an angular unconformity: (1) an initial period of sedimentation during which the older strata are deposited in a near-horizontal position, (2) a subsequent period of deformation during which the first sedimentary sequence is folded, (3) development of an erosional surface on the folded sequence of rock, and (4) a period of renewed sedimentation and the development of a younger sequence of sedimentary rocks on top of the old erosional surface.

At first, only the most obvious stratigraphic breaks were recognized as unconformities, but with more field observations, other, more subtle, discontinuities were recognized. In **Figure 8.3**, for example, foliated metamorphic and igneous rocks are overlain by flat-lying sedimentary strata. This relationship, in which plutonic igneous or metamorphic rocks are overlain by sedimentary rocks, is a **nonconformity**.

Why is a major unconformity implied where sandstone is deposited on top of granite?

(A) Sedimentation: A sequence of rocks is deposited over time.

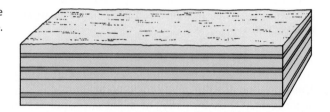

(B) Deformation: The sequence of rocks is deformed by mountain-building processes or by broad upwarps in Earth's crust, followed by erosion.

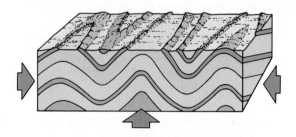

(C) Subsidence below sea level and renewed sedimentation.

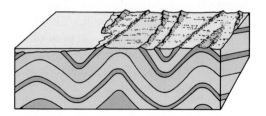

(D) A new sequence of rocks is deposited on the eroded surface of the older deformed rocks.

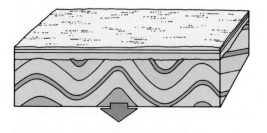

Figure 8.2 The geologic events implied from an angular unconformity represent a sequence of major events in the geologic processes operating within the area.

Figure 8.3 **A nonconformity** is an unconformity in which sedimentary rocks were deposited on the eroded surfaces of metamorphic or intrusive igneous rocks. The metamorphic rocks and the igneous dikes shown in this photograph of the inner gorge of the Grand Canyon were formed at great depths in the crust. Subsequent uplift and erosion exposed them at the surface. Younger sedimentary layers were then deposited on the eroded surface of the igneous and metamorphic terrain.

The deposition of sedimentary rocks on metamorphic or granitic rocks implies four major events: (1) the formation of an ancient sequence of rocks, (2) metamorphism or intrusion of granite, (3) uplift and erosion to remove the cover and expose the metamorphic rocks or granites at the surface, and (4) subsidence and deposition of younger sedimentary rocks on the eroded surface.

An example of another type of stratigraphic break is shown in **Figure 8.4**. Here the rock strata above and below the erosion surface are parallel. Erosion may strip off the top of the older sequence and may cut channels into the older beds, but there is no structural discordance between the older, eroded rock body and the younger, overlying rock. This type of discordance is called a **disconformity**.

An unconformity is best seen in a vertical section exposed in a canyon wall, road cut, or quarry, where it appears as an irregular line. An unconformity is not a line, however, but a buried erosional surface. The present surface of Earth is an example of what unconformable surfaces are like. Channels cut by streams are responsible for many irregularities, and resistant rocks protruding above the surrounding surface can cause local relief of several hundred meters. These buried erosion surfaces have been used in sequence stratigraphy to separate rock layers into packages that can be used to decipher past changes in sea level.

Relative Ages

Relative ages are determined by the chronologic order of a sequence of events. The most important methods of establishing relative ages are (1) superposition, (2) faunal succession, (3) crosscutting relations, and (4) inclusions.

An expanded conception of time is perhaps the main contribution of geology to the history of thought. Two different concepts of time, and hence two different but complementary methods of dating, are used in geology: relative time and numerical time. **Relative age** is simply determining the chronologic order of a sequence of events. Historians commonly use relative time when they place events in relation to ancient

Figure 8.4 Disconformities do not show angular discordance, but an erosion surface separates the two rock bodies. The channel in the central part of this exposure reveals that the lower shale units were deposited and then eroded before the upper units were deposited.

dynasties, reigns of kings and queens, or major events such as wars. Geologists do the same with geologic time. Earth's history is divided into eons, eras, periods, and epochs by markers such as the existence of certain fossils (for example, the "age of dinosaurs") or major physical events (for example, Appalachian mountain building), although the actual dates of the events may be unknown.

Relative dating implies that no quantitative or numerical length of time in days or years is deduced. An event can only be interpreted to have occurred earlier or later than another. This can be done by applying several principles of remarkable simplicity and universality.

The Principle of Superposition

The principle of **superposition** is the most basic guide in the relative dating of rock bodies. It states that in a sequence of undeformed sedimentary rock, the oldest beds are on the bottom and the higher layers are successively younger. The relative ages of rocks in a sequence of sedimentary beds can thus be determined from the order in which they were deposited.

In applying the principle of superposition, we make two assumptions: (1) Layers were essentially horizontal when they were deposited, and (2) the rocks have not been so severely deformed that the beds are overturned. (Many sedimentary rock sequences that have been overturned are generally easy to recognize by their sedimentary structures, such as cross-bedding, ripple marks, and mud cracks.)

Why is the principle of superposition so important in geologic interpretations?

The Principle of Faunal Succession

Fossils are the actual remains of ancient organisms, such as bones and shells, or the evidence of their presence, such as trails and tracks. Their abundance and diversity are truly amazing. Some rocks (such as coal, chalk, and certain limestones) are composed almost entirely of fossils, and others contain literally millions of specimens. Invertebrate marine forms are most common, but even large vertebrate fossils of mammals and reptiles are plentiful in many formations. For example, it is estimated that more than 50,000 fossil mammoths have been discovered in Siberia, and many more remain buried.

The principle of **faunal succession** states that groups of fossil animals and plants occur in the geologic record in a definite chronologic order. Consequently, a period of geologic time can be recognized by its characteristic fossils. Thus, in addition to superposition, a sequence of sedimentary rocks has another independent element that can be used to establish the chronologic order of events.

How can fossils be used to determine the relative age of a rock layer?

Even before Darwin developed the theory of evolution by natural selection, the principle of faunal succession was recognized by William Smith (1769–1839), a British surveyor and early geologist. Smith worked throughout much of southern England and carefully studied the fresh exposures of rocks in quarries, road cuts, and excavations. In a succession of interbedded sandstone and shale formations, he noted that several shales were very much alike, but the fossils they contained were not. Each shale layer had its own particular group (or assemblage) of fossils. By correlating types of fossils with rock layers, Smith developed a practical tool that enabled him to predict the location and properties of rocks beneath the surface. Soon after Smith announced that the fossil assemblages of England change systematically from the older beds to the younger, other investigators discovered the same to be true throughout the world.

By tracing the changing character of fossils in progressively younger rocks, you may appreciate the significance of these discoveries. The oldest rocks contain only traces of soft-bodied organisms. Progressively younger and younger sedimentary rocks contain marine invertebrates with shells, followed by the appearance of simple marine vertebrates like fish. Amphibians appear in even younger strata, followed in successively younger rocks by the first appearance of reptiles, birds, and mammals.

Today, the principle of faunal succession has been confirmed beyond doubt. It has been used extensively to find valuable natural resources, such as petroleum and mineral deposits. It is also the foundation for the standard geologic column, which divides geologic time into progressively shorter subdivisions called eons, eras, periods, epochs, and ages (see Figure 8.8).

The Principle of Crosscutting Relations

The relative age of certain events is also shown by the principle of **crosscutting relations**, which states that igneous intrusions and faults are younger than the rocks they cut (**Figure 8.5**). Crosscutting relations can be complex, however, and careful observation may be required to establish the correct sequence of events. The scale of crosscutting features is highly variable, ranging from large faults, with displacements of hundreds of kilometers, to small fractures, less than a millimeter long.

The Principle of Inclusion

The principle of **inclusion** states that a fragment of a rock incorporated or included in another is older than the host rock. The relative age of intrusive igneous rocks (with respect to the surrounding rock) is therefore commonly apparent if inclusions, or fragments, of surrounding rocks are included in the intrusion (**Figure 8.6**). As magma moves upward through the crust, it dislodges and engulfs large fragments of the surrounding material, which remain as unmelted foreign inclusions.

The principle of inclusion can also be applied to conglomerates in which large pebbles and boulders eroded from preexisting rocks have been transported and deposited in a new formation. The conglomerate is obviously younger than the formations from which the pebbles and cobbles were derived. In areas where superposition or

(A) Several generations of igneous dikes cut across the green metamorphic rock. The thick dike is the youngest because it cuts across all the other rock bodies. The green metamorphic rock is older than all of the dikes. Small fractures are younger than the dikes.

(B) This glacial moraine in the Sierra Nevada, California, is cut by a fault expressed as a low linear cliff (in shadow from left to right). The fault scarp is obviously younger than the moraine it cuts, indicating a tectonic event after deposition.

Figure 8.5 Crosscutting relations indicate the relative ages of rock bodies and geologic structures.

Figure 8.6 Inclusions of one rock in another provide a means of determining relative age. In this example fragments of granite are included in the basalt, clearly indicating that the granite is the older.

Figure 8.7 Ancient lava flows near St. George, Utah, now exist as long linear ridges with flat tops. When the lava was extruded, it flowed down stream valleys and displaced the stream to the margins of the flow. Subsequent erosion was then concentrated along the lava flow margins and developed a lava-capped ridge at the site of the original stream channel. These ridges are called inverted valleys. This flow is obviously older than other flows in the region that occupy the present drainage channels.

other methods do not show relative ages, a limit to the age of a conglomerate can be determined from the rock formation represented in its pebbles and cobbles.

Succession in Landscape Development

Surface features of Earth's crust are continually modified by erosion and commonly show the effects of successive events through time. Many landforms evolve through a series of stages, so a feature's relative age can be determined from the degree of erosion. This is especially obvious in volcanic features such as cinder cones and lava flows. These features are created during a period of volcanic activity and then subjected to the forces of erosion until they are completely destroyed or buried by erosional debris (**Figure 8.7**).

The Standard Geologic Column

Using the principles of relative dating, geologists have unraveled the chronologic sequence of rocks throughout broad regions of every continent and have constructed a standard geologic time scale that serves as a calendar for the history of Earth.

How are rock units on different continents correlated with one another?

Age **correlations** using fossils have made it possible to construct a diagram, called the **geologic column**, to show, in simple form, the major units of strata in Earth's crust (**Figure 8.8**). Most of the original geologic column was pieced together from strata studied in Europe during the mid-1800s. Major units of rock (such as the Cambrian, Ordovician, and Silurian) generally were named after geographic areas in Europe where they are well exposed. The rock units are distinguished from each other by major changes in rock type, unconformities, or different fossil assemblages. The original order of the units in the geologic column was based on the sequence of rock

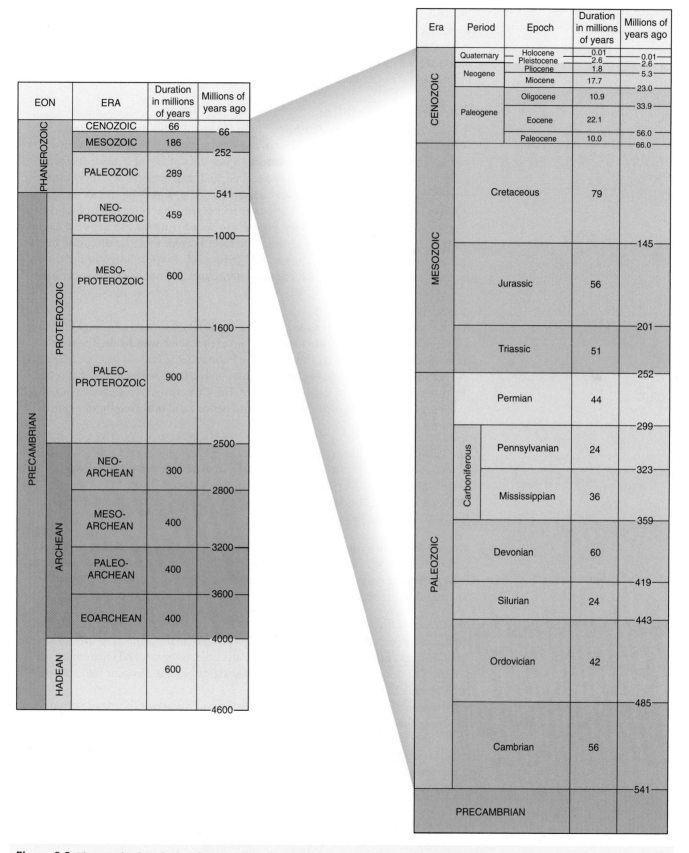

Figure 8.8 The standard geologic column was developed in Europe in the 1800s, based on the principles of superposition and faunal succession. Later, radiometric dates on igneous rocks provided a numerical scale for absolute ages of the geologic periods.

formations in their superposed order as they are found in Europe. In other areas of the world, rocks that contain the same fossil assemblages as a given part of the European succession are considered to be of the same age. The column is made of a hierarchy of relative time subdivisions of Earth's history. The longest units are called eons, and progressively shorter subdivisions are eras, periods, and epochs.

The Precambrian Eon

Precambrian time is represented by a group of ancient rocks that form a large volume of the continental crust. Most of these are igneous and intensely deformed metamorphic rocks. To produce these rocks, great thicknesses of sedimentary and volcanic rocks were strongly folded and faulted, and then intruded with granitic rock. Precambrian rocks contain only a very few fossils of the more primitive forms of life. Without common fossils, age correlations of rock layers from continent to continent are difficult, if not impossible. Major progress in deciphering the Precambrian has been made using radiometric ages for igneous and metamorphic rocks (as discussed in the next section). The **Hadean**, **Archean**, and **Proterozoic Eons**, subdivisions of the Precambrian, are defined based on these radiometric ages and not on fossil successions.

The Phanerozoic Eon

Rocks younger than the Precambrian are generally less deformed and contain many fossils, permitting geologists to identify and correlate them worldwide. The span of time since the end of the Precambrian is the **Phanerozoic Eon**—a term meaning "visible life" in Greek. The passage of time in this eon is chronicled by fossils that reveal the continual evolution of Earth's biosphere. The Phanerozoic is subdivided into three major eras, which are in turn subdivided into periods and into even shorter epochs (Figure 8.8).

The Paleozoic Era   The term Paleozoic means "ancient life." **Paleozoic** rocks contain numerous fossils of invertebrate (no backbone) marine organisms and later vertebrates, such as primitive fish and amphibians. The era is subdivided into periods (Figure 8.8) distinguished largely according to the sequence of sedimentary rock formations of Great Britain.

The Mesozoic Era   Mesozoic means "middle life." The term is used for a period of geologic time in which fossil reptiles, including dinosaurs, are present. Fossil invertebrates that were more like modern lifeforms dominate marine rocks. The **Mesozoic** Era includes three periods—the Triassic, Jurassic, and Cretaceous (Figure 8.8).

The Cenozoic Era   Cenozoic means "recent life." Fossils in these rocks include many types closely related to modern forms, including mammals, plants, and invertebrates. The **Cenozoic** Era has three periods—the Paleogene, Neogene, and Quaternary (Figure 8.8). Some geologists combine the Paleogene and the Neogene into a single time period called the Tertiary.

The geologic column by itself shows only the relative ages of the major periods in Earth's history. It tells us nothing about the length of time represented by a period. With the discovery of the radioactive decay of uranium and other elements, new tools for measuring geologic time were invented. They greatly enhanced our understanding of time and of the history of Earth and provided numerical benchmarks for the standard geologic column.

Radiometric Measurements of Time

Radiometric dating provides a method for directly measuring geologic time in terms of a specific number of years (numeric age). It has been used extensively during the last 50 years to provide a numerical time scale for the events in Earth's history.

Unlike relative time, which specifies only a chronologic sequence of events, **numerical age** is measured in hours, days, and years. In other words, numeric dating specifies quantitative (or absolute) lengths of time. Numeric time can be measured using any regularly recurring event, such as the swing of a pendulum or the rotation of Earth. You are very familiar with the concept of numeric dating in that it is the way in which you record your own age. It is one matter to reveal that you are older than someone else (relative age) and something quite different to say that you are a specific number of years old (numeric age). Using appropriate numerical dating techniques, geologists can estimate the time span between two geologic events.

Before Hutton and Lyell, few people even thought about the age of Earth. After Hutton presented his arguments for uniformitarianism, and Lyell further developed the concept, much interest was generated in the magnitude of geologic time, and scientists explored several ways to estimate Earth's age. Early attempts included estimates based on how long it would take for the ocean to become salty (about 100 million years) or how long it would take to accumulate the known thicknesses of fossil-bearing sedimentary strata (about 500 million years). By the end of the 1800s, many geologists had come to accept an age for the Earth of about 100 million years. This was based on Lord Kelvin's estimate of how long it would take the planet to cool from an initially molten state.

In short, before the 1900s there was no reliable method for measuring long periods of geologic time, and there seemed to be little hope of finding the secret of Earth's age. Each method showed that Earth was far older than many had supposed based on a reading of the Bible (6000 y), but the true dimensions of time remained elusive. Then, a major breakthrough occurred when Henri Becquerel (1852–1908), a French physicist, discovered natural **radioactivity** in 1896 and opened new vistas in many fields of science. Among the first to experiment with radioactive substances was the distinguished British physicist Lord Rutherford (1871–1937). After defining the structure of the atom, Rutherford was the first to suggest that radioactive decay could be used to calculate a numerical age for geologic events (Figure 8.8). The discovery of radioactive decay solved two problems at once. First it provided the means to calculate ages for igneous rocks. Second, the heat continually produced by radioactive decay explained why Earth was still hot billions of years after its formation.

Have scientists always thought the Earth was billions of years old?

Radioactive Decay

Many atoms spontaneously change into other kinds of atoms through radioactive decay. You will remember from our discussion of the nature of atoms that the number of protons in the nucleus is the determining characteristic of an element. If the number of protons in the nucleus changes, the atom becomes a different element with its own set of distinctive physical and chemical properties. This happens in minerals when potassium (K) decays to form the noble gas argon (Ar). A loss of a proton in the nucleus of potassium changes it into argon.

Of course, not all isotopes experience radioactive decay. To visualize the relationship between those isotopes that are radioactive and those that are not, refer to **Figure 8.9**. This chart shows the number of protons and the number of neutrons in the nuclei of most of the known isotopes. Remember, each isotope has a distinctive atomic weight—the sum of its protons and neutrons. Thus, one element can have many different isotopes, each with a different number of neutrons and different atomic weight, but the same number of protons. From this diagram, you can see that all of the stable isotopes lie in a diagonal band through the center of the diagram. Radioactive isotopes lie on either side. For example, you can see that potassium (K) has two stable isotopes (^{39}K and ^{41}K), one long-lived radioactive isotope (^{40}K) that is found in rocks and minerals, and several short-lived isotopes that do not occur naturally on Earth.

All decay reactions move an unstable isotope on a path that ultimately ends in a stable nuclear configuration in the central part of the diagram. There are several types of radioactive decay (**Figure 8.10**), but all involve changes in the number of protons

Why do some isotopes spontaneously decay to form other elements?

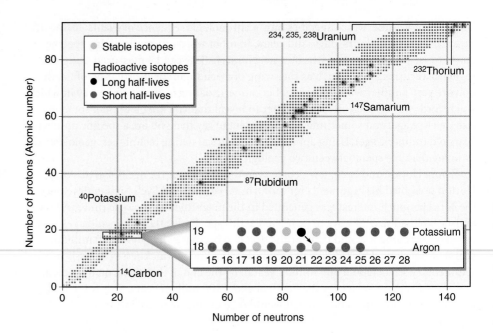

Figure 8.9 The chart of the nuclides shows the number of protons and neutrons in many of the almost 300 known isotopes. Most isotopes are radioactive and spontaneously decay to stable daughter isotopes that lie in the central band on the diagram. The elements that are commonly used in radiometric dating are labeled. The inset shows the decay of ^{40}K to ^{40}Ar. Some radioactive isotopes have such long half-lives that they can be found in rocks; others have half-lives of only fractions of a second and are not found naturally on Earth. Not all of these short-lived isotopes are shown.

and neutrons in the nucleus of an unstable atom. The atom that decays is the **parent isotope** and the product is the **daughter isotope**. Many decay reactions are composed of several separate steps in a long decay chain involving as many as a dozen intermediate isotopes.

Radiogenic heat is also produced by radioactive decay reactions (along with other particles not listed in Figure 8.10). In fact, radioactive decay of potassium, uranium, and thorium is one of Earth's principal sources of heat. Radiogenic heating explains why Earth is still hot after all these years of cooling.

Rates of Radioactive Decay

Each radioactive isotope disintegrates at its own distinctive rate. The rate of decay has been accurately and repeatedly measured for many radioactive isotopes. Unlike chemical reactions that only involve electrons, nuclear reactions are not affected by temperature or pressure, at least under the conditions found in the outer layers of Earth and other planets. Experimental measurements all show that decay follows a simple rule, under which half of the parent atoms present in a closed system decay to form daughter atoms in a fixed period of time known as a **half-life**. In other words, half the original atoms decay during one half-life. In a second half-life, half the remainder (or a quarter of the original atoms) decay. In a third half-life, half the remaining quarter decay, and so on (**Figure 8.11**).

Regular decay is the critical rhythm used in measuring radiometric dates. The time elapsed since the formation of a crystal containing a radioactive element can be calculated from the rate at which that particular element decays. For a simple decay reaction, the amount of the parent isotope remaining in the crystal can be compared with the amount of the daughter isotope.

Note that the fraction of atoms that decay during each half-life is always the same; one-half of the atoms change into another isotope. This means that the number of atoms that decay in a given time continually diminishes. Thus, the rate of decay actually declines over time. This declining rate is important in understanding radiometric dating, but it is also crucial for understanding why the amount of heat released by decay dramatically declines with the passage of time. Consequently, all of the planets have cooled with time.

The half-life for each unstable isotope is different. Most isotopes decay rapidly; that is, they have short half-lives and lose their radioactivity within a few days or years. Many of these short-lived isotopes may be formed naturally—in stellar interiors, for example. But they have such short half-lives compared with the age of the solar system that they have long since decayed away and are not found on Earth. However, other isotopes decay very slowly, with half-lives of hundreds of millions of years. These are the ones that can be used as atomic clocks for measuring long periods of time. The parent isotopes and their daughter products that are most useful for geologic dating are listed in **Table 8.1**.

Radiometric Dating

The use of radioactive decay to measure the passage of time numerically has dramatically improved our understanding of Earth and its evolution. **Radiometric dating** of rocks and minerals uses a variety of elements that decay in different ways (Table 8.1). Commonly, the same rock can be dated by several different techniques, giving geologists important cross checks on their interpretations and lending a firm foundation to our measurements of geologic time. Let us review several important systems for measuring radiometric dates.

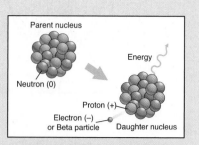

(A) Beta emission (Example: rubidim to strontium)

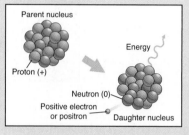

(B) Positron emission (Example: nitrogen to carbon)

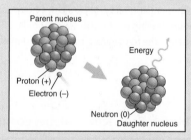

(C) Electron capture (Example: potassium to argon)

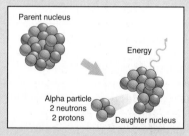

(D) Alpha emission (Example: uranium to thorium)

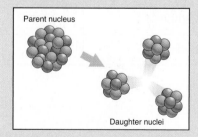

(E) Spontaneous fission (Example: uranium to various elements)

Figure 8.10 Several different modes of radioactive decay are important geologic clocks and important sources of heat.

Do all radioactive isotopes decay at the same rate?

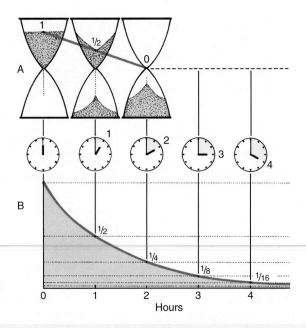

Figure 8.11 Rates of radioactive growth and decay are exponential. (A) Many processes are characterized by uniform, straight-line depletion, like sand moving through an hourglass. If half of the sand is gone in 1 hour, all of it will be gone in 2 hours. (B) Radioactive decay, in contrast, is exponential. If half is depleted in 1 hour, half of the remainder, or one-fourth, will be depleted in 2 hours, leaving one-fourth. Rates of radioactive decay are expressed in half-lives; the time required for half of the remaining amount to be depleted. In this case, the half-life is 1 hour.

Table 8.1 Radioactive Isotopes Commonly Used in Radiometric Dating

Parent Isotope	Daughter Isotope	Half-life (years)
Uranium-238	Lead-206	4.47 billion
Uranium-235	Lead-207	704 million
Thorium-232	Lead-208	14.0 billion
Samarium-147	Neodymium-143	106 billion
Rubidium-87	Strontium-87	49.6 billion
Potassium-40	Argon-40	1.28 billion
Carbon-14	Nitrogen-14	5730
Hydrogen-3	Helium-3	12.3

STATE OF THE ART   Single Grain Ages of Earth's Oldest Rocks

Many numerical dating techniques are based on the radioactive decay of one isotope into an isotope of a completely different element. One of the radioactive elements with a long half-life is uranium (Table 8.1). Uranium is found in fairly high concentrations in several minerals, including the mineral zircon ($ZrSiO_4$), where uranium can substitute for zirconium because of its similar size and charge. Zircon is found in very small quantities in many igneous and metamorphic rocks. Because zircon equilibrates very slowly with its environment, it is difficult to reset the radiometric clock. The parent uranium isotopes and daughter lead isotopes are held tightly in the mineral's atomic structure. Geologists have found that zircon grains are windows into the geologic past.

A new type of sophisticated instrument is used to open these windows. Called a SHRIMP (for Sensitive High Resolution Ion MicroProbe), these expensive tools allow scientists to get numerical dates for extremely small spots only a fraction

of a millimeter across. Thus, many dates can be obtained from a single grain (as shown in the photo). In contrast, many other dating techniques require the use of tens to hundreds of grains to get a single date.

The essence of this method is to bombard a polished grain surface with a beam of energetic ions (usually argon). These ions drill a small crater into the grain's surface and eject ionized atoms from the grain. These ions are swept into a mass spectrometer where the ratio of the parent isotope and the daughter isotope can be measured. For zircon, this means measuring the amount of uranium and its radiogenic daughter lead. By using this ratio and the known decay rate of uranium, a crystallization age for this small spot on the mineral can be calculated. After moving the crystal a small distance, another microhole is drilled and another age is calculated. The result is a map that shows the crystallization history of a single tiny grain of zircon (ages given in millions of years).

This technique shows that some gneisses have zircon grains that first crystallized in a magma, were then metamorphosed, and later still incorporated into a different hot magma without ever completely dissolving or even reaching equilibrium with the new molten environment. Each event is recorded by the crystallization of a rim (colored bands) of new zircon around the old core. In the photo you can see that the rim of the grain is about 80 million years younger than the core. Even more exciting results have come from dating zircons from the ancient rocks of the shields. The oldest rocks found on Earth so far are gneisses that crystallized about 4.0 billion years ago in what is now northern Canada. Even older grains of zircon, to as much as 4.4 billion years old, have been found in weakly metamorphosed quartzite from central Australia.

Courtesy of I.S. Williams, Australian National University

3976 4031 4031 4040 3924 4002

100 µm

Courtesy of I.S. Williams, Australian National University

Potassium-Argon (K-Ar) Clock   The potassium–argon method of radiometric dating is one of the most widely applied techniques. Because potassium is incorporated in common minerals, such as feldspars, micas, and amphiboles, this method is widely used to date igneous and metamorphic rocks. Although the half-life of ^{40}K is about 1.25 billion years (Table 8.1), rocks as young as a few hundred thousand years have been successfully dated. Fortunately, the potassium–argon clock is one of the easiest to understand. ^{40}Ar, the decay product of ^{40}K, is a noble gas and does not form chemical bonds with other ions in mineral structures. Thus, when a ^{40}K-bearing mineral crystallizes from hot magma, it contains very little or no ^{40}Ar. Once the rock cools, ^{40}Ar steadily accumulates as ^{40}K decays by electron capture. As long as the mineral is cool, less than a few hundred degrees Celsius, the ^{40}Ar will remain trapped within the mineral's crystalline structure, just as the grains of sand accumulate in the bottom of an hourglass. If this igneous rock is collected and the ^{40}K and ^{40}Ar contents are precisely measured, the crystallization age of the mineral can be calculated because the half-life of ^{40}K is accurately known.

What processes can reset the K–Ar clock to zero age?

It is important to keep in mind that a radiometric age is the time of crystallization of the mineral that is dated. These techniques tell us nothing about the time at which the elements themselves formed; most elements formed in the interiors of ancient stars billions of years before Earth was created. Moreover, this technique works best when used to date mineral grains that crystallized over very short periods of time and at the same time that the rock formed. Consequently, potassium–argon dating is best used to date metamorphic and igneous rocks, particularly quickly cooled volcanic rocks. It is nearly useless in dating clastic sedimentary rocks because argon may have been accumulating in the individual mineral grains for a very long time before the grains were transported and deposited to form a sedimentary rock.

To understand other potential complications in radiometric dating, imagine that a ^{40}K-bearing mineral grain leaks a little of the volatile ^{40}Ar already produced by radioactive decay. If we collect and analyze this grain, its calculated age will be lower than the real crystallization age of the mineral. In the same way, an underestimate of the passage of time would occur if we used an hourglass that had an undetected crack through which sand grains in the lower chamber escaped from the once-closed system. Leakage of ^{40}Ar can occur in real minerals if, for example, they are heated during metamorphic processes or if they are weathered by reactions with air and water at Earth's surface. Radiometric dating only works if the mineral has remained a closed system, with no loss or gain of the parent or daughter isotopes. Therefore, it is extremely important for the geologist to collect fresh, unaltered rocks for geochronologic work.

On the other hand, the fact that the potassium–argon clock can be partially or completely reset allows these techniques to be used to study the thermal histories of rocks, not just the time when they crystallized. A simple example is the metamorphism of an old granite to form gneiss. Here, all of the ^{40}Ar may be expelled from the igneous minerals as they become hot, effectively resetting the clock completely. Such a gneiss would then yield a K–Ar age that corresponds to the age of metamorphism. With increasingly sophisticated techniques, we are now able to look at the timing of even more subtle temperature changes, such as those that occur when sedimentary rocks are buried or when mountain belts are uplifted.

Carbon-14 Clock   Another important radiometric clock uses the decay of **carbon-14** (^{14}**C**), which has a half-life of 5730 years (**Figure 8.12**). If this isotope has such a short half-life, why is there any left at all? Carbon-14 is produced continually in Earth's upper atmosphere by cosmic rays—high-energy nuclear particles that arise from the Sun and elsewhere in space. Cosmic rays fragment the nuclei of gases in the atmosphere to produce free particles, including protons and neutrons. If one of these neutrons hits ^{14}N, with seven protons, the nucleus gives up a proton and unstable ^{14}C, with only six protons, is formed. The formation of ^{14}C is in balance with its decay, so the proportion of ^{14}C in atmospheric carbon dioxide (CO_2) is essentially constant.

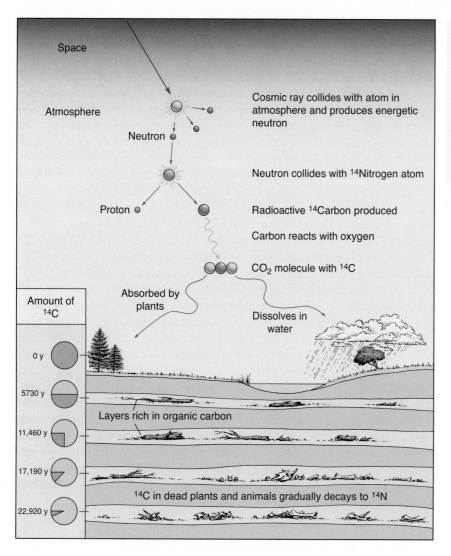

Figure 8.12 **The ^{14}C radiometric clock is** based on the production of ^{14}C from nitrogen in the atmosphere, the incorporation of the radioactive carbon into living plants and animals, and the regular decay of the ^{14}C after organisms die and stop absorbing carbon from the atmosphere. The short half-life of ^{14}C allows this system to be used to date events from a few hundred to as much as 40,000 years old.

The newly formed radioactive carbon becomes mixed with ordinary carbon atoms in atmospheric carbon dioxide. Plants use this radioactive carbon dioxide in photosynthesis, and animals also incorporate radioactive carbon by eating the plants. Both plants and animals thus maintain a fixed proportion of ^{14}C while they are alive. After death, however, no additional ^{14}C can replenish what is lost by radioactive decay. The ^{14}C steadily reverts to ^{14}N by beta decay. The time elapsed since an organism died can therefore be determined by measuring how much ^{14}C remains. The longer the time elapsed since death, the less ^{14}C. Because the isotope's half-life is 5730 years, the amount of ^{14}C remaining in organic matter older than 50,000 years is too small to be measured accurately. This method is therefore useful for dating very young geologic events involving organic matter and for dating archeological material.

There are several complications in using the ^{14}C system for age determinations. An important difficulty lies in the fact that the production rate of ^{14}C in the atmosphere has not been constant. Changes in the rate of cosmic-ray bombardment and in the strength of Earth's magnetic field, which can deflect cosmic rays, are probably the most important reasons why ^{14}C production changes. However, this error amounts to only 1% to 4%, when compared with ages obtained by other numeric dating techniques, such as tree ring dating. In addition, the short half-life limits ^{14}C dating to very young materials. Another possible drawback comes from carbon's mobility at Earth's surface and the fact that it may be added to or subtracted from an organism's remains after death.

Why can't we use carbon-14 to date old metamorphic rocks like those in the shields?

Problems in Radiometric Dating

Can all rocks be accurately dated using radiometric techniques?

The principles of radiometric dating may be simple enough, but the laboratory procedures are complex. Most radioactive elements are found in only trace quantities in rocks and minerals. They do not form their own minerals but are impurities in common rock-forming minerals. For example, uranium and thorium have concentrations of only a few parts per million in most rocks. Consequently, the principal difficulty lies in the precise measurement of minute amounts of parent and daughter isotopes. Modern instruments are capable of accurate measurements at the parts per billion level. The accuracy of a date also depends on the accuracy with which the half-life of the radioactive element is known. For example, measurements of the half-life for the decay of ^{235}uranium to ^{207}lead are considered accurate within 1%. Other errors are the result of the movement of parent and daughter isotopes after the radiometric clock has been started, as described previously for the K–Ar system. Isotopic systems can become open, or leaky, as a result of weathering and alteration, or by heating, burial, and uplift. By understanding these problems, one can minimize their effects by carefully collecting samples and using precise instruments in the laboratory.

Many rocks cannot be dated by radiometric methods. Some rocks are too young to have accumulated sufficient daughter products for analysis. Other rocks are too old and all of the parent isotopes have decayed away. Some rocks lack sufficient quantities of radioactive elements in the first place. For example, quartz does not accommodate radioactive elements in its crystal structure and is rarely useful for radiometric dating.

What kinds of rocks are most useful for radiometric dating?

To safeguard against these problems, geologists constantly check and recheck radiometric ages. Several radioactive isotopes are suitable for numerical dating, so one obvious test is to determine the age of a mineral or a rock by more than one method. If the results agree, the probability is high that the age is reliable. If the results differ significantly, additional methods must be used to decide which age, if either, is correct. Another independent check can be made by comparing the radiometric age with the relative age of the rocks, determined from such evidence as superposition and fossils. Through this system of tests and cross-checks, many very reliable radiometric ages have been determined.

Other Methods to Measure Numerical Time

Some natural processes produce records that can be used to calculate numerical ages. Some of the most important are tree rings, varves, and ice layers. Most of these methods can be used to understand the last few tens of thousands of years of Earth's history.

Tree Rings

The annual growth rings of trees found in temperate climates provide a simple way to measure the age of a tree, and is called **tree ring dating** (**Figure 8.13**). The thickness and texture of each ring are records of the environment—temperature, humidity, precipitation, insect infestations, and even fires will create distinctive patterns recorded in all of the trees in one area. Consequently, by comparing the pattern of rings from one tree to another tree, or even to ones that have died, the chronology of a forested region can be accurately deduced. By overlapping sections of rings from many different trees, scientists have been able to link together unbroken tree ring records that extend as far back as 8000 years in some localities.

Why can't we count tree rings to find out when dinosaurs lived on Earth?

Because the rings are so sensitive to the climate, tree ring dating is an ideal tool to date and explain many of Earth's most recent climate changes. Moreover, tree rings have been used to examine slumping and mass movement on hillsides and also disturbances caused by earthquakes. Pollutants from power plants, smelters, nuclear testing, and waste disposal sites are transported in wind and water and become incorporated in an annual ring. Consequently, both the extent and timing of these environmental perturbations are recorded as the tree grows.

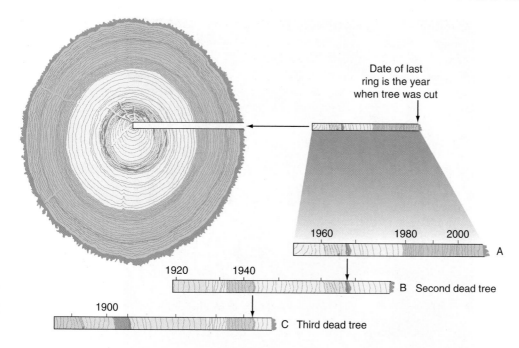

Figure 8.13 Tree rings can be used to determine the numerical ages of some very young geologic events. The patterns of annual growth rings can be matched from living to dead trees to form a record extending back several thousand years in some locations.

Varves

In some geologic settings, rhythmic changes in a sedimentary environment create a type of clock based on successions of thin sediment beds. Some of these rhythms correspond to the annual cycle of the seasons. For example, thin layers of clay, known as **varves**, accumulate in the still waters of some glacial lakes. Each graded layer represents one year and was created by seasonal variations in the amount of sediment-laden meltwater that flowed into a lake. By carefully counting these layers, geologists have pieced together a record that extends back about 20,000 years in the glaciated region around the Baltic Sea of northern Europe. Some varves can also be dated by ^{14}C techniques if they contain sufficient organic material.

In many other areas, thick sequences of laminated sedimentary rocks exist, but without knowledge of the numerical age of the first or last bed, we can only decipher the total length of time that passed during their deposition, not the numerical age of each layer. One famous section from the shales of the Green River Formation of Wyoming is built of layers 2 mm thick that accumulated into a sequence almost 1 km thick. If each bed represents a single year, then the shales were deposited over a period of about 5 million years, a result in accord with what is known about the radiometric age of the formation.

Ice Layers

A distinctive type of rhythmic sediment accumulates in Earth's high mountains and in its polar regions—ice sheets (**Figure 8.14**). Seasonal changes in the amount of snowfall and subsequent snowmelt create thin layers in glacial ice sheets. Drill cores obtained from both the Greenland and Antarctic ice sheets show that these layers can be counted to obtain a nearly continuous record almost 65,000 years long. Ice older than this is present in the ice sheets, but the annual layers have been destroyed by flow-age of the ice. Some of these cores are 3000 m long. Independent confirmation of the age of the ice layers can be obtained from radiometric dates of thin beds of volcanic ash preserved in the ice. Volcanoes in Iceland, Mount St. Helens in the United States, and even some volcanoes as far away as Mexico have erupted ash that is preserved in the Greenland ice cap.

What segment of geologic time can be dated using ice layers?

Figure 8.14 Annual layers of ice form in many glaciers. Many layers can be seen in this South American glacier. Layers of wind-blown dust deposited in the summer help define the layers. In some deep drill cores, these layers can be counted back for tens of thousands of years.

Courtesy of L. Thompson

These ice layers are rich storehouses of climatic information. Besides the information in their thickness, the isotopic and chemical compositions of the ice tell when changes in the environment occurred. Contaminants in the ice reveal the widespread changes that accompanied the industrial revolution, including the introduction of lead and copper from smelting operations and the buildup of carbon dioxide trapped in bubbles. Moreover, by carefully examining the oxygen isotopic composition of the ice, scientists can calculate the temperature of the polar regions.

Calibration of the Geologic Time Scale

Numerical ages of many geologic events have been determined from thousands of specimens throughout the world. These, in combination with the standard geologic column, provide a radiometric time scale from which the numerical age of other rock units and geologic events can be estimated.

Unfortunately, radiometric dates cannot be used to determine the age of every rock. Many sedimentary rocks do not contain minerals suitable for accurately dating the time of deposition. In most cases, a radiometric date of a sedimentary rock indicates when the mineral grains within the rock crystallized—not the time when the sediment was deposited. Radiometric dating can be used effectively to determine when igneous rocks crystallized or when new minerals formed in metamorphic rocks because, in these cases, the mineral and the rock formed together. The problem in developing a reliable numeric time scale is the placement of the radiometric dates of igneous and metamorphic rocks in their proper positions in the relative time scale established by relative dating of sedimentary rocks.

Layered Volcanic Rocks

The best reference points for the radiometric time scale are probably volcanic ash falls and lava flows. Both are deposited instantaneously, as far as geologic time is concerned. Because they commonly are interbedded with fossiliferous sediments, their exact positions in the geologic column can be determined.

For example, thin beds of volcanic ash are common in some richly fossiliferous shales and limestones of the western interior of the United States. Here, several different fossil species permit us to determine precisely the position of thin units in which the fossils occur within the geologic column. Comparisons with fossils found on other continents are also possible, so that their Late Cretaceous age is well established. The sequence of interbedded volcanic ash layers can be dated very precisely by radiometric means. Therefore, a numerical age for each successive fossil zone can be established. In this case, individual fossil zones have radiometric ages known within about 1% and provide tight constraints on the numerical ages of several important parts of the geologic time scale.

In other formations, sedimentary rocks lack fossils that are diagnostic of a certain period of geologic time. For example, the Morrison Formation of the western United States is famous for its abundant and distinctive dinosaur fauna, including stegosaurs, allosaurs, and apatosaurs, but it has few fossils that identify its exact position in the standard geologic column. Estimates of its age ranged from Jurassic to Cretaceous. However, volcanic ash beds are common in this unit, and careful radiometric dating has established that the Morrison Formation and its dinosaur remains range from 155 to 148 million years old. They are thus Late Jurassic in age (Figure 8.8).

How are igneous rocks used to calibrate the relative time scale?

Bracketed Intrusions

As shown in **Figure 8.15**, molten rock can cool within Earth's crust without ever breaking out to the surface. Subsequent erosion may expose this rock at the surface. Later, younger sediments may be deposited on top. In some cases, the entire sequence of events takes only a few million years. In others, a much longer time may pass. The relative age of the igneous rock is bracketed between the ages of the older sediments (1) and the younger sediments (4). Such a rock body is therefore known as a bracketed intrusion. Unfortunately, the span of time between 1 and 4 is commonly too long to permit the relative age of the intrusion to be useful in detailed geochronology. Radiometric dating of such rocks does establish the time of major igneous events, however.

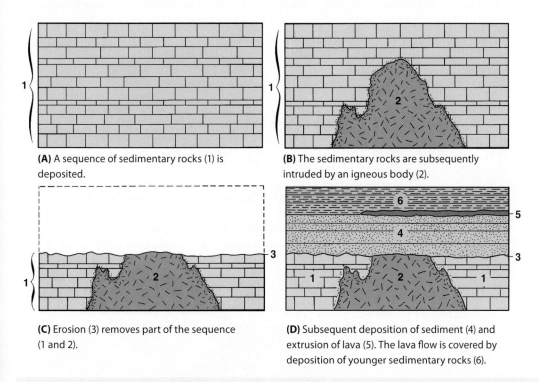

(A) A sequence of sedimentary rocks (1) is deposited.

(B) The sedimentary rocks are subsequently intruded by an igneous body (2).

(C) Erosion (3) removes part of the sequence (1 and 2).

(D) Subsequent deposition of sediment (4) and extrusion of lava (5). The lava flow is covered by deposition of younger sedimentary rocks (6).

Figure 8.15 Radiometric dating of igneous rocks can be used in developing a numerical time scale, but the relationship between rock bodies must be considered. In these diagrams, major geologic events are numbered. Radiometric dating of the intrusion (2) indicates that the granite cooled 355 million years ago and that the lava flow (5) cooled 288 million years ago. These two dates place certain constraints on the absolute age of all units. (1) is older than 355 million years; (3) and (4) formed between 355 and 288 million years ago; and (6) is younger than 288 million years.

GeoLogic   The Story of the Grand Canyon Revealed in Stone

Paleozoic rocks

Vishnu Schist

The Grand Canyon of Arizona exposes rocks that are as much as 1.7 billion years old. Relative and radiometric dates were used to reconstruct its geologic history. The major events that formed the rocks and landscapes of the Grand Canyon are shown in the block

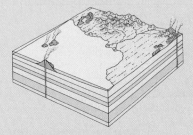

(A) Deposition of Vishnu sediment and associated island arc volcanic rocks (1.84 to 1.66 billion years ago).

(B) Mountain building and metamorphism form Vishnu Schist, including episodes of granite intrusion as the arcs accrete to growing North America (1.7 to 1.4 billion years ago).

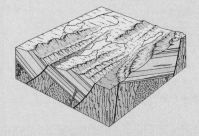

(E) Normal faulting and tilting as the ancient continent splits apart by rifting (800 to 700 million years ago).

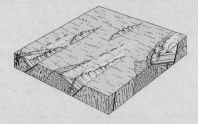

(F) Erosion and formation of Great Unconformity (700 to 540 million years ago).

Paleozoic rocks

Great Unconformity

Grand Canyon Supergroup

diagrams below. This sequence of geologic events was deciphered using superposition and crosscutting relations. The absolute ages of many of the events have been established by radiometric dating. The major rock units are labeled on the photograph.

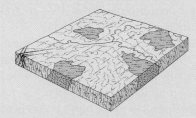

(C) Uplift and erosion remove the upper 20 km of rock (1.4 to 1.2 billion years ago).

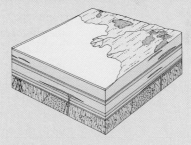

(D) Deposition of Unkar Group and capping basaltic lavas (parts of the Grand Canyon Supergroup) (1.2 to 1.0 billion years ago).

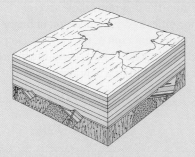

(G) Deposition of Paleozoic (blue), Mesozoic (green) and Cenozoic (yellow) rocks (550 to 20 million years ago).

(H) Uplift and erosion of Grand Canyon by Colorado River (20 million years to present).

The Geologic Time Scale

The currently accepted **geologic time scale** is based on the standard geologic column, established by faunal succession and superposition, plus the numerical radiometric dates of rocks that can be placed precisely in the column. Each dating system provides a cross-check on the other because one is based on relative time and the other on numerical time. Agreement between the two systems is remarkable and discrepancies are few. In a sense, the radiometric dates act as the scale on a ruler, providing reference markers between which interpolation can be made. Enough dates have been established so that the time span of each geologic period can be estimated with considerable confidence. The age of a rock can be determined by finding its location in the geologic column and interpolating between the nearest radiometric time marks.

From this radiometric time scale, we can make several general conclusions about the history of Earth and geologic time:

1. Present evidence indicates that Earth formed about 4.5 to 4.6 billion years ago. The oldest rocks on Earth are a little more than 4 billion years old.
2. The Precambrian constitutes more than 80% of geologic time.
3. Phanerozoic time began about 540 million years ago. Rocks deposited since Precambrian time can be correlated worldwide by means of fossils, and the dates of many important events during their formation can be determined from radiometric dating.
4. Some major events in Earth's history are difficult to place in their relative positions on the geologic column but can be dated by radiometric methods.

Deciphering Geologic History

Using the principles of relative and numerical dating, geologists can deduce the details of the geologic history of a region.

One of the best areas in the world to visualize the sequence of events recorded in the geologic record is the Grand Canyon of the Colorado River in Arizona (see the GeoLogic essay). Here, many rock formations are completely exposed, and the major events they record are exceptionally clear. Three major groups of rocks are shown in the photograph and are separated by major unconformities. They are (1) the Vishnu Schist and the granitic rocks that intrude it, which are exposed in the deep, rugged inner gorge; (2) the Grand Canyon Supergroup, a succession of sedimentary rocks nearly 4000 m thick, all tilted about 15°; and (3) the Paleozoic sedimentary rocks, which are 1500 m thick and are essentially horizontal. They form a series of alternating steep cliffs and gentle slopes extending from the top of the Grand Canyon Supergroup up to the canyon rim. Above the Paleozoic rocks to the north is another thick series of Mesozoic and Cenozoic strata (not visible in the photograph), which forms the Grand Staircase in southern Utah.

To interpret the events recorded by these rocks, we need only think a moment about how each major rock formation formed and what is implied by the relationship of one rock body to another. The major events are shown graphically in the series of block diagrams in the GeoLogic feature. The diagrams show several kinds of crosscutting relationships, as well as unconformities and superposition of major rock bodies. Study each figure carefully. Refer to the photograph of the canyon. Do you recognize each rock body? Do you understand the reasons for each event illustrated in the cross sections? If not, try again, and your reward will be a new awareness of time.

Magnitude of Geologic Time

The magnitude of geologic time is easier to comprehend when compared to some tangible linear time line.

Geologists speak of billions of years as casually as politicians speak of billions of dollars. A billion dollars is difficult to comprehend; indeed, it does not seem like real money. Great time spans are also difficult for most people to comprehend and do not seem to be real time. Today, we accept that our planet is 4.6 billion years old, but it is not easy to deal with this astonishing number. Yet, we should not accept it without trying to understand what it means. Without an expanded conception of time, extremely slow geologic processes, considered only in terms of human experience, have little meaning.

To appreciate the magnitude of geologic time, we will abandon numbers for the moment and refer instead to something tangible and familiar. In **Figure 8.16**, the length of a football field represents the lapse of time from the beginning of Earth's history to the present. A numeric time scale and the standard geologic periods are shown on the left. Precambrian time constitutes the greatest portion of Earth's history (87 yd). The Paleozoic and later periods are equivalent to only the last 13 yd. To show events with which most people are familiar, the upper end of the scale must be enlarged. The first abundant fossils occur at the 13-yd line. The great coal swamps are at about the 6-yd line. The dinosaurs became extinct about a yard from the goal line at the beginning of the Cenozoic, and the last ice age occurred an inch from the goal line. Recorded history corresponds to less than the width of a blade of grass.

What are some ways to envision the magnitude of geologic time?

> **It has not been easy for man to face time. Some, in recoiling from the fearsome prospects of time's abyss, have toppled backward into the abyss of ignorance.**
>
> **(Albritton, *The Abyss of Time*, 1980)**

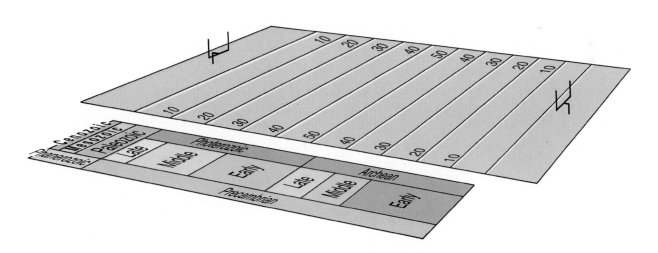

Figure 8.16 If the length of geologic time is compared to a football field, Precambrian time represents the first 87 yd, and all events since the beginning of the Paleozoic are compressed into the last 13 yd. Dinosaurs first appeared 5 yd from the goal line. The glacial epoch occurred in the last inch, and historic time is so short that it cannot be represented on this figure.

Key Terms

angular unconformity (p. 205)
Archean (p. 212)
carbon-14 (p. 218)
catastrophism (p. 203)
Cenozoic (p. 212)
correlation (p. 210)
crosscutting relations (p. 208)
daughter isotope (p. 214)

disconformity (p. 206)
faunal succession (p. 208)
geologic column (p. 210)
geologic time scale (p. 226)
Hadean (p. 212)
half-life (p. 214)
inclusion (p. 208)
Mesozoic (p. 212)

nonconformity (p. 205)
numerical age (p. 213)
Paleozoic (p. 212)
parent isotope (p. 214)
Phanerozoic (p. 212)
Precambrian (p. 212)
Proterozoic (p. 212)
radioactivity (p. 213)

radiometric dating (p. 215)
relative age (p. 206)
superposition (p. 207)
tree ring dating (p. 220)
uniformitarianism (p. 202)
varve (p. 221)

Review Questions

1. Explain the modern concept of uniformitarianism.
2. Explain the concept of relative dating.
3. Explain how the following principles are used in determining the relative age of rock bodies: (a) superposition, (b) faunal succession, (c) crosscutting relations, and (d) inclusions.
4. Discuss the sequence of events illustrated in Figure 8.2.
5. What is the standard geologic column? How did it originate?
6. Explain the meaning of half-life in radioactive decay.
7. Examine Figure 8.10 and give the simple nuclear change for each type of decay shown.
8. How is the numerical age of a rock determined?
9. Give examples of the important limitations on radiometric dating. In what situation could a rock be too old to obtain an accurate age? Too young? In what situations might there be insufficient radioactive elements?
10. What kinds of rocks—igneous, sedimentary, or metamorphic—are best for radiometric dating? Why?
11. Can you think of a simple test of the accuracy of radiometric dating using several minerals found in one volcanic rock?
12. Why are most rocks given relative ages based on their position in the standard geologic column rather than assigned a definite numerical age, even though accurate methods of radiometric dating are well established?
13. How old are the oldest rocks on Earth? How old is Earth thought to be?

Part II

Earth's Hydrologic System

Earth's hydrologic system is composed of all the paths through which water moves around and through the outer layers of the planet. In this part of the book, we explore not only the complex paths that water follows on the continents, but we will also consider the nature and dynamics of the atmosphere and the ocean. The sources of energy that drive these systems are the same–a combination of heat energy from the Sun and gravitational energy from the Earth. Solar heat, created in a vast nuclear furnace suspended in space, causes evaporation of seawater, makes some air masses more buoyant than adjacent ones, and speeds up chemical reactions–among a host of other effects. The relentless effect of gravity causes stream water, hillsides, glaciers, and air masses to move in a ceaseless effort to reach equilibrium. The surface features of the Earth, which supply us with so much scenic beauty, are all sculpted by the hydrologic system.

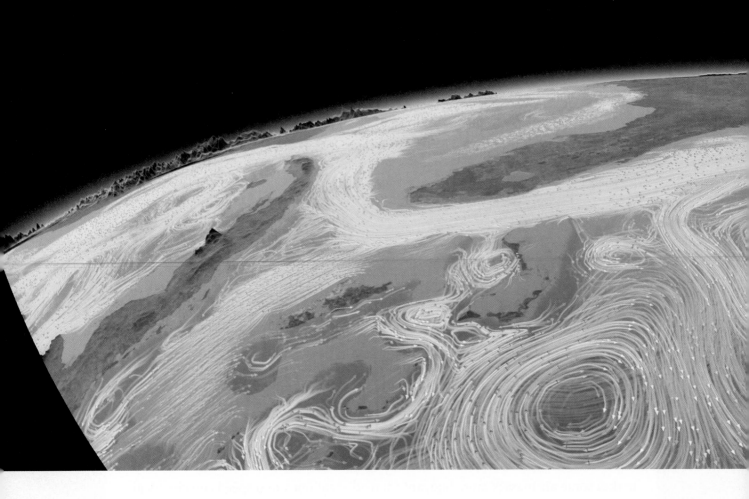

9 The Climate System: Earth's Atmosphere and Oceans

Earth's surface fluids form an enormous interconnected system of moving air and water that creates and controls the hydrologic system and the entire planet's climate. Every day we find our activities affected by the ceaseless movement of the atmosphere and oceans, as well as its constantly changing temperature and water vapor content.

We schedule our daily activities by the weather and on a longer term we look forward to the march of the seasons. They repeatedly bring alternating warmth and cold and drive the annual cycle of life. On an even longer scale, however, we may be completely unaware of variations in our overall climate. But, despite our ignorance, climate change over hundreds or thousands of years is just as relentless as any daily change in weather. Detecting a change in the climate is a daunting task, given the large daily, seasonal, and even annual variations in temperature and precipitation

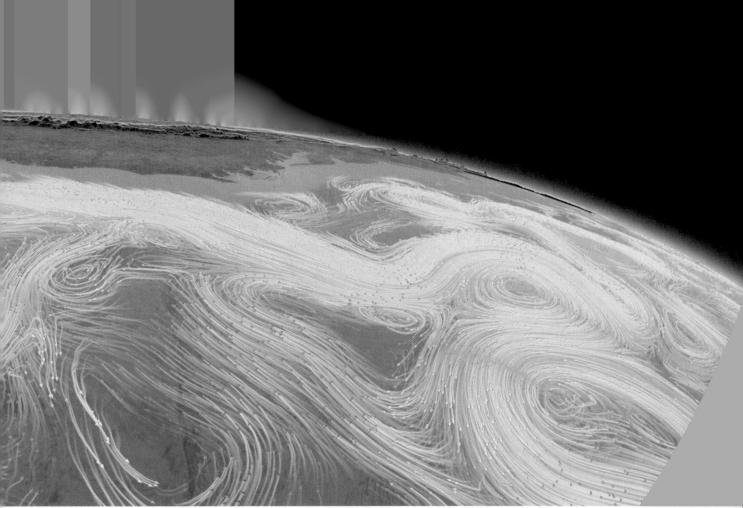

Courtesy of Goddard Space Flight Center/NASA

and the lack of good weather records before the 1800s. Moreover, natural events such as volcanic eruptions can cause short-term climate change.

Nonetheless, the record of climate change is extremely clear on the face of Planet Earth. Its sediments, rocks, and even its landscapes record the sometimes dramatic—but often slow and ponderous—changes in climate.

Climate significantly affects almost all geologic systems, because climate controls the operation of the hydrologic system. One of the most fundamental elements of Earth's climate system is the slow circulation of its great ocean currents. The Gulf Stream, shown above by a series of arrows showing flow direction, is one of the most important. Winds and continents shape this great current that carries warm water from the Gulf of Mexico and Caribbean Sea past the eastern shores of North America and then into the North Atlantic. The current moves as a series of interconnected swirls of warm water. Climate also controls Earth's tremendous river systems. The wind and endless crashing waves, the massive ice cap on Antarctica, the vast deserts, and even the soil in which we grow our food, all owe their existence to climatic controls. Consequently, climate profoundly affects all forms of life. In turn, we humans now can affect climate by altering Earth's surface and by introducing pollutants, such as carbon dioxide, into the atmosphere.

In this chapter, we focus on the two most important components of Earth's climate system: the oceans and atmosphere. To understand this system, you must have a basic understanding of the origin, composition, structure, and flow patterns of air and seawater. We close the chapter with a few notes on how humans affect the climate.

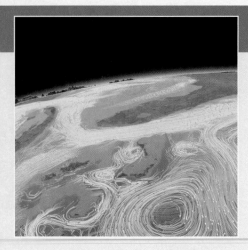

Major Concepts

1. Earth's climate system is driven by solar heat and the interactions of the oceans, the atmosphere, and their circulation patterns.
2. The atmosphere is the envelope of gases that surrounds Earth. It consists mostly of nitrogen and oxygen. Latitudinal variations in humidity and temperature are caused by the uneven distribution of solar radiation, and therefore heat, on Earth's surface. These differences drive a vast convective circulation system in the atmosphere.
3. Earth's ocean consists of liquid water, capped at the poles with sea ice. A strong vertical temperature gradient in ocean waters creates a thin, warm surface layer and a thick mass of cold deep water. The most important dissolved constituents in seawater are salt ($NaCl$) and calcium carbonate ($CaCO_3$).
4. A global circulation pattern involving surface and deep waters mixes the entire ocean. Circulation of the oceans is driven by the wind, by seawater density differences (caused by variations in salinity and temperature), and by coastal upwelling.
5. Global climate change can be caused by changes in solar radiation intensity, by volcanism, by the development of new mountain belts, by changes in the composition of the atmosphere (especially its carbon dioxide content), and to some extent by the tectonic position of the continents.
6. Concerns about global warming are based on increases in atmospheric carbon dioxide caused by the burning of fossil fuels.

Composition and Structure of the Atmosphere

> Among the planets of the Solar System, Earth's atmosphere is unique because it is rich in nitrogen and oxygen. Temperature variations divide it into several layers; the most important layer for geologic development of the surface is the troposphere.

Seen from space, the brilliant white swirling clouds of Earth's atmosphere are perhaps its most conspicuous feature. Although this tenuous envelope of gas is an insignificantly small fraction of the planet's mass (less than 0.01%), it is tremendously important because it is an extremely dynamic, open system. The atmosphere moves easily and rapidly and reacts chemically with surface materials. As a result, the atmosphere plays a part in the evolution of most features of the landscape.

What makes Earth's atmosphere unique?

The atmosphere transports heat energy from the tropics to the polar regions and moderates the far greater temperature extremes that would otherwise exist. Water from the oceans is evaporated and carried over the continents by wind. Above the land, it may precipitate to form rivers, glaciers, and systems of groundwater. Over vast desert areas, the flow of the wind drives the movement of sand. Winds also supply the energy that drives ocean surface currents, which transport heat, salt, and nutrients in addition to water. The wind drives the waves that modify our shorelines. Chemical reaction of the atmosphere with minerals drives weathering processes and creates soils and ore deposits. Consequently, an understanding of the composition and flow of the atmosphere is fundamental to understanding many of Earth's geologic systems.

Composition of the Atmosphere

Earth's atmosphere consists mainly of only a few gases. Just three constituents—oxygen, nitrogen, and argon—make up 99.9% of the atmosphere (**Table 9.1**). Earth's atmosphere is unique compared with the atmospheres of the eight other planets in our solar system (**Figure 9.1**). Ours is the only atmosphere with large amounts of oxygen. Oxygen is extremely important because it is necessary for most forms of animal life. Moreover, atmospheric oxygen is constantly reacting with minerals that originally

Table 9.1 Composition of Earth's Atmosphere

Component	Chemical Formula	Concentration (volume)
Nitrogen	N_2	78.0%
Oxygen	O_2	21.0%
Argon	Ar	0.9%
Carbon dioxide	CO_2	353 ppm
Neon	Ne	18 ppm
Helium	He	5 ppm
Methane	CH_4	2 ppm
Krypton	Kr	1 ppm
Water vapor	H_2O	About 1%

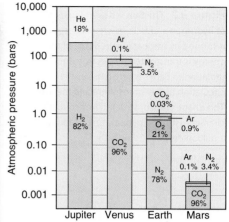

Figure 9.1 **Earth's atmosphere** is dramatically different in its composition and pressure from those of other planets. It is dominated by nitrogen and oxygen, whereas the atmospheres of other inner planets (Mars and Venus) are mostly carbon dioxide. The much larger outer planets, such as Jupiter, have thick atmospheres of hydrogen and helium.

formed deep inside Earth to make new minerals in equilibrium with the atmosphere. These reactions are one of the reasons for the continuing change at Earth's surface.

The major gases, along with the inert gases (helium, neon, and krypton), are all found in nearly constant proportions. But other constituents, most notably water, have concentrations that vary from place to place and from time to time. Water vapor (**humidity**) can vary from a bone-dry 0.01% to an extremely humid 3% of the atmosphere. Short-term variations in water content are hallmarks of an active hydrologic system. Another constituent—carbon dioxide—varies on a much longer time scale. Today it has an abundance of about 0.03%. Over the last few decades, carbon dioxide has been increasing in abundance because it is produced by the burning of fossil fuels.

The major atmospheric gases are nearly transparent to incoming solar radiation, and they do little to affect the heat balance at Earth's surface. However, some minor gases absorb certain wavelengths of light and thus help to heat the atmosphere. In fact, without these absorbent gases, Earth's surface temperature would be 30°C lower than it is today. Earth would be a frozen wasteland. The most important gases for the absorption of solar energy make up less than 1% of the atmosphere. In order of importance, they are water vapor, carbon dioxide, ozone, methane, and various nitrous oxides. Because the amount of carbon dioxide is increasing, this variation carries important implications for Earth's future climates.

Thermal Structure of the Atmosphere

The most widely recognized climate variable is temperature. Almost all of the heat in the atmosphere and oceans originates from nuclear fusion in the distant Sun (**Figure 9.2**). This energy is transmitted 150 million kilometers to Earth by radiation and heats our planet's surface. The global average temperature of the air just above the surface is 15°C (59°F). However, the range of surface temperatures across the globe is rather wide (–90°C to 58°C). Temperature also changes vertically above Earth's surface (**Figure 9.3**) and divides the atmosphere into several layers.

The Troposphere In the lowermost layer of the atmosphere, the temperature decreases with altitude at a rate of about 6.5°C per kilometer. In other words, if you climb 1000 m up the side of a mountain, it will be 6.5°C cooler than it was at the bottom. In reality, the rate of temperature change varies from place to place and even from season to season.

We live in this lowermost layer, called the **troposphere**. It is marked by turbulent movement of the air (wind) and wide variations in humidity and temperature. It is the zone where all the phenomena related to weather occur. From a geologic viewpoint,

What causes the atmosphere to divide into more or less distinct layers?

Figure 9.2 The Sun is a seething mass of hydrogen and helium where energy is formed by nuclear fusion. Some of this energy is transmitted by electromagnetic radiation to Earth, where it drives the circulation of the atmosphere and the ocean. This image was constructed from radiation characteristic of a temperature of about 1 million degrees Celsius.

Courtesy of the TRACE Project, Lockheed Martin Solar and Astrophysics Laboratory, and NASA

Figure 9.3 The main layers of Earth's atmosphere are defined according to their temperature gradients (left). Earth's weather systems develop in the troposphere. This profile shows mean temperatures at 15° N. The atmospheric pressure decreases regularly with height (right).

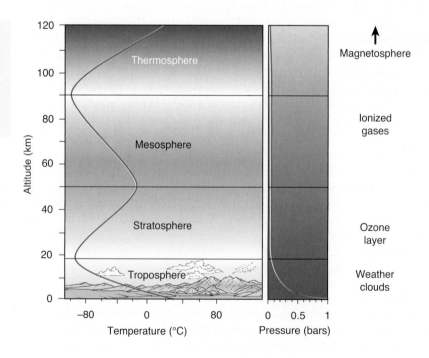

the most significant part of the atmosphere is the troposphere. This layer, or zone, contains about 80% of the atmosphere's mass and practically all of its water vapor and clouds. It is the zone in which evaporation, condensation, and precipitation occur, in which storm systems develop, and in which decay of solid surface rock takes place.

The Stratosphere   In the layer above the troposphere, the temperature change reverses and temperatures increase with altitude (Figure 9.3). This layer is known as the **stratosphere**. Temperatures increase to nearly the same level as at the surface, apparently because solar energy is absorbed by molecules of **ozone** (O_3) that are concentrated in the stratosphere. Because of this reversal of the temperature gradient, the lower two layers of the atmosphere do not readily mix. Consequently, gases and particulates move through the troposphere rapidly but only slowly across the boundary into the stratosphere. Once in the stratosphere, small particles may remain suspended for a very long time. No rain occurs in the stratosphere to remove particles that reach this height. Moreover, the stratosphere is quiet, and turbulence is uncommon. Because the density of the air in the stratosphere decreases with height, the stratosphere does not mix readily and it is stratified, or layered.

Although only a minor constituent in the atmosphere, ozone gas plays several important roles. Most of the ozone is concentrated in the stratosphere, where it forms the **ozone layer**. In addition to warming the stratosphere, ozone in the upper atmosphere also absorbs harmful ultraviolet (UV) radiation from the Sun. Thus, it forms a radiation shield for many kinds of plants and animals, including humans. Excessive ultraviolet radiation has been linked to skin cancer.

Unlike the other gases in the atmosphere, ozone is not released from the planet's interior, nor is it created by plants on the surface. Instead, stratospheric ozone is produced when sunlight breaks the bonds in an O_2 molecule to form atomic oxygen (O), which then reacts with another O_2 molecule to form O_3. Ozone concentrations reach a maximum of about 10 ppm (parts per million) at altitudes of 20 to 25 km. That is a very low concentration for any gas, but it is sufficient to create the "ozone shield."

Lower in the atmosphere, ozone plays a completely different role and is created by entirely different mechanisms. Ozone in the troposphere, a pungent gas, is strongly reactive and oxidizing. As a result, it is corrosive and constitutes a health hazard near the surface, where it is an important pollutant.

The Upper Layers of the Atmosphere   Above the stratosphere, the temperature decreases again in what is called the **mesosphere**, or "middle sphere" (coincidentally, the same term applied to the middle of Earth's interior). At an altitude of about 90 km, the temperature change reverses again to form the **thermosphere**. Here the temperature increases as ultraviolet energy from the Sun is absorbed by the molecules in the atmosphere (Figure 9.3). In fact, the gases become so warm that some molecular bonds are broken and charged ions of oxygen and nitrogen form.

Beyond the atmosphere is the **magnetosphere**, a tear-drop shaped zone where electrically charged particles streaming from the Sun are trapped by Earth's magnetic field. This part of the atmosphere is another powerful shield against damaging radiation that comes from outer space.

Atmospheric Pressure

The air about us seems so tenuous as to be almost weightless. In fact, the density of air at sea level is only about 1/800 the density of liquid water. Nonetheless, these air molecules exert a pressure that is just over 1 bar at sea level (1 bar = 1 kg/cm^2 or 14.7 $lb/in.^2$) at sea level. Perhaps you have seen what happens to a sealed aluminum can when the air inside is pumped out; the can collapses, crushed under the weight of the overlying air. Our own bodies are the same way; were it not for internal fluid pressures, we would be crushed flat.

Atmospheric pressure is greatest at sea level and drops rapidly with increasing altitude (Figure 9.3). At an elevation of about 5.6 km, atmospheric pressure is about 0.5 bars, and half of all the gas molecules in the atmosphere lie below this level. The

Which layer of the atmosphere has the greatest turbulent motion?

atmospheric pressure is cut in half again for each additional increase of 5.6 km in altitude. At an elevation of 8.8 km, the height of Mount Everest, the air is so "thin" that it is difficult for a human to get enough oxygen in each breath to survive. At 80 km above the surface, the pressure is only about 10 to 6 bars. Nonetheless, at 500 km above the surface, there is still a trace of an atmosphere, although it is extremely tenuous.

Atmospheric Water Vapor

What part of the atmosphere contains the greatest amount of water vapor?

One of the most important variables in the atmosphere is its water vapor content (**Figure 9.4**). Water is removed from the ocean by evaporation and rapidly carried aloft as water vapor in the turbulent troposphere. The amount of water that passes through the atmosphere each day is staggering. Water vapor is also important because it has a warming influence on the atmosphere—the so-called **greenhouse effect**. Moreover, the water vapor that condenses to form clouds controls the amount of solar energy that is reflected away from Earth, for clouds are highly reflective.

It is important to realize that cold air can hold much less water vapor than warm air. As a result, the percentage of water vapor in the atmosphere at the poles is almost 10 times less than at the equator. Likewise, the water vapor high in the atmosphere is much less than near the surface.

Precipitation occurs when the air becomes oversaturated with water vapor. Oversaturation occurs when vapor is no longer the stable form of water but must be joined by liquid as well. The vapor condenses to form small droplets of liquid water (or ice), which fall to the surface. Precipitation is generally caused by cooling of the air. Commonly air masses cool as they rise into the colder upper troposphere. Ascending air may be caused by winds that force air over high mountains or by the buoyancy of plumes of warm air. The distribution of precipitation on Earth's surface is shown in **Figure 9.5**. Most precipitation occurs along the equator. The least precipitation falls in the deserts north and south of the equator and also in the polar zones.

Figure 9.4 Water vapor is a minor but extremely important part of Earth's atmosphere. This global view of the water vapor (white) was constructed by images taken in infrared light and reveals the relative concentrations of water vapor and its transport paths in the atmosphere. You can see that water is almost everywhere in the atmosphere, but that high concentrations occur in swirling storm systems.

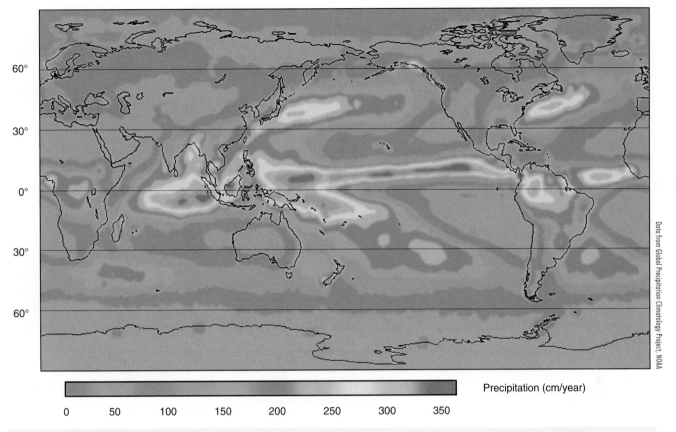

Figure 9.5 Precipitation is greatest near the equator, where warm, moisture-laden air rises, then cools at high altitude, and becomes supersaturated with water that falls as rain. This phenomenon causes the tropical rain forests. Dry regions lie in subtropical belts north and south of the equator, because here dry air descends, becomes heated, and can then absorb more water vapor. Such conditions cause evaporation to predominate over precipitation and a desert climate to exist.

Energy and Motion of the Atmosphere

Motion of the atmosphere is driven by the uneven distribution of solar energy. Solar heating is greatest in equatorial regions and causes water in the oceans to evaporate and the moist air to rise. The warm, humid air forms a belt of equatorial clouds and heavy rainfall. It is bordered in the middle latitudes by high-pressure zones that are cloud-free and contain dry descending air. Temperate and polar zones are marked by separately convecting masses of air.

Solar Radiation and Heat Balance

Much of the Sun's radiation that reaches Earth is reflected immediately back into space. Fully 30% of the energy is reflected from bright areas such as clouds, oceans, ice fields, and snow-covered plains. The solar energy that is not reflected is absorbed by the atmosphere and by the surface. As the surface warms, it radiates heat back toward space and warms the lower atmosphere in the process.

It has long been known that the amount of solar radiation absorbed by Earth decreases with the distance from the equator. This variation causes the pronounced temperature difference between the warm equatorial regions and the frigid polar regions. Measurements of the average surface temperature clearly show this global pattern (**Figure 9.6**).

The strikingly systematic temperature pattern north and south of the equator is the result of several factors. First, because Earth is a sphere, the angle at which the Sun's rays hit its surface varies from nearly vertical at the equator to nearly horizontal at the poles (**Figure 9.7**). Consequently, much less energy is received per square kilometer at the poles because the same amount of incoming radiation is spread over a larger area because of the angle. The same energy is concentrated in a much smaller area at the equator.

What is the fundamental control of climate?

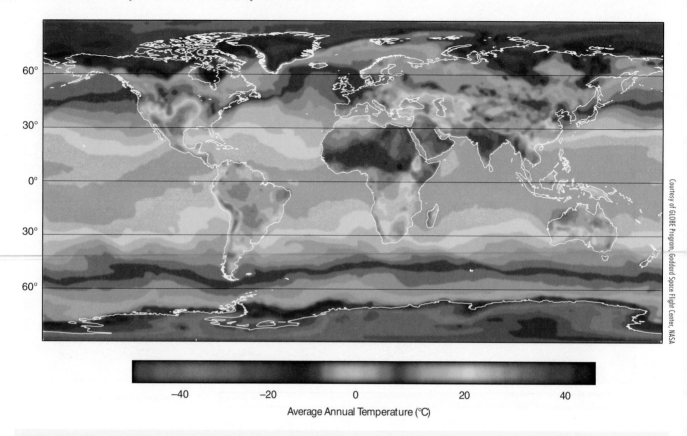

Courtesy of GLOBE Program, Goddard Space Flight Center, NASA

Average Annual Temperature (°C)

Figure 9.6 The mean surface temperature for June as obtained from National Oceanic and Atmospheric Administration weather satellites. Temperature increases from purple to blue to yellow to red.

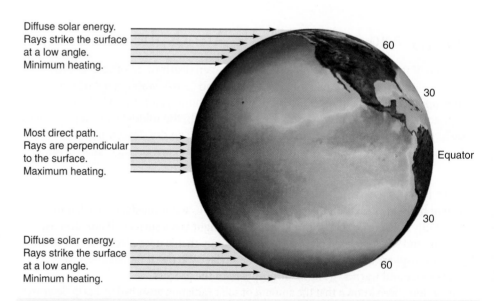

Diffuse solar energy. Rays strike the surface at a low angle. Minimum heating.

Most direct path. Rays are perpendicular to the surface. Maximum heating.

Diffuse solar energy. Rays strike the surface at a low angle. Minimum heating.

Figure 9.7 The Sun's energy is unevenly distributed across the surface of the nearly spherical Earth. The amount of solar energy per unit area varies with the angle at which the Sun's rays strike Earth's surface. At low latitudes, near the equator, the Sun's rays are nearly perpendicular and much more heat is received per unit area. At higher latitudes, where the angle is smaller, the same amount of energy is spread over a larger elliptical area and warms the surface less. Moreover, sunlight must travel through a much greater thickness of atmosphere near the poles than at the equator. This greater thickness also diminishes the amount of heat that reaches the surface.

In addition, in the polar regions, the low-angle Sun's rays travel through a much greater thickness of atmosphere, where more absorption and reflection occur. The result is a reduction in the energy received at the poles.

Still another critical factor affecting the heat distribution at Earth's surface is the length of the day. Because Earth's spin axis is tilted at an angle of 23.5° with respect to the plane of its orbit, the length of the day varies with the seasons. During the winter, the days are shorter because the spin axis is tilted away from the Sun. In the extreme polar regions, no sunlight falls on the surface for weeks. Thus, little solar heating occurs. During the summer at the poles, the "midnight Sun" does not set because the spin axis is now leaning toward the Sun. Nonetheless, the sunlight falls on the surface at such a low angle that little heating occurs.

In contrast at the equator, the Sun's rays strike the surface at a high angle, sunlight passes through less atmosphere, and the hours of sunlight change much less with the seasons. Consequently, the equatorial regions are efficiently heated.

Global Circulation of the Atmosphere

Systems not in equilibrium tend to change in a direction to reach equilibrium. The global circulation of the atmosphere is an attempt to reach equilibrium by equalizing the temperature differences between the poles and equator. The resulting flow pattern of the atmosphere is critically important for Earth's climates. This movement is the wind; it helps drive the circulation of the oceans as well as the atmosphere and, thus, the hydrologic system as a whole.

You have felt the wind blow, but you might not have noticed its systematic nature. Even from space, seeing that the atmosphere is in constant motion is easy; the circulation patterns are dramatically revealed by the shape and orientation of the clouds and the distribution of water vapor (Figure 9.4). At first glance, these circulation patterns may appear confused, but upon close examination, we find that they are well organized. If we smooth out the details of local weather systems, the global atmospheric circulation becomes apparent. For example, there is considerable symmetry in the flow patterns of the Northern and Southern Hemispheres.

The general circulation of the troposphere is depicted in **Figure 9.8**. If the circulation were due solely to solar heating, hot air would rise at the equator and flow toward the poles. As this air cooled, it would sink at the poles and then return to the equator by flowing across the surface. All surface winds would simply flow straight from the poles to the equator.

However, winds on the spinning Earth are deflected by the planet's rotation. This **Coriolis effect** is an illustration of Newton's first law of motion: a body in motion keeps its speed and direction unless acted on by an outside force. This "inertial" force divides atmospheric circulation into several latitudinal zones. Thus, the atmosphere flows in three separate loops, as shown in Figure 9.8. The tropical, temperate, and polar cells are spiraling convection cells that stretch around the planet.

The Global Patterns of Water Movement

Once you can visualize temperature variations across the globe and the flow of the atmosphere, you can also understand many facets of water movement in the hydrologic system. Combine the atmospheric pattern with two simple rules and we can explain the global transport patterns for water in the hydrologic system: (1) evaporation rate increases with temperature; (2) warm air holds more water vapor than cold air. Consequently, evaporation is high near Earth's equator and low at the poles (**Figure 9.9**). But precipitation rates are also high at the equator.

A Circulation Model   Over the equatorial oceans, hot, moist air rises because of the low density of the warm air. As it rises and cools, the moisture condenses. This condensation produces intense tropical rains, which fuel the growth of tropical rain forests in South America, Africa, and Indonesia. The Coriolis effect deflects surface winds in this climate zone, creating the trade winds that converge toward the equator creating the **intertropical convergence zone** (Figure 9.8).

Why is heat distributed unevenly across Earth's surface?

What are the main patterns of the atmosphere's global circulation?

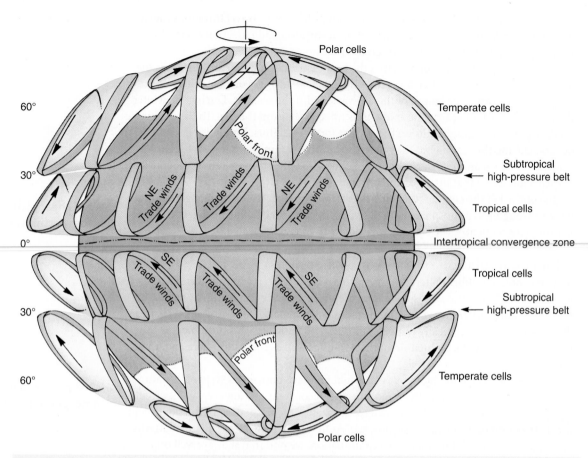

Figure 9.8 Atmospheric circulation and prevailing wind patterns are generated by the uneven distribution of solar radiation in combination with Earth's rotation. In the equatorial regions, air is intensely heated; the heating reduces its density, and the air rises. At higher altitudes, this air cools, becomes denser, and descends, forming the subtropical high-pressure belts (deserts) on either side of the equator. Near the surface, this air then moves back toward the equator to complete the cycle, causing trade winds. In the Northern Hemisphere, this air is deflected by Earth's rotation to flow southwestward. (In the Southern Hemisphere, flow is northwestward.) Temperate cells form a complementary spiral, creating strong west-to-east winds. Cold polar air tends to wedge itself toward the lower latitudes and forms polar fronts.

Why are there deserts in the mid-latitudes and polar regions?

As the rain is removed over the tropical regions, the rising air becomes much drier. At the top of the troposphere, the air splits into two convection paths, some flowing northward and some southward (Figure 9.8). As the dry air flows poleward, it cools and becomes denser until it begins to descend toward the surface. This dry air reaches the surface at about 30° north and south of the equator (Figure 9.8). As the dry, cool air descends, it warms, and its capacity to hold and absorb water vapor increases. As a result, evaporation exceeds precipitation (Figure 9.9). This is a very important factor in the movement of water from the oceans to the continents. Very little rain falls from this dry air. The low precipitation and high evaporation rates combine to cause the subtropical belts of deserts centered between 15° and 30° latitude (Figure 9.5). These deserts form a fundamental climatic zone on our planet.

Large temperate convection cells mark Earth's mid-latitudes between 30° and about 50°. In the Northern Hemisphere, the Coriolis force deflects the north-flowing air to the right to create the prevailing westerly winds of this zone. Consequently, most storm systems sweep from west to east in the temperate zone, as you probably have observed on weather maps for the United States. Mild, moist winds blow frequent cyclonic storm systems to the western sides of the continents. Precipitation is higher than in the desert zones, and temperatures are moderate. (Note that the directional patterns are reversed in the Southern Hemisphere.)

The polar front is another important element of mid-latitude climates (Figure 9.8). Here warmer air masses from the temperate cell rise over cold air masses that convect

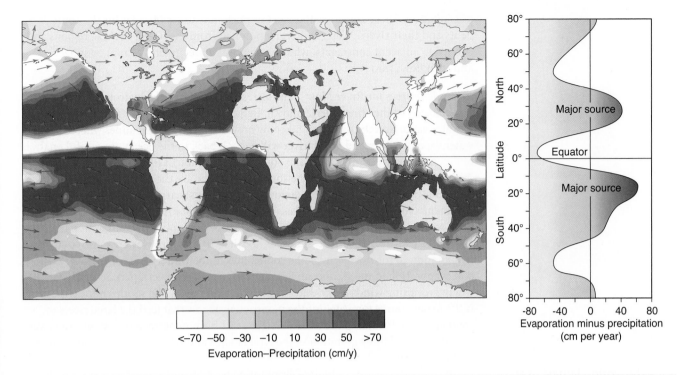

Figure 9.9 **The sources of continental precipitation** are where evaporation exceeds precipitation back into the ocean (dark blue). In these areas, great volumes of water vapor are left in the atmosphere and are blown onto the continents to fall as rain or snow. Light areas are where precipitation dominates, returning water from the atmosphere immediately back into the ocean. Arrows show the direction winds blow the water vapor. The graph on the right shows that the two major sources of water for the rivers on the continents are the oceans between 10° and 40° north and south of the equator. The equatorial oceans and the high-latitude oceans are not major sources of continental precipitation.

Data from: Gimeno, Luis, Andreas Stohl, Ricardo M. Trigo, Francina Dominguez, Kei Yoshimura, Lisan Yu, Anita Drumond, Ana María Durán-Quesada, and Raquel Nieto. "Oceanic and Terrestrial Sources of Continental Precipitation." *Reviews of Geophysics* 50 (2012). doi: 10.1029/2012RG000389.

separately in the polar regions. Near the surface, the air turns back toward the equator, completing the circulation of the temperate cell. A zone of unstable air, storm activity, and abundant precipitation is created at the polar front. When the warm, moist air flowing from the south cools, it drops its moisture as rain or snow. The irregular shifting position of the polar front is an important variable in weather conditions on the continents. A very fast-moving stream of cold air—the polar **jet stream**—marks the boundary between the two air masses.

The polar air that moves toward the equator becomes warm and rises and at high altitude flows northward to the pole. On its way to the pole, it cools and eventually sinks to complete the polar cell. Here, transport of water is limited because of low evaporation rates over the oceans—many of them are ice-covered—and because of the small amount of water vapor carried in cold air.

Evaporation-Precipitation Balance   One of the major factors influencing the patterns of water movement on a global scale is the balance between evaporation and precipitation in the oceans. Water vapor that forms by evaporation over the oceans is the major source of water that eventually falls as rain onto the continents. The graph in Figure 9.9 shows where annual evaporation is less than precipitation (on the left) and where evaporation exceeds precipitation (on the right). The major sources of water vapor that can blow onto the continents are areas of the ocean where evaporation exceeds precipitation. Where precipitation exceeds evaporation no vapor is left over to blow onto the continents, and these are not the major sources of water in Earth's rivers.

If you study Figure 9.9, you will see that the flow of water from the oceans to the atmosphere is almost symmetrical with respect to the equator. Near the equator, there is a comparatively narrow zone where evaporation *is less than* precipitation. Flanking this tropical zone are two broad areas where evaporation *exceeds* precipitation. These zones are the most important sources of water for the continents and their rivers. Water is literally pumped from the ocean in these regions and carried to adjacent continents, where it eventually precipitates as rain or snow and then flows into rivers.

Rainfall Sources for River Flow   The terrestrial branch of the hydrologic system is an extension of the atmospheric branch. Regions of excess evaporation are the sources of water vapor and the direction taken by prevailing winds determines where that water vapor ends up and which river will carry it back to the ocean. By careful analysis of Figure 9.9, you can find the major sources of water for Earth's large river systems. Areas of intense evaporation in the southern Atlantic are the major sources for the Amazon River. Prevailing winds (arrows) blowing west carry the water vapor over the continent. Northern Europe receives most of its precipitation from the North Atlantic from winds that blow to the east. Southern Europe is supplied by evaporation from the Mediterranean Sea. The sources of water for the Mississippi River are more complex. Evaporation in the northern Pacific contributes to the western part of the river basin and evaporation in the Caribbean Sea contributes to the eastern part of the basin where prevailing winds carry water vapor north and west. The southern Indian Ocean is the major source for water for the monsoonal storms that feed the large rivers of southern Asia, including the Indus, Ganges, and Brahmaputra. Surprisingly, the rivers of China are fed by evaporation in the Arabian Sea and transported by winds blowing to the northeast.

In contrast, the area of intense evaporation west of South America is not a major source of continental water. Prevailing winds carry most of the vapor into the central Pacific, where the water precipitates as rain and falls back into the oceans. Likewise, the water vapor from the South Atlantic does not supply rivers in adjacent Africa because winds carry it westward. Indeed, more than 90% of the water that evaporates from the ocean returns directly to the ocean as rain without falling on a continent.

Areas on land where evaporation exceeds precipitation are the world's major deserts (Sahara, Arabian, and Australian, for example; see Figure 9.21). Here, few rivers flow during the entire year.

Why are the world's largest river systems in the tropics?

Monsoons: An Important Regional Part of the Climate System

Monsoons occur in some parts of the world where a wet season is followed by a dry season as prevailing wind directions reverse direction. This is a type of regional, rather than global, weather pattern that requires special circumstances—circumstances largely controlled by plate tectonics. To understand this variation, we will examine the strong monsoons of southern Asia **(Figure 9.10)**.

During the winter, the Central Asian highlands, including the Tibetan Plateau, become very cold, especially compared to the adjacent lowlands near the equator. A high pressure zone develops, and cold, dry air masses sweep south pushing the intertropical convergence zone over the Indian Ocean. This forms the monsoonal dry season that lasts for several months. During the summer the situation reverses and the highlands of Central Asia heat up creating a strong upwelling and a low pressure zone. This draws the intertropical convergence zone north and along with it, warm, wet air from the ocean blows in with the trade winds. As this wet air flows over India and Bangladesh, in particular, heavy rains are triggered—especially as the air is forced over the high Himalaya front. Note the geological conditions that create the monsoons. First, a large continent is positioned near the equator. Second, continental collision created the mountainous highlands. These two factors combined with the global circulation of the atmosphere are largely responsible for this regional weather pattern.

Monsoonal rains help feed almost half of the world's population. Changes in the timing, strength, and location of the monsoon rains can, therefore, be devastating. For example, some of Earth's largest floods come during the monsoon months in this region. Every few years another season of torrential rains, floods, and landslides kill hundreds of people and displace millions.

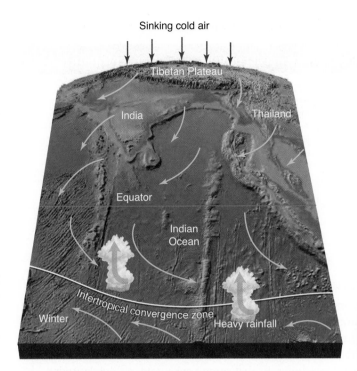

Sinking cold air

Rising warm air

(A) In winter dry seasons, high-pressures develop over the cool highlands of central Asia and outward flowing winds drive the intertropical convergence zone south of the equator.

(B) In summer wet seasons, the highlands heat up and the vertical flow of the atmosphere draws the intertropical convergence zone to the north. As the warm wet air moves over the Himalayas, the air cools and heavy rains fall over much of India and Bangladesh.

Figure 9.10 **Monsoons** cause wet and dry seasons in some regions.

Composition and Structure of the Oceans

Oceans are the great reservoirs of water in the hydrologic system and affect essentially every phase of Earth's dynamics. There are two principal layers of oceanic water: (1) a thin upper layer of warm, well-stirred water and (2) a thick mass of deeper, colder water that is relatively calm and slow-moving.

The importance of the oceans is hard to exaggerate. They influence practically every phase of Earth's dynamics. Consider a few facts about the oceans. Most of Earth's water resides in the seas, which cover 70% of the surface and contain about 97% of Earth's water. Some scientists compare the ocean to a huge boiler in which water is constantly changing from liquid to vapor and cycling through the hydrologic system. Indeed, the oceans are the major source of the water vapor that eventually precipitates onto the continents as snow or rain. The great capacity of the oceans to store heat moderates seasonal temperature changes and slows the rate of long-term climate change. Together with the atmosphere, the oceans help moderate temperature differences from the equator to the poles. Because gases can dissolve in seawater, the oceans play a major role in the composition of the atmosphere. Some of these dissolved gases eventually precipitate to form carbonate minerals in limestones. Seawater cycles through the oceanic crust, transforming hot, dry rock into wet, cold rock during seafloor metamorphism. Finally, the oceans were the womb and the cradle of life. Every element of the biosphere is directly or indirectly tied to the ocean. Everywhere you look, even in the heart of a large continent far from the seashore, you can see and feel the effects of the ocean.

Composition of Seawater

The oceans are not pure water, as you can easily taste. They contain many different kinds of dissolved salts, the importance of which is far-reaching. The major dissolved constituent of seawater is common table salt (sodium chloride or NaCl). The waters

Table 9.2 Major Dissolved Components in Earth's Ocean

Component	Chemical Formula	Concentration (g/kg)
Chloride	Cl^-	19.4
Sodium	Na^+	10.8
Magnesium	Mg^+	1.3
Sulfate	SO_4^{2-}	2.7
Calcium	Ca^{2+}	0.4
Potassium	K^+	0.4
Bicarbonate	HCO_3^-	0.1

of the open ocean have about 30 g of dissolved salt per kg of water. Other dissolved constituents add up to another 5 g per kg (**Table 9.2**).

Salinity is a measure of all of the dissolved salts in seawater. It varies with the amount of freshwater input from rivers or melting glaciers and with the rate of evaporation. In subtropical regions, the salinity of surface water is high because intense evaporation leaves the water rich in salts that cannot evaporate. At high latitudes where the temperature is lower, the evaporation rate is much lower and fresh rainwater makes the surface waters low in salinity.

However, there is little variation in salinity with depth in the ocean. Salinity is greater just below a surface layer of sea ice; the ice rejects the dissolved constituents, thus enriching them in the liquid beneath the floating ice. Highly saline waters are denser than fresher waters. These differences in salinity, along with temperature, help drive the circulation and flow of seawater, which is critical for daily weather patterns, climate control, and the movement of nutrients in the ocean.

The ocean plays another important role because it controls the composition of the atmosphere by exchanging gases, especially carbon dioxide. In turn, some dissolved carbon is removed by the precipitation of calcium carbonate—in the shells of living creatures, for example. Some dissolved carbon is converted to organic carbon and deposited in marine shales.

Thermal Structure of the Oceans

Water has one of the highest heat capacities of any substance. Consequently, the waters in the ocean have a tremendous capacity to store, transport, and release heat. Because of this fact, ocean temperature is important, and it powerfully affects Earth's weather and climates.

The average temperature of the global ocean is 3.6°C, but it varies widely. For example, ocean temperature generally decreases with depth (**Figure 9.11**). Near the surface, seawater is nearly the same temperature as the atmosphere. At great depths, the temperature is very nearly freezing, regardless of latitude on the globe. Consequently, the difference in temperature between the surface and the bottom water is small at the poles, and the temperature gradient is low.

Surface Water The ocean is layered because of temperature differences, just as the atmosphere is. The ocean has only two important layers (Figure 9.11). A thin upper layer (**surface water**) is generally warm and has a lower density, and a thick mass of cold water below has a higher density. The upper 100 m or so of the oceans are well mixed, stirred by winds, waves, and surface currents. This surface water communicates extensively with the atmosphere, freely exchanging constituents such as water vapor and carbon dioxide, and is relatively warm and oxygen-rich. In this surface layer, temperature and composition change very little with depth. It is a zone of turbulent mixing and, in this sense, is comparable to the troposphere.

What path does cold polar water follow to get to the equator?

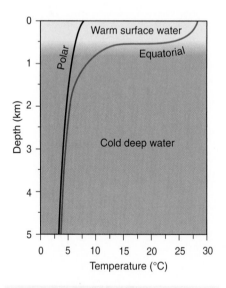

Figure 9.11 Ocean temperatures vary with depth and with latitude. In most cases, the ocean is stratified into warm surface water and cooler (and therefore denser) deep water. Except in polar regions, ocean water becomes markedly cooler with depth. Surface waters are warmest near the equator, but the temperature at 1 km depth varies little with latitude. Mixing is vigorous only in the surface layer.

Deep Water   Below the surface layer, the temperature rapidly drops through a transition zone to the cold **deep water** of the oceans. Deep water has an almost uniform temperature that changes very little north or south of the equator (Figure 9.11). This zone is very thick and contains most of the ocean water. Deep water moves very slowly. Moreover, the density difference between the two layers makes it very difficult for the dense cold water below to rise and mix with the lower-density warm surface layers. Thus, the stratification of the ocean is very stable. Deep water mixes slowly with the surface layer and is almost completely isolated from the atmosphere.

Sea Ice   Over large regions of the polar oceans, the temperature is so low that solid ice is the stable form of water. Sea ice plays a pivotal role in Earth's climate by increasing the amount of solar energy that is reflected back into space. Ice reflects much more solar energy than does the darker seawater. At any time, this **sea ice** covers as much as 15% of Earth's surface. The extent of ice changes with the season, growing to cover more of the ocean in the winter and shrinking to expose liquid during the summer (**Figure 9.12**). However, even in the present relatively warm climate, sea ice never completely disappears from the Arctic Ocean or from the fringes of the Antarctic continent. Sea ice is permanently present on about 7% of the ocean.

Sea ice is thin, usually no more than 4 m thick, because it is a good insulator that floats on top of the warmer water below. Fresh water freezes at 0°C, but seawater freezes at about −2°C, because of the salt it contains. The layer of sea ice grows thicker as ice crystallizes from the underlying water. It may also become thicker if snow falls onto the top of the ice sheet. Moreover, while sea ice crystallizes, distinctly dense and salty water forms beneath it as a complement to the salt-free ice. When the ice melts again in the spring, the fresh water from the ice dilutes the salty seawater.

Why are the ocean waters layered?

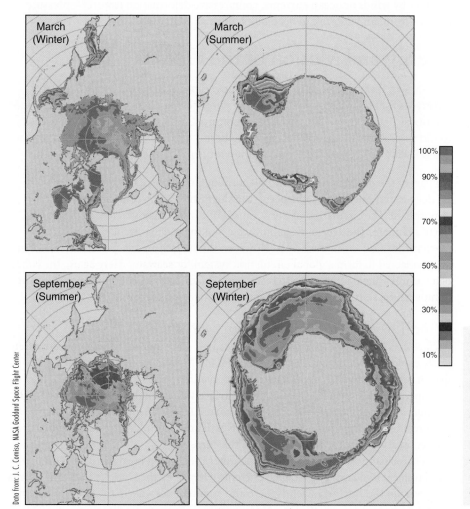

Data from: J. C. Comiso, NASA Goddard Space Flight Center

Figure 9.12   The extent of sea ice waxes and wanes with the seasons. The shading shows the percentage of area covered by ice. In the Northern Hemisphere, sea ice is at a maximum in March (at winter's end) and then declines through the summer (September). In the Southern Hemisphere, the variations in sea ice are similar, but at opposite times of the year, because the seasons are reversed. As sea ice declines in the north, it is expanding in the south, and vice versa.

Figure 9.13 **Sea ice** covers as much as 6% of the sea surface today. It slows the loss of heat from the oceans and reflects sunlight. Both affect Earth's climate. Note how the cracks in the ice have frozen over here in the Ross Sea, Antarctica.

© Armin Rose/ShutterStock, Inc.

If it is so cold in the polar regions, why don't the oceans freeze completely?

The average thickness of sea ice in the Arctic Ocean is 2 to 3 m; around Antarctica the ice is thinner, averaging about 1 to 2 m thick. Sea ice is quite smooth on a regional scale, but it is typically broken into a series of smaller blocks, or floes, of varying thickness (**Figure 9.13**). The ice is repeatedly fractured by its movement, which exposes seawater that then freezes between the ice floes. Where plates of sea ice are driven together by winds or ocean currents, compressive deformation features—pressure ridges—form, and the ice can become as much as 10 m thick below these ridges.

Energy and Motion of the Oceans

The warm surface layer of the ocean is moved principally by wind-generated currents that form circular patterns. The deep oceans circulate because of changes in density caused by salinity and temperature. Global oceanic circulation carries cold surface water deep into the North Atlantic, around Africa, and into the Indian and Pacific oceans. Surface currents then return the water to the North Atlantic.

No part of the ocean is completely still, although movement of water in the abyssal deep is extremely slow—the ocean is a sluggish beast by comparison with the atmosphere. The circulation of the oceans is one of the major factors in developing Earth's climate. The deep and shallow waters of the oceans circulate by distinctive mechanisms and at different rates. However, the surface and deep-water circulation are connected to form a global circulation system for seawater. Here again, the concept of a natural system attempting to reach equilibrium with its changing environment (temperature, salinity, wind pressure) is the critical motive force.

Wind-Driven Circulation of Surface Waters

Surface currents have been known and measured since ancient times and were extensively exploited by early navigators (**Figure 9.14**). Movement in the surface layer is driven primarily by the wind. In turn, the prevailing winds are caused by the uneven heating of Earth's surface. Circulation of surface waters might be best understood by considering the Pacific Ocean, which is bordered by continents on the east and west (Figure 9.14). Strong equatorial currents (North and South Equatorial Currents) are pushed westward by the trade winds. As the currents encounter the western land masses (Asia and Australia), some of the water is deflected northward and some southward to form two large ring-shaped currents. Helped by the Coriolis force, the flow is clockwise in the Northern Hemisphere and counterclockwise in the Southern Hemisphere. Other rings form at each pole and orbit in the opposite direction.

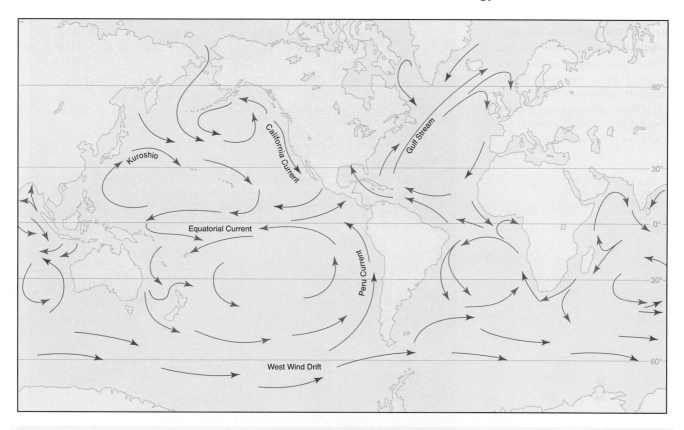

Figure 9.14 Surface currents of the oceans are driven by prevailing winds, which are in turn caused by the uneven heating of Earth's surface illustrated in Figure 9.6. Most of the currents have crudely circular patterns. Warm (red) and cold (blue) currents are shown.

Surface currents in other areas of the world ocean are more complex than in the Pacific because of the shapes and arrangements of land masses and the configuration of the ocean floor, but the basic pattern is still obvious (Figure 9.14). Large circular patterns dominate surface currents in the Atlantic, Pacific, and Indian oceans. Note that each circular current flow has a strong, narrow, poleward current on its west side and a weaker current on its east side. The most pronounced of these strong western-margin currents in the Northern Hemisphere are the Gulf Stream in the Atlantic and the Kuroshio current in the Pacific. Because these currents carry water from the south to the north, they are much warmer than the surrounding waters. The speed of these surface currents may exceed 2 m/sec. The return eastern-margin flow from the mid-latitudes to the equator is much slower and occurs over a broader area. Western boundary currents also occur in the Southern Hemisphere along the shores of South America and Africa. These currents are not as strong as those in the Northern Hemisphere.

The northward-flowing Gulf Stream is part of this global pattern. It is obvious as a warm anomaly in the North Atlantic (**Figure 9.15**). As it flows northward, it mixes with colder waters in turbulent swirls, or eddies, and eventually loses its identity. The warm waters carried by the Gulf Stream are very important for moderating the climate of northern Europe. The eastern Atlantic Ocean is much warmer at the surface than the western Atlantic. Consequently, western European winters are milder than their counterparts in eastern North America at the same latitude. Compare the climates of New York City and Lisbon, Portugal—two cities found at the same latitude but on opposite sides of the Atlantic. Some of the heat carried by the Gulf Stream is picked up by the Norwegian Current and carried even farther into polar regions.

Cold surface currents flow toward the equator along the eastern Pacific Ocean—the California Current along North America and the Peru Current off South America (Figure 9.14). These cool waters lower the air temperature along their shorelines, compared with continental regions at similar latitudes.

What determines the direction of surface currents?

Figure 9.15 The Gulf Stream is a warm north-flowing current in the North Atlantic. It is revealed in this computer model where reds are warm waters; blues and violets are cooler. Note the warm waters flow from the Gulf of Mexico along the Atlantic coast northward to the cool Arctic waters. The large swirls are eddies in the Gulf Stream.

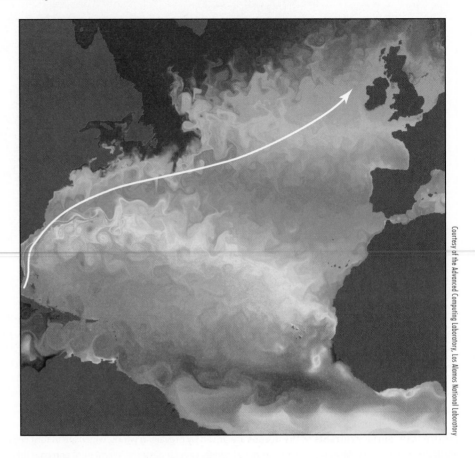

Courtesy of the Advanced Computing Laboratory, Los Alamos National Laboratory

Density-Driven Circulation of the Deep Ocean

At great depth, the oceans are not directly affected by the winds. Instead, the slow circulation of water in the deep ocean is caused by changes in water density. The major causes of this density variation in seawater are its temperature and salinity. This slow but vastly important movement is called **thermohaline circulation** (in Greek, *therme* is heat and *hals* is salt or sea). The movement is so slow that measuring it is difficult. Nevertheless, the patterns have been deduced from the distribution of minor dissolved constituents, principally salt and dissolved gases such as oxygen, and by careful measurement of water temperature at depth. The decay of the naturally formed radioactive isotope of carbon (^{14}C) also can be used to estimate the flow velocities for these density currents.

What causes the deep ocean water to move?

From these measurements, we infer that water from the surface sinks to great depths in the polar oceans (**Figure 9.16**). For example, cold, high-salinity water forms beneath the sea ice around Antarctica. This dense Antarctic water drops to the seafloor and then slowly flows along the base of the Atlantic Ocean, eventually reaching as far north as 40° N (Figure 9.16). Another large mass of cold deep water is formed in the North Atlantic, where frigid air cools surface waters. The water sinks and flows southward above the even denser Antarctic bottom water. This water rises back to the surface near 60° S. Cold water formed in mid-latitudes of the Southern Hemisphere wedges itself between the warm surface water and the North Atlantic deep water.

It takes about 1000 years for a complete cycle of surface water to become deep water and then surface water again. Most chemical and thermal properties of the oceans should not change rapidly because of this slow movement of deep ocean water. Thus, the oceans act to slow the rate of climate change. The transport of heat by the slowly moving waters of the oceans is one of the major factors in controlling Earth's climate.

Deep-ocean waters are generally rich in dissolved nutrients. They have spent considerable time at depth, where falling detritus can be dissolved and where no organisms exist to consume the nutrients because of the low amount of light. Where deep-ocean water wells upward into sun-drenched shallow waters, these nutrients become important for the marine food chain.

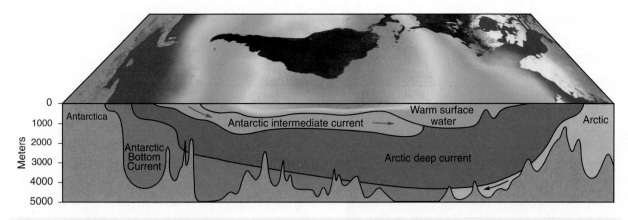

Figure 9.16 Deep circulation of the ocean is driven by density differences caused by temperature differences and to a lesser extent by differences in salinity. Cold bottom water in the Atlantic Ocean tumbles down the margins of Antarctica and flows northward, reaching as far as 40° N of the equator. Cold surface water in the north also sinks toward the ocean floor and flows southward.

Coastal Upwelling

Along the shorelines of many continents, strong **coastal upwelling** of deep-ocean water is an important part of ocean circulation. Because of the Coriolis effect, winds blowing toward the equator and along a coast cause seawater to move to the right of the wind direction in the Northern Hemisphere and to the left of the wind direction in the Southern Hemisphere (**Figure 9.17**). For appropriately oriented coasts, this movement of surface water away from the shore causes cold deep water to flow upward to take its place. This deep water is rich in nutrients and nourishes rich blooms of plankton, which in turn are food for a wide variety of sea animals. Some of the ocean's richest fisheries are found in these waters. Upwellings off the California coast and the western coasts of South America and Africa are excellent examples of this phenomenon. Ancient upwellings have created important deposits of phosphates (used for fertilizer) and organic materials that have contributed to the formation of oil.

A temporary change in surface currents can have an extreme impact on coastal upwelling. For example, occasionally the normally strong trade winds weaken. Their weakening allows warm currents to approach the western shore of South America, where surface waters are normally cold. This phenomenon, called **El Niño** (The Child) because it occurs around Christmas, disrupts the upwelling of cold nutrient-rich water (**Figure 9.18**). Consequently, the phytoplankton population diminishes and the fish population almost

What is El Niño and why is it important in oceanic circulation?

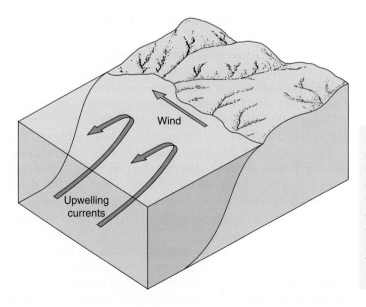

Figure 9.17 Upwelling ocean waters are common along the margins of the continents. In the Southern Hemisphere, coastal upwelling occurs in response to winds that blow northward. As a result, the Coriolis force drives surface water away from the shore. In response, deep waters upwell and bring nutrient-rich waters to the surface. This upwelling nourishes plankton and other organisms that feed on it.

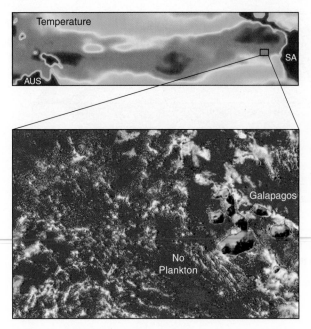

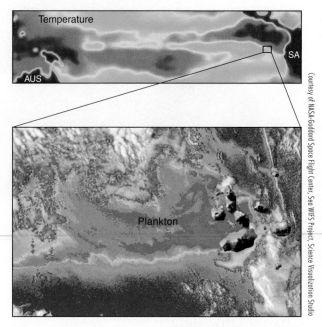

(A) Warm water (red in the upper map) accumulates against the western shore of South America (SA), inhibits deep upwelling, and reduces the amount of phytoplankton (red in the lower map) in the Pacific Ocean.

(B) As conditions return to normal, colder surface waters (green in upper map) return near shore of South America, upwelling resumes, and phytoplankton grows across the equatorial Pacific and especially near the western shore of the Galapagos Islands.

Figure 9.18 El Niño marks an anomalous buildup of warm water along the west coast of South America. This shuts off upwelling, nutrient-rich deep waters and stifles phytoplankton production.

disappears. Bird populations diminish, and fishermen are put out of work. Large El Niño disturbances also change precipitation patterns worldwide, causing flooding in North and South America and drought in areas as distant as India, Indonesia, and Australia.

Global Pattern of Oceanic Circulation

The waters of the entire ocean are mixed slowly but surely because of the wind-driven circulation of the shallow ocean, coastal upwelling, and the density-driven circulation that involves the deep ocean. All of these phenomena are shown in **Figure 9.19**. As ocean waters move by surface flow, by sinking, or by rising, other waters must move in to take their place. The surface flow is easy to observe and therefore is better understood, but the deep-water flow is much harder to decipher. Thus, the simple picture we elaborate below is based partly on observation and partly on theoretical models.

To visualize the global flow of the ocean, let us trace the long, slow journey of a parcel of water starting in the North Atlantic (**Figure 9.20**). As this water equilibrates with the cold air, it also becomes cold and dense and sinks as much as 2 km to the ocean floor. Once cold water fills the deep basin there, it surmounts the Iceland Ridge and plunges southward as an intense bottom current on the western side of the North Atlantic. Salty, dense, but warm water spills from the Mediterranean and mixes with the North Atlantic Deep Water. At any instant, the rate of water flowing along the floor of the Atlantic is estimated to be 80 times the volume of the Amazon River.

This deep current continues southward until it meets a strong northward- and eastward-flowing current of Antarctic Bottom Water in the South Atlantic. The two currents appear to merge and the water flows eastward into the deep Indian Ocean and ultimately northward into the deep Pacific Ocean. The water that originated in the North Atlantic has now been removed from contact with the atmosphere for several hundred years.

In the Pacific, much of this deep water warms as it moves toward the equator and slowly buoys to the surface. There it mixes with surface waters and begins a return flow to the Atlantic. But now, much of the journey is at the surface of the ocean. Wind-driven surface currents sweep it through the maze of Indonesian islands and into the Indian

How long does it take for North Atlantic Deep Water to make one complete circuit through the oceans?

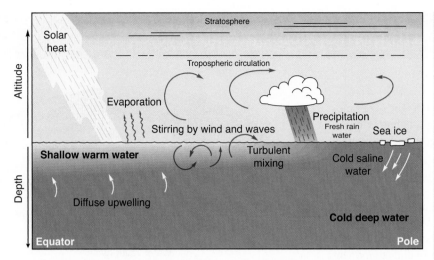

Figure 9.19 The ocean and atmosphere form a simple system of interacting water and air, but many processes are involved in the flow and mixing. Variation in solar heating causes winds, which drive the surface circulation of the ocean. Density-driven circulation vertically mixes the shallow and deep layers of the ocean. These density variations are caused by differences in temperature or salinity. Only the upper oceanic layer is well mixed by turbulence. Mixing with deep water is much less efficient and occurs mainly by the sinking of cold polar waters into the deep ocean. Diffuse upwelling throughout the ocean basins and coastal upwellings return deep water to the surface.

Ocean and then southward around Africa. The surface currents along the southern shore of Africa produce broad, spinning eddies that eventually feed the water into the Atlantic.

Beyond this area, northward transport is hindered by strong surface currents parallel to the equator (Figure 9.14); gigantic swirling eddies eventually spin much of the water across the equator and into the North Atlantic. Water is then swept northwestward across the Atlantic to near the Caribbean where, if caught in the Gulf Stream, it finally returns to the North Atlantic.

Of course, this long flow path is not like a pipe, conducting every molecule of water along the same path. As deep water traverses the ocean floor, there is very slow diffuse upwelling. More intense and localized upwelling of deep water also occurs along shorelines. Some water in the Pacific takes a shortcut to the North Atlantic by flowing through the Bering Sea, into the Arctic Ocean, and then southward (Figure 9.20).

The formation of North Atlantic deep water may have important implications for the global climate. Changes in the flow pattern, perhaps induced by shifting continents and swelling ridges, may have caused some of the climate change that led to the ice ages. In the future, greenhouse warming may also affect the rate at which North Atlantic deep water is formed. This effect could change oceanic circulation and have complicated global effects on precipitation.

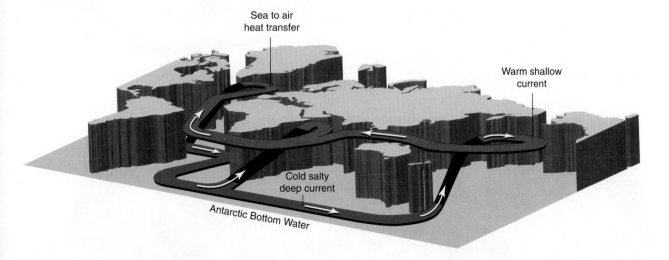

Figure 9.20 The global pattern of ocean circulation can be likened to a huge conveyor belt that carries surface water to great depths and then back again. Deep-water circulation (blue arrows) originates in the North Atlantic by the sinking of cold surface waters north of Iceland. This water flows southward at depth along the western side of the ocean basin and into the South Atlantic Ocean. Along the shores of Antarctica, it is joined by more cold sinking water and then flows eastward into the deep basins of the Indian and Pacific oceans. Diffuse upwelling in all of the oceans returns some of this water to the surface. In addition, a warm surface current from the Pacific (red arrows) may return water to the North Atlantic.

STATE OF THE ART   Earth from Space: Satellite Eyes

A multitude of satellites orbiting far above Earth's surface are equipped with a diverse array of sensors. One of the most straightforward applications is to take digital photographs of the surface in many different wavelengths and at different scales. The highest resolution satellite images available today show objects only 1 or 2 m across. (Examples of satellite photographs are found elsewhere in the text.)

Some of these satellites are in orbits that allow them to see the entire surface of Earth each and every day. As a result, they become monitors of the daily changes in Earth's hydrosphere and atmosphere—the fundamental components of our changing global climate. Movies showing motion of the hydrologic system can be made from these time-lapse images. Imagine the huge numbers of observation stations that would be needed to replace the global coverage provided by one of these satellites in a single day. Short- and long-range weather forecasts rely heavily on information derived from these satellites.

Sensors on satellites such as the TOPEX/Poseidon and Terra satellites measure the amount of water vapor in the atmosphere, wind speeds, or even wave height (to right). The map at the bottom of the page shows the elevation of the sea surface caused by differential heating and the effects of winds. Elsewhere in this book, you have seen global maps made from satellite data of precipitation (Figure 9.5), temperature (Figure 9.6), vegetation, and even the amount of ozone in the atmosphere (see GeoLogic later in this chapter).

Other satellites measure sulfur or other components in the air that emanate from volcanoes. Some are equipped with lasers for measuring the elevation of the surface below or with radar to construct images or topographic maps of the terrain beneath a dense forest canopy, through clouds, or even through a sheet of sand.

Satellites have literally become our eyes in the sky. These dramatic images have immensely changed our view of Earth and helped us see it as one global system.

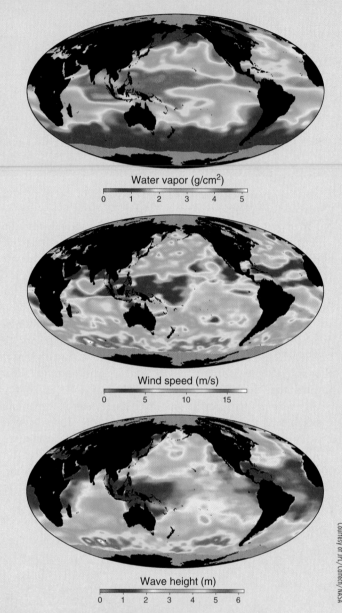

Water vapor (g/cm²)
0   1   2   3   4   5

Wind speed (m/s)
0   5   10   15

Wave height (m)
0   1   2   3   4   5   6

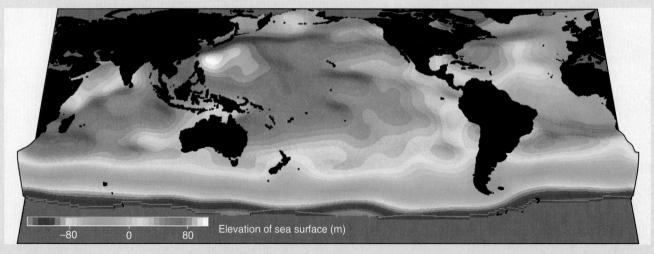

−80   0   80   Elevation of sea surface (m)

Climate Zones

Earth's climates vary with latitude. From equator to each pole, there are four fundamental climate zones: hot and wet (tropical); hot and dry (desert); moderately warm to cool and humid (temperate); and cold and dry (polar).

Climate is tremendously important in geologic processes that shape Earth's surface. As we have seen, sedimentary rocks form in immediate contact with and because of movement of the ocean or the atmosphere. These sediments record many details of ancient climates. In addition, the rates and styles of many surface processes are affected by climate. As a consequence, it is important to understand a few of the major climate zones.

Because of the systematic differences in the amount of solar radiation on the surface, combined with variations in the amount of precipitation, several major climatic zones surround the globe at different latitudes (**Figure 9.21**). One might expect these zones to be consistent around the globe, depending only on the radiation received from the Sun. However, local climate variations are widespread, resulting from secondary agents, such as mountain rain shadows, cold or warm surface currents, and elevation.

Tropical Climates

Moisture-laden warm air flows toward the equator from both hemispheres, as you saw in Figure 9.8. In these **tropical climates** (Figure 9.21), the annual average temperature exceeds 20°C. Precipitation rates as high as 2 m/y are not uncommon in these regions, but much rain usually falls in a wet summer season. Tropical cyclones bring much of this moisture to the continents.

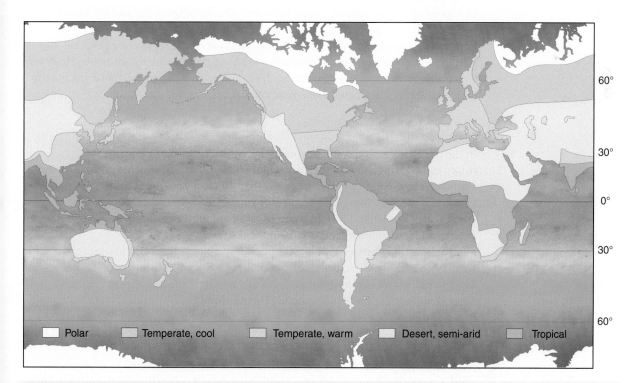

| Polar | Temperate, cool | Temperate, warm | Desert, semi-arid | Tropical |

Figure 9.21 A world climate map shows the systematic variations in temperature and precipitation across the globe. Four distinctive climates have important geologic implications. Near the equators lie the tropical climates with their high temperatures and precipitation, much of it falling from tropical cyclones. The world's great deserts lie in broad bands of desert climate north and south of the tropics. Temperate climates dominate the mid-latitudes. Nearer the poles, cold polar climates with varying amounts of precipitation are characteristic. Temperatures in the ocean range from systematically warm (red) near the equator to cold (blue) at the poles.

These climatic conditions are often recorded by distinctive rock types. For example, rain forests developed in the tropics may produce such lush vegetation that coal deposits form from the accumulation of dead plants. Moreover, large tropical rivers produce large deltas at their mouths. Elsewhere, thick red soils form as rocks are decomposed by deep weathering caused by the high temperatures and abundant precipitation. Nutrients are leached from the soil by the abundant precipitation.

Desert Climates

Where the precipitation is less than the evaporation rate, **desert climates** develop. Globally this situation occurs in the subtropical high-pressure zones, causing the great deserts of North Africa and the Middle East (Figures 9.8 and 9.21). Prevailing easterly winds and the high mountains of western North America are responsible for the deserts in that region. Small deserts can be produced locally in the rain shadows of high mountain belts.

Can global climate zones develop perpendicular to the equator?

Temperatures and precipitation rates vary greatly in the world's deserts. Portions of the Sahara in North Africa have an average summer high temperature of 38°C, a winter low of 16°C, and an annual precipitation of only about 2 cm/yr. Some deserts are cooler, and rain may not fall for decades. For example, the Atacama Desert of Chile has a summer average high of only 20°C and a winter minimum of 13°C. Although the Atacama is cooler than the Sahara, the average annual precipitation is extremely low—only 0.2 cm/yr.

Geologic processes in deserts include the deposition of evaporites and the development of large sand seas. Large rivers are rare, and closed river basins with no outlets to the sea are common. Decomposition of rock by weathering is slow because of a lack of water.

Temperate Climates

In regions of **temperate climates**, between latitudes of about 35° and 60° both north and south of the equator, annual temperatures range from less than 0°C to 25°C (Figure 9.21). Precipitation falls throughout the year. These regions are generally too cold to support coral reefs in the oceans or rich vegetation on the continents. However, large rivers form and deposit sediment. Rich soils are created by weathering, and temperatures are high enough for intensive agricultural development.

Polar Climates

In the region north and south of about 60°, temperatures are so low that water is frozen solid during much of the year (Figure 9.21). On the continents in these **polar climates**, the average temperature is less than 10°C all year and is below freezing for most of the year. Moreover, because of atmospheric circulation patterns, precipitation is also low (Figure 9.5). In terms of lower amounts of rainfall, these are polar deserts, and wind-blown sand occurs in many areas.

The differences between geologic processes in polar regions and in other climatic zones of the world can be explained by the presence of ice. Glaciers may form and devour continental landscapes that elsewhere are dominated by stream valleys. Beyond the reach of glaciers, seasonally frozen ground may overlie a layer of permanently frozen soil and enhance the downslope movement of material. Weathering is minimal in polar regions, and vegetation is sparse. Polar oceans are far too cold to support organic reefs and are often covered by sea ice.

At times in geologic history, polar climates have expanded much farther southward. Conversely, long periods of warmth have caused polar climates to collapse. Such global changes in climate are the focus of the next section.

Climate Change

Climate change on a continent can be caused by its slow tectonic movement through climate zones. Alternatively, the climate may change on a global scale. Heating by the greenhouse effect is one of many important factors that cause global climate change.

Nothing is as constant as change. This cliché applies to all of Earth's dynamic systems, but especially to the climate system. Climate fluctuations are nothing new. The general features of past climate change are clear in many sedimentary rock records. Past climate changes have run the gamut from minor changes in precipitation over a local area to dramatic changes that engulfed the entire planet. Future climate changes are just as inevitable, but they are very difficult to predict. Forecasting climate change requires understanding of Earth's vast climate web and its myriad interacting elements, which weave together in a complex and often chaotic fashion. We cannot isolate one piece of the system from the rest. Predictions of future change may improve when we better understand past changes in Earth's long climate history.

The amount of temperature change necessary to alter the climate significantly is not large. In fact, huge temperature changes have not occurred on Earth. Even its oldest rocks include metamorphosed sedimentary rocks, demonstrating that Earth had cooled enough for liquid water to exist continuously since at least four billion years ago. Apparently, the surface temperature has remained between water's freezing and boiling points for a very long time. This nearly constant temperature range on Earth's surface is indeed a remarkable fact.

Just as remarkable, perhaps, is how much change is wrought by small variations in the global temperature. From geologic and geochemical evidence, it has been suggested that temperatures during the last ice age, which reached their lowest point about 20,000 years ago, were only about 3°C to 5°C lower than today's balmy average of 20°C. Yet this small change was sufficient to create ice sheets that covered much of the Northern Hemisphere with a layer of ice 3 km thick.

In the past, two types of climate change often had dramatic geological and biological results: (1) regional climate change due to continents moving into different climate zones and (2) global climate change.

What causes climate change?

Continental Drift and Climate Change

Because the continents are carried about by the slow movements of the lithosphere, they commonly move from one climate zone to another. As a result, important changes in the climate at a single location can occur over a long span of time. For example, consider the climate zones that would be encountered by a continent drifting slowly from the southern polar regions to the north (**Figure 9.22**). At the South Pole, the continent would likely be covered with glacial ice, producing glacial deposits and destroying river systems. As the continent moved northward into the temperate zone, the glacial deposits would be succeeded by deposits from large rivers or shallow seas, depending on sea level.

Continued movement northward would bring the continent into the subtropical high-pressure belt marked by deserts. Stream sediments would be overlain by deposits of evaporites and wind-blown sand. Continuing movement into the tropics would bring greater precipitation, and therefore large river deposits interlayered with coal might lie atop the desert sands. Shallow marine deposits in this region would include coral reefs. With continued northward movement, this sequence would reverse, with tropical deposits on the bottom, progressively overlain by desert sands, then temperate river deposits that were capped by glacial deposits. This hypothetical trip would have taken about 400 million years at typical plate velocities.

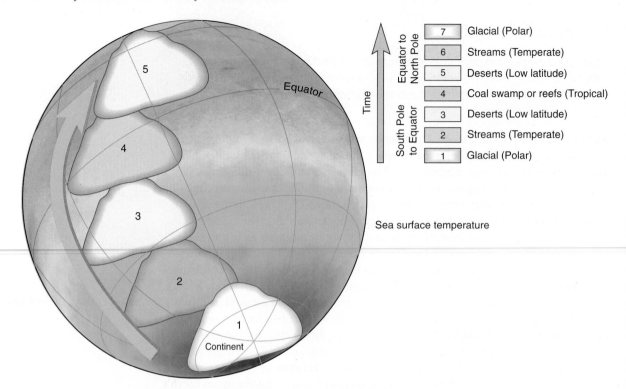

Figure 9.22 Continental drift causes local climate to change with the passage of time. A sequence of climate-controlled rocks develops as this hypothetical continent moves from the South to the North Pole. Polar climates produce glacial deposits; temperate climates may produce river deposits; desert climates produce wind-blown sand deposits and evaporites; and tropics yield thick deposits of deltaic sands, coals, or coral reefs in shallow-marine settings. A reverse version of this sequence accumulates as the continent moves through the climate zones of the Northern Hemisphere.

Obviously, such a simple succession of rocks is unlikely to form because plate tectonics rarely carries a continent continuously in a north-south direction. Moreover, plate collisions and mountain-building events would alter the simple sequence. However, the most important conclusion from this simple model is that profound change can occur in one area simply by shifting the continent through different climate zones. Climate—and the rock record it produces—changes continuously. Despite the evidence for climate change in this theoretical example, Earth's global climate remained unchanged.

Human Disturbance of the Climate System

Changes in the atmosphere–ocean system also may be induced by human activities. There are now so many people—over 7 billion—and our activities are so pervasive that humans have become an important part of the physical and chemical evolution of the entire planet. The changes we have made are not limited to the construction of buildings, dams, and highways, or to the destruction of natural vegetation in forests and plains. We are also changing the composition of the atmosphere and ocean in ways that may affect the global climate and even our own fate as a species.

Global climate change is caused by fundamental change in one of the major climatic factors and must be clearly distinguished from the movement of a continent from one climate zone to another. Instead, global climate change involves the expansion and contraction of entire climatic zones across Earth's surface. For example, during the past ice ages, polar climates extended as far south as the northern tier of the United States and into central Europe, about 1000 kilometers or so. In the geologic past, temperate climates were found far to the north of their present positions.

Obviously, global climate change and continental drift occur simultaneously, and separating their effects is not an easy task.

The most important global climate parameter is temperature. Consequently, the major controls on climate change are changes that affect the global temperature or its distribution on the surface. These include the energy output of the Sun, the composition of Earth's atmosphere, the reflectivity of Earth and its atmosphere, the ocean circulation patterns as continents move, blocking of sunlight by particles in the atmosphere, or even changes in Earth's spin or orbit. (Some of these global changes are described in more detail elsewhere in the text, where we discuss the cause of the ice ages.) Because these temperature factors are constantly changing, climate at one spot may change as a climate zone shrinks or expands.

The Greenhouse Effect Global climate change caused by human activities is a major political and scientific issue. Claims and counterclaims abound in the political debate about the reality of an increase in abundance of **greenhouse gases** in the atmosphere, about the source of the greenhouse gases in the atmosphere, and about a global temperature increase. From a scientific point of view, the facts are less contentious. Carbon dioxide in the atmosphere absorbs heat radiated from the surface and traps it in the troposphere. We have already described how certain gases absorb radiation at specific wavelengths depending upon the atoms in the molecule and the nature of the bonds that hold the atoms together. Gases that absorb energy and thus increase the atmosphere's temperature are called greenhouse gases. Carbon dioxide concentrations have been increasing since about 1800 as shown in **Figure 9.23**. Since 1958, regular sampling has shown that the concentration of carbon dioxide has increased from about 315 ppm to 390 ppm. Moreover, by carefully extracting bubbles of gas trapped in glacial ice, we have extended our carbon dioxide measurements back several hundred years. In the 1700s, the carbon dioxide content of the atmosphere was fairly constant at about 275 ppm, but it has increased to almost 400 ppm over the last 250 years.

The increased carbon dioxide is coming from fossil fuels, not volcanoes or other natural sources. This is revealed by the isotopic composition of the carbon in the air. Coal, oil, gasoline (all fossil fuels) burn to release carbon dioxide as a by-product which is isotopically distinct from CO_2 released from natural sources. The use of fossil fuels increased dramatically since the early 1800s because the population increased and because of the industrial revolution; the increased use correlates with the higher amounts of CO_2 in the atmosphere.

There is also a broad scientific consensus that the global temperature has increased by about 0.8°C over the past 100 years (Figure 9.23B). This is a classic case of trying to determine the relationships between two correlated observations (increasing carbon dioxide and increasing temperature). Does one cause the other? If so, which is cause and which is effect? Or are they completely unrelated and their correlation is simply coincidental? In this case, most atmospheric scientists have concluded that the two factors are actually related to one another because the effect (temperature increase) is in the direction predicted by the change in the causative factor (increase in carbon dioxide). As Earth warms, sea level is also rising—partly because warm water expands, but also because glacial ice is melting and returning to the oceans. A recent estimate of the magnitude of this effect was that sea level could rise by 7 m in 1000 y, if the temperature rises by 3°C over Greenland and melts all of its glaciers. Looking even farther into the future, the Earth may warm by as much as 8°C by the year 2400 if we continue on our current course.

Ocean Acidification Another troubling effect of the increasing CO_2 in the atmosphere is the increasing acidity of the oceans. About 30 to 40% of the CO_2 released to the atmosphere accumulates in the oceans where it reacts with water to make a weak acid, H_2CO_3 (carbonic acid). Carbonate minerals go into solution at lower pH (high acidity).

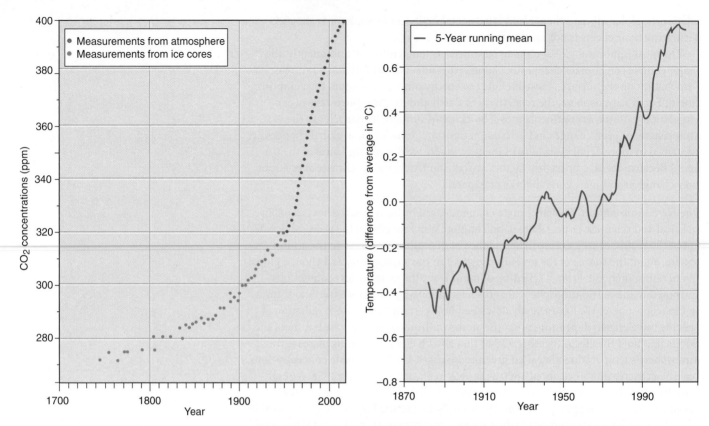

(A) Carbon dioxide concentrations in the atmosphere have risen dramatically since the mid-1700s, corresponding to the use of fossil fuels. CO_2 measurements were made on air bubbles trapped in ice layers from Antarctica (blue dots) combined with direct measurements of the atmosphere's composition (red dots) during the last 65 years.

(B) Average global temperature appears to be on the rise as well. The 5-year running average temperature deviates from the average temperature (calculated for 1951–1980). Annual variations are much more extreme, and comparisons from year to year are of little use for gauging global change. Taken at face value, these measurements indicate that the global temperature has risen by as much as 0.8°C over the last century.

Figure 9.23 Global temperatures and carbon dioxide concentrations in the atmosphere are rising.

The pH of the ocean is 8.1; if current trends continue it may drop to 7.8. That may not seem like much of a change, but the lower pH makes it difficult for corals, plankton, and other sea creatures to precipitate skeletal minerals. And this is only one example of many biologic effects. The rate of acidity increase exceeds that which has been seen for any past warming event.

Questions abound. Will it be possible for the biosphere to keep up with these rapid changes? Will Earth return to the greenhouse conditions of the Cretaceous period when temperate climates extended as far north as Alaska? How much will sea level rise? How acidic will the oceans become? How will the production of food and other biological resources be affected by this impending climate shift? Will a natural return to glacial conditions dampen the effects of the carbon dioxide increase? Only time will tell.

At this point, it is worth looking back over the atmosphere–ocean system in its entirety (**Figure 9.24**). The global climate system is based on energy delivered by the Sun and spread unequally on the surface. Because heat is concentrated at the equator and diffuses at the poles, a planet-wide circulation system is established in the atmosphere. Winds set up by the flow of the atmosphere drive the circulation of shallow ocean water, interference is supplied by the continents. Deep circulation of seawater is driven by density differences caused by heating and salt content. In turn, the movement of the waters in the ocean moderates the temperature differences around the globe. The climate zones are the ultimate result of these variations and have had an obvious impact on all aspects of the hydrologic system and its interaction with the lithosphere. Their role in the evolution of life and especially humans cannot be underestimated.

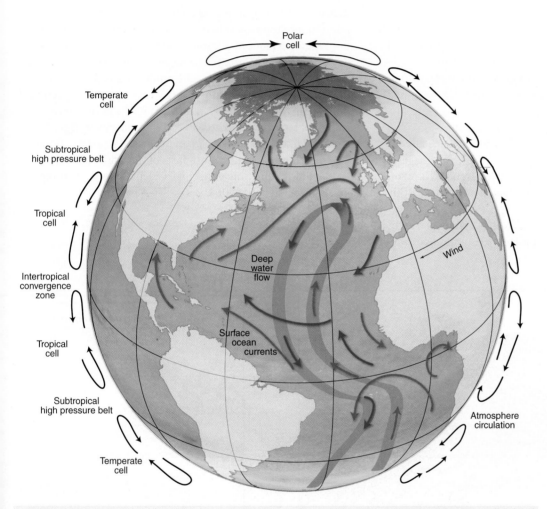

Figure 9.24 Earth's climate system is based on energy delivered by the Sun and spread unequally around the surface. This, in conjunction with variations in buoyancy and gravity-driven flow, creates vast circulation paths in the atmosphere and in the ocean. In turn, the convection moderates the temperature differences around the planet and shapes the biosphere.

GeoLogic   The Ozone Hole

The recent history of the atmosphere–ocean system involves the chemical changes caused by humans. Both the atmosphere and the ocean are becoming polluted with the products and by-products of our modern industrial and agricultural practices. Another example is the "hole" detected in the ozone layer.

Observations

1. Ozone forms in the stratosphere where sunlight breaks the bonds in an O_2 molecule to form atomic oxygen (O), which then reacts with another O_2 molecule to form O_3.
2. Ozone concentrations reach a maximum of about 10 ppm (parts per million) at altitudes of 20 to 25 km forming the ozone layer.
3. Ozone absorbs ultraviolet rays from the Sun.
4. An ozone hole over both poles has been detected by satellite measurements in Dobson units (DU). The image on the left shows the ozone hole over Antarctica. The ozone hole comes and goes with the seasons and is deepest in the winter (September).
5. The amount of ultraviolet radiation (a carcinogen at high dosages) that reaches the surface waxes and wanes with the size of the ozone hole.

6. Chlorinated fluorocarbons (CFCs, formerly used as refrigerants and as propellants in aerosol cans) are known to react with and destroy ozone molecules in the stratosphere.

Interpretations

The "hole" in the ozone layer is a region where the ozone abundance is lower. Apparently, even the tiny amounts of CFCs released into the atmosphere have modified the ozone balance, decreased the amount of ozone, and increased the area of the depleted ozone to create the ozone hole. The hole is deepest in the winter because natural ozone production is slower when the atmosphere is cooler. During this season ozone concentrations are about 70% than they should be. The lower concentration of stratospheric ozone means less ultraviolet light is absorbed and, consequently, more reaches the surface. Because CFCs are now banned in most parts of the world, the ozone hole is slowly returning to normal. Calculations suggest the hole may persist until 2068.

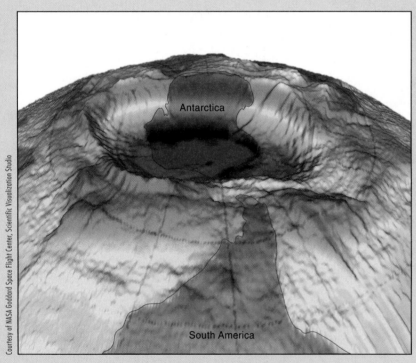

Courtesy of NASA Goddard Space Flight Center, Scientific Visualization Studio

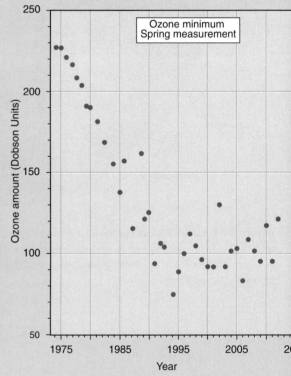

Key Terms

coastal upwelling (p. 251)
Coriolis effect (p. 241)
deep water (p. 247)
desert climate (p. 256)
El Niño (p. 251)
greenhouse effect (p. 238)
greenhouse gas (p. 259)

humidity (p. 235)
intertropical convergence
   zone (p. 241)
jet stream (p. 243)
magnetosphere (p. 237)
mesosphere (p. 237)
ozone (p. 237)

ozone hole (p. 237)
ozone layer (p. 237)
polar climate (p. 256)
precipitation (p. 238)
salinity (p. 246)
sea ice (p. 247)
stratosphere (p. 237)

surface water (p. 246)
temperate climate (p. 256)
thermohaline circulation
   (p. 250)
thermosphere (p. 237)
tropical climate (p. 255)
troposphere (p. 235)

Review Questions

1. Explain how weather and climate are different.
2. What causes the layering of Earth's atmosphere?
3. Why does the temperature gradient in the atmosphere reverse at the boundary between the troposphere and the stratosphere?
4. Describe the general temperature variations across Earth's surface. What causes these pronounced latitudinal changes in temperature?
5. Venus's cloudy atmosphere reflects away so much solar energy that its surface receives less than Earth. What then causes the very high (470°C) surface temperature on Venus?
6. Why does the pressure of the atmosphere decrease at high elevations?
7. Describe the global flow patterns of the atmosphere.
8. Why are the tropical rain forests found along the equator? Why does a band of deserts encircle Earth at 30° north and south of the equator?
9. Outline the role ozone plays in the atmosphere.
10. How is ozone formed in the stratosphere? What causes its destruction?
11. Outline the two major driving forces for ocean circulation.
12. Why is there so little mixing between the waters of the deep ocean and the waters in its surface layers?
13. Where is ocean water densest? What causes the density variations?
14. Why would the development of a warm surface layer associated with the El Niño event off the coast of South America inhibit coastal upwelling?
15. What are the major constituents of seawater? What is the origin of the dissolved ions?
16. Do you think Earth's climate has always been the way it is today? What evidence supports your conclusion?
17. What could cause the climate to change?
18. How is Earth's atmosphere different from any other in our solar system? What would you conclude if you discovered a planet with an oxygen-rich atmosphere?
19. What are the potential human-caused changes in Earth's climate system?
20. Why are wind velocities and wave heights so strongly correlated in the maps in the State of the Art section?

10 Weathering

Weathering involves the breakdown of rocks at Earth's surface by physical processes and chemical reactions with air and water. Erosion is the transport of this material.

A new building gradually deteriorates. The paint chips and peels, wood dries and splits, and even brick, building stone, and cement eventually decay and crumble. Left alone, most buildings decompose into a pile of rubble within a few hundred years. This process of natural decay is called weathering. Weathering is a general term describing all of the changes that result from the exposure of rock materials to the atmosphere.

The spires and columns of Bryce Canyon National Park in Utah vividly show how weathering modifies a rock body. Once a solid mass of sedimentary rock, these spires and columns were largely created by the gradual decomposition of rock by reactions with water and air. The loose material fell downslope and was eventually carried away by streams. Here, weathering is controlled by vertical joints and by differences in the various layers of colorful sedimentary strata. The intersecting joint systems produce a series of columns that are modified during weathering

into an infinite variety of forms by different bedding characteristics. Take a moment to study this photograph. Can you see that the columns are aligned in rows parallel to joint systems? Can you recognize certain horizons in the sedimentary strata that weather much more rapidly than others? Can you see joint systems being enlarged by weathering to separate the rocks into columns? Weathering has produced this remarkable landscape, but the effects of weathering can be seen everywhere.

From a geologic point of view, weathering is important because it transforms the solid bedrock into small, decomposed fragments and prepares those fragments for removal by the agents of erosion.

It would be difficult to overemphasize the importance of weathering to humans. Without weathering, Earth would be forbidding indeed. The continents would be bare, hard rock, for no soil cover could develop; consequently, Earth would be devoid of plant and animal life. In addition to producing the soil on which agriculture depends, weathering produces some other very practical products. Sand, gravel, and clay deposits are the indirect results of weathering. Practically all aluminum ore, most iron ore, and some copper ore are formed and concentrated by weathering. Consequently, it is important for us to understand this important component of Earth's systems.

Major Concepts

1. Weathering is the breakdown and alteration of rocks at Earth's surface through physical and chemical reactions with the atmosphere and the hydrosphere.
2. Physical weathering is the mechanical fragmentation of rocks from stress acting on them. Ice wedging may be the most important type.
3. Chemical weathering involves chemical reactions with minerals that progressively decompose the solid rock. The major types of chemical weathering are dissolution, acid hydrolysis, and oxidation.
4. Joints and fractures facilitate weathering because they permit water and gases in the atmosphere to attack a rock body at considerable depth. They also greatly increase the surface area on which chemical reactions can occur.
5. The major products of weathering are spheroidal rock forms, a blanket of regolith, and dissolved ions. Soil is the upper part of the regolith—a mixture of clay minerals, weathered rock particles, and organic matter.
6. Climate and rock type greatly influence the type and rate of weathering. On the other hand, weathering helps control the amount of carbon dioxide in the atmosphere, and thus climate.

The Nature of Weathering

To appreciate how geologic processes erode the surface of Earth, and how the landscape evolves, one first needs to understand the nature of weathering—the disintegration and decomposition of rocks. By definition, weathering is different from erosion. Weathering involves only the breakdown of rock, whereas erosion involves the removal of debris produced by the breakdown. In reality, however, weathering and erosion are intimately involved with one another. Weathering disintegrates solid rock and produces loose debris. Erosion by running water, wind, and ice removes the debris and exposes fresh rock, which is then weathered, and the cycle continues. The results of weathering are seen everywhere, from the debris along hill slopes to decomposed monuments of antiquity (**Figure 10.1**).

Like metamorphism, its counterpart deep within the crust, weathering reflects adjustments of rocks exposed to a new environment. Minerals in rocks are in equilibrium with where they originate (in terms of temperature, pressure, and chemical environment, for example). If they are exposed to a different environment the elements in minerals will slowly adjust to different forms that are stable under the new conditions. In weathering, rocks adjust and are altered to forms more stable at low pressure, low and fluctuating temperatures, and the chemical environment with abundant water that prevails at Earth's surface. Thus, metamorphic rocks and igneous intrusions are generally most susceptible to weathering.

Weathering, then, involves a multitude of physical, chemical, and biological processes, but two main types of weathering are recognized: (1) **physical weathering** and (2) **chemical weathering**. Physical (or mechanical) weathering breaks the rock mass into small particles. It is strictly a physical process involving no change in chemical composition. Chemical weathering alters the rock by chemical reactions between elements in the atmosphere and those in the rocks. Most geologists believe that chemical weathering is most important in terms of total amount of rock breakdown. In most places, however, the two processes work together, each facilitating the other, so that the final product results from a combination of the two processes.

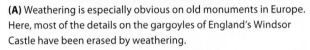

(A) Weathering is especially obvious on old monuments in Europe. Here, most of the details on the gargoyles of England's Windsor Castle have been erased by weathering.

(B) Weathering is apparent from the fallen debris on many slopes. Here the resistant sandstone butte is shrinking as weathering separates fragments that fall and accumulate at the base of the cliff.

Figure 10.1 **The effects of weathering** are seen whenever rocks are exposed. These photographs show typical examples.

Physical Weathering

> Physical weathering is the breakdown of rock into small fragments by physical processes without a change in chemical composition. No chemical elements are added to, or subtracted from, the rock. The most important types of physical weathering are ice wedging and sheeting, or unloading.

Ice Wedging

Figure 10.2 includes a simple diagram showing how **ice wedging** breaks a rock mass into small fragments. Water from rain or melting snow easily penetrates cracks, bedding planes, and other openings in the rock. As it freezes, it expands about 9%, exerting great pressure on the rock walls, similar to the pressure produced by driving a wedge into a crack. Eventually, the fractured blocks and bedding planes are pried free from the parent material. The stress generated each time the water freezes is approximately 110 kg/cm^2, roughly equivalent to that produced by dropping a 98-kg ball of iron (about the size of a large sledgehammer) from a height of 3 m. Stress is exerted with each freeze, so that, over a period of time, the rock is literally hammered apart.

Ice wedging occurs under the following conditions: (1) when there is an adequate supply of moisture; (2) where preexisting fractures, cracks, or other voids into which water can enter occur within the rock; and (3) where temperatures frequently rise and fall beyond the freezing point. Temperature fluctuation above and below the freezing point is especially important because pressure is applied with each freeze. In areas where freezing and thawing occur many times a year, ice wedging is far more effective than in exceptionally cold areas, where water is permanently frozen. Ice wedging thus

How does physical weathering break down a mass of solid rock into small fragments?

© Alexey Lebedev/ShutterStock, Inc.

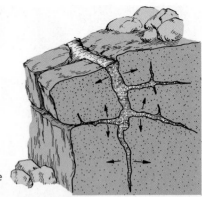

(A) Ice wedging occurs when water seeps into fractures and expands as it freezes. The expanding wedge forces the rock apart and produces loose, angular fragments that move downslope by gravity and accumulate at the base of the cliff as talus cones.

Courtesy of U.S. Department of Agriculture

(B) The effects of ice wedging in the Teton Range in Wyoming are seen in both the rugged surface of the mountain peaks and the accumulation of fragmented debris at the base of the cliff. Massive granite cut by numerous fractures forms the mountain range. Ice wedging, controlled in part by the fractures, produces the sharp, angular texture of the mountain peaks. The debris derived from ice wedging has accumulated in conical slopes near the base of the cliff.

Figure 10.2   Ice wedging is an important type of physical weathering in areas where temperatures rise above and fall below the freezing point.

occurs most frequently above the timberline. It is especially active on the steep slopes above valley glaciers, where meltwater produced during the warm summer days seeps into cracks and joints and freezes during the night (Figure 10.2).

The process of ice wedging has been known for years, and more than 100 years ago, ice wedging was used in some quarrying operations. Workers would drill a series of holes along the line of a desired cut and fill them with water. The expansion accompanying freezing would split the rock apart almost as cleanly as modern methods.

In arid regions, the growth of salt crystals in pores and cracks can also pry apart rock. The crystals grow by evaporation of the salt-laden spray that accumulates in the fractures of rocks exposed along the shores of salty lakes or the sea. This process is vividly expressed in the shattering of fence posts near the shore of the Great Salt Lake (**Figure 10.3**).

Figure 10.3 Growth of salt crystals has shattered these fence posts near the shore of the Great Salt Lake, Utah. Salty groundwater seeps into the wood, and, as it evaporates, salt crystals grow, expand, and break apart the wood fibers.

Sheeting

Rocks formed deep within Earth's crust are under great confining pressure from the weight of thousands of meters of overlying rocks. As this overlying cover is removed by erosion, the confining pressure is released, and the buried rock body tends to expand. The internal stresses, set up by expansion, can cause large fractures, or expansion joints, parallel to Earth's surface (**Figure 10.4**). The result is known as **sheeting**. It can be observed directly in quarries, where the removal of large blocks is sometimes followed by the rapid, almost explosive expansion of the quarry floor. A sheet of rock several centimeters thick may burst up, and at the same time, numerous new parallel fractures will appear deeper in the rock body. The same process occasionally causes rock bursts in mines and tunnels, when the confining pressure is released during the tunneling operation. It can also be seen in many valley walls and in excavations for roads, where rock slumping, due to sheeting, can cause serious highway problems.

Other Types of Physical Weathering

Animals and plants play a variety of relatively minor roles in physical weathering. Burrowing animals, such as rodents, mechanically mix the soil and loose rock particles, a process that facilitates further breakdown by chemical means. Pressure from growing roots widens cracks and contributes to the rock breakdown. Lichens can live on the surface of bare rock and extract nutrients from its minerals by ion exchange; the presence of lichens, therefore, results in both physical and chemical alteration of the minerals. These processes may seem trivial, but the work of innumerable plants and animals over a long period of time adds significantly to the disintegration of the rock. Although dismissed by some geologists, recent evidence is mounting that thermal expansion and contraction of the rock caused by daily or seasonal temperature changes may also be an effective process of physical weathering.

What is the difference between sheeting and stratification?

Figure 10.4 **Sheeting in granite** of the Sierra Nevada occurs as erosion removes the overlying rock cover and reduces the confining pressure. The bedrock expands, and large fractures develop parallel to the surface. Ice wedging may subsequently enlarge the fractures.

Talus

The products of physical weathering are best seen in high mountain country, where ice wedging dominates and produces a large volume of angular rock fragments. This material commonly accumulates in a pile at the base of the cliffs from which it was derived. Because most cliffs are notched by steep valleys and narrow ravines, the fragments dislodged from the high valley walls are funneled through the ravines to the base of the cliff, where they accumulate in cone-shaped deposits known as **talus cones** (**Figure 10.5**).

Talus cones are built up by isolated blocks loosened by physical weathering. The blocks commonly fall separately, as almost any mountain climber can testify, but large masses of the material on steep slopes may be moved by an avalanche. Earthquakes may also suddenly activate large numbers of blocks loosened by many seasons of ice wedging.

In the example shown in Figure 10.5, all of the talus has accumulated since the last ice age, which terminated 10,000 to 15,000 years ago. This is a considerable amount of material produced by physical weathering alone.

Chemical Weathering

Chemical weathering is the breakdown of minerals by chemical reactions with the atmosphere or hydrosphere. The three main types of chemical reactions are (1) dissolution, (2) hydrolysis, and (3) oxidation.

During chemical weathering, rocks are decomposed, the internal structure of the minerals is destroyed, and new minerals are created. Thus, there is a significant change in the chemical composition and physical appearance of the rock.

Dissolution

Dissolution is a process whereby a mineral passes completely into solution, like salt dissolving in water. Some minerals dissolve directly in water and the ions are **leached**, or flushed away. Halite (salt) is perhaps the best-known example. It is

Courtesy of L. F. Hintze

Figure 10.5 **Talus cones** are piles of rock debris that accumulate at the base of a cliff as the result of rockfall. Most rock fragments in talus cones are produced by ice wedging as here in the Canadian Rockies.

extremely soluble, surviving at Earth's surface only in the most arid regions. Gypsum is less soluble than halite but is also easily dissolved by surface water. Few, if any, large outcrops of these minerals occur in humid regions. This kind of dissolution happens because water is one of the most effective and universal solvents known. The structure of the water molecule requires the two hydrogen atoms to be positioned on the same side of the larger oxygen atom. The molecule thus has a concentration of positive charges on the side with the two hydrogen atoms, balanced by a negative charge on the opposite side. As a result, the water molecule is polar and behaves as a tiny magnet. It acts to loosen the bonds of the ions at the surface of minerals with which it comes into contact. Because of the polarity of the water molecule, practically all minerals are soluble to some extent in water, but those with ionic, rather than covalent, bonds are more easily dissolved.

Acid Hydrolysis

The most common dissolution reactions involve slightly acidic water. Carbonic acid (H_2CO_3) is common in natural environments and forms when water combines with carbon dioxides. This reaction takes place in the atmosphere and in the root zones of plants where carbon dioxide is released into the soil. In addition, bacteria in the soil combine oxygen with decaying organic materials to make carbonic acid. Consequently, water seeping through organic remains becomes more and more acidic and its effectiveness as a weathering agent continually increases. Other acids are also produced by plant activity and by bacterial decay of plant and animal remains. The result

What are the products of chemical weathering?

is seen dramatically in regions such as the Great Lakes area, where rivers flow through bogs and marshes and the organic acids stain the water yellowish brown. Human activities have also produced acids that contaminate surface waters, including sulfuric acid and nitric acid in acid rain and sulfuric acid from mining coal or sulfide minerals. The effects of these acids are seen in the corrosion of buildings and acidification of lakes and rivers and occasionally in the destruction of their biota.

Hydrolysis is a chemical reaction wherein water and another substance both decompose into ions; the OH^- ion groups with one of the fragments and the H^+ ion with another fragment. As you examine the following reactions, observe how the H^+ and the OH^- ions are derived from splitting water molecules. Hydrolysis can occur in pure water, but in the natural world it usually accompanies reactions with acids; thus, this kind of reaction between a mineral and an acid is usually called *acid hydrolysis*.

What are the major chemical reactions in weathering?

To simplify what is a far more complex series of reactions, we will illustrate weathering reactions involving calcite and carbonic acid. In pure water, calcite is not very soluble. But water with carbonic acid is capable of dissolving much more calcite than is pure water. Carbonic acid forms when rainwater combines with carbon dioxide in the atmosphere or the soil by the reaction:

$$\underset{\text{(water)}}{H_2O} \quad + \quad \underset{\text{(carbon dioxide)}}{CO_2} \quad = \quad \underset{\text{(carbonic acid)}}{H_2CO_3}$$

This acid may then react with calcite to form calcium and bicarbonate ions in solution. This reaction may be expressed as follows:

$$\underset{\text{(calcite)}}{CaCO_3} \quad + \quad \underset{\text{(carbonic acid)}}{H_2CO_3} \quad = \quad \underset{\text{(calcium bicarbonate)}}{Ca^{2+} + 2HCO_3^-}$$

Some silicate minerals may also dissolve, although not as readily as calcite. For example, pyroxene will slowly dissolve when it is in contact with acidic waters according to the following reaction:

$$\underset{\text{(pyroxene)}}{MgSiO_3} + \underset{\text{(water)}}{H_2O} + \underset{\text{(carbonic acid)}}{2H_2CO_3} = \underset{\text{(ions)}}{Mg^{2+} + 2HCO_3^-} + \underset{\substack{\text{(silicic} \\ \text{acid)}}}{H_4SiO_4}$$

Acid Hydrolysis and Secondary Minerals

Another important kind of hydrolysis reaction involves the formation of new minerals, in addition to the dissolved ions. You might think of this as a kind of partial solution, with some ions going into solution and being carried away at the same time as a new mineral forms. During chemical weathering, these new minerals are almost all *hydrated*—that is, they have water in their structures. The water is not merely absorbed, as by a sponge, but actually incorporated as OH^- ions into the atomic structure of the new mineral (**Figure 10.6**). Most silicate minerals, especially those containing aluminum, do not simply dissolve in water. Instead, they react to form new minerals and free ions.

How are natural acids formed?

A good example of the production of secondary minerals is the chemical weathering of feldspar. As you recall from previous chapters, feldspar is an abundant mineral in a great many igneous, metamorphic, and sedimentary rocks. It is therefore important to understand how feldspars weather and decompose to make clay minerals. In turn, these clay minerals are transported and deposited to form the most abundant sedimentary rock, shale (or, strictly speaking, mudrocks).

If plagioclase feldspar, the most common silicate mineral in the crust, comes in contact with water containing carbonic acid, the following general reaction takes place:

$$\underset{\text{(Na-plagioclase)}}{2NaAlSi_3O_8} \quad + \quad \underset{\text{(carbonic acid)}}{2H_2CO_3} \quad + \quad \underset{\text{(water)}}{9H_2O} \quad =$$

$$\underset{\text{(dissolved components)}}{2Na^+ \quad + \quad 2HCO_3^- \quad + \quad 4H_4SiO_4} \quad + \quad \underset{\text{(clay mineral)}}{Al_2Si_2O_5(OH)_4}$$

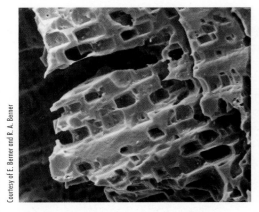

Courtesy of E. Berner and R. A. Berner

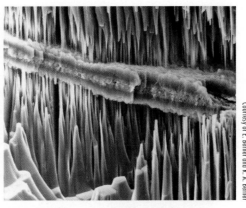

Courtesy of E. Berner and R. A. Berner

(A) Weathering creates small rectangular pits in plagioclase feldspar. The shapes of the pits are controlled by the internal arrangement of ions in the mineral's framework structure.

(B) Weathering corrodes amphibole to make long needles or cones of residual material controlled by the chain structure of the ions in the mineral. A thin vein of clay (purple) formed at the expense of the amphibole.

Figure 10.6 Chemical weathering of minerals can be seen with an electron microscope. The magnification is about 3000 times.

This reaction is simplified; it actually takes several steps to form clay. This clay mineral does not contain sodium, which was present in the original feldspar. The new mineral also has a new crystal structure, consisting of sheets of silicate tetrahedra that form submicroscopic crystals. The Na^+ ion is dissolved in the water. Silica is also released from the minerals and goes into solution in the water as a weak acid (H_4SiO_4). The water may then carry the dissolved components away from the site of reaction. For example, the sodium eventually accumulates in ocean water as dissolved salt. As a result of weathering reactions like this, the shales that form from the accumulation of such clays are poor in sodium, compared with their igneous precursors.

Potassium and calcium feldspars go through similar weathering reactions to produce clays, but K^+ ions are largely retained in the soil by absorption and become important nutrients for plants. When the plants die, the potassium is returned to the soil. Many of the Ca^{2+} ions released by weathering are transported to the oceans, but they eventually react with dissolved CO_3^{2-} to form carbonate minerals. These two processes leave seawater rich only in Na^+.

The effects of chemical weathering of minerals are clearly seen with a scanning electron microscope and are truly remarkable (Figure 10.6). Rectangular etch pits develop on weaknesses in the crystal structure of plagioclase, shown here at a magnification of 3000 times. As weathering proceeds, the pits grow and merge to destroy the fabric of the rock containing the feldspar. A macroscopic example of the effects of chemical weathering can be seen in the fragments of ancient granite columns that were partly buried in the mud of the Nile floodplain (**Figure 10.7**). Weathering destroyed the delicate carvings in the monuments as feldspar converted to clay minerals.

Oxidation

Oxidation is the chemical combination of oxygen, in the atmosphere or dissolved in water, with one mineral to form a completely different mineral in which at least one of the elements has a higher oxidation state (higher ionic charge). Of the elements that have variable charges, iron is the most important in weathering reactions on Earth. In most silicates, iron is present as Fe^{2+}, but in the presence of Earth's modern oxygen-rich atmosphere, Fe^{3+} is the favored oxidation state. Therefore, oxidation is especially important in the weathering of minerals that have a high iron content, such as olivine, pyroxene, and amphibole. Oxidation of silicates is commonly accompanied by hydrolysis and partial solution. In the case of olivine, the reaction is as follows:

Figure 10.7 A column of granite from a temple in Lower Egypt fell over and its right side was partially buried for hundreds of years. Moisture in the soil facilitated hydrolysis and altered much of the feldspar to clay. The left side of the column was exposed only to the dry atmosphere and remained fresh and unaltered for more than 2000 years.

$$2Fe_2SiO_4 \;+\; 4H_2O \;+\; O_2 \;=\; 2Fe_2O_3 \;+\; 2H_4SiO_4$$

(olivine)   (water)   (oxygen)   (hematite)   (silicic acid)

In this reaction, the iron in silicate minerals unites with oxygen to form the mineral hematite (Fe_2O_3). Hematite is deep red, and if it is dispersed in sandstone or shale, it imparts a red color to the entire rock. Limonite [$FeO(OH)$] is another common weathering product. It is formed by oxidation combined with a reaction with water.

Concluding Notes

Why is seawater so rich in sodium, calcium, and magnesium?

By carefully examining the reactions above and **Tables 10.1** and **10.2**, you should be able to detect a general pattern for chemical weathering. Most alkali (e.g., Na and K) and alkaline earth (e.g., Ca and Mg) elements are removed into solution by weathering reactions (Table 10.1) and eventually become enriched in seawater (Table 10.2). On the other hand, the solid mineral residue becomes enriched in Al, Si—incorporated in clays—and Fe— incorporated in oxides. These minerals are stable in the surface environment and relatively insoluble.

Figure 10.8 lists common minerals in order of their susceptibility to chemical weathering. This is a powerful example of how the materials in Earth's systems are constantly changing toward equilibrium. Minerals at the top of the list weather easily and rapidly; those at the bottom weather slowly and are resistant to change. Note that the order shown for igneous silicate minerals corresponds to their typical temperature of formation. Thus, olivine weathers more readily than plagioclase and plagioclase more readily than muscovite. Moreover, these minerals commonly weather to form minerals found low on the list—minerals that are stable at the cool, wet, oxygen-rich surface of the planet.

Inasmuch as feldspars and other silicate minerals that weather into clay constitute a large percentage of igneous and metamorphic rocks, an enormous amount of clay has been produced by the weathering of these minerals throughout geologic time. It has been calculated that sediment and sedimentary rocks have an average thickness of 1 km throughout the ocean basins, 2.5 km on the continental shelves, and 1.5 km on the continents. Because clay makes up about one-third of all sedimentary rocks, the total amount of clay would form a layer over 1 km thick if spread uniformly over the entire surface of Earth.

We have considered physical and chemical weathering as separate processes, but in nature they are inseparable because many types of weathering processes are usually involved in the weathering of any outcrop. Mechanical fracturing of a rock increases the surface area, where chemical reactions take place, and permits deeper penetration of reactive fluids that cause chemical decomposition. Chemical decay in turn facilitates physical disintegration. One process may dominate in a given area, depending on the climate and rock composition, but physical and chemical weathering processes generally attack the rock at the same time.

Table 10.1 Weathering Reactions for Common Minerals

Original Mineral	General Formula	Weathering Reactions	Dissolved Ions	Residual Minerals
Gypsum	$CaSO_4 \cdot 2H_2O$	Dissolution by water	Ca, SO_4	
Halite	NaCl	Dissolution by water	Na, Cl	
Olivine	$(Mg,Fe)_2SiO_4$	Oxidation Dissolution by acid	 Mg, Fe	Fe oxides
Pyroxene	$Ca(Mg,Fe)Si_2O_6$	Oxidation Dissolution in acid	 Mg, Fe, Ca	Fe oxides
Amphibole	$NaCa(Mg,Fe)_5AlSi_7O_{22}(OH)_2$	Oxidation Partial solution by acid	 Na, Ca, Mg	Fe oxides Clay
Plagioclase	$NaAlSi_3O_8$ to $CaAl_2Si_2O_8$	Partial solution by acid	Na, Ca	Clay
K-feldspar	$KAlSi_3O_8$	Partial solution by acid	K	Clay
Muscovite	$KAl_3Si_3O_{10}(OH)_2$	Partial solution by acid	K	Clay
Biotite	$K(Mg,Fe)_3AlSi_3O_{10}(OH)_2$	Oxidation Partial solution by acid	 K, Mg	Fe oxides Clay
Quartz	SiO_2	Resists dissolution		Quartz
Calcite	$CaCO_3$	Dissolution by acid	Ca	
Dolomite	$CaMg(CO_3)_2$	Dissolution by acid	Mg, Ca	
Pyrite	FeS_2	Oxidation	SO_4	Fe oxides

Table 10.2 Contribution of Weathering to Water Compositions (in parts per million)

Component	Rain Water	River Water	Groundwater	Ocean Water
HCO_3^-	0.5	58	93	28
SO_4^{2-}	2	11	32	905
Cl^-	1	7	5	19,400
Ca^{2+}	1	15	27	412
Mg^{2+}	0.5	44	6	1,290
Na^+	1	6	9	10,800
K^+	0.5	2	1	380
Fe^{2+}		0.6	1.6	0.1
Si^{4+}		6	18	2

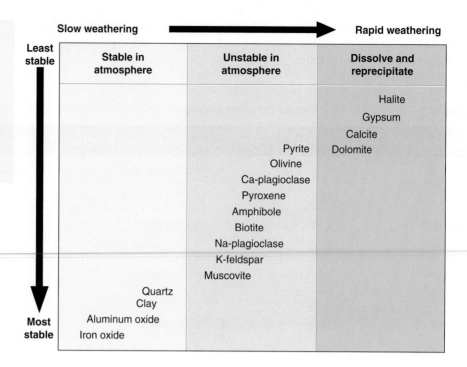

Figure 10.8 Relative susceptibility to weathering varies widely among common minerals found at Earth's surface. Minerals at the top of the diagram react to form minerals near the bottom that are stable at low temperatures and pressures and in the presence of abundant water and oxygen. The ultimate weathering products of many rocks are clays, quartz, and oxides of aluminum and iron.

Weathering of Major Rock Types

The weathering of rocks is influenced by a number of variables, such as the mineral composition, the texture of the rock, and the climate in which weathering occurs. Differential weathering is a result of differences in the rates of weathering.

Weathering is influenced by so many factors that it is difficult to make a meaningful generalization concerning the weathering of specific rock types. Limestone, for example, may weather and erode into a soil-covered valley in a humid climate, whereas the same formation forms a cliff in an arid climate. Similarly, well-cemented quartz sandstone may be extremely resistant to weathering, whereas sandstone with high clay contents is likely to be soft and weak and weather rapidly.

Mineral composition is of prime importance. Some minerals, such as quartz, are very stable and remain essentially unaltered for long periods of time; others, such as olivine and the feldspars, are very unstable and begin to decompose almost immediately (Figure 10.8). The texture of the rock is also very significant because of its influence on **porosity** and **permeability**, which govern the ease with which water can enter the rock and attack the mineral grains. Precipitation and temperature are the chief climatic controls, but weathering will be influenced not only by the total rainfall, but also by the distribution of precipitation through time, percent of runoff, and rate of evaporation. Therefore, a given rock will respond to weathering in a variety of ways, but in general the major rock groups weather in a characteristic fashion.

Granite is a massive homogeneous rock composed of feldspar, quartz, and mica, with minor amounts of other minerals. It forms at high temperatures, in considerable depth, and under great pressure, so it is out of physical and chemical equilibrium when exposed at Earth's surface. For example, the release of pressure resulting from the erosion of the overlying rocks produces expansion joints, which aid in the development of exfoliation.

Feldspars weather rapidly by chemical reaction with water and are altered to various clay minerals. Calcium plagioclase is least resistant, followed by potassium feldspars. Mica weathers somewhat more slowly than most feldspars but is easily attacked along its cleavage planes by water, and oxidation of iron and ion exchange are common. Micas alter, with little change in structure, to chlorite and clay minerals.

In contrast, quartz is very resistant to both chemical and physical weathering and remains essentially unaltered as the other minerals are decomposed. Therefore, it constitutes the most significant particle or fragment produced by the weathering of granite.

Basalt is a fine-grained rock composed mostly of feldspar, olivine, and pyroxene. The surface of a basalt flow is generally vesicular and very porous, and the interior of the rock body is commonly broken by a system of columnar joints. Therefore, flows are highly permeable and susceptible to decomposition. Olivine, pyroxene, and calcium plagioclase—the most common minerals in basalt—are all highly susceptible to chemical weathering. Quartz is not present in basalt, so most minerals in this rock are eventually converted to clay and iron oxides. The ultimate weathering product is a red or brown soil.

Sandstone is composed mostly of quartz grains, with varying amounts of small rock fragments, feldspar, and clay minerals. The quartz in sandstone is highly resistant to chemical weathering, so chemical decomposition of the rock consists largely of an attack on the cement. The major cementing materials in sandstones are calcite, iron oxide, and quartz.

Limestone is composed mostly of the mineral calcite, although it generally contains some clay and other impurities. It is the most soluble of the common rock types, and (except in extremely dry climates) solution is the dominant weathering process. In most limestone regions in humid climates, solution activity enlarges joints and bedding planes and forms a network of caverns and caves; the limestone formations in such regions typically form valleys. In arid regions, where solution activity is at a minimum, limestone forms cliffs.

Shale commonly weathers faster than most other rocks because it is fine-grained and soft. Because it contains a high proportion of clay, it has the ability to absorb and expel large amounts of water.

Why do various rock types weather in different ways?

Differential Weathering

As can be seen from the preceding brief descriptions, different rock masses, or different sections of the same rock, weather at different rates. This variation is known as **differential weathering**. It occurs on a broad scale, from the great sandstone ridges of the Appalachian Mountains to delicate etching of thin layers in sedimentary rock. The more-resistant zones stand out as ridges, and the weaker zones form depressions. Differential weathering can lead to the formation of unusual shapes and forms, such as the spindles and pinnacles in Bryce Canyon (see the chapter-opening photograph) or pits and caverns on a rock face. Differential erosion on dikes of igneous rocks can form trenches or walls, depending on whether the dike is harder or softer than the surrounding rock.

Differential weathering can be seen everywhere a rock is exposed. Study the photo of Bryce Canyon and you will notice that each layer has its own weathering characteristics. The white layers erode most rapidly and tend to form slender columns. The thicker beds of sand are more resistant, whereas the interbeds of siltstone and shale weather rapidly. Thus, the horizontal layers are etched into ridges and furrows, which are responsible for much of the beauty in this scene.

Why do rocks weather at different rates?

Products of Weathering

The major products of weathering are (1) rock bodies modified into spherical shapes; (2) a blanket of loose, decayed rock debris, known as regolith, of which soil is an important part; and (3) ions in solution.

Geometry of Weathered Rock Fragments

The breakdown of rocks and the shapes of most rock fragments are inherited from patterns of joints, bedding, cleavage, and other planes of structural weakness in the parent rock material. The best way to appreciate how joints, bedding planes, and other planes of weakness influence the geometry of rock fragmentation is to compare and contrast outcrops of several rock types and consider the shape of the fragmented material that weathering has produced (**Figure 10.9**).

Importance of Fractures and Joints   Almost all rocks are broken into a system of fractures that greatly influence the weathering of rock bodies in two ways. First, they effectively cut large blocks of rock into smaller ones, thereby increasing the surface area where chemical reactions take place. The importance of joints in weathering processes can be appreciated by considering the amount of new surface area produced by jointing. Consider, for example, a cube of rock that measures 10 m on each side (**Figure 10.10**). If only the upper surface of the cube were exposed and the rock were not jointed, weathering could attack only the exposed top surface of 100 m². If the block were bounded by intersecting joints 10 m apart, however, the surface area exposed to weathering processes would be 600 m². If three additional joints cut the cube into eight smaller cubes, the surface exposed to weathering would be 1200 m². If joints 1 m apart cut the rock, 6000 m² of rock surface would be exposed. Obviously, a highly jointed rock body weathers much more rapidly than a solid one. The breakdown of a rock along a system of jointing planes is known as **joint-block separation**. Figure 10.9A shows a basalt flow that broke into hexagonal columns as it cooled.

Besides providing a larger surface area for chemical decomposition, joints also act as a system of channels through which water can more readily penetrate a rock body. Joints thus permit physical and chemical weathering processes to attack the rock from several sides, even hundreds of meters below the surface.

Spheroidal Weathering

In the weathering process, there is a universal tendency for rounded (or spherical) surfaces to form on a decaying rock body regardless of the original shapes of the rock fragments. The sphere is the geometric form that has the least amount of surface area per unit of volume. A rounded shape is produced because weathering attacks an exposed rock from all sides at once, and decomposition is most rapid along the corners and edges of the rock (**Figure 10.11**). As the decomposed material falls off, the corners become rounded, and the block eventually is reduced to an ellipsoid or a sphere. Once the block attains this shape, it simply becomes smaller with further weathering. This process is known as **spheroidal weathering**.

Examples of spheroidal weathering can be seen in almost any exposure of rock (**Figure 10.12**). It can also be seen in the rounded blocks of ancient buildings and monuments. The original blocks had sharp corners and were fitted together with precision. The edges are now completely decomposed, and each block has assumed an ellipsoidal or spherical shape. In nature, spheroidal weathering is produced both at the surface and at some depth.

Exfoliation is a special type of spheroidal weathering in which the rock breaks apart by separation along a series of concentric shells or layers that look like cabbage leaves (Figure 10.9E). The layers, essentially parallel to each other and to the surface, develop by both chemical and physical means. Exfoliation may involve sheeting in rocks such as granite; if they are brought to the surface after deep burial, they have a tendency to expand upward and outward as the overlying rock is removed. In cold climates, ice wedging along the sheeting joints helps to remove successive layers gradually. The increase in volume of mineral grains associated with the chemical weathering of feldspar might also promote exfoliation. Exfoliation causes massive rocks, such as granite, to develop a spherical form characterized by a series of concentric layers ranging from boulders to a mountain.

Regolith

The results of weathering can be seen from the driest deserts and the frozen wastelands, to the warm, humid tropics. The most obvious product of weathering is a blanket of loose, decayed rock debris known as **regolith**, which forms a discontinuous cover over the solid, unaltered bedrock below it. The term *regolith* comes from the Greek work *rego*, meaning "blanket" (blanket rock). It is a layer of soft, disaggregated

© Chris Geszvain/ShutterStock, Inc.

(A) Joint-block separation results when prominent fractures divide the rock into small blocks. The Devil's Post Pile in California is an excellent example, where columnar joints control the geometric patterns of rock breakup.

(B) Bedding-plane separation occurs along a bedding zone of weakness in sedimentary rocks and causes the rock to break up into slabs. Foliation in metamorphic rock causes a similar type of weathering.

(C) Jointing is commonly the major type of structural weakness in granite and related rocks and causes the rock to break up into large blocks. Spheroidal weathering then rounds the edges of the fragments.

(D) Granular disintegration in granite is common, producing crumbly spheroidal boulders. The disintegrated material consists of feldspars weathered to clay and quartz grains. The dissolution of calcite cement in sandstone also causes granular breakdown.

(E) Exfoliation occurs when the solid rock mass comes apart in a series of shells or plates that roughly conform to the shape of the outer surface. Exfoliation can occur on a very large scale, such as on this dome in Yosemite National Park, California, or on a very small scale, with the individual plates being only a millimeter or less thick.

(F) Shattering occurs when a rock is subjected to severe stress that ruptures the rock into sharp, irregular, angular blocks. Ice wedging shatters rock outcrops in nature. Repeated cycles of heating and cooling may also cause shattering. Blasting bedrock with explosives produces shattering artificially.

Figure 10.9 The geometric patterns of rock disintegration depend on the composition, texture, and structure (especially layers and joints) of the parent rock body.

rock material formed in place by the decomposition and disintegration of the bedrock that lies beneath it. Within the regolith, the individual grains or small groups of mineral particles are easily separated, one from the other. The thickness of the regolith ranges from a few centimeters to hundreds of meters, depending on the climate, type of rock, and length of time that weathering processes have been operating. The transition from bedrock to regolith can be seen in road cuts and stream valleys.

Gravel, sand, silt, and mud deposited by streams, wind, and glaciers are sometimes referred to as transported regolith, in order to distinguish them from the residual

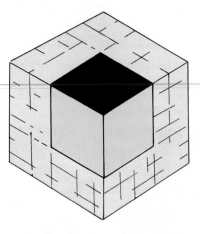

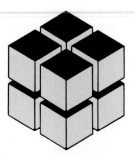

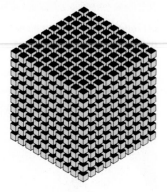

(A) A surface of bedrock, 10 m long and 10 m wide, with no joints, exposes a total area of 100 m² to weathering processes. A set of joints 10 m apart would expose more rock to the atmosphere and would increase the surface area vulnerable to weathering to 600 m².

(B) Three additional joints, dividing the block into eight cubes, would increase the surface area to 1200 m².

(C) If joints 1 m apart cut the rock, the surface area exposed to weathering would be increased to 6000 m².

Figure 10.10 A system of joints cutting a rock body greatly increases the surface area exposed to weathering.

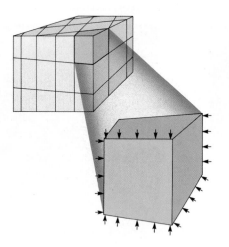

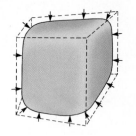

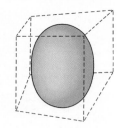

(A) Joint systems cut a rock body into angular blocks.

(B) On each block, weathering proceeds inward from the joint face.

(C) The corners of the block are soon completely decomposed, so the weathered rock assumes a spherical or ellipsoidal shape.

Figure 10.11 Spheroidal weathering occurs because the edges and corners of a joint block are easily decomposed.

Figure 10.12 Spheroidal weathering shapes natural outcrops like these jointed and weathered granite blocks from near Prescott, Arizona.

regolith produced by weathering. Many types of transported regolith, or surficial deposits, have been identified, and are associated with rivers, glaciers, and wind.

A regional view of the regolith and its relationship to bedrock is given in **Figure 10.13**. The photograph shows exposures of bedrock limited to certain areas of resistant limestone and sandstone strata, which form discontinuous cliffs along the upper part of the mountain front. On the steep canyon walls, little soil is retained, and bedrock is exposed from the base to the top of the canyon. The sketch in Figure 10.13B was made from the photograph and outlines the rock outcrops. In Figure 10.13C, the outcropping bedrock is not shown, so the regolith appears as a thin, discontinuous blanket with "holes" where bedrock is exposed. Sediment fills the valley in the foreground, but the regolith there is not shown in the diagram. If you carefully study the pattern of exposed bedrock areas Figure 10.13B, you can see that the strata are warped into broad folds, shown in Figure 10.13D, which form the internal structure of the mountains.

Soil

The uppermost layer of the regolith is the **soil**. It is composed chiefly of small particles of rock, new minerals formed by weathering, plus varying amounts of decomposed organic matter. Soil is so widely distributed and so economically important that it has acquired a variety of definitions, and you should be aware that the term, as used by engineers, geologists, farmers, and soil scientists, has somewhat different definitions.

The transition from the upper surface of the soil down to fresh bedrock is a **soil profile**, which shows a rather systematic sequence of layers, or **horizons**, distinguished by composition, color, and texture. These are shown in **Figure 10.14** for a humid temperate climate. The **A horizon** is the topsoil layer, which often is visibly divided into three layers: A0 is a thin surface layer of leaf mold, especially obvious on forest floors; A1 is a humus-rich, dark layer; and A2 is a light, bleached layer. The **B horizon** is the subsoil, which contains fine clays and colloids washed down from the topsoil. It is largely a zone of accumulation and commonly is reddish in color. The **C horizon** is a zone of partly disintegrated and decomposed bedrock. The individual rock fragments are often weathered, spheroidal boulders that may be completely decomposed. The C horizon grades downward into fresh, unaltered bedrock.

Several special kinds of soils are worth noting here. The major soil orders are described in **Table 10.3** and their distribution is shown in **Figure 10.15**. Some soils are noteworthy because they are ore deposits. For example, aluminum does not migrate far during weathering, and, in fact, it may be concentrated as a residual

What is the ultimate origin of soil?

(A) The Wasatch Range in central Utah displays contrasting areas of bedrock and regolith.

(B) Outcrops of bedrock appear in cliffs and canyons. Slopes are covered with regolith.

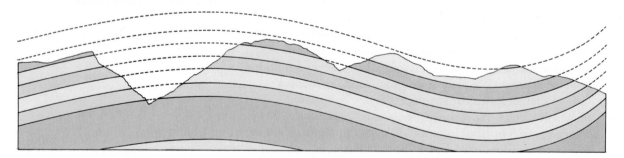

(C) The discontinuous blanket of regolith almost completely covers some formations, while others are exposed as discontinuous cliffs. Outcrops of bedrock form "holes" in the regolith cover.

(D) The structure of the bedrock consists of rock layers warped into broad folds, some of which are cut by canyons. Compare with (A).

Figure 10.13 The relationship between bedrock and regolith is depicted in the photograph and diagrams.

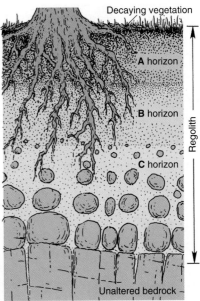

Figure 10.14 **A soil profile** for a temperate climate shows the transition from bedrock to regolith through a sequence of layers, or horizons, consisting of successively smaller fragments capped by a dark layer of decomposed organic material.

Table 10.3 Major Types of Soil

Soil Order	General Characteristics	Typical Geographic or Geologic Setting
Andisols	Soils that form by weathering of volcanic parent material.	Common in volcanic areas like Japan, Alaska, and Pacific Northwest
Entisols	Only slightly weathered with minimal development of soil horizons. Similar to parent material.	Young surfaces, slopes, and sand dunes
Aridosols	Thin A and B horizons with little organic matter. Carbonate deposits (caliche) common in B horizon.	Arid regions
Gelisols	Soils with a dark organic surface layer underlain by permafrost.	Arctic regions of North America and Russia
Histosols	Dark soils dominated by decomposed organic matter.	Typically in poorly drained low-lying areas
Inceptisols	Well-developed A horizon, but little clay; still retain weatherable minerals.	Relatively young surfaces or resistant bedrock
Mollisols	Dark organic-rich A horizon. Nutrient-rich parent material.	Common in grasslands of stable platforms in temperate climates
Alfisols	Thin A horizon and clay and nutrient-rich B horizon.	Semi-arid to humid climates; good agricultural soil
Spodosols	Strongly leached A horizon with aluminum and iron-rich B horizon.	Coniferous forests in cool temperate climates
Vertisols	Clay-rich soils that shrink when dry and swell when wet.	Tropical regions with wet and dry seasons
Oxisols	Deeply weathered with oxidized iron apparent in B horizon. Red or yellow.	Humid tropical to subtropical regions of low relief
Ultisols	Highly weathered clay-rich B horizon with high concentrations of aluminum and low in nutrients.	Warm, humid regions; with fertilizer they can be used for agriculture

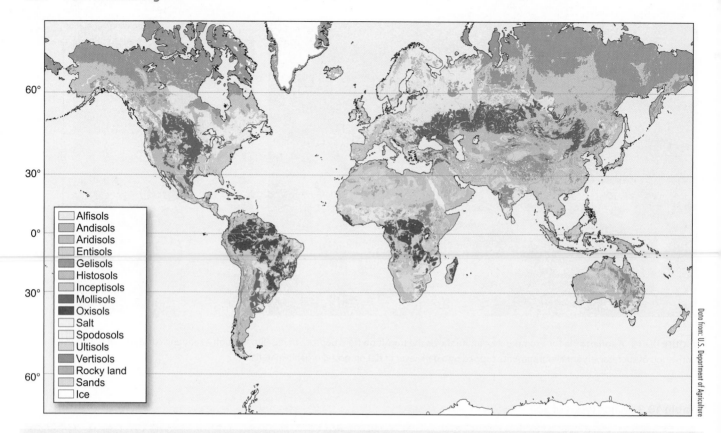

Alfisols
Andisols
Aridisols
Entisols
Gelisols
Histosols
Inceptisols
Mollisols
Oxisols
Salt
Spodosols
Ultisols
Vertisols
Rocky land
Sands
Ice

Data from: U.S. Department of Agriculture

Figure 10.15 **Principal soil types** of the world are shown on this map. The various soils are classified based on obvious physical characteristics including organic matter and clay content, the degree of nutrient depletion, soil chemistry, and origin or age of the soil. See Table 10.3 for a description of the major features of each type of soil. Note the strong control of climate on soil types.

deposit as other elements are removed. The enrichment may be great enough to form aluminum ore—bauxite. Iron is also insoluble in the presence of dissolved oxygen and also accumulates in soils. Intensely weathered soils are typically red because of the secondary iron oxide minerals formed by weathering. Such residual laterites may also be enriched in other insoluble elements such as nickel and make ore deposits. (see Figure 10.18).

As do so many other aspects of weathering, the type and thickness of soil depend on a number of factors, the most important of which are parent rock composition, topography, and time. The mineral composition of the bedrock strongly influences the type of soil because the bedrock provides the chemical elements and mineral grains from which the soil develops. Pure quartzite, for example, contains 99% quartz and is extremely deficient in minerals that can weather to make clays. Its decomposition produces a thin, infertile soil poor in plant nutrients (**Figure 10.16A**).

Topography affects soil development because it influences the amount and rate of erosion and the nature of drainage. Flat, poorly drained lowlands develop a bog-type soil, rich in decomposed vegetation and saturated with water, whereas steep slopes permit rapid removal of regolith and inhibit the accumulation of weathered materials. Well-drained uplands are conducive to thick, well-developed soils (Figure 10.16B).

Time is important in soil development in that it takes time for physical and chemical processes to break down the bedrock. In Figure 10.16C, the young lava flow has a very thin, patchy soil, whereas the older flow has had time for a thick soil layer to develop.

Ions in Solution

The ions dissolved in water are almost invisible products of weathering. A chemical analysis of rainwater compared with river water illustrates the effectiveness of chemical weathering in dissolving and transporting many elements (Table 10.2). Fresh rainwater contains relatively little dissolved mineral matter, but surface water soon dissolves the more soluble minerals in the rock and transports ions away in solution (**Figure 10.17**). In general, Na, K, Ca, and Mg are the most soluble ions. Each year the rivers of the world carry about 4 million metric tons of dissolved materials to the oceans (Table 10.2). It is not surprising, then, that seawater contains 3.5% (by weight) dissolved salts, most of which were derived from the continents by chemical weathering.

The major source of ions in solution is carbonate rock. About 45% of the dissolved material in rivers is derived from carbonates, even though they constitute only about 16% of the continental area exposed to weathering. Evaporites are also made of very soluble minerals—salts of potassium, sodium, magnesium, chlorine, and sulfate. Despite the fact that they make up only about 1% of the area of the continents, fully 18% of the ions in solution in rivers appear to be derived from evaporites. If we consider only one constituent, about 80% of the chloride in the Amazon River comes from halite dissolution, not from the weathering of silicate minerals. Other rocks are much less soluble than carbonates and evaporites. Silicate rocks account for more than 80% of the land area but only about 35% of the dissolved constituents of river water. The effect of susceptibility of rock types to chemical weathering is the important factor here.

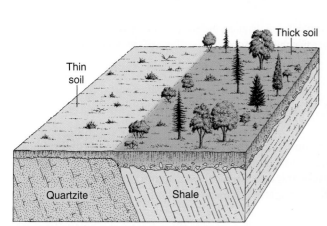

(A) The influence of rock type is illustrated by the difference between a layer of quartzite and a layer of shale. Quartzite resists chemical decomposition, so the soils produced from it are thin and poorly developed. Shale is much more susceptible to chemical weathering and forms thicker soils.

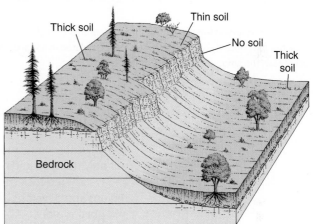

(B) The influence of topography is apparent from the contrast between slope soils and valley soils. Thick soils can form on flat or gently sloping surfaces, but steep slopes permit only thin soils to develop.

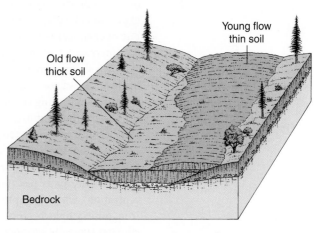

Figure 10.16 Topography, rock type, and time also affect the thickness of soil.

(C) The influence of time can be seen in areas of volcanism. Thick soils have developed on old lava flows, in contrast to thin soils on younger flows.

© Krzysztof Wiktor/ShutterStock, Inc.

Figure 10.17 Dissolved ions are one of the major products of chemical weathering. Here, along the margins of a salt pan in Death Valley, California, the salts once dissolved in river water accumulated in a shallow lake that then evaporated. The volume of salt in these evaporite deposits reveals the amount of dissolved materials.

Climate and Weathering

Climate is the single most important factor influencing weathering. It determines not only the type and rate of weathering, but also the characteristics of regolith and weathered rock surfaces. Intense chemical weathering occurs in hot, humid regions and develops a thick regolith. Chemical weathering is minimal in deserts and polar regions.

Climate is of major importance in weathering because rainfall, temperature, and seasonal changes all directly affect the style and rates of weathering. The influence climate has on weathering is apparent in the striking contrasts of the soil in the tropics, deserts, and polar regions (**Figure 10.18**).

In physical weathering, perhaps the most important temperature changes are the ones that produce continual cycles of freezing and thawing that result in repeated expansion of water ice in the rock and soil, and thus mechanical fragmentation. The rate of chemical reactions (and biological activity) also tends to increase as temperature increases. Commonly, a 10°C increase in temperature doubles reaction rates.

Most chemical reactions, such as hydrolysis, dissolution, and oxidation, require the presence of water, so the total amount of precipitation in an area is clearly a major factor in weathering (Figure 10.15). But many factors such as the intensity of rain, seasonal variations, infiltration, runoff, and the rate of evaporation combine to influence weathering and weathered products in a given region.

The extent and style of chemical weathering are not controlled entirely by temperature and total water supply; weathering may be greatly affected by other conditions. For example, many reactions are controlled by the water's acidity (hydrogen ion concentration), which is expressed as the pH value, ranging from 1 (acid) to 14 (alkaline). Iron, for example, becomes 100,000 times as soluble at pH 6 as it is at pH 8.5. Tropical climates typically support lush vegetation that yields organic acids. As a result, forested areas experience higher rates of chemical weathering than in otherwise similar areas that lack such growth. Some studies suggest weathering to be eight times as

(A) Arctic weathering profile. Thin soils, with partly decomposed rock fragments, develop in polar regions. Physical weathering is dominant.

(B) Tropical weathering profile. Thick red (oxidized) soils develop in tropical regions. Chemical weathering dominates.

Figure 10.18 Climate is one of the important controls on the type of soil found in an area.

high as in forested areas as compared to nonforested areas, other things being equal. Moreover, high temperatures and abundant water can also increase the rate of bacterial activity, important in the production of acid.

The relative importance of various types of weathering under different climatic conditions (temperature and rainfall) is shown in **Figure 10.19**. High temperature and high precipitation cause intense chemical weathering. Physical weathering dominates in regions of low temperature and low rainfall. Perhaps the best way to appreciate the influence of climate on weathering is to consider variations in the types and thicknesses of soils from the equator to the poles, as shown in **Figure 10.20**. This diagram summarizes the relationships between the amount of chemical weathering and variations in precipitation and temperature.

In humid, *tropical climates*, extreme chemical weathering rapidly develops thick soils to depths greater than 70 m (Figure 10.20). In central Brazil, the zone of decayed rock is more than 150 m thick. Under such conditions, the feldspars are completely altered to clays, and all soluble minerals are leached out. Only the most insoluble materials (such as silica, aluminum, and oxidized iron) remain in the thick, red soil (Figure 10.18). These are not good soils for continued agriculture, because plant nutrients are leached away by the abundant rainfall. In Figure 10.15, you can see that **oxisols** dominate the tropical zones in both Africa and South America. The high temperatures in tropical zones speed chemical reactions, so chemical decomposition is very rapid. Frost action, of course, is essentially nonexistent in the tropics, except on the tops of high mountains.

In low-latitude *deserts*, north and south of the tropical rain forests, chemical weathering is minimal because of the lack of precipitation. Moreover, organic matter is not abundant in these **aridosols**. Consequently, the soil is thin, and exposures of fresh, unaltered bedrock are common. Physical weathering is evident, however, in the fresh, angular rock debris that litters most slopes.

In *temperate regions*, precipitation ranges from humid to subarid, and temperatures range from cool to warm. Both chemical and physical processes operate, and the soil and regolith develop to depths of several meters. The agricultural breadbaskets of the world are not in the tropics where soil is thickest, but in these temperate zones. Here, soils are moderately thick and have retained their nutrients. Moreover, temperate soils are commonly enhanced by added deposits of wind-blown dust. **Mollisols** (from Latin *mollis* for soft) with their organic-rich A horizons dominate the temperate zones (Figure 10.15).

Figure 10.19 The relative importance of various types of weathering depends on temperature and rainfall. This diagram shows that strong chemical weathering occurs where both temperature and precipitation are high. Physical weathering is strongest where the mean annual temperature is between −10° and 10°C and precipitation is between 25 and 100 cm. Weathering is at a minimum where annual precipitation is below 25 cm.

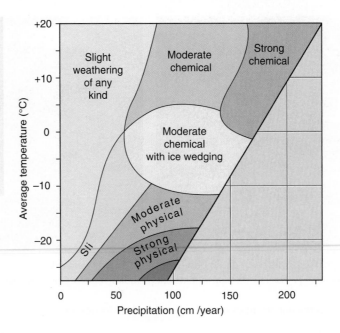

Figure 10.20 Climate controls the type and extent of weathering because of the combined effects of precipitation, temperature, and vegetation. (Other variables are also involved, such as those shown in Figure 10.16.) Weathering is most pronounced in the tropics, where precipitation, temperature, and vegetation reach a maximum. Conversely, a minimum of weathering is found in deserts and polar regions, where these factors are minimal. Compare this cross section with the map in Figure 10.15.

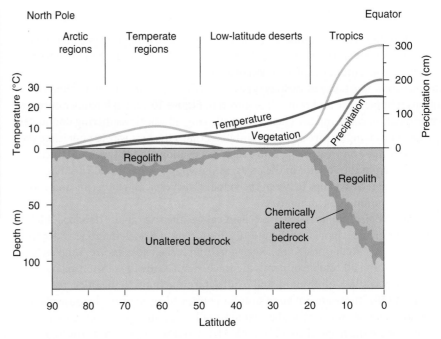

Why are thin soils produced in both polar and desert regions?

In the *polar regions*, weathering is largely physical. Temperatures are too low for much chemical weathering, so the soil typically is thin and unproductive **gelisols** (from Latin *gelare* to freeze). They are composed mostly of angular, unaltered rock fragments (Figure 10.18). In permafrost zones (areas where water in the pore spaces of soil and rock is permanently frozen), the surface layer melts during the summer but freezes again in the winter. This unique condition produces polygonal ground patterns, which result from thermal contractions and the differential thawing and freezing.

Not only does the climate affect weathering rates, but the climate is also affected by weathering. This interplay of chemical reactions involving carbon forms one of Earth's great chemical cycles. We showed earlier that CO_2 can be dissolved in water to form the bicarbonate ion (HCO_3^-). When this water comes in contact with calcium-bearing silicate minerals calcium carbonate (calcite or aragonite) can precipitate. This important reaction removes carbon dioxide from the atmosphere. Because CO_2 is an important greenhouse gas, the reduction of this key component eventually reduces the atmosphere's ability to absorb heat and the planet cools. You can see that this will form a complex feedback cycle and that the lower temperatures will result in lower reaction rates. In fact, there may be a link between the uplift of the Himalaya mountains in the middle Cenozoic and a global decrease in the amount of CO_2 in the atmosphere. The reaction to form carbonates may have caused the lower temperatures that have prevailed in the second half of the Cenozoic.

Conversely, carbon dioxide gas can be released to the atmosphere by other weathering reactions. For example, CO_2 gas can be produced by weathering of rocks that have organic materials like black shales. In essence, the organic carbon reacts with oxygen in the atmosphere or oxidized groundwater to create CO_2. In this case, atmospheric carbon dioxide and temperatures could rise.

Rates of Weathering

The rate at which weathering processes decompose and break down a solid rock body depends on three main factors: (1) susceptibility of the constituent minerals to weathering, (2) climate, and (3) the amount of surface exposed to the atmosphere.

A consideration of the rate at which weathering proceeds is a good way to review its controlling factors. Rates of weathering can be calculated by measuring the amount of decay on rock surfaces of known age. Tombstones, ancient buildings, and monuments, for example, provide datable rock surfaces for estimating rates of weathering. These studies show that in some climates, several centimeters of rock can be decomposed in a few decades, whereas the same rock remains unaltered in other climates.

In places, rates of weathering have been measured on volcanic ash and basaltic lava flows that have been dated by radiometric means. For example, on the subtropical island of St. Vincent in the West Indies, a volcanic ash deposited 4000 years ago has weathered to produce a layer of clay soil 2 m thick. Soils have also formed on the ash deposits resulting from the 1883 eruption of the equatorial volcano Krakatau. Measurements made 45 years after the Krakatau eruption showed a new soil nearly 50 cm thick. Thinking on a grander scale, the rate of dissolution of rock may be comparable to that of erosion caused by running water. For example, more material is dissolving from the Hawaiian island of Oahu than is being moved by erosion. The dissolved ions are carried to the ocean by both rivers and groundwater. Of course, rates of weathering are much lower in polar climates and in the arid deserts.

The Egyptian pyramids provide an interesting example of rates of weathering in arid climates. The Great Pyramid of Cheops, near Cairo, was originally faced with polished, well-fitted blocks of travertine limestone. These blocks protected the rock in the pyramid core from weathering until the outer, polished layers were removed, about 1000 years ago, to build mosques in Cairo. Since then, without the rock facing, weathering has attacked all four main rock types used in the construction of the pyramid. The least weathered rock in the pyramid is granite, which today remains essentially unweathered. Also resistant is a hard, gray limestone, which still retains marks of the quarry tools used to carve its final shape. The shaly limestone and fossiliferous limestone used for other blocks, however, have weathered rapidly. Individual building blocks have weathered to spherical forms. Many of these blocks have a zone of decayed minerals as deep as 20 cm. Most of the weathered debris remains as talus on individual tiers and around the base of the pyramids. During the last 1000 years, the volume of weathered debris produced from the pyramid has been calculated to be 50,000 m^3.

Near Saqqara, Egypt, on older pyramids built nearly 4600 years ago, deeper weathering has occurred, with talus debris nearly covering the steps. Some of the smaller pyramids in the area are completely covered by their own weathered debris.

In Mexico, pyramids built nearly 2000 years ago have also been deeply weathered. Some were completely covered by their own weathered debris (see the *GeoLogic* essay on the next page).

GeoLogic   Weathering of Mexican Pyramids

When Europeans first entered the highland valleys of central Mexico, they found tree-covered mounds dotting the landscape. Eventually, archeologists discovered that these mounds were not natural at all, but were the weathered remnants of the buildings of a once great city. Teotihuacan was at its zenith about 400 AD but the city was abandoned and partially destroyed about 700 AD. In the photograph, you can see that the classic Mesoamerican architecture appears to be wonderfully preserved. But that is a popular misconception. The temples with their staircases, platforms and altars were restored after careful archeological work. Study the mounds to the left; the low rounded conical hills covered with grass and a few trees. These are the unrestored segments of the Teotihuacan complex.

Observations

1. After Teotihuacan was abandoned some 1300 years ago the buildings and monuments were left to decay as a result of weathering.
2. A regolith, in some places more than 1 m thick, topped the ruined structures. It consisted of decomposed rock fragments, clay, and organic material from plants that colonized the newly created soil.
3. Once angular building stones are now spheroidal, crumbly, and ill-fitting.
4. Numerous blocks of rock are scattered over the weathered structures.
5. Most of the delicate carvings in the solid rock are destroyed, only being preserved where collapse or overhangs had protected them from the attack by water and air.
6. Some of the temples have now been partially reconstructed, revealing their former grandeur. New stone was quarried and carefully placed to recreate the ancient buildings. But so extensive was the weathering that it was unclear how many steps ascended to the top of a temple or even how many wide platforms break the profile.

Interpretations

Considering the location, landscape, and present system of slopes, could the pyramids have been covered by deposition of sediment from running water, wind, or lakes? The principal conclusion you can draw from these observations is that the monuments were not covered with sediment transported in from some other area, but simply covered in their own weathered debris. Weathering is a rapid process capable of producing a regolith in only a few hundred years in a temperate climate. The products of weathering are spheroidal boulders, decomposed rock fragments, and soil.

Key Terms

A horizon (p. 281)
aridosols (p. 287)
B horizon (p. 281)
C horizon (p. 281)
chemical weathering (p. 266)
differential weathering
  (p. 277)
dissolution (p. 270)

exfoliation (p. 278)
gelisols (p. 288)
granular disintegration
  (p. 279)
horizon (p. 281)
hydrolysis (p. 272)
ice wedging (p. 267)
joint-block separation (p. 278)

leach (p. 270)
mollisols (p. 287)
oxidation (p. 273)
oxisols (p. 287)
permeability (p. 276)
physical weathering (p. 266)
porosity (p. 276)
regolith (p. 278)

shattering (p. 279)
sheeting (p. 269)
soil (p. 281)
soil profile (p. 281)
spheroidal weathering
  (p. 278)
talus cone (p. 270)

Review Questions

1. List five ways in which the effects of weathering are expressed in natural outcrops of bedrock.

2. Discuss the processes involved in the most important types of physical weathering.

3. Which rock types are most susceptible to chemical weathering?

4. Discuss the chemical reactions involved in chemical weathering of feldspar.

5. Explain why joints are important in weathering processes.

6. How does soil originate?

7. Why are mudrocks (dominated by clay minerals) the most common sedimentary rocks?

8. Explain how rock types, topography, and time influence the types of soil produced by weathering.

9. Why is quartz sand so abundant in clastic sedimentary rocks?

10. What is spheroidal weathering?

11. Draw a schematic diagram showing variations in weathering from arctic regions to the tropics.

12. How do the pyramids of Egypt and Mexico provide information on rates of weathering?

13. Put these rock types in the order by which they would weather the fastest: granite, rock salt, gabbro, sandstone, mica schist, shale, serpentinite, and gypsum. Assume that the climate is humid and temperate.

14. Which rock type, limestone, granite, or basalt, would weather to form the most clay? Which would be most susceptible to oxidation?

15. Compare the soil map in Figure 10.15 with a climate map. Comment on the distribution of different soil types in each of the major climates.

11 Slope Systems

Most of Earth's surface is not perfectly horizontal; sloping surfaces are everywhere. On Earth, most slopes are formed by stream erosion and are related to the walls of stream valleys. Some, however, are the result of tectonic activity such as faulting; others result from wave erosion, extrusion of lava, glaciation, and even impact of meteorites. Regardless of a slope's origin, gravity has a universal tendency to pull materials on it to a lower gravitational potential. Consequently, the downslope transfer of material through the direct action of gravity—mass movement—is extremely common.

Slope failures can be rapid and devastating, as in great landslides on steep cliffs, or they can be imperceptibly slow, as in the creep of soil down the gentle slope of a grass-covered field. Slope failures occur on all planetary surfaces—modifying impact crater rims on the Moon and on Mercury, enlarging the huge canyons on Mars, in addition to shaping stream valleys, sea cliffs, and mountain fronts on Earth. The inexorable force of gravity works in concert with weathering to slowly create new landscapes. In all cases, the net effect of mass movement is the transportation of loose rock material from hillsides onto low-lying areas.

As geologic hazards, slope failures and related subsidence features pose a constant threat to people in many regions of the world. Moreover, landslides are also hazards in open pit mines where steep slopes sometimes fail. The panorama here shows a large landslide from the wall of the Bingham Canyon Copper Mine that slipped in 2013. Fortunately, geologists and engineers at the mine were closely monitoring the deformation using a sophisticated radar system and predicted that the wall would fail. The visitors center was evacuated, mining equipment was moved to safer places, and mining activity stopped hours before the wall collapsed. Nonetheless, this may have been the most costly landslide in the history of the United States (nearly $1 billion) and required months of work before the mine was brought back to full production. El Niño rains cause millions of dollars of damage in California as steep oceanfront properties slump into the Pacific Ocean. In Central America, hillsides, drenched by torrential rains from hurricanes, occasionally give way and destroy whole villages. Many lives have been lost in the region as a result of these debris flows and other kinds of slope failures. Consequently, assessment of slope stability is an important job for many geologists.

In this chapter, we consider the various types of mass movement, some rapid and others extremely slow. We also consider the fundamental factors that determine if a slope is stable or unstable and susceptible to failure. The great overriding theme in this chapter is that all slopes are mobile and constantly changing under the continuous pull of gravity.

© Rio Tinto Kennecott

Major Concepts

1. Mass movement is the downslope transfer of material through the direct action of gravity. It is a major geologic process operating on all slopes.
2. The most important factors influencing slope failures are saturation of slope material with water, earthquakes, over steepening of slopes, freezing and thawing, and the strength of the materials in the slope.
3. The major types of mass movement are creep, debris flows, slumps, rock falls, and subsidence.
4. Creep is the very slow downslope movement of soil and rock, produced primarily by the expansion and contraction of the surface materials.
5. Debris flows are mixtures of rock fragments and water that flow rapidly downslope as a viscous fluid. A lahar is a special type of debris flow composed of volcanic materials.
6. Slumps or landslides are a type of mass movement in which the material moves as a unit or block along definite slippage planes.
7. Rockfalls include the free-fall of a single fragment ranging from small grains upward to huge blocks that may break up and become fluidized as they flow.
8. Subsidence is essentially vertical motion caused by collapse into voids or as a result of compaction of loose materials.
9. Slopes are open dynamic systems in which regolith and near-surface bedrock move downslope toward the main stream, where they are removed through the drainage system.

Factors Influencing Mass Movement

> Gravity is the driving force for the downslope movement of material, but several factors are important in causing movement to occur. The most important are (1) saturation of material with water; (2) vibrations from earthquakes; (3) over steepening of slopes by undercutting; (4) alternating freezing and thawing; and (5) strength of the slope materials.

What factors influence mass movement?

Gravity pulls continuously downward on all materials everywhere on Earth's surface. Bedrock is usually so strong that it remains fixed in place, but if a slope becomes too steep or if fractures form, masses of bedrock may break free and move downslope. Soil and regolith, in contrast, are held together poorly and are much more susceptible to downslope movements. There is abundant evidence that, on most slopes, at least a small amount of downhill movement is occurring all of the time.

Gravity is the driving force behind all slope processes. Mass movement is not limited to stream valleys but occurs on all slopes, including sea cliffs and fault-block mountain fronts, on the ocean floor, and even on the slopes of craters on other planets. The force of gravity is continuous, of course, but it can move material only when it exceeds the **cohesive strength** of the surface material. As the products of weathering accumulate on a hill slope, the dry, loose rock fragments will tend to accumulate at a nearly uniform slope angle inclined at what geologists call the **angle of repose**. This angle is the steepest slope at which loose material, such as talus, will remain at rest without rolling farther downslope. This is the inclination of a slope at equilibrium. The angle of repose is commonly about 30° for dry sand, but it varies depending on the size, shape, and sorting of the fragments and the amount of moisture between the grains. Rock slopes also have a profile of equilibrium. Some rocks, such as sandstone, form steep cliffs, whereas soft shale forms gentle slopes. That is why in many areas, alternating layers of sandstone and shale develop a profile of alternating cliffs and slopes. The profile of equilibrium on any rock surface is the preferred profile, and any time it is modified (naturally or by humans), it will readjust to the original profile.

Some factors that influence mass movement can be understood by considering the forces acting on a rock fragment that rests on a slope (**Figure 11.1**). The weight of the object (the force caused by gravity) is directed vertically downward, here 1 kg. The force directed down this hill slope is only 0.5 kg [1 kg · sin(30°)]. On a stable slope, the frictional cohesion of the object to the slope is greater than this downhill force, and the rock fragment remains stationary. Any factor that either weakens the cohesion of the object with the surface or increases the downslope force may initiate downslope movement. Such factors include (1) saturation of the material with water, (2) vibrations from earthquakes, (3) alternating expansion and contraction of the regolith, (4) the undercutting of slopes by streams or waves, and (5) modification of slopes by humans, including the removal of vegetation.

Figure 11.1 Forces acting on a rock on a hill slope determine if it will move downslope. The force of gravity is vertical, but it can be separated into one component that is parallel to the surface and another that is perpendicular to the slope. Consequently, the force directed downslope depends on the weight of the object and the angle of the slope. If the downhill force exceeds the forces of friction that resist movement, the rock will start to move.

Water is an important factor in mass movement because it lubricates the unconsolidated material on slopes (reduces cohesion) and adds weight to the mass (increases downhill force), thereby promoting mobility and downslope movement. Heavy rainfalls, whether prolonged over many days or in a single storm, are particularly effective in triggering mass movement.

Earthquakes, with their initial shock and aftershocks, can loosen fragments of rocks on steep slopes, overcoming the cohesion of the slope, and set the regolith in motion. In many areas, more damage is caused by mass movement than by the earthquake itself. For example, an earthquake in Guatemala in 1976 set off more than 10,000 mass movements. Most occurred on steep slopes, but some were on gentle slopes where water-saturated regolith was mobilized and turned into debris flows.

Undercutting of slopes is a fundamental cause of slope failures in that it creates steep gravitationally unstable surfaces. Natural undercutting is caused by streams eroding their banks, or by waves cutting cliffs on a shoreline. Home and road construction commonly undercut natural slopes that were at the angle of repose. The new steeper slope is unstable and susceptible to failure. Almost everyone has seen the evidence of slope failure on a steep road cut.

A significant factor in mass movement has been the modification of natural slopes to suit humans. Since prehistoric times, farming and deforestation have brought changes in vegetation cover, soil, and drainage. In more recent times, large scale engineering works have modified coastlines, river systems, and landforms on an even larger scale. All of these changes by humans result in new and artificial surfaces imposed on existing geologic systems that had attained some degree of equilibrium. They commonly provoke unforeseen reactions that cause widespread damage. A dramatic example is the deforestation of large areas in Madagascar that resulted in tens of thousands of major slope failures and accelerated erosion (**Figure 11.2**).

The strength of the materials in the slope is obviously important. We have alluded to this already in the discussion about frictional cohesion of a rock to the hillside. But the rocks and regolith that make up the slope will also control its failure. Consider the strength of a well-cemented sandstone compared to a shale. Shales are weak and fail easily. Fractures and bedding planes also impart weakness to a rock body.

Another example of slope failure related to human activities is the landslide that occurred at Vajont Dam in northern Italy. This, the worst dam disaster in history, resulted from a huge landslide into the Vajont Reservoir on October 9, 1963 (**Figure 11.3**). The dam was built to provide electricity for the city of Venice and was the highest dam in the world at the time. As the reservoir behind the dam was filled, a landslide on the steep canyon walls moved slowly downhill over a 3-year period. The

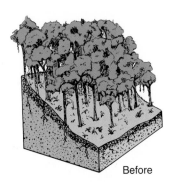

Before

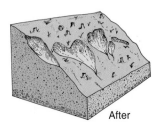

After

Figure 11.2 Accelerated mass movement resulting from deforestation in Madagascar is vividly shown in these diagrams and photograph. As a consequence of removing the forests, numerous landslides formed and deep gullies eroded into the hillsides and fed sediment into adjacent streams. The ocean around Madagascar is colored red by the silt eroded from the recently deforested areas.

rate of creep had been as much as 7 cm per week, until a month before the catastrophe; then, it increased to 25 cm per day. On October 1, animals grazing on the slopes sensed the danger and moved away. Finally, on the day before the slide, the rate of creep was about 40 cm per day. Engineers expected a small landslide and did not realize, until the day before the disaster, that a large area of the mountain slope was moving en masse downhill. When the slide broke loose, more than 270 million cubic meters of rock rushed down the hill slope at nearly 100 km/h and splashed into the reservoir. It produced a wave of water more than 100 m high that swept over the dam and rushed down the valley, destroying everything in its path for many kilometers downstream. The entire catastrophic event, including the slide and flood, lasted only 7 minutes, but it took approximately 2000 lives and caused untold property damage.

(A) Longarone before the flood.

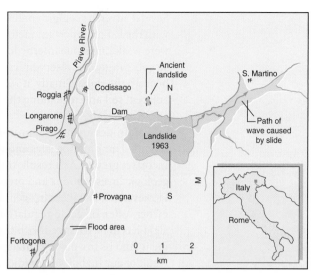

(B) Longarone is downstream from a dam with a larger reservoir. In 1963, a large landslide into the reservoir created a huge wave that swept away part of the town.

(C) Longarone after the landslide drove flood waters into the town.

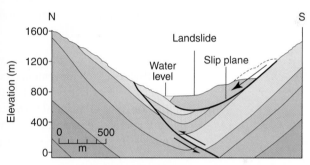

(D) Cross section through the Vajont landslide shows the role played by a tilted succession of weak sedimentary rocks.

(E) Vajont dam today. The catastrophic landslide fell from the southern (right) wall of the canyon. Older landslides scar the northern wall as well.

Figure 11.3 The Vajont Dam disaster is illustrated in this map, cross section, and photographs. The map shows the location of the landslide and the area covered by the resulting flood. The photographs show part of the town of Longarone, Italy, before and after the flood.

Several adverse geologic conditions contributed to the slide. Compare these conditions with the list of important factors just discussed. The rocks in the steep mountainside were weak limestone interbedded with thin layers of clay. The beds were inclined steeply toward the reservoir, creating inherent planes of weakness in the bedrock beneath the slope. Finally, it was the rising water level in the reservoir, saturating the adjacent soil and rock along the banks, which reduced the slope's cohesiveness and caused the slide.

Landslides are natural processes and constantly occur, but as slopes are artificially modified for building sites and roads, the magnitude and frequency of mass movement increases greatly. As a result, millions of dollars in property are lost each year. Careful geologic investigations and proper land-use planning could greatly reduce these losses. However, it also takes stringent laws that are enforced to make a real difference. Land-owners often object to regulations that limit their ability to sell their property. This right has to be balanced with the need to protect the buyers and to protect the government (and its tax-paying citizens) from the need to pay for the eventual damages. As a result of our inability to predict all landslides, millions of dollars in property are lost each year. Careful geologic investigations and proper land-use planning could greatly reduce these losses.

Types of Mass Movement

Many types of mass movement can be recognized on the basis of the behavior of the material and the mechanics of movement. The most important are (1) creep, (2) debris flows, (3) slumps or landslides, and (4) subsidence.

Mass movements include all types of slope failure. Because of their potential for destruction, such movements have been studied extensively by engineers and geologists. As a result, they have been classified in various ways, depending on the type of motion, type of material involved, and rate of movement (**Figure 11.4**). As you read through these descriptions you will see that slope failures—described by how they move—involve flowing, sliding, falling, and subsiding.

Creep

Creep is an extremely slow, almost imperceptible downslope movement of soil and rock debris that results from the constant minor rearrangement of the constituent particles (Figure 11.4A). The motion is so slow that observing it directly generally is difficult, but it is expressed in a variety of ways. On weakly consolidated, grass-covered slopes, evidence of creep can be seen as bulges or low, wavelike swells in the soil. In road cuts and stream banks, creep can be expressed by the bending of steeply dipping strata in a downslope direction or by the movement of blocks of a distinctive rock type downslope from their outcrop (**Figure 11.5**). Additional signs of creep include curved tree trunks and tilted posts, deformed roads and fence lines, and damage to retaining walls. The slow movement of large blocks of bedrock (block slides) can be considered a type of creep.

How is creep expressed on a hillside?

Many factors combine to cause creep, but the heaving process that results from the alternating expansion and contraction of the loose rock fragments in the regolith is probably the most important. The heaving process is accomplished in two principal ways: (1) by wetting and drying and (2) by freezing and thawing. In both instances, the regolith expands and shifts upward perpendicular to the hill slope. When it contracts, it settles back vertically under the force of gravity. With each cycle of expansion and contraction, each particle of rock comes to rest slightly downslope from its original position (**Figure 11.6**). Repeated expansion and contraction cause the particles to move downslope in a zigzag path. Freeze-thaw cycles are most numerous in regions where the temperature regularly crosses the freezing point. Therefore, creep is facilitated by cold climates. Cycles of wetting and drying will occur in greatest number where heavy precipitation alternates with periods of desiccation.

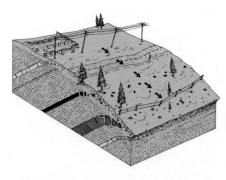

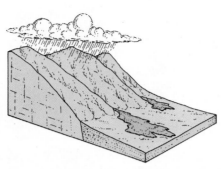

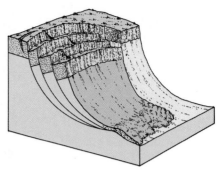

(A) Creep is the slow downslope migration of soil and loose rock fragments resulting from a variety of processes, including frost heaving.

(B) A debris flow is the rapid flow of a mixture of rock fragments, soil, mud, and water. The mixture generally contains a large proportion of mud and water.

(C) A slump or landslide is the slow or moderately rapid movement of a coherent body of rock along a curved rupture surface.

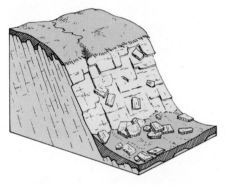

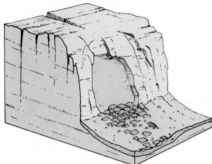

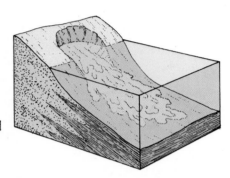

(D) A rockslide is the rapid downslope movement of rock material along a bedding plane, joint, or other plane of structural weakness.

(E) A rockfall is the free-fall of rock from steep cliffs.

(F) Subaqueous slope failures form on steep submarine slopes involving clasts of all sizes.

Figure 11.4 **Mass movement** takes various forms, all of which produce slope retreat and enlarge valleys. Examples of various types of mass movement are illustrated in the diagrams.

Many other factors also contribute to creep. Growing plants exert a wedgelike pressure between rock particles in the soil and thus cause them to be displaced downslope. Burrowing organisms also displace particles, and with each movement, however slight, the force of gravity pulls the particles downslope. In addition, creep can be facilitated by undercutting from rain runoff and streams, increased loads of rainwater and snow, and disturbance of slope surfaces by earthquakes and construction by humans.

Studies in various parts of the world show that the rate of creep is highly variable, but some general patterns have been discovered. On moderately steep slopes (10° to 15°), a rate of 1 to 2 mm/yr is common in humid temperate regions. In semiarid regions with cold winters, creep reaches an average of 5 to 10 mm/yr.

Solifluction (soil flowage) is a special type of creep. It is common in polar regions, where groundwater in the pore spaces of soil and rock is permanently frozen (**Figure 11.7**). The layer of permanently frozen ground is called the **permafrost** layer. It ranges from less than a meter to several hundred meters thick and occupies some 20% of the world's land. The presence of permafrost presents some special conditions for the downslope movement of regolith. During the spring and early summer, the ground begins to thaw from the surface downward. Because the meltwater cannot percolate downward into the impermeable permafrost layer, the upper zone of soil becomes completely saturated, and large areas of the regolith will flow slowly down even the gentlest slopes. These hillsides are covered by lobes of moving debris and look like melted wax on a candle. Solifluction can also occur in temperate regions in nonfrozen soil if a sufficient amount of water accumulates in the upper soil. Once liquefied, the soil can start moving downslope.

In what parts of the world would you most likely find solifluction?

Figure 11.5 Creep, the slow downslope movement of soil and rocky debris, is a common phenomenon on slopes. This photo shows how creep downhill bent the upper edges of these stratified rocks to the right.

Another type of slow downslope movement occurs in cold regions. **Rock glaciers** are long, tonguelike masses of angular rock debris and ice that resemble glaciers in general outline and form. The surface of a rock glacier is typically furrowed by a series of parallel flow ridges similar to those in an advancing lava flow (**Figure 11.8**). Evidence of movement includes concentric ridges within the body, the rock glacier's lobate form, and its steep front. Measurements show that rock glaciers move downslope at rates ranging from 5 cm a day to 1 m a year. Rock glaciers commonly occur at the heads of glaciated valleys and are fed by a continuous supply of rock fragments produced by ice wedging on the cirque wall. Excavations into rock glaciers reveal ice in the pore spaces between the rock fragments. Presumably, the ice is responsible for much of the flow movement. With a continuous supply of rock fragments from above, the constantly increasing weight causes the ice in the pore spaces to flow. Favorable conditions for the development of rock glaciers thus include steep cliffs

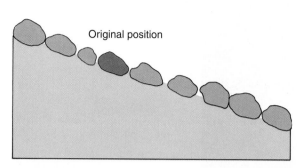

(A) Water seeps into the pore spaces between fragments of loose rock debris.

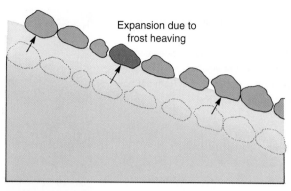

Expansion due to frost heaving

(B) As the water freezes and expands, the soil and rock fragments are lifted perpendicular to the ground surface.

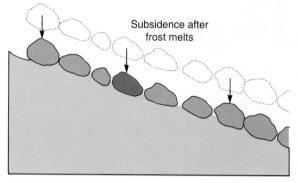

Subsidence after frost melts

(C) As the ice melts, gravity pulls the particles down vertically, displacing them slightly downhill.

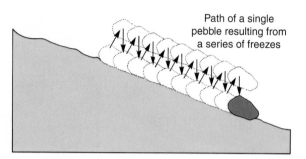

Path of a single pebble resulting from a series of freezes

(D) Repeated freezing and thawing cause a significant net displacement downslope.

Figure 11.6 Creep can result from repeated expansion and contraction of the regolith. With each cycle, there is a net downslope displacement of all loose material.

Figure 11.7 Solifluction is a major type of mass movement in cold polar regions and in some high mountains. On this hill in Alaska, the water-saturated regolith moves slowly downslope like a viscous fluid to form a series of lobate terraces.

A rock glacier in the Canadian Rockies, British Columbia, illustrates many features of this type of mass movement. Note the long tongues of moving rock debris and the wrinkled surface resulting from this flow of rock fragments and ice.

Courtesy of L. F. Hintze

and cold climates. The steep cliffs supply coarse rock debris with large spaces between fragments in which ice can form; the cold climates keep the ice frozen. Some rock glaciers may be debris-covered, formerly active glaciers.

Debris Flows

Debris flows consist of mixtures of rock fragments, mud, and water that rapidly flow downslope as viscous fluids (**Figure 11.9**). They are commonly mislabeled mudslides. Debris flows vary in size and rate of flow, depending on water content, slope angle, and available debris. Movement can range from flow like that of freshly mixed concrete to one that is similar to the flow of muddy water. Debris flows are a common type of mass movement that generally occur during intense rainfall or sudden thaw. Water penetrates and soaks into the regolith and adds weight to the hillside mass because it replaces the air in the open spaces between the fragments. The added weight and lubricating effect of water are keys to the origin of debris flows. Consequently, they commonly begin on steep hill slopes as soil slumps that liquefy and then flow downhill at speeds as great as 50 km/hr. The reason for the high velocity of flow is the presence of large amounts of water. Water acts as a lubricant by decreasing the friction between grains. Therefore, on a given slope, the more water present, the greater the speed of the flow. Many are more than 80 km long. If they reach a mountain front, they spread out in the shape of a large lobe, or fan. Because of their high densities, debris flows can transport huge boulders, cars, and buildings. Obviously, the consequences of a debris flow can be catastrophic if human habitation lies in its path.

Debris flows are common in many wet climates but they also form in arid and semiarid regions. They typically originate in steep-sided gullies where there is abundant loose, weathered debris (Figure 11.9). Recently burned hillsides are common sites for debris flows. Many disastrous "landslides" that occur during wet El Niño years in southern California are debris flows that formed on fire ravaged slopes.

Debris flows can be shed from mine dumps where unstable slopes of mine waste accumulate at the angle of repose for dry rock fragments. Perhaps the best known, and certainly one of the most tragic, was the disastrous flow of 1966 that destroyed

Why do debris flows move so quickly and so far from their sources?

(A) Debris flows in Coast Ranges of California have stripped away vegetation and exposed the light-colored sedimentary rocks.

Courtesy of U.S. Department of Agriculture

(B) Debris flow from the Lemhi Range, Idaho, forms a broad lobe at the mouth of a narrow canyon.

Figure 11.9 Debris flows are rapidly moving slurries of rock fragments, mud, and water.

the village school at Aberfan, South Wales, with the loss of 140 lives. The debris flow had its origin in a large dump from a coal mine that was 40 m high. During a period of heavy rain, the debris became saturated with water, the slope failed, and the mass of mine debris flowed down into the valley, covering a school and adjacent buildings.

Mudflows are a variety of debris flows that consist of small silt and clay-sized particles. Their water content can be as much as 30%. They may move over slopes as gentle as 5°, and have been known to move houses and barns from their foundations. A special type of mudflow occurs in the St. Lawrence valley in eastern Canada and in various parts of Scandinavia. In both regions, marine mud deposited near the margins of receding glaciers has a remarkable property known as "sensitive clay" or "quick clay." The sediment particles are loosely packed and consequently have a high water content. With only a slight disturbance, the material can become liquefied, or "quick" (transformed from a weak solid to a viscous fluid). The material can flow rapidly, even on very gentle slopes, once this change takes place. Disastrous flows of quick clay have affected several settlements along the valley of the St. Lawrence. Large masses of clay may become liquefied completely and flow as fast as a river.

How are lahars different from other kinds of debris flows?

A volcanic debris flow is called by its Indonesian name **lahar** (**Figure 11.10**). Many occur because the abundant loose pyroclastic material that accumulates on the flanks of a steep-sided volcano is inherently unstable. Dramatic lahars have occurred when newly erupted pyroclastic material became saturated by rain, by melting snow warmed by volcanic heat, or by water expelled from a crater lake. Lahars are especially dangerous because they travel at high velocities and can flow for great distances. The explosive eruption of Mount St. Helens, for example, triggered several large mudflows that flowed

© Yann Arthus-Bertrand/Corbis

Figure 11.10 **These light-colored lahars** from the Mt. Pinatubo volcano in the Philippines filled river valleys with flowing volcanic debris moving at speeds of over 30 km/hr. Most of the Pinatubo lahars were triggered by torrential rains from typhoons that mobilized loose volcanic ash on the flanks of the volcano. Fifty thousand homes were destroyed, but only a few hundred deaths were reported because of the advance warning provided by volcanologists.

many kilometers down the Toutle River. Other lahars from prehistoric eruptions of Mount Rainier in Washington traveled more than 80 km. On November 13, 1985, a lahar raced down the slopes of the ice-capped Andean volcano called Nevado del Ruiz at speeds of more than 150 km/hr. The lahar roared down the Lagunillas River valley, completely destroying the city of Armero, Colombia, 50 km away. It buried more than 25,000 people—90% of the city. This lahar was a watery mass of boulders and mud 40 m deep traveling 40 km/hr through town. Nothing could escape it; humans and livestock were engulfed and swept away by the slurry. Because it took more than an hour for the flow to reach Armero, a single telephone call from an observer or an electronic monitor nearer the mountain could have averted the human tragedy.

Landslides and Slumps

Although the vague term **landslide** has been applied to almost any kind of slope failure, true landslides involve movement along a well-defined slippage plane. Landslides, therefore, differ from creep and debris flows in their mechanics of movement. A landslide block moves as a unit (or series of units) along a definite fracture (or system of fractures), with much of the material moving as a large **slump block** (**Figure 11.11**). The detached block leaves behind a distinct curved incision, or scar. The slippage plane is typically spoon-shaped. As the block moves downward and outward, it commonly rotates so that bedding or other identifiable surfaces are tilted backward toward the source (Figure 11.11). In the lower part of the slump block, part of the displaced material may move as a debris flow. Several slippage planes commonly develop in the same slide, so the top of the slump block is broken into a series of steps, or small terraces. The characteristic scar, tilting of bedding or other surfaces, and jumbled, poorly drained small hills formed by previous slides serve to identify terrains that are prone to landslides.

Landslides are common phenomena and occur on a small scale nearly everywhere. Large slides are less numerous, but they commonly develop on steep slopes of weak shale. They can move in a matter of seconds or slip gradually over a period of weeks and months.

Many landslides come to rest on valley floors and often dam the streams flowing through the valleys, forming lakes behind them. Such lakes are temporary because the impounded water soon overflows the barrier and rapidly erodes through the unconsolidated rock debris. This sequence may result in catastrophic flooding downstream as the lake is almost instantly drained. Many landslides are started by earthquakes. The 2008 Sichuan earthquake in China triggered 15,000 landslides, rock falls, and debris flows that killed 20,000 people.

The term **rockslide** is used to denote the rapid movement of a large block of rock along a bedding plane, joint, or other plane of structural weakness (Figure 11.4D). A large block may move en masse for a short distance, but generally there is some disintegration as the body moves downslope and breaks into smaller blocks of rubble. A rockslide may grade into a rockfall or into a landslide, with the entire mass moving as a coherent unit. Joint systems are critical in the development of rockslides because they are continuous fractures through massive rock that ultimately weaken the structure and lead to failure. Once the stress exceeds the cohesive strength along any plane in a rock, mass movement will be initiated. The failure tends to be progressive because weakening along one joint will direct additional stress onto others.

Rockslides usually occur on steep mountain fronts, but they can develop on slopes with gradients as low as 15°. They are among the most catastrophic of all forms of mass movement. Sometimes millions of metric tons of rock plunge down the side of a mountain in a few seconds.

Rock Falls and Avalanches

Rockfalls include the free-fall of a single fragment ranging from a small grain upward to huge blocks (Figure 11.4E). Over time, great quantities of small to moderate-sized fragments (a few centimeters to a few meters long) shower down from the face of a cliff and accumulate at the base as talus. Ice wedging is a major process in dislodging the fragments.

Are all rapidly moving slope failures saturated with water like debris flows?

Some rockfalls are much larger; a whole hillside may break off from the face of a mountain and evolve into a flowing mass of particles called a **rock avalanche** (**Figure 11.12**). Even though it flows like a fluid, a rock avalanche is dry. The huge rock avalanche that buried the town of Frank, Alberta, Canada, in 1903 was among the largest of rock avalanches in the world. A gigantic wedge of limestone 400 m high, 1200 m wide, and 160 m thick crashed down from Turtle Mountain at 4:10 a.m.

(A) This photograph shows the upper part of a landslide in California, where homes were displaced along the curved rupture surface of the slump block.

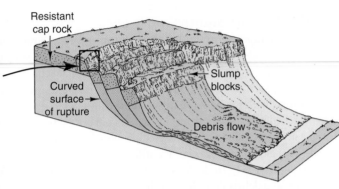

(B) Landslides occur along rather well-defined slippage surfaces. Large blocks slump and rotate downslope, and many grade into debris flows at their lower margins.

(C) This landslide along a hillside in Hong Kong destroyed four major roads and many buildings. As development continues, so will devastation from landslides.

(D) A landslide into a harbor in Hong Kong. Note the trucks and cars in the upper part of the slide.

Figure 11.11 Landslides are a significant kind of mass movement that have discrete planes or rupture surfaces along which rocks slip or slide downhill. They are often extremely destructive where human development encroaches on hill slopes as seen in these three examples.

on April 29 and destroyed much of the town. Seven million tons of rock fell off the mountain face and swept over the valley floor, burying many homes, mines, railways, and 3200 acres of farmland to a depth of 30 m. The fall occurred in 100 seconds, dammed the Crowsnest River, and created a small lake at the base of Turtle Mountain.

A fascinating aspect of some large rock avalanches is their capacity to move rapidly and spread out over vast distances, seemingly violating the laws governing friction. Most dry rockfalls move horizontally less than twice the distance they fall, but some show remarkable mobility, traveling roughly 10 to 100 times as far as they fall. One of the more spectacular examples is the Blackhawk "slide" in the Mojave Desert about 135 km east of Los Angeles (**Figure 11.13**). This rock avalanche occurred about 17,000 years ago when a large part of Blackhawk Mountain collapsed and a mass of rock debris fell about 1.5 km down the mountainside and spread out 9 km beyond the mountain front. The great avalanche came to rest as a huge lobe on the almost flat surface of the valley floor. The velocity of movement is estimated to have been 120 km/hr.

© Robert Harding Picture Library Ltd/Alamy

Figure 11.12 The Frank rockfall, Alberta, Canada, involved a huge mass of rock that broke away from the mountain face, completely burying the mining town of Frank, Alberta. Among the factors that contributed to this slope failure were the steepness of the mountain front, the dip of the bedding planes parallel to the mountain face, and underlying weak shale and coal beds. Mining activity may have triggered the movement.

Figure 11.13 The Blackhawk slide in southern California is a good example of a long rock avalanche. The vertical fall was only 1.5 km, but the flow spread out 9 km beyond the mountain front.

STATE OF THE ART   Landslide Hazard Maps and GIS

Landslides and other types of slope failures are some of the most damaging of all geologic hazards, even though they are not as dramatic as earthquakes or volcanic eruptions. Each year more than two billion dollars of damage results from landslides in the United States alone. As a consequence, some governments have devoted significant efforts to understanding slope failures and to preventing some of the damage. Only a few things can be done to stop natural landslides, but much can be done to prevent people from living and developing properties where landslides are a threat.

One tool used in this effort is the development of landslide hazard maps. Such maps show the relative risk of a landslide. Modern hazard maps are constructed using *geographic information systems (GIS)*. Geographic information systems use "layers" of computerized maps. Each spot on a map layer represents a specific property, for example, rainfall, elevation, angle of slope, or any other parameter that can be measured (or calculated) in an area.

What types of information would be useful in predicting where landslides might occur in the future? By reviewing the information in this chapter, you should be able to come up with a list of important factors. Four very important factors are slope, precipitation, past occurrence of landslides, and rock type.

Elevation above sea level is not important, but the slope of a hillside is very important. The steeper the slope, the greater the chance for landslides. The slope can be calculated from a topographic map and is recorded on a grid for each spot in the map area. Precipitation is also important. Water adds to the weight of the material in a slope, reduces its cohesiveness and lubricates flowing materials. Precipitation is measured at many meteorological stations and estimated between the stations so that it can also be represented as a grid of values on a map.

A hazard value is assigned to each spot on the map for each of these separate factors. Then they are, in effect, added together to make a composite map of the overall landslide hazard, like the one shown here. Locate your area and compare the relative hazard with your understanding of local topography and precipitation.

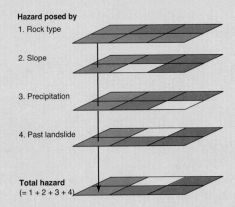

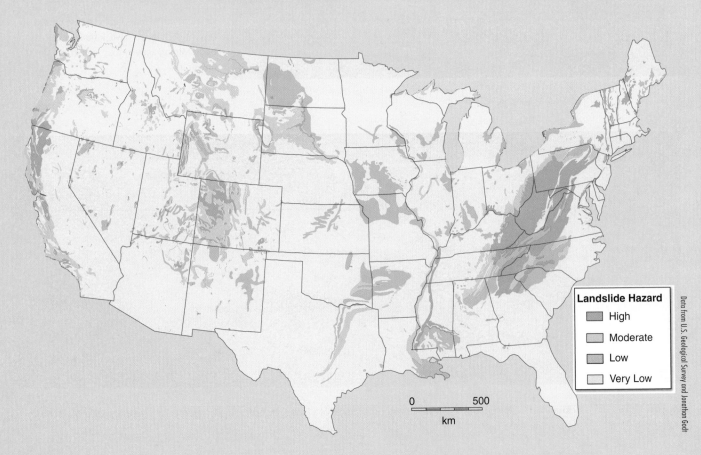

Landslide Hazard
- High
- Moderate
- Low
- Very Low

0   500 km

Several theories have been proposed to explain how a dry mass of rock debris can flow like a fluid. Some consider that the debris moves over a cushion of compressed air beneath it. Others think that the vibration of the fragments lowers the friction between them and allows them to move more freely. This acoustic fluidization permits the mass of debris to flow like a liquid. Laboratory experiments show that vibrating sand has mechanical properties quite different from ordinary sand and can be "fluidized" by the vibrations.

Land Subsidence

Subsidence is the downward movement of earth material lying at or near the surface. It differs from other types of mass movement in that movement is essentially vertical; there is little or no horizontal component. The primary force producing subsidence, of course, is gravity, but before gravity can act, other processes must operate to create space into which the earth can sink. The formation of caves by dissolution of rock by groundwater is a major cause of subsidence. The natural burning of combustible materials, such as peat and coal, in the subsurface also removes support for overlying rock, as does the melting of isolated blocks of glacial ice covered with glacial sediment. When swampy areas, such as Florida's Everglades, are drained, bacteria can oxidize the organic materials to make water and carbon dioxide gas. The loss of carbon in the soil can cause subsidence of several centimeters per year. Lava tubes are also areas of potential subsidence.

Examples of subsidence that result from human activity are varied and numerous. Where subsurface mining has removed large quantities of rock, subsidence into the abandoned workings may be so widespread that entire towns are abandoned. Subsidence may also follow the removal of fluids such as water, oil, or gas from the subsurface. A notable example is the subsidence of buildings in Mexico City because of the excessive pumping of groundwater from aquifers below them. Subsidence like this also occurred in arid southern Arizona. Since 1900, a tremendous amount of groundwater was pumped from wells to provide for the rapidly expanding metropolitan areas and for agriculture. In some areas 500 times the amount of water that naturally replenished the groundwater system was withdrawn and water levels in some wells dropped by 150 m. Fissures ruptured the surface when the water was removed and the loose sediment compacted unevenly. The fissures typically form in linear swarms, with individual fissures as much as 1 m wide. A total area of more than 7500 km^2 was affected. Since 1985, the problem has lessened because water has been imported to the arid valleys of southern Arizona through a system of canals. This imported water reduced the demand for groundwater and has also been used to artificially recharge the groundwater system. In Long Beach, California, pumping from the Wilmington oil field caused the surface to subside 10 m in 30 years. Pipelines, bridges, roads, and harbor facilities had to be modified to counter the effect of subsidence. The injection of water into the petroleum reservoir rock has now reduced subsidence in the area by raising the fluid pressure in the subsurface rock.

Perhaps the most devastating type of subsidence results from the expansion and contraction of clay-rich soils. The process may at first seem harmless enough, but the damage is very expensive. When dry, expansive soils are hard and strong, they are almost like rock, but when water is added, they expand and soften. Some clay-rich soil will expand more than 15 times its dry volume. Expansion of one and a half times the dry volume is common. When fully saturated with water, these soils lose much of their strength and become soft and slippery, much like lubricating grease. Upon drying, they shrink, causing the structures built upon them to collapse or buckle (**Figure 11.14**).

The shrinking and swelling of soils inflict enormous loss—of homes, commercial buildings, roads, and pipelines. The average annual loss in the United States is more than $2.3 billion—more than twice the loss from floods, hurricanes, tornadoes, and earthquakes combined. The damage from expansive soils is not sensational and draws little attention because it happens to individuals one by one throughout the country.

Figure 11.14 The swelling and contraction of clay minerals as the wet and dry affects roads and buildings, causing millions of dollars of damage each year. Repeated swelling and contraction of clays have caused this road in Colorado to buckle into waves like a roller coaster.

Courtesy of D. C. Noe, Colorado Geologic Survey

Subaqueous Mass Movement

Subaqueous mass movements affect large areas of the seafloor and are probably as common as those on land (Figure 11.4F). They are especially active near deltas and convergent continental margins, where sediment accumulates rapidly and slopes are steep. Weak, water-saturated sediment may slide or flow downhill if a slope becomes unstable. Landslides with slump blocks are common along the continental slope (**Figure 11.15**). Sand flows and turbidity currents commonly move farther downslope and out across the abyssal plain, where they can damage submarine cables and other installations and are potential dangers to offshore oil fields. Submarine mass movements are often triggered by earthquakes or large storms, but the giant Storegga slide off the coast of Norway may have been set off by the breakdown of an icy cement called methane hydrate.

Other spectacular submarine landslides have scarred the flanks of volcanic islands and seamounts, such as Hawaii (**Figure 11.16**). Recent detailed mapping of the ocean floor around the Hawaiian island chain shows that 17 giant submarine landslides have occurred off the Hawaiian Islands and 40 more surround the submerged volcanoes that extend from Kauai to Midway. These are among the largest landslides on our planet. One occurred when the northern part of the island of Oahu collapsed, sending debris 235 km out across the deep ocean floor. The horizontal extent of this landslide was more than 30 times the height of its fall. Huge blocks of debris, one the size of Manhattan Island, were swept across the ocean floor. Such an event must have created a tremendous tsunami, much larger than any created by an earthquake. The tsunamis created by giant submarine landslides in Hawaii swept debris onto slopes 300 m above sea level! Fragments of coral reefs are embedded in mountainsides 365 m above sea level. Elsewhere, large scour marks and rip-up channels are found on other islands, indicating wave erosion far beyond anything witnessed in historic times.

The process of giant submarine landslides and their associated tsunamis has been going on for many millions of years as volcanic islands grow from the seafloor and then collapse, with huge sections of the islands peeling off every 100,000 to 200,000 years. A long fault marked by a prominent escarpment has formed as a 300-km segment of the island of Hawaii slowly moves seaward. The block is moving at a rate of more than 10 cm/yr (Figure 11.16). The giant landslides around Hawaii are almost certainly driven by the active magmatic systems inside the volcanoes. A magma chamber inflates when new magma enters from the mantle, and extrusion follows. Every day, lava flows dump millions of tons of new rock along the edge of the island. Eventually the stress and weight of the new land will trigger another great landslide

Where are subaqueous landslides most likely to occur?

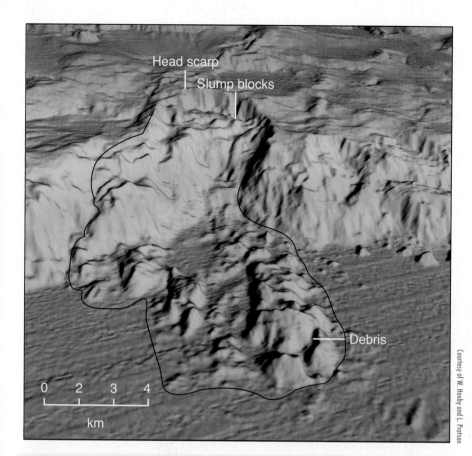

Courtesy of W. Haxby and L. Pratson

Figure 11.15 A subaqueous landslide of the coast of central Oregon formed when tectonic activity made the slope too steep, as seen in this shaded relief map. A trail of debris extends downslope from the arcuate head of the landslide. The crescent-shaped head of the slope failure is 6 km across. Some of the blocks are hundreds of meters high and kilometers across. The collapse may have been triggered by an earthquake on the underlying subduction zone.

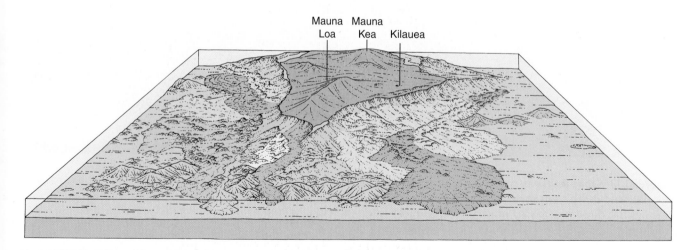

Figure 11.16 Vast subaqueous slumps and flows flank the island of Hawaii and others in the seamount chain. Each major lobe is outlined. The names of the major shield volcanoes are also given.

and a catastrophic tsunami. Geologists agree that if the south side of the island breaks free suddenly, the results will be one of the worst disasters in recorded history, far beyond anything ever witnessed by humankind. The force of this catastrophic landslide crashing into the ocean could trigger a tsunami that could travel across the ocean at the speed of a jet plane inundating the coasts of the Pacific Rim. The south flank of the island of Hawaii is unstable. It is going to slide into the sea, and there is nothing we can do to stop it.

Slow-forming, but nonetheless spectacular, subsidence features also form on the ocean floor. Many develop as a result of the flow of salt interlayered with marine sediments (**Figure 11.17**). Subsidence bowls several kilometers across dot the seafloor west of the Mississippi delta. They developed while salt diapirs welled upward. The horizontal and vertical displacement of the salt caused adjacent areas to collapse.

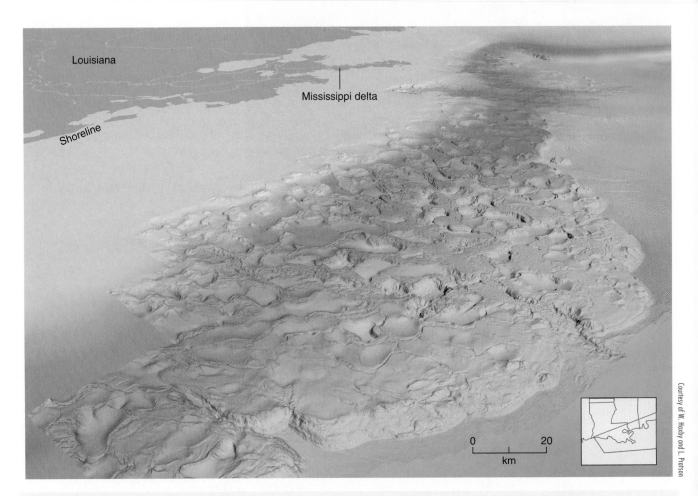

Louisiana

Mississippi delta

Shoreline

0   20
km

Figure 11.17 Subaqueous subsidence occurred in thick deposits of sediment and salt near the delta of the Mississippi River. This pockmarked, Moonlike seascape was created by the shifting flow of the salt. The movement of salt created the high scarp on the right side of the map. Elsewhere, salt diapirs mushroomed into domes that eventually collapsed to make the bowl-shaped depressions as the salt flowed away.

GeoLogic   Slope Systems

If you carefully study this photograph and the accompanying sketch, you will soon see what geologists see as they look at a hillside. What makes the difference in slope angle? Why are the rocks in the cliff so jointed, whereas those below are not? What processes shaped the form of the cliff and slope? Abundant angular boulders litter the slope. Where did they come from? Do you see any systematic changes in particle size away from the cliff? Why is there a network of small gullies even in this arid region? Finally, does the stream at the foot of the hill play a role in the evolution of the slope?

Observations

1. The cliff is receding by rockfalls.
2. Talus produced by frost action and rockfalls accumulates at the base of the cliff.
3. Larger blocks of rock occur on the higher slopes below the cliff.
4. Small rocks are found progressively downslope.
5. Near the river the slope is covered by mostly fine-grained material (sand) and a few very small blocks.
6. Sediments carried by the river are mostly sand-size particles (note sand bars) and mud (note muddy water).

Interpretations

The basic elements of a **slope system** are shown in the diagram. Rockfalls from the cliff are produced by weathering and frost action along joints in the sandstone formation. Coarse boulders accumulate at the base of the cliff. They are then transformed into smaller and smaller particles by mechanical and chemical weathering and move downslope by creep and other types of mass movement. Some debris may be collected by minor tributaries on the slope and move into the main stream by running water. The remainder of the fine-grained debris continues to move down-slope by gravity (mostly by creep). It eventually enters the main stream and is carried away as sand, salt, and mud.

A slope is perhaps best thought of as an open and very dynamic system where gravity is the major source of energy. The system has inputs of rock material produced by physical and chemical weathering, and an output of fine rock fragments into a stream as shown above. Mass movement, together with weathering, and erosion of small gullies transport regolith and loose rock material downslope to a stream.

Key Terms

angle of repose (p. 294)
cohesive strength (p. 294)
creep (p. 298)
debris flow (p. 302)
lahar (p. 304)

landslide (p. 305)
mass movement (p. 298)
mudflow (p. 304)
permafrost (p. 299)
rock avalanche (p. 306)

rockfall (p. 305)
rock glacier (p. 300)
rockslide (p. 305)
slope system (p. 313)
slump block (p. 305)

solifluction (p. 299)
subaqueous mass movement
  (p. 310)
subsidence (p. 309)

Review Questions

1. List the factors that affect mass movement on a slope.
2. How would you decide if a building lot was safe with regard to slope failure?
3. Calculate the downhill force on a boulder weighing 2.5 kg that rests on a slope inclined at 27°.
4. Why does deforestation cause unstable slopes and accelerate mass movement?
5. Describe four types of rapid mass movement.
6. List the types of mass movement that are dominantly slow.
7. List five ways in which creep is expressed on a hill slope.
8. What factors promote debris flows?
9. Speculate about the variety of ways that a lahar might be generated.
10. What causes creep?
11. What is solifluction? In what climates is it common?
12. How does subsidence differ from other kinds of mass movement?
13. What kinds of slope failures occur below sea level? What causes them?
14. Explain why slopes are considered open systems.

12 River Systems

A river system is a network of connecting channels through which water, precipitated on the surface, is collected and funneled back to the ocean. At any given time, about 1300 km³ of water flows in the world's rivers. As it moves, it picks up weathered rock debris and carries it to the oceans. Rivers are the dominant agents of erosion on our planet. No matter where you go, rivers have played some role in shaping the surface.

The tremendous waterfalls of the Iguassú River, shown in the panorama above, illustrate both the beauty and power of rivers. The river rises in a coastal mountain range hardly 50 km from the Atlantic Ocean. From its headwaters, it flows 700 km inland along the border of Argentina and Brazil before it joins the great Paraná River and empties into the Atlantic Ocean. Shortly before it merges with the Paraná, the river plunges over a series of high escarpments created by erosion of the flood basalts of the Paraná basin. With a torrential roar, the falls plunge over cliffs that are 3 km long and greater than 70 m high. More than 273 separate falls can be counted along the rugged cliffs.

But river systems also formed the gentle rolling hills of Ohio and Kansas, the levees and back-swamps of Louisiana, the ridges and valleys of the Appalachian Mountains, and the erosional details of the flat coastal plains. No other geologic agent is so universally important in the origin and evolution of the surface upon which we live.

Rivers are ideal examples of natural systems. The energy that drives the flow of water ultimately comes from the Sun and from gravity. Moreover, it is easy to see that a river is an open system with multiple sources for water and for the sediment it carries. Water can come from melting snow and ice, from direct precipitation, from groundwater, and of course from other rivers. Sediment is added to the system by erosion and solution and eventually it leaves the system when it is deposited far from its source. This majestic system interacts with other parts of the hydrologic system and, as we shall see, is modified by the tectonic system in diverse ways. Rivers respond to climate change as well as to the motion of the continents.

In this chapter, we discuss the effects and controls on rivers as natural systems: how water flows and how it carries and eventually deposits sediment. We also consider how the entire river system responds to changes.

Major Concepts

1. Running water is part of Earth's hydrologic system and is the most important agent of erosion. Stream valleys are the most abundant and widespread landforms on the continents.
2. A river system consists of a main channel and all of the tributaries that flow into it. It can be divided into three subsystems: (a) a collecting system, (b) a transporting system, and (c) a dispersing system.
3. The most important variables in stream flow are (a) discharge, (b) gradient, (c) velocity, (d) sediment load, and (e) base level.
4. The variables in a stream constantly adjust toward a state of equilibrium.
5. Rivers erode by (a) removal of regolith, (b) downcutting of the stream channel by abrasion, and (c) headward erosion.
6. As a river develops a low gradient, it deposits part of its load on point bars, on natural levees, and across the surface of its floodplain.
7. Most of a river's sediment is deposited where the river empties into a lake or ocean. This deposition commonly builds a delta at the river's mouth. In arid regions, many streams deposit their loads as alluvial fans at the base of steep slopes.
8. The origin and evolution of the world's major rivers are controlled by the tectonic and hydrologic systems.

Geologic Importance of Running Water

Running water is by far the most important agent of erosion. Other agents, such as groundwater, glaciers, and wind, are locally dominant but affect only limited parts of Earth's surface.

An attempt to appreciate the significance of streams and stream valleys in Earth's regional landscape presents a problem of perspective. Viewed from the ground, Earth's stream valleys may appear to be only irregular depressions between rolling hills and plains. Viewed from space, however, stream valleys are seen to dominate most continental landscapes of Earth.

What is the most common landform on Earth's surface?

The ubiquitous stream valleys on Earth's surface, and the importance of running water as the major agent of erosion, can best be appreciated by taking a broad, regional view of the continents and their major river systems (**Figure 12.1**). As the topographic maps in **Figure 12.2** show, the surface, throughout broad regions of the continents, is little more than a complex of valleys created by stream erosion. Even in the desert, where it sometimes does not rain for decades, networks of dry stream valleys commonly are major landforms. No other landform on the continents is as abundant and significant. Look at a satellite photograph of Earth. Is any part of the terrain not influenced by stream erosion?

Major Characteristics of River Systems

A river system consists of a main channel and all of the tributaries that flow into it. It can be divided into three subsystems: (1) a collecting system, (2) a transporting system, and (3) a dispersing system.

Although rivers and the valleys through which they flow are the most familiar of all landforms, it is difficult to define precisely the word *river* because of the great variety of physical characteristics rivers exhibit. There are big rivers, such as the Mississippi, Amazon, and Nile, and there are little rivers, streams, creeks, or brooks. Some rivers in arid regions flow only after a heavy rain and then dry up, whereas rivers in the Arctic are frozen two-thirds of the year. From the viewpoint of geology, it is perhaps most useful to consider a river not as a natural channel through which water flows, but as a system. A **river system**, or **drainage basin**, consists of a main channel and all of the tributaries that flow into it (**Figure 12.3**). It is bounded by a **divide** (ridge), beyond which water is drained by another system. Within a river system, the surface of the ground slopes toward the network of tributaries, so the drainage system acts as a funneling mechanism for removing surface runoff and weathered rock debris.

Figure 12.1 Rivers drain most of the continents, but their distribution and patterns are controlled by climate and plate tectonics. For example, the role of climate is evident in that there are few rivers in the mid-latitude deserts, large rivers in the tropics, and no rivers in the coldest polar areas. Short rivers drain convergent margins (like western North and South America) and long rivers drain the stable platforms (like central North America and Russia) and some shields (like the South American shield).

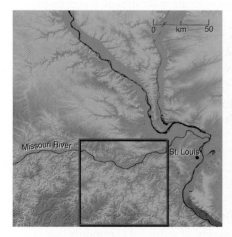

(A) A map of an area in Missouri shows the regional patterns of the valleys formed by the Missouri river system near St. Louis. The area is about 200 km across.

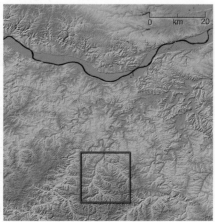

(B) A detailed view of the area reveals an intricate network of streams and valleys within the tributary regions of the large streams.

(C) At even higher resolution, many smaller streams and valleys in the main drainage system are revealed.

Courtesy of JPL/NIMA/NASA

Figure 12.2 Erosion by running water is the dominant process in the formation of the landscape and stream valleys are apparent at all scales.

Figure 12.3 The major parts of a river system are characterized by different geologic processes. The tributaries in the headwaters constitute a subsystem that collects water and sediment and funnels them into a main trunk stream. Erosion is dominant in this headwater area. The main trunk stream is a transporting subsystem. Both erosion and deposition can occur in this area. The lower end of the river is a dispersing subsystem, where most sediment is deposited in a delta or an alluvial fan and water is dispersed into the ocean. Deposition is the dominant process in this part of the river.

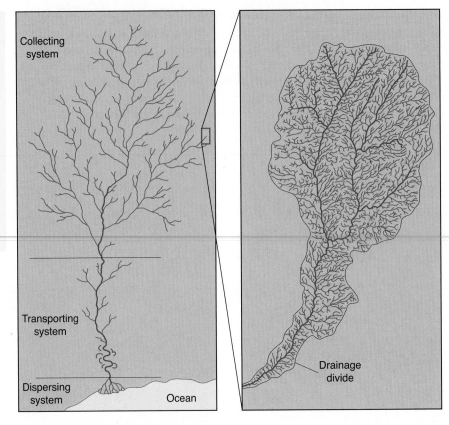

A map of a typical river system is shown in Figure 12.3. Three subsystems—a collecting system, a transporting system, and a dispersing system—can be identified. Although the boundaries between the three subsystems are somewhat gradational, the distinguishing characteristics of each subsystem on a regional scale are readily apparent.

The Collecting System

A river's **collecting system** consists of the network of **tributaries** in the headwater region that collect and funnel water and sediment to the main stream. It commonly has a **dendritic** (treelike) **drainage pattern**, with numerous branches that extend upslope toward the divide. Indeed, one of the collecting system's most remarkable characteristics is the intricate network of tributaries, shown in the enlargement in Figure 12.3. This map was made by plotting all visible streams shown on an aerial photograph. It is not, however, the entire system. Each of the smallest tributaries shown on the map has its own system of smaller and smaller tributaries, so the total number becomes astronomical. From the details in Figure 12.2, it is apparent that most of the land's surface is part of some drainage basin.

The Transporting System

The **transporting system** is the main trunk stream, which functions as a channel through which water and sediment flow from the collecting area toward the ocean. Although the major process is transportation, this subsystem also collects additional water and sediment. Deposition of sediment commonly occurs where the channel meanders back and forth and when the river overflows its banks during a flood stage. Erosion, deposition, and transportation thus occur, but the main process in this part of a river is the movement of water and sediment.

The Dispersing System

The **dispersing system** consists of a network of **distributaries** at the mouth of a river, where sediment and water are dispersed into an ocean, a lake, or a dry basin. The major processes are the deposition of the coarse sediment load and the dispersal of fine-grained material and river waters into the basin.

Order in Stream Systems

It is apparent from Figure 12.3 and **Figure 12.4** that a stream does not occur as a separate, independent entity. Every stream, every river, and every gully and ravine are part of a drainage system, with each tributary intimately related to the stream into which it flows and to the streams that flow into it. Every stream has tributaries, and every tributary has smaller tributaries, extending down to the smallest gully. Studies of drainage systems show that when a stream system develops freely on a homogeneous surface, definite mathematical ratios characterize the relationships between the tributaries and the size and gradient of the stream and of the stream valley. Some of the more important relationships and generalizations are the following:

1. The number of stream segments (tributaries) decreases downstream in a mathematical progression.
2. The length of tributaries becomes progressively greater downstream.
3. The gradient, or slope, of tributaries decreases exponentially downstream.
4. The stream channels become progressively deeper and wider downstream.
5. The size of the valley is proportional to the size of the stream and increases downstream.

These relationships are the basis for the conclusion that streams erode the valleys through which they flow.

If valleys were ready-made by some process other than stream erosion, such as faulting or other earth movements, these relationships would be "infinitely improbable." You can easily confirm the high degree of order in streams by studying Figures 12.2 and 12.3. Does each tributary have a steeper gradient than the stream into which it flows? Does each tributary flow smoothly into a larger stream without an abrupt change in gradient? Are the tributary valleys smaller than the valleys into which they drain?

Geologists have studied stream erosion in great detail over the last century, and they have been able to observe and measure many aspects of stream development and erosion by running water. The origin of valleys by erosion is well established, and running water is clearly the most significant agent of erosion on Earth's surface.

How do we know that streams erode the valleys through which they flow?

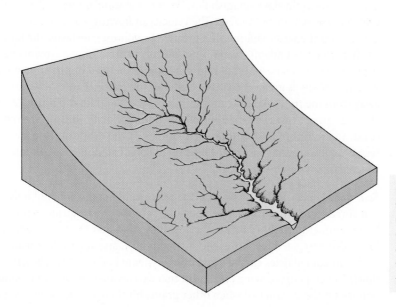

Figure 12.4 **The characteristics of a river change systematically** downstream. The gradient decreases downstream, and the channel becomes larger. Other downstream changes include an increase in the volume of water and an increase in the size of the valley through which the stream flows.

The Dynamics of Stream Flow

Rivers are highly complex systems influenced by several variables. As is the case with so many natural systems, if one variable is changed, it produces a change in the others. The most important variables are (1) discharge, (2) gradient, (3) velocity, (4) sediment load, and (5) base level.

Anyone who has watched the fascinating flow of water in a river realizes that the process is complex. The water moves down the stream channel through the force of gravity, and the velocity of flow increases with the slope, or gradient, of the streambed. In fact, the flow of water in natural streams depends on several factors, the most important of which are discussed below. These variables are intimately related and a change in one causes change in others.

Factors Influencing Stream Flow

Discharge   The amount of water passing a given point during a specific interval of time is called **discharge**. It is usually measured in cubic meters per second. The discharges of most of the world's major drainage systems have been monitored by gauging stations for years. The water for a river system comes from both surface runoff and seepage of groundwater into the stream's channels. Groundwater seepage is important because it can maintain the flow of water throughout the year.

Stream Gradient   Certainly one of the most obvious factors controlling stream flow is the **gradient**, or slope, of the stream channel. The gradient of a stream is steepest in the headwaters and decreases downslope. The **longitudinal profile** (a cross section of a stream from its headwaters to its mouth) is a smooth, concave, upward curve that becomes very flat at the lower end of the stream (Figure 12.4). The gradient usually is expressed in the number of meters the stream descends for each kilometer of flow. The headwater streams that drain the Rocky Mountains can have gradients of more than 50 m/km; the lower reaches of the Mississippi River have a gradient of only 1 or 2 cm/km.

Velocity   Streams flow downhill with velocities that range from a few centimeters per second to as much as 10 meters per second (about 35 km/hr). The velocity of flowing water is proportional to the gradient of the stream channel. Steep gradients produce rapid flow, which commonly occurs in high-mountain streams. Where slopes are very steep, waterfalls and rapids develop, and the velocity approaches that of free fall. Low gradients result in slow, sluggish flow. Where a stream enters a lake or an ocean, its velocity is soon reduced to zero. The velocity of flowing water in a given channel also depends on the water volume. The greater the volume, the faster the flow.

The velocity of flowing water is not uniform throughout a stream channel. It depends on the shape and roughness of the channel and on the stream pattern. The velocity usually is greatest near the center of the channel and above the deepest part, away from the frictional drag of the channel walls and floor (**Figure 12.5**). As the channel curves, however, the zone of maximum velocity shifts to the outside of the bend, and a zone of minimum velocity forms on the inside of the curve. This flow pattern is an important cause of the lateral erosion of stream channels and of the migration of stream patterns.

Sediment Load   Running water is the major cause of erosion, not only because it can abrade and erode its channel, but also because of its enormous power to transport loose sediment produced by weathering. Flowing water is a fluid medium by which loose, disaggregated regolith is picked up and transported to the ocean.

Sediment particles can be lifted from a stream bed by hydraulic lift—just as air flowing over a curved wing creates lift that carries an airplane aloft. Some grains bounce off the stream bed when other grains hit them and knock them into the

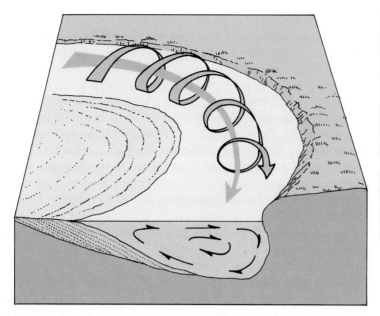

Figure 12.5 Flow of water around a meander bend in a river follows a corkscrew pattern. Water on the outside of the bend is forced to flow faster than that on the inside of the curve. This difference in velocity, together with normal frictional drag on the channel walls, produces a corkscrew pattern. As a result, erosion occurs on the outside bank, where velocity is greatest, and deposition occurs on the inside of the bend, where velocity is at a minimum. Erosion on the outside of the meander bend and deposition on the inside cause the stream channel to migrate laterally.

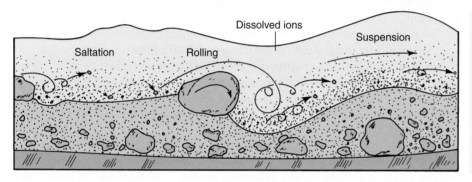

Figure 12.6 Movement of the sediment load in a stream is accomplished in a variety of ways. Mud is carried in suspension. Particles that are too large to remain in suspension are moved by sliding, rolling, and saltation. Some ions are dissolved and carried in solution. Increases in discharge, due to heavy rainfall or spring snowmelt, can flush out all of the loose sand and gravel, so the bedrock is eroded by abrasion.

flowing water. In addition, water has a very low viscosity, many times less than that of flowing lava. As a result, its usual flow cannot be described by smooth, simple, streamlines. Instead, it is **turbulent** with many secondary eddies and swirls in addition to the main downstream current. One part of the turbulent flow is vertical and tends to keep small grains suspended in the stream flow.

Once within a stream, sediment is transported in three ways (**Figure 12.6**):

1. Fine particles are moved in suspension (suspended load).
2. Coarse particles are moved by traction (rolling, sliding, and saltation) along the streambed (bed load).
3. Dissolved material is carried in solution (dissolved load).

The **suspended load** is the most obvious, and generally the largest, fraction of material moved by a river. In most major streams, silt and clay-sized particles remain in suspension most of the time and move downstream at the velocity of the flowing water, to be deposited in an ocean, in a lake, or on a floodplain.

Particles of sediment too large to remain in suspension collect on the stream bottom and form **bed load**, or traction load. These particles move by sliding, rolling, and **saltating** (short leaps). The bed load moves only if there is sufficient velocity to move the large particles. Part of the bed load can suddenly move in suspension, or part of the suspended load can settle. The bed load can constitute 50% of the total load in some rivers, but it usually ranges from 7% to 10% of the total sediment load. The movement of the bed load is one of the major tools of stream abrasion because as the sand and gravel move, they abrade (wear away) the sides and bottom of the stream channel. In some rivers, the grinding action of the bed load can be heard as large boulders are moved along the river's bottom.

The **dissolved load** is matter transported as chemical ions and is essentially invisible. All streams carry some dissolved material, which is derived principally from the groundwater that emerges from seeps and springs along the riverbanks. The most abundant materials in solution are calcium and bicarbonate ions, but sodium, magnesium, chloride, ferric, and sulfate ions are also common. Various amounts of organic matter are present, and some streams are brown with organic acids derived from the decay of plant material. Flow velocity, which is so important to the transportation of the suspended and traction loads, has little effect on a river's ability to carry dissolved material. Once mineral matter is dissolved, it remains in solution, regardless of velocity, and is precipitated and deposited only if the chemistry of the water changes. Chemical analysis shows that most rivers carry a dissolved load of less than a thousand parts per million. Although these amounts of dissolved material seem small, they are far from trivial. Sampling shows that 5% to 50% of all the material carried to the ocean is in solution. For example, in the Mississippi River the dissolved load is about 30% of the total sediment load.

Velocity is an important control on a stream's ability to erode, transport, and deposit sediment. The **capacity** of a stream is the amount or weight of sediment it carries. Stream capacity increases to a third or fourth power of flow velocity; that is, if the velocity is doubled, the stream can move from 8 to 16 times as much sediment. Another measure of sediment load is its **competence**—the size of the largest particle the stream is able to carry. Competence also increases with velocity.

The results of experimental studies show that a minimum or **threshold velocity** is required to move grains of a certain size (**Figure 12.7**). The graph shows that at low velocities only small grains can be transported. Higher velocities will generally move larger particles. On the other hand, where the stream's velocity is low, a significant part of the sediment load is deposited along the channel or on the floodplain. Sediment may also be deposited where the velocity is reduced, such as when a river enters a lake or the ocean.

Base Level   The **base level** of a stream is the lowest level to which the stream can erode its channel. The base level is, in effect, the elevation of the stream's mouth, where the stream enters an ocean, a lake, or another stream. A tributary cannot erode lower than the level of the stream into which it flows. Similarly, a lake controls the level of erosion for the entire course of the river that drains into it. The levels of tributary junctions and lakes are temporary base levels: Lakes can be filled with sediment or drained, and streams can then be established across the former lake bed. For all practical purposes, the ultimate base level is sea level because the energy of a river is quickly reduced to zero as it enters the ocean. Therefore, base level is an extremely important control on the extent of stream erosion, and a drop in base level commonly creates the basal unconformity of a sedimentary sequence.

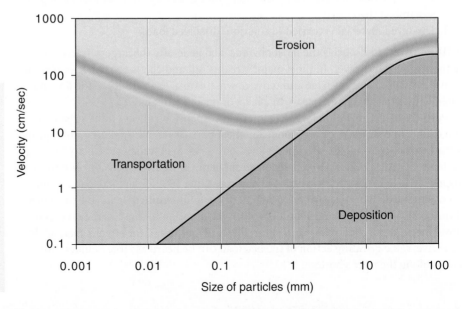

Figure 12.7 The threshold velocity for sediment transport shows the minimum velocities at which a stream can pick up and move a particle of a given size. This threshold velocity is represented by a zone, not by a line, because of variations resulting from stream depth, particle shape, and density. The lower curve indicates the velocity at which a particle of a given size settles out and is deposited. Note that fine particles stay in suspension at velocities much lower than those required to lift them from the surface of the streambed.

Equilibrium Gradients in River Systems

A river system functions as a unified whole: Any change in one part of the system affects the other parts. The major factors that determine stream flow constantly change toward a balance, or equilibrium, so that the gradient of the stream is adjusted to accommodate the volume of the water available, the channel's characteristics, and the velocity necessary to transport the sediment load.

We have repeatedly emphasized the fact that any one part of a stream does not occur as a separate, independent entity. One of the most important characteristics of a river system is that it functions as a unified whole: Any change in one part of the system affects the other parts. The major factors that determine stream flow (discharge, velocity, channel shape, gradient, base level, and load) constantly change. A change in any of these factors causes compensating adjustments in another factor to restore balance or equilibrium in the entire drainage system. A river is in equilibrium if its channel form and gradient are balanced so that neither erosion nor deposition occurs. Rivers are constantly adjusting to approach this ideal condition. This adjustment is important in understanding the natural evolution of the landscape. It also has practical considerations: If we are going to continually manipulate rivers to suit our needs, we should know how river systems respond to changes.

The concept of equilibrium in a river system can be appreciated by considering a hypothetical stream in which equilibrium has been established. In **Figure 12.8A**, the variables in the stream system are in balance, so neither erosion nor deposition occur along the stream's profile. There is just enough water to transport the available sediment down the existing slope. Such a stream is in equilibrium and is known as a **graded stream**. In Figure 12.8B, the stream's profile is displaced by a fault that creates a waterfall. The increased gradient across the fault greatly increases the stream's velocity at that point, so rapid erosion occurs, and the waterfall (or the rapid) begins to migrate upstream. The eroded sediment added to the stream segment on the dropped fault block is more than the stream can transport because the system was already in equilibrium before faulting occurred. The river therefore deposits part of its load at

What is the profile of equilibrium in a river system?

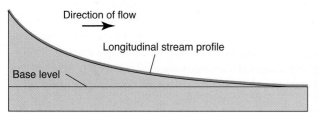

(A) Initially, when the stream profile is at equilibrium, the velocity, load, gradient, and volume of water are in balance. Neither erosion nor deposition occurs.

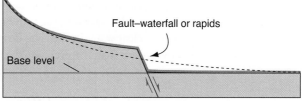

(B) Faulting disrupts equilibrium by decreasing the gradient downstream and increasing the gradient at the fault line.

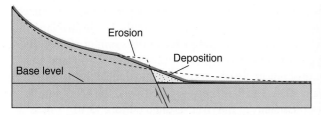

(C) Erosion proceeds upstream from the fault and deposition occurs downstream and a new stream profile starts to develop.

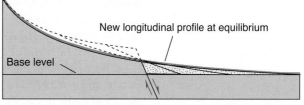

(D) Erosion and deposition eventually develop a new stream profile at which the velocity, load, gradient, and volume of water will be in balance so that neither erosion nor deposition occurs.

Figure 12.8 Adjustment of a stream to reestablish equilibrium is shown by profile changes after disruption by faulting. Erosion and deposition develop a new stream profile at which the velocity, load, gradient, and volume of water will be in balance so that neither erosion nor deposition occurs.

Figure 12.9 The volume of sediment transported by a stream is illustrated by the Mono Reservoir in California, which has been completely filled with sand and mud.

that point, thus building up the channel gradient (the yellow areas in Figure 12.8C–D) until a new profile of equilibrium is established.

An example of the adjustments just described occurred in Cabin Creek, a small tributary of the Madison River, north of the Hebgen Dam in Montana. In 1959, during the Hebgen Lake earthquake, a 3-m fault scarp formed across the creek. By June 1960, erosion by Cabin Creek had erased the waterfall at the cliff formed by the fault, and only a small rapid was left. By 1965, the rapid was completely removed, and equilibrium was reestablished.

Equilibrium in a river system is also illustrated by the results of dam construction. In the reservoir behind a dam, the gradient is reduced to zero. Hence, where the stream enters the reservoir, its sediment load is deposited as a delta and as layers of silt and mud over the reservoir floor (**Figure 12.9**). Because most sediment is trapped in the reservoir, the water released downstream has practically no sediment load. The clear water in the river downstream of the dam is therefore capable of much more erosion than the previous river, which carried a sediment load adjusted to its gradient. As a result, extensive scour and erosion commonly result downstream from a new dam.

The Nile and River Equilibrium

The Aswan High Dam on the Nile River in Egypt provides a good example of the many consequences of modifying a river system that has approached equilibrium. For centuries, the Nile River has been the main source of life in Egypt. The Nile's principal headwaters are located in the high plateaus of Ethiopia. Once a year, for approximately a month, the Nile used to rise to flood stage and cover much of the fertile farmland in the Nile Delta area. The Aswan High Dam was completed in the summer of 1970. It was intended to provide Egypt with water to irrigate 1 million acres of arid land and to generate 10 billion kilowatts of power, which, in turn, was to double the national income and permit industrialization. The dam, however, destroyed the Nile's equilibrium, and many unforeseen adjustments in the river resulted (**Figure 12.10**). This is what happened.

The Nile is not only the source of water for the delta; it is also the source of sediment. When the dam was finished and began to trap sediment in a reservoir (Lake Nasser), the physical and biological balance in the delta area was destroyed. Without the annual "gift of the Nile," the delta coastline is now exposed to the full force of marine currents, and wave erosion is eating away at the delta front. Some parts of the delta are receding several meters a year.

© Yann Arthus-Bertrand

Figure 12.10 **The Nile River** has been dramatically affected by the construction of the Aswan High Dam. Without the annual flood, the river stagnates and is commonly overgrown with vegetation.

The sediment previously carried by the Nile was an important link in the aquatic food chain, nourishing marine life in front of the delta. The recent lack of Nile sediment has reduced plankton and organic carbon to a third of the former levels. This change either killed off or drove away sardines, mackerel, clams, and crustaceans. The annual harvest of 16,000 metric tons of sardines and a fifth of the fish catch have been lost.

The sediment of the Nile also naturally fertilized the floodplain. Without this annual addition of soil nutrients, Egypt's 1 million cultivated acres need artificial fertilizer.

The water discharged from the reservoir is clear, free of most of its sediment load. Without its load, the discharged water flows swiftly downstream and is vigorously eroding the channel bank. This scouring process has already destroyed three old barrier dams and more than 500 bridges built since 1953. Ten new barrier dams must be built between Aswan and the ocean at a cost equal to one-fourth the cost of the Aswan Dam itself.

The annual Nile flood was also important to the area's ecology because it washed away salts that accumulated in the arid soil. Soil salinity has already increased, not only in the delta, but throughout the middle and upper Nile areas. Unless costly corrective measures are taken, millions of acres will become unproductive. Also, the control of the river has resulted in stagnation of the channels and overgrowth of vegetation (Figure 12.10).

The change in the river system has permitted double cropping, but this eliminated periods of dryness. The dry seasons previously helped limit the population of schistosomiasis, a blood parasite carried by snails that infects the intestinal and urinary tracts of humans. In some areas of Egypt, 30% of the people had the infection, but recent drug treatments have reduced that to about 2%.

Problems have also occurred in the lake behind the dam. The lake was to have reached a maximum level in 1970, but it might actually take 200 years to fill. More than 15 million cubic meters of water annually seep underground into the porous Nubian Sandstone, which lines 480 km of the lake's western bank. The sandstone is capable of absorbing an almost unlimited quantity of water. Moreover, the lake is in one of the hottest and driest places on Earth, and the rate of evaporation is staggering. A high rate was expected, but additional losses from transpiration by plants growing along the lakeshore and increased evaporation caused by high winds have brought the total loss of water from the lake to nearly double the expected rate. This loss equals half the total amount of water that once was "wasted," flowing unused to the ocean.

What changes occur when a dam is built on a river?

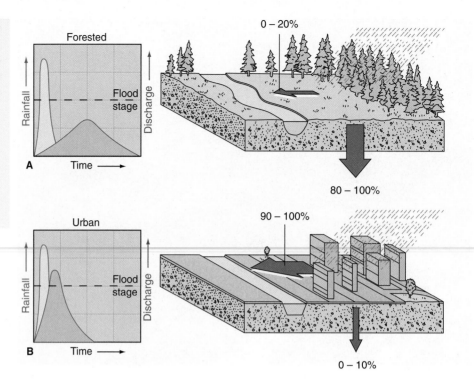

Effects of Urbanization on River Equilibrium

Another way that equilibrium in river systems has been disrupted is through urbanization. The construction of cities may at first seem unrelated to the modification of river systems, but a city significantly changes the surface runoff, and the resulting changes in river dynamics are becoming serious and costly. Water that falls to Earth as precipitation usually follows several paths in the hydrologic system. In general, from 54% to 97% returns to the air directly by evaporation and transpiration; the remaining water collects in stream systems as surface runoff or infiltrates the ground and moves slowly through the subsurface toward the ocean. Under natural conditions, 80% to 100% of the surface runoff infiltrates into the subsurface. Urbanization disrupts each of these paths in the normal hydrologic system. It changes the nature of the terrain and consequently affects the rates and percentages of runoff and infiltration. Roads, sidewalks, and roofs of buildings render a large percentage of the surface impervious to infiltration. Not only does the volume of surface runoff increase, but runoff is much faster because water is channeled through gutters, storm drains, and sewers. As a result, flooding increases in intensity and frequency (**Figure 12.11**).

Processes of Stream Erosion

River systems erode the landscape by three main processes: (1) removal of regolith, (2) downcutting of the stream channel by abrasion, and (3) headward erosion.

Erosion of the land is one of the major effects of the hydrologic system. It has occurred on all continents throughout all of geologic time and will continue as long as the system operates and land is exposed above sea level (**Figure 12.12**). Evidence of erosion is ubiquitous and varied. We see it in the development of gullies on farmlands and in the cutting of great canyons. We see it in the thick layers of sedimentary rocks that cover large parts of the continents and bear witness to erosion and deposition in past ages. But exactly how does a river system erode the land? How can a relatively small stream such as the Colorado River erode the Grand Canyon, which is more than 2 km deep and 25 km

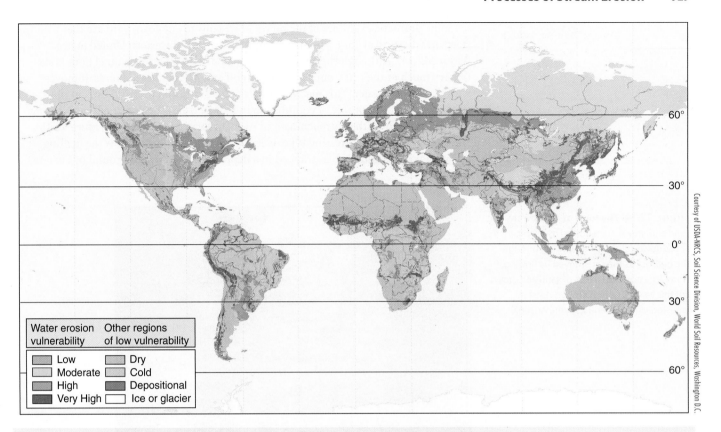

Courtesy of USDA-NRCS, Soil Science Division, World Soil Resources, Washington D.C.

Water erosion vulnerability | **Other regions of low vulnerability**

Low
Moderate
High
Very High

Dry
Cold
Depositional
Ice or glacier

Figure 12.12 Water erosion vulnerability varies across the globe in a systematic way and is closely related to climate and topography. Two large regions are climatically controlled: The Arctic regions (blue), which have been glaciated and/or influenced by permafrost, and the great desert regions (gray) that have low vulnerability simply because there isn't enough water for significant stream erosion. In much of the desert regions deposition by wind and intermittent streams dominates. Areas of very high vulnerability are mountainous regions with high rainfall.

wide? What processes are involved in erosion? How do river systems evolve? Answers to these basic questions have eluded scientists until recently, and even today some details remain controversial. However, we now know that erosion by running water and the evolution of river systems are accomplished by three basic processes: (1) removal of regolith, (2) downcutting of the stream channel by abrasion, and (3) headward erosion.

Removal of Regolith

One of the most important processes of erosion is the removal and transport of rock debris (regolith) produced by weathering. The process is simple but important. Loose rock debris created by weathering is washed downslope into the drainage system and is transported as sediment load in streams and rivers. In addition, soluble material is carried in solution. The net result is that the blanket of regolith created by weathering is continually being removed and transported to the sea by stream action. As it is removed, however, it is also continually being regenerated by the weathering of the fresh bedrock below. Measurements of the amount of sediment carried by rivers suggest that about 6 cm/1000 yr are removed from the continents.

Downcutting of Stream Channels

Downcutting is a fundamental process of erosion in all stream channels, whether small hillsides, gullies, or great canyons of major rivers. The process is accomplished by the **abrasion** of the channel floor by sand and gravel as they are swept downstream by the flowing water. It is similar in many respects to the action of a wire saw used in quarries to cut and shape large blocks of stone (**Figure 12.13**). An abrasive such as garnet, corundum, or quartz, dragged across a rock by a wire, can cut through a stone block with remarkable speed.

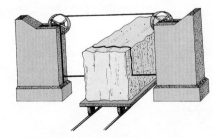

Figure 12.13 Sketch of a wire saw commonly used in quarry operations. The wire is pulled across the rock, dragging abrasives as it moves. When the rock is raised (or the wire is lowered), the abrasives, dragged across the rock by the wire, slice through the block.

Some dramatic examples of the power of streams to cut downward are the steep, nearly vertical gorges in many canyons in the southwestern United States (**Figure 12.14**). Although the bed load of sand and gravel on the channel floor is stationary much of the time, during spring runoff and periodic flash floods, it moves with the flowing water. This material is an effective abrasion tool and can cut the stream channel to a profile of equilibrium in a short time. The power of downcutting is also expressed by the entrenchment of rivers to form deep canyons (**Figure 12.15**).

An effective and interesting type of abrasion of the channel floor is the drilling action of pebbles and cobbles trapped in a depression and swirled around by currents.

Figure 12.14 **The tools of erosion** are sand and gravel. Transported by a river, they act as powerful abrasives, cutting through the bedrock as they are moved by flowing water. The abrasive action of sand and gravel cut this vertical gorge through resistant limestone in the Grand Canyon, Arizona.

Figure 12.15 **Entrenched meanders** of the Colorado River resulted from downcutting of the river channel more than 300 m.

The rotational movement of the sand, gravel, and boulders acts like a drill and cuts deep holes known as **potholes**. As the pebbles and cobbles are worn away, new ones take their place and continue to drill into the bedrock of the stream channel. Some potholes are several meters in diameter and more than 5 m in depth (**Figure 12.16**).

As the pebbles and cobbles are carried by flowing water, they themselves are worn down by striking one another and the channel bottom. Their corners and edges are chipped off, and the particles become smaller, smoother, and more rounded. Large boulders that have fallen into a stream and are transported only during a flood are thus slowly broken and worn down to smaller fragments. Ultimately, they are washed away as grains of sand.

Another important factor in the downcutting of a stream channel is the upstream migration of waterfalls and rapids. Here again, the process is simple but important. It can be appreciated by considering the erosion of Niagara Falls (**Figure 12.17**). The increased velocity of the falling water sets up strong turbulence at the base of the falls, causing rapid erosion of the underlying weak, nonresistant rock layers. The cliff is gradually undermined, and the falls retreat upstream. During the last 12,000 years, Niagara Falls has migrated headward more than 11.5 km.

Slope retreat also plays an important role in shaping stream valleys. Mass movement and the erosion of small tributaries reduce steep valley walls to gentle slopes. In the process, much new sediment enters the river system.

© vitmark/ShutterStock, Inc

Figure 12.16 Potholes are eroded in a streambed by sand, pebbles, and cobbles whirled around by eddies. These potholes on the floor of the Blyde River Canyon in South Africa are about 3 m across.

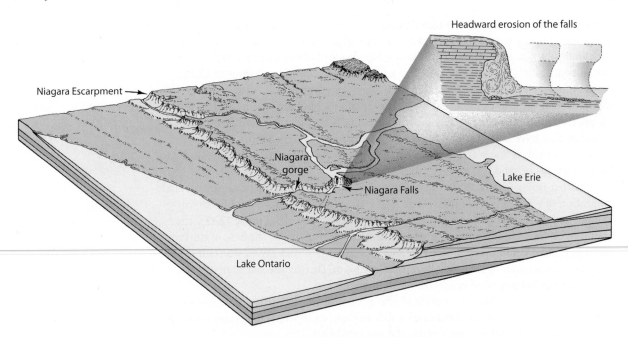

Niagara Escarpment

Headward erosion of the falls

Niagara gorge

Niagara Falls

Lake Erie

Lake Ontario

(A) The Niagara River originated as the last glacier receded from the area and water flowed from Lake Erie to Lake Ontario over the Niagara Escarpment. Erosion causes the waterfalls to migrate upstream at an average rate of about 1 m/yr.

© R Rusak/ShutterStock, Inc.

(B) Niagara Falls presents a spectacular scene as large volumes of water fall vertically over the cliffs of limestone. The falls are 70 m high and have migrated headward more than 11.5 km in the last 12,300 years.

Figure 12.17 **Retreat of Niagara Falls upstream** occurs as hydraulic action undercuts the weak shale below the limestone.

Headward Erosion

In the process of stream erosion and valley evolution, streams have a universal tendency to erode headward, or upslope, and to increase the lengths of their valleys until they reach the divide. **Headward erosion** can be analyzed by referring to **Figure 12.18**. The reason erosion is more vigorous at the head of a valley than on its sides is apparent from the relationships between the valley and the regional slope. Above the head of a valley, water flows down the regional slope as sheets (sheet flow), but the water starts to converge to a point where a definite stream channel begins. As the water is concentrated into a channel, its velocity and erosive power increase far beyond those of the slower-moving sheet of water on the surrounding ungullied surface. The additional volume and velocity of the channel water erode the head of the valley much faster than sheet flow erodes ungullied slopes or the valley walls. In addition, groundwater moves toward the valley, so the head of the valley is a favorable location for the development of springs and seeps. These, in turn, help to undercut overlying resistant rock and cause headward erosion to occur much faster than retreat of the valley walls. The head of the valley is thus extended upslope.

Stream Piracy   With the universal tendency for headward erosion, the tributaries of one stream can extend upslope and intersect the middle course of another stream, thus diverting the headwater of one stream to the other. This process, known as **stream piracy**, is illustrated in **Figure 12.19**. Stream piracy is most likely to occur if headward erosion of one stream is favored by a steeper gradient or by a course in more easily eroded rocks. Some of the most spectacular examples occur in the folded Appalachian Mountains, where nonresistant shale and limestone are interbedded with resistant sandstone formations. The process of stream capture and the evolution of the region's drainage system are shown in the series of diagrams in **Figure 12.20**. The original streams flowed in a dendritic pattern (a branching, treelike pattern) on horizontal sedimentary layers that once covered the folds. As uplift occurred, erosion removed the horizontal sedimentary rocks, and the dendritic drainage pattern became superposed, or placed on, the folded rock beneath. The **superposed stream** thus cuts across weak and resistant rocks alike. As the major stream cuts a valley across the folded rocks, new tributaries rapidly erode headward along the nonresistant formations. By headward erosion, these new streams progressively capture the superposed tributaries and change the dendritic drainage pattern to a **trellis drainage pattern** (a pattern in which the tributaries join the main stream at right angles).

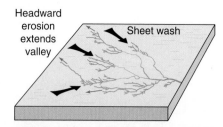

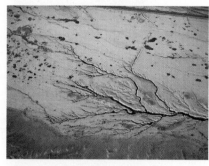

Figure 12.18 **Headward erosion** is constantly extending the drainage upslope so that the network of tributaries is enlarged and consumes the flat, undissected upland. Water flows as a sheet down the undissected regional slope. As it converges toward the head of a tributary valley, its velocity and volume are greatly increased, so its ability to erode also increases. The tributary valley is thus eroded headward, up the regional slope.

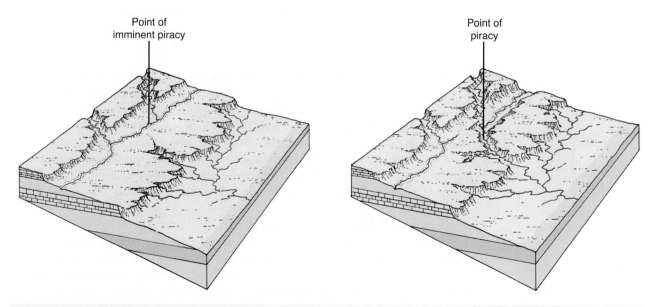

Figure 12.19 **Stream piracy** occurs where a tributary with a high gradient rapidly erodes headward and captures a tributary of another stream.

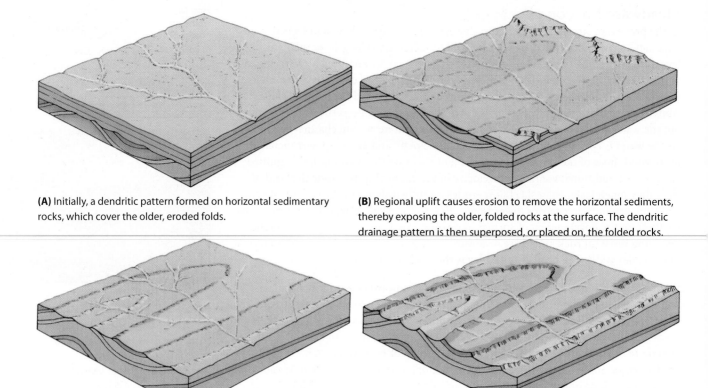

(A) Initially, a dendritic pattern formed on horizontal sedimentary rocks, which cover the older, eroded folds.

(B) Regional uplift causes erosion to remove the horizontal sediments, thereby exposing the older, folded rocks at the surface. The dendritic drainage pattern is then superposed, or placed on, the folded rocks.

(C) Main streams cut across both resistant and nonresistant rock by channel abrasion.

(D) Rapid headward erosion along exposures of weak rocks results in stream capture and modification of the dendritic pattern to a trellis pattern.

Figure 12.20   A dendritic drainage pattern superposed on a series of folded rocks evolves into a trellis pattern as headward erosion proceeds along nonresistant rock formations.

Extensive stream piracy and development of a trellis drainage pattern can be seen almost any place where folded rocks are exposed at the surface. In the folded Appalachian Mountains, the major streams that flow to the Atlantic (such as the Susquehanna and the Potomac) are all superposed across the folded strata. Their tributaries, however, flow along the nonresistant rocks parallel to the geologic structure and have captured many superposed tributary streams.

Another example of stream piracy is the Pecos River in New Mexico. By extending itself headward to the north along the weak shale and limestone, which crop out in a north–south zone parallel to the Rocky Mountain front, it has captured a series of eastward-flowing streams that once extended from the Rockies across the Great Plains. The original eastward drainage (shown in **Figure 12.21A**) resulted from the uplift of the Rocky Mountains. Now the headwaters of most of the original streams have been captured by the Pecos River. Water that once would have flowed across the Llano Estacado (the High Plains of Texas) now flows down the Pecos valley (Figure 12.21B).

Extension of Drainage Systems Downslope

In addition to downcutting and headward erosion, a drainage system can grow in length simply by extending its course downslope as sea level falls or as the landmass rises. This process is probably fundamental in determining the original course of many major streams, especially in the interior lowlands, where the oceans once

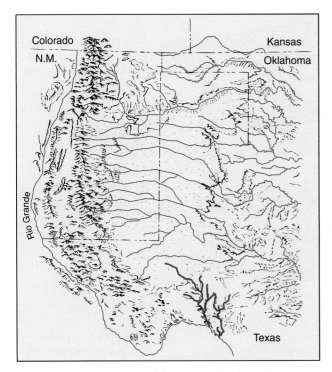

(A) Prior to the development of the Pecos valley, drainage is believed to have been eastward from the Rocky Mountains across the Great Plains.

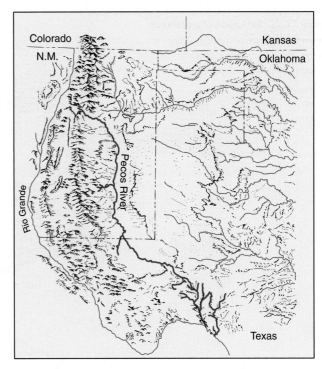

(B) Headward erosion of the Pecos River northward along the nonresistant rocks of the Pecos plains captured the headwaters of the eastward-flowing streams.

Figure 12.21 **The Pecos River** evolved as headward erosion extended the drainage network northward and captured the eastward-flowing streams.

covered much of the continents and then slowly withdrew. During a regression of the sea, drainage systems were extended down the newly exposed slopes. Later, they were modified by headward erosion and stream piracy.

Tidal channels (major channels formed by tidal currents extending from offshore well into the tidal flat) along coastal plains are examples of the beginning of a new segment of a drainage system as a result of a fall in sea level. The pattern of land and tidal drainage is characteristically dendritic because the material on which this drainage is established consists of recently deposited horizontal sediments. If the slope is pronounced, however, the tributaries, as well as the major streams, flow parallel for a long distance. If sea level were to drop, the streams would continue to flow downslope, following the courses established by tidal channels, as is shown in **Figure 12.22**. Major streams would extend their drainage patterns over the deltas that they deposited. Most of the streams in the Gulf and Atlantic coastal plains originated in this way. If sea level is falling, the youngest parts of a river are therefore near the shoreline and upslope, where headward erosion develops new channels.

The Grand Canyon: A Model of Stream Erosion

Because the evolution of a drainage system may require tens of millions of years, we can study the origin of stream valleys only indirectly. One approach is to study the interaction of downcutting and slope retreat by means of a computer model of the Colorado River's erosion of the rock sequence in the Grand Canyon area. Variables of a drainage system that affect various rock formations, such as rates of downcutting and slope retreat, were analyzed. This study produced hundreds of computer-calculated profiles of the Grand Canyon, showing changes that have occurred between the time the Colorado River began cutting through the Colorado Plateau and the present.

Figure 12.22 Extension of a drainage system downslope occurs as a shoreline recedes. This downslope extension commonly results in a dendritic pattern.

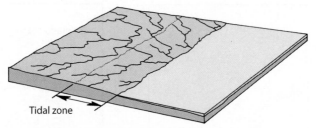

(A) In the original position of the shore, tidal channels develop between high tide and low tide.

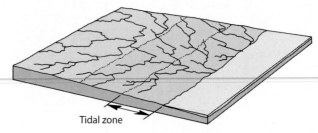

(B) As the sea level falls and the shoreline recedes, tidal channels become part of the permanent drainage system.

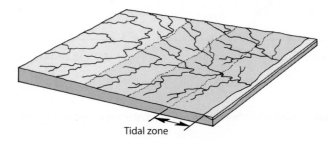

(C) With each successive retreat of the shoreline, new tidal channels develop and drainage is extended farther downslope. A dendritic drainage pattern typically is produced on the homogeneous tidal-flat material.

Although this model cannot be verified directly, you can get a glimpse of the stages of canyon development by studying the canyon longitudinally (**Figure 12.23**). Upstream, near Lees Ferry, the river is just beginning to cut through the Kaibab Limestone. Here, uplift has been minimal, and the entire sequence of strata exposed farther downstream in the Grand Canyon is below the surface. The river cuts only a narrow gorge in the Kaibab Limestone, which forms the upper rim of the Grand Canyon downstream. Farther downstream (near the bottom of the photograph), uplift permitted the river to cut much deeper into the rock sequence, and the sequence of profiles across the canyon is similar to the one developed by the computer model. Evidence of the evolution of slope morphology from the canyon itself thus supports the findings of the computer model.

Processes of Stream Deposition

In the lower parts of a river system (transporting and dispersing systems), the gradient of a river is very low. As a result, the stream's velocity is reduced, and deposition of much of the sediment load occurs, to create: (1) floodplains, (2) alluvial valleys, (3) deltas, and (4) alluvial fans.

The fact that rivers transport and deposit huge volumes of sediment is apparent in the practical problems of the silting of reservoirs and the maintenance of navigable channels and harbors. Most large rivers are always muddy; in some rivers, the weight sediment sometimes exceeds the weight of water. Sediment is deposited when the velocity of the current falls below the minimum velocity required to keep the particles of a certain size in motion (Figure 12.7). Thus, if a river carrying silt, sand, and gravel is slowed by a gentler gradient, or by entering a lake or the sea, the coarsest particles of

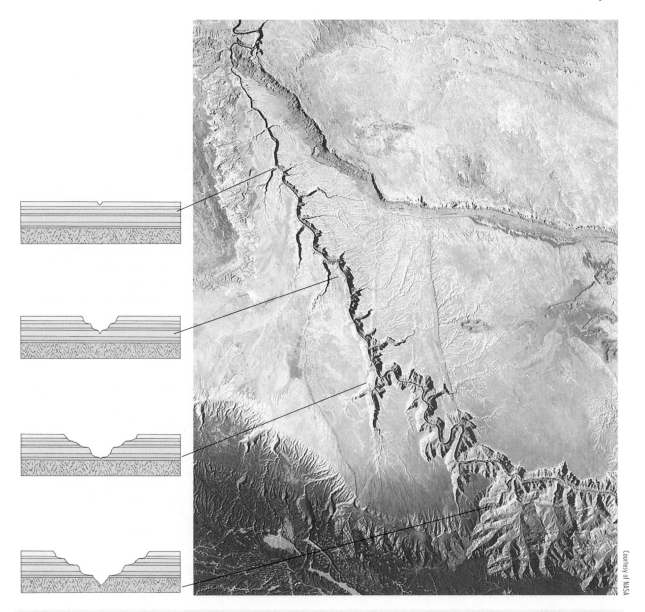

Figure 12.23 The effects of erosion of the eastern Grand Canyon are seen in this space photograph. The river flows from Lees Ferry, in the upper left, toward the lower right. At Lees Ferry, the river is just beginning to cut through the sedimentary rock sequence and has produced a profile like the one shown in the computer model. Downstream, uplift has permitted the river to cut deeper, and it has produced a sequence of profiles of alternating cliffs and slopes formed on resistant and nonresistant rock bodies.

the load are deposited first, and progressively finer particles are deposited as the velocity of the current continues to decrease. Deposition of the sediment load in the lower transporting and dispersal segments of a river creates prominent and distinctive landforms. Foremost among these are the great floodplains and alluvial valleys. Farther downstream, where the river enters the sea, most of its load is deposited as huge deltas.

Floodplains

On the gentle slopes of shields and stable platforms, most stream valleys are covered with large quantities of sediment that make up a flat surface over which the stream flows. This surface is called the **floodplain**, and during high floods it may be completely covered with water. Rivers that flow across floodplains are characterized by channels that either meander in sinuous loops or braid in interweaving multiple channels. These differences in channel configurations reflect variations in the type of sediment load and fluctuations in the volume of water. A schematic diagram showing the features commonly developed on a meandering river floodplain is shown in **Figure 12.24**. It serves as a simple graphic model of floodplain sedimentation.

What major geologic processes operate on a floodplain?

Meanders and Point Bars   All rivers naturally tend to flow in a sinuous pattern, even if the slope is relatively steep, because water flow is turbulent, and any bend or irregularity in the channel deflects the flow of water to the opposite bank. The force of the water striking the stream bank causes erosion and undercutting, which initiate a small bend in the river channel. In time, as the current continues to impinge on the outside of the channel, the bend grows larger and is accentuated, and a small curve ultimately grows into a large **meander** (**Figure 12.25**). On the inside of the meander, velocity is at a minimum, so some of the sediment load is deposited. This type of deposit occurs on the point of the meander bend and is known as a **point bar**. The two major processes around a meander bend—erosion on the outside and deposition on the inside—cause meander loops to migrate laterally.

Because the valley surface slopes downstream, erosion is more effective on the downstream side of the meander bend; thus, the meander also migrates slowly down the valley (Figure 12.25). As a meander bend becomes accentuated, it develops an almost complete circle. Eventually, the river channel cuts across the meander loop and follows a more direct course downslope. The meander cutoff forms a short but

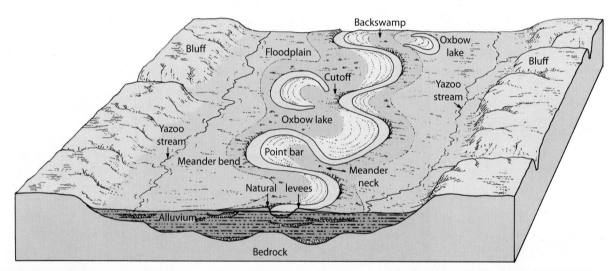

Figure 12.24 The major features of a floodplain include meanders, point bars, oxbow lakes, natural levees, backswamps, and stream channels. A stream flowing around a meander bend erodes the outside curve and deposits sediment on the inside curve to form a point bar. The meander bend migrates laterally and is ultimately cut off, to form an oxbow lake. Natural levees build up the banks of the stream, and backswamps develop on the lower surfaces of the floodplain. Yazoo streams have difficulty entering the main stream because of the high natural levees and thus flow parallel to it.

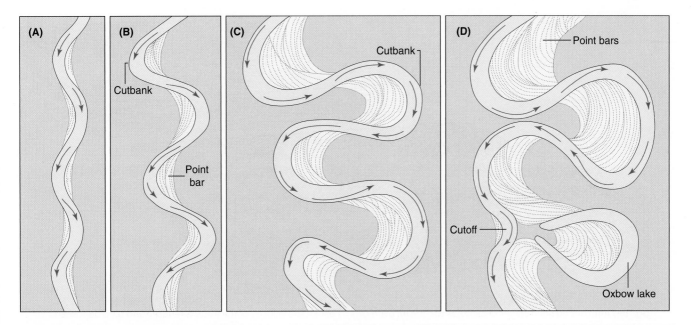

Figure 12.25 Stream meanders evolve because erosion occurs on the outside of a curved stream channel, where velocity is greatest, and deposition occurs on the inside of the curve, where velocity is low. An irregularity deflects stream flow (A) to the opposite bank and erosion begins. This starts the development of a meander loop (B–D). At the same time, sediment is deposited on the inside of the bend, forming point bars. The meander enlarges and migrates laterally (C–D). Continued growth of the meander bends ultimately cuts off the channel and forms an oxbow lake (D).

sharp increase in stream gradient, causing the river to completely abandon the old meander loop, which remains as a crescent-shaped lake known as an **oxbow lake** (Figure 12.25).

Natural Levees   Another key process operating on a floodplain is the development of high embankments, called **natural levees**, on both sides of the river. Natural levees form when a river overflows its banks during flood stage and the water is no longer confined to a channel but flows over the land surface in a broad sheet. This unchanneled flow significantly reduces the water's velocity, and some of the suspended sediment settles out. The coarsest material is deposited close to the channel, where it builds up a high embankment. Natural levees grow with each flood. Some grow high enough so that the river channel is higher than the surrounding area (**Figure 12.26**).

Backswamps   As a result of the growth and development of natural levees, much of the floodplain may be lower than the river flowing across it. This area, known as the **backswamp**, is poorly drained and commonly is the site of marshes and swamps. Tributary streams in the backswamp are unable to flow up the slope of the natural levees, so they are forced either to empty into the backswamp or to flow as **yazoo streams**, streams that run parallel to the main stream for many kilometers. Strangely enough, then the highest parts of the floodplain may be along the natural levees immediately adjacent to the river.

The lower Mississippi River is well known for its floodplain features (**Figure 12.27**). Between Cairo, Illinois, and the Gulf of Mexico, the Mississippi meanders over a broad floodplain, forming high natural levees, oxbow lakes, and backswamps. The dynamics of the river and the changes it can bring about by deposition are illustrated by the fact that, from 1765 to 1932, the river cut off 19 meanders between Cairo, Illinois, and Baton Rouge, Louisiana. Now the level of the Mississippi is controlled by dams and artificial levees, which have modified its hydrology, much as the Nile and Colorado rivers have been artificially manipulated.

How can a river build its own levees?

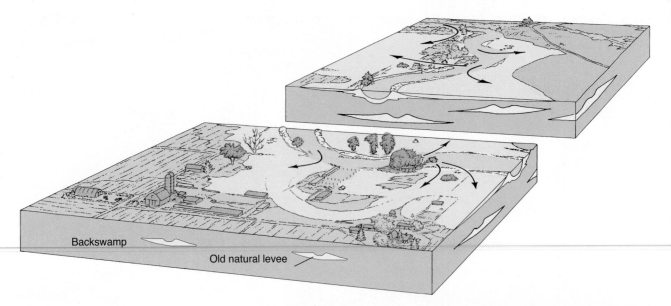

Backswamp

Old natural levee

(A) They form during flood stages because, as the stream overflows its banks, the velocity of the water is reduced and silt is deposited.

© Vladimir Melnikov/ShutterStock, Inc.

(B) As the levees grow higher, the stream channel also rises, and thus the river can be higher than the surrounding floodplain.

Figure 12.26 **Natural levees** are wedge-shaped deposits of fine sand, silt, and mud that taper away from the stream banks toward the backswamp.

Braided Streams If streams are supplied with more sediment than they can carry, they deposit the excess material on the channel floor as sand and gravel bars. These deposits may force a stream to split into two or more channels, so the stream pattern forms an interlacing network of braided channels and islands (**Figure 12.28**). The **braided stream** pattern is best developed in rivers that carry coarse sand and gravel and fluctuate greatly in the volume of water they discharge. These conditions commonly occur in arid or semiarid regions, where the amount of water in a stream varies greatly from season to season, or from storm to storm. Melting ice caps and glaciers also produce favorable conditions for braided streams because the streams in front of the melting ice cannot transport the exceptionally large load of sediment deposited by the glaciers. As a result, deposition occurs in mid-channel and new channels develop. For example, meltwater from the Nabesna glacier of southeast Alaska created the braided stream in Figure 12.28. Moreover, the cold climate near glaciers causes most rivers to freeze during the winter, so the volume of water discharged fluctuates from almost nothing in the winter to spring floods. Compare the channel pattern in this photograph with the meandering channels on the Mississippi River floodplain shown in Figure 12.27.

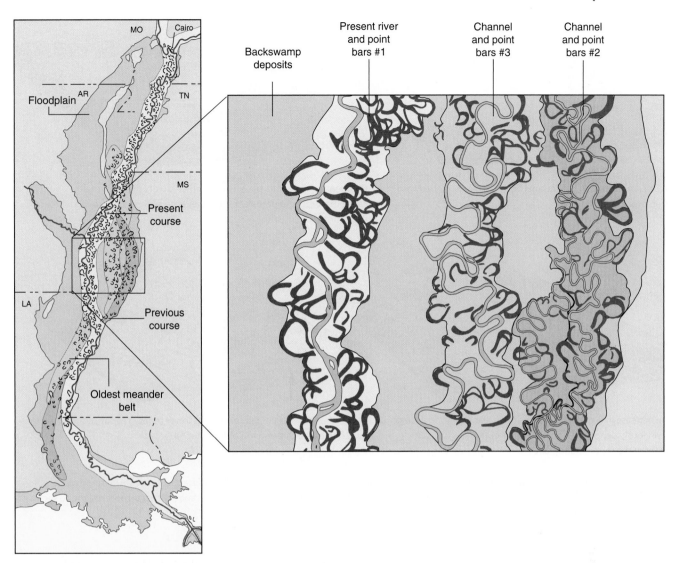

Figure 12.27 labels: MO, Cairo, AR, TN, MS, LA, Floodplain, Present course, Previous course, Oldest meander belt, Backswamp deposits, Present river and point bars #1, Channel and point bars #3, Channel and point bars #2

Figure 12.27 The floodplain of the Mississippi River extends from Cairo, Illinois, to the sea. It is more than 1000 km long and 200 km wide. The main meander belt has shifted several times during the last few thousand years. Progressively older meander belts are shown in blue, brown, and tan.

Alluvial Valleys

Many streams fill part of their valleys with sediment during one part of their history and then cut through the sediment fill during a subsequent period. This fluctuation in stream processes commonly produces **stream terraces**. Deposition can be initiated by any change that reduces a stream's capacity to transport sediment. These changes include (1) a reduction in discharge (as a result of climatic change or of a loss of water volume due to stream piracy), (2) a change in gradient (caused by a rise in base level or by regional tilting), and (3) an increase in sediment load.

The basic steps in the evolution of stream terraces are shown in **Figure 12.29**. In Figure 12.29A, a stream cuts a valley by downcutting and **slope retreat**. In Figure 12.29B, changes, such as regional tilting of the land or rising of base level, cause the stream to deposit part of its sediment load and build up a floodplain, which forms a broad, flat valley floor. In Figure 12.29C, subsequent changes (such as uplift or increased runoff) cause renewed downcutting into the easily eroded floodplain deposits, so a single set of terraces develops on both sides of the river. Further erosion can produce additional terraces Figure 12.29D by the lateral shifting of the meandering stream.

During the last ice age, the hydrology of most rivers changed significantly and produced stream terraces in many river systems. Stream runoff was increased greatly

Figure 12.28 A braided stream pattern commonly results if a river is supplied with more sediment than it can carry, as at the front of a glacier.

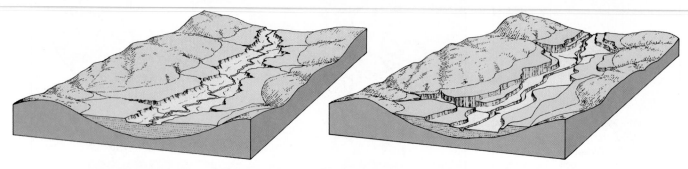

(A) A stream cuts a valley by normal downcutting and headward erosion processes.

(B) Changes in climate, base level, or other factors that reduce flow energy cause the stream to partially fill its valley with sediments, forming a broad, flat floor.

(C) An increase in flow energy causes the stream to erode through the previously deposited alluvium. A pair of terraces is left as a remnant of the former floodplain.

(D) The stream shifts laterally and forms lower terraces as subsequent changes cause it to erode through the older valley fill.

Figure 12.29 The evolution of stream terraces involves the deposition of sediment in a stream valley, subsequent change in the stream's gradient, and renewed downcutting. These changes can be initiated by various factors that affect a stream's capacity to transport sediment, such as changes in climate, changes in base level, or regional uplift.

by the melting ice, and large quantities of sediment deposited by the glaciers were reworked by the streams, many of which became overloaded. In addition, the climatic changes accompanying the ice age caused a general worldwide increase in precipitation. As a result, many streams filled part of their valleys with sediment that they are now cutting through to form stream terraces (**Figure 12.30**).

Deltas

As a river enters a lake or the ocean, its velocity suddenly diminishes, and most of its sediment load is deposited to form a **delta**. The growth of a delta can be complex, especially for large rivers depositing huge volumes of sediment. However, three major processes are fundamental to the formation and growth of a delta: formation of distributaries, splay development, and avulsion.

Distributaries   The diagrams in **Figure 12.31** illustrate the development of distributaries. As a river enters the ocean (or a lake) and the flowing water is no longer confined to a channel, the currents flare out, rapidly losing velocity and flow energy. The coarse material carried by the stream is deposited in two specific areas: (1) along the margins of quiet water on either side of the main channel (deposits build up subaqueous natural levees) and (2) in the channel at the river mouth, where there is a sudden loss of velocity (deposits build a bar at the mouth of the channel). These two deposits effectively create two smaller channels (distributary channels), which can build seaward for some distance. The process is then repeated, and each new distributary is divided into two smaller distributaries. In this manner, a system of branching distributaries builds seaward in a fan-shaped pattern.

Figure 12.30 Stream terraces along the Pahsimeroi River, Idaho, were formed by recent recurrent uplift. More than seven well-defined terraces can be identified in this area.

Splays   Figure 12.31 shows how the area between distributaries is filled with sediment. A local break in the levee, a **crevasse**, forms during periods of high runoff and diverts a significant volume of water and sediment from the main stream. The escaping water spreads out and deposits its sediment to form a **splay**, which is essentially a small delta, with small distributaries and systems of subsplays.

The sediment deposited by distributaries and splays is vulnerable to erosion and transportation by marine waves and tides. The growth of a delta is therefore influenced by the balance between the rate of input of sediment by the river and the rate of erosion by marine processes. If waves or tides are strong, the development of distributaries is limited, and the sediment is reworked into bars, beaches, and tidal flats.

Avulsion   A major phenomenon in the construction of deltas is the shifting of a river's entire course in an action called **avulsion**. Distributaries cannot extend indefinitely into the ocean because the river's gradient and capacity to flow gradually decrease. The river, therefore, is eventually diverted to a new course, which has a higher gradient. This diversion generally happens during a flood. The river breaks through its natural levee, far inland from the active distributaries of the delta, and develops a new course to the ocean. The new channel shifts the site of sedimentation to a different area, and the abandoned segment of the delta is attacked by wave and current action. The new, active delta builds seaward, developing distributaries and splays, until eventually it also is abandoned, and another site of active sedimentation is formed. The shifting back and forth of the main river channel is thus a major way in which sediment is dispersed and a delta grows (**Figure 12.32**).

Types of Deltas   Several types of deltas are illustrated in **Figure 12.33**. Each shows a different balance between the forces of stream deposition and the forces reworking the sediment (waves and tides). In the Mississippi Delta, processes of river deposition

(A) Natural levee breached. Part of stream flow diverted to backswamp.

(B) Reduced velocity causes deposition of sediment in a fan-shaped splay.

(C) Growth of splay by development of small distributaries and subsplays.

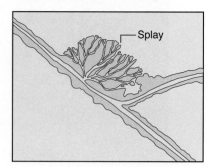

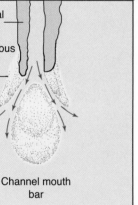

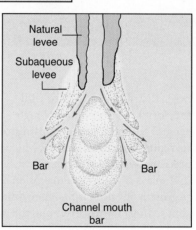

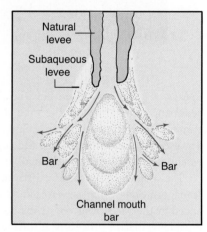

(D) A bar forms at the mouth of the river channel. Subaqueous levees form.

(E) Stream flow is channeled between bar and natural levees.

(F) Process is repeated, forming branching distributaries.

Figure 12.31 Distributaries and splays play key roles in the growth of a delta. These sequential diagrams show the evolution of splays (A–C) and distributaries (D–F).

dominate (Figure 12.33A). The delta is fed by the extensive Mississippi River system, which drains a large part of North America and discharges an annual sediment load of approximately 454 million metric tons. The river is confined to its channel throughout most of its course, except during high floods. Most of the sediment reaches the ocean through two or three main distributary channels and has rapidly extended the delta far into the Gulf of Mexico. This extension is known as a **bird-foot delta**.

Seven major subdeltas have been constructed by the Mississippi River during the last 5000 years as repeated avulsion occurred in the region between Baton Rouge and New Orleans. These are shown in Figure 12.32. The oldest lobe (1) was abandoned

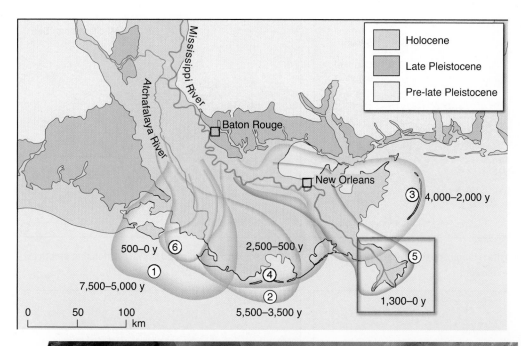

Courtesy of GSFC/METI/ERSDAC/JAROS/NASA, and U.S./Japan ASTER Science Team

Figure 12.32 The history of the Mississippi Delta involves repeated avulsion of the main channel, which has formed seven subdeltas. Most of the sediment is deposited in a small sector of the delta front. A major break in the natural levee upstream eventually diverts the entire flow to some other sector, and the process is repeated. Wave action then erodes the inactive bird-foot deltas. Subdeltas are indicated by numbers (1–6) according to age. One of the active distributary systems (5) has built a major bird-foot delta in the last 1300 years, the details of which are shown in this satellite photograph.

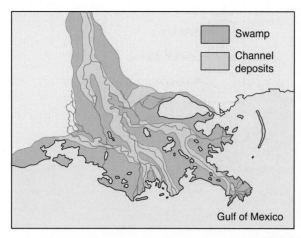

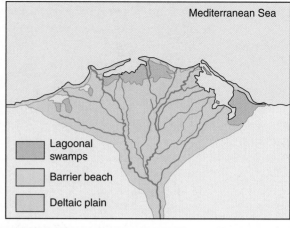

(A) The Mississippi Delta is dominated by fluvial processes that produce a bird-foot extension.

(B) The Nile Delta is dominated by wave action that produces an arcuate delta front.

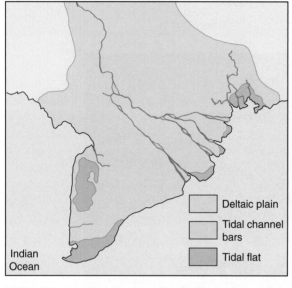

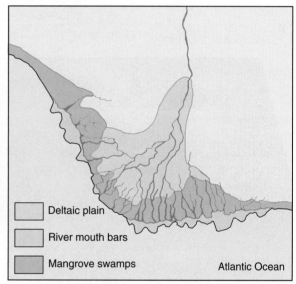

(C) The Mekong Delta is dominated by tidal forces that produce wide distributary channels.

(D) The Niger Delta has formed where stream deposition, wave action, and tidal forces are about equal. An arcuate delta front and wide distributary channels are thus produced.

Figure 12.33　The shape of a delta depends on the balance between fluvial and marine processes. The most important processes include the ability of waves and tides to rework the sediment in the delta. The Mississippi, Nile, Mekong, and Niger deltas are dominated by different processes.

approximately 4000 years ago and since then has been eroded back and inundated. Only small remnants remain exposed today. The successive lobes, or subdeltas (2 to 7), have been modified to various degrees. The abandoned channels of the Mississippi are well preserved and can be recognized on satellite photographs. The currently active delta lobe (7) has been constructed during the last 500 years. River studies show that the present bird-foot delta has been extended as far as the balance of natural forces permits. Without continued human intervention, the Mississippi will shift to the present course of the Atchafalaya River.

The Nile Delta (Figure 12.33B) differs from the Mississippi Delta in several ways. Instead of being confined to one channel, the Nile begins to split up into distributaries at Cairo, Egypt, more than 160 km inland, and fans out over the entire delta. Before construction of the Aswan High Dam, the Nile's annual flood briefly covered much of the delta each year and deposited a new layer of silty mud. Two of the large distributaries have built major lobes extending beyond the general front of the delta, but strong wave action in the Mediterranean redistributes the sediment at the delta front.

The reworked sediment forms a series of arcuate barrier bars, which close off segments of the ocean to form lagoons. The lagoons in turn form a subenvironment, which soon becomes filled with fine sediment. The difference between the Nile Delta and the Mississippi Delta is due largely to dissimilar balances between the influx of sediment, which builds bird-foot deltas, and the strength of wave action, which redistributes sediment to form barrier bars.

The Mekong Delta, along the southern coast of Vietnam, is dominated by tidal currents that redistribute sediment in the river channels and along the delta fronts (Figure 12.33C). The distributaries branch into two main courses near Phnom Penh, about 500 km inland. Sediment carried by the river is reworked by tidal current, forming broad distributary channels, a distinctive feature of tide dominated deltas.

The Niger Delta (Figure 12.33D) is a good example of a delta in which the important energy systems are nearly in equilibrium. Stream deposition, wave action, and tidal currents are more evenly balanced there than in the other delta types, so the Niger Delta is remarkably symmetrical.

Subsidence of Deltas A glance at the delta map (see Figure 12.38) should remind you that many of these large deltas are densely populated and intensely farmed. And yet many of them appear to be slowly sinking into the sea. Why are they sinking if sediment is continually delivered to the delta from the drainage basin? The answer is revealed by carefully considering what is happening in our modern world: (1) global sea levels are rising, (2) oil and gas are being removed from resource-rich deltas, (3) sediment is being trapped upstream behind dams before it can reach the delta, and (4) as distributaries became increasingly managed and channelized, natural avulsion has been eliminated. As a result, sediment is delivered to the very end of a channel and is not building up the surface of the delta (Figure 12.32).

The Mississippi delta provides us with a good example. Over the past few hundred years, 25% of its deltaic wetlands have been covered by the sea. Diversion of the river and its sediment load to subsiding parts of the delta could help build new land, but continued drowning may be inescapable because sea level is rising three times faster than sediment is being deposited.

Over the last few decades, this has become obvious as more and more of the delta surface becomes inundated with seawater as the delta continues to subside under its own weight.

Alluvial Fans

An **alluvial fan** is a stream deposit that accumulates in a dry basin at the base of a mountain front. Such areas are usually arid and have a large quantity of loose, weathered rock debris on the surface; when rain falls, the streams then transport huge volumes of sediment. Many or most of the streams are intermittent and run for only a short time each year or just during storms. Deposition of the load carried by the streams results from the sudden decrease in velocity as a stream emerges from the steep slopes of the upland and flows across the adjacent basin with its gentle gradient. The channel soon becomes clogged with sediment, and the stream is forced to seek a new course. In this manner, the stream shifts from side to side and builds up an arcuate, fan-shaped deposit (**Figure 12.34**). Debris flows are also common on alluvial fans. Their coarse unsorted deposits are found interlayered with the stream gravels and sands. As several fans build basinward at the mouths of adjacent canyons, they eventually merge to form broad slopes of alluvium at the base of the mountain range (**Figure 12.35**).

Although alluvial fans and deltas are somewhat similar, they differ in mode of origin and internal structure. In deltas, sediment is deposited in a body of water. The level of the ocean or lake effectively forms the upper limit to which the delta can be built. In contrast, a fan is deposited in a dry basin, and its upper surface is not limited by water level. Cobbles and gravel in alluvial fans are commonly less rounded than in other stream deposits. The coarse-grained, unweathered, poorly sorted sands and gravels of an alluvial fan also contrast with the fine sand, silt, and mud that predominate in a delta.

Figure 12.34 Alluvial fans form in arid regions where streams enter dry basins and deposit their sediment load as the stream gradient becomes smaller. This fan is in Death Valley, California.

Figure 12.35 Alluvial slopes develop as fans grow and merge. This photograph of the eastern slope of the Sierra Nevada of eastern California shows large alluvial slopes covering much of the dry basin.

Floods

Flooding is the overflow of water from the stream channel onto adjacent land that is usually dry. It is a natural process in all river systems and has occurred throughout all of geologic time.

Flooding of rivers is not a rare event, but it is often a seasonal occurrence corresponding to prolonged rain or rapid snowmelt. Human populations have always concentrated on deltas and floodplains of major rivers. It is therefore no accident that more than 500 flood stories, like that of Noah in the Bible, from more than 250 peoples or tribes are well documented. Indeed, floods are the most frequent and lethal of all natural disasters.

Why do floods occur?

Flooding on Deltas and Floodplains

Deltas and floodplains are the regions most susceptible to flooding because in these areas flooding is a fundamental recurring geologic process. Indeed, the deltas and floodplains originate and grow through the process of flooding, and in these areas flooding is as natural as windstorms in a desert. Unfortunately, more than half of the world's population lives along riverbanks, deltas, and seacoasts, where devastating floods are a natural, common process.

Most rivers experience seasonal flooding in which waters overflow their banks and spread out over the floodplain. Exceptionally high water can cause extensive flooding over thousands of square kilometers (**Figure 12.36**). This type of flooding results because rivers that flow over the lowlands and deltas tend to build up high natural levees; the river channel is actually higher than the surrounding area. Ultimately, a river may break through its levee and develop a new course to the sea. Such a breakthrough usually occurs in the delta region of a river where numerous distributaries form over a period of time.

The flood in the upper Mississippi River Basin in 1993 was the greatest flood disaster in U.S. history. Property damage exceeded $10 billion, and millions of acres of productive farmland were under water for weeks. To understand the details of this event, take a moment and study the satellite images of the region around St. Louis, Missouri (Figure 12.36). The geologic setting of the upper Mississippi River Basin is strikingly different from that of the lower basin. In the upper basin, the Mississippi River and its major tributaries (the Missouri and Illinois) flow through relatively deep, narrow valleys throughout much of their course, and their floodplains are like long narrow trenches bounded by steep bluffs. In the lower basin, the floodplain is much wider and is able to accommodate high water. The narrow floodplain in the upper basin was completely covered with floodwater for much of its length, whereas there was no serious flooding in the lower Mississippi River Basin.

Weather in the upper Mississippi River Basin during the latter part of 1992 and continuing into 1993 was highly unusual. Heavy rainfall began in September 1992 and continued for eight months. In some areas, more than three times the "normal" annual rainfall had occurred by June 1993. Soil moisture was therefore at saturation point for essentially the entire region, and reservoirs were at or near maximum capacity. Following this unusually wet spring, excessive precipitation persisted through June and July. Eighty percent of the upper basin received more than 200% normal rainfall for July, and 30% of the area received more than 400%. The flood occurred because the soil throughout the region was saturated, and there was literally no storage capacity in the ground for the incredible amounts of rainfall during the summer.

The floodplains of the lower Mississippi were inundated in 2011. Heavy precipitation during the preceding fall and winter persisted through the spring causing streams to be at high levels. Finally, heavy rains over the basins of the Ohio River and the central Mississippi drove the river over its banks from Memphis, Tennessee, south to the delta. In a type of controlled avulsion, the Army Corps of Engineers opened a

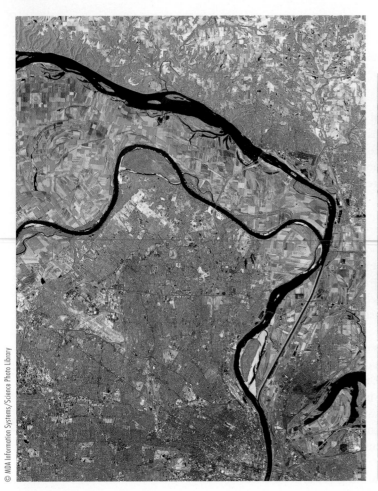

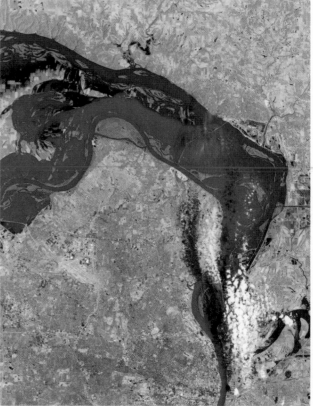

(A) The region around St. Louis, Missouri, on July 4, 1988, when the area was experiencing a drought.

(B) The same area on July 18, 1993, at the peak of the flood, when the Mississippi River was 5 m above flood level.

Figure 12.36 **The 1993 flood** in the Mississippi River Basin covered vast areas of the floodplain.

floodgate into the Atchafalaya Basin (Figure 12.32) to divert some of the flow. It was the main distributary for the delta until about 4,000 years ago when avulsion shifted the main channel eastward to the vicinity of New Orleans.

Recorded history bears grim witness to the destruction flooding can bring. Bangladesh, a country built almost entirely on the huge delta of the Ganges and Brahmaputra rivers, experienced its worst flood of many in 1991, in which more than 60% of the country (140,000 km^2) was under water. More than 138,000 people died and tens of millions were uprooted and displaced (**Figure 12.37**). A storm surge from a hurricane caused much of the damage. On the nearly flat, low lying delta no topographic barriers hindered the surge. Floods are also common along the floodplains of these vast river systems, with thousands killed and millions left homeless in the floods of 1987, 1988, 1998, 2004, and again in 2012.

A similar story is told for other great deltas of the world. The Huang He (Yellow River) in China, for example, periodically overflows its natural levees, causing destruction and misery. In 1887, floodwaters covered more than 130,000 km^2 of the delta's surface, with an estimated loss of life exceeding 1,000,000. Perhaps Earth's most deadly natural disaster, a huge 1931 flood of this river left as many as 4 million people dead. In 1991, 200,000 km^2 were flooded; 2000 people died and 1,000,000 were made homeless. China's other major river, the Yangtze, had four major floods in the 1900s, with the loss of life ranging from 3600 to 145,000 people in each flood. The Chinese people

Figure 12.37 Floods of the Ganges River of Bangladesh are caused by the monsoons, which occur between June and September each year. Floods inundate the low-lying delta of the river.

know all too well the process of flooding along their two great rivers, yet these same floodplains have nurtured Chinese civilization for centuries.

Flash Floods

Flash floods are local, sudden, short-lived floods in which great volumes of water rush downstream at high velocities. They frequently occur in the upper reaches of a river, especially in mountain valleys. Flash floods are a major process in developing alluvial fans. Ordinarily, they are caused by brief but heavy rainfall (a cloudburst) that transforms even a dry streambed into a rushing torrent of water and mud. Flash floods are especially likely to occur in regions that have narrow, deeply incised valleys where the river channel is so restricted that an exceptionally high wave of floodwater develops and rushes downstream with tremendous force. Disaster can strike with lightning speed. In the narrow canyons of Zion National Park in Utah, it is not uncommon for a summer storm to cause the river level to rise 30 m.

The flash flood on the Big Thompson River that drains part of the Colorado Rockies near Denver is a classic example. Spawned by 25 to 30 cm of rainfall from a violent cloudburst on the night of July 31–August 1, 1976, the downpour transformed a small mountain stream into a raging torrent of muddy water. The wall of water swept down the canyon, demolishing nearly everything in its path, including canyon highways, bridges, homes, and commercial buildings. At least 150 people perished and property damage exceeded $50 million.

It is clear that the greatest cause of flood damage is the choice (or necessity) of humans to build near rivers. Flooding is a natural geologic process that has become a hazard to humans only since they have built and developed communities in mountain valleys, floodplains, and deltas.

Rivers, Climates, and Plate Tectonics

> The evolution of the major rivers of the world is influenced directly and indirectly by plate tectonics and by climate zones.

In previous sections of this chapter, we considered river systems on a local basis: how they erode, transport, and deposit material. The major rivers of the world have other features, of a much larger scale, that are related to the global patterns of the hydrologic and tectonic.

Climate and River Systems

How is the distribution of major rivers related to climate?

A glance at a drainage map of the world may give a first impression that the drainage of the continents is haphazard and unsystematic. Rivers appear to flow in any direction, in an almost unlimited variety of patterns. Upon further study, however, some system becomes apparent in the locations of the major rivers, their tributaries, and the patterns they form.

Figure 12.38 shows the locations of the world's major rivers and the relative sizes of their deltas. The major rivers are found on broad, gently sloping platforms that lie near the equator—where there is maximum rainfall. Likewise, the sizes of the deltas formed at the mouths of the major rivers of the world are controlled by the size of the drainage basins, elevation of the land, and climate (which controls the amount of surface runoff). Maximum sediment load occurs in large rivers that drain mountainous topography in a humid climate. The world's largest deltas are built by the Amazon, Tigris-Euphrates, Ganges, Mekong, and Hwang Ho. In addition, there is a common association of many rivers with a submarine canyon and a huge submarine fan built out onto the abyssal plains in the deep-ocean basin. A submarine fan, like a delta built at a river mouth, is an indication of the vast amount of erosion accomplished by the work of a river system.

It is also apparent from Figure 12.38 that large areas of the continents do not have major river systems. Arid and semiarid low-latitude deserts, such as the Sahara in

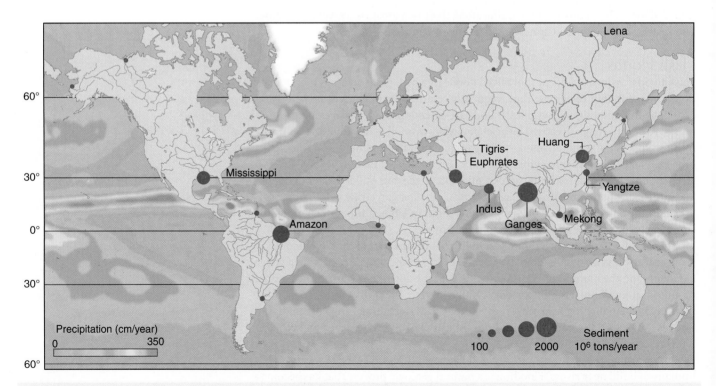

Figure 12.38 The world's largest rivers transport vast volumes of sediment, most of which is deposited as huge deltas. The size of the deltas is partially controlled by climate. Most of the large rivers are in the tropics or originate in the tropics where precipitation is highest (red highest to purple lowest precipitation). The size of the circles shows the annual discharge of sediment for the largest drainage systems of the world.

North Africa, the Kalahari in South Africa, and the great desert in Australia, are the most obvious examples (see elsewhere in the text). Sand seas cover much of them.

Low surface runoff is not confined to the low-latitude deserts, however. Large river systems are not found in the polar regions of North America and Europe. There are several reasons for this. First, these areas were covered with glaciers during the ice age, which ended only a few thousand years ago. Previous drainage systems were obliterated as the glaciers expanded over the region, and there has not been sufficient time for new integrated drainage systems to develop. Second, large tracts of the polar regions are, in fact, arid and have low precipitation (see Figure 12.1).

Humid areas underlain by porous limestone have poor drainage because solution activity develops a network of subterranean caverns and enlarged fractures, which divert the drainage to the subsurface. Many of these areas have no integrated drainage systems, despite their humid climate. Parts of Kentucky, Florida, and Mexico's Yucatan Peninsula are in this category. Large rivers do not develop on tropical islands because the catchment areas are too small even though rainfall is very high.

River Systems and Plate Tectonics

Plate tectonics is a fundamental factor in the origin and evolution of Earth's major river systems. As shown in **Figure 12.39**, tectonics can influence rivers in a variety of ways. The most obvious is that tectonism creates the principal relief of continents such as mountain belts along plate margins and continental tilt. The convergence of the Pacific plates with North and South America produced a long linear mountain belt with an eastward continental tilt. The result is that most of the drainage of the Americas is a simple pattern away from the convergent margin toward the passive margins of the Atlantic Ocean (purple arrows). The Amazon River of South America and the pre-glacial drainage of North America are classic examples of this type of tectonic control. Perhaps this was the most common drainage pattern in the geologic past.

Even more impressive are the mountains and highlands extending from France to the South China Sea, a distance of more than 13,000 km. This highland resulted from India and Africa impinging on Eurasia. This created a subradial drainage of the great rivers of Asia (purple arrows on Figure 12.39). In fact, seven of the ten largest rivers originate in the Himalayan orogenic belt.

Associated with continent-to-continent collision are depressions or basins, downwarps of the crust, parallel to the mountain belt. Drainage commonly develops in the downwarp parallel to the axis of the mountain belt (red arrows on Figure 12.39). The Ganges, Indus, Tigris-Euphrates, and Danube are good examples (see Figure 12.38 for their locations).

Continental rifting is one of the most direct and obvious ways in which drainage can be modified by plate tectonics. The most recent rivers generated by rifting are those associated with the African Rift Valleys and the Red Sea (black arrows). Older rift-generated drainages flow away from the escarpments formed by the rifts that broke up an older supercontinent, Pangaea. The rift-generated rivers commonly flow across basalt extruded along the rift system. The great escarpments along the west coast of India, southern and western Africa, and the eastern coast of South America, as well as eastern Australia were formed during the breakup of Pangaea and rivers flowing down the shoulders of the rifts are among the oldest rivers in the world.

Modification of Basic River Pattern

Many exceptions and modifications to this basic pattern are influenced by tectonics. Continental rifting effectively beheads or dismembers a previously established river system. Also, if rifting or subsidence occurs in the shield or platform, it will tend to focus and orient the trunk system of the drainage. Examples are the lower Niger, Amazon, Parana, and lower Mississippi rivers.

Indirectly, the tectonics of a continent influence drainage patterns because folded rocks produced by crustal deformation create zones of alternating hard and soft rock, parallel to the trend of the mountain belt. Headward erosion follows the zones of weakness and modifies the pattern so that large segments of a river flow parallel to the

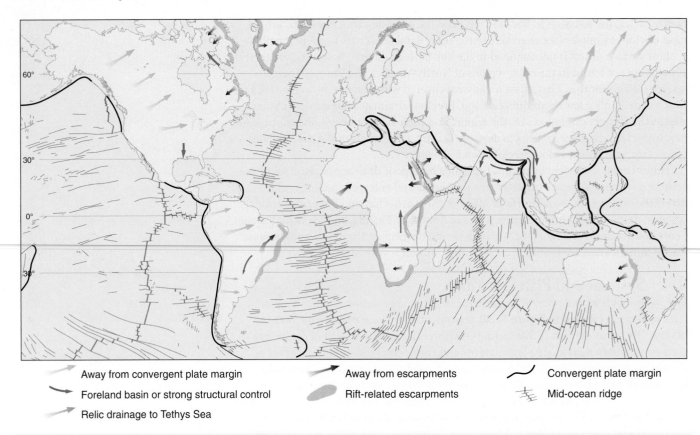

Away from convergent plate margin	Away from escarpments	Convergent plate margin
Foreland basin or strong structural control	Rift-related escarpments	Mid-ocean ridge
Relic drainage to Tethys Sea		

Figure 12.39 The relationship between river systems and plate tectonics is shown on this map of plate boundaries and the flow direction of the major rivers. Convergent plate boundaries (purple arrows) have had the greatest influence on the drainage of Eurasia and the Americas where drainage is away from the mountain belt on the convergent margin. Downwarped basins and tightly folded mountain belts create drainage parallel to the orogenic belt (red arrows). Divergent plate margins commonly create drainage away from the rift system (black arrows).

structural trends of the folded mountain belt. The Mekong (Vietnam) and Irrawaddy (Burma) rivers in Southeast Asia are examples.

Volcanic activity is another method by which the tectonic system modifies a river's drainage pattern. Extrusion of flood basalts can obliterate the preexisting drainage system. A new pattern is then established on the volcanic surface or along the margins of the flows.

The drifting of a continent into a new climatic zone is yet another way in which tectonic activity modifies a river system. As a continent drifts into the low latitudes, precipitation is greatly reduced, and wind-blown sand can completely cover large parts of the previously established drainage. Proof of this type of modification was recently discovered in radar images of parts of the Sahara, made during a flight of the space shuttle *Columbia* (**Figure 12.40**). These images show a large and extensive ancient drainage system now buried beneath the sand.

The drifting of continents into cold climatic zones may cause similar destruction of a substantial part of the drainage system if glaciers develop and cover the continent. Continental glaciation obliterates the drainage system beneath it and forces the major rivers to establish a new course along the margins of the ice. After the continental ice sheet retreats, a new and complex drainage pattern is integrated through a system of overflowing ponds and lakes.

Age of Rivers

Do rivers have histories? How is the history of a river related to plate tectonics?

The age and history of a river system are fundamental questions, but the answers may be difficult to determine. We may consider the time of the origin of a river to be the earliest date at which a continuous system drained the region in question. We may consider a river to date from the last marine regression, the last significant tectonic

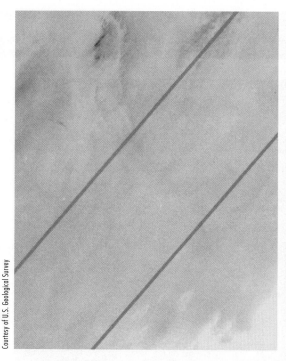 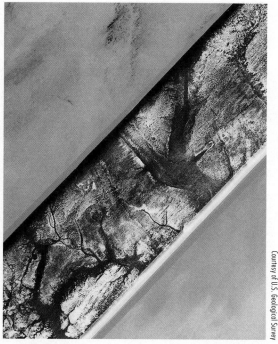

Courtesy of U.S. Geological Survey

Courtesy of U.S. Geological Survey

Figure 12.40 Ancient river systems in the Sahara Desert are now buried under sand but are revealed by radar imagery made during a flight of the space shuttle. On a satellite image, the present sand desert surface is yellowish orange (top) and gives no hint of a river drainage. The black-and-white radar image cutting diagonally across the picture on the right covers an area about 50 km wide and 300 km long. The largest valley on the image is as wide as the present Nile River valley and represents millions of years of erosion when the Sahara had a much wetter climate.

uplift, the termination of lava extrusion, or the waning of an ice sheet. All produce new surfaces upon which a drainage system may evolve. A river could be terminated by a new marine invasion, new tectonism, glaciation, volcanic extrusion, or expansion of a sand sea, but it is not always that simple. In fact, some of today's major river systems may date back to the early Cenozoic Era, 40 to 50 million years ago.

However, various parts of a river system originate and evolve at different times and in different ways, so we cannot establish a precise time when an entire river system originated. Very few rivers (and certainly no major ones) begin or end without some relationship to the drainage system that preceded them. Instead, a drainage system continually evolves by headward erosion and stream capture, adjustment to the structure of the underlying rocks, and modifications related to marine transgressions, continental glaciation, desert sand, and continental rifting. As the system continually evolves, each period of its history inherits something from the preceding conditions. The reason rivers continue to evolve is that the hydrologic system is continuous. Uplift of a mountain belt cannot divert or change the course of a river because a river has the capacity to downcut its channel much faster than uplift occurs. A river's history is a history of the landscape over which it flows.

In spite of the complexity and longevity of river systems, a few age estimates have been made. The Colorado River, which carved the Grand Canyon, probably started its current path about 5 million years ago. And the Amazon River of South America probably started as a transcontinental river about 11.5 million years ago and reached its modern shape and size only about 2.5 million years ago.

In all probability, the great Amazon is not the largest river the world has seen. Larger rivers probably drained Pangaea, the ancient landmass that existed about 200 million years ago, before the present continents were outlined and drifted apart. The ancestral Congo, for example, could have flowed across South America before rifting.

GeoLogic   Russia's Lena River Delta

The delta of the Lena River in northern Russia clearly shows many of the processes of delta building in this false color image. Sediment picked up by a huge tributary system flows as much as 4000 km before it is deposited here as a large fan-shaped delta.

Observations

1. The river freezes over during the winter, but during spring high runoff transports large volumes of sediment.
2. The seasonal fluctuation in stream runoff plus the load of coarse gravel develop braided stream channels.
3. Channel bars are abundant and numerous distributary bars develop where the river channel approaches the sea.
4. When the major river channel shifts part of the delta becomes inactive and no additional sediment is deposited. Numerous small lakes associated with permafrost activity form on the abandoned part of the delta (on the west).
5. The Arctic Ocean is frozen during the long winter so wave action along the delta front is limited.

Interpretations

The Lena Delta is dominated by fluvial processes. As the river encounters the ocean, its velocity slows and the clastic sediment it carries drops out to make channel mouth bars that split the channel and create a branching system of distributaries. Avulsion moves the course of the major stream and, as a result, the focal point of deposition moves back and forth across the delta. Today, the eastern part of the Lena delta is most active and the northwestern part has been abandoned. The inactive part of the delta has been reshaped over into a series of irregular ponds by the repeated freezing and thawing, swelling and collapsing, of the water-saturated sediment. Unseen in this vertical view, is the thick (1 to 5 km) wedge of clastic sediment that has accumulated on the continental margin. Deltas are major sedimentary environments that contribute to the continuing growth of the continents.

Key Terms

abrasion (p. 329)
alluvial fan (p. 347)
avulsion (p. 343)
backswamp (p. 339)
base level (p. 324)
bed load (p. 323)
bird-foot delta (p. 344)
braided stream (p. 340)
capacity (p. 324)
collecting system (p. 320)
competence (p. 324)
crevasse (p. 343)

delta (p. 342)
dendritic drainage pattern
  (p. 320)
discharge (p. 322)
dispersing system (p. 321)
dissolved load (p. 324)
distributary (p. 321)
divide (p. 319)
downcutting (p. 329)
drainage basin (p. 319)
floodplain (p. 337)
graded stream (p. 325)

gradient (p. 322)
headward erosion (p. 333)
longitudinal profile (p. 322)
meander (p. 338)
natural levee (p. 339)
oxbow lake (p. 339)
point bar (p. 338)
pothole (p. 331)
river system (p. 319)
saltating (p. 323)
slope retreat (p. 341)
splay (p. 343)

stream piracy (p. 333)
stream terrace (p. 341)
superposed stream (p. 333)
suspended load (p. 323)
threshold velocity (p. 324)
transporting system (p. 320)
trellis drainage pattern
  (p. 333)
tributary (p. 320)
turbulent flow (p. 323)
yazoo stream (p. 339)

Review Questions

1. Explain the reasons for concluding that stream action (running water) is the most important process of erosion on Earth.
2. Describe and illustrate the three major subsystems of a river.
3. Draw a diagram showing the general nature of transportation of (a) bed load, (b) suspended load, and (c) dissolved load.
4. Explain the role of flow velocity in the transportation and deposition of stream sediment.
5. Explain the concept of equilibrium in river systems, and cite several examples of how streams adjust to attain equilibrium.
6. How does urbanization affect surface runoff?
7. Explain how a stream cuts a valley through solid bedrock.
8. What is headward erosion? Why does it occur?
9. Explain the process of stream piracy, and cite examples of how it modifies a drainage system.
10. How does a stream system grow longer?
11. Name and describe the important landforms associated with floodplain deposits.

12. Describe the steps involved in the growth of a stream meander and the formation of an oxbow lake.
13. How does a point bar develop?
14. Explain the origin of natural levees.
15. What conditions are conducive to the development of braided streams?
16. Describe and illustrate the steps in the development of stream terraces.
17. Explain how a delta is built where a stream enters a lake or the sea.
18. Outline the history of the Mississippi Delta. Why is it subsiding today?
19. Make a series of sketches to show the form of a delta in which (a) fluvial processes dominate, (b) wave processes dominate, and (c) tidal processes dominate.
20. Explain how an alluvial fan is built.
21. What role does climate play in shaping river systems?
22. Contrast the nature of a river system that flows from a mountain belt toward a convergent plate margin and one that flows toward a passive continental margin.

13 Groundwater Systems

The movement of water in the pore spaces of rocks beneath Earth's surface is a geologic process that is not easily observed and therefore not readily appreciated; however, groundwater is an integral part of the hydrologic system and a vital natural resource. Groundwater is not rare or unusual. It is distributed everywhere beneath the surface of Earth. It occurs not only in humid areas, but also beneath desert regions, under the frozen polar regions, and in high mountain ranges. In many areas, the amount of water seeping into the ground equals or exceeds the surface runoff.

In many ways, groundwater systems are like river systems. For each there is a collecting system, transporting system, and a dispersal or discharge system. For groundwater, the collection area is the zone of recharge where surface water enters the subsurface. The path followed by the groundwater is controlled by permeability of the rocks in an aquifer. Along the flow path, groundwater, like river water, picks up materials and transports them. In groundwater systems, most of the transported materials are carried away as ions dissolved in water. Finally, like a river, a groundwater system has a discharge zone where the water comes back to the surface or enters a lake, river, or ocean. In areas dominated by groundwater dissolution, unique landscapes evolve. There are no integrated river systems; most of the water that falls on the surface as rain enters the groundwater system. One such area, shown above, is the exotic landscape of southern China. Here, groundwater has dissolved large volumes of limestone, leaving a maze of residual towers

and conical hills. Every hill contains a labyrinth of interconnected caves. There are no stream valleys on the hillsides: only grooves, channels, and ridges produced by solution activity. The water commonly collects in small lakes and ponds reflecting the high water table. Much of the missing material was carried away as ions dissolved in water that eventually reaches the ocean.

In this chapter we study groundwater systems: how water moves through various types of pore spaces in the rock and how it forms karst topography, caves, and cave deposits. You will see that groundwater is related to surface drainage and how it erodes and deposits material to change loose sand into sandstone and fallen tree trunks into petrified wood. We then consider how we use groundwater, this most precious natural resource, and how we attempt to cope with the environmental problems that result when we modify and manipulate this part of the hydrologic system.

Major Concepts

1. Groundwater is an integral part of the hydrologic system, and it is intimately related to surface water drainage.
2. The movement of groundwater is controlled largely by the porosity and permeability of the rocks through which it flows.
3. The water table is the upper surface of the zone of saturation.
4. Groundwater moves slowly through the pore spaces in rock.
5. The natural discharge of groundwater is generally into springs, streams, marshes, and lakes.
6. Aquifers are saturated permeable rocks; they may be confined between impermeable layers or unconfined and open to the surface.
7. Erosion by groundwater produces karst topography, which is characterized by caves, sinkholes, solution valleys, and disappearing streams. Precipitation of minerals from groundwater creates deposits in caves and along fractures and cements many kinds of clastic sedimentary rocks.
8. Alteration of the groundwater can produce many unforeseen problems, such as pollution, subsidence, collapse, and disruption of ecosystems.

Groundwater Systems

Two physical properties of a rock largely control the amount and movement of groundwater. One is porosity, the percentage of the total volume of the rock consisting of voids. The other is permeability, the capacity of a rock to transmit fluids.

Groundwater is not stagnant and motionless. Rather, it is a dynamic part of the hydrologic system, in constant motion, and is intimately related to surface drainage. Gravity is the principal driving force for the flow of groundwater. Moreover, like other parts of the hydrologic system (rivers and glaciers), the groundwater system is an open system (**Figure 13.1**). Water enters the system when surface water infiltrates the ground (**recharge**); water moves through the system by percolating through the pore spaces of rock and ultimately leaves the system by seeping into streams, springs or lakes (**discharge**). Along this flow path, groundwater does geologic work—mostly as a result of solution or precipitation of rock. The characteristics of the material through which the water moves are fundamental controls on the groundwater system.

Porosity

How can a rock be highly porous and still have low permeability?

Water can infiltrate the subsurface because solid bedrock—as well as loose soil, sand, and gravel—contains **pore spaces**. There are four main types of pore spaces, or voids, in rocks (**Figure 13.2**): (1) spaces between mineral grains, (2) fractures, (3) solution cavities, and (4) vesicles. In sand and gravel deposits, pore space can constitute from 12% to 45% of the total volume. If several grain sizes are abundant and the smaller grains fill the space between larger grains, or if a significant amount of cementing material fills the spaces between grains, the **porosity** is greatly reduced. All rocks are cut by fractures, and in some dense rocks (such as granite), fractures are the only significant pore spaces (Figure 13.2). Solution activity, especially in limestone, commonly removes soluble material, forming pits and holes. Some limestones thus have high porosity. As water moves along joints and bedding planes in limestone, solution activity enlarges fractures in the rock and develops passageways that may grow to become caves. In basalts and other volcanic rocks, vesicles formed by trapped gas bubbles significantly affect porosity. Vesicles commonly are concentrated near the top of a lava flow and form zones of very high porosity; these zones can be interconnected by columnar joints or through the voids in cinders and rubble at the top and base of the flow.

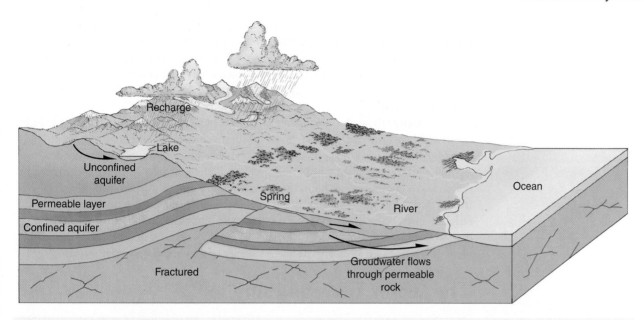

Figure 13.1 **The groundwater system** is an open system of water flowing below the surface but still under the influence of gravity. Water enters this system in recharge zones, generally flows slowly through connected pores in soil and rock and eventually is discharged through springs, rivers, or directly into the ocean. As water moves through the system, it dissolves soluble rocks or locally, carries the ions in solution, and locally deposits minerals along fractures and in caves.

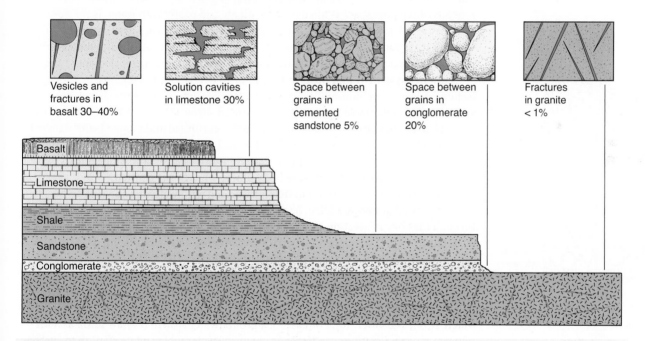

Figure 13.2 **Various types of pore spaces** in rocks permit the flow of groundwater.

Permeability

Permeability, the capacity of a rock to transmit a fluid, is measured in units called darcies, after the French engineer who did early experiments to understand groundwater flow. Permeability varies with the fluid's viscosity, the **hydrostatic pressure**, the size of openings, and particularly the degree to which the openings are interconnected. If the pore spaces are very small, a rock can have high porosity but low permeability because it is difficult for water to move through small openings.

Rocks that commonly have high permeability are conglomerates, sandstones, basalt, and certain limestones. Permeability in sandstones and conglomerates is high because of the relatively large, interconnected pore spaces between the grains. Basalt is permeable because it is often extensively fractured by columnar jointing and because the tops of most flows are vesicular. Fractured limestones are also permeable, as are limestones in which solution activity has created many small cavities. Rocks that have low permeability are shale, unfractured granite, quartzite, and other dense, crystalline metamorphic rocks.

Water moves through the available pore spaces following a tortuous path as the flow twists and turns through the tiny voids. Whatever the permeability of the rock, groundwater flows slowly and the flow is laminar. Thus, the flow of groundwater contrasts sharply with the turbulent flow of rivers. Whereas the flow velocity of water in rivers is measured in kilometers per hour, the flow velocity of groundwater commonly ranges from 1 m/day to 1 m/yr. The rate of percolation in exceptionally permeable material is only 250 m/day. Some water takes more than a million years to move from recharge to discharge zone. Only in special cases, such as the flow of water in caves, does the movement of groundwater even approach the velocity of slow-moving surface streams.

The Water Table and Aquifers

The water table is the upper surface of the zone of saturation. Aquifers are saturated permeable rocks; they may be open or confined.

What is the general configuration of the water table?

As water seeps into the ground, gravity pulls it downward through two zones of soil and rock. In the upper zone, the pore spaces in the rocks are only partly filled with water, and the water forms thin films, clinging to grains by surface tension. This zone, in which pore space is filled partly with air and partly with water, is the **zone of aeration**. Below a certain level, all of the openings in the rock are completely filled with water (**Figure 13.3**). This area is called the **zone of saturation**. The **water table**, which is the upper surface of the zone of saturation, is an important element in the groundwater system. It may be only a meter or so deep in humid regions, but it might be hundreds or even thousands of meters below the surface in deserts. In swamps and lakes, the water table is essentially at the land surface (**Figure 13.4**). Although the water table cannot be observed directly, it has been studied and mapped with data collected from wells, springs, and surface drainage. In addition, the movement of groundwater has been studied by means of radioactive isotopes, dyes, and other tracers, so extensive knowledge of this invisible body of water has been acquired.

A permeable zone or formation that is saturated with water is known as an **aquifer** (**Figure 13.5**). Aquifers are filled, or recharged, as surface water seeps downward through the zone of aeration. An **unconfined aquifer** is connected to the surface by open pore spaces through which it can be recharged, as shown in Figure 13.4. In an unconfined aquifer, the amount of water is indicated by the height of the water table. Saturated zones in surficial deposits of sand and gravel are commonly unconfined aquifers. They can easily be contaminated by fluids at the surface. At considerable depths, all pore spaces in the rocks are closed by high pressure, and there is no free water. This is the lower limit, or base, of a groundwater system (**Figure 13.6**). Several important generalizations can be made about the water table and its relation to surface topography and surface drainage (Figure 13.4). In general, the water table tends to mimic the surface topography. In flat country, the water table is flat. In areas of rolling hills, it rises and falls with the surface of the land. The reason is that groundwater moves very slowly, so the water table rises in the areas beneath the hills during periods of greater precipitation but takes a long time to flatten out during droughts.

In humid areas, the water table is at the surface in lakes, swamps, and most streams, and water moves in the subsurface toward these areas, following the general paths shown in Figure 13.4. In arid regions, however, most streams lie above the water table, so they lose much of their water through seepage. Where impermeable layers (such as shale) occur within the zone of aeration, the groundwater is trapped above the general water table, forming a local **perched water table**. If a perched water table extends to the side of a valley, springs and seeps occur.

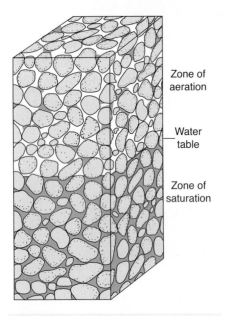

Zone of aeration

Water table

Zone of saturation

Figure 13.3 The water table is the upper surface of the zone of saturation. Water seeps into the ground through pore spaces in rock and soil. It passes first through the zone of aeration, in which the pore spaces are occupied by both air and water, and then into the zone of saturation, in which all of the pore spaces are filled with water. The depth of the water table varies with climate and amount of precipitation.

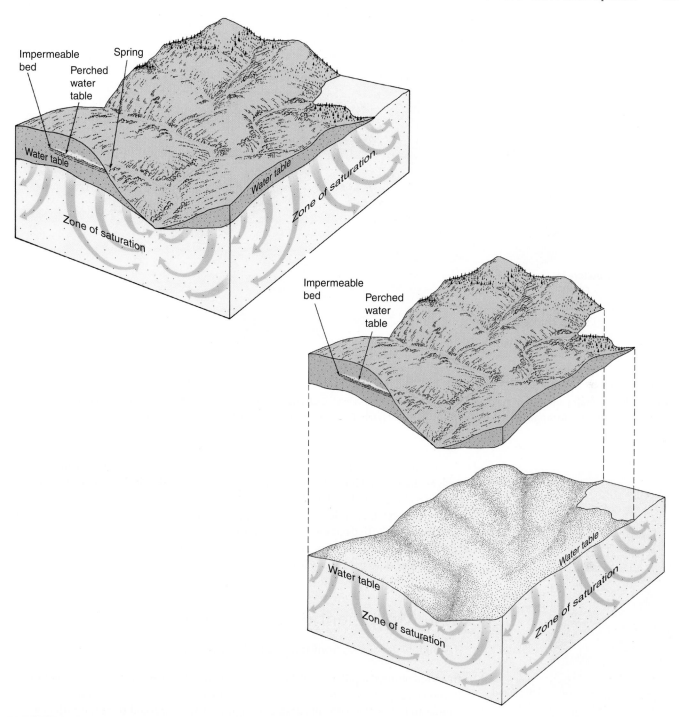

Figure 13.4 **The movement of groundwater** in an unconfined aquifer is directed toward areas of least pressure. In the idealized system depicted here, groundwater moves downward to the water table (by the pull of gravity) and then moves toward areas of lower pressure. The configuration of the water table has a strong influence on the direction of movement. In most areas, the water table is a subdued replica of the topography. The enlarged view shows high and low areas in the water table much like the hills and valleys on the surface. Differences in the height of the water table cause differences in the pressure on water in the zone of saturation. Water thus moves downward, beneath the high areas of the water table (because of the higher pressure), and upward, beneath the low areas. Water can also flow across the surface of the water table. It commonly seeps into streams, lakes, and swamps, where the water table is near the surface. In humid areas, as represented here, the surface of a lake may represent the water table (right side of block diagram). In this case, a stream gains water from groundwater. A line of springs and seeps commonly occurs where an impermeable rock layer creates a perched water table.

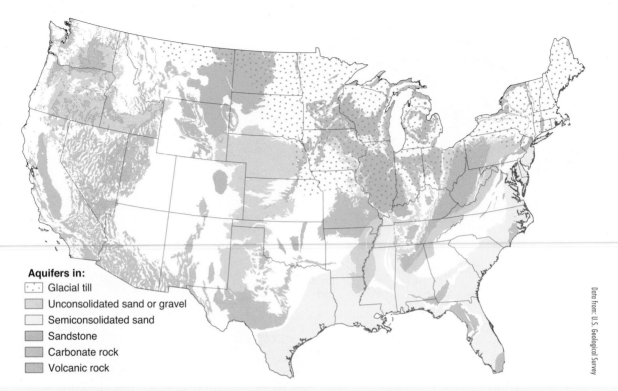

Aquifers in:
- Glacial till
- Unconsolidated sand or gravel
- Semiconsolidated sand
- Sandstone
- Carbonate rock
- Volcanic rock

Data from: U.S. Geological Survey

Figure 13.5 **The major aquifers** of the United States are shown on this map. Each aquifer consists of permeable rocks. Much of the country's drinking water and water for irrigation is extracted from these reservoirs. For some aquifers, extraction is faster than recharge.

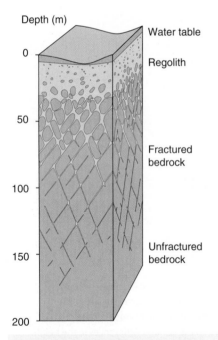

Figure 13.6 **The base of an unconfined groundwater reservoir** is not an abrupt surface like the water table. Most of the groundwater reservoir is in porous regolith and bedrock. Different rock types have substantially different porosities and permeabilities. Open pores gradually close with depth, so the base of the reservoir varies from place to place.

Confined aquifers are permeable rock units enclosed within impermeable strata such as shales (see Figure 13.10). Confined aquifers typically lie deep below the surface but are recharged in highlands where the permeable rocks are exposed. A sandstone aquifer that extends under much of the upper midwestern states of Iowa and Missouri is recharged hundreds of kilometers to the north in Wisconsin, where the permeable rocks outcrop at the surface. The water pressure in a confined aquifer can reach high levels. Water does not readily leak into a confined aquifer by moving directly downward from the surface.

The Movement of Groundwater

Groundwater moves from zones of high pressure to zones of lower pressure.

In an unconfined aquifer, the difference in elevation between parts of the water table is known as the **hydraulic head**. These pressure differences cause the water in the aquifer to follow the paths illustrated in Figure 13.4. If we could trace the path of a particle of water, we would find that gravity slowly pulls it through the zone of aeration to the water table. When a particle encounters the water table, it continues to move downward, by the pull of gravity, along curved paths from areas where the water table is high toward areas where it is low (lakes, streams, and swamps). The explanation for this seemingly indirect flow is that the water table is not a solid surface like the ground surface. Water at any given point below the water table beneath a hill is under greater pressure than water at the same elevation below the lower water table in a valley. Groundwater therefore moves downward and toward points of less pressure.

Although these paths of groundwater movement may seem indirect, they conform to the laws of fluid physics and have been mapped in many areas by tracing the movement of dye injected into the system. The movement of the dye reveals a continuous, slow circulation of groundwater, from infiltration at the surface to seepage into streams, rivers, and lakes.

Natural and Artificial Discharge

Natural discharge of the groundwater reservoir occurs wherever the water table intersects the surface of the ground. In general, such places are inconspicuous, typically occurring in the channels of streams and on the floors and banks of marshes and lakes. This discharge is the major link between groundwater reservoirs and other parts of the hydrologic system.

The natural discharge of groundwater into streams, lakes, and marshes is the major link between groundwater reservoirs and other parts of the hydrologic system (**Figure 13.7**). If it were not for groundwater discharge, many permanent streams would be dry during parts of the year. Most natural discharge is into streams and lakes and, therefore, usually goes unnoticed. It is detected and measured directly by comparing the volume of precipitation with the volume of surface runoff.

Artificial discharge results from the extraction of water from wells, which are made by simply digging or drilling holes into the zone of saturation. Many thousands of wells have been drilled, so in some areas artificial discharge has modified the groundwater system. Indeed, in some areas, more water is removed from the groundwater system by artificial discharge than is added by natural recharge, and the level of the water table drops.

Natural Discharge

Several geologic conditions that produce natural discharge in the form of **seeps** and **springs** are shown in Figure 13.7. If permeable beds alternate with impermeable layers (Figure 13.7A), the groundwater is forced to move laterally to the outcrop of the permeable bed. Conditions such as this usually are found in mesas and plateaus where permeable sandstones are interbedded with impermeable shales. The spring line commonly is marked by a line of vegetation. Figure 13.7B shows a limestone terrain in which springs occur where the base of the cavernous limestone outcrops. Kentucky's Mammoth Cave area is a good example. Figure 13.7C shows springs along a fault that produces an avenue of greater permeability. Faults frequently displace strata for significant distances; thus, impermeable beds, which block the flow of groundwater, may be displaced along a fault so that they are positioned against permeable rocks. The water then moves up along the fault plane and forms springs along the fault line. Many of the great spas of Europe depend upon springs originating on faults. Figure 13.7D shows lava formations that outcrop along the sides of a canyon. Springs develop because groundwater migrates readily through the layers of vesicular and jointed basalt. Note that surface drainage disappears as the water flows over the lava plain.

An excellent example of this last type of discharge is found in the Thousand Springs area of southern Idaho, where numerous springs occur along the sides of the Snake River Canyon (**Figure 13.8**). This region is a vast lava plain extending across the entire southern part of the state. It was built up by innumerable flows of basaltic lava, with some interbedded sand and gravel deposited in streams and lakes that occupied the region during the intervals between volcanic eruptions. The porosity and permeability of these basaltic rocks are remarkably high. Porosity is produced by columnar joints and vesicular texture in the basalt and by pore space in zones of rubble at the tops and bottoms of the basalt flows. In addition, porosity is naturally high in lava tubes and in layers of unconsolidated coarse sand and gravel between some flows. In terms of permeability, the rock sequence is almost like a sieve. The Snake River lies near the southern margin of the lava-covered plain, so tributary streams coming from the mountains to the north are forced to flow across the plain before they can join the Snake (Figure 13.8A). Only one river actually completes the short journey. The rest end after flowing a short distance across the plain; they lose their entire volume of water by seepage into the subsurface. Two of the main would-be tributaries are known as Big Lost and Little Lost Rivers. The groundwater returns to the surface in a series of spectacular springs approximately 200 km downstream. The largest and best known

What geologic conditions produce natural springs?

(A) A line of springs develops on valley walls where impermeable beds cause groundwater in permeable layers to migrate laterally and eventually to seep out at the surface. They are commonly marked by an abnormal growth of vegetation.

(B) Springs form along valley slopes where cavernous limestone permits the free flow of groundwater to the surface.

(C) Many faults displace rocks so that impermeable beds are placed next to permeable beds. A spring line commonly results as groundwater migrates upward along a fault.

(D) Surface water readily seeps into vesicular and jointed basalt flows. It then migrates laterally and forms springs where basalt units are exposed in canyon walls.

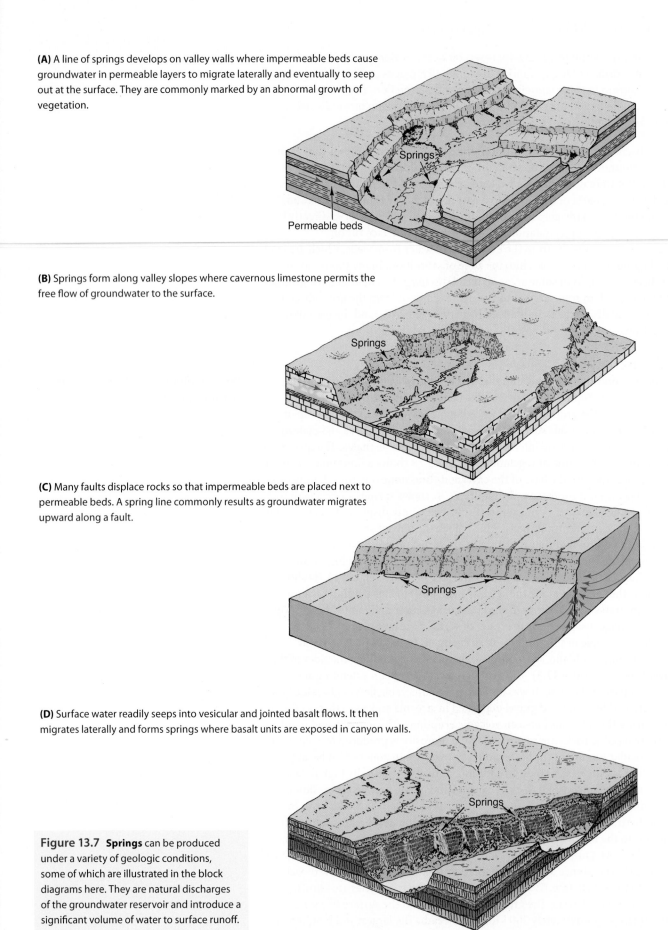

Figure 13.7 Springs can be produced under a variety of geologic conditions, some of which are illustrated in the block diagrams here. They are natural discharges of the groundwater reservoir and introduce a significant volume of water to surface runoff.

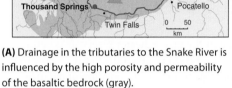

(A) Drainage in the tributaries to the Snake River is influenced by the high porosity and permeability of the basaltic bedrock (gray).

(B) The springs issue from the north wall of the canyon and are fed by water that flowed underground.

Figure 13.8 Thousand Springs along the Snake River Canyon, Idaho, are fed by water that seeps into the basaltic lava plain about 200 km to the northeast. The major tributaries coming from the north lose their entire volume of water by seepage into the subsurface and the rivers simply end. Much of the groundwater reappears to form the spectacular Thousand Springs.

are the Thousand Springs just west of Twin Falls, Idaho (Figure 13.8B). These springs clearly show the tremendous movement of groundwater as they discharge about 1500 m³/sec (nearly 37,000 gal/sec). The visible springs issue from a layer of vesicular basalt 50 m above the river. However, the volume of water that seeps into the Snake River in a less-spectacular fashion below the banks is no doubt many times as great. So much water comes from the Thousand Springs area that an electric power plant has been built on the site to use the energy.

Wells—Artificial Discharge

Ordinary wells are made simply by digging or drilling holes through the zone of aeration into the zone of saturation, as shown in **Figure 13.9**. Water then flows out of the pores into the well, filling it to the level of the water table. When a well is pumped, the water table is drawn down around the well in the shape of a cone, known as the **cone of depression**. If water is withdrawn faster than it can be replenished, the cone of depression continues to grow, and the well ultimately goes dry. The cone of depression around large wells, such as those used by cities and industrial plants, can be many hundreds of meters in diameter. All wells within the cone of depression are affected (Figure 13.9). This undesirable condition has been the cause of "water wars," fought physically and in the courts. Because groundwater is not fixed in one place, as mineral deposits are, it is difficult to determine who owns it. Many disputes are now being arbitrated using computer models that simulate subsurface conditions such as permeability, direction of flow, and level of the water table. The models predict what changes will occur in the groundwater system if given amounts of water are drawn out of a well over specified times.

Extensive pumping can lower the general surface of the water table. This effect has had serious consequences in some metropolitan areas in the southwestern United States, such as Phoenix, Arizona, where the water table has fallen hundreds of meters.

What is the cone of depression in the water table? How is it produced?

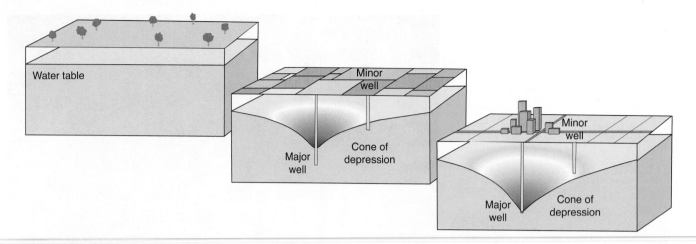

Figure 13.9 A cone of depression in a water table results if water is withdrawn from a well faster than it can be replenished. The cone can extend for hundreds of meters around large, deep wells and effectively lower the water table over a large area. Shallow wells nearby then run dry because they lie above the lowered water table.

The supply of groundwater is limited. Although the groundwater reservoir is being continually replenished by precipitation, the migration of groundwater is so slow that it can take hundreds of years to raise a water table to its former position of balance with the hydrologic system.

Confined Aquifers

> Water in a confined aquifer lies between impermeable beds and is under pressure, like water in a pipe. Where a well or fracture intersects the aquifer, water rises in the opening to reach the potentiometric surface and may produce a flowing (or artesian) well.

A very important type of groundwater occurs in a confined aquifer where a permeable rock body is confined between impermeable beds, where it is under pressure. A well drilled into such an aquifer will commonly allow water to rise and flow freely without pumping. The name **artesian** was originally applied to flowing wells in a French province along the English Channel, where this condition is common. The necessary geologic conditions for artesian water, illustrated in **Figure 13.10**, include the following:

1. The rock sequence must contain interbedded permeable and impermeable strata to create a confined aquifer. This sequence occurs commonly in nature as interbedded sandstone and shale. Permeable beds form the aquifers.
2. The rocks must be tilted and exposed in an elevated area where water can infiltrate into the aquifer.
3. Sufficient precipitation and surface drainage must occur in the outcrop area to keep the aquifer filled.

Water confined in an aquifer behaves much as water does in a pipe. Hydrostatic pressure builds up, so where a well or fracture intersects the bed, water rises in the opening. The height to which water in a confined aquifer rises is shown by the colored line in Figure 13.10. The imaginary surface defined by this level is called the **potentiometric surface**. You might expect it to be a horizontal surface, but actually, a potentiometric surface slopes away from the recharge area. The mineral grains in the aquifer provide resistance to flow, lowering the water pressure. Some pressure is also lost through minor leaks in the underground plumbing system. If a well were drilled at location A or C in Figure 13.10, water would rise in the well, but it would not flow to

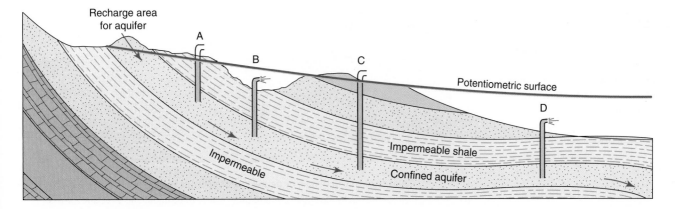

Figure 13.10 **The necessary geologic conditions for a flowing well** include (1) a permeable bed (aquifer, blue) confined between impermeable layers, (2) rocks tilted so the aquifer can receive infiltration from surface waters, and (3) adequate infiltration to fill the aquifer and create hydrostatic pressure. Consequently, water rises in all of the wells (A, B, C, D) to a level called the potentiometric surface. Flowing (or artesian) wells occur only when the top of the well is below the potentiometric surface and require no pumping.

the surface because the potentiometric surface is below the ground surface. Water in a well at location B or D, where the potentiometric surface is above the ground, would flow to the surface. Nonetheless, the water in all of these wells is under pressure and rises above the top of the aquifer.

Confined aquifers are common in most areas underlain by sedimentary rocks because the necessary geologic conditions are present in various ways. One of the better-known confined aquifers underlies the Great Plains states (**Figure 13.11A**). The sequence of interbedded sandstones, shales, and limestones is nearly horizontal throughout most of Kansas, Nebraska, and the Dakotas, but it is upwarped along the eastern front of the Rockies and the margins of the Black Hills. Several sandstone formations are important aquifers. Water is confined in them under hydrostatic pressure. The recharge area is along the foothills of the Rockies.

Figure 13.11B illustrates another type of confined aquifer in the inclined strata of the Atlantic and Gulf Coast plains of North America. The rock sequence consists of permeable sandstone and limestone beds, alternating with impermeable clay. Surface water, flowing toward the coast, seeps into the beds where they are exposed at the surface. It then moves slowly down the dip of the permeable strata.

A third example is from the western United States, where the arid climate makes confined aquifers an important resource (Figure 13.11C). In this region, the subsurface rocks in an intermontane basin consist of sand and gravel deposited in ancient alluvial fans. Farther into the basin, these deposits are interbedded with layers of clay and silt deposited in playa lakes. These fine-grained strata act as confining layers bounding permeable layers of sand and gravel. Water seeping into the fan deposits becomes confined as it moves away from the mountain front.

Confined aquifers also underlie some of the world's great desert regions. Natural discharge from them is largely responsible for oases. Part of the Sahara system is shown in Figure 13.11D. Oases occur where water from a confined aquifer is brought to the surface by fractures or folds or where the desert floor is eroded down to the top of the aquifer.

Note that each example shown in Figure 13.11 has the basic geologic conditions necessary for artesian water: (1) There is a sequence of interbedded permeable and impermeable strata, and (2) the sequence of strata is tilted so that the strata are exposed in an elevated area, enabling surface water to infiltrate into the aquifer. The main difference in each area is in the details of the rock structure and sequence of strata.

What produces flowing wells?

Figure 13.11 Confined aquifers develop under a variety of geologic conditions, some of which are illustrated in these block diagrams. The main difference is in the geometry of the rock structures in each area. The potentiometric surface is shown with a dashed red line.

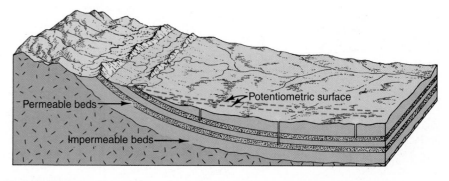

(A) The Great Plains states are underlain by permeable Cretaceous sandstones that are warped up along the front of the Rocky Mountains, where they receive infiltration. This structure forms a widespread confined aquifer in Kansas, Nebraska, and the Dakotas.

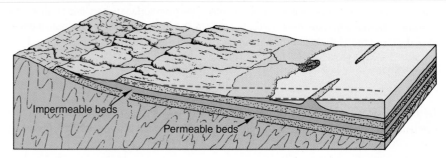

(B) Under the Atlantic and Gulf Coast states Tertiary and Cretaceous rocks dip uniformly toward the ocean. Water enters permeable beds where they are exposed and becomes confined down the dip to form a large confined aquifer.

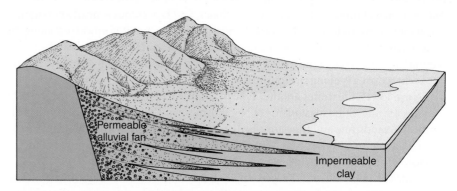

(C) The intermontane basins in the western United States contain permeable sand and gravel (deposited in alluvial fans) interfingered with impermeable clay deposits (deposited in playa lakes). Water seeps into the lenses of buried fan deposits and is confined by the clay to form a confined system.

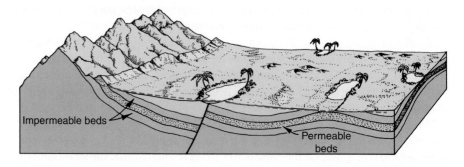

(D) The Sahara Desert of North Africa is underlain mostly by gently warped permeable beds that receive water where they are exposed at the base of the Atlas Mountains. Oases form when water from the confined aquifer discharges through fractures or where the aquifer is exposed by erosion.

Thermal Springs and Geysers

In areas of recent igneous activity, rocks associated with old magma chambers can remain hot for hundreds of thousands of years. Groundwater migrating through these areas of hot rocks becomes heated and, when discharged to the surface, produces thermal springs and geysers.

The most spectacular manifestation of groundwater is in the areas of thermal springs and geysers, where scalding water and steam commonly erupt high into the air. Geysers and thermal springs are usually the results of groundwater migrating through areas of hot, but not molten, igneous rocks.

The three most famous regions of hot springs and **geysers** (a hot spring that intermittently erupts jets of hot water and steam) are Yellowstone National Park of Wyoming, Iceland, and New Zealand. All are regions of recent volcanic activity, so the rock temperatures just below the surface are quite high. Although no two geysers are alike, all require certain conditions for their development:

1. A body of hot rocks must lie relatively close to the surface.
2. A system of fractures must extend downward from the surface.
3. A relatively large supply of groundwater must be present.

Eruptions of Geysers

Geyser eruptions occur when groundwater pressure in fractures, caverns, or porous rock builds to a critical point at which the temperature-pressure balance is such that a small change will cause the water to convert instantly into steam (**Figure 13.12**). Because the water at the base of the fracture is under greater pressure than the water above, the deeper water must be heated to a higher temperature before it boils. Eventually, a slight increase in temperature or a decrease in pressure (resulting from the liberation of dissolved gases) causes the deeper water to boil. The expanding steam throws water from the underground chambers high into the air. After the pressure is released, the caverns refill with water and the process is repeated.

This process accounts for the periodic eruption of many geysers. The interval between eruptions is the amount of time required for water to percolate into the fracture and be heated to the critical temperature. Geysers such as Old Faithful in Yellowstone National Park erupt at definite intervals because the rocks are permeable and the "plumbing system" refills rapidly. Other geysers, which require more time for water to percolate into the chambers, erupt at irregular intervals because the water supply over a longer period of time can fluctuate.

Hot water migrates upward through the surrounding rock without losing much heat or energy and emerges at the surface as thermal springs, sometimes at boiling temperatures. These waters are loaded with chemicals dissolved from the rocks through which they flow. Where they reach the surface, they quickly cool and precipitate various minerals in beautiful splashes of color (**Figure 13.13**).

Why do geysers erupt in cycles?

Geothermal Energy

The thermal energy of groundwater, or **geothermal energy**, offers an attractive source of energy for human use. At present, it is used in various ways in areas of the United States, Mexico, Italy, Japan, and Iceland.

In Iceland, geothermal energy has been used successfully since 1928. Wells are drilled in geothermal areas, and the steam and hot water are piped to storage tanks and then pumped to homes and municipal buildings for heating and hot water. The cost of this direct heating is only about 60% of that of fuel-oil heating and about 75% of the cost of the cheapest method of electrical heating. Steam from geothermal energy is also used to run electric generators, producing an easily transported form of energy.

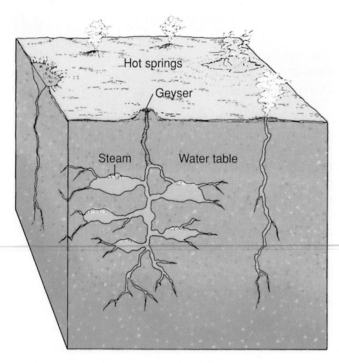

(A) Groundwater circulating through hot rocks in an area of recent volcanic activity collects in caverns and fractures. As temperature rises the water boils and steam bubbles rise, grow in size and number, and may accumulate in restricted parts of the geyser tube.

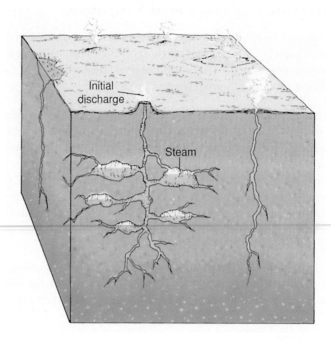

(B) The expanding steam forces water upward until it is discharged at the surface vent. The deeper part of the geyser system becomes ready for the major eruption.

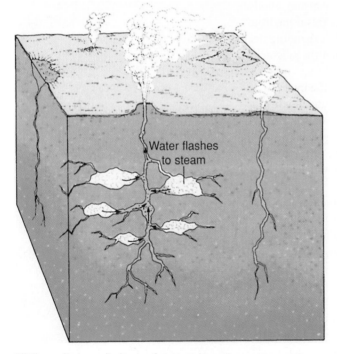

(C) The preliminary discharge of water reduces the pressure on the water lower down. Consequently, water from the side chambers and pore spaces begins to flash into steam, forcing the water in the geyser system to erupt.

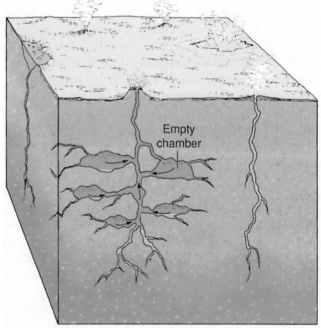

(D) Eruption ceases when the pressure from the steam is spent and the geyser tubes are empty. The system then begins to fill with groundwater again, and the eruption cycle starts anew.

Figure 13.12 The origin of geysers is depicted in this series of diagrams. A geyser can develop only if (1) a body of hot rock lies relatively close to the surface, (2) a system of irregular fractures extends down from the surface, and (3) there is a constant supply of groundwater. Hot springs and mud pots develop where groundwater has freer access to the surface.

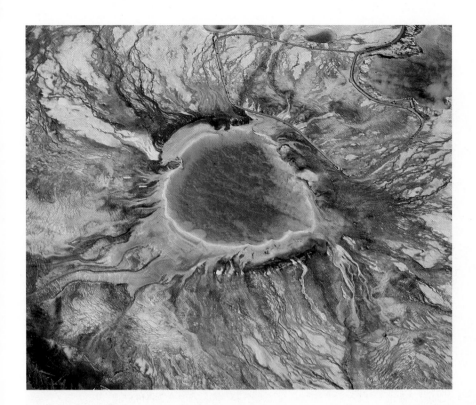

Figure 13.13 Hot springs such as Prismatic Springs in Yellowstone National Park are common where groundwater is heated in regions with young volcanism or deep faulting. The brilliantly colored water is a result of chemical reactions between the wall rocks and the hot groundwater. Orange hues show where overflows carry the alteration products. A walkway in the upper right part of the photo shows the scale.

Erosion by Groundwater

Slow-moving groundwater can dissolve huge quantities of soluble rock and carry it away in solution. In some areas, it is the dominant agent of erosion and produces karst topography, which is characterized by sinkholes, solution valleys, and disappearing streams.

Groundwater can accomplish erosion on an enormous scale, but unlike streams, groundwater erodes only by dissolving soluble rocks such as limestone, dolostone, rock salt, and gypsum. It then transports the dissolved mineral matter and either discharges it into other parts of the hydrologic system or deposits it in the pore spaces within the rock. Groundwater erosion starts with water percolating through joints, faults, and bedding planes and dissolving the soluble rock (**Figure 13.14**). In time, the fractures enlarge to form a subterranean network of caves that can extend for many kilometers. The caves grow larger until ultimately the roof collapses, and a craterlike depression, or **sinkhole**, is produced. Solution activity then enlarges the sinkhole to form a solution valley, which continues to grow until the soluble rock is removed completely.

The most important acid in groundwater is carbonic acid (H_2CO_3). This acid forms readily as carbon dioxide in the atmosphere and soil dissolves in water. Most of the carbon dioxide comes from gases in the soil, where plants have enriched carbon dioxide to as much as 10%; in the atmosphere, carbon dioxide makes up only 0.03%. Sulfuric acid—formed from sulfur compounds abundant in organic sediments such as coal, peat, and liquid petroleum—is also commonly present in groundwater. Together with various more-complex organic acids generated in the soil, these dilute acids react with the minerals in rocks and remove them in solutions. These processes of acid production and mineral dissolution can be represented by two chemical reactions:

$$\underset{\text{(water)}}{H_2O} + \underset{\text{(carbon dioxide)}}{CO_2} = \underset{\text{(carbonic acid)}}{H_2CO_3}$$

$$\underset{\text{(acid)}}{H_2CO_3} + \underset{\text{(calcite)}}{2CaCO_3} = \underset{\text{(dissolved calcium bicarbonate)}}{2Ca^{2+} + 2HCO_3^-}$$

Figure 13.14 The importance of fractures in the evolution of a cave system is revealed in this photograph of the Redwall Limestone exposed on the walls of the Grand Canyon. Dissolution was enhanced along vertical fractures and on nearly horizontal bedding planes.

The rate of erosion of limestone terrains by the chemical action of groundwater can be measured in several different ways. One is to measure precisely the weight of small limestone tablets and place them in different climatic conditions. Once the weight loss of the limestone during a specific period of time is known, the average rate at which limestone terrains are being lowered, by chemical processes alone, can be calculated. Other methods include measuring the amount of mineral matter in water dripping through caves and the dissolved mineral matter carried by a river system in limestone regions. Recently, precise measurements of dissolution rates have been obtained with a microerosion meter, an instrument capable of measuring the erosion of a rock surface to the nearest 0.005 mm. The results of these measurements indicate that in temperate regions the landscape is being lowered at an average rate of 10 mm/1000 yr. In areas of greater rainfall, rates may be as high as 300 mm/1000 yr. These averages may seem small, but they indicate that in some areas erosion by groundwater can be greater than the average erosion of a surface by running water. It is clear from these measurements, and from the characteristics of limestone terrains, that groundwater accomplishes erosion on a grand scale.

The permeability and porosity created by dissolution of limestone can create important reservoirs for oil and natural gas accumulation. In fact, about 50% of all petroleum comes from carbonate rocks.

Rock salt and gypsum, the most soluble rocks, are eroded rapidly by solution activity and can cause overlying layers to collapse. They are relatively rare, however, and are not widely distributed on any of the continents.

Caves

Perhaps the best way to appreciate the significance of solution activity is to consider the nature of a cave system and the amount of rock removed by solution. Shallow groundwater dissolves carbon dioxide and forms a weak acid. The slightly acidic water then percolates through the fractures and bedding planes, slowly dissolving the limestone and enlarging the openings (**Figure 13.15**). In the zone of aeration, the general direction of groundwater motion is downward, toward the water table. The water then moves toward a natural outlet, such as a river system, and as it moves, it dissolves the limestone. In time, a main subterranean channel is developed that transports the solution to the main streams. If the water table drops (usually by downcutting of the river), water in the main subterranean channel begins once again to seep downward to a new level. Eventually, the old horizontal channel drains, and water dripping from the old channel ceiling begins to deposit calcite, in time creating major deposits in the open cave. The origin and evolution of **caves** are shown in Figure 13.15.

As caves grow larger, they become unstable and tend to collapse. The fallen rubble occupies about one-third more volume than it did as intact rock on the cave ceiling. Consequently, small caves may completely fill because of debris falling from the ceiling, but larger caverns tend to migrate upward as their roofs collapse and bury their floors. Larger or shallower caves ultimately break through to the surface to become sinkholes. A map of a cave system may show long, winding corridors, with branched openings that enlarge into chambers, or a maze of interlacing passageways and channels, controlled by intersecting joint systems. Where a sequence of limestone formations occurs, several levels of cave networks may exist. Kentucky's Mammoth Cave, for example, has more than 50 km of continuous subterranean passages on several different levels (**Figure 13.16**).

Karst Topography

Karst topography is a distinctive type of terrain resulting largely from erosion by groundwater (**Figure 13.17**). *Karst* is a German word for the Kras Plateau in Slovenia where this landscape is common. In contrast to a landscape formed by surface streams, which is characterized by an intricate network of stream valleys, karst topography lacks a well-integrated drainage system. Sinkholes are generally numerous and, in many karst regions, they dominate the landscape. Where sinkholes grow and enlarge, they merge and form elongate or irregular closed depressions known as **solution valleys**. Small streams commonly flow on the surface for only a short distance and then disappear down a sinkhole, becoming **disappearing streams**. There the water moves slowly through a system of caverns and caves, sometimes as sluggish underground streams. Springs, which are common in karst areas, return water to the surface drainage.

In tropical areas, where dissolution is at a maximum because of the abundance of water from heavy rainfall, a particular type of karst topography, known as **tower karst**, develops. In China, this terrain is called a peak forest. Tower karst is characterized by steep, cone-shaped hills rather than sinkholes and solution valleys (see Figure 13.20). The towers are largely residual landforms left after most of the rock has been removed by solution activity along fracture systems, collapse of caverns, and enlargement of solution valleys. They are the remnants of a once-continuous layer of rock that covered the area.

Thus, in detail, karst topography is highly diverse, ranging from fantastic tower landscapes to the low relief of a plane pitted with small depressions. What is common to all karst terrains is that their landforms are caused by the unusually great solubility of certain rock types. Humid climate is a very important factor in developing karst topography. The more water moving through the system, the more solution activity will occur. Karst topography is, therefore, largely restricted to humid and temperate climatic zones. In desert regions, where little rain falls, extensive karst topography will not develop.

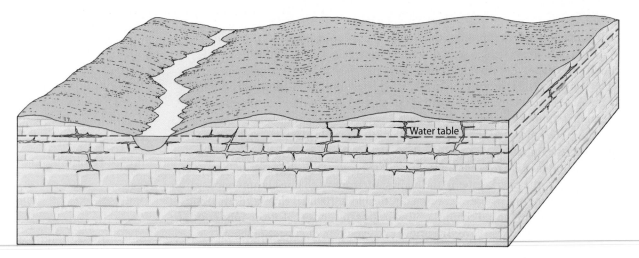

(A) In the early stages, water seeps through the fractures and bedding planes in limestone. The groundwater seeps downward to the water table and then moves toward the surface streams. Soluble minerals are dissolved and the flow paths become enlarged.

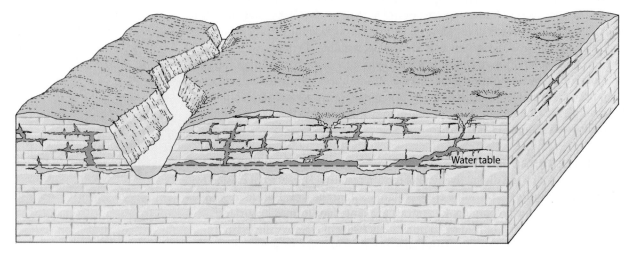

(B) As the surface streams erode the valley floor, the water table drops. The surface water seeping through the zone of aeration enlarges the existing joints and caves. Movement of water toward the surface stream develops a main system of horizontal caverns.

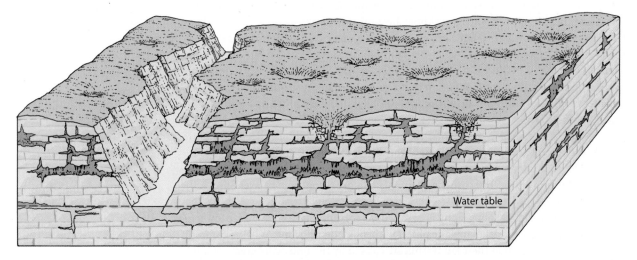

(C) As the river erodes a deeper valley, the water in the main underground channel seeks a new path to the lower river level. A new, lower system of horizontal caverns develops. The older, higher caverns may continue to enlarge and ultimately collapse to form sinkholes, or they may fill with fallen rubble or cave deposits.

Figure 13.15 The evolution of a cave system is shown schematically in these diagrams.

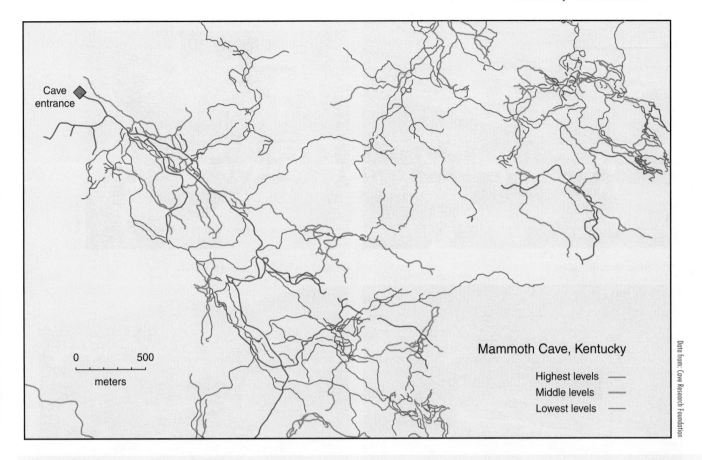

Figure 13.16 **A map of Mammoth Cave**, Kentucky, shows the complexity of cavern systems in limestone regions. The major northwest-southeast lineaments show how joints have controlled groundwater erosion. Solution activity along the joints has produced the network of caverns. Various cave systems develop in different layers of limestone.

A simplified model of the evolution of karst topography is shown in the block diagrams in **Figure 13.18**. Initially, water follows surface drainage until a large river cuts a deep valley below the limestone layers. Groundwater then moves through the joints and bedding surfaces in the limestone and emerges at the riverbanks. As time goes on, the passageways become larger and caverns develop. Surface waters disappear into solution depressions. The roofs of caves collapse, so many sinkholes are produced (Figures 13.17B and 13.18A). Springs commonly occur along the margins of major stream valleys. Sinkholes proliferate and grow in size as the limestone formation is dissolved away. The cavernous terrain of central Kentucky, for example, is marked by more than 60,000 holes. As solution processes continue, sinkholes increase in number and size. Some merge to form larger depressions with irregular outlines. This process ultimately develops solution valleys (Figure 13.18B). Most of the original surface is finally dissolved, with only scattered hills remaining (Figure 13.18C). When the soluble bedrock has been removed by groundwater solution, normal surface drainage patterns reappear.

Karst Regions of the World   The development of karst topography is not random but is developed in specific regions where limestone formations are exposed near the surface and where there is adequate precipitation for vigorous chemical reactions to occur. High temperatures are favorable because heat enhances the dissolution process. The map in **Figure 13.19** shows the major karst regions of the world, and, although it appears that karst topography is scattered across the globe, close examination of this map will reveal why karst landscapes develop in specific areas and not in others.

Courtesy of John S. Shelton

Courtesy of USGS

(A) Sinkhole karst, Kentucky.

(B) Sinkhole in a karst terrain in Florida.

(C) In some karst regions, streams disappear into subsurface caverns like this one in China.

(D) Groundwater solution enlarged these fractures in limestones in New Zealand.

Figure 13.17 Karst topography includes a wide variety of landforms ranging in size from small solution pits, to sinkholes and caves, to residual towers. These solution features are distinctive and stand out in striking contrast to landscapes formed by running water, glaciers, or the wind.

Only about 20% of Earth's surface has major limestone sequences exposed at the surface; the development of karst topography is limited to those areas. The continental shields do not favor karst development, nor do terrains dominated by granite, basalt, sandstone, and the like. As can be seen in Figure 13.19, karst topography is not well developed in desert regions, where there is inadequate rainfall, although limestone occurs at the surface. Karst landforms are also rare in regions recently covered by continental glaciers, such as those in Canada and northern Europe.

The best-developed and most-extensive karst topography is in humid and tropical regions. The Alpine Mountain belt extending through the Pyrenees and southern Europe and across Croatia, Bosnia, Yugoslavia, and Turkey has thick sequences of deformed limestone on which many classic karst landforms have developed. In the United States about 15% of the area is favorable for karst development. The best-known regions are the sinkhole country of Kentucky, Indiana, and Florida. These regions consist of plains pockmarked by innumerable small, isolated depressions (Figure 13.17A). Surface streams of any significant length are extremely rare. The limestone regions of Puerto Rico, Jamaica, Cuba, and Mexico's Yucatan Peninsula have extensive karst regions because of widespread limestone and heavy rainfall. Little-known karst

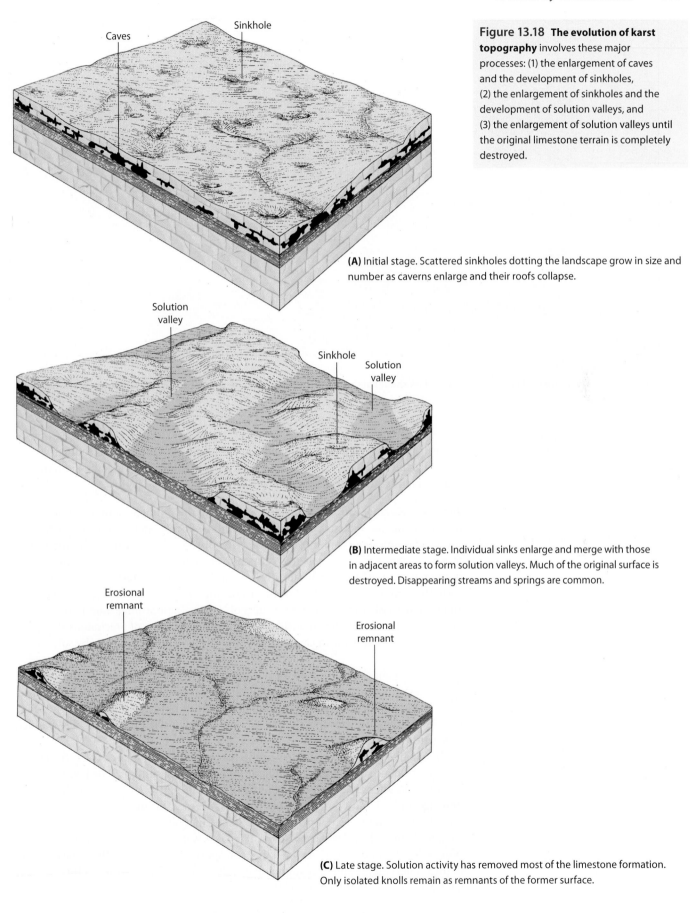

Caves

Sinkhole

Figure 13.18 The evolution of karst topography involves these major processes: (1) the enlargement of caves and the development of sinkholes, (2) the enlargement of sinkholes and the development of solution valleys, and (3) the enlargement of solution valleys until the original limestone terrain is completely destroyed.

(A) Initial stage. Scattered sinkholes dotting the landscape grow in size and number as caverns enlarge and their roofs collapse.

Solution valley

Sinkhole

Solution valley

(B) Intermediate stage. Individual sinks enlarge and merge with those in adjacent areas to form solution valleys. Much of the original surface is destroyed. Disappearing streams and springs are common.

Erosional remnant

Erosional remnant

(C) Late stage. Solution activity has removed most of the limestone formation. Only isolated knolls remain as remnants of the former surface.

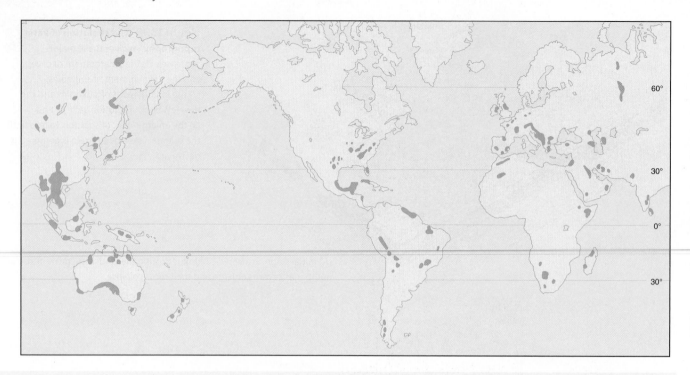

Figure 13.19   The major areas of karst topography (dark shade) of the world are restricted to regions where outcrops of limestone occur in humid climatic conditions.

regions in other parts of the world include the large Nullarbor Plain in southern Australia and the karst regions of Southeast Asia, particularly in Malaysia and the Indonesian islands of Sumatra and Java.

The Tower Karst of China   In striking contrast to the sinkhole plains of Indiana and Kentucky, the tower karst topography of southern China presents some of the most spectacular limestone scenery on Earth (see **Figure 13.20**). Here, an area of thousands of square kilometers, once covered by thick layers of limestone, is in an advanced stage of dissection by groundwater. The region consists of a "forest" of hills that rise abruptly from the surrounding terrain. These hills are remnants between sinkholes and solution basins and stand like clusters of towers. These strange mountains, shaped like upended loaves of French bread, form an intricate system of precipitous slopes and overhanging cliffs, with caves, arches, and strange landforms made by solution activity.

Classical Chinese art is noted for portraying these bizarre and exotic landforms, which appear unreal to the foreign eye. Western artists believed that the Chinese masters who painted these landforms were impressionists, but anyone fortunate enough to visit the region realizes that the artists were not visionaries; the shapes they painted were nature's own.

The small-scale solution features in this area are almost as impressive as the larger features. Evidence of solution activity is everywhere. Every hill is riddled with caves, caverns, solution pits, and voids that form an intricate network for subterranean drainage. Indeed, most outcrops look like Swiss cheese. This maze of cavities is where the action is, and in a way, it performs the same function as the network of gullies and streams do in surface drainage systems. It collects water and funnels it through the system. As the water moves, it erodes and transports rock material. Thus, the cavities in the towers enlarge and ultimately collapse. Examples of the features in this amazing landscape are shown in Figure 13.20.

(A) Residual towers reveal the extent of groundwater solution. Caves are common in such towers.

(B) Surface grooves, pits, and sharp ridges caused by dissolution of limestone in surface waters.

(C) Dissolution is expressed vividly by the abundant cavities so that the limestone outcrops resemble Swiss cheese.

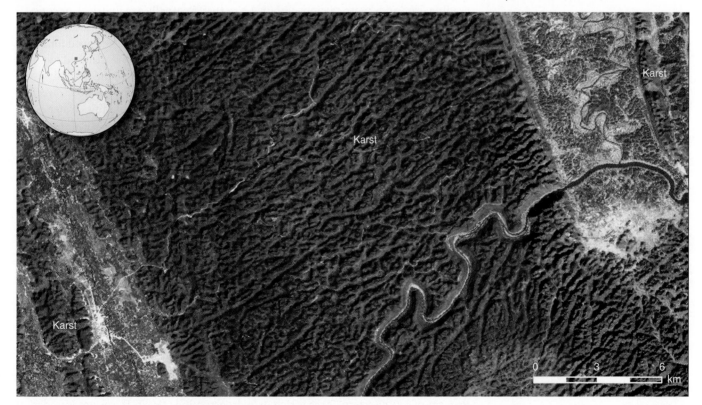

(D) Karst terrain. Karst processes form distinctive regional landscapes as shown on this map of southern China. The fractures that control groundwater flow become etched into the landscape as linear valleys.

Figure 13.20 Karst features in southern China are well developed because of the high rainfall and thick layers of limestone exposed at the surface and include large residual towers, as well as small scale features.

Deposition by Groundwater

The mineral matter dissolved by groundwater can be deposited in a variety of ways. The most spectacular deposits are stalactites and stalagmites, which are found in caves. Less obvious are the deposits in permeable rocks such as sandstone and conglomerates. Here, groundwater commonly deposits mineral matter as cement between grains.

The chemical processes that cause groundwater to dissolve soluble material are easily reversed, and the minerals are precipitated in the pore spaces, voids, and caves within the rock. The change from solution to precipitation is commonly caused by a lowering of the water table. The main solution processes occur in the zone of saturation; precipitation occurs in the zone of aeration after the caves and pore spaces are drained. This process can be understood by examining what happens in an air-filled cave. Groundwater dripping from a cave's roof contains more CO_2 than the surrounding air. In an attempt to reach equilibrium with the air, CO_2 diffuses out of the water droplet. This diffusion reduces the amount of carbonic acid as well as the amount of calcite that can be dissolved in the water. As a result, the water becomes saturated with $CaCO_3$ and it precipitates. You can visualize this process by examining the chemical reactions shown previously. Both reactions are driven to the left as a result of the escape of CO_2 from the groundwater.

The deposits formed in caves are some of nature's fancywork; the endless variety of cave deposits is familiar to almost everyone. They originate in a variety of ways and are collectively called **speleothems**. One formative process is shown in **Figure 13.21**. As water enters a cave (usually from a fracture in the ceiling), carbon dioxide escapes during evaporation, and a small amount of calcium carbonate crystallizes. Each succeeding drop adds more calcium carbonate, so that eventually a cylindrical, or cone-shaped, projection is built downward from the ceiling. Many beautiful and strange forms result, some of which appear in **Figures 13.22** and **13.23**. Icicle-shaped forms growing down from the ceiling are **stalactites**. These commonly are matched by deposits growing up from the floor, known as **stalagmites**, because the water dripping from a stalactite precipitates additional calcium carbonate onto the floor directly below. Many stalactites and stalagmites eventually unite to form columns. Water percolating from a fracture along a slanting ceiling may form a thin, vertical sheet of rock known as drapery because of its shape. Pools of water on the cave floor flow from one place to another, and as they evaporate, calcium carbonate is deposited on the floor, forming terraces made of travertine—a layered cave or hotspring rock composed of calcium carbonate.

Lechuguilla Cave

What features are unique in Lechuguilla Cave?

One of the best-kept secrets of the National Park Service is Lechuguilla Cave, in the Guadalupe Mountains near Carlsbad, New Mexico. It was first explored in 1986 and was found to be the deepest known cave in the United States, with a vertical range of 475 m. Within this cave are some of the most spectacular cave "formations" in the world. As a result, it is commonly referred to as the "Jewel of the Underground."

Lechuguilla Cave is remarkable, not only for its size and beauty, but also for its strange origin. Most of the world's caves are dissolved from limestone as groundwater with carbonic acid (H_2CO_3) moves downward by the pull of gravity. Lechuguilla developed in an entirely different way. Sulfuric acid (H_2SO_4), rather than carbonic acid, played the dominant role in its formation. Water associated with the vast petroleum deposits of southwestern Texas is rich in hydrogen sulfide (H_2S). This water is under pressure and seeps up, from the trapped oil and gas, through fractures in the rock. Eventually it reaches the shallow groundwater reservoir; there, the hydrogen sulfide combines with oxygen in the fresh groundwater to form sulfuric acid (H_2SO_4). The rising acidic water dissolves the cave system and, in places, excess hydrogen sulfide gas may even dissolve caverns above the water table.

The unique role played by sulfuric acid in the formation of Lechuguilla Cave has produced an enchanting array of sulfate cave deposits, most notably gypsum ($CaSO_4 \cdot 2H_2O$). Some blocks are as large as houses; elsewhere, crystals are as fine as hair. The hallmarks of Lechuguilla are dozens of unique stalactites hanging from the ceiling like monstrous, grotesque chandeliers (Figure 13.23). Some are more than 6 m long, with downward-twisting trunks ending in branching, clawlike arms of transparent crystals. But the gypsum deposits of Lechuguilla come in many other strange forms. Long, slender, glasslike needles, some as much as 6 m long, extend downward from the ceiling. Other exotic cave deposits, both large and small, occur in great profusion and display, in a most spectacular way—the results of fluid and gases migrating beneath the surface.

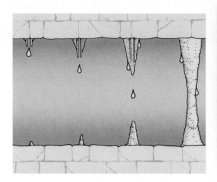

(A) Diagram showing, left to right, the evolution of stalactites, stalagmites, and columns.

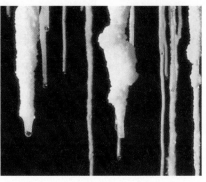

Courtesy of David Herron

(B) Long, slender stalactites (soda straws) grow as a drop of water suspended at the end loses carbon dioxide and evaporates.

Figure 13.21 **Stalactites** originate on the ceilings of caves. Water seeps through a crack and loses carbon dioxide as it partially evaporates. Consequently, a small ring of calcite is deposited around the crack. The ring grows into a tube, which commonly acquires a tapering shape as water seeps from adjacent areas and flows down its outer surface.

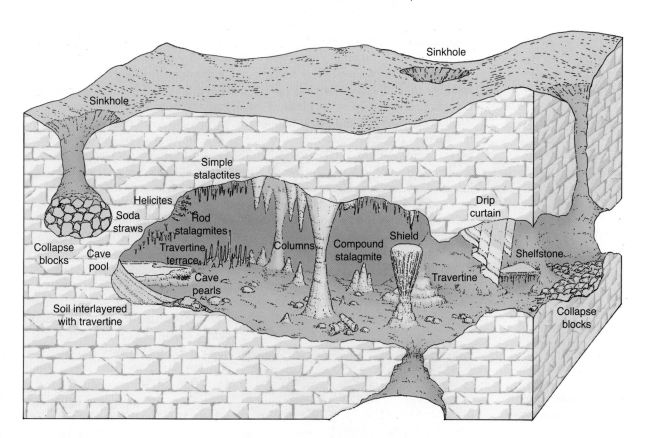

Figure 13.22 **Many varieties of cave deposits** are shown in this idealized diagram. Most are composed of calcite deposited by water that seeps into the open cave and then loses carbon dioxide as the water evaporates.

(A) Stalactites and stalagmites formed from calcium carbonate in Lechuguilla Cave.

(B) Gypsum stalactites form coarsely crystalline, clawlike branches. These "chandeliers" are up to 6 m long and are thought to be the world's largest.

Figure 13.23   Deposits in Lechuguilla Cave, New Mexico.

Although cave deposits are spectacular expressions of deposition by groundwater, they are trivial compared with the amount of material deposited in the pore spaces of rock. In sandstones and conglomerates, precipitation of silica and calcium carbonate cement the loose grains into a hard, strong rock body. In some formations, the cementing minerals deposited by groundwater may exceed 20% of the volume of the original rock (**Figure 13.24**).

Mineral precipitation by groundwater action is a slow process, in some cases involving the slow removal—one at a time—of atoms or molecules of organic matter and their simultaneous replacement by other mineral ions carried by the groundwater. One example of this process is petrified wood. Perhaps the best-known deposit of petrified wood is the Petrified Forest National Park in eastern Arizona. Here, great accumulations of petrified logs, buried in ancient river sediments, are now being uncovered by weathering and erosion (**Figure 13.25**). The Petrified Forest is not really a forest at all but a great collection of driftwood. This driftwood washed down from adjacent highlands about 230 million years ago and accumulated as logjams in ancient river bars and floodplains. It was subsequently covered with hundreds of meters of younger sediments. While the driftwood was covered with sediment, groundwater percolating through the strata replaced the cellular structure of the wood with silica. This process transformed the wood to agate, a variety of silicon dioxide (SiO_2).

Geodes are another common example of the result of the action of groundwater deposition. A geode is a roughly spherical, hollow rock mass with its central cavity lined with mineral crystals. Geodes are common in limestones, but they also occur in preexisting voids in shales and in silica-rich volcanic rocks. The formation of geodes

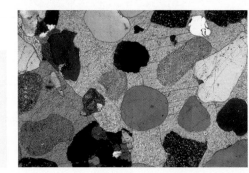

Figure 13.24   Calcite deposited by groundwater cements the rounded quartz sand grains together, as shown in this thin ⸺tone. The quartz grains are ⸺ of gray and the calcite is pink ⸺kles. The area shown is 3 mm

can be explained as a two-stage development. First, a cavity is formed in the rock by groundwater solution activity. Then, under different conditions, the mineral matter carried by groundwater is precipitated on the walls of the rock cavity. Quartz, calcite, and fluorite are the most common minerals precipitated. They accumulate very slowly and form perfect crystals, pointing toward the center of the cavity. Subsequent

Figure 13.25 **Petrified trees** litter the area, piled like giant jackstraws about a rolling landscape on the Petrified Forest Member of the Chinle Formation, Arizona.

Figure 13.26 **Mammoth Hot Springs**, Yellowstone National Park, was formed by the deposition of travertine ($CaCO_3$) as the warm spring water evaporated and lost carbon dioxide.

erosion removes the material around the geode cavity, but the mineral-lined walls of the cavity are resistant, so geodes remain; they are found as boulderlike remnants, left from the weathering of the parent rock material. In a sense, a geode is a fossil cavity.

Another expression of deposition by groundwater is the mineral deposit formed around springs. Mammoth Hot Springs in Yellowstone National Park is one of the most spectacular examples (**Figure 13.26**).

Groundwater Resources

Groundwater is a valuable resource that is being exploited at an ever-increasing rate. Ancient groundwater systems have also produced valuable mineral resources.

Groundwater is of major importance to civilization because it is the largest reserve of drinkable water in many regions. This probably constitutes the most important use of groundwater in our modern society for both urban and rural peoples in arid and semi-arid regions. Many cities in the western United States derive a substantial portion of their municipal water from wells. Another major use of groundwater is in agricultural irrigation. Although most irrigation water is derived by diverting rivers, increasingly irrigation water in the United States is derived directly from groundwater reserves. Increased use of drinking and irrigation water has caused the water levels in many aquifers to drop. In developing countries, groundwater resources are often the controlling factor for development.

In addition, some important metallic ore deposits are formed by groundwater processes. Groundwater can carry dissolved metals from one place to another, concentrating it in specific areas. Some groundwaters are heated by deep circulation or igneous intrusions; the heating usually increases their ability to carry dissolved ions. For example, deposits of fluorite, lead, and zinc in the upper Mississippi Valley formed as groundwater carried dissolved metals to shallow levels and concentrated them in deposits of sulfide minerals. Deposits of uranium, vanadium, and copper in sedimentary rocks are also formed by the movement of groundwater. Many nonmetallic resources are also controlled by groundwater systems. Oil is sometimes trapped in paleokarst deposits formed in ancient limestones. The Yates field of western Texas is developed in a buried karst terrain and contains the highest-yielding wells in the United States because of the high permeability and porosity of such partially dissolved rock units. We have already mentioned the use of heated groundwaters to produce electricity and space heating. Even many building stones are derived from groundwater deposits. These include travertine and onyx marble. Many valuable gemstones form from groundwaters: for example, opal, agate, and onyx (forms of amorphous or cryptocrystalline silica) and emerald.

Alteration of Groundwater Systems

A variety of problems resulting from human activities alter the groundwater system. Important problems are: (1) changes in the chemical composition of groundwater (pollution), (2) saltwater encroachment, (3) changes in the position of the water table, and (4) subsidence.

Groundwater is an integral part of the hydrologic system and is intimately related to other parts of the system. As we have seen in this chapter, its source is precipitation and infiltration from surface runoff. Its natural discharge is into streams and lakes and, ultimately, the sea. With time, a balance, or equilibrium, among precipitation, surface runoff, infiltration, and discharge is established. These in turn approach equilibrium with surface conditions, such as slope angles, soil cover, and vegetation. When any one of these interrelated factors is changed or modified, the others respond to reestablish equilibrium.

Changes in Composition

The composition of groundwater can be changed by increases in the concentration of dissolved solids in surface water. The soil is like a filtration system through which groundwater moves. Obviously, any concentration of chemicals or waste creates local pockets that potentially can contaminate the groundwater reservoir. Material that is leached (dissolved by percolating groundwater) from waste disposal sites, for example, includes both chemical and biological contaminants. Upon entering the groundwater flow system, the contaminants move according to the hydraulics of that system. The character and concentration of the pollutants depend partly on the length of time the infiltrated water is in contact with the waste deposit, partly on the volume of infiltrated water, and partly on the solubility of waste involved. In humid areas, where the water table is shallow and in constant contact with refuse, leaching continually produces maximum potential for pollution.

How does human activity alter the groundwater system?

Figure 13.27 illustrates four geologic environments in which waste disposal affects the groundwater system. In the environment shown in Figure 13.27A, the near-surface material is permeable and essentially homogeneous. Dissolved pollutants percolate

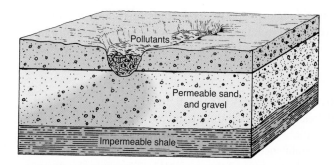

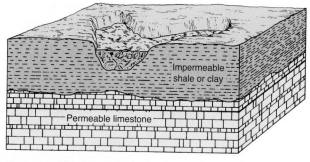

(A) A permeable layer of sand and gravel overlying an impermeable shale creates a potential pollution problem because contaminants are free to move with groundwater.

(B) An impermeable shale (or clay) confines pollutants and prevents significant infiltration into the groundwater system in the limestone below.

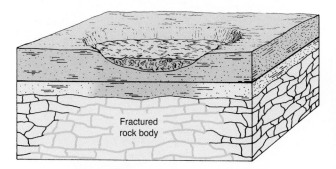

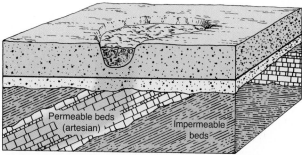

(C) A fractured rock body provides a zone where pollutants can move readily in the general direction of groundwater flow.

(D) An inclined, permeable aquifer below a disposal site permits pollutants to enter a confined aquifer and move down the dip of the beds, so that they contaminate the system.

Figure 13.27 **The effects of waste disposal or leaking storage tanks on a groundwater system** depend on the geologic setting. In many cases, water seeping through the disposal site enters and pollutes the groundwater system.

downward through the zone of aeration and, upon reaching the water table, enter the groundwater flow system. The flowing pollutants ultimately become part of the surface drainage system. As shown in Figure 13.27B, an impermeable layer of shale confines pollutants and prevents their free movement in the groundwater system. As a result, the pollutants are restricted and inhibited from moving freely through the groundwater system. Figure 13.27C illustrates a disposal site above a fractured rock body. Upon reaching the fractured rock, the contaminants can move more readily in the general direction of the groundwater flow. Dispersion of the contaminants is limited, however, because of the restriction of flow to the fractures. Figure 13.27D illustrates a critical condition in which a waste disposal site was constructed in highly permeable sand and gravel, above an inclined aquifer. Here, pollutants move down past the water table and enter the aquifer as recharge. If the waste disposal site is directly above the aquifer, as shown in the diagram, most of the pollutants will enter the aquifer and contaminate the groundwater system.

Saltwater Encroachment

On an island or a peninsula, where permeable rocks are in contact with the ocean, a lens-shaped body of fresh groundwater is buoyed up by the denser saltwater below, as is illustrated in **Figure 13.28A**. The fresh water literally floats on the saltwater and is in a state of balance with it. If excessive pumping develops a large cone of depression in the water table, the pressure of the fresh water, on the saltwater directly below the well, is decreased, and a large cone of **saltwater encroachment** develops below the well, as is shown in Figure 13.28B. Continued excessive pumping causes the cone of saltwater to extend up the well and contaminate the fresh water. It is then necessary to stop pumping for a long time to allow the water table to rise to its former position and depress the cone of saltwater. Restoration of the balance between the freshwater lens and the underlying saltwater can be hastened if fresh water is pumped down an adjacent well (Figure 13.28C).

Changes in the Position of the Water Table

The water table is intimately related to surface runoff, the configuration of the landscape, and the ecological conditions at the surface. The balance between the water table and surface conditions, established over thousands or millions of years, can be completely upset by changes in the position of the water table. Two examples illustrate some of the many potential ecological problems.

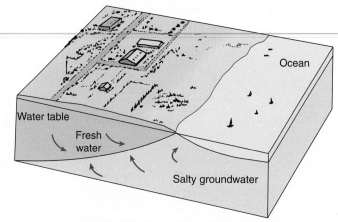

(A) A lens of fresh groundwater beneath the land is buoyed up by denser saltwater below.

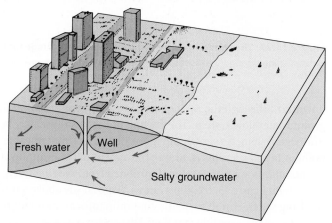

(B) Excessive pumping causes a cone of depression in the water table on top of the freshwater lens and a cone of saltwater encroachment at the base of the freshwater lens.

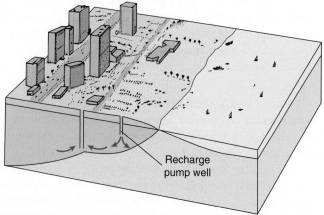

Figure 13.28 The relationship between fresh water and saltwater on an island or a peninsula is affected by the withdrawal of water from wells. Excessive pumping causes a cone of saltwater encroachment, which limits the usefulness of the well.

(C) Fresh water pumped down an adjacent well can raise the water table around the well and lower the interface between the fresh water and the saltwater.

STATE OF THE ART   Computer Models of Flowing Groundwater

Deep below your feet, groundwater is probably inching its way through a permeable aquifer. We are largely unaware of this vast resource, but groundwater is used extensively for city water supplies and for irrigating croplands. Although we rely on this water supply to be pure, groundwater invariably carries dissolved ions—in most cases of little consequence to people. However, in some places groundwater has become contaminated with bacteria, arsenic, gasoline, and a host of other pollutants. If such contaminated groundwaters are pumped out of wells and enter municipal water supplies, they could have serious consequences.

Short of eliminating the use of groundwater altogether, what can be done to protect humans from pollutants moving invisibly in groundwater? Contaminated groundwater does not simply diffuse outward away from the source of pollution. Instead, it follows a path determined by the permeability and structure of the aquifer (Figure 13.27). Around some waste sites, monitor wells are drilled to see where the contaminated water is flowing. Drilling is costly and time-consuming. An alternative is to create a computer model of the flow path of the contaminated groundwater. If such models were accurate, precise predictions about the direction, depth, or rate of flow could be made quickly and inexpensively and the risk of pumping water from contaminated wells avoided.

However, any computer calculation depends heavily on the quality of the starting conditions used in the model. What would you need to know to construct such a model? Of prime importance is a thorough understanding of the geologic and hydrogeologic characteristics of the rocks through which the water is flowing. How deep is the water table? What is its slope? What is the hydraulic head? Are the rocks permeable or impermeable? Do they have fractures or bedding planes along which fluids flow? What is the subsurface configuration of the pressure gradients? How much precipitation is there and how is the aquifer recharged? Are the wastes soluble in water? Will they react with minerals along the flow path and become neutralized? And so on.

Given the complexity of natural rocks, many of these questions can only be approximately answered. Within these limitations, however, reasonable models of groundwater flow can be constructed. In the three-dimensional figure below, you can see a colored plume of contaminated groundwater emanating from an industrial facility that leaked chlorinated hydrocarbons. The contaminant is soluble in water and has leaked into the groundwater system. The contaminated water then moved down through the aquifer. The colored bands show the concentration in the contaminated plume. You can see that the core of the plume, near its source, is more polluted than the rest. Such a computer model would allow city planners to avoid the plume and help environmental geologists devise ways to shut off continued contamination, pump the contaminated water out, or otherwise mitigate the damage.

3D model and image provide courtesy of ctech.com

In southern Florida, fresh water from Lake Okeechobee has flowed for the past 5000 years as an almost imperceptible "river" only a few centimeters deep and 64 km wide. This sheet of shallow water created the swampy Everglades. The movement of the water was not confined to channels. It flowed as a sheet, in a great curving swath for more than 160 km (**Figure 13.29**). The surface of the Everglades slopes southward only 2 cm/km, but this gradient was enough to keep the water moving slowly to the coast and to prevent saltwater from invading the Everglades and the subsurface aquifers along the coast. In effect, the water table in the swamp was at the surface, and the ecology of the Everglades was in balance with the water table.

Today, many canals have been constructed to drain swamp areas for farmland, to help control flooding, and to supply fresh water to the coastal megalopolis (Figure 13.29). The canals diverted the natural flow of water across the swamp, in effect lowering the water table, in some places as much as 0.5 m below sea level. This change in position of the water table produced many unforeseen and often unfortunate results. As the water table was lowered, saltwater encroachment occurred in wells all along the coast. Some cities had to move their wells far inland to obtain fresh water.

The most visible effects, however, involve the ecology of the swamp. In the past, the high water table could maintain a marsh during periods of natural drought. Now the surface is dry during droughts. Forest fires ignite the dry organic muck, which burns like peat, smoldering long after the surface fires die out. This effectively destroys the ecology of the swamp. The lowering of the water table also caused the muck to compact, so that it subsided as much as 2 m in places. In addition, muck exposed to the air oxidizes and disappears at a rate of about 2.5 cm/yr. Once the muck is gone from the swamp, only nature can replace it.

Raising the water table can also modify many surface processes. An example is found in the environmental changes caused by irrigation in Washington's Pasco Basin.

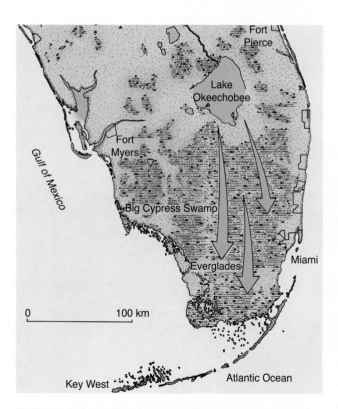

(A) Natural drainage of southern Florida in 1871 spread southward from Lake Okeechobee in a broad sheet only a few centimeters deep. This sheet maintained swampy conditions in the Everglades and established a water table very close to the surface.

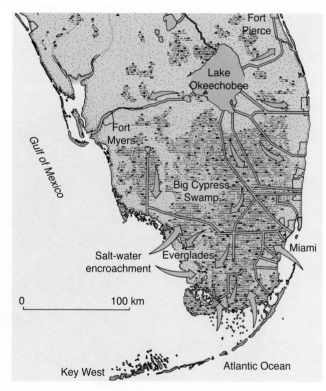

(B) Canals diverted the natural flow of surface water across the Everglades. The water table was lowered, the swamp was destroyed in some areas, and saltwater encroached in wells along the coast.

Figure 13.29 Modification of the natural drainage system of southern Florida.

This area, which lies in the rain shadow of the Cascade Mountains, receives only 15 to 25 cm of precipitation a year. In recent years, extensive irrigation has caused the water table to rise, introducing many changes in the surface conditions. Today, from 100 to 150 cm of water is applied each year to the ground by irrigation, which simulates the effect of a large climatic change. The higher water table has rapidly developed large springs along the sides of river valleys. The springs are now permanent, reflecting saturation of much of the ground. Erosion is accelerated, and many farms and roads have been damaged severely. Landslides present the most serious problems. Slopes that were stable under arid conditions are now unstable because they are partly saturated from the high water table and from the formation of perched water bodies.

In many areas, it is imperative that we modify the environment by reclaiming land or by irrigation; unless we are careful, however, the detrimental effects of our modifications may outweigh the advantages. Before we modify an environment, we must attempt to understand the many consequences of altering natural systems.

Subsidence

Surface **subsidence** related to groundwater can result from natural Earth processes, such as the development of sinkholes in a karst area or from the artificial withdrawal of fluids. An ever-present hazard in limestone terrains is the collapse of subterranean caverns and the formation of sinkholes (Figure 13.17B). Buildings and roads have frequently been damaged by sudden collapses into previously undiscovered caverns below. In the United States, important karst regions appear in central Tennessee, Kentucky, southern Indiana, Alabama, Florida, and Texas (Figure 13.19). The problem of potential collapse is difficult to solve. Important construction in karst regions should be preceded by test borings to determine whether subterranean cavernous zones are present. Geophysical studies using ground-penetrating radar and seismic investigations can be used to detect some shallow caverns and soil cavities. Wet concrete can be pumped down into caves and solution cavities, but such remedies can be very expensive.

Compaction and subsidence also present serious problems in areas of recently deposited sediments. In New Orleans, for example, large areas of the city are now 4 m below sea level, a drop due largely to the pumping of groundwater. As a result, the Mississippi River flows some 5 m above parts of the city, and rainwater must be pumped out of the city at considerable cost. Also, as the surface subsides, waterlines and sewers are damaged.

Where groundwater, oil, or gas is withdrawn from the subsurface, significant subsidence can also occur, damaging construction, water supply lines, sewers, and roads. Long Beach, California, has subsided 9 m as the result of 40 years of oil production from the Wilmington oil field. This subsidence resulted in almost $100 million worth of damage to wells, pipelines, transportation facilities, and harbor installations. Parts of Houston, Texas, have subsided as much as 1.5 m as a result of the withdrawal of groundwater.

Probably the most spectacular example of subsidence is Mexico City, which is built on a former lake bed. The subsurface formations are water-saturated clay, sand, and volcanic ash. The sediment compacts as groundwater is pumped for domestic and industrial use, and slow subsidence is widespread. The opera house (weighing 54,000 metric tons) has settled more than 3 m, and half of the first floor is now below ground level. Other large structures are noticeably tilted (**Figure 13.30**).

Another type of groundwater problem is shown in **Figure 13.31**. In western Wyoming, a dam for storing irrigation water was built in the tilted strata of the Madison Limestone Formation. The limestone, however, was so porous and permeable that all of the water that was supposed to be stored in the reservoir seeped into the subsurface and was lost. The reservoir never filled, and the project was abandoned.

We are using and altering the groundwater system at an ever-increasing rate.

Approximately 20% of all water used in the United States is pumped from the subsurface. This amounts to more than 83 billion gallons a day, almost three times as much as in 1950. How much groundwater will be needed in the future? What effect will pumping groundwater have on the environment?

Figure 13.30 **Subsidence of buildings in Mexico City** resulted from compaction after groundwater was pumped from unconsolidated sediment beneath the city. Subsidence has caused this building to tilt and sink more than 2 m.

Figure 13.31 **A dam constructed on permeable limestone** in western Wyoming never functioned because the surface water seeped into the subsurface. The dam lies at the beginning of the gorge. The light-colored sediment behind the dam marks the fraction of the reservoir that formed before water was lost through seepage.

GeoLogic   Cave Systems of Guilin, China

One of the most exotic landscapes in the world is the tower karst in southern China. Here, groundwater is a powerful agent of erosion and forms vast cave systems largely hidden from view. Openings to many caves occur on hillsides but they only give a hint to the extent of the cave system within the mountains. What were the major controlling factors in their formation?

Observations

1. Limestone, a soluble rock, is the dominant rock type in this region.
2. The hills are unusual steep-sided, conical mounds.
3. Intersecting sets of fractures cut the rocks and bedding planes are etched into the rocks.
4. The hills are remnants of the once extensive limestone layer.
5. The hillside is riddled with openings to caves—some go through the entire hill.

6. The abundance of water and vegetation show that the climate is humid.

Interpretations

Taken together, these observations lead to the logical conclusion that this landscape was shaped largely by groundwater solution. In this humid climate, surface water seeped into intersecting joints and enlarged them by solution activity. This separated the once continuous limestone layer into numerous conical towers. Caves and smaller openings on the face of the hills indicate extensive cavern systems inside the hills. In many cases, cave exploration has verified this conclusion. The caverns were also dissolved by groundwater percolating along fractures and bedding planes. The diagram illustrates how a geologist might view this area. Without abundant water none of these processes would be possible and the limestone would resist weathering and erosion.

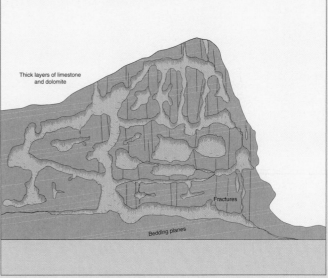

Key Terms———————————————————————————

aquifer (p. 362)

artesian water (p. 368)

cave (p. 375)

cone of depression (p. 367)

confined aquifer (p. 364)

disappearing stream (p. 375)

discharge (p. 360)

geothermal energy (p. 371)

geyser (p. 371)

groundwater (p. 360)

hydraulic head (p. 364)

hydrostatic pressure (p. 361)

karst topography (p. 375)

leach (p. 386)

perched water table (p. 362)

permeability (p. 361)

pore spaces (p. 360)

porosity (p. 360)

potentiometric surface (p. 368)

recharge (p. 360)

saltwater encroachment (p. 387)

seep (p. 365)

sinkhole (p. 373)

solution valley (p. 375)

speleothem (p. 382)

spring (p. 365)

stalactite (p. 382)

stalagmite (p. 382)

subsidence (p. 391)

tower karst (p. 375)

unconfined aquifer (p. 362)

water table (p. 362)

zone of aeration (p. 362)

zone of saturation (p. 362)

Review Questions

1. Define porosity and permeability.
2. Describe and illustrate the major types of pores, or voids, in rocks.
3. What rock types are generally impermeable or nearly impermeable?
4. Describe the major zones of subsurface water, and explain how water moves through each zone.
5. Contrast the geologic conditions that form confined and unconfined aquifers.
6. What is a potentiometric surface? How does it control the movement of water in confined aquifers?
7. Explain some ways in which springs originate.
8. What effects are produced in the water table by excessive and rapid pumping?
9. Explain how a flowing (artesian) well develops.
10. Explain the origin of geysers.
11. What is the source of heat for hot springs and geysers?
12. Describe the evolution of a landscape in which groundwater is the dominant agent of erosion.
13. Explain how stalagmites and stalactites originate.
14. What important resources are related to groundwater systems?
15. Describe the forms and processes of groundwater pollution.
16. Describe the relationship between salty groundwater and fresh groundwater beneath an island or a peninsula.
17. What undesirable effects can result from withdrawing an excessive amount of groundwater from wells close to the ocean?
18. Explain how the alteration of the natural drainage system in southern Florida has affected the Everglades.
19. How can subsidence of the land result from the withdrawal of groundwater? Give examples.

14 Glacier Systems

No event in recent geologic history has had as profound an effect on Earth as the last great ice age. Its impact extended far beyond the margins of the ice itself and influenced almost every aspect of the physical and biological world. For example, the present sites of many northern cities, such as Chicago, Detroit, Montreal, and Toronto, were buried beneath thousands of meters of glacial ice as recently as 15,000 to 20,000 years ago.

Much of the magnificent mountain scenery in Canada's Yukon Territory shown above was sculpted by valley glaciers, many of which still exist. The Kaskawulsh Glacier and its tributaries have cut the deep U-shaped valleys, carved horns, and transported the eroded debris downhill to be carried away by melt waters and river systems. As we will see in this chapter, the great continental glaciers, which covered much of North America and Europe, had an even more profound effect upon the landscape.

When glaciation occurs, many geologic processes are interrupted or modified significantly. Much precipitation becomes trapped in glaciers instead of flowing immediately back to the ocean.

Consequently, sea level drops and the hydrology of streams is greatly altered. As gigantic ice sheets advance over continents, they obliterate preexisting drainage networks. The moving ice scours and erodes the landscape and deposits the debris near its margins, covering the preexisting topography. The crust of Earth is pushed down by the weight of the ice, and meltwater commonly collects and forms lakes along the ice margins. As the glaciers melt, new drainage systems are established to accommodate the large volume of meltwater. Far beyond the margins of the glaciers, stream systems are modified by changing climatic patterns. Even in arid regions, the imprint of climatic changes associated with glaciation is seen in the development of large lakes in closed basins.

We now know that glacial epochs have come and gone repeatedly over the last few million years. Today, the planet basks in the relative warmth of an interglacial period, but it has been cyclically plunged into cold episodes. Will there be another ice age?

In this chapter, we study how glaciers operate as systems of flowing ice and how they modify the landscape. We then consider the causes of an ice age, which remain tantalizing questions still partly unanswered.

Major Concepts

1. Glaciers are systems of flowing ice that form where more snow accumulates each year than melts.
2. As ice flows, it erodes the surface of the land by abrasion and plucking. Sediment is transported by the glacier and deposited where the ice melts. In the process, the landscape is greatly modified.
3. The two major types of glaciers—continental and valley glaciers—produce distinctive erosional and depositional landforms.
4. The Pleistocene ice age began 2 to 3 million years ago. During the ice age, there were several glacial and interglacial epochs. The last glacial maximum was about 18,000 years ago and glaciers have been receding since then.
5. The major effects of an ice age include glacial erosion and deposition, modification of drainage systems, creation of numerous lakes, the fall of sea level, isostatic adjustments of the lithosphere, and migration and selective extinction of plant and animal species.
6. Periods of glaciation have been rare events in Earth's history. The causes of glacial episodes are not completely understood, but they may be related to several simultaneously occurring factors, such as astronomical cycles, plate tectonics, and ocean currents.

Glacier Systems

A glacier is an open system of flowing ice. Water enters the system as snow, which is transformed into ice by compaction and recrystallization. The ice then flows through the system, under the pressure of its own weight, and leaves the system by evaporation and melting. The balance between the rate of accumulation and the rate of melting determines the size of the glacial system.

Glacial Ice

A glacier is a natural body of ice formed by the accumulation, compaction, and recrystallization of snow that is thick enough to flow. It is a dynamic system involving the accumulation and transportation of ice. The movement of the ice is a critical factor. A mass of ice must move or flow to be considered a glacier. Bodies of ground ice, formed by the freezing of groundwater within perennially frozen ground, are not glaciers, nor is the relatively thin sheet of frozen seawater known as sea ice, which is so abundant in the polar regions. Perennial snowfields that do not move are also not considered glaciers. Glacial ice is really a type of metamorphic rock that begins as sediment (an aggregate of mineral particles, or snow) and is then metamorphosed by compaction and recrystallization into glacial ice.

The essential parts of a glacial system are (1) the **zone of accumulation**, where there is a net gain of ice, and (2) the **zone of ablation**, where ice leaves the system by melting, calving (shedding of large blocks of ice from a glacier edge, usually into a body of water), and evaporating (**Figure 14.1**). The boundary between these zones is the **snow line**. In the zone of accumulation, snow is transformed into glacial ice. Freshly fallen snow consists of delicate hexagonal ice crystals or needles, with as much as 90% of their total volume as empty space (**Figure 14.2**). As snow accumulates, the ice at the points of the snowflakes melts from the pressure of snow buildup and migrates toward the center of the flake, eventually forming an elliptical granule of recrystallized ice approximately 1 mm in diameter. The accumulation of these particles packed together is called firn, or névé. With repeated annual deposits, the loosely packed névé granules are compressed by the weight of the overlying snow. Meltwater, which results from daily temperature fluctuations and the pressure of the overlying

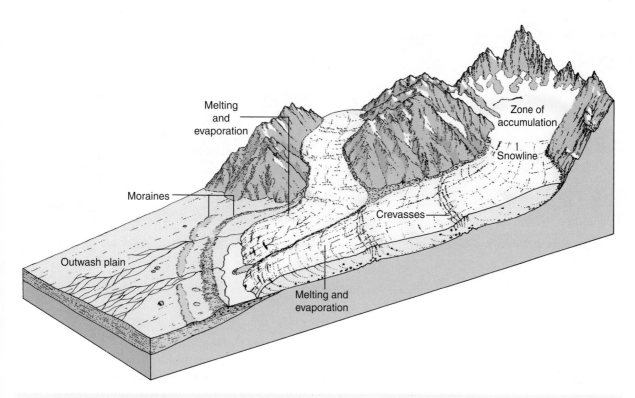

Figure 14.1 **A glacial system** is an open system of ice that flows under the pull of gravity. Snow enters the system by precipitation and is transformed into ice. The ice flows outward from the zone of accumulation under the pressure of its own weight. The ice leaves the system by evaporating and melting in the zone of ablation. The boundary between these zones is approximated by the snow line. As the ice moves through the system, it erodes and deposits sediment at the end of the glacier.

Figure 14.2 **Snow crystals** change to granular ice by partial thawing and refreezing along their delicate edges. Burial produces compaction and recrystallization, cementing all the grains into glacial ice.

Scanning electron microscope images courtesy of E. Erbe, U.S. Department of Agriculture

snow, seeps through the pore spaces between the grains; when it freezes, it adds to the recrystallization process. Most of the air in the pore spaces is driven out. When the ice reaches a thickness of approximately 60 m, it can no longer support its own weight and yields to plastic flow.

Types of Glaciers

There are two main types of glaciers: (1) valley glaciers and (2) continental glaciers (**Figure 14.3**). **Valley glaciers** are ice streams that are confined to the canyons and valleys of mountainous terrains. They originate in snowfields at the mountain crest and flow down the canyons. A valley glacier that emerges from the mountain front and spreads out as a large lobe at the foot of the mountain is commonly called a piedmont

Courtesy of M. F. Sheridan

(A) Continental glaciers are huge ice sheets that cover a large part of a continent such as Antarctica, shown here.

(B) Valley glaciers are streams of ice that flow down canyons and the valleys of mountainous terrains.

Figure 14.3   Types of glaciers can be recognized by their geometry and relationship to topography.

glacier. **Continental glaciers** are huge sheets of ice that spread out over a large part of the continent. They are commonly more than 3000 m thick and completely cover the underlying terrain, except for the peaks of the highest mountains. These huge sheets of ice generally flow outward in all directions from one or several central regions of accumulation. Flow directions may be influenced by subglacial topography such as highlands and mountain ranges. In rugged terrain, the direction and rate of ice flow may be greatly influenced by mountain ranges, and the ice is funneled through mountain passes in large streams called outlet glaciers. Antarctica and Greenland are present-day examples of continental glaciers with a maximum thickness approaching 5000 m.

Glacier Flow

Ice is a brittle substance and, when struck sharply, it will fracture and break; however, like many substances that are normally regarded as solids, ice will flow if adequate stress is applied over an extended period of time. Gravity is the fundamental force that causes ice to flow, and where an accumulation of ice exceeds a depth of about 60 m, depending on temperature, slope angle, and so on, flow is initiated. Ice has a much higher viscosity, or resistance to flow, than liquid water. As a result, its flow is not turbulent, but rather the flow of glacial ice is laminar. That is, the planes of flow are parallel. The planes curve but do not intersect or cross.

Several mechanisms have been observed by which ice undergoes solid-state flow (**Figure 14.4**). In ice composed of loose individual granules, the shifting and rotation of grains can produce a flowlike movement similar to that of sand being poured from a bucket. In glaciers, the ice crystals are packed tightly together, so this type of flow is minimal. Stress exerted on intergrown crystals causes them to melt at points where pressure is concentrated. The water is then moved to areas of lower pressure and refreezes. Another mechanism involves minor displacement along a series of parallel slip planes within individual ice crystals. Thin layers of ice move past each other as a sheared deck of cards would.

The ice in most glaciers normally moves too slowly for us to see at any given moment, but a simple experiment can demonstrate the movement of glacial ice. We need only lay a series of boulders in a straight line across a glacier from wall to wall, and within a year or two the line will no longer be straight (**Figure 14.5**). This was done in the Alps at least as early as the 1800s. More sophisticated measurements show that various parts of glaciers move at different rates. One method to observe this difference is to drill a vertical hole through the glacier, insert a flexible pipe, and then survey the pipe's position and inclination over a period of several years (Figure 14.5).

(A) Rotation of grains

(B) Melting and freezing

(C) Internal slipping

Figure 14.4   Mechanisms of ice flow involve (A) rotation of grains, (B) melting and freezing, and (C) internal slipping within the ice mass.

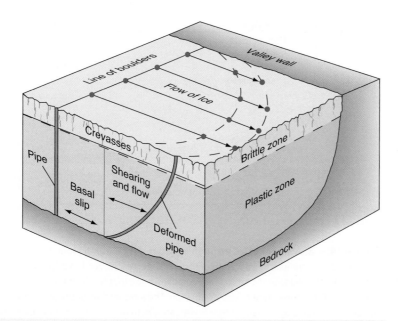

Figure 14.5 **Ice flow** in a glacier can be shown by laying a series of rocks in a straight line across the glacier and observing their position in subsequent years. The displaced boulders show that the center of the glacier, where ice is thickest, moves most rapidly. A glacier moves by basal slip across the underlying bedrock and by internal flow. The upper part of the glacier moves faster than the basal part, which drags against bedrock. The upper part of the glacier is brittle and as a result has many fractures or crevasses.

The results show that the pipe not only moves down the valley, but it also bends into a curve, with greater movement at the top than at the bottom. In addition, the total downslope movement is greater than that accounted for by bending. This result indicates that along with internal flow, the glacier also slips over the underlying bedrock surface. Up to 95% of the movement in a glacier can occur by basal slip. Basal slip is accomplished by melting and freezing of the ice near the contact with the bedrock. Indeed, a relatively warm glacier is not everywhere in firm contact with its bed, but it is locally separated from the bedrock below by lubricating pockets of water.

Direction and Amount of Movement   The movement of glacial ice can best be understood by considering what happens within the zone of accumulation and the zone of ablation of a valley glacier that is in a steady state; that is, the size of the glacier is neither shrinking nor expanding. Each year, a wedge-shaped layer of snow, thickest at the head and thinnest at the snow line, is added to the surface of accumulation (**Figure 14.6**). A similar wedge of ice, thickest at the end of the glacier and thinnest at the snow line, is removed by melting. If a glacier is at equilibrium, the volume of water represented by these two wedges must be the same. It cannot keep getting thicker at the head and thinner at the snout. It adjusts to the accumulation and removal of ice by changing the inclination of the direction of flow (Figure 14.6). In the zone of accumulation, the direction of movement is inclined downward, with respect to the surface of the glacier. The degree of downward inclination decreases from the head of the glacier to the snow line. At the snow line, the direction of movement is parallel to the surface of the glacier. In the zone of ablation, the movement is upward, toward the surface, with upward inclination increasing from the snow line to the snout. As shown in Figure 14.6, the ice at the head of the glacier flows downward through the glacier. This same ice is near the base of the glacier when it passes the snow line. It then flows upward and laterally to the snout.

Extending and Compressing Flow   The movement of glacial ice is not uniform. The vector lines in Figure 14.6 show that velocities of ice flow in the zone of accumulation increase progressively from the head to the snow line. Here, the ice is under tension

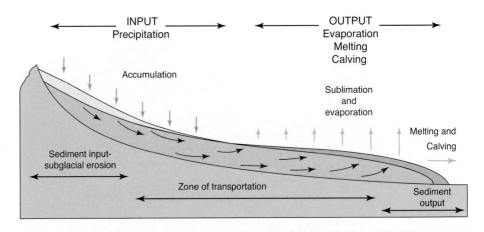

Figure 14.6 **Movement of ice through a glacier** is shown in this longitudinal cross section. In the zone of accumulation, a wedge of snow is added each year. It is thickest at the head and thins to a feather edge at the snow line. A similar wedge that is thin at the snow line and thickest at the end of the glacier is removed by ablation. The internal flow of the glacial ice resulting from this accumulation and wastage is shown with arrows. In the zone of accumulation, the ice moves downward. Near the snow line, movement is horizontal. Near the lower end of the glacier, the movement of the ice is upward.

and is constantly pulling away from upvalley ice. This is the condition of **extending flow**. Below the snow line, velocities progressively decrease; therefore, upvalley ice is continually pushing against downvalley ice. This is a condition of **compressing flow**. Where bedrock slopes steepen, glacier velocities increase and extending flow prevails; where the bedrock slopes are gentle, velocities decrease and compressing flow occurs (**Figure 14.7**). Where glaciers descend over extremely steep slopes, the ice descends with high velocities, creating a veritable icefall. These are zones of extreme extending flow, and the ice is greatly thinned and completely broken by numerous deep crevasses (see Figure 14.14B). The flow velocity in an icefall can exceed 10 times that of the glacier elsewhere along its course. At the base of an icefall, conditions are reversed; flow decreases rapidly, compressing flow dominates, and the glacier thickens.

In continental glaciers, the flow is radially outward from the zone of accumulation or from broad domes of maximum thickness. Movement, however, is strongly influenced by the configuration of the subglacial landscape. Preglacial valleys will channelize the flow of ice and greatly influence both direction and rate of flow. Domes or highlands, in contrast, will act as barriers and inhibit flow.

How fast do glaciers flow?

Velocity Variations   The flow of glacial ice, like that of running water in streams and rivers, is not constant, but varies significantly with time and place. Ice flow in a glacier may seem extremely slow compared with the flow of water in rivers, but the movement is continuous, and over the years, vast quantities of ice can move through a glacier. Measurements show that some of Switzerland's large valley glaciers move as much as 180 m/yr. Smaller glaciers move from 90 to 150 m/yr. Some of the most rapid rates have been measured on the outlet glaciers of Greenland, where ice is funneled through mountain passes at a speed of 8 km/yr. From these and other measurements, flow rates of a few centimeters per day appear common, and velocities of 3 m/day are exceptional.

Surging Glaciers   An extremely rapid flow of glacial ice, with velocities more than 100 times normal, is referred to as a *glacial surge*. Flow is extremely rapid, with daily advances of more than 90 m. One of the most rapid surges on record was observed in the Kutiak Glacier in Asia, where the glacier advanced 12 km in 3 months. In fall 1993, the Bering Glacier in Alaska surged 225 m in a day. Most glaciers occur in remote areas, so few surges were well documented in the past; today, however, satellite

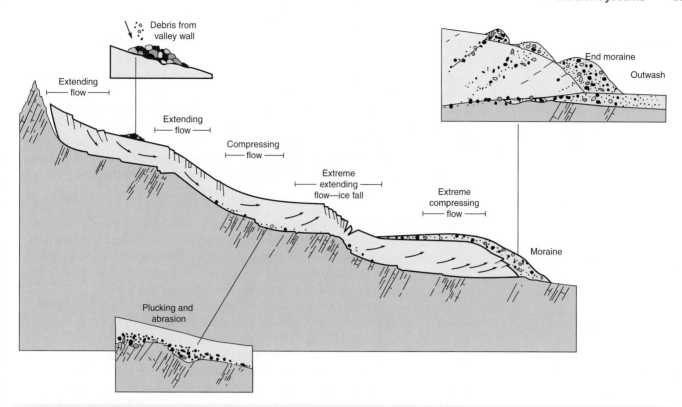

Figure 14.7 Extending and compressing flow result from variations in velocity. Where velocity increases, a glacier is extended (stretched) and thinned. Where velocity decreases, the glacier is compressed and thickened. Fast-flowing ice over a steep surface results in extreme extending flow. Gentle gradients produce compressing flow. As the ice moves, it erodes and transports sediment in various ways. Rockfalls from the valley walls add sediment to the surface of the ice. Plucking and abrasion erode bedrock at the base of the glacier. Near the terminus, this sediment load is transported upward by the flowing ice and deposited as the ice ablates.

imaging monitors the flow velocities of glaciers throughout the world. Preliminary studies indicate that glacier surges are much more common than previously thought.

It is believed that glacial surge results from basal slip, as water gradually accumulates in small, interconnected cavities under the ice. Under such conditions, basal hydraulic pressure could increase to the point at which the glacier is locally raised a centimeter or two off its bed. Raising of the glacier could greatly increase basal slip and initiate a surge. Glacial surges may also result from sudden slippage along the base of a glacier, caused by the buildup of extreme stress upstream. Stagnant or slow-moving ice near the terminus can act as a dam for the faster-moving ice upstream. If this damming happens, stress builds up behind the slow-moving ice, and a surge occurs when a critical point is reached. Surges can also be caused by a sudden addition of mass to the glaciers, such as a large avalanche or landslide on its surface. Glacier surge is more than a feature of academic interest. Where glaciers reach the sea, surging glaciers can create many times more icebergs than normal, which constitute hazards in shipping lanes. Greenland, for example, discharges on the order of 10,000 icebergs each year into the Atlantic Ocean. If this amount were suddenly increased 100 times, it could have a significant effect on shipping lanes.

Crevasses

The most obvious and abundant structures in a glacier are **crevasses**—large cracks opened by the fracturing of a brittle upper layer of ice as the underlying ice continues to flow (Figure 14.5). Crevasses are nearly vertical and may be more than 30 m deep and thousands of meters long. Crevasses are tensional fractures produced by differential motion in the ice (see Figure 14.14B). Almost any part of a glacier involved in differential flow velocities is likely to develop crevasses transverse to the direction of ice flow. Marginal crevasses are present in almost all valley glaciers, along their lateral

Why do crevasses form only in the upper part of glaciers?

margins, because the ice drags along the valley walls (see Figure 14.11). These crevasses are usually short and point upstream. Transverse crevasses form at right angles to the direction of flow, where flexing of the ice occurs as the glacier moves over bumps or ridges on the bedrock floor. Similarly, icefalls are intensely crevassed by the greatly accelerated rate of flow as the ice moves down a steep slope (Figure 14.7). Longitudinal crevasses develop at the terminus of a glacier, where the ice stream spreads out, setting up tensional stresses at right angles to the flow direction. Radial crevasses are similar but form a radial pattern where the ice spreads out in a lobate pattern. Crevasses allow geologists to study the interior of a glacier, but they are extremely hazardous because they may become bridged over with snow, forming veritable ice death traps.

Ablation

The zone of ablation is where ice leaves the system by melting, evaporating, and calving (Figure 14.6). Melting, of course, is a major process. It is influenced by many complex factors, such as cloud cover, air temperature, rain, dust, and dirt on the surface of the glacier. Surface rock debris can significantly influence melting because the darker rock absorbs much more solar radiation than the lighter ice and snow.

Anyone who has visited a glacier during the summer is impressed with the large amount of meltwater. Melting occurs not only at the end of a glacier (see Figure 14.14C), but over its entire surface. When meltwater is abundant, it percolates into the crevasses and pore spaces between the ice grains, creating a zone of saturation within the glacier. A water table is thus created in the glacier and is commonly seen a few meters below the surface in many crevasses. Near the snow line, a thin layer of snow covers the impervious glacial ice below, and the concentration of meltwater may create snow swamps. Water derived from these swamps collects into a surface drainage system in which streams may cut steep-walled channels tens of meters deep. Velocity in these streams may be abnormally fast because the smooth ice surface of the channel offers minimal resistance to flow.

A surface stream may disappear down a large cylindrical hole in the ice and into a system of subglacial tunnels (see Figure 14.11). Subglacial tunnels are largest and most numerous near the end of a glacier. Some are tightly confined and may be completely full of meltwater and operate under pressure like domestic water pipes. Where the water is brought back to the surface, it may emerge with enough force to form a geyserlike eruption. The amount of water lost by melting is apparent in the expanse of braided streams in the outwash plain beyond the glacier.

Calving occurs primarily where the glacier enters the sea and is broken into large fragments that float away as icebergs and ultimately melt (see Figure 14.15). Most of the ice on the entire Antarctic continent reaches the sea; some of it extends over the ocean's surface as a floating ice shelf. Calving is thus a major form of wastage for Antarctic glaciers, as huge tabular icebergs break away from the shores and drift northward. Calving is also a major process of wastage in valley glaciers that reach the sea.

In the zone of ablation, only a small volume of ice changes from the solid state directly to the vapor state. This accounts for less than 1% of the total ablation. In most glaciers, melting and calving are the dominant processes of ablation.

Glacial Equilibrium

Does ice flow through a glacier even though the end of the glacier is receding?

Glaciers are open systems and have much in common with other gravity flow systems, such as rivers and groundwater. Water enters the system primarily in the upper parts of the glacier, where snow accumulates and is transformed into ice. The ice then flows out of the zone of accumulation. At the glacier's lower end, or terminus, ice leaves the system by melting, calving, and evaporating. For most glaciers, ice accumulation dominates during winter when snowfall is greatest, and ablation is highest during spring and summer. The annual difference between accumulation and ablation on a glacier is the net mass balance. If more snow is added in the zone of accumulation than is lost by melting or evaporation at the end of the glacier, the ice mass increases and the glacial system expands. If the accumulation of ice is less than ablation, there is a net loss of mass and the size of the glacial system is reduced. If accumulation and ablation are in balance, the mass of ice remains constant, the size of the system remains constant, and the terminus of the ice remains stationary. It is important to understand that the margins of a glacier constitute the boundaries of a system of flowing ice, much as the banks and mouth of a river constitute the boundaries of a river system. Ice within the glacier continually flows toward the terminus, or terminal margins, regardless of whether the terminal margins are advancing, retreating, or stationary.

The behavior of a glacial system (the size of the mass of ice) is determined by the balance between the rate of input and the rate of output of ice. The two major variables in this balance are temperature and precipitation. A glacier can grow or shrink with an unchanging rate of precipitation if the temperature varies enough to increase or decrease the rate of melting (rate of output). The size of a glacier in no way represents the amount of ice that has moved through the system, just as the length of a river does not represent the volume of water that has flowed through it. Size simply shows the amount of ice currently in the system.

An example, from the last ice age, illustrates this point. A glacial valley 20 km long in the Rocky Mountains was eroded 600 m deeper than the original stream valley. This large amount of erosion was not accomplished by 20 km of ice moving down the valley. It was the result of many thousands of kilometers of ice flowing through the valley. If the ice occupied the valley during each glacial epoch and moved 0.3 m/day, a total of approximately 72,000 km of ice would have moved down the valley. Yet, the glacier was never more than 20 km long. The enormous abrasion caused by such a long stream of ice would be able to wear down the valley to a depth of 600 m.

Erosion

Continental ice sheets and valley glaciers are powerful agents of erosion. An ice sheet may erode its base at a rate of 0.1 to 0.35 mm/yr. The North American ice sheet may have eroded as much as 1 to 2 m of bedrock during the last glacial cycle and tens of meters during the entire series of glacial advances. Because of the cold temperature accompanying glaciation, ice wedging contributes to the process. Wherever hills or mountains stand higher than the surface of the glacier, intense ice wedging occurs, loosening blocks of rock that then roll onto the surface of the glacier, which carries them away. Indeed, ice wedging is responsible for much of the detailed form of the sharp, jagged peaks that characterize glaciated mountains (Figure 14.3).

Glacial plucking is the lifting out and removal of fragments of bedrock by the moving ice (Figure 14.7). It is one of the most effective ways in which a glacier erodes the land. The process involves ice wedging. Beneath the glacier, meltwater seeps into joints or fractures, where it freezes and expands, wedging loose blocks of rock. The loosened blocks freeze to the bottom of the glacier and are plucked, or quarried, from the bedrock, becoming incorporated in the moving ice. The process is especially effective where the bedrock is cut by numerous joints and where the surface of the bedrock is unsupported on the downstream side.

Abrasion is essentially a filing process. The angular blocks plucked and quarried by the moving ice freeze firmly into the glacier; thus firmly gripped, they are ground against the bedrock over which the glacier moves (Figure 14.7). The process is similar to the rasping action of a file or sandpaper. Rivers, wind, and waves do not have this ability to grasp and use rock fragments as a rasp. The process is a trademark of glaciers. Aided by the pressure of the overlying ice, the angular blocks are very effective agents of erosion, capable of wearing away large quantities of bedrock. The fragments become abraded and worn down as they grind against the bedrock surface. As a result, glacial boulders usually develop flat surfaces that are deeply scratched.

Evidence of the distinctive abrasive and quarrying action of glaciers can be seen on most bedrock surfaces over which glacial ice has moved. Hills of bedrock (**roches moutonnées**) commonly are streamlined by glacial abrasion. Their upstream sides typically are rounded, smoothed, striated, and locally polished by abrasion, while their downstream sides are made steep and rugged by glacial plucking (**Figure 14.8**).

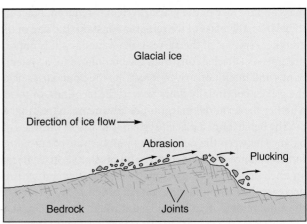

Figure 14.8 A roche moutonnée, like this one in a fjord in Norway, is an erosional feature that forms as ice moves over bedrock, eroding it into a streamlined shape by a combination of abrasion and plucking. Abrasion produces a smooth, grooved, and striated surface on one flank; plucking develops a jagged irregular cliff on the downstream flank. Roches moutonnées range from small knobs a few meters high to major domes more than 200 m high.

Roches moutonnées range in size from small knobs a few meters long to large hills that rise 50 to 200 m above the surrounding landscape. Many are elongate in the direction of the ice flow and are best developed in resistant but jointed rock. *Whalebacks* are similar to roches moutonnées but are smooth, elongate features that are typically grooved and polished. They lack the rugged downstream face produced by glacial plucking and perhaps are more common than roches moutonnées.

Glacial striations are numerous, small scratches, a fraction of a millimeter deep and several tens of centimeters long. They are formed by angular sand-sized particles dragged across the rock surface by the flowing ice. The striations are parallel to the direction of ice movement. When used in conjunction with other features like moraines and drumlins, they reveal the direction that the ice flowed. Glacial polish is produced where very fine debris is incorporated into the basal ice. Glacial grooves are similar to striations but are larger, longer, and deeper. They are distinctively linear and U-shaped, with a smooth base, walls, and rounded edges (**Figure 14.9**). Typical grooves are 10 to 20 cm wide and 50 to 100 m long. Exceptional grooves, 30 m deep and 12 km long, are found in northwestern Canada.

Figure 14.9 Glacial grooves and striations result from the abrasive action of a glacier and (when used in conjunction with other features) clearly show the direction in which the ice moved across this landscape in southern Ontario.

Valley Glacier Systems

Erosion and deposition by valley glaciers produce many distinctive land-forms, the most important of which are: (1) U-shaped valleys, (2) cirques, (3) hanging valleys, (4) horns, (5) moraines, and (6) outwash plains.

Valley glaciers are responsible for some of the most rugged and scenic mountainous terrain on Earth. The Alps, the Sierra Nevadas, the Rockies, and the Himalayas were all greatly modified by glaciers during the last ice age, and the shapes of their valleys, peaks, and divides retain the unmistakable imprint of erosion by ice.

As a result, they have been studied for many years, and their general characteristics are well understood. Valley glaciers are long, narrow streams of ice that originate in the snowfields of high mountain ranges and flow down preexisting stream valleys (**Figure 14.10**). They range from a few hundred meters to more than a hundred kilometers in length. In many ways, they resemble river systems. They receive an input of water (in the form of snow) in the higher reaches of mountains, and they have a system of tributaries leading to a main trunk system. Their flow direction is controlled by the valley the glacier occupies, and as the ice moves, it erodes and modifies the landscape over which it flows. Unlike a stream of liquid water, the ice may be as much as 1000 m deep and flow slowly, perhaps a fraction of a meter a day. As the ice moves, it picks up rock and debris along its margins from abrasion, forming a marginal zone of dirty ice. In addition, the mass movement of rock debris from the valley walls above the glacier

Courtesy of NASA/USGS

Figure 14.10 Valley glaciers on Bylot Island just off the northern end of Canada's Baffin Island originate in the snowfields that almost completely cover the mountain peaks. Note that the snow line extends down almost to sea level. The main glaciers extend down from the highland as tongues of ice (blue). Note that glaciers, like river systems, consist of a main trunk stream and an intricate system of branching tributaries.

contributes to the rock debris along the ice margin. Below the snow line, the melting of the dirty ice concentrates the debris into a linear band along the side of the glacier that is called a **lateral moraine** (see Figure 14.14A). Where a tributary glacier joins the main stream, the two adjacent lateral moraines merge to form a **medial moraine**. Remember that the debris in a **moraine** represents only the outcropping of a band of dirty ice normally extending from the surface to the floor of the glacier (**Figure 14.11**). Thus, most valley glaciers are composed of multiple ice streams from tributary glaciers, separated by zones of dirty ice underlying the moraines at the surface. Downstream, the glacier undergoes progressive melting, and the morainal ridges become higher and broader.

If the floors of two merging glaciers join at the same level, the ice streams merge side by side, each extending from the surface to the floor of the valley, separated by a zone of dirty ice and debris. If a tributary glacier enters the main stream above the floor of the main glacier, the tributary glacier does not extend down to the floor of the main stream but rests above it.

The idealized diagrams in **Figure 14.12** and the photographs in **Figures 14.13** and **14.14** illustrate the major erosional landforms resulting from valley glaciation. Figure 14.12 permits a comparison and contrast of landscapes formed only by running water with those that have been modified by valley glaciers. Figure 14.12A shows the typical topography of a mountain region being eroded by streams. A relatively thick mantle of soil and weathered rock debris covers the slopes. The valleys are V-shaped, in cross section, and have many bends at tributary junctions so that ridges and divides between tributaries appear to overlap if you look up the valley. In Figure 14.12B, the valleys are shown occupied by glaciers. The growing glaciers expand down the tributary valleys and merge to form a major glacier.

A valley glacier commonly fills more than half of the valley depth, and as it moves, it modifies the former V-shaped stream valley into a broad U-shaped, or troughlike form. The head of the glacier is enlarged by plucking and grows headward toward the mountain crest to form a **cirque** (Figure 14.12). Where two or more cirques approach the summit crest, they sculpt the mountain crest into a sharp, pyramid-shaped peak, called a **horn**. The projecting ridges and divides, between glacial valleys, are subjected to rigorous ice wedging, abrasion, and mass movement. In contrast to the rounded topography developed by stream erosion, these processes produce sharp, angular crests and divides, called **arêtes**.

What unique landforms result from sediment deposited by valley glaciers?

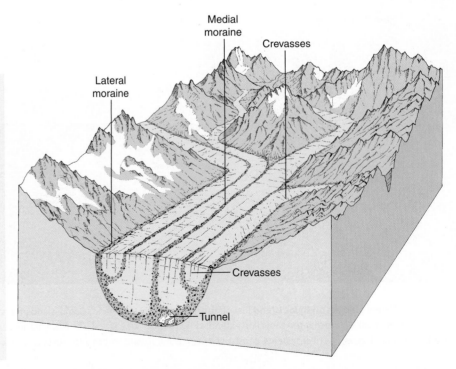

Figure 14.11 The internal structure of a valley glacier consists of the merging of different ice streams. Each tributary is separated from the others by morainal debris or dirty ice. Crevasses develop in the upper, brittle part of the glacier by differential motion in the ice. Common types of crevasses in valley glaciers include marginal crevasses due to drag along the valley walls, transverse crevasses that form as the glacier moves over irregularities on the bedrock floor, and longitudinal crevasses that develop at the end of a glacier where the ice stream begins to spread out. Meltwater may accumulate in marginal lakes or flow in short streams before seeping into the interior of the ice. Much of the meltwater moves through tunnels in or at the base of the ice.

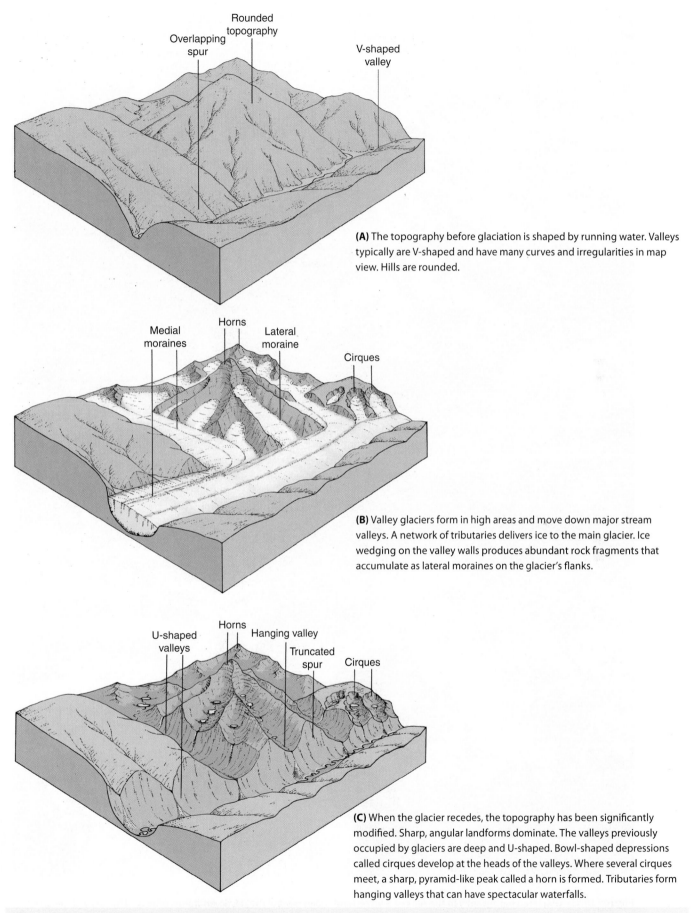

(A) The topography before glaciation is shaped by running water. Valleys typically are V-shaped and have many curves and irregularities in map view. Hills are rounded.

(B) Valley glaciers form in high areas and move down major stream valleys. A network of tributaries delivers ice to the main glacier. Ice wedging on the valley walls produces abundant rock fragments that accumulate as lateral moraines on the glacier's flanks.

(C) When the glacier recedes, the topography has been significantly modified. Sharp, angular landforms dominate. The valleys previously occupied by glaciers are deep and U-shaped. Bowl-shaped depressions called cirques develop at the heads of the valleys. Where several cirques meet, a sharp, pyramid-like peak called a horn is formed. Tributaries form hanging valleys that can have spectacular waterfalls.

Figure 14.12 Landforms produced by valley glaciers constitute some of the most spectacular scenery in the world. In these diagrams, an idealized landscape formed by stream erosion is shown as it might appear before, during, and after glaciation.

Figure 14.13 Glaciated topography in the Alps of Switzerland shows most of the classic landforms produced by valley glaciers. Note the major U-shaped valley and its tributaries. Cirques and horns dominate the landscape in the background, where remnants of valley glaciers still exist. Compare the landforms shown here with those in Figure 14.12C.

(A) Lateral moraines form on the margins of valley glaciers. Where two glaciers merge, the lateral moraines merge to form a medial moraine.

(B) Crevasses are large cracks in the brittle upper part of a glacier. These transverse crevasses formed perpendicular to the flow direction as the glacier moved over bumps on the valley floor.

(C) The outwash plain is covered by braided streams emanating from the melting Nabesna glacier of the Yukon Territory.

(D) A terminal moraine forms a high arcuate ridge after a valley glacier receded in California's Sierra Nevada range.

(E) Extending flow forms crevasses and may occur where a glacier descends over a steep slope. Compressing flow is common at the end of the glacier, where stagnant ice prevents its forward motion.

(F) Glacial erosion features, such as the horns shown here, dominate glaciated mountain regions. The Grand Teton, Wyoming, was shaped by erosion during the ice age.

Figure 14.14 Valley glaciers produce distinctive erosion and deposition features.

What unique landforms result from erosion by valley glaciers?

Note that where tributaries enter the main glacier, the upper surfaces of the glaciers are at the same level. The main glacier, however, is much thicker, and it therefore erodes its valley to a greater depth than that of the tributary valleys. When the glaciers recede from the area, the floors of the tributary valleys will be higher than the floor of the main valley; the tributary valleys are therefore known as **hanging valleys**.

Part of a valley glacier's load consists of rock fragments that avalanche down the steep valley sides and accumulate along the glacier margins. Frost action is especially active in the cold climate of valley glaciers and produces large quantities of angular rock fragments. This material is transported along the surface of the glacial margins, forming conspicuous lateral moraines (Figure 14.11). Where a tributary glacier enters the main valley, the lateral moraine of the tributary glacier merges with a lateral moraine of the main glacier to form a medial moraine in the central part of the main glacier. In addition to transporting the load near its base, a valley glacier thus acts as a conveyor belt and transports a large quantity of surface sediment to the terminus. At the terminus, ice leaves the system through melting and evaporation, and the load is deposited as an **end moraine**. End moraines commonly block the ends of the valleys, so meltwater from the ice accumulates and forms ponds and lakes (Figure 14.14D). Downstream from the glacier, meltwater reworks the glacial sediments and redeposits them to form an outwash plain (Figure 14.14C).

Figures 14.12C and 14.13 show regions after glaciers have disappeared. The most conspicuous and magnificent landforms developed by valley glaciers are the long, straight, U-shaped valleys, or troughs. Many are several hundred meters deep and tens of kilometers long. The heads of glacial valleys terminate in large amphitheater-shaped or bowl-like cirques, which commonly contain small lakes.

The landforms that develop at the terminus of a valley glacier are illustrated in Figures 14.1 and 14.14. The **terminal moraine** characteristically extends in a broad arc, conforming to the shape of the terminus of the ice. It commonly traps meltwater and forms a temporary lake. If periods of stabilization occur during the recession of ice, **recessional moraines** may form behind the terminal moraine.

The great volume of meltwater released at the terminus of a glacier reworks much of the previously deposited moraine and redeposits the material beyond the glacier in an **outwash plain** (Figures 14.1 and 14.14C). Outwash sediment has all of the characteristics of stream deposits, and the sediment is typically rounded, sorted, and stratified.

Continental Glacier Systems

Continental glaciers greatly modify the entire landscape they cover. The flowing ice removes the soil and commonly erodes several meters of the underlying bedrock. Material is transported long distances and deposited near the ice margins, producing depositional landforms such as moraines, drumlins, eskers, kettles, lake sediment, and outwash plains. The preexisting drainage is disrupted or obliterated, so numerous lakes form after the ice melts.

In terms of their effect on the landscape and on Earth's hydrologic system, continental glaciers are by far the most important type of glacial system. These large ice sheets form in some of the most rigorous and inhospitable climates on Earth. Nonetheless, teams of scientists from various countries use modern technology to study existing continental glaciers in Canada, Greenland, and Antarctica. From these studies, we can construct a reasonably accurate model of an idealized continental glacial system and analyze how it operates (**Figure 14.15**).

What are the major elements in a continental glacier system?

The basic elements of a continental glacier are much the same as those in a valley glacier. Both systems have a zone of accumulation, where there is a net gain of ice from snowfall. The ice flows out from the zone of accumulation to the zone of ablation, where it leaves the system through melting, evaporation, and calving. A continental glacier is a roughly circular or elliptical plate of ice, rarely more than 3000 m thick. Ice does not have the strength to support the weight of an appreciably thicker

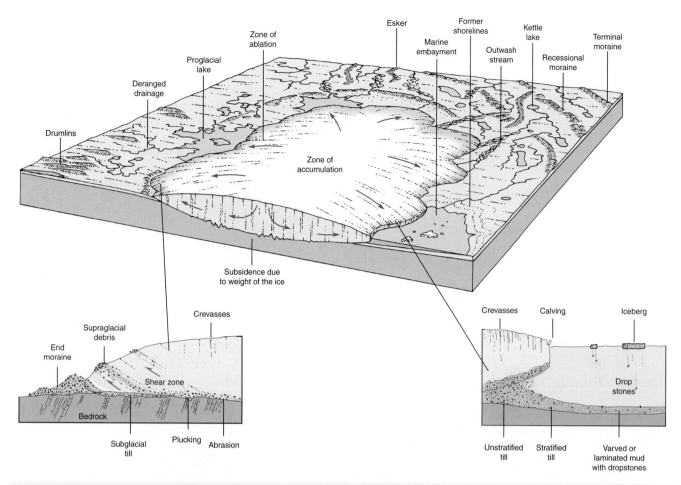

Figure 14.15 A continental glacier system covers a large part of a continent and causes significant changes across the entire landscape. The weight of the ice depresses the ground surface, so the land commonly slopes toward the glacier, and lakes form in the depressions along ice margins, or an arm of the ocean may invade the depression. The preexisting river systems are greatly modified, and streams that flow toward the ice margins are impounded to form lakes. The glacier advances more rapidly into lowlands, so the margins are not straight but are typically irregular or lobate. As the system expands and contracts, ridges of sediment are deposited along the margins, and a variety of erosional and depositional landforms develop beneath the ice. The balance between the rate of accumulation and rate of melting determines the size of the glacier.

accumulation. If more ice is added by increased precipitation, the glacier simply flows out from the centers of accumulation more quickly.

The weight of such a huge ice mass causes Earth's crust to subside, so the surface of the land commonly slopes toward the glacier. Subsidence creates lowland along the ice margin, which traps meltwater to form large lakes. If the margin of the glacier is near the coast, an arm of the sea may flood the depression.

Preexisting drainage systems are modified or completely obliterated. Rivers that flow toward the ice margins are impounded, forming lakes, which may overflow and develop a new river channel parallel to the ice margin. Drainage systems covered by the glacier are destroyed. Thus, when the ice melts, no established, integrated drainage system exists, so numerous lakes form in the natural depressions.

The margins of continental glaciers commonly form large lobes. These develop because the ice moves most rapidly into preexisting lowlands. The sediment deposited at the ice margins form arcuate or lobate terminal moraines. Erosion is at a minimum where the ice extends into the near polar regions because ice is frozen to the land surface. Most of the movement occurs in the middle of the sheet rather than at its bottom.

The Barnes Ice Cap of Baffin Island, Canada (**Figure 14.16**), is one of the last remnants of the glacier that covered much of Canada and parts of the northern United States only 14,000 years ago. This example illustrates the relationship between the continental

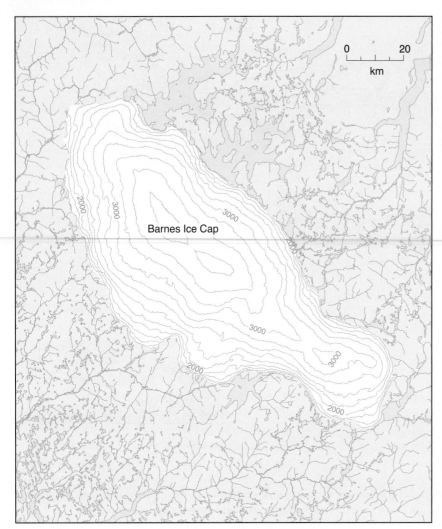

Figure 14.16 The Barnes Ice Cap, Baffin Island, Canada, is a remnant of the last continental glacier that covered large parts of North America and shows many features typically produced by continental glaciation. Isostatic adjustment of the crust causes the surface of the land to slope toward the ice, so lakes form along the ice margins. Drainage coming from the north is blocked by the ice and also contributes to lake formation near the ice margins. Irregularities in the surface over which the ice flows cause the ice margins to be uneven, or lobate.

glacier and the regional landforms. As shown on the map, the glacier is elliptical, with irregular, or lobate, margins. The ice is thickest in the central part and thins toward the edges. The presence of the glacier has caused an isostatic subsidence of the crust, so the land slopes toward the ice margins. In addition, the glacier has completely disrupted the former drainage system. Meltwater has therefore accumulated along the ice margins, forming a group of lakes. A photograph of the southern margins of the ice cap (**Figure 14.17**) shows the large, gently arched surface of the glacier, sediment deposited along the ice contact, and stream channels formed by meltwater on the glacier's surface.

The ice cap that covers nearly 80% of Greenland is much larger than the remnant on Baffin Island. In cross section, the glacier is shaped like a drop of water on a table (**Figure 14.18**). Its upper surface is a broad, almost flat-topped arch and is typically smooth and featureless. The base of the glacier is relatively flat. The Greenland glacier is more than 3000 m thick in its central part, but it thins toward the margins. The zone of accumulation is in the central part of the island, where the ice sheet is nourished by snowstorms moving from west to east. The snowline lies from 50 to 250 km inland; thus, the area of ablation constitutes only a narrow belt along the glacial margins.

What are outlet glaciers?

In rugged terrain, especially in areas close to the margins, the direction of ice movement is greatly influenced by mountain ranges, and the ice moves through mountain passes in large streams of **outlet glaciers** (**Figure 14.19**). These resemble valley glaciers in that they are confined by the topography. Pressure builds up in the ice behind a mountain range and forces outlet glaciers through mountain passes at relatively high speeds. Measurements in Greenland show that the main ice mass advances at approximately 10 to 30 cm/day. Outlet glaciers, however, can move as fast as 1 m/hr. In some places, you can actually see the ice move.

Figure 14.17 The margins of the Barnes Ice Cap are marked by ridges of sediment deposited as the ice melts and the glacier retreats. The glacier's upper surface is gently arched, and meltwater has formed small meandering streams. The landscape of the Great Lakes region must have appeared something like this 20,000 years ago.

The glacier of Antarctica is similar to that of Greenland in that it covers essentially the entire land mass (**Figure 14.20**). Antarctica, however, is much larger than Greenland, and its glacier contains more than 90% of Earth's ice. Much of the glacier is more than 3000 m thick, and its weight has depressed large parts of the continent's surface below sea level. Parts of Antarctica (mostly near the continental margins) are mountainous, with the higher peaks and ranges protruding above the ice. In the mountains, outlet glaciers funnel ice from the interior to the coast.

In addition to the continental glacier that blankets most of the land surface, Antarctica possesses two vast, fringing ice shelves and several smaller ones. These are not true glaciers but tabular bodies of ice that float on the ocean waters in the embayments of the Ross and Weddell seas. The shelves are several hundred meters thick and are fed by glaciers flowing out toward the edge of the Antarctic landmass. They are attached to the coast but calve off into the sea to form huge tabular icebergs that may exceed 100 km in length.

Many outlet glaciers flow through the valleys of the rugged Transantarctic Mountains onto the western edge of the Ross Ice Shelf (Figure 14.20). The largest of these is the Byrd Glacier, which is more than 20 km wide and 100 km long. The flow of ice is expressed by long ridges and furrows, parallel to the valley walls. Byrd Glacier is one of the fastest-moving glaciers in Antarctica, flowing at a rate of 750 to 800 m/yr.

Both Greenland and Antarctica are surrounded by water, so there is an ample supply of moisture to feed their glaciers. In contrast, Siberia is cold enough for glaciers to exist, but it lacks sufficient precipitation for ice to accumulate, so it has no glaciers.

Perhaps the best way to approach the study of landforms produced by continental glaciers is to study the photograph of an ice cap in Iceland (**Figure 14.21**) and the block diagrams of an ice sheet margin (**Figure 14.22**). From viewpoints on the ground, the landforms developed by continental glaciers are relatively inconspicuous and

Figure 14.18 The Greenland Ice Sheet
covers nearly 80% of the island. In this map, the thickness of the glacier is shown by contour lines in meters. The upper surface of the glacier is a broad, almost flat-topped arch and is typically smooth and featureless. The arrows show the direction of ice flow. Note from the cross section that the central part of Greenland has been depressed below sea level by the weight of the ice. Glaciers in Iceland are smaller and discontinuous; they have formed on volcanoes that are higher than their surroundings.

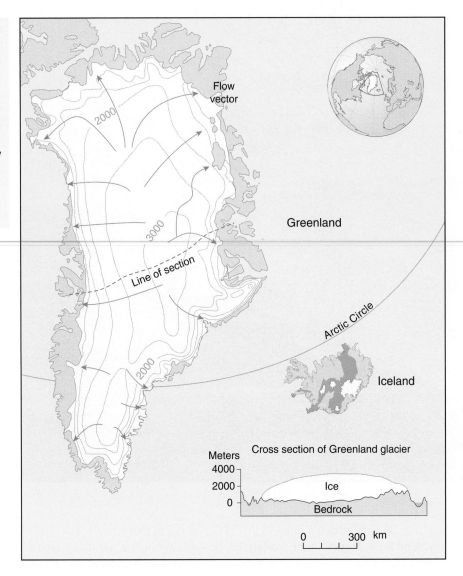

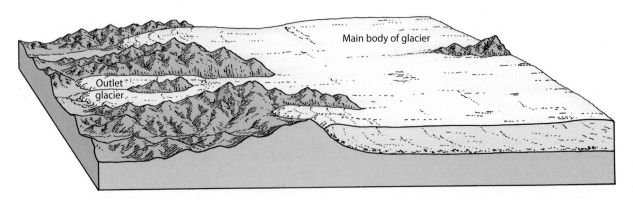

Figure 14.19 Outlet glaciers are segments of a continental glacier that advance rapidly through a mountain pass. A mountain range is a physical barrier to the movement of a continental glacier, and great pressure builds up in the ice behind the range. This pressure causes the ice in an outlet glacier to move very rapidly, in comparison with the main body ice.

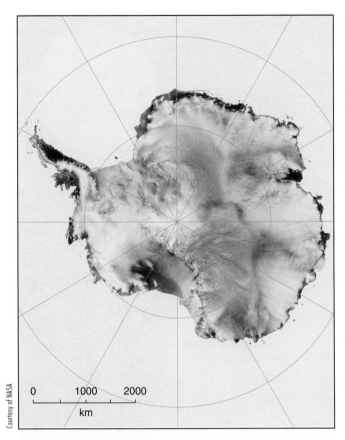

(A) Satellite image of Antarctica shows the vast extent of a continental glacier. Two large ice shelves and several smaller ones float on seawater. The ice shelves appear smooth and flat.

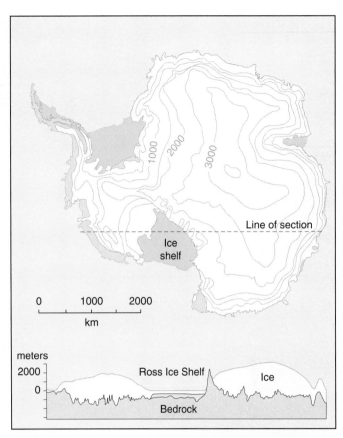

(B) The tremendous weight of the Antarctic ice sheet has depressed large parts of the continent below sea level. The higher peaks of Antarctica's mountain ranges protrude above the glacier as "islands of rock" in a sea of ice. Elevation contours in meters.

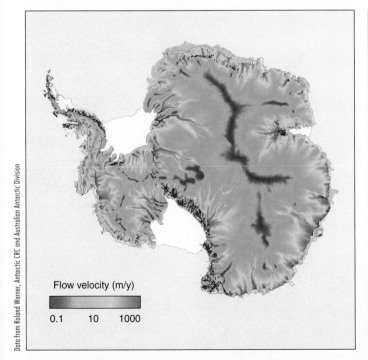

(C) The ice moves as much as several hundred meters per year as shown in this computer model. "Drainage basins" are bounded by slow ice. Fast streams of ice are concentrated on the margins but extend deep into the interior.

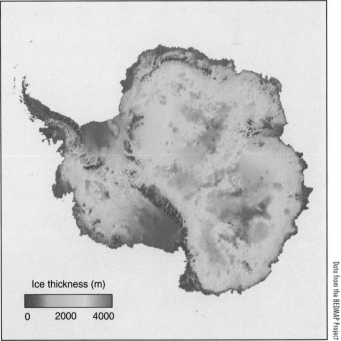

(D) The Antarctic ice sheet is as much as 3000 m thick but tapers toward the margins of the continent.

Figure 14.20 Antarctica is buried by Earth's largest mass of ice. The maps show its topography, velocity, and thickness of the glacier.

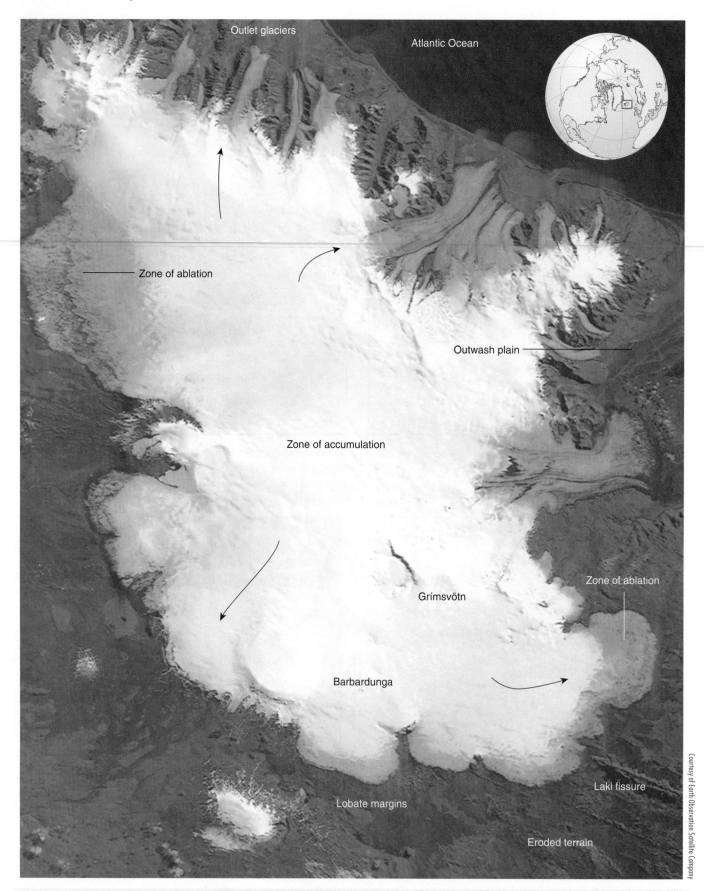

Outlet glaciers

Atlantic Ocean

Zone of ablation

Outwash plain

Zone of accumulation

Grímsvötn

Zone of ablation

Barbardunga

Laki fissure

Lobate margins

Eroded terrain

Figure 14.21 Iceland's Vatnajokull glacier completely buries the underlying surface, including two large shield volcanoes, Grímsvötn and Barbardunga. Beyond the snow line, large lobes form the margins of the ice cap. In the more-rugged terrain, toward the top, outlet glaciers advance through the valleys toward the sea. Sediment, carried by braided rivers on the outwash plains, tints the near-shore water. A subglacial eruption of Grímsvötn in 1996 caused a huge subglacial flood to burst from the glacier and flow down the outwash plain to the ocean, destroying bridges and roads.

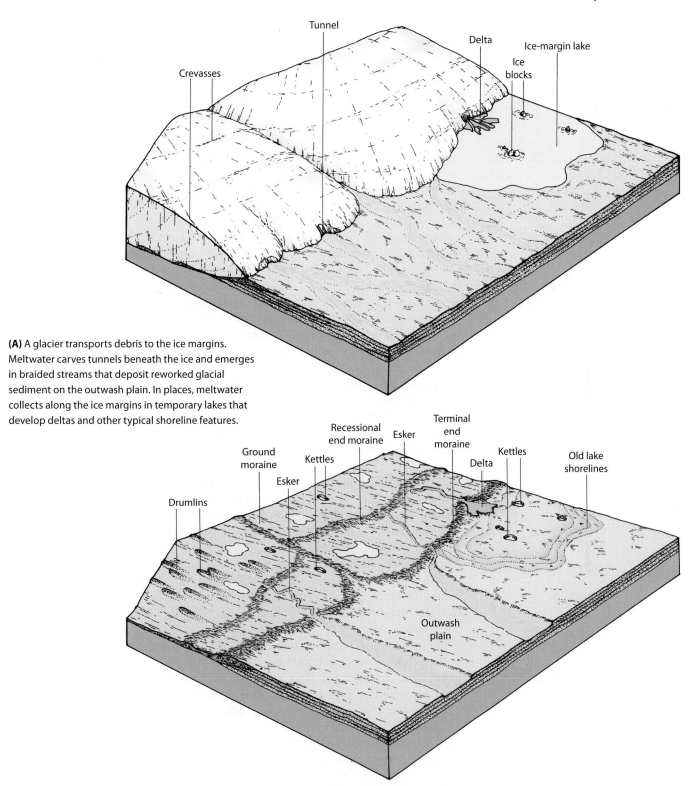

(A) A glacier transports debris to the ice margins. Meltwater carves tunnels beneath the ice and emerges in braided streams that deposit reworked glacial sediment on the outwash plain. In places, meltwater collects along the ice margins in temporary lakes that develop deltas and other typical shoreline features.

(B) After the ice has receded, the hummocky hills of a terminal moraine stretch in an arcuate line, conforming to the original shape of the ice margins at the farthest advance of the glacier. The retreating glacier leaves behind unsorted debris in ground moraines, and recessional moraines mark the positions of the ice margin where the glacier paused during its retreat. Hills of ground moraine can be reshaped by a subsequent advance of ice, forming drumlins. Sinuous eskers remain where sediment was deposited by subglacial streams, and sediment reworked by meltwater forms outwash plain and lake deposits. Where ice blocks were stranded by the receding glacier and partly buried under debris, the melting of the ice produces kettles.

Figure 14.22 Landforms developed by continental glaciers commonly are related to the position of the ice margin or the direction of the flow.

not nearly as spectacular as those produced by valley glaciers. Regionally, however, continental glaciation modifies the entire landscape, producing many important and distinctive surface features (**Figure 14.23**). Debris (**till**) transported by the glacier accumulates at the ice margin as a terminal moraine. Beneath the ice is a variable thickness of till, transported by the glacier and deposited as a **ground moraine**. This material, together with outwash plain sediment, can be reshaped by subsequent advances of ice to produce streamlined hills, called **drumlins**. The upstream end of a drumlin is blunt and steeper than the tail, so the form resembles a raindrop. The long axis is oriented parallel to the direction of ice movement. Drumlins are usually found in groups or swarms containing as many as 10,000 individuals. Excellent drumlin fields are found in Ireland, England, Canada, Michigan, Wisconsin, New England, New York, and western Washington. Some of the islands in Boston Harbor are drumlins, as is Bunker Hill, a famous landmark in U.S. history.

Streams of meltwater flow in tunnels within (and beneath) the ice and carry a large bed load, which is ultimately deposited to form a long, sinuous ridge known as an **esker** (Figure 14.23D). Debris-laden meltwater forms braided streams that flow from the glacier, over the outwash plain, where they deposit much of their load. During the retreat of the glaciers, meltwater forms subglacial channels and tunnels, which open into the outwash plain. Temporary lakes can develop where meltwater is trapped along the edges of the glacier, and deltas and other shoreline features form along the lake margins. Deposits on the lake bottom typically are stratified in a series of alternating light and dark layers known as **varves** (Figure 14.23B). The coarse, light-colored material accumulates during spring and summer runoff. During the winter, when the lake is frozen over, no new sediment enters the lake, and the fine mud settles out of suspension to form the thin, dark layers.

Ice blocks, left behind by the retreating glacier front, can be partly or completely buried in the outwash plain or in moraines. Where an isolated block of debris-covered ice melts, a depression known as a **kettle** is formed.

Figure 14.22B shows the area after the glacier has disappeared completely. The end moraine appears as a belt of hummocky hills, which mark the former position of the ice. The size of the moraine reflects the duration of a stable ice front. Continental moraines can be large. For example, the Bloomington Moraine in central Illinois is 25 to 30 km wide and more than 300 km long but only 20 to 60 m high. From the ground, it probably would not be recognized by an untrained observer as anything more than a series of hills. Mapped over a large area, however, it can be seen to have an arcuate pattern, conforming to the lobate margin of the glacier. Many small depressions occur throughout the moraine, some of which may be filled with water, forming small lakes and ponds.

Scattered across the surface of the glaciated regions of North America and Europe are large fields of boulders known as **erratics** (**Figure 14.24**). Many midwestern U.S. erratics are composed of igneous or metamorphic rocks and are completely different from the underlying bedrock of sandstone, limestone, and shale. They could come only from the interior of Canada, hundreds of kilometers to the north. Some erratics are incorporated in the body of glacial sediment, whereas others lie free on the ground. Most erratics are small but many exceed 3 m in diameter, and others are enormous, weighing thousands of tons like the one shown in Figure 14.24.

Pleistocene Glaciation

The Pleistocene ice age was one of the most significant events in recent Earth history. The major effects of the ice age were: (1) glacial erosion and deposition over large parts of the continents that modified river systems, (2) creation of millions of lakes, (3) changes in sea level, (4) pluvial lakes developed far from the ice margins, (5) isostatic adjustment of the crust, (6) abnormal winds, (7) impact on the oceans, (8) catastrophic flooding, and (9) modifications of biologic communities.

(A) Glacial till resting on horizontal limestone in Iowa is responsible for much of the rich farmland in that area.

(B) Varves are annual layers of sediment accumulated in glacial lakes. Large boulders dropped from melting icebergs (dropstones) accumulate contemporaneously.

(C) Moraines form a distinctive topography of rolling hills and numerous closed depressions.

(D) Eskers form long, sinuous ridges composed of sand and gravel deposited by streams that flowed beneath the glacier.

(E) Drumlins are streamlined hills that were shaped by the movement of the glacier and show the direction in which the ice flowed.

(F) The outwash plain forms from meltwaters from the glacier and is characterized by fluvial sediments deposited by braided streams.

Courtesy of D. Easterbrook

Figure 14.23 Glacial landforms reveal the former presence of vast continental glaciers.

Figure 14.24 Erratics are large boulders transported by glaciers and then dropped far from their point of origin. This isolated block was carried 300 km by glacial ice and now lies near Okotok, Alberta, Canada. In some areas, diamond and other ore deposits have been found by tracing distinctive erratics back to their bedrock sources.

The cycles of glacial and interglacial periods, which began between 2 and 3 million years ago, constitute one of the most significant events in the recent history of Earth. During this time, the normal hydrologic system was completely interrupted throughout large areas of the world and was considerably modified in others. The evidence of such an event in the recent past is overwhelmingly abundant. Over the last century, extensive field observations have provided incontestable evidence that continental glaciers covered large parts of Europe, North America, and Siberia (**Figure 14.25**). The most recent cycle of glacial retreat started only about 20,000 years ago (**Figure 14.26**). A detailed map of glacial features in the northeastern United States is given in **Figure 14.27**. These maps were compiled after many years of fieldwork by hundreds of geologists who mapped the location and orientation of drumlins, eskers, moraines, striations, and glacial stream channels. These maps revealed the extent of the ice sheet, the direction of flow, and the locations of systems of meltwater channels, and they allowed us to decipher a history of multiple advances and retreats of the ice.

Several periods of Pleistocene glaciation in the United States are recorded by broad sheets of till and complex moraines, separated by ancient soils and layers of wind-blown silt. Striations, drumlins, eskers, and other glacial features show that almost all of Canada, the mountain areas of Alaska, and the eastern and central United States, down to the Missouri and Ohio rivers, were covered with ice (Figure 14.27). There were three main zones of accumulation, the largest of which was centered over Hudson Bay. Ice advanced radially from there, northward to the Arctic islands and southward into the Great Lakes area. A smaller center was located in the Labrador Peninsula. Ice spread southward from this center into what are now the New England states. In the Canadian Rockies, to the west, valley glaciers coalesced into ice caps. These grew into a single ice sheet, which then moved westward to the Pacific shores and eastward down the Rocky Mountain foothills, until it merged with the large sheet from Hudson Bay.

Throughout much of central Canada, the glaciers eroded 120 m of regolith and solid bedrock. This material was transported to the glacial margins and accumulated as ground moraine, end moraines, and outwash in a broad belt from Ohio to Montana (Figure 14.27). In places, the glacial debris is more than 300 m thick, but the average thickness is about 15 m. Meltwater carried sediment down the Mississippi and other rivers and deposited it on the continental shelf. Much of the fine-grained sediment was transported and redeposited by wind.

What evidence indicates multiple cycles of advance and retreat of the glaciers during the ice age?

Even before the theory of worldwide glaciation was generally accepted, many observers recognized that more than a single advance and retreat of the ice had occurred during the **Pleistocene Epoch**. Extensive evidence now shows that a number of periods of growth and retreat of continental glaciers occurred during the ice age. The interglacial periods of warm climate are represented by buried soil profiles, peat beds, and lake and stream deposits separating the unsorted, unstratified deposits of glacial debris.

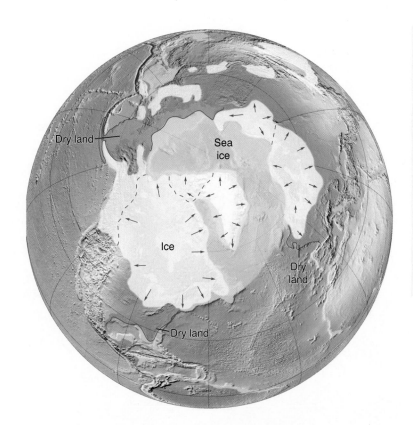

Figure 14.25 Pleistocene glaciers covered large areas in North America, Europe, and Asia, as well as many high mountain regions. Parts of Alaska and Siberia were not glaciated because those areas were too dry. They were cold enough, but not enough precipitation fell for glaciers to develop. With the accumulation of so much ice on the land, the shoreline moved as much as 200 km seaward along the Atlantic coast of the United States. The drop in sea level formed a broad land bridge between Siberia and North America (gray). Sea ice covered most of the Arctic Ocean and extended into the North Atlantic well south of Iceland (light blue). Ocean temperatures dropped by as much as 10°C.

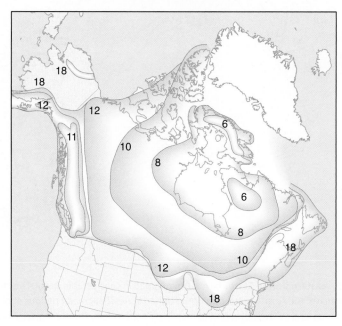

Figure 14.26 Successive positions of the ice front during the recession of the last ice sheet have been mapped from data collected by geologists in Canada and the United States. Contours indicate the position and age of the ice front in thousands of years before the present.

In North America, radiometric dating shows that the ice first began to advance between 2 and 3 million years ago, and after repeated cycles of advancing and receding, the last ice sheet began to retreat about 20,000 years ago. Remnants of these last glaciers, now occupying about 10% of the world's land surface, still exist in Greenland and Antarctica.

The Effects of the Pleistocene Glaciation

The presence of so much ice upon the continents had a profound effect upon almost every aspect of Earth's hydrologic system. The most obvious effects, of course, are the spectacular mountain scenery and other continental landscapes fashioned by both

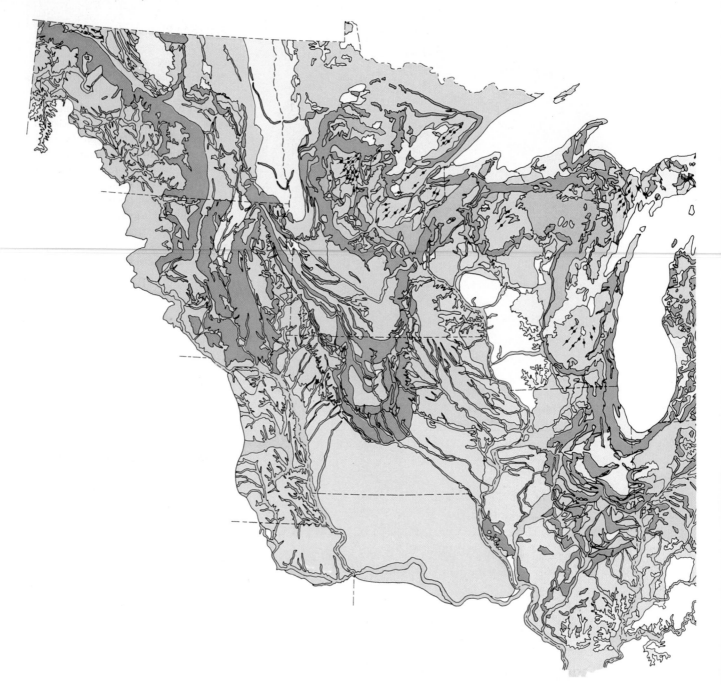

Figure 14.27 **The major glacial features in the eastern United States** have been mapped in considerable detail. They tell a story of repeated advance and retreat of the great ice sheets over much of the north-central part of the United States. Four major periods of glaciation are recorded; each had multiple cycles of expansion and contraction, and all are shown on this map in different colors. The older glacial deposits (Nebraskan, Kansan, and Illinoian) have been modified by erosion and partly covered by the most recent deposits of Wisconsin age (shown in various shades of green). The only part of this area not covered by ice was in southern Wisconsin. It was protected from ice advance by the Wisconsin highlands to the north, which diverted lobes of ice around the area.

The moraines (darker tones) indicate the former positions of the ice margins. Two major lobes of ice moved into the lowlands of the Great Lakes area and another large lobe moved south from the Dakotas and Minnesota and into Iowa. These lobes scoured out the basins, which were filled with water for large glacial lakes (shown in blue) much larger than the present Great Lakes. In New England, the ice moved southeastward beyond the present coastline and deposited moraines out on the continental shelf; they are now covered with water. Long Island and Cape Cod are the northernmost remnants of this morainal system. Eskers formed in subglacial streams are shown in linear patterns of orange and are especially abundant in Maine. Spillways or meltwater channels, shown in yellow, reveal the drainage system that carried off the meltwater from the ice. Note the major drumlin fields in New York, Wisconsin, and Minnesota, which indicate the direction of ice movement. Isostatic depression of the crust due to the weight of the ice permitted the sea to invade the coast of Maine and large parts of the St. Lawrence lowlands in southern Canada.

Compiled from glacial map of the United States east of the Rocky Mountains, The Geological Society of America

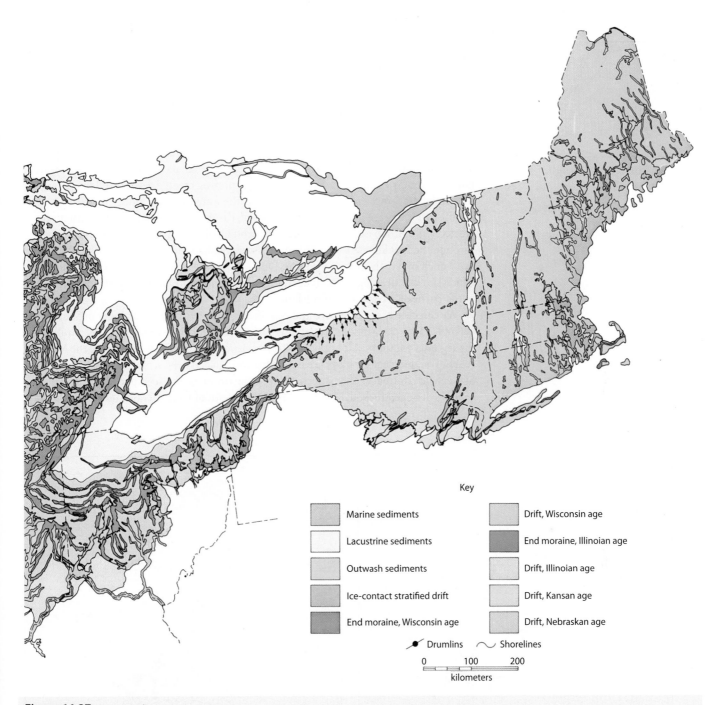

Key

Marine sediments	Drift, Wisconsin age
Lacustrine sediments	End moraine, Illinoian age
Outwash sediments	Drift, Illinoian age
Ice-contact stratified drift	Drift, Kansan age
End moraine, Wisconsin age	Drift, Nebraskan age

Drumlins   Shorelines

0   100   200
kilometers

Figure 14.27 *(continued)*

glacial erosion and deposition instead of by running water. Entirely new landscapes covering millions of square kilometers were formed in a relatively short period of geologic time. In addition, the vast bodies of glacial ice affected the Earth well beyond the glacier margins. Directly or indirectly, the effects of glaciation were felt in every part of the globe.

Modification of Drainage Systems   Before glaciation, the landscape of North America was eroded mainly by running water. Well-integrated drainage systems collected runoff and transported it to the ocean. Much of North America was drained by rivers flowing northeastward into Canada because the regional slope throughout the north-central part of the continent was to the northeast. The preglacial drainage patterns are not known in detail. Various features of the present system, however, together with segments of ancient stream channels now mostly buried by glacial sediments, suggest a pattern similar to that shown in **Figure 14.28**. Before glaciation, the major tributaries of the upper Missouri and Ohio rivers were part of a northeastward-flowing drainage system. This system also included the major rivers draining the Canadian Rockies, such as the Saskatchewan, Athabasca, Peace, and Liard Rivers. It emptied into the Arctic Ocean, probably through Lancaster Sound and Baffin Bay, and an eastern drainage was out of the St. Lawrence River.

As the glaciers spread over the northern part of the continent, they effectively buried the trunk streams of the major drainage systems, damming up the northward-flowing tributaries along the ice front. This damming created a series of lakes along the glacial margins. As the lakes overflowed, the water drained along the ice front and established the present courses of the Missouri and Ohio rivers. A similar situation created Lake Athabasca, Great Slave Lake, and Great Bear Lake, and their drainage through the Mackenzie River. This process established the present drainage pattern over much of North America (Figure 14.28). Compare this diagram with Figure 14.16, which shows a drainage system currently undergoing similar modifications as a result of the Barnes Ice Cap on Baffin Island.

We can clearly see extensive and convincing evidence of these changes in South Dakota. There, the Missouri River flows in a deep, trenchlike valley, roughly parallel to the regional contours. All important tributaries enter from the west. East of the Missouri River, preglacial valleys are now filled with glacial debris, marking the remnants of preglacial drainage. The pattern of preglacial drainage is also supported by recent discoveries of huge, thick, deltaic deposits in the mouth of Lancaster Sound and in Baffin

(A) Before the ice age, drainage of central North America was northeastward, from the northern and central Rocky Mountains into the St. Lawrence Bay, Hudson Bay, and the Arctic area. The area eventually covered by ice is shown with the light shading.

(B) Present drainage patterns show major modifications. Preglacial drainage was impounded against the glacial margins and developed new outlets to the ocean through the Missouri, Ohio, and Mackenzie rivers. The drainage system beneath the ice was obliterated. The present drainage in most of Canada is deranged, consisting of numerous lakes, swamps, and unintegrated meandering streams.

Figure 14.28 Glacial modification of North American drainage was extensive and created thousands of lakes.

Bay, Canada. These deposits are difficult or impossible to explain as results of the present drainage pattern because no major drainage system currently empties into those areas.

Beyond the margins of the ice, the hydrology of many streams and rivers was profoundly affected, either by the increased flow from meltwater or by the greater precipitation associated with the glacial epoch. With the appearance of the modern Ohio and Missouri rivers, water that formerly emptied into the Arctic and Atlantic oceans was diverted to the Gulf of Mexico through the Mississippi River. Other streams became overloaded and their valleys partly filled with sediment. Still others became more effective agents of downcutting, as a result of glacial sediment, and their valleys deepened. Although the history of each river is complex, the general effect of glaciation on rivers was to produce thick alluvial fill in their valleys; the fill is now being eroded to form stream terraces.

Lakes   Pleistocene glaciation created more lakes than all other geologic processes combined. The reason is obvious if we recall that a continental glacier completely disrupts the preglacial drainage system. The surface over which the glacier moved was scoured and eroded by the ice, leaving myriad closed, undrained depressions in the bedrock. These depressions filled with water and became lakes (**Figure 14.29**).

Farther south, in the north-central United States, lakes formed in a different manner. There, the surface was covered by glacial deposits of ground moraine and end moraines. Throughout Michigan, Wisconsin, and Minnesota, these deposits formed closed depressions that soon filled with water to form tens of thousands of lakes. Many of these lakes still exist. Others have been drained or filled with sediment, leaving a record of their former existence in peat bogs, lake silts, and abandoned shorelines.

Exceptionally large lakes were created along the glacial margins. We can envision their formation with the help of the basic model of continental glaciation shown in Figure 14.15. The ice on both North America and Europe was about 3000 m thick near the centers of maximum accumulation, but it tapered toward the glacier margins. Crustal subsidence was greatest beneath the thickest accumulation of ice. In parts of Canada and Scandinavia, the crust was depressed more than 600 m. As the ice melted, rebound of the crust lagged behind and the regional slope toward the ice persisted. This slope formed basins that have lasted for thousands of years. These basins became lakes or were invaded by the ocean. The Great Lakes of North America and the Baltic Sea of northern Europe were formed primarily in this way.

Although the origin of the Great Lakes is extremely complex, the major elements of their history are known and are illustrated in the four diagrams in **Figure 14.30**. The preglacial topography of the Great Lakes region was influenced greatly by the structure and character of the rocks exposed at the surface. A geologic map of this region shows that the major structural feature is the Michigan Basin, which exposed a broad, circular belt of weak Devonian shale and Silurian salt formations, surrounded by the more-resistant Silurian limestone. Preglacial erosion undoubtedly formed a wide valley or lowland along the shale, and escarpments developed on the resistant limestone.

As the glaciers moved southward into this area, large lobes of ice advanced down the great valleys, eroding them into broad, deep basins. Lakes Michigan, Huron, and Erie were scoured from the belt of weak Devonian shale by these lobes of ice. Figure 14.30A shows the Great Lakes area as it probably appeared when glaciers began to recede about 20,000 years ago. Meltwater flowed away from the glacier margin to the south. As the glaciers receded, lower land was uncovered, and meltwater became impounded in front of the ice margins to form the ancestral Great Lakes (Figure 14.30B). Drainage was still to the south through various ancient channels that joined the Mississippi River. As deglaciation continued, an eastern outlet was established (Figure 14.30C). Finally, as the ice receded farther (Figure 14.30D), a new outlet was developed through Lake Iroquois and into the St. Lawrence estuary. Niagara Falls came into existence at this time, when water from Lake Erie flowed across the Niagara Escarpment into Lake Iroquois (now Lake Ontario). The exposed sequence of rock consists of a resistant limestone formation underlain by a weak shale layer. Undercutting of the shale below the limestone causes the falls to retreat upstream.

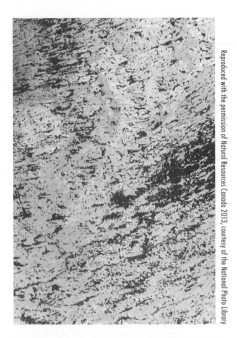

Figure 14.29 Lakes created by continental glaciation in the shield area of North America were photographed from a height of approximately 900 km. More lakes were created by glaciation than by all other geologic processes combined.

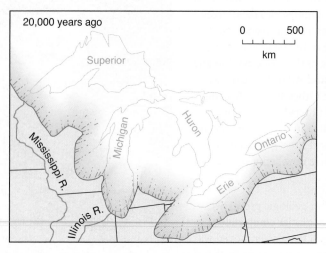

(A) Approximately 20,000 years ago, the ice front extended beyond the present Great Lakes. The ice advanced into lowlands surrounding the Michigan Basin, with large lobes extending down from the present sites of Lakes Erie and Michigan.

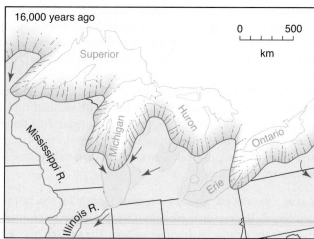

(B) The ancestral Great Lakes appeared about 16,000 years ago, as the ice receded. The northern margins of the lakes were against the retreating ice. Drainage was to the south, to the Mississippi River. Drainage through the St. Lawrence River was blocked.

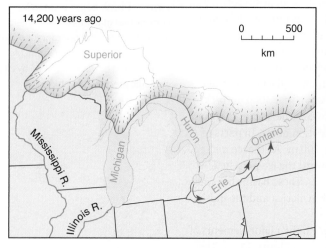

(C) As the ice front continued to retreat, an eastern outlet developed as the lakes connected across northern Michigan. Niagara River, and eventually the great falls, developed as water from early Lake Erie flowed over the Niagara Escarpment into ancient Lake Iroquois. Meltwater drained to the ocean through the Hudson River in what is now New York.

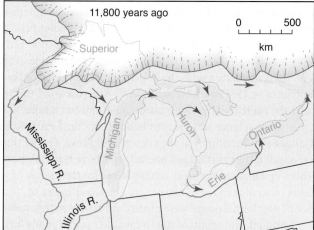

(D) By about 11,800 years ago, the ice had receded enough to allow water to drain eastward to the Atlantic through the St. Lawrence River. Lake Erie was very small at this time. The western lakes were still mostly buried by ice and drainage from them was to the south. The lakes began to assume their present outlines about 10,000 years ago.

Figure 14.30 The evolution of the Great Lakes can be traced from their origin along the ice margins about 20,000 years ago. The sequence of events and modifications of the landscape are inferred from numerous studies of glacial features in the Great Lakes area.

Data from: J.A. Clark et al., 2012

To the northwest, another group of lakes formed in much the same way, but they have since been reduced to small remnants of their former selves. The largest of these marginal lakes, known as Lake Agassiz, covered the broad, flat region of Manitoba, in northwestern Minnesota, and the eastern part of North Dakota (**Figure 14.31**). It drained into the Mississippi River and then, at lower stages, developed outlets into Lake Superior. Later, when the ice dam retreated, it drained into Hudson Bay. Remnants of this vast lake include Lake Winnipeg, Lake Manitoba, and Lake of the Woods. The sediments deposited on the floor of Lake Agassiz provided much of the rich soil for the wheatlands of North Dakota, Manitoba, and the Red River Valley of Minnesota. Even now, ancient shorelines of Lake Agassiz remain, marking its former margins.

Northward, along the margin of the Canadian Shield, Lake Athabasca, Great Slave Lake, and Great Bear Lake are remnants of the other great ice-marginal lakes. In northern Europe, the recession of the Scandinavian ice sheet caused similar depressions along the ice margins, and the large lakes that were thus produced ultimately connected with the ocean to form the Baltic Sea.

Changes of Sea Level   One of the most important effects of Pleistocene glaciation was the repeated worldwide rise and fall of sea level, a phenomenon that corresponded to the retreat and advance of the glaciers. During a glacial period, water that normally returned to the ocean by runoff became locked on the land as ice, and sea level was lowered. When the glaciers melted, sea level rose again. The amount of change in sea level can be calculated because the area of maximum ice coverage is known in considerable detail, and the thickness of the ice can be estimated from the known volumes of ice in the glaciers of Antarctica and Greenland. The Antarctic ice sheet alone contains enough water to raise sea level throughout the world by about 70 m.

The dates of sea level changes are well documented by radiocarbon dates from terrestrial organic matter and from near-shore marine organisms obtained by drilling and dredging off the continental shelf. These dates show that about 35,000 years ago, the sea was near its present position. Gradually, it receded. By 18,000 years ago, it had dropped nearly 137 m. It then rose rather rapidly to within 6 m of its present level. The fall in sea level caused the Atlantic shoreline to recede between 100 and 200 km, exposing vast areas of the continental shelf (Figure 14.25). Early humans probably inhabited large parts of the shelf that are now more than 100 m below sea level.

The glaciers extended far across the exposed shelf of the New England coast, as is evidenced by unsorted morainal debris and the remains of mastodons dredged from the seafloor in those areas. In the oceans off the central and southern Atlantic states, depth soundings reveal drainage systems and eroded stream valleys that extended across the shelf. Great Britain was connected to the European continent during glacial maximums (Figure 14.25). Moreover, Asia was connected to North America by a land bridge across the Bering Strait allowing humans to migrate on dry land, arriving 17,000 to 15,000 years ago. Paleoindians may have used the ice-free corridor in western Canada if they came by land (Figure 14.26) or they may have traveled along coastal waterways.

Pluvial Lakes   The climatic conditions that caused glaciation had an indirect effect on arid and semiarid regions far removed from the large ice sheets. The increased precipitation that fed the glaciers also increased the runoff of major rivers and intermittent streams, resulting in the growth and development of large **pluvial lakes** (Latin *pluvia*, "rain") in numerous isolated basins in nonglaciated areas throughout the world. Most pluvial lakes developed in relatively arid regions where, prior to the glacial epoch, there was insufficient rain to establish an integrated, through-flowing drainage system to the sea. Instead, stream runoff in those areas flowed into closed basins and formed playa lakes. With increased rainfall, the playa lakes enlarged and sometimes overflowed. They developed a variety of shoreline features—wave-built terraces, bars, spits, and deltas—now recognized as high-water marks in many desert basins. Pluvial lakes were most extensive during glacial intervals. During interglacial stages, when less precipitation fell, the pluvial lakes shrank to form small salt flats or dry, dusty playas.

The greatest concentration of pluvial lakes in North America was in the northern part of the Basin and Range Province of western Utah and Nevada. The fault-block structure there has produced more than 140 closed basins, many of which show evidence of former lakes or former high-water levels of existing lakes. The distribution of the former lakes is shown in **Figure 14.32**. Lake Bonneville was the largest, by far, and occupied a number of coalescent intermontane basins. Remnants of this great body of fresh water are Great Salt Lake and Utah Lake. At its maximum extent, Lake Bonneville was about the size of Lake Michigan, covering an area of 50,000 km², and was 300 m deep. The principal rivers entered the lake from the high Wasatch Range, to the east. They built large deltas, shoreline terraces, and other coastal features that are now high above the valley floors along the mountain front (**Figure 14.33**).

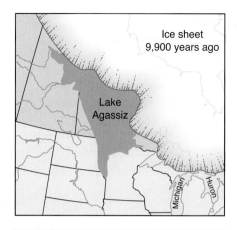

Figure 14.31 **Lake Agassiz** was the largest glacial lake in North America. Its former shorelines are now marked by beach ridges, spits, and bars. The dry lake bed now forms the fertile soils of Manitoba and North Dakota. Remnants of this former glacial lake include Lake Winnipeg, Lake Manitoba, and Lake of the Woods.

Figure 14.32 Pluvial lakes were formed in the closed basins of the western United States as a result of climatic changes associated with the glacial epoch. Most are now dry lake beds because of the arid climate. Former shorelines of the pluvial lakes are well marked along the basin margins. Lake Bonneville, in western Utah, was the largest. Great Salt Lake, Utah, is one of its remnants.

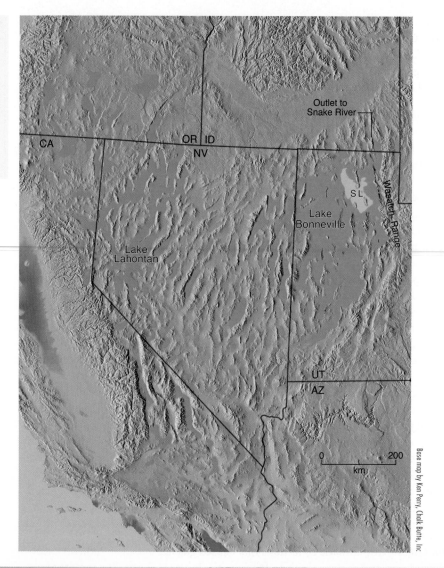

Figure 14.33 Shoreline features of Lake Bonneville include deltas, beaches, bars, spits, and wave-cut cliffs. Multiple shorelines were produced as Lake Bonneville dried up. This photograph shows shorelines on Fremont Island, Great Salt Lake, Utah. Note the wave-cut cliffs and terraces at the highest levels.

As the level of the lake rose to 300 m above the floor of the valley, it overflowed to the north into the Snake River and thence to the ocean. The outlet, established on unconsolidated alluvium, rapidly eroded down to bedrock, 100 m below the original pass. The level of the lake was then stabilized, fluctuating only with the pluvial epochs associated with glaciation. Some valley glaciers from the Wasatch Range extended down to the shoreline of the old lake, and some of their moraines were carved by wave action. This wave erosion shows conclusively that glaciation was contemporaneous with the high level of the lake. As the climate became drier, the lake dried up, leaving faint shorelines at lower levels.

Isostatic Adjustment   Major isostatic adjustments of the lithosphere during the Pleistocene glaciation were caused by the weight of the ice, which depressed the continents. In Canada, a large area around Hudson Bay was depressed below sea level, as was the area in Europe around the Baltic Sea. The land has been rebounding from these depressions ever since the ice melted. The area around Washington's Puget Sound rose at a rate of more than 10 cm/yr shortly after the ice disappeared 13,500 years ago, but the rate slowed to 2 cm/yr by 11,000 years ago. The total uplift in the area is about 150 m. The former seafloor around Hudson Bay has risen almost 600 m and is still rising at a maximum rate of about 3.5 cm/yr. The land must rise an additional 80 m before it regains its preglacial level and reestablishes isostatic balance. Some of these isostatic movements triggered large earthquakes in Scandinavia about 9000 years ago. These earthquakes are unique in that they are not associated with plate boundaries.

The tilting of Earth's crust, as it rebounds from the weight of the ice, can be measured by careful surveying and also by mapping the elevations of the shorelines of ancient lakes (**Figure 14.34**). The shorelines were level when they formed but were tilted as the crust rebounded from the unloading of the ice. In the Great Lakes region, old shorelines slope downward to the south, away from the centers of maximum ice accumulation, indicating a local rebound of 400 m or more.

Effects of Winds   The presence of ice over so much of the continents greatly modified patterns of atmospheric circulation. Winds near the glacial margins were strong and unusually persistent because of the abundance of dense, cold air coming off the glacier fields. These winds picked up and transported large quantities of loose, fine-grained sediment brought down by the glaciers. This dust accumulated as **loess** (wind-blown silt), sometimes hundreds of meters thick, forming an irregular blanket over much of the Missouri River valley, central Europe, and northern China.

Sand dunes were much more widespread and active in many areas during the Pleistocene. A good example is the Sand Hills region in western Nebraska, which covers an area of about 60,000 km^2. This region was a large, active dune field during the Pleistocene, but today the dunes are largely stabilized by a cover of grass.

The Oceans   Pleistocene glaciation affected to some extent the waters of all of the oceans. Besides changing the sea level so that shorelines were altered and much of the continental shelves were exposed, the glacial periods cooled the ocean waters by as much as 10°C. The lower temperatures affected the kind and distribution of marine life and also influenced seawater chemistry. Furthermore, patterns and strengths of oceanic currents were changed. Circulation was significantly restricted by glacially formed features such as the Bering Strait, extensive pack ice, and exposed shelves.

Floods of glacial meltwater may have altered the flow of the Gulf Stream. About 13,400 years ago, a dam on glacial Lake Iroquois (Figure 14.30) failed and cold water flowed down the Hudson Valley, across the exposed continental shelf, and into the Atlantic. The cooling effect reduced the flow of the Gulf Stream and may have caused an abrupt, but brief episode of cooling in the Northern Hemisphere.

As sea level fell and then rose at the end of the last ice age other dramatic events occurred. For example, during the last glacial maximum, the Black Sea became an isolated freshwater lake separated from the salty Mediterranean Sea by a dry strip of land where the Straits of Bosporus are today. But about 7500 years ago, the Mediterranean

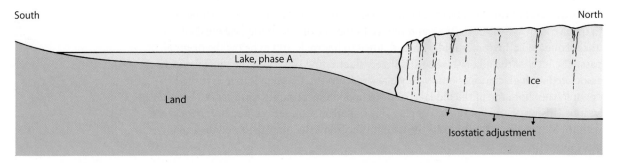

(A) When a lake develops along a glacier's margins, the shoreline features, such as beaches and bars, are horizontal.

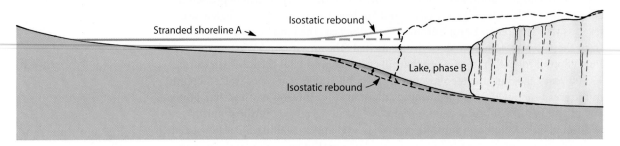

(B) As the ice recedes, isostatic rebound occurs. The shoreline features formed during phase A are tilted away from the ice. Younger horizontal shoreline features are formed by the lake during phase B.

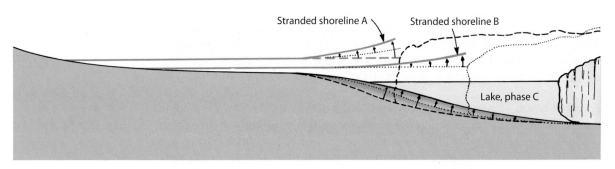

(C) Continued retreat of the ice causes further isostatic rebound and tilting of both shorelines A and B, which converge away from the glacier.

Figure 14.34 The tilted shorelines of glacial lakes can be used to measure the rate and extent of isostatic adjustment of the crust after the ice recedes.

Sea had risen sufficiently to spill over the barrier and into the Black Sea basin. The "flood" cut a 90 m deep trough and triggered a deluge that flooded the coastal area around the Black Sea.

Even the deep-ocean basins did not escape the influence of glaciation. Where glaciers entered the ocean, icebergs broke off and rafted their enclosed load of sediment out into the ocean. As the ice melted, debris ranging from huge boulders to fine clay settled on the deep-ocean floor, resulting in an unusual accumulation of coarse glacial boulders in fine oceanic mud. Ice-rafted sediment is most common in the Arctic, the Antarctic, the North Atlantic, and the northeastern Pacific.

In the warmer reaches of the oceans, the glacial and interglacial periods are recorded by alternating layers of red clay and small calcareous shells of microscopic organisms. The red mud accumulated during cold periods, when fewer organisms inhabited the colder water. During the warmer interglacial periods, life flourished, and layers of shells mixed with mud were deposited.

Channeled Scablands The continental glacier in western North America moved southward from Canada only a short distance into Washington, but it played an important role in producing a strange complex of interlaced deep channels, a type of topography found nowhere else on Earth. This area, the **Channeled Scablands**, covers much of eastern Washington and consists of a network of braided channels from 15 to 30 m deep. The term *scabland* is appropriately descriptive because, viewed from the air, the surface has the appearance of great wounds or scars (**Figure 14.35**). Many of the channels have steep walls and dry waterfalls or cataracts. In addition, there are sediment deposits with giant ripple marks and huge bars of sand and gravel (Figure 14.35). These features attest to extreme erosion by running water—a catastrophic flooding by normal standards—yet, today the area does not have enough rainfall to maintain a single permanent stream.

The scablands were eroded by the following process. A large lobe of ice advanced southward across the Columbia Plateau and temporarily blocked the Clark Fork River,

(A) Scabland channels west of Spokane, Washington.

(B) Giant ripple marks west of Spokane, Washington.

(C) Patterned ground formed by permafrost activity at the end of the glacial period.

Figure 14.35 **The Channeled Scablands** of Washington consist of a complex of deep channels cut into the basalt bedrock. The scabland topography is completely different from that produced by a normal drainage system. It is believed to have been produced by "catastrophic" flooding.

one of the major northward-flowing tributaries of the Columbia River (**Figure 14.36**). The impounded water backed up to form glacial Lake Missoula, a long, narrow lake extending diagonally across part of western Montana. Sediments deposited in this lake now partly fill the long, narrow valley. As the glacier receded, the ice dam failed, releasing a tremendous flood over the southwestward-sloping Columbia Plateau. The enormous discharge, barely diverted by the preexisting shallow valleys, spread over the basalt surface, scouring out channels and forming giant ripple marks, bars, and other sediment deposits. Estimates suggest that, during the flood, as much as 40 km³ of water per hour may have been discharged from Lake Missoula. Because the glaciers advanced several times into the region, such catastrophic flooding probably occurred many times, perhaps as far back as 2.5 million years ago. Lake Missoula formed each time the ice front advanced past the Clark Fork River and then flooded the Scablands with each recession of the ice and subsequent dam failure.

Biological Effects of the Ice Age   The severe climatic changes during the ice age had a drastic impact on most life forms. With each advance of the ice, large areas of the continents (the areas beneath the ice) became totally depopulated, and plants and animals retreating southward in front of the advancing glacier were under tremendous stress. The most severe stresses resulted from drastic climatic changes, reduced living space, and a curtailed food supply. As the glaciers advanced, most species were displaced, along with their environments, across distances of approximately 3200 km. As the ice retreated, some new living space became available in deglaciated areas, but the formerly exposed continental shelves were inundated by the rising sea. During the major glacial advances, when sea level was lower, new routes of migration opened from Asia to North America, because much of Alaska and Siberia were not glaciated (see Figure 14.25), and from Southeast Asia to the islands of Indonesia. Land plants were forced to migrate with the climatic zones in front of the glaciers. As the glaciers pushed cold-weather belts southward, displaced storm tracks and changes in precipitation affected even the tropics.

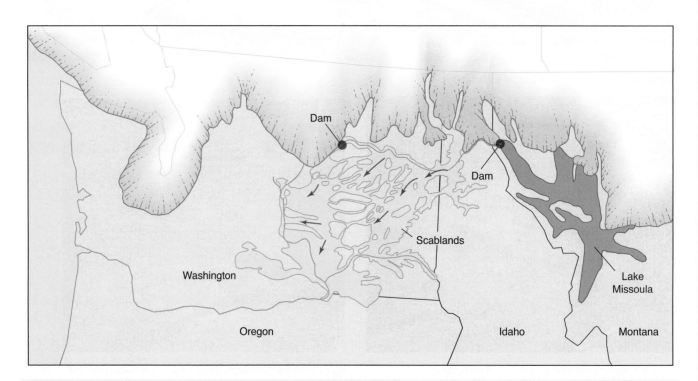

Figure 14.36 The origin of the Channeled Scablands is attributed to "catastrophic" flooding, on a magnitude apparently unique in Earth's history. The flood resulted when the ice dam that formed glacial Lake Missoula failed as the glacier receded. The ice sheet blocked the drainage of the northward-flowing Clark Fork River to form Lake Missoula, a long, deep lake in western Montana. As the glacier receded, the ice dam that formed Lake Missoula failed, and water from the lake quickly flowed across the Scablands, eroding deep channels. The glacier's repeated advance and retreat probably produced several ice dams that failed as the ice melted, each time causing catastrophic flooding.

Many life forms could not cope with the repeated and overwhelming environmental changes brought about by the cycles of advancing and retreating ice. Numerous species, particularly giant mammals, became extinct. During glaciation, the now-extinct imperial mammoth, 4.2 m high at the shoulders, roamed much of North America. The saber-toothed tiger became extinct about 14,000 years ago. Fossils of the giant beaver, as large as a black bear, and the giant ground sloth, which measured 6 m tall standing on its hind legs, have been found in Pleistocene sediments. In Africa, fossil sheep 2 m tall have been found, in addition to pigs as big as a present-day rhinoceros. In Australia, giant kangaroos and other marsupials thrived during the Pleistocene.

Records of Pre-Pleistocene Glaciation

Glaciation has been a rare event in Earth's history, but there is evidence of widespread glaciation during late Paleozoic time (200 to 300 million years ago) and during late Precambrian time (600 to 800 million years ago).

Before the great ice age, which began 2 to 3 million years ago in North America, Earth's climate was typically mild and uniform for long periods of time. This climatic history is implied by the types of fossil plants and animals and by the characteristics of sediments preserved in the stratigraphic record. There are, however, widespread glacial deposits—unsorted, unstratified debris containing striated and faceted cobbles and boulders—recording several major periods of ancient glaciation in various parts of the geologic record. These glacial deposits commonly rest on striated and polished bedrock, and they are associated with varved shales and with sandstones and conglomerates that are typical of outwash deposits. Such evidence implies several major periods of glaciation prior to the last ice age.

The best-documented record of pre-Pleistocene glaciation is found in late Paleozoic rocks (formed 200 to 300 million years ago) in South Africa, India, South America, Antarctica, and Australia. Exposures of ancient glacial deposits are large and numerous in these areas, many resting on a striated surface of older rock (**Figure 14.37**). Deposits of even older glacial sediment exist on every continent but South America.

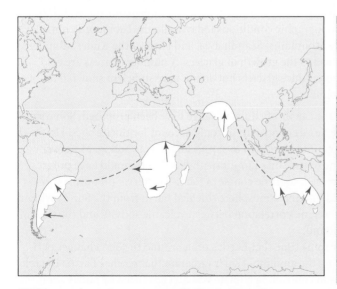

(A) This map shows the areas covered by ice during the late Pennsylvanian and the Permian periods and the direction of ice movement.

(B) The glacially striated bedrock in southern Australia has been exposed by erosion of overlying sedimentary strata. The striations originally formed in the late Paleozoic.

Figure 14.37 Late Paleozoic glaciation is well documented in southern continents by deposits of glacial sedimentary rocks, striated bedrock surfaces, and other glacial features.

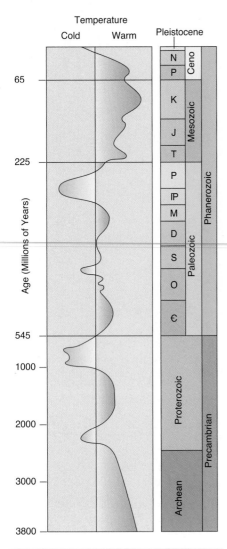

Figure 14.38 Several periods of glaciation have been identified in Earth's long history that may record changes in the surface temperature. The graph shows one estimate of relative temperature changes with time. The curve shows when temperatures were higher (to the right) or lower (to the left) than today.

These indicate that two other periods of widespread glaciation occurred during late Precambrian time (**Figure 14.38**).

Small bodies of glacial sediment from other geologic periods have been found in local areas, but they are not nearly as well documented or as widespread as the Precambrian and Late Paleozoic deposits. Glaciation, therefore, has been a relatively rare phenomenon and has not occurred in regular cycles throughout Earth's long history. Glacial epochs must require a special combination of conditions, which has occurred only a few times in the 4.5 billion years of Earth's history.

Causes of Glaciation

No completely satisfactory theory has been proposed to account for Earth's history of glaciation. The cause of glaciation may be related to several simultaneously occurring factors, such as astronomical cycles, a drop of atmospheric CO_2, the locations of the continents, and changes in ocean currents.

Although the history of Pleistocene glaciation is well established and the many effects of glaciation are clearly recognized, we do not know with complete certainty why Earth's climate changes and why glaciation takes place. For more than a century, geologists and climatologists have struggled with this problem, but it remains unsolved. An adequate theory of glaciation must account for the following facts:

1. During the last ice age, repeated advances of the ice in North America and northern Europe were separated by interglacial periods of warm climate (**Figure 14.39**).
2. Glaciation is an unusual event in Earth's history. Widespread glaciation also occurred at the end of the Paleozoic Era, 200 to 300 million years ago, and during late Precambrian time, approximately 700 million years ago (Figure 14.38).
3. Throughout most of Earth's history, the climate was milder and more uniform than it is now. Several lines of evidence suggest that the global average temperature was about 22°C throughout much of Earth's history. Today the global average is only about 14°C. A period of glaciation requires lowering of Earth's present average surface temperature by about 5°C.
4. Continental glaciers grow on elevated or polar land masses that are situated so that storms bring moist, cold air to them. Glaciers can move into lower latitudes, but they originate in highlands or in high latitudes. Greenland and Antarctica provide favorable topographic conditions today, as do the Labrador Peninsula, the northern Rocky Mountains, Scandinavia, and parts of the Andes Mountains.
5. Precipitation is critical to the growth of glaciers. A number of areas are cold enough at present to produce glaciers but do not have sufficient snowfall to develop glacial systems.

Many hypotheses for the causes of climate change have been proposed. Some suggest that variations in the Sun's energy output could account for the ice ages. However, glaciation is cyclical and cannot be related to simple long-term cooling. Moreover, our present understanding of the Sun's luminosity holds that it should have progressively increased, not decreased, over the course of Earth's history. Still others argue that volcanic dust injected into the atmosphere shielded Earth from the Sun's rays and initiated an ice age. However, no correlation between volcanic activity and the start of the last ice age has been found.

It has been known for some time that Earth's orbit around the Sun changes periodically, cyclically affecting the amount of solar radiation that reaches Earth. The role of Earth's orbital changes in controlling climate was first advanced by James Croll in the late 1800s. Later, Milutin Milankovitch, a Serbian geophysicist, elaborated on the theory and convincingly calculated that these irregularities in Earth's orbit could cause the climatic cycles now known as **Milankovitch cycles**. They are the result of

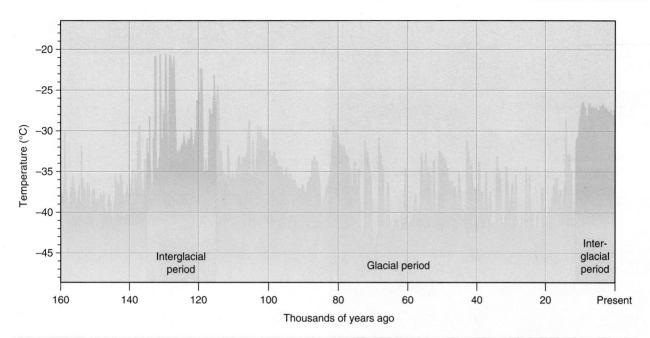

Figure 14.39 A record of climatic change during the last 160,000 years was assembled from studies of ice cores from Greenland's glacier. It shows that the normal pattern of change involves numerous rapid fluctuations in temperature—not only during glacial periods, but throughout interglacial periods as well. The stable warm temperature of the present interglacial period is distinctly abnormal.

the additive behavior of several types of cyclical changes in Earth's orbital properties. Changes in the eccentricity (a measure of the noncircularity) of Earth's orbit occur in a cycle about 96,000 years long. The inclination, or tilt, of Earth's axis varies periodically between 22° and 24.5°. The tilt of Earth's axis, of course, causes the seasons: The greater the tilt, the greater the contrast between summer and winter temperatures. Changes in the tilt occur in a cycle 41,000 years long. Also, Earth wobbles on its spin axis and completes one wobble, or *precession*, every 21,700 years. According to the Milankovitch theory, these astronomical factors cause a periodic cooling of Earth, with the coldest part in the cycle occurring about every 40,000 years (**Figure 14.40**). The main effect of the Milankovitch cycles is to change the contrast between the seasons and not to change the amount of solar heat delivered to Earth. These cycles within cycles predict that during maximum glacial advances, winter temperatures are milder but so too are summer temperatures. As a result, less ice is melted than is received and a glacier may build up.

Milankovitch worked out the ideas of climatic cycles in the 1920s and 1930s, but it was not until the 1970s that a sufficiently long and detailed chronology of the Pleistocene temperature changes was worked out to test the theory adequately. A correspondence between astronomical cycles and late Cenozoic climate fluctuations now seems clear. Furthermore, studies of deep-sea cores, and the fossils contained in them, indicate that the fluctuation of climate during the last few hundred thousand years is remarkably close to that predicted by Milankovitch.

A problem with this theory is that the astronomical cycles have been in existence for billions of years. We might expect that glaciation would have been a cyclic event throughout geologic time, instead of a rare occurrence (Figure 14.38). Other factors must also be involved that caused Earth's temperature to drop below a critical threshold. Once the temperature is low enough, Milankovitch cyclicity will act as an ice age pacemaker, forcing the planet into and out of glacial epochs.

An attractive theory holds that decreases in atmospheric carbon dioxide, an important greenhouse gas, started the long-term cooling trend that eventually led to glaciation. Recent studies of the carbon dioxide content of gas bubbles preserved in the

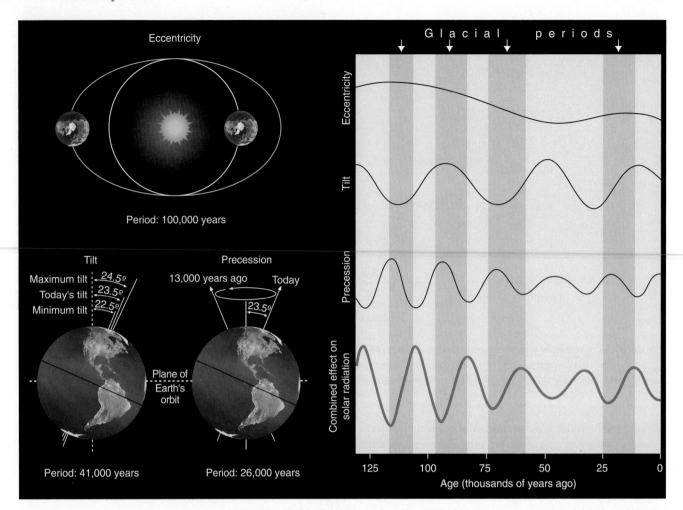

Figure 14.40 Milankovitch climate cycles are caused by periodic changes with time in Earth's orbital elements, including orbital eccentricity, obliquity or tilt of the spin axis, and precession or wobble of the spin axis. When all of these cycles are added together, they affect the seasonal differences in temperature on Earth. The total solar radiation at 65° N is shown as an example. The principal periods of glaciation as defined from the continents, seafloor sediments, and polar ice cores are also shown.

Greenland ice cap lend support to this idea. High carbon dioxide contents correspond to warm interglacial periods, and low carbon dioxide to glacial epochs. Similarly, conclusions drawn from our current understanding of the geochemical cycle of carbon indicate a greater than 10-fold decrease in atmospheric carbon dioxide since the middle of the Mesozoic Era (**Figure 14.41**). However, we must still ask: Is this decline the cause of global cooling or is it the result? What caused the carbon dioxide levels to decline?

Another important component of the cause for the long-term temperature drop (Figure 14.38) may be related to the positions of the continents, relative to the poles. This relation can control the circulation of the oceans and atmosphere, affecting how ocean currents carry heat to high latitudes. Here again, the theory of plate tectonics helps to explain how Earth's systems operate. Throughout most of geologic time, the North Pole appears to have been in a broad, open ocean that allowed major ocean currents to move unrestricted. Equatorial waters flowed into the polar regions, warming them with water from more temperate latitudes. This unrestricted circulation produced mild, uniform climates that persisted throughout most of geologic time.

Throughout the Cenozoic Era, the large North American and South American continental plates moved westward from the Eurasian plate. This drift culminated in the development of the Atlantic Ocean, trending north–south, with the North Pole in the small, nearly landlocked basin of the Arctic Ocean. The Isthmus of Panama developed at a convergent plate margin about 4 million years ago, and further separated oceanic circulation and created the Pacific and Atlantic oceans. Meanwhile, Antarctica drifted over the South Pole and a strong circumpolar current developed in the surrounding ocean. This current prevented Antarctica from exchanging heat with the tropics. By about 34 million years ago, alpine glaciers started to develop and then they spread to form a continental ice sheet that covered almost all of Antarctica by about 14 million years ago. The ages of volcanoes on glaciated basement rocks and evidence from deep-sea cores in the southern oceans, show that glaciation in the Antarctic began long before the Pleistocene and has continued ever since.

Plate tectonics may have also caused other important changes in Earth's climate. While the Arctic Ocean became enclosed and surrounded by continents, the Rocky Mountains and the Himalayas rose. These mountains may have altered the flow of the atmosphere and, according to computer models, could have created a colder climate in the Northern Hemisphere. Plate tectonics may even provide an explanation for the drop in the abundance of carbon dioxide during the last 60 million years. The process may have begun with the rise of the high Himalaya mountains. Uplift and erosion exposed large volumes of rock to weathering. As weathering attacked the silicates, many ions went into solution. When feldspar weathers, calcium ions go into solution and are carried by rivers to the oceans. Once in the oceans, calcium may have combined with dissolved carbon dioxide to make limestone, in essence fossilizing part of the atmosphere's carbon dioxide. Over the course of millions of years of weathering and carbonate deposition, the carbon dioxide levels may have dropped (Figure 14.41). As carbon dioxide was cleared from the air, the greenhouse effect was also diminished and Earth's climate cooled.

As can be seen, there are many variables in the interactions of climate, atmospheric composition, the circulation of the oceans, and plate tectonics that could have helped cause the ice ages. No single causative agent has been identified. Apparently, an ice age occurs because of several simultaneously occurring factors. The Pleistocene glaciation appears to have had two major underlying causes. One is related to a gradual long-term drop in global temperature. The most likely causes may be plate tectonic movements and a drop in atmospheric carbon dioxide. A second cause is needed to explain the waxing and waning of the glacial epochs on short time scales. Milankovitch cycles seem to provide appropriately timed changes.

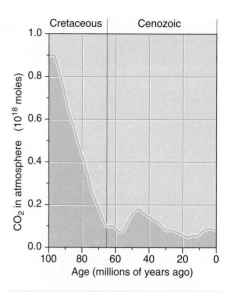

Figure 14.41 **The abundance of carbon dioxide in Earth's atmosphere** has declined dramatically during the last 100 million years. Loss of this important greenhouse gas may have allowed Earth to cool enough for glaciers to accumulate.

STATE OF THE ART   Mass Spectrometry and Earth's Climate

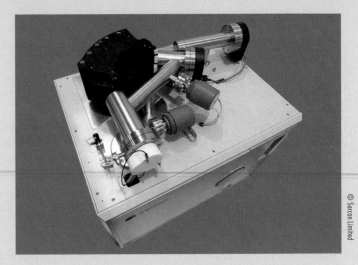

Many of the speculations about the cause of climate variations are rooted in highly specialized geochemical studies. Earth's changing climate has been tracked using tiny marine fossils and a sophisticated instrument called a mass spectrometer. Tiny marine animals secrete shells made of calcium carbonate ($CaCO_3$) by extracting ions of Ca^{+2} and CO_3^{-2} from seawater. The oxygen in the carbonate is in equilibrium with the oxygen isotopic ratio of the surrounding seawater. The isotopic composition of seawater is in turn controlled by the amount of ice on the continents. This is so because light isotopes of oxygen are preferentially evaporated from the ocean to enter the atmosphere and eventually fall as snow on the continents. Thus, during periods of glaciation, the light isotopes of oxygen are extracted and locked into ice. Consequently, the remaining seawater becomes richer in the heavy isotopes of oxygen during periods of glaciation. By analyzing the ratio of the heavy (^{18}O) to light (^{16}O) oxygen isotopes in the fossils, geologists are able to estimate the amount of ice on the continents—a neat trick indeed. The higher the $^{18}O/^{16}O$ ratio in the fossils, the higher the amount of ice on the continents and the lower the paleotemperature.

Samples containing these tiny fossils are collected from the seafloor when researchers on a ship drill cores through the uppermost layers of sediment. Paleontologists carefully extract the fossil shells from each layer. They are able to determine the age of each layer using the principle of faunal succession. In the laboratory, the shells from each layer are treated to release carbon dioxide gas. This gas is collected in a glass tube and then released into a mass spectrometer, which measures the ratio of the oxygen isotopes to one another.

As the name implies, a mass spectrometer is able to separate the isotopes in the gas sample according to their masses (atomic mass is the sum of the protons and neutrons). Specifically, it measures the ratio of the heavy to the light isotopes of oxygen. This ratio is exactly what we need to know to reconstruct Earth's glacial history.

But how does the mass spectrometer do this? In our case, the carbon dioxide gas is heated so that it ionizes—loses an electron and becomes charged. Once the ion is charged, it is accelerated by a magnetic field down a curved metal tube less than one meter long. Magnets along the tube bend the path of the individual ionized particles. The paths of the heavy isotopes are not as easily bent as the paths of the light isotopes, and separation occurs. Ion detectors are placed at the end of the tube and record the numbers of each ion that strike it. By comparing the counts on two detectors, the isotope ratio is calculated.

We now have the information needed to construct a climate variation curve. By plotting (1) the age of the fossil and (2) its isotopic composition, we create a curve that shows the changing amount of ice on the continents. Each bend to the right is a cool period with much ice on the continent; a bend to the left

is a warmer interglacial period with little ice on the continents. We can verify these climate curves by performing the same kind of analysis in widely separated parts of the ocean. What a story is told by the isotope ratios in these tiny fossils!

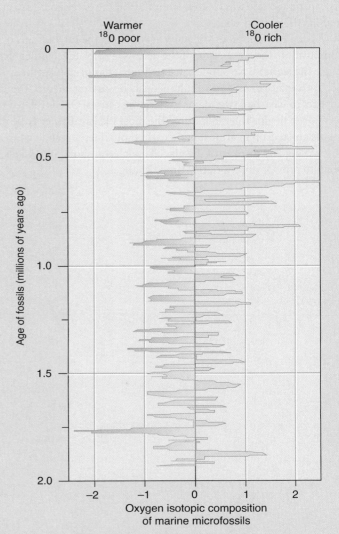

Oxygen isotopic composition of marine microfossils

When Will the Next Ice Age Start?

The next ice age seemed imminent when paleoclimatologists met in 1972 to discuss this question. The previous interglacial periods seemed to have lasted about 10,000 years each. Assuming that the present interglacial period would be just as long, they concluded that "it is likely that the present-day warm epoch will terminate relatively soon *if man does not intervene*." Since 1972 our understanding of the climate system has improved. We know now that not all interglacial periods are of the same length and that solar heating varies in an irregular fashion forced by Milankovitch orbital changes. But we also know that greenhouse gases are increasing in concentration with each passing year. What will be their effect on the onset of the next ice age? Predicting next week's weather is so hard you may think predicting temperatures thousands of years in the future is impossible. But based on the variations in solar heating and on the amount of carbon dioxide in the atmosphere, some calculations of future temperatures have been made (**Figure 14.42**). According to these estimates, the interglacial period we are in now may persist for another 50,000 years. That is, if carbon dioxide levels increase to 750 ppm. Right now, the concentration of carbon dioxide in the atmosphere is only about 400 ppm, but it is rising rapidly as people continue to burn fossil fuels. If carbon dioxide in the atmosphere drops instead, then the next glacial epoch may be only 15,000 years away.

In addition, recent studies of seafloor sediments and cores from glaciers around the world, especially the Greenland glacier, indicate that climatic change is anything but smooth. The long cores drilled from the ice yield information about atmospheric gases, temperature, precipitation, wind, and volcanic activity. Scientists can count the annual layers of ice, just as tree rings are counted, but the ice layers contain much more information than do tree rings. They can preserve elemental carbon from forest fires and datable layers of ash from volcanic eruptions. But the unique feature of ice cores is that they contain actual samples of Earth's ancient atmosphere. According to studies of the oxygen isotopic composition of the ice in these cores, the change from warm to frigid temperatures can occur in a decade or two. Moreover, the ice cores show that an ice age is not uniformly cold, nor are interglacial periods uniformly warm. Analysis of ice cores of the entire thickness of the Greenland Glacier shows that the climate over the last 250,000 years has changed frequently and abruptly (Figure 14.39). The present interglacial period (the last 8000 to 10,000 years) has been fairly stable and warm, but the previous one was interrupted by numerous frigid spells lasting hundreds of years. If the previous period was more typical than the present one, the period of stable climate in which humans flourished—inventing agriculture and thus civilization—may have been possible only because of a highly unusual period of stable temperature.

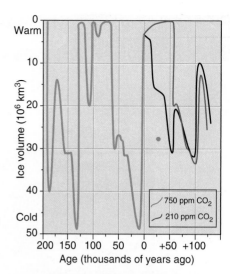

Figure 14.42 An exceptionally long interglacial period may be in store if these calculations are correct. Computer models using the amount of solar heating (as controlled by Milankovitch orbital variations) and the amount of carbon dioxide, predict that the next glacial period may begin about 50,000 years from now (red line assuming a concentration of 750 ppm CO_2) or as soon as 15,000 years from now (if CO_2 drops to 210 ppm, black line).

GeoLogic   Glaciation in the Sierra Nevada

John Muir (1838–1914) was one of the first to suggest that the Sierra Nevada range of eastern California had been extensively glaciated. Originally, his claims were met with skepticism, especially by J.D. Whitney, then the state geologist of California. How could the sunny state of California have sustained glaciers? Now more than 140 years later, few doubt Muir's claims. Look at this photograph to see what evidence you can find of a former period of glaciation.

Observations

1. A U-shaped mountain valley set amid a distinctive mountainous terrain marked by horns and hanging valleys.
2. Linear trails of sediment along the valley walls join arcuate ridges of sediment at the mouth of the canyon.
3. The sedimentary deposits are made of till—unstratified and poorly sorted.

Interpretations

The logical conclusion that glaciers shaped this landscape was made by comparing it with modern glaciated regions. Horns and hanging valleys flank U-shaped valleys that are still filled with flowing masses of glacial ice. Mounds of till have accumulated at the end of the glaciers. In Muir's words, *"The main lateral moraines that extend from the jaws of the amphitheater... are continued in straggling masses along the walls of the amphitheater, while separate boulders, hundreds of tons in weight, are*

left stranded here and there out in the middle of the channel." All that was missing was the ice.

Here, we have reconstructed an ancient glacier that carved this valley in the Sierra Nevada Mountains of California, then transported the material and dropped it in arcuate ridges at the terminus of the glacier. Such evidence helps to show that Earth is coming out of a frigid glacial epoch that peaked 18,000 years ago.

Key Terms

abrasion (p. 405)	erratic (p. 420)	lateral moraine (p. 408)	recessional moraine (p. 412)
arête (p. 408)	esker (p. 420)	loess (p. 431)	roche moutonnée (p. 405)
Channeled Scablands (p. 433)	extending flow (p. 402)	medial moraine (p. 408)	snow line (p. 398)
cirque (p. 408)	glacial plucking (p. 405)	Milankovitch cycles (p. 436)	terminal moraine (p. 412)
compressing flow (p. 402)	glacial striation (p. 406)	moraine (p. 408)	till (p. 420)
continental glacier (p. 400)	ground moraine (p. 420)	outlet glacier (p. 414)	valley glacier (p. 399)
crevasse (p. 403)	hanging valley (p. 412)	outwash plain (p. 412)	varve (p. 420)
drumlin (p. 420)	horn (p. 408)	Pleistocene Epoch (p. 422)	zone of ablation (p. 398)
end moraine (p. 412)	kettle (p. 420)	pluvial lake (p. 429)	zone of accumulation (p. 398)

Review Questions

1. Describe the processes by which snow is transformed into glacial ice.
2. Draw a cross section of a typical valley glacier, and explain how a valley glacial system operates.
3. Contrast compressing and extending flow in a glacier. What part of a glacier is dominated by each type of flow?
4. Which moves faster, the base of a glacier or the ice near the surface? Why? What evidence supports your answer?
5. Why is the flow of glacial ice laminar instead of turbulent as the flow of water in streams?
6. Sketch a model of a continental glacial system and explain how it operates.
7. Explain the processes by which glaciers erode the surface over which they flow.
8. Name and describe landforms produced by valley glaciers.
9. Draw a simplified map of North America showing the extent of the ice sheet during Pleistocene time.
10. Briefly describe the major effects, both direct and indirect, of Pleistocene glaciation.
11. Explain the origin of the Channeled Scablands.
12. Compare and contrast the origins of Lake Michigan and the Great Salt Lake.
13. List several hypotheses for the causes of continental glaciation.
14. Explain the origin of Hudson Bay.
15. Why did sea level change during each period of advance and retreat of the ice?
16. Explain the origin of the present course of the Missouri River.
17. Study Figure 14.16 and explain why the terminal moraines occur in a series of lobate patterns rather than in a straight line.
18. How do geologists measure isostatic adjustments of the crust that result from glaciation?
19. Why did a large number of lakes develop in the arid part of the western United States during each major advance of the ice during the Pleistocene ice age?
20. List the periods of major pre-Pleistocene glaciation that are well documented in the geologic record.

15 Shoreline Systems

Water in oceans and lakes is in constant motion. It moves by wind-generated waves, tides, tsunamis (seismic sea waves), and a variety of density currents. As it moves, it constantly modifies the shores of all the continents and islands of the world, reshaping coastlines with the ceaseless activity of waves and currents. Shoreline processes can change in intensity from day to day, and from season to season, but they never stop.

Shoreline systems are complex open systems where the principal source of energy is wind-generated waves. Ultimately, the wind's energy is derived from the Sun. Gravity is also an important source of energy in the system—its influence is felt in tides and near-shore currents. The materials in the system include the shore itself, sand (and other shoreline sediment), and seawater. The world's present shorelines, however, are not the result of present-day processes alone. The shapes of many coastlines are largely the result of processes other than marine and may owe their outlines to stream erosion or deposition, glaciation, volcanism, tectonism, or even the growth of plants and animals.

The panorama shows the southern coast of Australia and a group of famous sea stacks called the Twelve Apostles—even though only eight remain. As waves pound the shore, they erode the

cliffs of soft marine limestone to form caves, which evolve into arches, which in turn collapse, leaving the tall sea stacks—the highest is 45 m tall. It is apparent that the stacks were once part of a continuous sheet of limestone that extended much farther to the south than it now does. The wave erosion caused the cliffs to recede from the beach, leaving a wide, flat terrace or wave-cut platform in their wake. The platform is produced just below sea level where wave action is vigorous. The cliffs are eroding back at a rate of about 2 cm a year; eventually, new caves, arches, and stacks will form from them. The result of this process is evident onshore as well. The flat terrace above the cliffs is an ancient wave-cut platform that was lifted out of the sea by tectonic movements. During uplift, the shoreline gradually moved south. Thus, wave erosion, followed by tectonic uplift and renewed wave erosion, created this landscape.

Why is this shoreline so different from those of the Atlantic coast of the United States, or the coasts of tropical islands such as Tahiti? In this chapter, we consider these and other questions of coastal dynamics. Shorelines are especially important to our society because of the concentration of population on or near the coasts. In fact, over **twenty percent** of the world's population is within 100 km of the shoreline. To live in harmony with these rapidly changing environments, we must understand their histories and dynamics.

Major Concepts

1. Wind-generated waves provide most of the energy for shoreline processes.
2. Wave refraction concentrates energy on headlands and disperses it in bays.
3. Longshore drift, generated by waves advancing obliquely toward the shore, transports sediment parallel to the coast. It is one of the most important shoreline processes.
4. Erosion along a coast tends to develop sea cliffs by the undercutting action of waves and longshore currents. As a cliff recedes, a wave-cut platform develops, until equilibrium is established between wave energy and the shape of the coast.
5. Sediment transported by waves and longshore current is deposited in areas of low energy to form beaches, spits, and barrier islands.
6. Erosion and deposition along a coast tend to develop a straight or gently curving shoreline that is in equilibrium with the energy expended upon it.
7. Reefs grow in tropical climates and thrive only in shallow, clear marine waters. Fringing reefs around volcanic islands can evolve into atolls.
8. The worldwide rise in sea level, associated with the melting of the Pleistocene glaciers, drowned many coasts. Coasts are classified on the basis of either the process—subaerial or marine—that has been most significant in developing their configurations or their tectonic setting.
9. Tides are produced by the gravitational attraction of the Moon and locally exert a major influence on shorelines.
10. Tsunamis are waves generated by earthquakes, volcanic eruptions, and subaqueous landslides that disturb the seafloor.

Waves

Shorelines are dynamic systems involving the energy of waves and currents. Wind-generated waves provide most of the energy for erosion, transportation, and deposition of sediment. Waves approaching a shore are bent, or refracted, so that energy is concentrated on headlands and dispersed in bays.

Most shoreline processes are directly or indirectly the result of wave action. An understanding of wave phenomena is therefore fundamental to the study of shoreline processes. All waves move some form of energy from one place to another. This is true of sound waves, radio waves, and water waves. The most important types of ocean waves are generated by wind.

As wind moves over the open ocean, the turbulent air distorts the surface of the water. Gusts of wind depress the surface where they move downward; as they move upward, they cause a decrease in pressure, elevating the water's surface. These changes in atmospheric pressure produce an irregular, wavy surface in the ocean and transfer part of the wind's energy to the water. In a stormy area, waves are choppy and irregular, and wave systems of different sizes and orientations may be superposed on each other. As the waves move out from their place of origin, however, the shorter waves move more slowly and are left behind, and the wave patterns develop some measure of order.

What is the nature of the motion of water in wind-generated waves?

Wave Motion in Water

Water waves are described in the same terms as those applied to other wave phenomena. These are illustrated in **Figure 15.1**. The **wavelength** is the horizontal distance between adjacent **wave crests**. The **wave height** is the vertical distance between wave crest and wave trough. The time between the passage of two successive crests is the **wave period**.

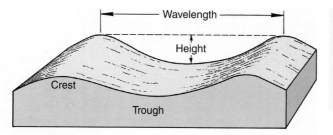

Figure 15.1 The morphology of a wave can be described in terms of its length (the distance from crest to crest), height (the vertical distance between crest and trough), and period (the time between the passage of two successive crests).

Wave motion can be observed easily by watching a floating object move upward (as the crest of a wave approaches) and then sink into the following trough. Viewed from the side, the object moves in a circular orbit with a diameter equal to the wave's height (**Figure 15.2**). Beneath the surface, this orbital motion dies out rapidly, becoming negligible at a depth equal to about one-half the wavelength. This level is known as the **wave base**. The motion of water in waves is therefore distinctly different from the motion in currents, in which water moves in a given direction and does not return to its original position.

A wave's energy depends on its length and height. The greater the wave's height, the greater the size of the orbit in which the water moves. The total energy of a wave can be represented by a column of water in orbital motion.

Breakers

Wave action produces little or no net forward motion of the water because the water moves in an orbital path as the wave advances. As a wave approaches shallow water, however, some important changes occur (**Figure 15.3**). First, the wavelength decreases because the wave base encounters the ocean bottom, and the resulting friction gradually slows the wave. Second, the wave height increases as the column of orbiting water encounters the seafloor. As the wave form becomes progressively higher and the velocity decreases, a critical point is reached at which the forward velocity of the orbit distorts the wave form. The wave crest then extends beyond the support range of the underlying column of water, and the wave collapses, or breaks. At this point, all of the water in the column moves forward, releasing its energy as a wall of moving, turbulent surf known as a **breaker**.

After a breaker collapses, the **swash** (a turbulent sheet of water) flows up the beach slope. The swash is a powerful surge that causes a landward movement of sand and gravel on the beach. After the force of the swash is dissipated against the slope of the beach, the water flows down the beach slope as **backwash**, although some seeps into the permeable sand and gravel.

In summary, waves are generated by the wind on the open ocean. The wave form moves out from the storm area, but the water itself moves in a circular orbit with little or no forward motion. As a wave approaches the shore, it breaks, and the energy of the forward-moving surf is expended on the shore, causing erosion, transportation, and deposition of sediment.

Wave Refraction

A key factor in shoreline processes is **wave refraction** because it influences the distribution of energy along the shore as well as the direction in which coastal water and sediment move. It occurs because the part of a wave in shallow water begins to drag the bottom and slows, whereas the segments of the same wave in deeper water move forward at normal velocity. As a result, the wave is bent, or refracted, so the crest line tends to become parallel to the shore. Wave refraction thus concentrates energy on headlands and disperses it in bays.

To appreciate the effect of wave refraction on the concentration and dispersion of energy, consider the energy in a single wave. In **Figure 15.4**, the unrefracted wave is divided into three equal parts (AB, BC, and CD), each having an equal amount of energy. As the wave moves toward the shore, segment BC, in front of the **headland**,

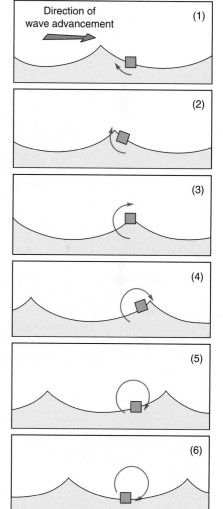

Figure 15.2 The motion of a water particle as a wave advances is shown by the movement of a floating object. As the wave advances (from left to right), the object is lifted up to the crest and then drops down to the trough (top). The wave form advances, but the water particles move in orbits, returning to their original position.

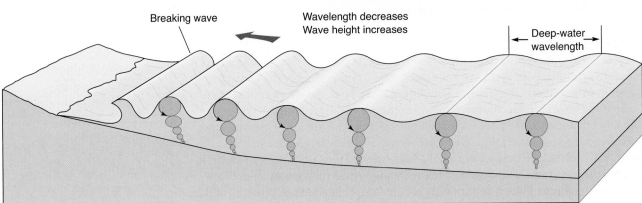

Figure 15.3 **A wave approaching the shore** undergoes several significant changes as the water in orbital motion encounters the seafloor. (1) The wavelength decreases because of frictional drag, and the waves become crowded together as they move closer to shore. Note also that the orbital motion of water in a wave decreases with depth and dies out at a depth equal to about half its wavelength. (2) The wave height increases as the column of water, moving in an orbit, stacks up on the shallow seafloor. (3) The wave becomes asymmetrical, because of increasing height and frictional drag on the seafloor, and ultimately breaks. The water then ceases to move in an orbit and rushes forward to the shore. The photograph shows the characteristic shape of a breaking wave.

How does wave refraction influence erosion and deposition along coasts?

first interacts with the shallow floor and is slowed down. Meanwhile, the rest of the wave (segments AB and CD) moves forward at normal velocity. This difference in velocity causes the crest line of the wave to bend as it advances shoreward. The wave energy between points B and C is concentrated on a relatively short segment (B′C′) of the headland, whereas the equal amounts of energy between A and B, and between C and D, are distributed over much greater distances (A′B′ and C′D′). Breaking waves are thus powerful erosional agents on the headlands but are relatively weak in bays, where they commonly deposit sediment to form beaches. Where major wave fronts are refracted around islands and headlands, the refraction patterns are obvious from the air (**Figure 15.5**).

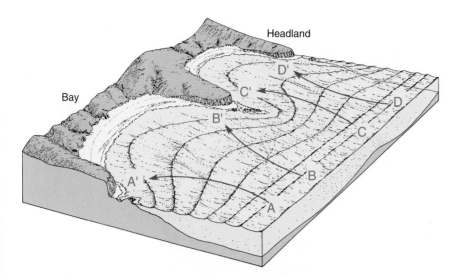

Figure 15.4 Wave refraction concentrates energy on headlands and disperses it across bays. Each segment of the unrefracted wave—AB, BC, and CD—is the same length and therefore has the same amount of energy as the other segments. As the wave approaches shore, segment BC encounters the seafloor sooner than AB or CD and moves more slowly. This difference in the velocities of the three segments causes the wave to bend, so that the energy contained in segment BC is concentrated on the headland (B′C′), while the energy contained in AB and CD is dispersed along the beach (A′B′ and C′D′).

Longshore Drift

Longshore drift is generated as waves strike the shore at an angle. Water and sediment move obliquely up the beach face but return directly down the beach, perpendicular to the shoreline. This movement results in a net transport parallel to the shore. As a result, an enormous amount of sediment is constantly moving parallel to the shore.

Longshore drift is one of the most important shoreline processes. It is generated as waves advance obliquely to the shore (**Figure 15.6**). As a wave strikes the shore at an angle of less than 90°, water and sediment moved by the breaker are transported obliquely up the beach, in the direction of the wave's advance. When the wave's energy is spent, the water and sediment return with the backwash, directly down the beach, perpendicular to the shore. The next wave moves the material obliquely up the shore again, and the backwash returns it again directly down the beach slope. A single grain of sand is thus moved in an endless series of small steps, with a resulting net transport parallel to the shore. This process is known as **beach drift**. A similar process, known as a **longshore current**, develops in the breaker zone; thus, longshore movement occurs in two zones. One is along the upper limits of wave action and is related to the surge and backwash of the waves. The other is in the surf and breaker zone, where material is transported in suspension and by saltation. The two processes work together, and their combined action is known as **longshore drift**.

Longshore drift results in the movement of an enormous volume of sediment. A beach can be thought of as a river of sand, moving by the action of beach drift. If the wave direction is constant, longshore drift occurs in one direction only. If waves approach the shore at different angles during different seasons, longshore drift is

Figure 15.5 Wave refraction around headlands and islands is clearly shown in aerial photographs taken along the coast of Oregon. The energy concentrated on the headlands has reduced some of them to offshore islands.

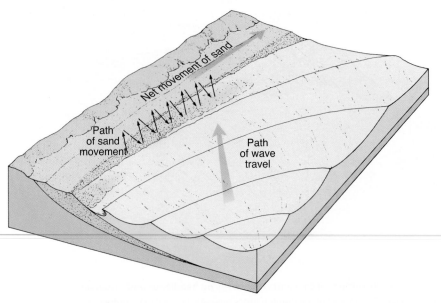

(A) As a wave breaks on the shore, sediment lifted by the surf is moved diagonally up the beach slope. The backwash then carries the particles back down the beach at a right angle to the shoreline. This action is repeated by each successive wave and transports the sediment along the coast in a zigzag pattern. Particles also are moved underwater in the breaker surf zone by this action.

Courtesy of U.S. Geological Survey

(B) Longshore drift can be seen in patterns of sediment in this aerial photograph. The sediment is moved parallel to the shore in a series of waves.

Figure 15.6 Longshore drift occurs where waves strike the beach at an oblique angle.

periodically reversed. Longshore currents can pile significant volumes of water on the beach, which return seaward through the breaker zone as a narrow **rip current**. These currents can be strong enough to be dangerous to swimmers.

Longshore Drift at Santa Barbara, California

One of the best ways to appreciate the process of longshore drift is to consider how it has influenced human affairs. A good example occurred near Santa Barbara, along the southern coast of California, where data have been collected over a considerable period. Santa Barbara is a picturesque coastal town at the base of the Santa Ynez Mountains. It is an important educational, agricultural, and recreational area, and the people there wanted a harbor that could accommodate deep-water vessels. Studies by the U.S. Army Corps of Engineers suggested that the site was unfavorable because of the strong longshore currents, which carry large volumes of sand to the south (**Figure 15.7A**). Rivers draining the mountains of the coastal ranges supply new sediment to the coast at a rate of 600 m³/day. Longshore drift continually moves the sand southward from beach to beach. The currents are so strong that boulders 0.6 m in diameter can be transported. Ultimately, the sand transported by longshore drift is delivered to the head of a submarine canyon and then moves down the canyon to the deep-sea floor. Despite reports advising against the project, a breakwater 460 m long was built and a deep-water harbor was constructed in 1925 at a cost of $750,000. This breakwater was not tied to the shore. Sand, moved by longshore drift, began to pour through the gap and fill the harbor, which was protected from wave refraction and longshore currents by the breakwater (Figure 15.7B). To stop the filling of the harbor, the town had to connect the breakwater to the shore. Sand then accumulated behind the breakwater, at its southern end. Soon a smooth, curving beach developed around

Why is a beach commonly called "a river of sand"?

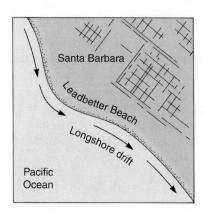

(A) The Santa Barbara coast had significant longshore drift before the breakwater was built.

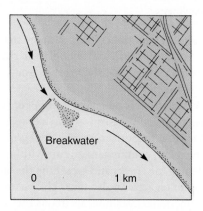

(B) The initial breakwater prevented the generation of longshore currents in the protected area behind it, and therefore the harbor filled with sand.

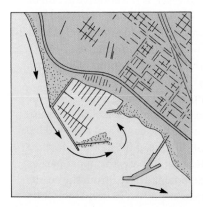

(C) After the breakwater was connected to the shore, longshore currents moved sand around the breakwater and filled the mouth of the harbor. Sand is now dredged from the harbor and pumped down the coast.

(D) Photo of Santa Barbara Harbor.

Courtesy of National Oceanic and Atmospheric Administration

Figure 15.7 **The effect of a breakwater on longshore drift** in Santa Barbara, California, is documented by a series of maps of the coast from 1925 to 1938.

the breakwater, and longshore drift carried sand around the breakwater and deposited it inside the harbor (Figure 15.7C). Two disastrous effects were produced: First, the harbor became so choked with sand that it could accommodate only vessels with very shallow draft; second, the beaches downcoast were deprived of their source of sand and began to erode. Within 12 years, more than $2 million worth of damage had been done to property down the coast from Santa Barbara, as the beach in some areas was cut back 75 m. The problem was solved by the installation of a dredge in the Santa Barbara harbor to pump out the sand and return it to the longshore drift system on the downcurrent side of the harbor. Most of the beaches have been partly replenished, but dredging is very expensive.

Erosion Along Coasts

Erosion along coasts results from the abrasive action of sand and gravel, moved by the waves and currents and, to a lesser extent, from solution and hydraulic action. The undercutting action of waves and currents typically produces sea cliffs. As a sea cliff recedes, a wave-cut platform develops. Minor erosional forms associated with the development of sea cliffs include sea caves, sea arches, and sea stacks.

Coastal regions are sculpted in many shapes and forms, such as rocky cliffs, low beaches, quiet bays, tidal flats, and marshes. The topography of a coast results from the same basic forces that shape other land surfaces: erosion, deposition, tectonic uplift, and subsidence.

Wave action is the major agent of erosion along coasts, and its power is awesome during storms. When a wave breaks against a sea cliff, the sheer impact of the water can exert a pressure exceeding 100 kg/m^2. Water is driven into every crack and crevice of the rocks, compressing the air within. The compressed air then acts as a wedge, widening the cracks and loosening the blocks.

Solution activity also takes place along the coast and is especially effective in eroding limestone. Even noncalcareous rocks can be weathered rapidly by solution activity because the chemical action of seawater is stronger than that of fresh water.

Sea Cliffs and Wave-Cut Platforms

The most effective process of erosion along coasts, however, is the abrasive action of sand and gravel moved by the waves. These tools of erosion operate like the bed load of a river. Instead of cutting a vertical channel, however, the sand and gravel moved by waves cut horizontally, forming wave-cut cliffs and wave-cut platforms.

Why are most shores undergoing vigorous erosion?

To understand the nature of wave erosion and the principal features of its forms, study the diagram and photographs in **Figure 15.8**. Where steeply sloping land descends beneath the water, waves act like a horizontal saw, cutting a notch into the bedrock at sea level. This undercutting produces an overhanging **sea cliff**, or **wave-cut cliff**, which ultimately collapses. The fallen debris is broken up and removed by wave action, and the process is repeated on the fresh surface of the new cliff face. As the sea cliff retreats, a **wave-cut platform** is produced at its base, the upper part of which commonly is visible near shore at low tide. Sediment derived from the erosion of the cliff, and transported by longshore drift, may be deposited in deeper water to form a **wave-built terrace**. Stream valleys that formerly reached the coast at sea level are shortened and left as **hanging valleys** when the cliff recedes.

As the platform is enlarged, the waves break progressively farther from shore, losing much of their energy by friction as they travel across the shallow platform. Wave action on the cliff is consequently greatly reduced. Beaches can then develop at the base of the cliff, and the cliff face is gradually worn down, mainly by weathering and mass movement. Because wave-cut platforms effectively dissipate wave energy, the size to which they can grow is limited. However, some volcanic islands have been truncated completely by wave action and slope retreat so that only a flat-topped platform is left near low tide.

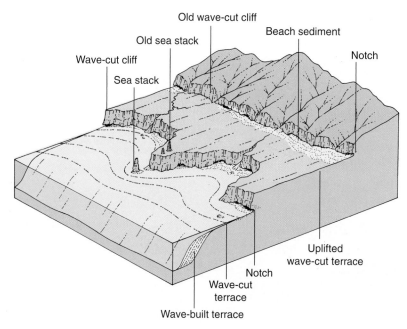

(A) An uplifted wave-cut platform and a new sea cliff and platform in the process of forming. As erosion continues, the cliff recedes to form a wave-cut platform. Some sediment eroded from the shore is deposited in deeper water as a wave-built terrace.

(C) Wave action operates like a horizontal saw cutting at the base of the cliff like this one in Oman.

(B) A wave-cut platform on the Washington coast.

Figure 15.8 A wave-cut platform is the fundamental landform produced by wave erosion.

Sea Caves, Sea Arches, and Sea Stacks

The rate at which a sea cliff erodes depends on the durability of the rock and the degree to which the coast is exposed to direct wave attack. Zones of weakness (such as outcrops with joint systems, fault planes, and beds of shale between harder sandstones) are loci of accelerated erosion. If a joint extends across a headland, wave action can hollow out an alcove, which may later enlarge to a **sea cave**. Because the headland commonly is subjected to erosion from two sides, caves excavated along a zone of weakness can join to form a **sea arch** (**Figure 15.9**). Eventually, the arch collapses,

Figure 15.9 The development of sea caves, sea arches, and sea stacks is associated with differential erosion of a headland.

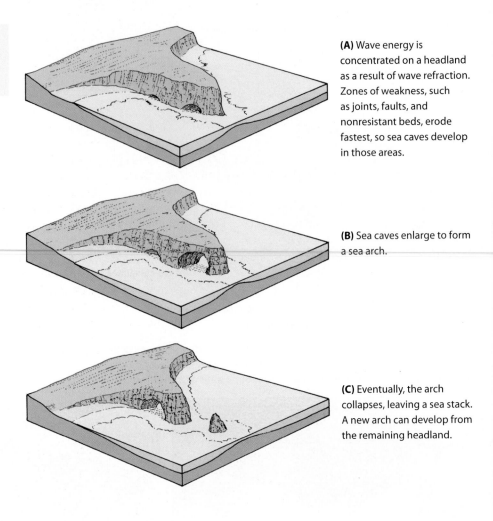

(A) Wave energy is concentrated on a headland as a result of wave refraction. Zones of weakness, such as joints, faults, and nonresistant beds, erode fastest, so sea caves develop in those areas.

(B) Sea caves enlarge to form a sea arch.

(C) Eventually, the arch collapses, leaving a sea stack. A new arch can develop from the remaining headland.

and an isolated pinnacle known as a **sea stack** is left in front of the cliff. An excellent example of the development of sea arches and stacks is shown in **Figure 15.10** in which the collapse of a sea arch is documented by photographs.

Summary of Coastal Erosion

Coastal erosion is a natural process that has altered the world's shorelines ever since the oceans were first formed at least 4 billion years ago. Every day, the surging action of waves, the movement of longshore currents, and the pounding of storms erode shorelines. In addition, sea level is constantly changing; with each rise or fall, a new coastline is formed and the process of reshaping the shore begins anew. Because of the recent rise in sea level, due to the melting of the glaciers, many shorelines of the world are several hundred meters higher than they were 20,000 years ago. Vigorous erosion by waves and currents will continue.

Can coastal erosion be stopped?

The rate at which wave action cuts away at the shore is extremely variable. It depends on the configuration of the coast, the size and strength of the waves, and the physical characteristics of the bedrock. In poorly consolidated material, such as glacial moraines, stream deposits, or sand dunes, the rate of cliff retreat may be as much as 30 m/yr, but rates of erosion along most coasts are much slower.

An interesting example of rates of coastal erosion over a longer time is documented by maps made by the ancient Romans when they conquered Britain. These maps show that in approximately 2000 years, parts of the British coast have been eroded back more than 5 km, and the sites of many villages and landmarks have been swept away. Other examples of rapid wave erosion are found on new volcanic islands, such as Surtsey, near Iceland. The newly formed volcanic ash that makes up such islands can be completely planed off by wave action in a matter of only a few decades.

(A) Sea arches and sea stack as they appeared in 1969.

(B) The same area in 1987, after collapse of arch.

Figure 15.10 Collapse of a sea arch along the coast of California. Between 1969 and 1987, wave erosion also eliminated a small sea stack.

The reality of coastal erosion is made painfully clear by the passion of Americans to live and vacation on the seashore. Development projects unwittingly put more and more people and property on the shore, an area that by its very nature is dynamic and mobile. In a remarkably short period of time, waves can erode high cliffs like those that surround most of the island of Hawaii (**Figure 15.11**). About 86% of California's coast is receding at an average rate of 0.15 to 0.75 m/yr (**Figure 15.12**). Parts of Monterey Bay lose as much as 2 to 3 m/yr. Cape Shoalwater, Washington, about 100 km west of Olympia, has been eroding at a rate of more than 30 m/yr. Parts of Chambers County, Texas, have lost 3 m of coast in nine months. In parts of North Carolina, erosion in one year has cut into beachfront property up to 25 m.

To combat these losses, people build sea walls and breakwaters, but these are local and temporary solutions at best. A sea wall, or jetty (a long concrete or rock structure that juts out into water to restrain waves and currents), may protect threatened property near it, but it often hastens erosion in other areas. There may be no simple answer. Carefully written zoning laws that limit coastal development may be the best option. In our battle with nature, retreat might be the ultimate solution.

© Bob Abraham/Getty Images

Figure 15.11 High sea cliffs of Hawaii indicate that the shoreline has receded tens of kilometers by wave action and slumping. Older islands in the chain have been completely planed off in a matter of only a few million years.

Figure 15.12 Erosion of sea cliffs is a major process along shorelines. This photo shows erosion of a sea cliff along the coast of California. In the last two decades, dozens of homes and hundreds of other structures have been lost or damaged because of sea cliff recession, particularly during storms triggered by El Niño oscillations.

© Mark Gibson/Photolibrary/Getty Images

Deposition Along Coasts

Sediment transported along the shore is deposited in areas of low wave energy and produces a variety of landforms, including beaches, spits, tombolos, and barrier islands.

A shoreline is a system that involves input of sediment from various sources, transportation of the sediment, and ultimate deposition. Much of the sediment is derived from the land and delivered to the sea by major rivers. The sediment is then transported by waves and longshore currents and is deposited in areas of low energy, where it builds a variety of landforms. Changes continue by both erosion and deposition until the coastline is smooth and straight, or gently curving.

Figure 15.13 shows some important elements in a coastal system. The primary sources of sediment, for beaches and assorted depositional features, are the rivers that drain the continents. Sediment from the rivers is transported along the shore by longshore drift and is deposited in areas of low energy. Erosion of headlands and sea cliffs is also a source of sediment. In tropical areas, the greatest source of sand commonly is shell debris, derived from wave erosion of near-shore coral reefs. Sediment can also leave the system by landward migration of coastal sand dunes and by transportation into deep areas of the ocean floor, where turbidites accumulate as submarine fans.

Beaches

What is the source of sand on beaches?

A **beach** is a shore built of unconsolidated sediment. Sand is the most common material, but some beaches are composed of cobbles and boulders and others of fine silt and clay. The physical characteristics of a beach (such as slope, composition, and shape) depend largely on wave energy, but the supply and size of available sediment particles are also important. Beaches composed of fine-grained material generally are flatter than those composed of coarse sand and gravel.

Spits

In areas where a straight shoreline is indented by **bays** or **estuaries**, longshore drift can extend the beach from the mainland to form a **spit**. A spit can grow far out across the bay as material is deposited at its end (**Figure 15.14**). Eventually, it may extend completely across the front of the bay, forming a **baymouth bar**.

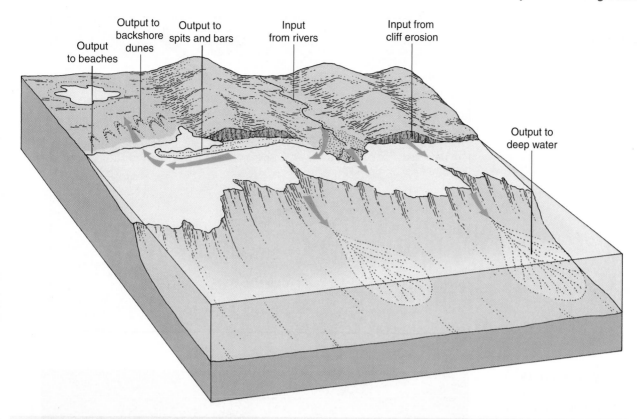

Figure 15.13 A shoreline is a dynamic system of moving sediment. Most of the sediment in a shoreline system is supplied by rivers bringing erosional debris from the continent and by the erosion of sea cliffs by wave action. This material is transported by longshore drift and can be deposited on growing beaches, spits, and bars. Some sediment, however, leaves the system either by transportation to deeper water by turbidites or by the landward migration of coastal sand dunes.

Tombolos

Beach deposits can also grow outward and connect the shore with an offshore island to form a **tombolo**. This feature commonly is produced by the island's effect on wave refraction and longshore drift (**Figure 15.15**). An island near a shore can cause wave refraction to such an extent that little or no wave energy strikes the shore behind it. Longshore drift, which moves sediment along the coast, is not generated in this wave shadow zone. Sediment carried by longshore currents is therefore deposited behind the island. The sediment deposit builds up and up and eventually forms a tombolo, a bar or beach connecting the shore to the island. Longshore currents then move uninterrupted along the shore and around the tombolo.

Barrier Islands

Barrier islands are long, offshore islands of sediment, trending parallel to the shore (**Figure 15.16**). Almost invariably, they form long shorelines next to gently sloping coastal plains, and they typically are separated from the mainland by a lagoon. Most barrier islands are cut by one or more tidal inlets. Many barrier islands develop from the growth of spits across an irregular shoreline (Figure 15.14).

Transportation and deposition along many coasts can be measured using historical monuments, maps, and sequences of aerial photography. In northern France, a dike built at the shoreline in 1597 is now more than 3 km inland from the present shore, indicating an average rate of spit migration of about 1 km/100 yr.

Other rates of spit migration, based on dated maps, include the western end of Fire Island (off the southern coast of Long Island, New York) and the Rockaway spit (western Long Island). The rate of lateral migration for both Long Island spits is 65 m/yr, or 6.5 times the rate of migration along the French coast.

(A) Waves strike a shore obliquely and cause longshore sediment transport. The sediment moves in a series of sand waves around the end of a beach and into a bay.

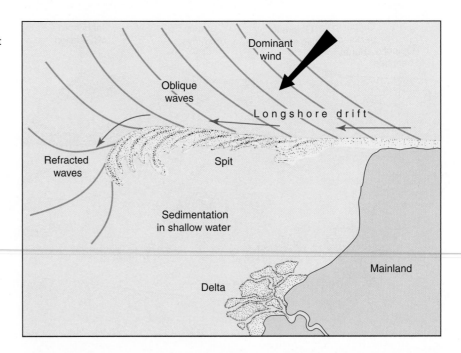

(B) A spit or barrier island can form by migration of a spit. Sediment moving along the shore is deposited as a spit in the deeper water near a bay. The spit grows parallel to the shore by longshore drift.

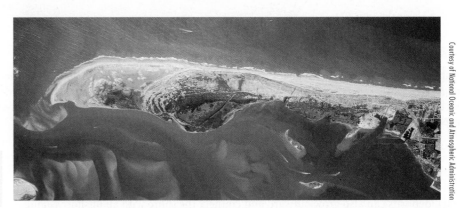

Courtesy of National Oceanic and Atmospheric Administration

Figure 15.14 Curved spits develop as longshore drift moves sediment around the point of a barrier beach.

How does a spit evolve into a barrier?

One of the best-documented examples of coastal modification is the barrier beach at Chatham, Massachusetts (**Figure 15.17**). Before 1987, the barrier beach known as Nauset, or North Beach, curved southward in a long, graceful arc and terminated as a spit south of the town of Chatham. On January 2, 1987, a storm and high tide cut a narrow slice (less than a half-meter deep and 5 m wide) through the barrier. By 1992, the break was more than 3 km wide and 8 m deep, and a new beach was connected to the mainland. Modification of the barrier has continued to the present. Major storms have created new breaches. The break in the protective barrier permitted extensive erosion on the mainland coast.

Evolution of Shorelines

Processes of shoreline erosion and deposition tend to develop long, straight, or gently curving coastlines. Headlands are eroded, and bays and estuaries are filled with sediment. The configuration of the shoreline evolves until wave energy is distributed equally along the coast, and neither large-scale erosion nor deposition occurs.

All of the coastlines throughout the world are constantly changing. In many areas, changes are rapid, and within only a few decades the local configuration of a shoreline can be significantly modified. Over a longer period, regional variations in coast

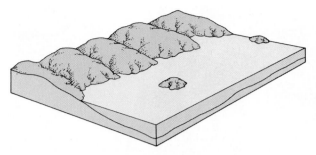

(A) An offshore island acts as a breakwater to incoming waves and creates a wave shadow along the coast behind it.

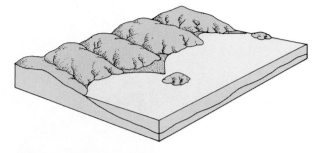

(B) Sediment moved by longshore drift is trapped in the shadow zone.

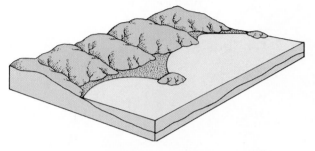

(C) The zone of sediment deposition eventually grows until it connects with the island. Longshore drift will then move sediment along the shore and around the tombolo.

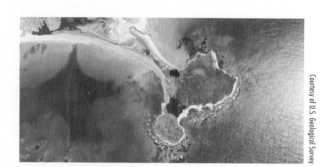

Courtesy of U.S. Geological Survey

(D) An aerial photograph of a tombolo.

Figure 15.15 A tombolo is a bar or beach that connects an island to the mainland. It forms because the island creates a wave shadow zone along the coast, in which longshore drift cannot occur.

configuration occur. This constant and rapid change in our coasts is due, in part, to the rise in sea level that accompanies the melting of glaciers. Other changes in the coast result from uplift or subsidence of the land or expansion or contraction of the sea. Thus, the shape of most coastlines is far from being at equilibrium with the wave energy expended upon them. The general trend is for headlands to be eroded and bays and estuaries to become filled with sediment. The change in the configuration of the shoreline is always in the direction so that energy is equally distributed along the shore, and neither large-scale erosion nor deposition occurs.

In such a condition, the energy of the waves and longshore drift is just sufficient to transport the sediment that is supplied. A shoreline with such a balance of forces is called a **shoreline of equilibrium**. As is the case with a stream profile of equilibrium, a delicate balance is maintained between the landforms and the geologic processes operating on them.

We can construct a simple conceptual model of a shoreline's evolution toward equilibrium and show the changes that would be expected to occur as erosion and deposition operate (**Figure 15.18**). Figure 15.18A shows an area originally shaped by stream erosion and subsequently partly drowned by rising sea level. River valleys are invaded by the sea to form irregular, branching bays, and some hilltops form peninsulas and islands. Next, as shown in Figure 15.18B, marine erosion begins to attack the shore. The islands and headlands are eroded into high wave-cut cliffs. As erosion proceeds (Figure 15.18C), the islands and headlands are worn back, and the sea cliffs increase in height. Minor features, such as sea caves, sea arches, and sea stacks, form by differential erosion in weak places in the bedrock. These are continually being formed and destroyed as the sea cliff recedes. A wave-cut platform develops, reducing wave energy, so a beach forms at the base of the cliff. In a more advanced stage of

What is the configuration of a shoreline of equilibrium?

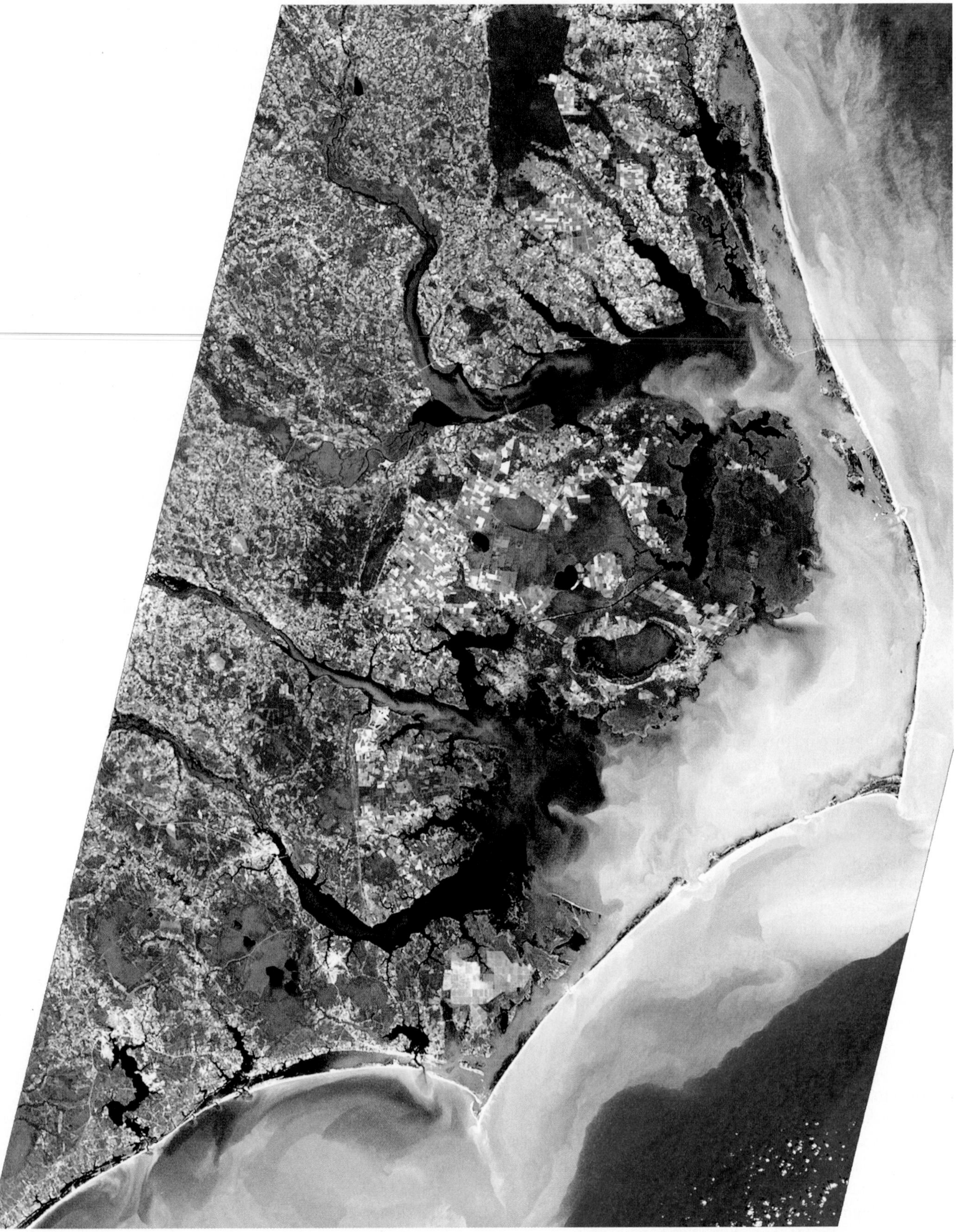

Courtesy of Goddard Space Flight Center/NASA

Figure 15.16 **A barrier island** along the Atlantic coast of the United States has a smooth seaward face, where wave action and longshore drift actively transport sediment. A tidal inlet may form a break in the island, and sediment transported through it is deposited as a tidal delta in the lagoon.

Figure 15.17 **Changes of the shoreline near Chatham, Massachusetts.**

(A) July 1984. Before the break, the barrier-spit acted as a shield protecting Chatham's shore and harbor.

(B) March 1987. Two months after the break, the inlet enlarged and exposed the mainland to erosion.

(C) October 2013. Twenty-six years after the initial breach, the barrier continues to be modified with new breaches and bars forming.

(A) A rise in sea level floods a landscape eroded by a river system and forms bays, headlands, and islands.

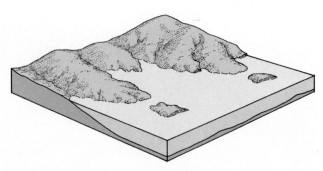

(B) Wave erosion cuts cliffs and, locally, sea stacks and arches on the islands and peninsulas.

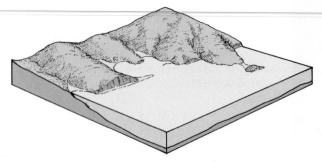

(C) Wave-cut cliffs recede and grow higher, and headlands erode to form new sea cliffs. Sediment accumulates, forming beaches and spits.

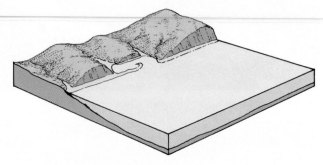

(D) Islands are completely eroded, beaches and spits enlarge, and lagoons form in the bays.

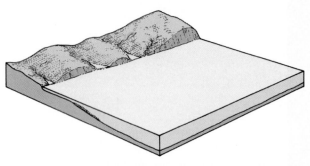

(E) A straight shoreline is produced by the additional retreat of the cliffs and by sedimentation in bays and lagoons. The large wave-cut platform then limits further erosion by wave action.

Figure 15.18 The evolution of a shoreline of equilibrium from an embayed coastline involves changes due to both erosion and deposition. Eventually, a smooth coastline is produced, and the forces acting on it are essentially at equilibrium; thus, neither erosion nor deposition occurs on a large scale.

development (Figure 15.18D), the islands are eroded away and bays become sealed off, partly by the growth of spits, forming lagoons. The shoreline then becomes straight and simple. In the final stages of marine development (Figure 15.18E), the shoreline is cut back beyond the limits of the bay. Sediment moves along the coast by longshore drift, but the wave-cut platform is so wide that it effectively eliminates further erosion of the cliff by wave action. The shoreline of equilibrium is straight and essentially in equilibrium with the energy acting on it. Further modification of the cliffs results from weathering, mass movement, and stream erosion.

Naturally, the development of a shoreline is also affected by special conditions of structure and topography and by fluctuations of sea level or tectonics. The process of erosion of the headlands by wave action and the straightening of the shoreline by both erosion and deposition follow the general sequence of this idealized model; however, actual shorelines rarely proceed through all these stages because fluctuations of sea level upset the previously established balance.

The development of a shoreline is interrupted in many areas by tectonic uplift, which abruptly elevates sea cliffs and wave-cut platforms above the level of the waves.

Photograph by J. Shelton

Figure 15.19 **A series of elevated beach terraces** resulted from tectonic uplift along the southern coast of California and the offshore islands. This photograph of San Clemente Island was taken with the Sun at a low angle, to emphasize the sequence of terraces.

When this happens, wave erosion begins at a new, lower level, and the **elevated marine terraces**, stranded high above sea level, are attacked and eventually obliterated by weathering and stream erosion (**Figure 15.19**).

Hurricanes, Storm Surges, and Shoreline Evolution

Coastal changes are particularly great during intense storms such as hurricanes, typhoons, or "northeasters," as they are called in New England. Hurricanes produce torrential rain, strong winds, and storm surges. In 2005, Hurricane Katrina combined all of these to damage the shorelines of the Gulf Coast of the United States; its cost exceeded $120 billion and almost 2000 people lost their lives. In the Mississippi Delta, levees collapsed and drowned large areas of the city of New Orleans—much of which is already below sea level (**Figure 15.20**).

Storm surges expend tremendous amounts of energy along coastal regions and produce considerable damage and rapid changes in coastal morphology. Intense storms, such as hurricanes and typhoons, are centered around strong low-pressure systems in the atmosphere that cause the sea surface to rise in a broad dome, while they depress the surface farther away. Such a buildup of water beneath the storm produces extensive flooding when it reaches shallow coastal areas. In addition, during the storm, the drag of the wind on the sea surface not only produces high waves but also creates currents that push the water in the direction of the wind. When the water reaches the coastal area, it impinges along the shore, resulting in abnormally high waves and tides (Figure 15.20). Storm surges may be as high as 5 to 7 m. The effect of storm surges on local coastal morphology and sediment transport is profound and extremely rapid. As a result, the erosion, transportation, and deposition of sediment cause extensive damage to life and property. Perhaps the deadliest known storm surges in the Bay of Bengal in 1876 took 100,000 lives, and another in 1970 killed an estimated 300,000 people.

Barrier islands are particularly vulnerable to the storm surges and the waves associated with hurricanes (Figure 15.20). On September 8, 1900, a storm surge hit the barrier island city of Galveston, Texas, killing 6000 people. Large bodies of sediment may be washed over a barrier island to form a washover fan in the adjacent bay. New surge channels (tidal inlets) may be opened, while previously formed inlets may be sealed. In some cases, the barriers are swept away completely. Beach and dune sand may be moved inland, rearranging the area's surface features. Consequently, barrier islands are temporary and poor places for people to live. This was forcefully illustrated by the effects of Hurricane Sandy on the barrier islands of the east coast of the United States in 2012 (**Figure 15.21**).

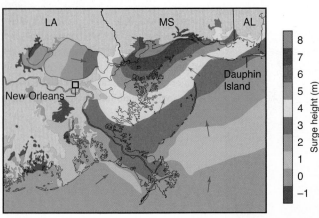 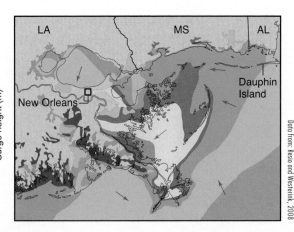

(A) Maps of the height of the storm surge show how they moved with the hurricane across the Gulf of Mississippi. The arrows show the wind directions and the lengths are scaled to the velocity of the wind. As Katrina made landfall, the surge was 6 m high and localized against the levees on a distributary of the Mississippi River delta. The surge spilled into the river and moved upriver to flood New Orleans.

(B) Five hours later the surge had moved against the coast of the state of Mississippi and had a height of 8 m causing massive flooding of low-lying coastal areas. Dauphin Island is one of the barrier islands along the eastern part of the shoreline shown here.

Dauphin Island, AL

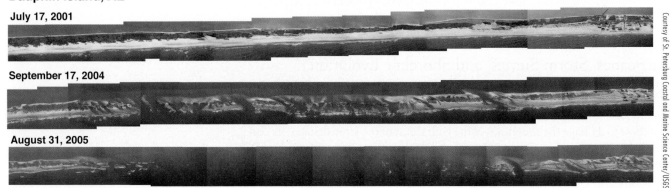

July 17, 2001

September 17, 2004

August 31, 2005

(C) Dauphin Island, a barrier island offshore Alabama, has changed dramatically over the years and was breached completely by erosion caused by the storm surge of Hurricane Katrina.

Figure 15.20 **Storm surges** are produced by storms with low atmospheric pressures and strong winds like those associated with Hurricane Katrina in 2005.

Reefs

Reefs form a unique type of coastal feature because they are biological in origin. Modern reefs are built by a complex community of corals, algae, sponges, and other marine invertebrates. Most reefs grow and thrive only in the warm, shallow waters of semitropical and tropical regions.

In many regions of the ocean, **coral** reefs grow and flourish to such an extent that they significantly modify, if not control, the configuration of a coastline. Reefs are especially important in the warm tropical waters of the South Pacific, where they are a major influence along the coasts of most islands (**Figure 15.22**). Reefs are constructed from invertebrate colonial animals that, instead of building separate isolated shells, build enormous "apartment houses" in which thousands of individuals live. When the animals die, the shell structure remains intact and subsequent generations build their apartments upon the abandoned homes of their predecessors; soon a reef develops into a "rocky" coast. The most impressive modern reef is Australia's Great Barrier Reef, which stretches a distance of 2000 km. Others are noted throughout the South Pacific

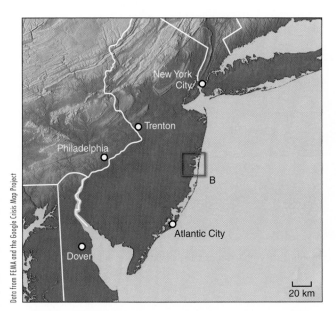

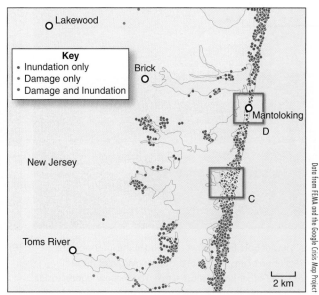

(A) Barrier islands protect much of the Atlantic shoreline of the United States including this area that extends from southern New Jersey to New York City.

(B) Each colored dot represents a home damaged by the hurricane near Tomsriver, New Jersey. You can see that the damage was localized on the barrier island but damage was minimal behind it on the mainland. The damage is not a surprise and has been predicted for decades.

Key
- Minor damage
- Modest damage
- Major damage

(C) This neighborhood map shows that the beachfront properties, the ones most valued for their ocean views, are also the ones most susceptible to storm damage.

(D) Hurricane Sandy caused a bridge to collapse and dramatically changed the shoreline of the barrier island near Mantoloking, New Jersey.

Figure 15.21 Storm surges can be extremely damaging to coastal buildings and cause rapid erosion as shown here for Hurricane Sandy in 2012.

(A) This aerial view of Bora Bora shows an island in the intermediate stage in the evolution of an atoll. Note the outer margin of the reef, where the growth of organisms is most active. The shallow lagoon inside the reef, light blue, is mostly calcareous sand formed by erosion of the reef. The remnant volcano in the center is highly dissected by stream erosion, indicating the elapse of a long period of time since the volcano was active.

(B) Reef mounds are visible through the shallow water in the lagoon.

(C) An underwater view of the reef, showing the community of organisms involved in reef construction.

Figure 15.22 A barrier reef surrounds a volcanic island in the Society Islands, French Polynesia.

as barrier reefs and atolls surrounding ancient volcanic islands. Indeed, coral reefs cover an area about the size of Italy.

Reef Ecology

How can an entire shoreline be formed by growth of organisms?

The marine life that forms a reef can flourish only under strict conditions of temperature, salinity, acidity, and water depth. Most modern coral reefs occur in warm tropical waters between the limits of 30° S latitude and 30° N latitude (**Figure 15.23**). Colonial corals need sunlight, and they cannot live in water deeper than about 75 m. They do not grow up from abyssal depths on the sea floor. Instead, they grow most luxuriantly just a few meters below sea level. Dirty water inhibits rapid, healthy growth because it cuts off sunlight, and the suspended mud chokes the organisms that filter feed. Corals are therefore absent or stunted near the mouths of large muddy rivers. They can survive only if the salinity of the water ranges from 27 to 40 parts per thousand; thus, a reef can be killed if a flood of fresh water from the land reduces the salinity. Coral reefs are remarkably flat on top, the upper surface is usually exposed at low tide but must be covered at high tide. Reefs can grow upward with rising sea level if the rate of rise is not excessive. They can also grow seaward over the flanks of reef

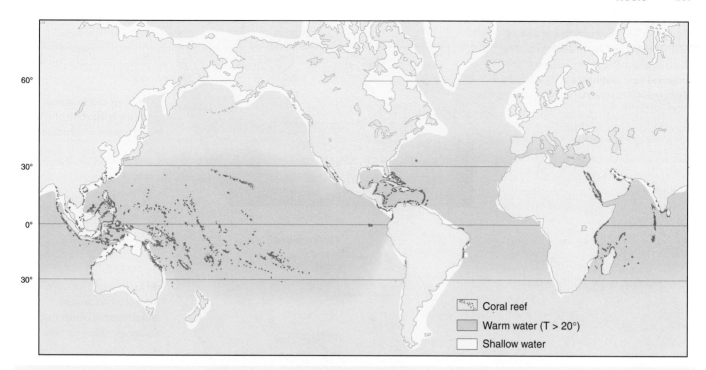

Figure 15.23 The distribution of coral reefs is restricted to low-latitude areas, where the average water temperature exceeds 20°C throughout the year. The water must also be shallow and clear. Reefs do not form where major rivers empty into the ocean, nor do they grow upward from the deep ocean floor of the abyssal plains. Reefs are widely developed throughout the Pacific and Indian Oceans.

debris. The fact that reefs form in such restricted environments makes them especially important as indicators of past climatic, geographic, and tectonic conditions. It also makes them especially sensitive to climate change. Increased carbon dioxide in the atmosphere and the consequent acidification of the oceans threaten many of the organisms that construct coral reefs. The lower pH makes carbonate minerals more soluble, which makes it difficult for many coral species to precipitate the carbonate that forms their skeletons.

Types of Reefs

Fringing reefs, generally ranging from 0.5 to 1 km wide, are attached to such land masses as the shores of volcanic islands (**Figures 15.22 and 15.24**) or continents. The corals grow seaward, toward their food supply. Because coral and other reef-building lifeforms, need sunlight to grow, reefs are usually absent near deltas and mouths of rivers, where the waters are muddy. Heavy sedimentation and high runoff also make some tropical coasts of continents unattractive to fringing reefs.

Barrier reefs are separated from the mainland by a lagoon, which can be more than 20 km wide. As seen from the air, the barrier reefs of islands in the South Pacific are marked by a zone of white breakers. At intervals, narrow gaps occur, through which excess shore and tidal water can exit. The finest example of this type is the Great Barrier Reef, which stretches for 2000 km along the northern shore of Australia, from 30 to 160 km off the Queensland coast.

Platform reefs grow in isolated oval patches in warm, shallow water on the continental shelf. They were apparently more abundant during past geologic periods of warmer climates. Most modern platform reefs seem to be randomly distributed, although some appear to be oriented in belts. The latter feature suggests that they were formed on submarine topographic highs, such as drowned shorelines.

Atolls are roughly circular reefs that rise from deep water, enclosing a shallow lagoon in which there is no exposed central land mass. The outer margin of an atoll is naturally the site of most vigorous coral growth. It commonly forms an overhanging rim, from which pieces of coral rock break off, accumulating as submarine talus on

Figure 15.24 The evolution of an atoll from a fringing reef was first recognized by Charles Darwin. The theory assumes that continued slow subsidence of the ocean floor allows the reef to continue growing upward.

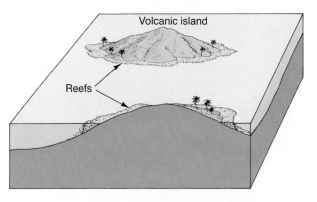

(A) A fringe reef begins to grow along the coast of a newly formed volcanic island.

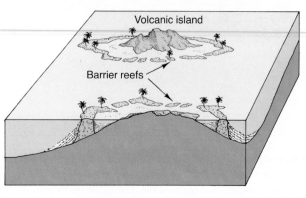

(B) As the island subsides, the reef grows upward and develops a barrier that separates the lagoon from open water.

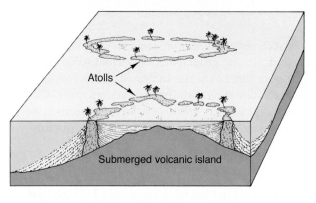

(C) Further subsidence completely submerges the island, but if subsidence is not too rapid, the reef continues to grow upward to form an atoll.

the slopes below. A cross-sectional view of a typical atoll shows that the lagoon floor is shallow and is composed of calcareous sand and silt with rubble derived from erosion of the outer side (see the foreground of Figure 15.24).

Atolls are by far the most common type of coral reef. More than 330 are known, of which all but 10 lie within the Indo-Pacific tropical area. Drilling into the coral of atolls tends to confirm the theory that atolls form on submerged volcanic islands. In one instance, coral extends down as much as 1400 m below sea level, where it rests on a basalt platform carved on an ancient volcanic island. Because coral cannot grow at that depth, it presumably grew upward as the volcanic island sank. A reef this thick probably accumulated over 40 or 50 million years.

Origin of Atolls

In 1842 Charles Darwin first proposed a theory to explain the origin of atolls. As is indicated in Figure 15.24, the theory is based on the continued relative subsidence of a volcanic island. Darwin suggested that coral reefs are originally established as fringing reefs along the shores of new volcanic islands. As the island gradually subsides, the coral reef grows upward along its outer margins. The rate of upward growth essentially keeps pace with subsidence. With continued subsidence, the area of the island becomes smaller, and the reef becomes a barrier reef. Ultimately, the island is completely submerged, and the upward growth of the reef forms an atoll. Erosional debris from the reef fills the enclosed area of the atoll to form a shallow lagoon.

Types of Coasts

On a global scale, coasts are classified on the basis of their tectonic setting. On a local scale, coasts are classified on the basis of the process most responsible for their configuration.

Coastal landscapes exist in an almost infinite variety of marvelous forms, ranging from the sandy barrier islands of the Atlantic, to the rocky shores of New England, to the swamps of southern Florida, to the rugged cliffs of California. Nearly all coasts are complex, both in the types of landforms and in their geologic history. All are dependent upon the landforms that preceded them, all are subject to the effects of changes in sea level that took place during the ice age, and all are influenced by the operation of present coastal processes. Despite these complexities, insight into the nature of shorelines can be gained by considering the processes responsible for their configuration. To do this effectively, we must consider coastal features on a regional scale and contrast them with features developed on a local scale.

Shoreline Classification Based on Plate Tectonics

On a regional basis there are fundamental reasons for similarities and differences along coasts, and it is not surprising that plate tectonics, the fundamental dynamic system of Earth, has a tremendous influence on the origin and evolution of coastlines. The broadest features of coasts, those extending for thousands of kilometers, are directly related to types of plate boundaries and may be classified as follows: (1) convergence coasts, (2) passive-margin coasts, and (3) marginal seacoasts.

A map showing the global distribution of the various types of coasts classified on the basis of plate tectonics is shown in **Figure 15.25**. This map highlights the relationship of various coastal types and the tectonic setting of the continents. You should refer to this map as you study the material that follows.

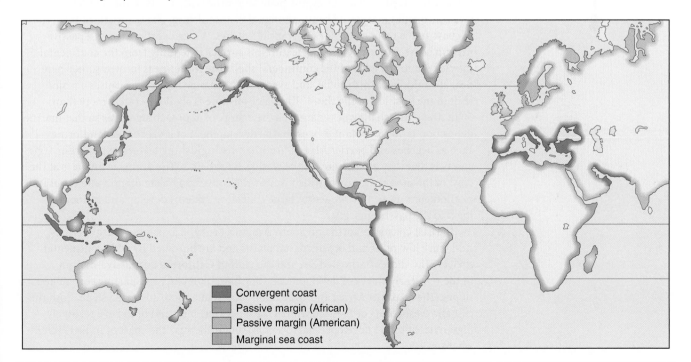

Convergent coast
Passive margin (African)
Passive margin (American)
Marginal sea coast

Figure 15.25 The tectonic classification of coasts is based on the tectonic setting of the continental margins. Passive margins characterize the eastern coasts of North and South America, Africa, and Australia. Mountainous coasts are typical of converging plate margins.

Convergence Coasts   Convergence coasts develop where one plate collides with another to form a subduction zone. They are all relatively straight and mountainous and are distinguished by rugged sea cliffs, raised marine terraces, and narrow continental shelves. Convergence coasts are regions of active seismicity, volcanism, and rapid tectonic uplift, all of which have profound influences on the characteristics of the coast. The high mountain ranges of convergence coasts have rivers that are short, steep, and straight because the drainage divide is at a high elevation near the coast. Consequently, the rivers do not transport enough sediment to build up large deltas; they deposit their sediment load into drowned river valleys or onto small open beaches. None of the world's 25 largest deltas occur on convergence coasts. Indeed, the only areas of significant sediment accumulation along convergence coasts lie in relatively small bays caused by drowning of stream valleys or depressions formed by faulting.

The steep gradients of the seafloor off convergence coasts descend to depths of hundreds of meters, and what sediment is transported by longshore currents is intercepted by submarine canyons and transported to the adjacent trenches, where they form submarine fans. The deep water close to shore also permits large waves to maintain their size because there is no shallow seafloor to interfere with wave motion and diminish their size. Thus, large waves strike the shore with high energy, resulting in rapid rates of erosion.

The west coasts of North and South America are excellent examples of continental convergence coasts, and the Aleutian Islands, Japan, and the Philippines are typical convergence coasts of island arcs.

How do tectonics influence the nature of shorelines?

Passive-Margin Coasts   Passive-margin coasts are initially formed by rifting and the movement of a continent away from a spreading ridge. The coastlines evolve through a series of stages and develop more diverse types. In the early stages of rifting, the separating land masses have high relief marked by steep cliffs. The drainage divide is close to the shore, so rivers and streams emptying into the sea are short, small, and carry little sediment. The coasts during this initial stage are very similar to convergence coasts. The cliffs formed by uplift and rifting are ultimately eroded and, as the continent moves away from the uparched oceanic ridge and as the lithosphere cools, the coast subsides. Passive-margin coasts then remain tectonically stable until they become involved with a converging plate or another rift. The tectonic stability of a passive continental margin is the fundamental distinguishing characteristic of this type of coast and is responsible for many local features. A cross section across a passive-margin coast shows a nearly flat, featureless surface extending from the continental interior out to the edge of the continental shelf. The shoreline is located somewhere near the middle of this profile, and even a slight change in sea level causes a major shift in the position of the shore. The water offshore is shallow—rarely more than 50 m deep. This shallowness dampens the energy of approaching waves so that marine deposition is an important process in the development of local coastal landforms, such as beaches and barrier islands. Another distinguishing feature of passive-margin coasts is that they are the sites of the world's large deltas. The drainage divide is at the crest of the mountain range, which is near the converging plate margin and thousands of kilometers away. Consequently, large collecting systems develop and funnel sediment to the passive margin.

The east coasts of North and South America are typical passive-margin coasts. Both have low relief and broad coastal plains and are bordered by wide continental shelves. The shores of the Red Sea and the Gulf of California are passive-margin coasts in the very early stages of development; they have high cliffs and narrow continental slopes. The coasts of Africa and Greenland represent a more advanced stage. Both the east and west coasts of these continents face ocean ridges, and both have relatively high relief but have developed narrow continental shelves. The coast of India is somewhat similar.

Marginal Sea Coasts   Some continental coasts are near converging plate boundaries but are removed from their influences by an offshore volcanic arc. Although they are near the plate margin and the subduction zone, they are far enough away to be unaffected by convergent tectonics. They thus behave more like passive-margin coasts. Major rivers commonly carry large quantities of sediment and build large deltas and other depositional features, such as beaches, bars, tidal flats, and marshes in the shallow seas, which are protected from vigorous wave action of the open ocean by the associated volcanic arc. For example, the South China Sea is protected from the open ocean by the Philippine island arc, and the Gulf of Mexico is protected by the island arc of the Caribbean.

Shoreline Classification Based on Local Geologic Processes

It should be emphasized that the tectonic classification of coasts is intended to apply only on a regional scale—distances of thousands of kilometers. On a smaller scale, lengths of 100 km or so, the local coastal features are highly diverse. The configurations of the shorelines are controlled by geologic processes involving erosion and deposition. The recent rise of sea level associated with the melting of the last glaciers has had a profound effect on all shorelines. The rise of sea level developed new shorelines that are only a few thousands of years old. Many of these coasts are dramatically out of equilibrium with shoreline erosion. Climate is also a major controlling factor influencing local coastal features because it controls glacial systems, the location of major deltas, and the growth of reefs.

On this smaller scale, two principal types of coasts are recognized: (1) coasts shaped mainly by terrestrial processes of erosion and deposition, and (2) coasts shaped mainly by marine processes.

Coasts Formed by Subaerial Processes

The configuration of many coasts is largely the result of subaerial geologic agents, such as streams, glaciers, volcanism, and earth movements. These processes produce highly irregular coastlines characterized by bays, estuaries, fjords, headlands, peninsulas, and offshore islands. The landforms can be either erosional or depositional, but they are only slightly modified by marine processes. Many of these coasts have experienced a relative rise in sea level. Some of the more common types are illustrated in **Figure 15.26**.

Stream Erosion Coasts   If an area eroded by running water is subsequently flooded by a rise in sea level, the landscape becomes partly drowned. Stream valleys become bays or estuaries, and hills become islands. The bays extend up the tributary valley system, forming a coastline with a dendritic pattern. Chesapeake Bay is a well-known example (Figure 15.26A).

Stream Deposition (Deltaic) Coasts   At the mouths of major rivers, fluvial deposition builds deltas out into the ocean. The deltas dominate the configuration of the coast. They can assume a variety of shapes and are locally modified by marine erosion and deposition (Figure 15.26B).

Glacial Erosion Coasts   Drowned glacial valleys, usually known as **fjords**, form some of the most rugged and scenic shorelines in the world. Fjords are characterized by long, troughlike bays that cut into mountainous coasts, extending inland as much as 100 km. In polar areas, glaciers still remain at the heads of many fjords. The walls of fjords are steep and straight. Hanging valleys with spectacular waterfalls are common (Figure 15.26C).

Glacial Deposition Coasts   Glacial deposition dominates some coastlines in the northern latitudes, where continental glaciers once extended beyond the present shoreline, over the continental shelf. The ice sheets left drumlins and moraines, to be drowned by the subsequent rise in sea level. Long Island, for example, is a partly submerged moraine. In Boston Harbor, partly submerged drumlins form elliptical islands.

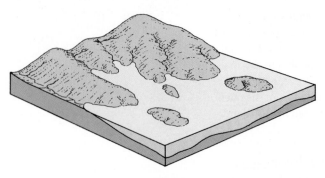

(A) Stream erosion produces an irregular, embayed coast with offshore islands.

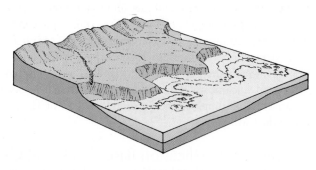

(B) Stream deposition produces deltaic coasts.

(C) Glacial erosion produces long, narrow, deep bays (drowned glacial valleys) called fjords.

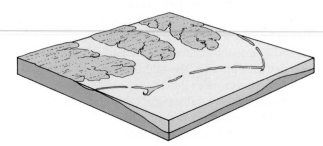

(D) Marine erosion produces wave-cut cliffs.

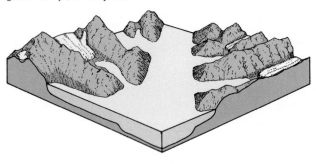

(E) Marine deposition produces barrier islands and beaches.

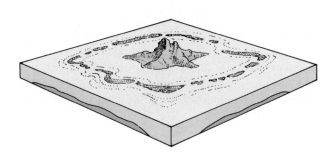

(F) The growth of coral reefs produces barrier reefs and atolls.

Figure 15.26   Classification of coasts is based on the dominant geologic process in developing their configuration.

Coasts Formed by Marine Processes

As marine erosion and deposition begin to act on a coastline, marine processes ultimately dominate and control the coastal configuration. These coasts are characterized by wave-cut cliffs, beaches, barrier islands, spits, and (in some cases) sediment deposited through the action of biological agents, such as marsh grass, mangroves, and coral reefs. Marine erosion and deposition smooth out and straighten shorelines and establish a balance between the energy of the waves and the configuration of the shore.

Wave Erosion Coasts   Wave erosion begins to modify the shoreline as soon as the landscape produced by other agents is submerged. Wave energy is concentrated on the headlands, and a wave-cut platform develops slightly below sea level. Ultimately, a straight cliff and a large wave-cut platform are created. The White Cliffs of Dover, England, are prime examples (Figure 15.26D).

Marine Deposition Coasts   Where abundant sediment is supplied by streams or ocean currents, marine deposits determine the characteristics of the coast. Barrier islands and beaches are the dominant features. The shoreline is modified as storm waves break over the barriers and transport sand inland. The barriers also increase in length and width as sand is added. The lagoons behind the barriers receive sediment and fresh water from streams; thus, they are often capable of supporting dense marsh vegetation. Gradually, a lagoon fills with stream sediment, with sand from the barrier bar (which enters through tidal deltas), and with plant debris from swamps. The barrier coasts of the southern Atlantic and Gulf Coast states are coasts of this type (Figure 15.26E).

Coasts Built by Organisms   Coral reefs develop a type of coast that is prominent in the islands of the southwestern Pacific. The reefs are built up to the surface by corals and algae, and they can ultimately evolve into an atoll. Another type of organic coast that is prevalent in the tropics is formed by intertwined root systems of mangrove trees, which grow in the water, particularly in shallow bays (Figure 15.26F).

Tides

> Tides are produced by the gravitational attraction of the Moon and the centrifugal force of the Earth–Moon system. They affect coasts in two major ways: (1) by initiating a rise and fall of the water level and (2) by generating tidal currents.

On most shorelines throughout the world, the sea advances and retreats in a regular rhythm twice in approximately 24 hours. These changes are the **tides**, and their cause has intrigued people for thousands of years. In the Mediterranean, tides are almost imperceptible; in the Bay of Fundy, they are more than 20 m high. Tides raised by the Moon have gradually slowed Earth's spin. Nine hundred million years ago, a "day" on Earth was only 18 hours long. Ignorance of tides has had an impact on history. Caesar's war galleys were devastated on the British shore because he failed to pull them high enough out of the water to avoid the returning tide. King John of England (1167–1216) was caught in a high tide, lost his treasure and part of his army, and was so enraged he died a week later. The origin of tides was not known until Isaac Newton (1642–1727) showed how tides arise from the gravitational attraction of the Moon and Earth.

The diagram in **Figure 15.27** illustrates, on a highly exaggerated scale, the principal forces that produce tides. The gravitational force exerted by the Moon tends to pull the oceans facing the Moon into a bulge. Another tidal bulge, on the side of Earth opposite the Moon, is caused by centrifugal force. Earth and the Moon rotate around a common center of mass, which lies approximately 4500 km from the center of Earth on a

What are the major effects of the rise and fall of tides?

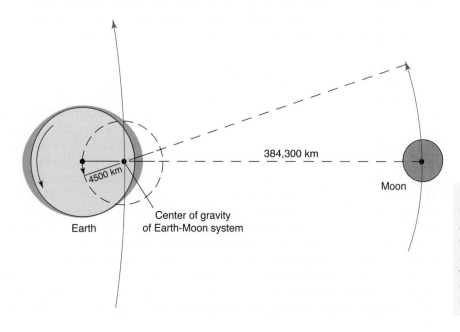

Figure 15.27 Ocean tides are caused by the gravitational attraction of the Moon and the centrifugal force of the Earth-Moon system. On the side of Earth facing the Moon, the gravitational attraction is greater, forming a tidal bulge in the ocean's water. On the other side of Earth, the centrifugal force is greater, causing another tidal bulge.

384,300 km

4500 km

Moon

Earth

Center of gravity of Earth-Moon system

(A) Low tide near Halls Harbor, Nova Scotia.

(B) High tide near Halls Harbor, Nova Scotia.

Figure 15.28 Tidal variations are extreme in some restricted inlets such as the Bay of Fundy in Nova Scotia.

line directed toward the Moon. The eccentric motion of Earth, as it revolves around the center of mass of the Earth–Moon system, creates a large centrifugal force, which forms the second tidal bulge. Earth rotates beneath the bulges, so the tides rise and fall twice every day.

The major effect of the rise and fall of tides is the transportation of sediment along the coast and over the adjacent shallow seafloor. Extremely high tides are produced in shallow seas where the rising water is funneled into bays and estuaries. For example, in the Bay of Fundy, between New Brunswick and Nova Scotia, the tide range (the difference in height between high tide and low tide) is as much as 21 m (**Figure 15.28**). Where fine-grained sediment is plentiful and the tide range is great, the configuration of the coast is greatly influenced by tides and tidal currents.

Tsunamis and Shorelines

Movement of the ocean floor by an earthquake, volcanic eruption, or submarine landslide may produce a wave called a tsunami, which has a long wavelength and travels across the open ocean at high speeds. As a tsunami approaches shore, its wavelength decreases and its wave height increases; therefore, a tsunami can be a formidable agent of destruction along shorelines.

Large waves, known as seismic sea waves or by the Japanese term **tsunamis**, originate from disturbances on the ocean floor. They are also commonly referred to as tidal waves, but they have no relationship with tides at all. Tsunamis can be caused by volcanic eruptions, submarine landslides, or even meteorite impact, but most result from earthquakes that displace the ocean floor. It is not surprising then that most tsunamis occur in the Pacific Ocean, which is circled by active volcanoes and intense seismicity, both of which result from a series of subduction zones surrounding the Pacific. For example, in 2011 a magnitude 9.0 Tohoku earthquake in a subduction zone east of Japan created a tsunami that was as much as 6 m high when it struck land. Nearly 20,000 people died from the tsunami alone (**Figure 15.29**). The tsunami affected about 700 km of shoreline in Japan and covered over 500 km² of land. The tsunami inundated the shoreline with debris-laden waves, destroyed entire towns, damaged a nuclear reactor, swept ships onto land, and cars out to sea. In some places, the tsunami crashed onshore as a breaking wave, but in many others it swept on land like a broad powerful river laden with debris—ships, docks, buildings, and cars in a muddy slurry. Nor did it come silently; witnesses said it sounded like a roaring jet or a freight train.

(A) A surge of seawater as much as 30 m high destroyed buildings along much of the densely populated shoreline in Minamisanriku.

(B) Specially built barriers were insufficient to stop the tsunami in Miyako, Japan.

Figure 15.29 Tohoku tsunami, caused by a subduction zone earthquake, struck the Japanese islands in 2011.

A tsunami differs from wind-produced ocean waves in that energy is transferred to the water by displacement of the seafloor during vertical faulting or by disturbances from volcanic eruptions or submarine landslides. When the seafloor is displaced rapidly, the entire body of water above it is affected. Whatever happens on the seafloor is reflected on the water surface above. Thus, the entire body of water, 5000 to 6000 m deep, participates in the wave motion. Consequently, where part of the ocean floor is uplifted or subsides, a bulge and its adjacent depression are produced on the ocean surface (**Figure 15.30**). The alternating swell and collapse may cover up to 10,000 km² and spread out across the ocean like ripples in a pond. In the open ocean, a tsunami is not a huge wall of water as many people might think; it is usually less than 5 m high, with a wavelength of up to 1000 km. Thus, the slope of the wave surface is very gentle (1 cm or so per kilometer). Such a wave in the open ocean is essentially invisible, because it is masked by the normal surface waves. Indeed, a passing tsunami would not even disturb a game of shuffleboard on a cruise ship.

To understand tsunamis, scientists have made measurements using sensitive pressure meters lying on the seafloor and constructed computer models. In 4000 to 5000 m of water, these instruments are capable of detecting changes in sea level of less than a millimeter. These studies show that a tsunami is not a single wave, an image held by many, but is more like a series of concentric waves—similar to that produced by a pebble thrown into a pond (**Figure 15.31**). The wave front travels at tremendous speed ranging from 500 to 800 km/hr, roughly the speed of a jetliner. It can thus travel across the entire ocean in a few hours. Only as a tsunami approaches the shore does it reveal its tremendous energy (Figure 15.30). The energy distributed in the thick

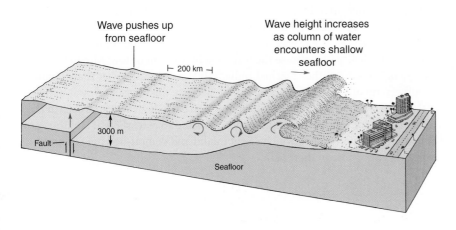

Wave pushes up from seafloor

Wave height increases as column of water encounters shallow seafloor

├─ 200 km ─┤

3000 m

Fault

Seafloor

Figure 15.30 Tsunamis are produced by disruption of the seafloor (by earthquakes, volcanic eruptions, or landslides), which causes a large column of water to move. Initially, the wave is not very high, but as the wave approaches the shore, the column piles up, dramatically increasing the wave's height. The wave may be as much as 30 m high when it strikes the shore.

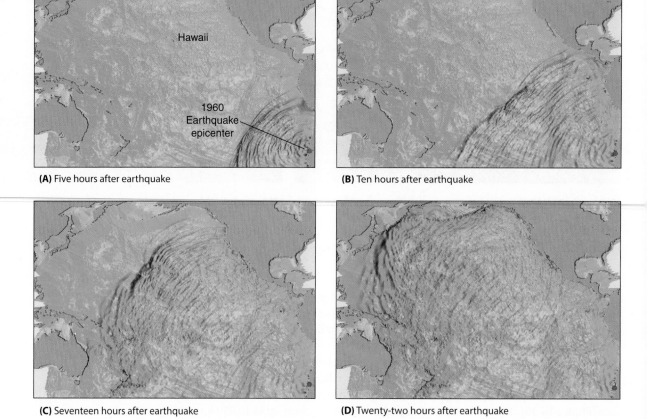

(A) Five hours after earthquake

(B) Ten hours after earthquake

(C) Seventeen hours after earthquake

(D) Twenty-two hours after earthquake

Figure 15.31 Computer model of a 1960 tsunami caused by a large earthquake along the subduction zone beneath Chile. The tsunami created a system of waves that rapidly spread across the entire Pacific Ocean. The wave front reached Japan about one day later. If you look carefully at the wave as it nears Japan, you will also see that tsunamis are reflected and refracted just like any other wave.

Courtesy of P. L. F. Liu, S. N. Seo, S. B. Yoon, C. Devine/Cornell University Center for Advanced Computing

column of water becomes concentrated in a progressively shorter column, resulting in a rapid increase in wave height at the surface. Waves that are fewer than 50 cm high in the deep ocean can build rapidly to heights exceeding 15 m in many cases and over 30 m in rare instances. They exert an enormous force against the shore and can inflict serious damage and great loss of life. For every meter along the coast, a tsunami can deliver more than 100,000 tons of water with a destructive power that is difficult to imagine. The earthquake-triggered tsunami that swept across the Indian Ocean in 2004 was one of the largest in historical times. The surge it created was as much as 30 m high and drove water inland for 2 km or more (**Figure 15.32**). Waves 1.5 m high were generated on shorelines as far away as South Africa to the west and across the Pacific in Mexico to the east. The waves moving across the open ocean were imperceptible to ships in their path because the wave height was 1 m or less. Moving at an average speed of 700 km/hr, they reached India, 1500 km away, in fewer than 2 hours and the southern tip of South Africa in 12 hours. Because the wavelength was about 200 km, the wave crests arrived about 15 minutes apart. As the waves approached the shores of the islands of Indonesia, their heights increased to at least 30 m and thus produced extremely destructive surf, which swept inland and demolished houses, trees, and almost everything else in their path.

© AFP/AFP/Getty Images

Figure 15.32 **The great Indian Ocean Tsunami** of 2004 devastated shorelines and was one of the deadliest known on Earth; over 230,000 people lost their lives in 15 countries.

(A) Drawdown of the sea may precede a tsunami as it did here in Thailand. You can see the white tsunami wave in the background as curious swimmers walk out to see what is going on. Fortunately, all of these Swedish vacationers survived.

Courtesy of DigitalGlobe, Inc

Before

After

Courtesy of DigitalGlobe, Inc

(B) These satellite views show the destruction of most of the buildings, bridges, and roads in the coastal city of Banda Aceh, Indonesia, by the tsunami.

Like all other waves, a tsunami consists of a crest and a trough. Commonly, the first sign that a tsunami is approaching is not an immense wall of water but the sudden withdrawal of the sea (Figure 15.32). Shorelines recede and harbors are emptied because the trough reaches the coast first. This seaward pull of the water from shore may extend out a great distance (over tens of kilometers), often with tragic results. When an earthquake and tsunami struck Lisbon in 1755, the withdrawal of the sea exposed the bottom of the city's harbor. This bizarre sight drew curious crowds, who drowned when the crest of the tsunami rushed in a few minutes later. Many died the same way when a tsunami, generated by an earthquake in the Aleutian trench, hit Hawaii in 1946.

Not all tsunamis are produced by violent earthquakes. For example, in 1896 in Japan, a mild earthquake barely felt on shore was followed by a large tsunami that drowned 22,000 people. Similarly, in Nicaragua in September 1992, no one felt the offshore quake that caused a destructive tsunami that swept coastal homes out to sea and killed 170 people. The reason for this seemingly anomalous condition is that some earthquakes release their energy very slowly, over a minute or more, rather than in a brief snap. This may happen if the boundary between the moving blocks of rock is lubricated. The seismic energy from such a quake moves Earth's surface in long undulations that humans do not feel.

Why is a tsunami so small that it is imperceptible in the open ocean, but may be more than 30 m high when it reaches the shore?

GeoLogic   Tidal Inlet, Eastern Canada

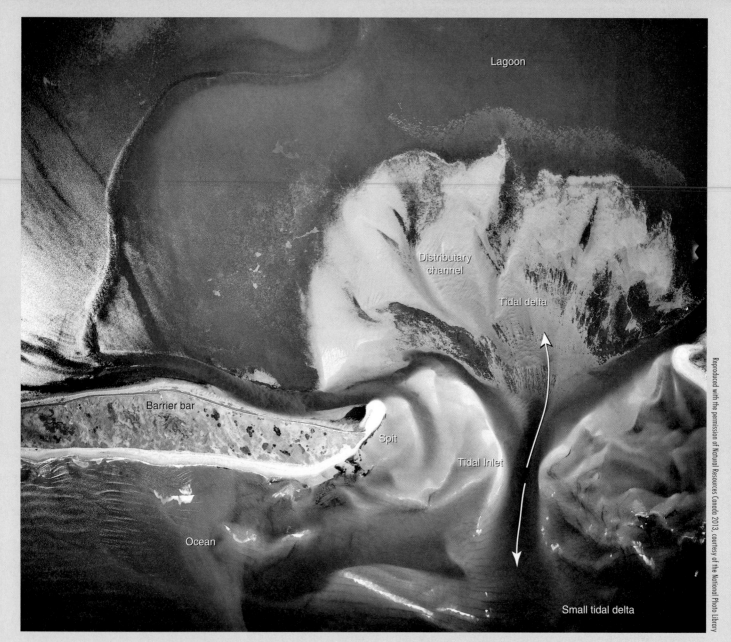

The shore is dynamic being constantly changed and reshaped by waves, currents and tides. Two major processes are active in this area, (1) longshore drift and (2) tides.

Observations

1. A barrier bar extends across the lower part of the photo, broken by a tidal inlet.
2. A lagoon (upper part of photo) separates the mainland from the barrier bar.
3. The open ocean is in the lower part of the area.

Interpretations

1. The curved spit indicates longshore drift from left to right.

2. A large tidal delta is built by incoming tides entering the quiet water of the lagoon through an inlet across the barrier bar.
3. A small tidal delta is built in the open ocean by outgoing tides.
4. Subaqueous sediment is moved by longshore currents forming subaqueous spits that restrict the tidal channel.
5. The source of sediment for the tidal delta is sediment transported by longshore currents.
6. The tidal delta has a series of distributary channels much like a river delta.
7. The tidal delta in the lagoon is much larger than that in the open ocean because wave-action in the open ocean redistributes the delta sediment and inhibits delta growth.

Key Terms

atoll (p. 467)
backwash (p. 447)
barrier island (p. 457)
barrier reef (p. 467)
bay (p. 456)
baymouth bar (p. 456)
beach (p. 456)
beach drift (p. 449)
breaker (p. 447)
coral (p. 464)
elevated marine terrace (p. 463)

estuary (p. 456)
fjord (p. 471)
fringing reef (p. 467)
hanging valley (p. 452)
headland (p. 447)
longshore current (p. 449)
longshore drift (p. 449)
platform reef (p. 467)
rip current (p. 450)
sea arch (p. 453)
sea cave (p. 453)

sea cliff (p. 452)
sea stack (p. 454)
shoreline of equilibrium (p. 459)
spit (p. 456)
storm surge (p. 463)
swash (p. 447)
tide (p. 473)
tombolo (p. 457)
tsunami (p. 474)
wave base (p. 447)

wave-built terrace (p. 452)
wave crest (p. 446)
wave-cut cliff (p. 452)
wave-cut platform (p. 452)
wave height (p. 446)
wavelength (p. 446)
wave period (p. 446)
wave refraction (p. 447)

Review Questions

1. Describe the motion of water in a wind-generated wave.
2. Explain how wave refraction alters the form of a coastline.
3. Explain the origin of longshore drift.
4. Describe the stages in the evolution of a sea cliff and wave-cut platform.
5. Name the major depositional landforms along a coast, and explain the origin of each.
6. What effect would the construction of dams on major rivers have on beaches along the coast?
7. How are elevated marine terraces formed?
8. What conditions are necessary for the formation of a coral reef?

9. Why are coral reefs so poorly developed along the shoreline of equatorial Brazil?
10. Explain the origin of atolls.
11. Describe six common types of shoreline.
12. What are the differences between a shoreline formed along a convergent continental margin and one on a passive continental margin?
13. Explain how ocean tides are generated.
14. Explain the origin of a tsunami.
15. What is the difference in the source of energy for a normal wave, a tide, and a tsunami?

16 Eolian Systems

Geologists once thought that wind, like running water and glaciers, was an effective agent of erosion, but even in the deserts, few major topographic features are the result of wind abrasion. Wind can, however, pick up and transport large quantities of loose sand and dust. As a result, shifting sand dunes or thick sheets of wind-blown dust dominate the landscape in many areas—especially in low-latitude deserts, where precipitation is low and evaporation is high. The great "sand seas" that completely cover large areas of north Africa's Sahara Desert, Saudi Arabia, and central Australia are the most spectacular examples of wind activity on Earth, but sand dunes also occur in many coastal areas and in smaller "rain-shadow" deserts.

Even in the arid regions of the western United States, sand dunes are locally significant landforms. The Great Sand Dunes National Monument in Colorado, the Algodones dunes in southern California, and the Sand Hills of Nebraska are good examples of relatively small dune fields. The panorama above is from the rain-shadow desert in Death Valley, California. It vividly illustrates a landscape where wind is the dominant geologic process and huge sand dunes are the most striking feature.

Wind activity is also important in forming large deposits of wind-blown dust called loess that blanket millions of square kilometers in the mid-latitude continents, including portions of China, the central United States, and central Eurasia. Wind-blown dust covers about one-tenth of the land surface. This fact is important because soils from these deposits are some of Earth's richest farmland and are the foundation for a large percentage of the world's food supply. Wind-blown dust is also carried far out over the seas, where it settles to the floor of the ocean and forms sediment.

The eolian system, like other parts of the hydrologic system, takes its energy from the Sun and the uneven heating of the planet. Like water, it is a fluid that flows readily. Also like water, the wind picks up and transports huge volumes of sediment. Deposition and erosion are associated with distinctive landforms just as they are for streams, groundwater, and flowing ice. Thus, your challenge as you read this chapter is to clearly comprehend the similarities in the different components of the hydrologic system, but also to come to grips with the significant differences between the role of moving water and moving air.

In this chapter, we consider eolian systems as geologic agents: how the wind erodes the surface, transports and deposits sediment, and forms unique features of the landscape.

Major Concepts

1. Wind is not an effective agent in eroding the landscape, but it can produce deflation basins and yardangs as well as small pits and grooves on rocks.
2. The major result of wind activity is the transportation of loose, unconsolidated fragments of sand and dust. Wind transports sand by saltation and surface creep. Dust is transported in suspension, and it can remain high in the atmosphere for long periods.
3. Sand dunes migrate as sand grains are blown up and over the windward side of the dune and accumulate on the lee slope. The internal structure of a dune consists of strata inclined in a downwind direction.
4. Various types of dunes form, depending on wind velocity, sand supply, and constancy of wind direction.
5. Wind-blown dust (loess) forms blanket deposits, which can mask the older landscape beneath them. The source of loess is desert dust or the fine rock debris deposited by glaciers. Some deep oceanic sediment is wind-blown dust from continents.
6. Desertification, the loss of farmable land on the margins of deserts, can be caused by human activity or by slight climatic fluctuations.

The Global Eolian System

The eolian system is a dynamic open system driven by heat from the Sun. The great deserts of the world, where the effects of the wind are most obvious, form in low-latitude regions in zones roughly 30° north and south of the equator. There wind lifts, transports, and eventually deposits loose sand and dust, but its ability to erode solid rock is limited.

Earth's eolian system is intimately tied to the hydrologic system and shares many common features with it. Like the hydrologic system, the eolian system is a manifestation of a moving fluid across the surface. In addition, the energy source for both is the same. The kinetic energy of the wind originates in the Sun and is radiated to Earth. The uneven heating of Earth's surface makes the atmosphere a vast convecting fluid that envelopes the entire planet. Prevailing wind patterns are determined by (1) variations of solar radiation with latitude, (2) the **Coriolis effect** (deflection due to Earth's rotation), (3) the configuration of continents and oceans, and (4) the location of mountain ranges.

The geologic effects of the wind are most obvious in deserts, where precipitation and runoff are low and vegetation is sparse. The locations of most deserts are controlled by the pattern of atmospheric convection. The world's great deserts, such as the Sahara and the deserts of Asia, are mostly in low-latitude belts (**Figure 16.1**). As described earlier, humid equatorial air, heated by solar radiation, rises because it is buoyant. As the air rises to higher altitudes, it cools and releases its moisture, which falls as tropical rains in the equatorial regions. This air is much drier as it continues to convect poleward. Eventually, the dry air descends to the surface near 30° to 35° north and south of the equator. As the air descends, it warms, so it rarely produces any precipitation. Consequently, evaporation of surface moisture, rather than precipitation, occurs in the low latitudes where the convecting air descends. The trade winds move this air back to the equator, where it is again heated, humidified, and rises to start the cycle again. Dry descending air currents also create polar deserts, where precipitation is low, but so is the temperature.

Other deserts lie in **rain shadows**, behind high mountain ranges that intercept moisture-laden air currents. As the air is forced to rise over the mountain range, it cools and precipitates its moisture. On the other side of the range, the dry descending

What controls the location of deserts?

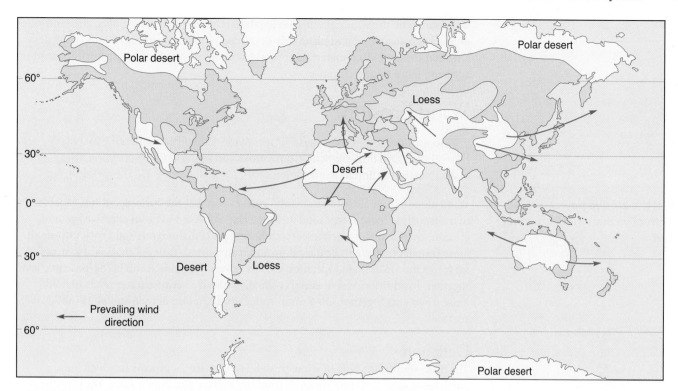

Figure 16.1 The major deserts of the world, the Sahara, the Arabian, the Kalahari, and the deserts of Australia, are near 30° north or south of the equator. These bands are under almost constant high atmospheric pressure where dry air subsides. Desert and near-desert areas cover nearly one-third of the land surface. Wind-blown dust (loess) accumulates downwind from major deserts or along former margins of Pleistocene glaciers. Prevailing winds (arrows) transport dust from the Sahara and Kalahari deserts to the Atlantic Ocean and from deserts of Australia to adjacent seas. Dry descending air currents also create polar deserts, where precipitation is low, but so is the temperature.

air is heated. The arid regions of Nevada and Utah lie in the rain shadow of the Sierra Nevada.

Wind action is most significant in desert areas, but it is not confined to them. Many coasts are modified by winds that pick up loose sand on the beach and transport it inland.

The eolian system is summarized in **Figure 16.2**. Weathering produces sediment particles in a range of sizes. Water transports some of this material downslope before it

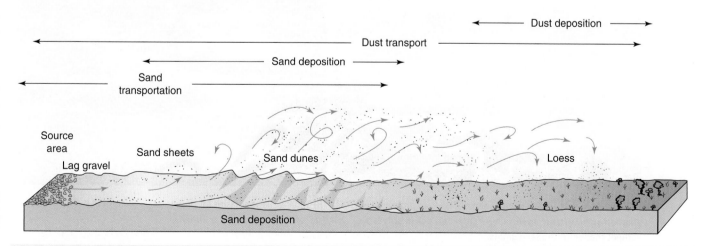

Figure 16.2 The eolian system is driven by energy from the Sun. Flowing air erodes, transports, and deposits fine sediment to form distinctive landforms and rock bodies. Sand sheets, sand dunes, and layers of loess are the major eolian deposits. Lags of coarse particles are left behind.

is picked up by the wind. The coarsest materials are left to form lag deposits; sand-size grains are transported and ultimately deposited to form sheets and dunes in the deserts; and the dust-sized materials are carried far away. Much of the dust is deposited in adjacent, more humid climates.

Wind Erosion

Wind erosion acts in two ways: (1) by deflation, the lifting and removal of loose sand and dust particles from Earth's surface, and (2) by abrasion, the sandblasting action of wind-blown sand.

Why is the erosive power of wind less effective than that of running water?

Wind alone can do little to erode solid rock exposed at the surface, but it is capable of transporting loose unconsolidated material. For wind to be an effective agent of erosion, chemical and mechanical weathering must disintegrate solid rock into small loose fragments that can be picked up and transported. A dry climate is also necessary; in a humid climate vegetation usually covers the surface and holds loose particles together. In addition, wet material is usually cohesive because water tends to hold loose fragments together. On a small scale, wind can also abrade and polish solid rock surfaces.

Deflation

The most significant type of wind erosion is **deflation**, a process in which loose particles of sand and dust are lifted from the surface and blown away. The turbulence of the wind is able to lift these fine materials. Deflation commonly occurs in semiarid regions where the protective cover of grass and shrubs has been removed by the activity of humans and animals. The results are broad shallow depressions called **deflation basins**. Deflation basins also commonly develop where calcium carbonate cement, in sandstone formations, is dissolved by groundwater, leaving loose sand grains that are picked up and transported by the wind (**Figure 16.3**). Large deflation basins, covering areas of several hundred square kilometers, are associated with the great desert areas of the world, particularly in North Africa near the Nile Delta.

Perhaps the best example of wind erosion in the United States is in the Great Plains, especially the High Plains of Colorado, Kansas, and Texas. In this area, innumerable

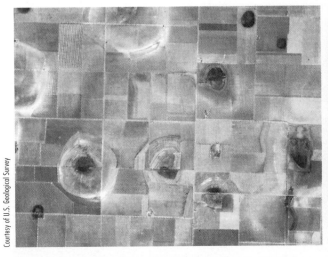

Courtesy of U.S. Geological Survey

(A) Deflation basins in the Great Plains of Texas are produced where solution activity in the layers of horizontal bedrock dissolves the cement that binds the sand grains.

(B) Small deflation basins in sandstone formations in the Colorado Plateau form in a similar manner.

Figure 16.3 Deflation basins are wind erosion features that range in size from small depressions to large basins several kilometers in diameter.

deflation basins—ranging from small dimples, 30 cm deep and 3 m in diameter, to larger basins, 15 to 20 m deep and more than a kilometer across—are scattered across the landscape (Figure 16.3). Many have permanent or intermittent lakes in them. Although some of these depressions may be the result of collapse and local subsidence, deflation has played a major role in their development. During wet periods, water collecting in small depressions will dissolve the calcareous cement in the horizontal sandstone that covers the area. Thus, many of the individual grains in the sandstone formation are loose and free to move about. During dry periods, wind will pick up the loose grains and blow them away. This process creates a larger basin, which collects more water, which in turn dissolves more cement, to produce more loose grains. The process is therefore self-perpetuating. Many of these depressions are enlarged by the activity of animals. Animals were especially influential in the time when great herds of buffalo thronged to the temporary ponds for water. After wading and wallowing, the herds would carry away mud on their bodies and destroy the surrounding vegetation, producing conditions that favored further wind erosion. The depressions have thus been referred to as "buffalo wallows."

In general, wind can move only sand and dust-sized particles, so deflation leaves concentrations of coarser material known as **lag deposits**, or **desert pavements** (**Figure 16.4**). These striking desert features of erosion (**Figures 16.5** and **16.6**) stand out in contrast to deposits in dune fields and playa lakes. Deflation occurs only where unconsolidated material is exposed at the surface. It does not occur where there are thick covers of vegetation or layers of gravel. The process is therefore limited to areas such as deserts, beaches, and barren fields.

Abrasion

Wind abrasion is essentially the same process as the artificial sandblasting used to clean building stone. Energy for abrasion comes from the kinetic energy of the wind. Wind-driven grains impact rock surfaces and small particles are knocked off the rock. Some effects of wind abrasion can be seen on the surface of the bedrock in most desert regions (**Figure 16.7**). In areas where soft, poorly consolidated rock is exposed, wind erosion can be both spectacular and distinctive. Some pebbles, known as **ventifacts** (literally meaning "wind-made"), are shaped and polished by the wind (**Figures 16.7** and **16.8**). Such pebbles are commonly distinguished by two or more flat faces that meet at sharp ridges and are generally well polished. Some faceted ventifacts are up to 3 m long. Other ventifacts have a variety of shapes. Some have surface irregularities, pits, and grooves, or U-shaped depressions with roughly parallel sides aligned with the wind direction (Figure 16.7).

Larger landforms produced by wind abrasion are less common, but in some desert regions, distinctive linear ridges, called **yardangs**, are produced by wind erosion. These features were first discovered in China's Taklimakan Desert. The name is derived from the Turkistani word *yar*, meaning "ridge" or "bank." Typical yardangs

What are the major features produced by wind erosion?

Figure 16.5 **Lag gravels** consist of angular clasts that range from fragments 30 cm or so across to small pebbles. Once the clasts become concentrated like this, they protect the surface from further erosion.

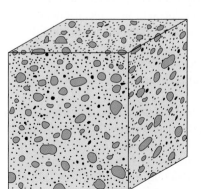

Time 1: Original gravel is dispersed

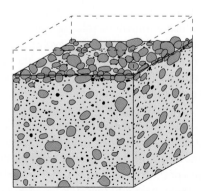

Time 2: Deflation removes fine grains

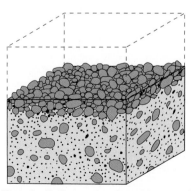

Time 3: Deflation develops lag gravel

Figure 16.4 **Desert pavement** results as wind selectively removes sand and fine sediment, leaving the coarser gravels to form a lag deposit. The protective cover of lag gravel acts as an armor limiting future deflation.

Figure 16.6 **Lag gravel** and desert pavement, not sand dunes, are major parts of the landscape in many desert regions like the Sahara Desert in Morocco. Desert pavement forms where small particles are swept away by the wind.

Courtesy of U.S. Geological Survey

(A) Grooves eroded on bedrock near Palm Springs, California.

(B) Ventifact shaped by wind abrasion into flat surfaces.

Figure 16.7 **Wind abrasion** is a process very much like sandblasting. Grooves and polished surfaces are apparent on cobbles and on bedrock in most of the world's deserts.

have the form of an inverted boat hull (**Figure 16.9**) and commonly occur in clusters, oriented parallel to the prevailing wind that formed them. Theoretically, they can be formed in any rock type, but they are best developed in soft, unconsolidated, fine-grained sediment or volcanic ash that is easily sculpted but is cohesive enough to retain steep slopes. Yardangs evolve into streamlined shapes that offer minimum resistance to the moving air. In a way, they are analogs to drumlins, which are shaped by moving ice. This shape may involve the combination of erosion with deposition to sculpt the flanks or the end of the yardang.

Yardangs are generally restricted to the most arid parts of deserts, which are relatively sand-poor and are areas where vegetation and soil are minimal. There is some indication that the Sphinx in Egypt was constructed out of a yardang.

Some of the most spectacular wind erosional features on Earth are the great yardangs of the Tibesti area of Chad in northern Africa (**Figure 16.10**). There, ridges almost 150 m high and several kilometers long are carved by the wind out of nonresistant sediments. The ridges are separated by troughs 100 m or more wide. There are no stream erosion channels between the yardangs, and no evidence of water erosion can be seen on the floor of the yardang field.

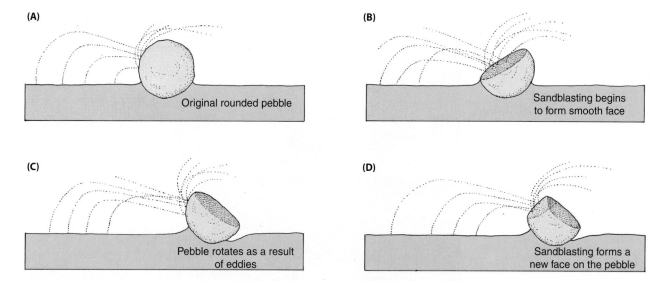

(A) Original rounded pebble

(B) Sandblasting begins to form smooth face

(C) Pebble rotates as a result of eddies

(D) Sandblasting forms a new face on the pebble

Figure 16.8 **Ventifacts** are pebbles shaped and polished by wind action. They commonly have two or more well-polished sides formed as the sandblasting action of wind reshapes one side and then another. In (B), a facet (smooth face) is formed by sandblasting on the side facing the prevailing wind. Removal of the sand by deflation causes the pebble to rotate so that another faceted and polished surface is produced on another side of the pebble (D).

Photograph by Alan L. Mayo

Figure 16.9 **Yardangs** are elongate ridges formed by wind erosion of relatively soft material. Note the streamlined shape.

Transportation of Sediment by Wind

Wind transports sand by saltation and surface creep. Silt and dust-sized particles are carried in suspension.

Movement of Sand

Although both wind and water transport sediment, the mechanics of motion involved are somewhat different because the viscosity of water is much greater than that of air. A sediment grain can be picked up by the wind when the forces acting to move the grain overcome the forces resisting movement. The main forces resisting motion are the weight of the grain and its cohesion to other grains. Wind blowing over the surface creates aerodynamic lift and drag on a grain. **Lift** is caused by the air flowing over the grain, creating a zone of low air pressure over the grain. (This is the same lift generated above the top of an airplane's wing.) The low pressure causes the grain to be

Courtesy of NASA

Figure 16.10 Yardangs and windstreaks in the Tibesti area of Chad are carved out of horizontally bedded nonresistant sedimentary rocks as regional winds are diverted around the eastern side of a large shield volcano.

"sucked" into the air flow. **Drag** is caused by the impact of air molecules on the grain's surface. High-speed photography shows that a particle begins to shake and then lifts off, spinning into the air. A critical wind velocity must be reached before a grain of a certain size will begin to move. For sand-sized grains with diameters of about 0.1 mm, this critical wind velocity is only about 20 cm/sec (0.7 km/hr). Once entrained into the moving wind, a sediment particle moves in a variety of waves, which are similar to those in which sediment moves in water (**Figure 16.11**).

Saltation results from impact and elastic bounce (Figure 16.11). When a grain falls to the surface, it collides with other sand grains. The impact causes one or more grains to bounce into the air, where they are driven forward by the drag of the stronger wind above the surface. Gravity soon pulls them back, and the grains strike the ground at angles generally ranging from 10° to 15°. If the sand is moving over solid rock, the grains bounce back into the air. If the surface is loose sand, the impact of a falling grain can knock several grains into the air, setting up a chain reaction, which eventually sets in motion the entire sand surface. Saltation normally lifts sand grains less than 1 cm above the ground, but heights as much as a meter are not unusual. The forward velocity of saltating sand grains is usually about one-half the wind speed.

Some grains that are too large to be ejected into the flowing air move by surface **creep** (rolling and sliding). These large grains are moved by the impact of saltating grains and the drag of the wind, but they do not lose contact with the bed. Approximately one-fourth of the sand moved by a sandstorm travels by rolling and sliding. Particles with a diameter greater than 2 mm are rarely moved by wind.

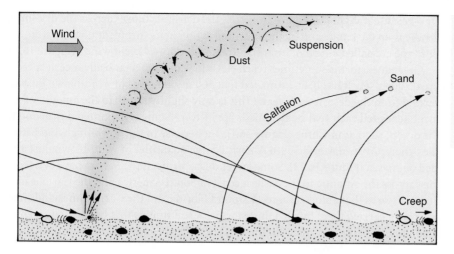

Figure 16.11 **The transportation of sediment by wind** is accomplished by surface creep, saltation, and suspension. Coarse grains move by impact from other grains and slide or roll (surface creep). Medium grains move by skipping or bouncing (saltation). Fine silt and clay move in suspension.

Movement of Dust

If you have ever been in a dust storm, you know that the wind also carries fine sediment in another way. Small grains of dust (silt- and clay-sized particles with diameters of less than 0.06 mm) are carried in **suspension** by turbulence in the wind flow. Such particles are lifted high into the atmosphere and are carried great distances before they settle back to earth. Dust storms are major processes in deserts. They can transport thousands of tons of sediment hundreds of kilometers (**Figure 16.12**). Dust storms are a major dynamic process, and they subtly, but constantly, change the surface. Throughout human history, dust storms have been a major cause of soil erosion. References to dust storms were recorded in 1150 BCE in China and in biblical times in the Middle East.

Dust storms are commonly initiated by the downdraft of cool air from a cumulonimbus cloud. When such a cloud develops to the point that rain begins to fall from it, the rain cools the air as it falls. Because the cool air is denser than the surrounding air, it descends in a downdraft. As the heavy, cooled air reaches the ground, it is deflected forward and moves in a large tongue-shaped pattern. It flows across the ground as a density current—a body of moving air that is heavier than the surrounding

© Povilho/iStockphoto

Figure 16.12 **A dust storm** in the Blue Nile area (Sudan, Africa) results when cool air descends and moves laterally over the surface as a density current. As the dense, cool air moves across the surface, it sweeps up dust and sand in its turbulent flow, creating a dust storm, or haboob. Eolian dust from such storms is an important component of deep-ocean sediment.

air—because it is cooler (Figure 16.12). As the dense air moves across the dry surface, it sweeps up dust and sand by the churning action of its turbulent flow. Dust storms of this type are called *haboobs*, from the Arabic word for "violent wind" (Figure 16.12).

Great dust storms sometimes reach elevations of 2500 m and advance at speeds of up to 200 m/sec. It has been estimated that 500 million tons of wind-blown dust are carried from the deserts each year. (This is only slightly less than the amount of sediment deposited each year by the Mississippi River.) Some is deposited downwind from the desert, such as in China (Figure 16.1), but because of the prevailing wind pattern in the Sahara, Australia, and South America, large quantities of wind-blown dust are carried out to sea (**Figure 16.13**). Some larger dust storms in the Sahara have even carried dust across the Atlantic to the eastern coast of South America and the Caribbean Sea. Some diseases that afflict coral reefs around Florida and the Caribbean islands may have been carried there by such dust storms. A soil fungus that has caused an epidemic among some of the organisms in the coral reefs has been traced to central Africa.

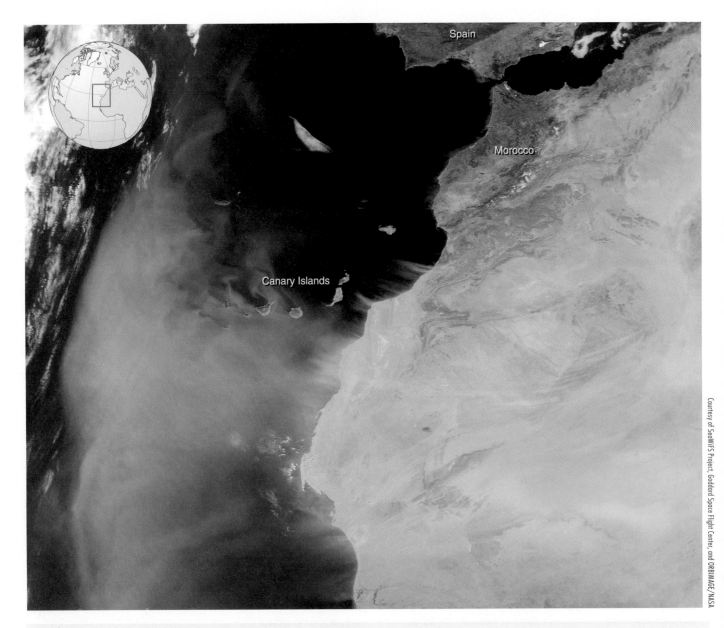

Courtesy of SeaWiFS Project, Goddard Space Flight Center, and ORBIMAGE/NASA

Figure 16.13 Dust storms in the Sahara Desert, as seen from a satellite. This storm formed by dissipating thunderstorms and then picked up loose sediment and moved it westward. The Canary Islands disturb the flow of the dust-laden winds. Storms like this can carry sediment into the Atlantic Ocean and as far west as the Americas. Eolian dust from such storms is an important component of deep-ocean sediment.

Deposits of Wind-Blown Sand

Wind-blown sand accumulates to form sand sheets, ripples, and dunes. Different kinds of sand dunes result from variations in sand supply, wind direction, and velocity. The most significant types of dunes are (1) transverse dunes, (2) barchan dunes, (3) longitudinal dunes, (4) star dunes, and (5) parabolic dunes.

Where wind velocity decreases, moving sand grains may become deposited to form a variety of sedimentary bodies. About 40% of these deposits are gently undulating, nearly flat **sand sheets** (see Figure 16.19). Grains that are too big to move by saltation are the principal constituent, and many sand sheets grade into sand dunes. The active part of a sand sheet is only a few centimeters thick, but the sand sheet may cover a very large area. For example, much of southern Egypt and northern Sudan are covered by a featureless sand sheet.

Saltating grains form ripples that are perpendicular to the wind direction. The distance between each ripple is about the same as the average jump made by each saltating grain. These ripples are usually only a centimeter or so high. Ripples form on the surfaces of many dunes.

How can wind move an entire sand dune?

Sand dunes are the most commonly recognized deposits of sand. Dunes migrate relentlessly downwind and may completely modify the landscape, damaging or obliterating almost anything in their path. Forests have been entombed by advancing dunes, streams diverted, and villages completely covered (**Figure 16.14**). Examples of such migration occurred in England and France, where entire towns were overwhelmed by advancing dunes so that nothing was seen but the church spire. Then the dunes marched on, leaving behind a devastated countryside of dead trees and collapsed buildings. But why do dunes form, and why do some grow so large? How do they move?

In many respects, **dunes** are similar to ripple marks (formed by either air or water) and to the large sand waves or sandbars found in many streams and in shallow-marine water. Many dunes originate where an obstacle such as a large rock, a clump of vegetation, or a fence post creates a zone of quieter air behind it (**Figure 16.15**). As sand is blown up or around the obstruction and into the protected area (the wind shadow), its velocity is reduced and deposition occurs. Once a small dune is formed, it acts as a barrier itself, disrupting the flow of air and causing continued deposition downwind. Dunes range in size from 30 cm to as much as 500 m high and 1 km wide.

What is the difference in the way wind and water transport sediment?

The movement of sand in a typical dune is diagrammed in **Figure 16.16**. A typical dune is asymmetrical, with a gently inclined windward slope and a steeper downwind

(A) Dunes near Cairo, Egypt, encroach upon an apartment complex.

(B) Dune fields in northern Canada's polar desert migrate over a forest.

Figure 16.14 Migrating sand dunes may obliterate or damage almost anything in their path. These dunes in North Africa and North America are gradually inundating buildings and forests.

Figure 16.15 **Sand dunes commonly originate in wind shadows**. Any obstacle that diverts the wind, such as a bush or a fence post, creates eddies and reduces wind velocity. Wind-blown sand is deposited in protected areas, and eventually enough sand accumulates in the wind-shadow area to form a dune. The dune itself then acts as a barrier, making its own wind shadow, and thus causes additional accumulation of sand.

Windward face

Wind

Migration of the dune

Slipface

Figure 16.16 **A sand dune migrates** as sand grains move up the slope of the dune and accumulate in a protected area on the downwind face. The dune slowly moves grain by grain. As the grains accumulate on the downwind slope, they produce a series of layers (cross-beds) inclined in a downwind direction.

slope—the **lee slope**, or **slip face**. The steep slip face of the dune shows the direction of the prevailing wind. Dunes migrate grain by grain as wind transports sand by saltation and surface creep up the windward slope. The wind continues upward past the crest of the dune, creating divergent airflow and eddies just over the lee slope. Beyond the crest, the sand drops out of the wind stream and accumulates on the slip face. When the sand grains exceed the angle of repose, they spill down the slip face in small landslides or avalanches. As more sand is transported from the windward slope and accumulates on the lee slope, the dune migrates downwind. The internal structure of a migrating dune consists of **cross-beds** formed as the saltating grains accumulate on the inclined downwind slope of the dune (Figure 16.16). Strata formed on the lee slope are therefore inclined downwind. Geologists map the directions of ancient winds by measuring the directions in which the cross-strata of wind-blown sandstone are inclined (**Figure 16.17**).

Quartz is the most common mineral in wind-blown sand because of its hardness, strength, and resistance to weathering. In some areas, gypsum also forms sand-sized grains that can saltate and accumulate in dunes.

Why do sand dunes assume different shapes?

Types of Sand Dunes

Sand dunes vary greatly in size and shape and, although they form bleak and barren wastelands, they impart a sense of stark beauty and hidden mysteries to the areas they cover.

Transverse dunes typically develop where there is a large supply of sand and a constant wind direction (**Figure 16.18A**). These dunes cover large areas and develop wavelike forms, with sinuous ridges and troughs, perpendicular to the prevailing wind. Transverse dunes commonly form in deserts where exposed ancient sandstone formations provide an ample supply of sand. They usually cover large areas known as

Figure 16.17 Cross-bedding in the Navajo Sandstone in Zion National Park, Utah, is evidence that the rock formed in an ancient desert. The inclination of the strata shows that the wind blew from north to south (from left to right, in the photograph) for most of the time during which this formation was being deposited.

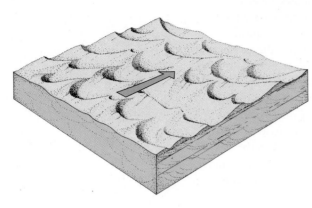

(A) Transverse dunes develop where the wind direction is constant and the sand supply is large.

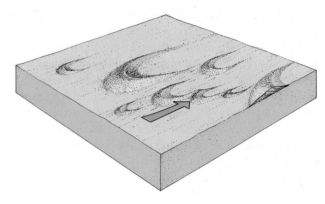

(B) Barchan dunes develop where the wind direction is constant but the sand supply is limited.

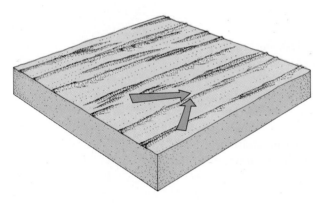

(C) Linear or seif dunes are formed by converging winds in an area with a limited sand supply.

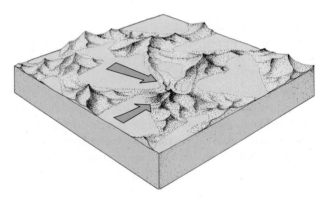

(D) Star dunes develop where the wind direction is variable.

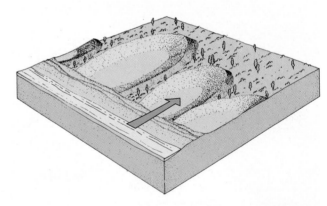

(E) Parabolic dunes are formed by strong onshore winds.

Figure 16.18 Each of the major types of sand dunes represents a unique balance between sand supply, wind velocity, and variability in wind direction. The arrows show the major wind directions.

sand seas, so called because the wavelike dunes produce a surface resembling that of a stormy sea (**Figure 16.19**).

Barchan dunes form where the supply of sand is limited and winds of moderate velocity blow in a constant direction (Figure 16.18B). These crescent-shaped dunes are typically small, isolated dunes from 1 to 50 m high. The tips (or horns) of a barchan point downwind, and sand grains are swept around them as well as up and over the crest. With a constant wind direction, beautiful symmetric crescents form. With shifts in wind direction, however, one horn can become larger than the other. Although

Figure 16.19 **Sand seas in south Africa** have been mapped with the use of satellite photography. A sand sea covers much of Namibia where some of Earth's highest sand dunes are found. Some are as much as 300 m high. A variety of dune types are formed as a result of variations in wind velocity and direction, supply of sand, and the nature of the surface over which the sand moves. This image shows an area about 100 km across.

barchans typically are isolated dunes, they may be arranged in a chainlike fashion, extending downwind from the source of sand.

Linear dunes or **seif dunes** (Arabic, "sword") are long, parallel ridges of sand, elongate in a direction parallel to the vector resulting from two slightly different wind directions (Figure 16.18C). They develop where strong prevailing winds converge and blow in a constant direction over an area having a limited supply of sand. Many linear dunes are less than 4 m high, but they can extend downwind for several kilometers. In larger desert areas, they can grow to 100 m high and 120 km long, and they are usually spaced from 0.5 to 3 km apart. Linear dunes occupy a vast area of central Australia called the Sand Ridge Desert. They are especially well developed in some desert regions of North Africa and the Arabian Peninsula.

Star dunes are mounds of sand having a high central point, from which three or four arms, or ridges, radiate (Figure 16.18D). This type of dune is typical in parts of North Africa and Saudi Arabia. The internal structure of these dunes suggests that they were formed by winds blowing in three or more directions. Some are 500 m high.

Parabolic dunes (blowouts) typically develop along coastlines where vegetation partly covers a ridge of wind-blown sand, transported landward from the beach (Figure 16.18E). Where vegetation is absent, small deflation basins are produced by strong onshore winds. These blowout depressions grow larger as more sand is exposed and removed. Usually, the sand piles up on the lee slope of the shallow deflation hollow, forming a crescent-shaped ridge. In map view, a parabolic dune is similar to a barchan, but the tips of the parabolic dune point upwind and are fixed in place by vegetation. Because of their form, parabolic dunes are also called hairpin dunes.

Sand Seas

Although Earth is commonly called the water planet, several continents have vast areas where precipitation is rare, and the surface is covered with wind-blown sand. Some of these areas are so vast they are known as **sand seas**, or **ergs** (Figure 16.19). It has been calculated that 99.8% of all wind-blown sand is in the great sand seas of the world. The largest are in Africa, Asia, and Australia. In Africa, about 800,000 km^2 (or one-ninth of the entire area of the Sahara) is covered by stable or active sand dunes. One-third of Saudi Arabia, approximately 1,000,000 km^2, is covered by eolian sand, and in the vast Rub' al Khali (Empty Quarter), dunes may be more than 200 m high and cover about 400,000 km^2. The Australian sand seas are mainly in the western and central portions of that continent (Figure 16.1). Dunes cover most of the area (about 60%) of a typical sand sea; the rest consists of sand sheets.

Ancient Sand Seas   Sand deposits from ancient deserts are found in many parts of the world, including those where deserts do not exist today. These ancient sand seas are distinctive sedimentary deposits with a texture (cross-bedded) and composition (clean quartz sand) that reveals their origin. They range in age from a few thousand to more than 500 million years old and provide an important record of climatic change and movement of tectonic plates.

Some ancient sand seas indicate that climatic conditions have changed considerably over vast areas. For example, the Sand Hills of Nebraska are an inactive dune field covering some 57,000 km^2. The dunes were active during the last ice age, which ended about 15,000 years ago. Barchan and transverse types are well preserved; some are 120 m high. Because of climate change during the last interglacial period, they are now covered with grass and do not migrate.

Other sand seas developed when moving continents drifted through the low-latitude zones where deserts form. Such is the case for those in western North America. Ancient sand seas were especially numerous and widely distributed throughout the western United States during late Paleozoic and Mesozoic time. More than eight formations of eolian sand have been recognized, and much of the spectacular scenery of the Colorado Plateau is developed in these colorful strata (**Figure 16.20**). The formations are typically white, buff, or pale red, and they erode into steep cliffs. Many national parks and monuments including Zion, Capitol Reef, Canyonlands, Arches, Canyon de Chelly, and the Grand Canyon expose formations developed in ancient sand seas. Ancient wind directions can easily be determined by the dip direction of the cross-strata. It is thus possible to map patterns of ancient winds when the sandstone formations were being deposited. These great accumulations of sand record the passage of North America through the dry desert latitudes during the late Paleozoic and Mesozoic Eras.

Figure 16.20 Ancient sand seas are evidenced by thick deposits of wind-blown sand in the Colorado Plateau and adjacent regions of the western United States. During the early Mesozoic Era, this region was a vast dry plain where eolian sands accumulated in a subsiding basin. This photograph of the White Cliffs in the Grand Staircase is typical of the deposits of more than five major sand seas now preserved in the rock record.

Deposits of Wind-Blown Dust: Loess

Loess is a deposit of wind-blown dust (silt and clay) that accumulates slowly and ultimately blankets large areas. It covers one-tenth of the world's present land surface. The dust is derived either from nearby deserts or from rock flour originating near recently glaciated regions. Wind-blown dust settles in the oceans and is an important source of deep-marine mud.

Wind-blown dust may accumulate in thick deposits called **loess** (German, "loose") that blankets many regions throughout the world. Loess is distinctive sediment. It is composed mostly of silt-size grains, but smaller clay-size grains are also common. Loess is typically yellowish brown and is composed of small angular grains of quartz, feldspar, and clay. Although it is loose, friable, and porous, it typically erodes into vertical walls that do not crumble unless they are disturbed. As the dust accumulates, successive generations of grass roots (now represented by narrow tubes partly filled with calcium carbonate) make it sufficiently coherent to stand up in vertical cliffs (**Figure 16.21**).

Dust is dispersed high into the atmosphere and carried great distances by the wind (**Figure 16.22**). Therefore, loess is widespread beyond desert areas and, unlike sand dunes, loess deposits blanket the landscape, covering hills and valleys alike. The map in Figure 16.1 shows the distribution of loess. Loess covers as much as one-tenth of Earth's land surface and is particularly widespread in semiarid regions along the margins of the great deserts between latitude 24° to 55° north and 30° to 45° south. Equatorial regions are free from loess because as soon as dust accumulates, it is washed away by heavy rainfall. Areas formerly covered by continental glaciers are also free from loess because glaciated terrains are new surfaces that were covered with ice until only a few thousand years ago. Loess is not widespread in the major desert regions because dust is swept out of the deserts and deposited in adjacent areas.

As can be seen in Figure 16.22, the global dispersion of wind-blown dust is related to the prevailing wind patterns in the major desert regions of the world. Wind carries dust from the deserts of central Asia to the south and east, where it is deposited over vast areas of China. The greatest deposits of loess are in the Shansi and adjacent provinces of China, where it locally exceeds 300 m in thickness. Over immense areas, many meters of dust have accumulated, completely burying the entire landscape (**Figure 16.23**). From the deserts of North Africa, dust is carried by the prevailing

What geologic features are produced by wind-blown dust?

Figure 16.21 **Loess deposits in central China** cover vast areas and are exceptionally thick. They illustrate the typical properties of loess, including fine grain size, sequences of buried soils, buff color, and steep cliff faces.

Courtesy of J. Herman, Laboratory for Atmospheres, Goddard Space Flight Center

Figure 16.22 **Dust sources and distribution paths** are shown for April 2002. They are based on satellite measurements of aerosols in the atmosphere. Most dust is derived from the deserts of Arabia and north Africa and then blown toward the Atlantic by prevailing winds. Dust plumes are also common north and south of the Himalaya Mountains.

winds to the west, where it settles out over the Atlantic Ocean. Oceanographic research has identified wind-blown dust that forms distinctive widespread layers on the seafloor. Indeed, wind-blown dust is an important source of deep-marine mud. Some dust from the Sahara is blown northward and is trapped in the Mediterranean Sea, but occasionally some reaches southern Europe and has been known to turn white Alpine snow into delicate shades of pink and brown.

Most of the great loess deposits of North America and Europe are not related to desert dust but are considered to have their sources in the outwash plains of continental glaciers. Rock debris pulverized and transported by glaciers is ultimately deposited

Figure 16.23 **Young loess deposits in northern China** are as much as 100 m thick and are widely used for agriculture. Stream erosion has cut deep valleys, but there is still a tendency to stand in vertical cliffs. Inasmuch as the loess forms an excellent soil, it can be terraced for optimum agricultural production.

as glacial outwash. This sediment, commonly called rock flour, was then picked up by strong thermal winds moving down off the glacier. These winds carried the fine particles aloft and then deposited them as a blanket of loess beyond the margins of the ice. In the central United States, most of the loess is in the uplands of the Mississippi drainage system and is well exposed as vertical cliffs along the riverbanks. Near rivers where floodplain muds provide a ready source of dust, the loess is 30 m or more thick, but it thins out away from the river channels. The greatest accumulations are east of the floodplains; their thickness is controlled by the prevailing winds.

In Eurasia, a long belt of loess derived from glacial material stretches from France to China (Figure 16.1). Beginning as local thin patches in France and Germany, the deposits become thicker and more extensive as they are traced across Russia and Turkestan.

Loess has played an important role in human history, and it continues to be highly significant today. Loess forms some of the most productive soils in the world; the steppes of eastern Europe (the Ukraine and Russia) and the plains of the midwestern United States are all blanketed by loess (Figure 16.1). This correlation is not by chance. Loess is not deposited in dry desert regions but in adjacent areas where there is enough precipitation to support agriculture. It is transported, not decomposed, soil and is rich in nutrients because they have not been leached away. Moreover, because it is transported soil, it is commonly thicker and not as easily stripped away as relatively thin weathered soils.

Long before the Qin (Ch'in) dynasty in 221 BCE, farming was initiated on the loess plains of northern China. This exceptional soil, blown in from the Gobi Desert during the ice age, shaped the origins of China. On this rich land along the middle Yellow and Wei rivers, the first Chinese culture developed in the fifth millennium BCE. The fertility of the soil when irrigated led to the development of the remarkable water control work of ancient China. The loess in China is up to 335 m thick and covers an area of more than 400,000 km^2. This soft, loose fine material is easily eroded and is carved into a remarkable maze of dendritic stream valleys (**Figure 16.24**). In this region, everything is yellow: the land, the homes, and the water. It is loess that gives the name to the Yellow River, whose sediment load is simply loess remobilized by water. In fact, the Yellow River is more like a thin mudflow than like running water. The loess in the Yellow River is ultimately delivered to the ocean, where much stays in suspension for a long time, imparting its characteristic color to the Yellow Sea.

Figure 16.24 Loess plateaus of China are highly dissected into an intricate network of tributaries to the Yellow River, so-named because of the abundance of fine sediment in the river.

Desertification

Desertification (or land degradation) occurs naturally or as a result of the activities of people. Exploitation and overgrazing of sensitive lands adjacent to natural deserts can cause a desert to expand.

The great deserts of the world formed by natural processes over long periods, as continents migrated into dry climates produced in low-latitude, high-pressure zones. A desert may expand and shrink in response to short-term cyclic climatic fluctuations. The margins of deserts, therefore, have always been transitional or gradational to the adjacent, more humid environments.

There are two distinct forms of desert expansion. One is a natural process due to climatic change or the migration of moving sand. It takes place on the immediate edges of existing deserts, and when the distribution of rain over the desert and surrounding areas shifts, the desert expands or contracts. Some desert expansion takes place over thousands of years.

The second form of desertification results from human-induced breakdown of soils in the zones adjacent to deserts. Poor cultivation practices, overgrazing, and deforestation are the major causes. Sparse vegetation inhibits wind erosion, but when it is destroyed, the desert expands. Along the desert margins, human activity is commonly superimposed upon the natural processes that cause the expansion and contraction of the deserts. Grazing livestock, the compaction of soil by hooves, and even the collection of firewood by humans can reduce the plant cover. The soil ultimately degrades, and, in many cases, the desert expands. This process, in which productive land becomes unproductive, is called **land degradation** or **desertification**. If the general climatic trend is toward increasing aridity, desertification can occur with remarkable speed. This type of desertification results when the number of people and livestock exceeds the capacity of the rainfall to supply their food.

Desertification does not occur in a broad, even swath that can easily be mapped along the desert fringe. Deserts advance erratically, forming patches on their borders, and areas far from the desert may quickly degrade into barren rock and sand. Desertification presents an enormous problem for human existence. About one-third of Earth's land is arid or semiarid, but only about one-half of this area is so dry that it cannot support human life. More than 600 million people live in the dry areas, and about 80 million live on land that is nearly useless because of desertification. The most severe problems are in Africa and Asia. The United Nations estimates that more than 11 billion acres (35% of the world's cropland) show signs of human-induced degradation.

The Sahel, a semiarid zone south of the Sahara Desert, is commonly used as an example of desertification. It extends across the entire continent of Africa and includes Ethiopia, the Sudan, Mauritania, and all of the small countries in between. It is a transition zone, 800 to 1000 km wide, separating the Sahara to the north from the well-watered grasslands of central Africa. The area receives rainfall from the seasonal northward shift of moist equatorial air masses, but it suffers a long, hot, rainless season that is the result of dry northeasterly trade winds blowing southward out of the Sahara. A slight shift in the wind patterns causes significant changes in climate and weather. An example of these changes can be seen in the shrinking shoreline of Lake Chad (**Figure 16.25**). This natural situation, combined with overgrazing, removal of trees for firewood, and an annual population growth rate of 2.5% to 3.0%, have stressed the Sahel's sensitive environment. Drought and human suffering go hand in hand in this fragile area.

Clearly, neither shipping emergency food nor drilling more and deeper wells are the answers to desertification. Reaction to disaster must be replaced by predisaster planning, based on a clear understanding of an area's basic geologic systems and of the kinds of changes that occur naturally.

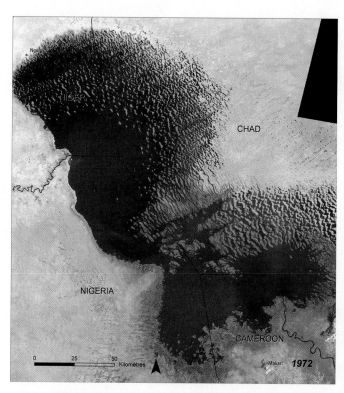

(A) Lake Chad in 1972.

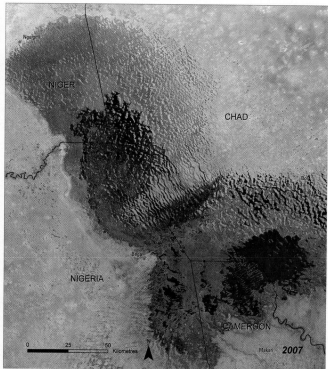

(B) Lake Chad in 2007.

Figure 16.25 Lake Chad is a shallow freshwater lake between the Sahara Desert and the tropics. Climatic variations cause the lake to expand or shrink. During the ice age, the lake covered 1,000,000 km². In 1972 (A) the lake covered only about 20,000 km². The lake shrank to 1200 km² during the drought of the Sahel (B). Fishing villages once on the shore are now stranded many kilometers from water.

© UNEP, n.d., Lake Chad. Environmental Change Hotspots. Division of Early Warning and Assessment (DEWA). United Nations Environment Programme (UNEP)

GeoLogic   Mars: The Eolian Planet

Mars is the fourth planet from the Sun. Its surface temperature and atmospheric pressure are both much lower than on Earth. So low, that liquid water is not stable at the present time. It either vaporizes or freezes. How has this modified the surface of Mars?

Observations

1. Occasionally global dust storms envelope the entire planet, obscuring the peaks of volcanoes that are 20 km high.
2. Winds comparable to those in a strong hurricane on Earth rage for several months at a time.
3. Windstreaks caused by erosion and deposition are found near many impact craters.
4. Dunes are found in almost all regions on Mars. The largest sand sea encircles the polar ice cap and is comparable in size to the entire Sahara Desert. Long transverse dunes break up into isolated barchans on the margin of the erg.
5. Yardangs have developed by abrasion of young deposits that are weakly consolidated and long narrow spines project from low plateaus, The ridges are narrow and keel-like, and the ends taper sharply. Their alignment shows the dominant wind direction.
6. Irregular blankets of loess mantle most of the planet and are interlayered with ice in the polar ice caps.
7. On the surface, landers have photographed small sand dunes. The blocky, angular rocks that mantle the area resemble desert pavement. Some of the rocks are fluted and shaped into ventifacts, as if eroded by the wind.

Interpretations

These features collectively show that Mars is a cold desert where the wind is the dominant active process. Everywhere you look the surface is shaped by the wind. The wind is constantly moving and redepositing loose surface material. If not for its liquid water and abundant plant life, Earth too would be an eolian planet. Which processes do you think are more common in our solar system—those driven by wind or those caused by running liquid water?

Courtesy of JPL/Malin Space Science Systems/NASA

Sand dunes like these barchans are very common on Mars.

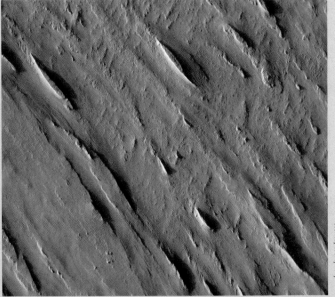

Courtesy of JPL/ASU/NASA

Yardangs emanate from incompletely stripped hills.

Courtesy of JPL/NASA

The surface of Mars is dominated by eolian erosion and deposition. Small dunes have formed from saltating sand. The wind has also fluted some pebbles and scoured away loose sediment leaving a lag of pebbles and boulders.

Key Terms

barchan dune (p. 494)
blowout (p. 496)
Coriolis effect (p. 482)
creep (p. 488)
cross-bed (p. 493)
deflation (p. 484)
deflation basin (p. 484)
desertification (p. 500)

desert pavement (p. 485)
drag (p. 488)
dune (p. 491)
erg (p. 496)
lag deposit (p. 485)
land degradation (p. 500)
lee slope (p. 493)
lift (p. 487)

linear dune (p. 495)
loess (p. 497)
parabolic dune (p. 496)
rain shadow (p. 482)
saltation (p. 488)
sand sea (p. 496)
sand sheet (p. 491)
seif dune (p. 495)

slip face (p. 493)
star dune (p. 496)
suspension (p. 489)
transverse dune (p. 493)
ventifact (p. 485)
yardang (p. 485)

Review Questions

1. Describe the processes involved in wind erosion.
2. What controls the distribution of the major desert regions of Earth?
3. What landforms are produced by wind erosion and deflation?
4. Explain the origin of ventifacts.
5. Explain the origin of desert pavements.
6. Draw a simple diagram showing how sand is transported by wind.
7. Why is wind an effective agent in sorting sand and dust?
8. Describe how a sand dune forms and how it migrates.

9. List the five major types of dunes, and state the conditions under which each type forms (wind direction and velocity, sand supply, and the characteristics of the surface over which the sand moves).
10. What is the origin of loess?
11. Where are the major areas of loess deposits in the world today?
12. What changes in the Australian desert would occur if that continent were to drift 2000 km (approximately 20° of latitude) northward?

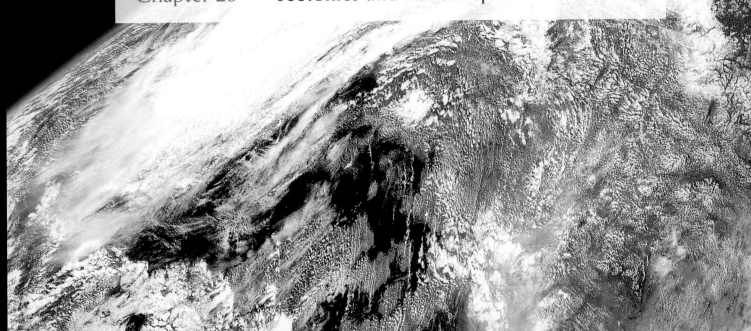

Part III

Earth's Tectonic System

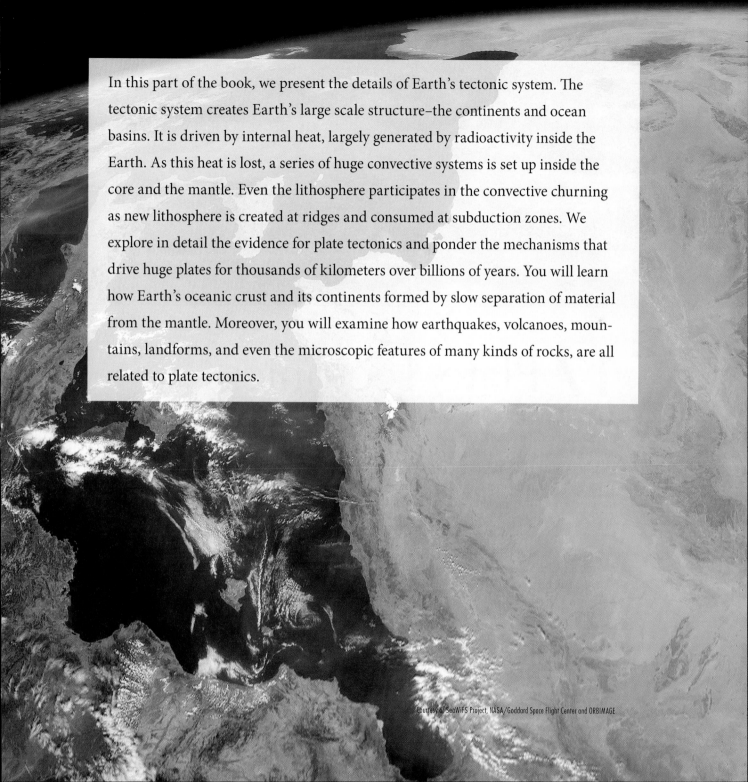

In this part of the book, we present the details of Earth's tectonic system. The tectonic system creates Earth's large scale structure–the continents and ocean basins. It is driven by internal heat, largely generated by radioactivity inside the Earth. As this heat is lost, a series of huge convective systems is set up inside the core and the mantle. Even the lithosphere participates in the convective churning as new lithosphere is created at ridges and consumed at subduction zones. We explore in detail the evidence for plate tectonics and ponder the mechanisms that drive huge plates for thousands of kilometers over billions of years. You will learn how Earth's oceanic crust and its continents formed by slow separation of material from the mantle. Moreover, you will examine how earthquakes, volcanoes, mountains, landforms, and even the microscopic features of many kinds of rocks, are all related to plate tectonics.

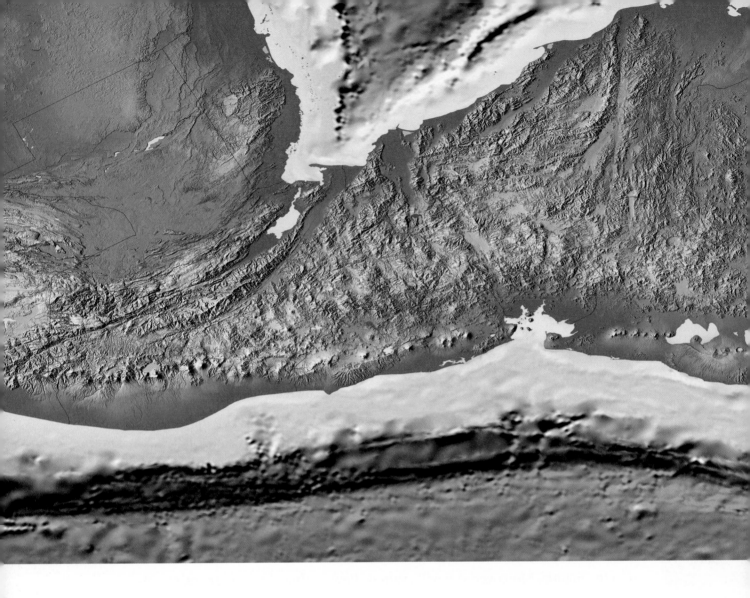

17 Plate Tectonics

An outstanding example of the surface expression of plate movement is dramatically shown in this shaded relief map of Central America created with data collected by astronauts on the Space Shuttle. The map spans a distance of almost 2000 km. A long linear trench lies parallel to the shore and marks the zone where the oceanic Cocos plate bends and dives back into the mantle. The area between the trench and the shore is underlain by strongly deformed sediment scraped off the downgoing plate on land. The system of ridges and valleys, high plateaus, grabens, and volcanoes is part of the great Cordilleran mountain chain formed by subduction of oceanic lithosphere beneath the American continents. The mountains furrowed with arcuate ridges (on the far right and left) are eroded fold and thrust belts formed by compression. On the left, you can see a series of long linear troughs and rugged ridges that cut diagonally across the region; this is a transform plate boundary that connects the Caribbean trench with the Middle American trench. Strike-slip movement has sheared continental and oceanic plates past one another. Huge ash-flow calderas are filled by lakes and smaller andesite volcanoes are aligned parallel to the trench. Magma is generated at a depth of about 100 km before rising, intruding the crust to form vast batholiths, or erupting explosively. These volcanoes and innumerable earthquakes along the faults pose a direct threat to millions of people. All of these tectonic features have been extensively eroded by rivers

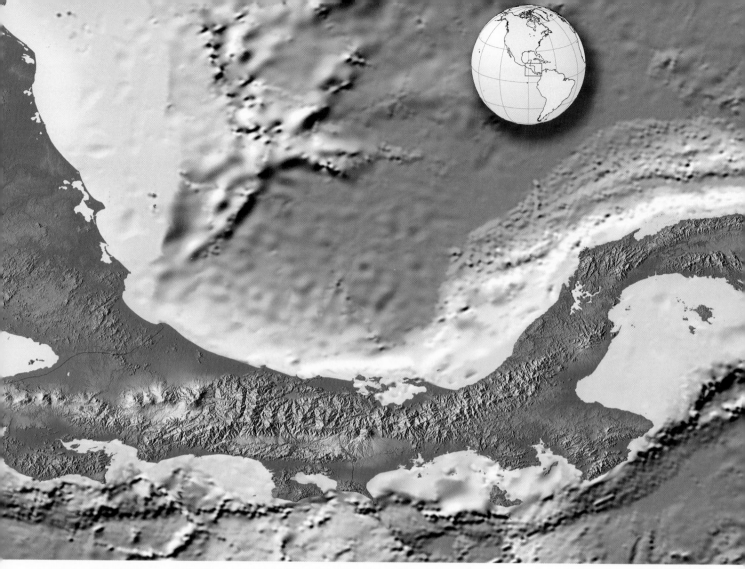

to form short but delicate dendritic patterns. Wave action has shaped the shoreline and helped to create the wide continental shelf. Elsewhere coral reefs have shaped the Caribbean coastline.

Plate tectonics has done much more than explain the deformation of these mountains in Central America. It explains the San Andreas fault system and its relationship to the Gulf of California and how the Cascade Mountains are related to the far-off mid-ocean ridge that traverses the Pacific. It explains many aspects of the interrelationships of volcanoes, earthquakes, climate change, and even of the evolution of life itself. In brief, it provides a single unifying theory of Earth's dynamics. Essentially everything about our planet is related either directly or indirectly to plate tectonics.

How did scientists develop such a revolutionary theory? A few decades ago, most geologists believed that continents and ocean basins were fixed, permanent features on Earth, and the theory of continental drift was considered a radical idea. What brought about the remarkable change in the entire science of geology? In this chapter we consider how the theory of plate tectonics developed and the evidence upon which it is based. We then consider the nature of the lithospheric plates, what causes them to move, and how we measure the rates and direction of plate motion.

Courtesy of JPL/NIMA/NASA

Major Concepts

1. The theory of continental drift was proposed in the early 1900s and was supported by a variety of geologic evidence. Lack of knowledge of the nature of the oceanic crust, however, prevented a complete theory of Earth's dynamics from being developed.

2. A major breakthrough in the development of the plate tectonics theory occurred in the early 1960s, when the topography of the ocean floors was mapped and magnetic and seismic characteristics of the oceanic crust were determined.

3. Most tectonic activity occurs along plate boundaries. Divergent plate boundaries are zones where the plates split and spread apart. Convergent plate boundaries are zones where plates collide. Transform fault boundaries are zones where plates slide horizontally past each other.

4. The direction of the relative motion of plates is indicated by (a) the trend of the oceanic ridge and associated transform faults, (b) seismic data, (c) magnetic stripes on the seafloor, and (d) the ages of chains of volcanic islands and seamounts. The motion of a plate can be described in terms of rotation around a pole.

5. Heat from the mantle (generated by radioactivity) and from the core is probably the fundamental cause of Earth's internal convection.

6. The major forces acting on plates are (a) slab-pull, (b) ridge-push, (c) basal drag, and (d) friction along transform faults and in subduction zones. The most important forces that make the plates move are probably slab-pull and ridge-push.

Continental Drift

> The theory of continental drift was proposed in the early 1900s and was supported by a variety of impressive geologic data. Lack of an understanding of the nature of the oceanic crust, however, prevented the development of a complete theory of Earth's dynamics.

The theory of plate tectonics wrought a sweeping change in our understanding of Earth and the forces that shape it. Some scientists consider this conceptual change as profound as those that occurred when Darwin reorganized biology in the nineteenth century or when Copernicus, in the sixteenth century, determined that Earth is not the center of the universe.

What evidence indicates that continents split and then drift apart?

The predecessor of the plate tectonic theory, the concept of **continental drift**, is an old idea. Soon after the first reliable world maps were made, scientists noted that the continents, particularly Africa and South America, would fit together like a jigsaw puzzle, if they could be moved. Antonio Snider-Pelligrini, a Frenchman, was one of the first to study the idea in some depth. In his book *Creation and Its Mysteries Revealed* (1858), he showed how the continents looked before they separated (**Figure 17.1**). He cited fossil evidence in North America and Europe but based his reasoning on the catastrophe of Noah's flood. The idea seemed too far-fetched for science or the general public, so it was forgotten, not to be revived for 50 years. The theory was first considered seriously in 1908, when American geologist Frank B. Taylor pointed out several geologic facts that could be explained by continental drift.

However, Alfred Wegener, a German meteorologist, was the first to exhaustively investigate the idea of continental drift and to convince others to take it seriously. In his book *The Origin of the Continents and Oceans* (1915), Wegener based his theory not only on the shapes of the continents, but also on geologic evidence, such as similarities in the fossils found in Brazil and Africa. He drew a series of maps showing three stages

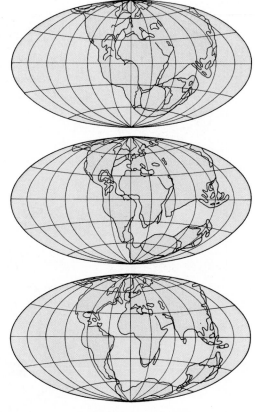

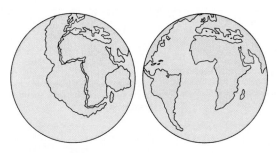

(A) Maps made by Antonio Snider-Pelligrini in 1858.

Figure 17.1 Continental drift was illustrated as early as 1858 by Antonio Snider-Pelligrini when he published these maps **(A)** in his book Creation and Its Mysteries Revealed. The idea seemed too far-fetched to the public and the scientific communities of the time and was forgotten, not to be revived for 50 years. Wegener published his series of maps **(B)** in 1915. His evidence, most of which was quite valid, was drawn from all of the sciences. Wegener called the original land mass Pangaea ("all lands") and believed that the continents somehow plowed through the oceanic crust as they drifted.

(B) Maps made by Alfred Wegener in 1915.

in the drifting process, beginning with an original large land mass, which he called **Pangaea** (meaning "all lands") (Figure 17.1). Wegener believed that the continents, composed of less dense silicic rock, somehow plowed through the denser rocks of the ocean floor, driven by forces related to the rotation of Earth.

Most geologists and geophysicists rejected Wegener's theory, although many scientific observations supporting it were known at the time. A few noted scholars, however, seriously considered the theory. Alexander L. du Toit, from South Africa, compared the landforms and fossils of Africa and South America and further expounded the theory in his book *Our Wandering Continents* (1937). Arthur Holmes, of England, later developed it in his textbook *Principles of Physical Geology* (1944). The early arguments concerning the breakup of the supercontinent Pangaea and the theory of continental drift were supported by some important and imposing evidence, most of which resulted from regional geologic studies as outlined below.

Paleontological Evidence

The striking similarity of certain fossils found on the continents on both sides of the Atlantic is difficult to explain unless the continents were once connected. The fossil record indicates that a new species appears at one point and disperses outward from there. Floating and swimming organisms could migrate in the ocean, from the shore of one continent to another, but the Atlantic Ocean would present an insurmountable obstacle for the migration of land-dwelling animals, such as reptiles and insects, and certain land plants. Consider the profound implications of the following examples (**Figure 17.2**).

Fossils of *Glossopteris*, a fernlike plant, have been found in rocks of the same age from South America, South Africa, Australia, India, and Antarctica. Mature seeds of this plant were several millimeters in diameter, too large to have been dispersed across the ocean by winds. The simultaneous presence of *Glossopteris* on all of the southern continents, therefore, is strong supporting evidence that the continents were once connected.

Are marine fossils important in supporting the theory of continental drift?

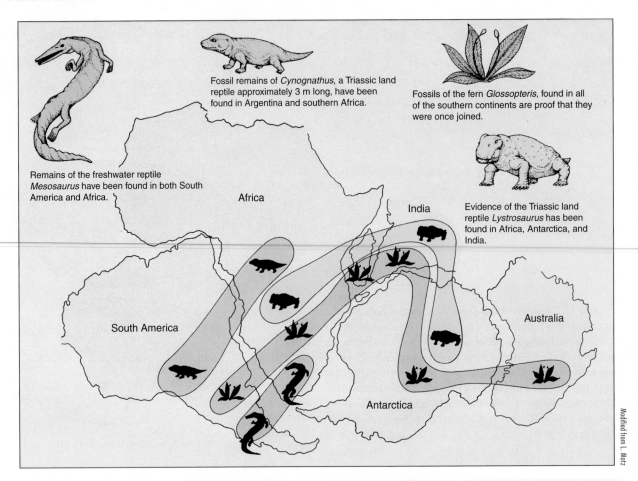

Remains of the freshwater reptile *Mesosaurus* have been found in both South America and Africa.

Fossil remains of *Cynognathus*, a Triassic land reptile approximately 3 m long, have been found in Argentina and southern Africa.

Fossils of the fern *Glossopteris*, found in all of the southern continents are proof that they were once joined.

Evidence of the Triassic land reptile *Lystrosaurus* has been found in Africa, Antarctica, and India.

Africa

India

South America

Australia

Antarctica

Modified from L. Motz

Figure 17.2   Paleontologic evidence of continental drift can be appreciated by considering the distribution of some fossil plants and animals found in South America, Africa, Madagascar, India, Antarctica, and Australia. *Mesosaurus*, a Permian freshwater reptile, is found in both Brazil and South Africa. *Glossopteris*, a fossil fern, is found on all of the southern continents in the zone shown on the map. *Lystrosaurus*, a Triassic land reptile, is found in South Africa, South America, India, and Antarctica. *Cynognathus*, an older Triassic reptile, is found in Argentina and South Africa.

The distribution of Paleozoic and Mesozoic reptiles provides similar evidence; fossils of several species have been found in the now-separated southern continents. An example is a mammal-like reptile belonging to the genus *Lystrosaurus*. This creature was strictly a land dweller. Its fossils are found in abundance in South Africa, South America, Asia, and in Antarctica. This genus thus inhabited all of the southern continents except Australia during the same geologic period. Clearly, these reptiles could not have swum thousands of kilometers across the Atlantic and Antarctic oceans, so some previous connection of the continents must be postulated. A former land bridge between the continents could explain the distribution of *Lystrosaurus* in distant parts of the world. Surveys of the ocean floor show no evidence for such a submerged land bridge like today's Central America.

Evidence from Structure and Rock Type

Several geologic features end abruptly at the coast of one continent and reappear on the facing continent across the Atlantic (**Figure 17.3**). The folded mountain ranges at the Cape of Good Hope, at the southern tip of Africa, trend from east to west and terminate sharply at the coast. An equivalent structure, of the same age and style of deformation, appears near Buenos Aires, Argentina (Figure 17.3). The folded Appalachian Mountains are another excellent example. The deformed structures of the mountain belt extend northeastward across the eastern United States and through

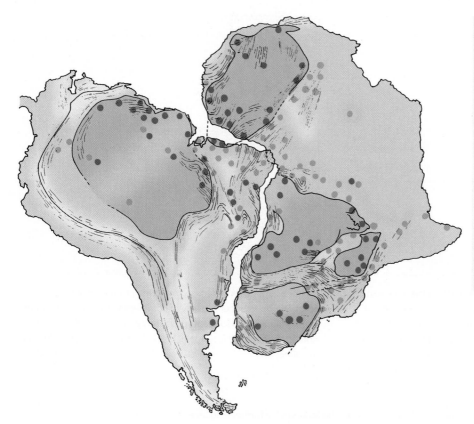

Figure 17.3 **South America and Africa fit together**, not only in outline, but also in rock types and geologic structure. The green areas represent the shields of metamorphic and igneous rocks, formed at least 2 billion years ago. Structural trends such as fold axes are shown by dashed lines. The gray areas represent younger rock, much of which has been deformed by mountain building. Most of the deformation occurred from 450 million to 650 million years ago. Several fragments of the African shield are stranded along the northern coast of Brazil. Green dots represent rocks that are more than 2 billion years old. Orange dots represent younger Precambrian rocks.

Data from: P. M. Hurley

Newfoundland, terminating abruptly at the ocean. The mountain belt with a similar age, rock sequence, fossils, and structural style reappears on the coasts of Ireland, Scotland, and Norway.

Other examples could be cited, but the important point is that the continents on both sides of the Atlantic fit together, not only in outline, but in rock type and structure. They are related much like matching pieces of a torn newspaper (**Figure 17.4**). The jagged edges fit, and the printed lines (structure and rock types) join in a coherent unit. One important point needs emphasis. The geologic similarities on opposite sides of the South Atlantic are found only in rocks older than the Cretaceous Period, which began about 145 million years ago. The southern continents are believed to have split and begun drifting apart in Jurassic time, about 200 million years ago.

Evidence from Glaciation

During the latter part of the Paleozoic Era (about 300 million years ago), glaciers covered large portions of the continents in the Southern Hemisphere. The deposits left by these ancient glaciers are distinct and can be readily recognized, and they cannot be mistaken for other types of sediment. In addition, striations and grooves on the underlying rock show the direction in which the ice moved (**Figure 17.5A**). Except for Antarctica, all of the continents in the Southern Hemisphere now lie close to the equator, far removed from a latitude that could produce glaciation. In contrast, the present-day continents in the Northern Hemisphere show no trace of glaciation during this time. In fact, fossil plants in North America and Europe indicate a tropical climate in those areas. This evidence is difficult to explain in the context of immovable continents because the climatic belts are determined by latitude.

Even more difficult to explain is the direction in which the glaciers moved. Regional mapping of striations and grooves indicates that in South America, India, and Australia, the ice accumulated in the oceans and moved inland. Such movement

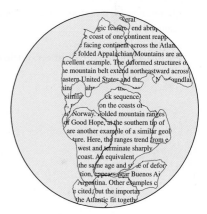

Figure 17.4 **When reconstructed, the continents fit together** like a jigsaw puzzle or pieces of a torn newspaper. Not only do the outlines of the torn pieces fit together, but the printing on them (analogous to the ages and structural features of the continents) also matches across their edges.

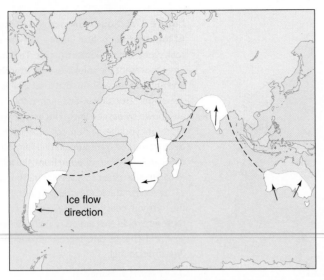

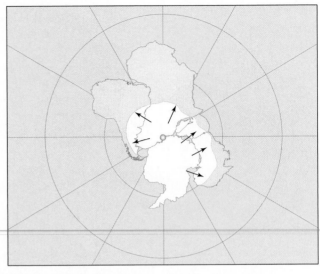

(A) Late Paleozoic glacial deposits are found only in the Southern Hemisphere and India, areas now close to the tropics. The present-day cold latitudes in the Northern Hemisphere show no evidence of glaciation at this time. Arrows show the direction of ice movement was from the sea toward the land. This flow direction is impossible; glaciers flow from centers of accumulation on the continents *outward toward* the sea.

(B) If the continents were restored to their former positions according to Wegener's theory of continental drift, and if the former South Pole were located approximately where South Africa and Antarctica meet, the location of late Paleozoic glacial deposits and the directions in which the ice flowed would be explained nicely.

Figure 17.5 The distribution and flow direction of late Paleozoic glaciers provide further evidence of continental drift.

of ice would be impossible unless there was a land mass where the oceans now exist. Glaciers do not form in the ocean. If glaciers could form in the sea, a large permanent glacier would exist in the Arctic Ocean. Instead, glaciers originate on land and move *toward* the edge of a continent.

However, if the continents were grouped together as Wegener proposed, the glaciated areas would have made up a neat package near the South Pole (Figure 17.5B), and Paleozoic glaciation could be explained nicely. The pattern of glaciation was considered strong evidence of continental drift, and many geologists who worked in the Southern Hemisphere became ardent supporters of the theory because they could see the evidence with their own eyes.

Evidence from Other Paleoclimatic Records

Why is the distribution of Paleozoic glacial features such powerful evidence for continental drift?

Other evidence of striking climatic changes recorded in the geologic record tends to support the drift theory. Great coal deposits in Antarctica show that abundant plant life once flourished on that continent, now covered with ice more than a kilometer thick.

On the other continents, thick deposits of salt, formations of wind-blown sandstone, and extensive fossil coral reefs provide additional clues that permit us to reconstruct the climatic zones of the past. The paleoclimatic patterns shown by these rocks are baffling with the continents in their present positions, but if the continents are grouped together in their predrift positions, the patterns are easily explained (**Figure 17.6**).

The evidence for the theory of continental drift was considered and debated for years. Wegener was criticized for failing to explain what forces would permit continents of granite to plow through oceans of rock. The idea of a moving lithosphere was yet to come. In the absence of a reasonable mechanism for drift, there was little further development of the theory until after World War II. An explosion of knowledge then provided renewed support for the drift hypothesis and also led to the discovery of a possible mechanism.

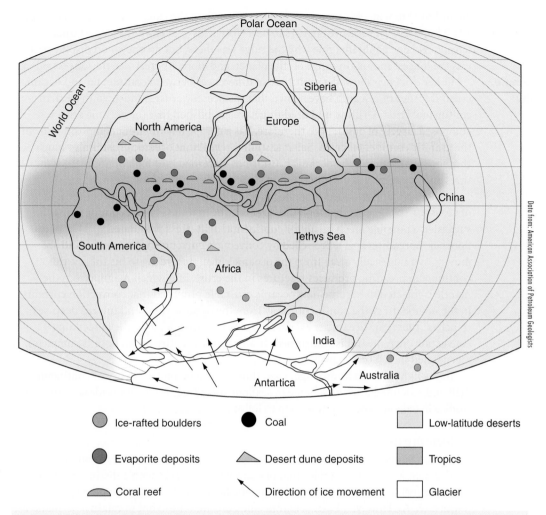

Figure 17.6 Paleoclimatic evidence for continental drift includes deposits of coal, desert sandstone, rock salt, wind-blown sand, gypsum, and glacial deposits about 300 million years ago near the end of the Paleozoic Era. Each indicates a specific climatic condition at the time of its formation. The distribution of these deposits is best explained if we assume that the continents were grouped together at the end of the Paleozoic Era, as shown in this diagram.

Development of the Theory of Plate Tectonics

The plate tectonics theory was developed during the early 1960s, when new instruments permitted scientists to map the topography of the ocean floor and to study its geologic and paleomagnetic characteristics.

Although the theory of continental drift was supported by some convincing evidence, the data on which it was originally based came only from the continents because, before the 1950s, there was no effective means of studying the ocean floor. Before 1950, therefore, geologists faced an almost total absence of data about the geology of three-fourths of Earth's surface; then, in the 1950s and 1960s, new technology resulted in a burst of new data and new ideas about the geology of the ocean floor and about paleomagnetism.

Geology of the Ocean Floor

In the 1950s and 1960s, newly developed echo-sounding devices enabled marine geologists and geophysicists to map in considerable detail the topography of the ocean floor. When the results of these studies were compiled, they revealed that several ocean basins are divided by a great ridge, approximately 65,000 km long and

The plate tectonics theory is simple, clear, and straightforward. Why wasn't it developed earlier?

about 1500 km wide. Moreover, at the crest of much of the ridge is a central valley, from 1 to 3 km deep. This feature appears to be a **rift valley** that is splitting apart under tension. No one could imagine why the ridge was there, but no one could dispute that it was the longest mountain range on the planet, and along its crest was the longest valley.

Other evidence showed a multitude of differences between continental and oceanic crust. Decades of research have shown that the oceanic crust is much younger than continental crust. Drilling and dredging have established that the oceanic crust is composed largely of basalt and, therefore, has a completely different composition from the granitic continental crust. Seismic studies reveal that oceanic crust is also much thinner. Furthermore, the oceanic crust is not deformed into folded mountain structures and apparently is not subjected to strong compressional forces.

In 1960 H. H. Hess, a noted geologist from Princeton University, proposed a theory of **seafloor spreading** that took into account the new data from echo soundings and suggested a possible mechanism for continental drift. Hess postulated that the ocean floors are spreading apart, propelled by convection currents in the mantle, and are moving symmetrically away from the oceanic ridge. According to his theory, this continuous spreading produces fractures in the crust, into which magma from the mantle is injected to become new oceanic crust. He proposed that convection currents in the mantle carry the continents away from the oceanic ridge and toward deep-sea trenches. There, the oceanic crust descends into the mantle, with the descending convection current, and is reabsorbed. In this way, the entire ocean floor is completely regenerated in 200 or 300 million years.

In the light of fresh knowledge, Hess thus elaborated on the theory of continental drift and redefined it in the scheme of seafloor spreading. A test of his ideas, using new studies in paleomagnetism, was soon to follow.

Paleomagnetism

Like most planets, Earth has an internally generated magnetic field. In many ways, Earth's magnetic field resembles that of a simple bar magnet with a distinct north and south magnetic pole. The axis of the magnetic field is inclined 11° from the spin axis (**Figure 17.7A**). However, Earth's mantle and core are far too hot to retain a permanent magnetic field. Earth's magnetism, therefore, must be constantly generated electromagnetically. Geophysicists still lack a complete understanding of how the field forms. The electromagnetic, or dynamo, theory postulates that the outer core of liquid iron convects and the motion generates electrical currents that establish a magnetic field (Figure 17.7B).

The study of rock magnetism developed during the 1950s with the perfection of new, highly sensitive magnetometers. Certain rocks, such as basalt, are fairly rich in iron and become weakly magnetized by Earth's magnetic field as they cool. In a sense, the mineral grains in the rock become "fossil" magnets that show the orientation of Earth's magnetic field at the time when the minerals crystallized and cooled; they thus preserve a record of **paleomagnetism**. Similarly, the iron-oxide grains in some red sandstones become oriented in Earth's magnetic field as the sediment is deposited, so some sedimentary rocks also can show the orientation of the paleomagnetic fields. These rocks therefore retain an imprint of Earth's magnetic field at the time of their formation.

Apparent Polar Wander   Some of the first paleomagnetic studies were conducted in Europe. Paleomagnetism in these rocks of widely different ages appears to show that Earth's north magnetic pole has steadily changed its position. As illustrated in **Figure 17.8A**, the north magnetic pole appears to have slowly migrated northward and westward to its present position. The change in position was systematic, not random. Migration of the magnetic pole was found from paleomagnetic work in North America, and although the path of migration was systematically different, it paralleled that of the European shift. Soon, paleomagnetic results collected from the southern continents were reported. Again, a systematic change in the position of the magnetic pole through time was documented—but with different paths for different continents.

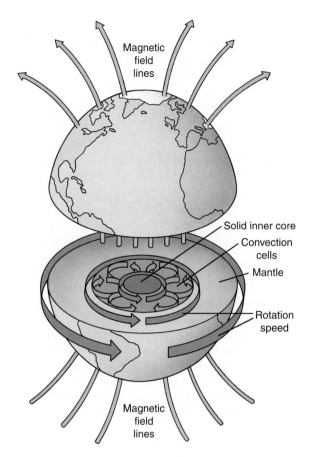

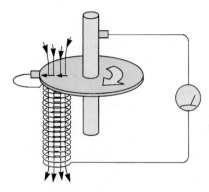

Figure 17.7 **Earth's magnetic field** is like that of a simple bar magnet because it has a north and a south pole. The temperature in the core and mantle, however, is far too high for permanent magnetism. Earth's magnetism must therefore be generated electromagnetically.

(A) Lines of force in Earth's magnetic field are shown by arrows. If a magnetic needle were free to move in space, it would be deflected by Earth's magnetic field. Close to the equator, the needle would be horizontal and would point toward the poles. At the magnetic poles, the needle would be vertical. Field lines are shown for a reverse polarity time period.

(B) Theoretically, convection in Earth's core can generate an electrical current (in a manner similar to the operation of a dynamo), which produces a magnetic field.

It is impossible that there were numerous magnetic poles migrating systematically and eventually merging. The most logical explanation is that there has always been only one magnetic pole, which has remained fixed, while the continents moved with respect to it. Consequently, students of paleomagnetism became leading proponents of the theory of continental drift.

The results of paleomagnetic studies make sense if the continents were once arranged as shown in Figure 17.8B and then drifted to their present positions. This discovery brought renewed interest in the theory of continental drift and lent support to the conclusion that the Atlantic Ocean opened relatively recently.

Patterns of Magnetic Reversals on the Seafloor Studies of the magnetic properties of numerous layers of volcanic rock, from many parts of the world, demonstrate that the polarity of Earth's magnetic field has reversed many times over its history. Epochs of **normal polarity** (that is, periods when the magnetic field was oriented as it is today with the north magnetic pole in the north and close to its present location) have been followed by periods during which the locations of the north magnetic pole and the south magnetic pole were reversed. At least a dozen magnetic reversals have occurred in the last 4.0 million years (**Figure 17.9**). The present period of normal polarity began about 780,000 years ago. It was preceded by a major period of **reversed polarity**, which began about 2.5 million years ago. That period of generally reversed polarity contained two short episodes of normal polarity. The major intervals of alternating polarity (about 1 million years apart) are termed **polarity chrons**.

How does apparent polar wandering support the theory of plate tectonics?

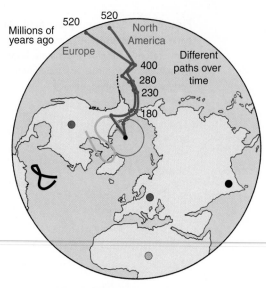

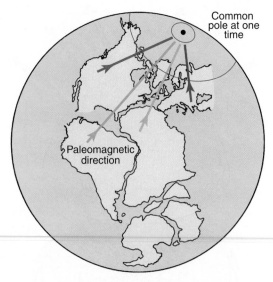

(A) The magnetic properties of rocks in North America show that the north magnetic pole has apparently migrated in a sinuous path over the last several hundred million years (red). Evidence from other continents shows similar migration, but along different paths. How could different continents show different paths of polar migration? The paleomagnetic evidence implies that, if the continents had remained fixed, different continents would have had different magnetic poles at the same time, but that would be impossible.

(B) The question can be answered if the magnetic pole has remained fixed while the continents have drifted. If, for example, Europe (blue) and North America (red) were previously joined, the paleomagnetic field preserved in their rocks would indicate a single pole location until they drifted apart. The sequence of rocks on each continent would then show the pole taking a different path of migration to its present position.

Figure 17.8 Apparent changes in the locations of the magnetic poles in the geologic past are shown by paleomagnetic studies of rocks.

How do patterns of magnetic reversals support the plate tectonics theory?

The pattern of alternating polarities has been clearly defined, and evidence of the occurrence of polarity reversals has been found in widely separated places. From the sequence of **magnetic anomalies** and their radiometric ages, a reliable chronology of **magnetic reversals** has been established for the last 4 million years (Figure 17.9). The paleomagnetic time scale is gradually being extended back in time.

In 1963 Fred Vine and D. H. Matthews saw a way to use paleomagnetism to test the idea of seafloor spreading put forth by Hess. If seafloor spreading has occurred, they suggested, it should be recorded in the magnetism of the basalts in the oceanic crust. (The same idea was developed independently by L. W. Morley.) If Earth's magnetic field reversed intermittently, new basalt forming at the crest of the oceanic ridge would be magnetized according to the polarity at the time it cooled. As the ocean floor spreads, a symmetrical series of magnetic stripes, with alternating normal and reversed polarities, would be preserved in the crust along either side of the oceanic ridge. Subsequent investigations have conclusively proved this theory.

To understand the origin of these magnetic patterns better, consider how the seafloor could have evolved during the last few million years. **Figure 17.10A** shows the seafloor as it is considered to have been about 2.75 million years ago, during the Gauss normal polarity chron (named for German mathematician Karl Friedrich Gauss). Basalt was injected into dikes below the ocean ridge or was extruded over the seafloor as submarine flows. As it crystallized and cooled, it became magnetized in the direction of the existing (normal) magnetic field, and thus, basalt extruded along the oceanic ridge formed a zone of new crust with normal magnetic polarity. As the seafloor spread, this zone of crust split and migrated away from the ridge but remained parallel to it. About 2.5 million years ago, Earth's magnetic polarity reversed. New crust generated at the oceanic ridge was then magnetized in this new direction

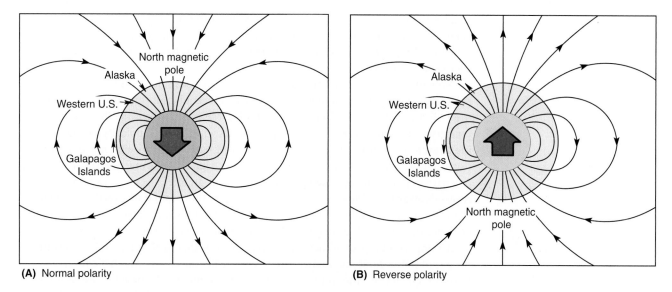

(A) Normal polarity   (B) Reverse polarity

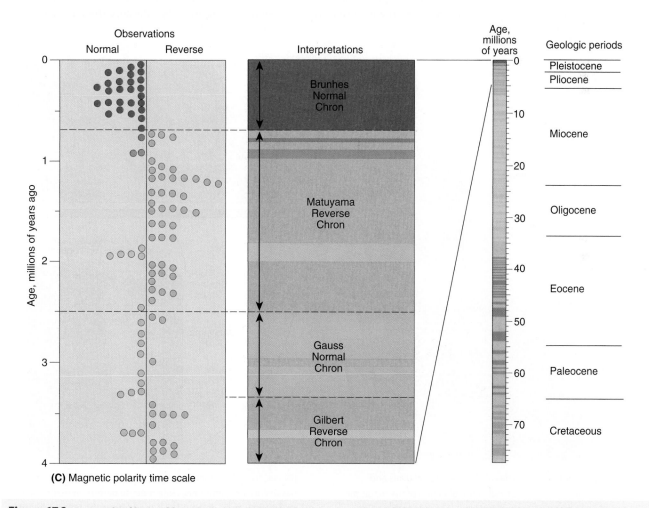

(C) Magnetic polarity time scale

Figure 17.9 **Reversals of lines of force in Earth's magnetic field** are documented by paleomagnetic studies of numerous rock samples from throughout the world. Lines of force with normal polarity are shown in (A). With reverse polarity (B), the lines of force are oriented in the opposite direction. (C) shows the patterns of changing polarity with time. The pattern of change during a period of 1 or 2 million years is distinctive, and it can be used to help establish the age of a rock sequence.

Data from: A. Cox, G. B. Dalrymple, and R. R. Doell

(Figure 17.10B), producing a zone of crust with reverse polarity. When the polarity changed to normal again, the newest crustal material was magnetized in the normal direction. In this way, the sequence of polarity reversals became imprinted like a bar code on the oceanic crust.

Note that the patterns of magnetic stripes on the ocean floor, on either side of the ridge, match the patterns found in a sequence of recent basalts on the continents (Figure 17.10A and B; see also Figure 17.9C); that is, the crest of the ridge shows normal polarity and is flanked by a broad stripe of rocks with reversed polarity (formed during a reversed chron) and containing two narrow bands of rocks with normal polarity (formed during normal chrons). Then follows a stripe with normal polarity, containing one narrow band with reversed polarity, and so on. In brief, the patterns of magnetic reversals away from the ridge crest are the same as those found in a vertical sequence of rocks on the continents, from youngest to oldest. These data provide compelling evidence that the seafloor is spreading and that continents drift.

An important aspect of these reversal patterns is that they enable us to determine the age of the seafloor and to measure rates of plate movement. Magnetic reversals in rock sequences on the continents have been radiometrically dated. These studies show that the present normal polarity has existed for the last 700,000 years and was preceded by the pattern of reversals shown in Figure 17.9C. Because the same pattern

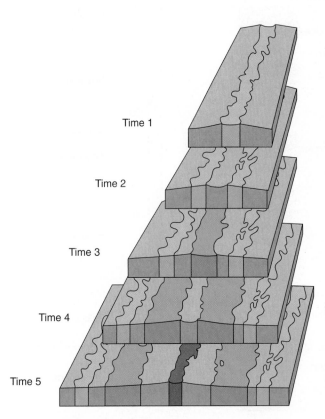

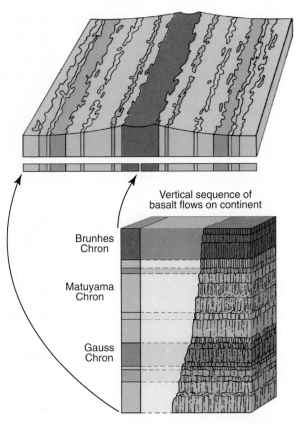

(A) As magma cools and solidifies along the ridge in dikes and flows (top), it becomes magnetized in the direction of the magnetic field existing at that time (normal polarity). As seafloor spreading continues, the magnetized crust formed during earlier periods separates into two blocks. Each block is transported laterally away from the ridge, as though on a conveyor belt. New crust, formed at the ridge, becomes magnetized in the opposite direction.

(B) Patterns of magnetic reversals in a vertical sequence of basalts on the continents. Note that the pattern of magnetic reversals away from the ridge is the same as the pattern found in this sequence of basalt flows. The youngest (upper) continental rocks correlate with the youngest oceanic crust (at the center of the oceanic ridge).

Figure 17.10 Specific patterns of magnetism are preserved in the newly formed crust generated at the oceanic ridge as the lithosphere moves laterally. The patterns of magnetic reversals away from the ridge are identical to the patterns of magnetic reversals in a vertical sequence of rocks on the continents.

exists in the oceanic crust, we can assign provisional ages to the magnetic anomalies on the ocean floor based on known ages of continental rocks. Magnetic surveys have now determined patterns of magnetic reversal for most of the ocean floor, and from these patterns, the age of various segments of the seafloor has been established (**Figure 17.11**). These studies show that most of the deep seafloor was formed during the Cenozoic Era (the last 65 million years). It now seems probable that very little or none of the present ocean basin was formed before the Jurassic. From the pattern of magnetic reversals, the rate of seafloor spreading appears to range from 1 to 17 cm/yr.

Evidence from Sediment on the Ocean Floor

To many geologists, some of the most convincing evidence for the plate tectonics theory comes from recent drilling in the sediment on the ocean floor. The Deep-Sea Drilling Project is truly a remarkable example of scientific exploration. It began in 1968 with the *Glomar Challenger*, a special ship designed by a California offshore drilling company. The *Challenger* can lower more than 6000 m of drilling pipe into the open ocean, bore a hole in the seafloor, and bring up bottom cores and samples. The project was funded by the National Science Foundation and is under the direction of the Scripps Institution of Oceanography. Since 1968, the *Challenger* and its successors have drilled hundreds of holes in the seafloor and penetrated more than a kilometer into the oceanic crust. These drilling projects have provided considerable data in support of the theory of plate tectonics.

Deep-sea drilling confirms the conclusions drawn from paleomagnetic studies by providing samples of the fossils that accumulated on different portions of the ocean

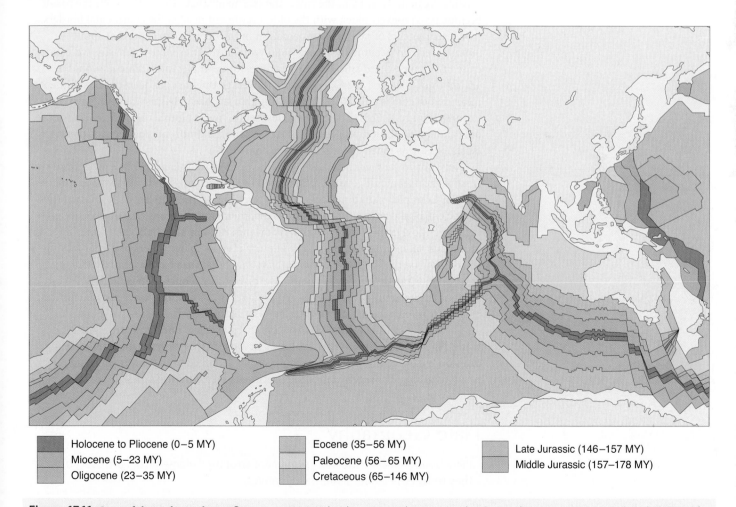

Holocene to Pliocene (0–5 MY)	Eocene (35–56 MY)	Late Jurassic (146–157 MY)
Miocene (5–23 MY)	Paleocene (56–65 MY)	Middle Jurassic (157–178 MY)
Oligocene (23–35 MY)	Cretaceous (65–146 MY)	

Figure 17.11 Ages of the rocks on the seafloor are symmetrical with respect to the oceanic ridge. By correlating magnetic reversals with the age of rocks found on the continents, we can estimate the age of the seafloor. The youngest crust is along the crest of the ridge. Away from the ridge, the crust is progressively older. The oldest oceanic crust is found in the Pacific Ocean and is less than 200 million years old.

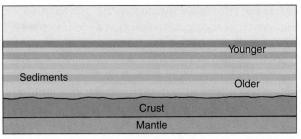

(A) With no seafloor spreading, the entire ocean floor would be covered with a thick sequence of oceanic sediment, with alternating polarity preserving a record of Earth's magnetism since Precambrian time.

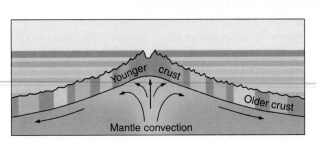

(B) With seafloor spreading, the blanket of oceanic sediment thins progressively toward the crest of the oceanic ridge and is almost nonexistent on the ridge. The edge of each layer of magnetized sediment lies upon basaltic crust, which was generated at the spreading ridge during the same time interval the sediment was deposited.

Figure 17.12 The thickness of sediment and magnetic reversals on the oceanic ridge confirms the theory of seafloor spreading.

floor. As is predicted by the plate tectonics theory, the *youngest sediment resting on the basalt* of the ocean floor is found near the oceanic ridge (**Figure 17.12**), where new crust is being created. Away from the ridge, the sediments that lie directly above the basalt become progressively older, with the oldest sediment nearest the continental borders.

Measurements of rates of sedimentation in the open ocean show that about 3 mm of red clay accumulates every 1000 years. If the present ocean basins were old enough to have existed since Cambrian time, for example, the sediments would be 1.5 km thick (Figure 17.12A); however, the average thickness of deep-ocean sediments measured to date is only 300 m, suggesting that the ocean basins are young geologic features (Figure 17.12B). In fact, the oldest sediments yet found on any ocean floor are only about 200 million years old. In contrast, the metamorphic rocks of the continental shields are as much as 3.8 billion years old.

Not only do the thickness and age of the deepest sediments increase away from the crest of the oceanic ridge, but certain types of sediment also indicate seafloor spreading. For example, plankton thrive in the upwelling, warm, nutrient-rich water of the Pacific equatorial zone. As the creatures die, their tiny skeletons rain down unceasingly to build a layer of soft, white chalk on the seafloor. The chalk can form only in the equatorial belt, as plankton do not flourish in the colder waters of higher latitudes; yet, drilling by the *Glomar Challenger* has shown that the chalk layer on the Pacific floor extends north of today's equator. The only logical conclusion is that the Pacific seafloor has been migrating northward for at least 100 million years, carrying its load of chalk formed anciently when the plate was farther south.

The theory of plate tectonics is now firmly established and accepted as the fundamental theory of Earth's dynamics. It was first used to explain the meaning of features on the ocean floor. Now the emphasis has switched to the continents, and most previous geologic observations of the continents are being reexamined in light of plate tectonics theory.

Plate Geography

Plate boundaries are the most significant structural elements of Earth because they reflect the planet's internal dynamics.

How does the geography of tectonic plates differ from classical physical geography?

The shorelines of the continents are major geographic features but have little significance from the standpoint of Earth's tectonics. Plate boundaries are the planet's most significant geologic elements, and to understand plate tectonics, you must learn a new geography: the geography of plate boundaries. This should not be difficult because plate boundaries generally are marked by major topographic features. You only need to focus your attention on Earth's structural features, rather than on the boundaries between land and ocean.

The new geography of **tectonic plates** is illustrated in **Figure 17.13**. Earth's outer rigid layer—the lithosphere—is divided into a mosaic of seven major plates and several smaller subplates. The major plates are outlined by oceanic ridges, trenches, and young mountain systems. These include the Pacific, Eurasian, North American, South American, African, Australian, and Antarctic plates. Remember, the continents are not moving separately from the oceanic crust, but both are parts of moving lithospheric plates that extend into the mantle.

The largest is the Pacific plate, which is composed almost entirely of oceanic crust and covers about one-fifth of Earth's surface. The other large plates contain both continental crust and oceanic crust. No major plate is composed entirely of continental crust. Smaller plates include the Philippine, Arabian, Juan de Fuca, Cocos, Nazca, Caribbean, and Scotia plates, plus others that have not been defined precisely.

Individual plates are not permanent features. They are in constant motion and continually change in size and shape. Plates that do not contain continental crust can be completely consumed in a subduction zone. Even plate margins are not fixed. A plate can change its shape by splitting along new lines, by welding itself to another plate, or by the accretion of new oceanic crust along its passive margin. The movement and modification of a plate margin can change its size and shape across the entire plate.

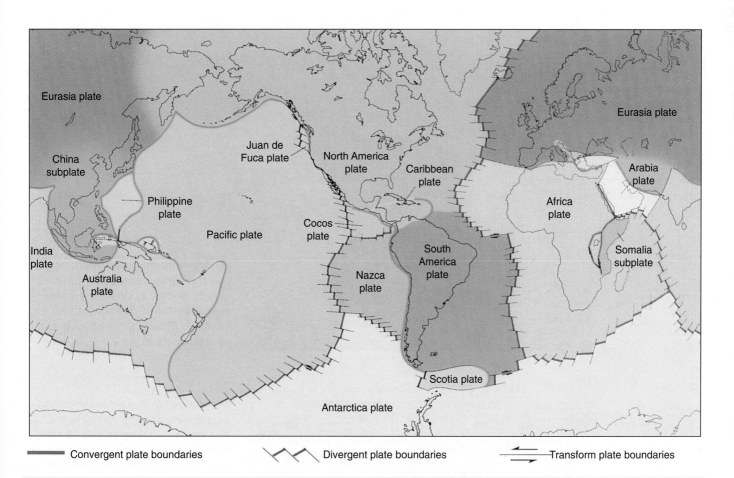

Figure 17.13 **The major tectonic plates** are delineated by the major tectonic features of the globe: (1) the oceanic ridge, (2) deep-sea trenches, and (3) young mountain belts. Plate boundaries are outlined by earthquake belts and volcanic activity. Most plates (such as the North American, African, and Australian) contain both continental and oceanic crust. The Pacific, Cocos, and Nazca plates contain predominantly oceanic crust.

Plate Boundaries

Three kinds of plate boundaries are recognized and define three fundamental kinds of deformation and geologic activity: (1) divergent plate boundaries—zones of tension, where plates split and spread apart, (2) convergent plate boundaries—zones where plates collide and one plate moves down into the mantle, and (3) transform fault boundaries—zones of shearing, where plates slide past each other without diverging or converging.

How are plate boundaries expressed at the surface?

Each tectonic plate is rigid and moves as a single mechanical unit—that is, if one part moves, the entire plate moves. It can be warped or flexed slightly as it moves, but relatively little change occurs in the middle of a plate. Nearly all major tectonic activity occurs along the plate boundaries, and thus, geologists and students of geology focus their attention on the plate margins, the ones that are active as well as the ancient plate boundaries preserved on the continents (**Figure 17.14**).

Divergent Plate Boundaries

A **divergent plate boundary** forms where a plate splits and is pulled apart. Except for a few rift zones in Africa and western North America, essentially all present divergent plate margins are submerged beneath the sea. Where a zone of spreading extends into a continent, rifting occurs, and the continent splits (**Figure 17.15**). The separate

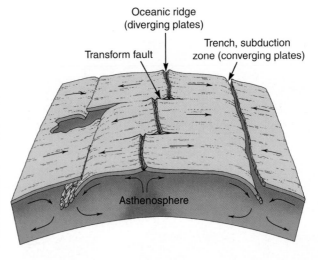

Oceanic ridge (diverging plates)

Transform fault

Trench, subduction zone (converging plates)

Asthenosphere

Figure 17.14 Types of plate margins are depicted in this idealized diagram. Constructive margins (divergent plate boundaries) occur along the oceanic ridge, where plates move apart. Destructive margins (convergent plate boundaries) occur along the deep trenches. Margins with no change in seafloor area during displacement occur along transform faults.

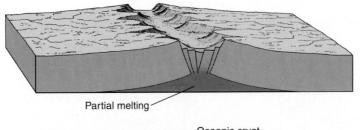

Partial melting

Oceanic crust

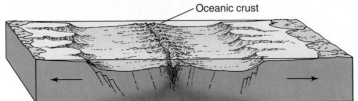

(A) Continental rifting begins when the crust is uparched and stretched, so that block faulting occurs. Continental sediment accumulates in the depressions of the downfaulted blocks, and basaltic magma is injected into the rift system.

(B) As the continents separate, new oceanic crust and new lithosphere are formed in the rift zone, and the ocean basin becomes wider. Remnants of continental sediment can be preserved in the down-dropped blocks of the new continental margin.

Figure 17.15 Divergent plate boundaries are found in the ocean basins and continents. The mid-ocean ridge is one type of divergent plate boundary and has abundant normal faults, shallow earthquakes, and basaltic magmatism.

continental fragments drift apart with the diverging plates, so a new and continually enlarging ocean basin is formed at the site of the initial rift zone. Divergent plate boundaries are thus characterized by tensional stresses that produce normal faults along the margins of the separating plates. Basaltic magma, derived from the partial melting of the mantle, is injected into the fissures or extruded as fissure eruptions. The magma then cools and becomes part of the moving plates.

Divergent plate boundaries are some of the most active volcanic areas on Earth; they are, however, generally characterized by unspectacular, quiet fissure eruptions, most of which are concealed beneath the sea. The importance of volcanism along this zone is underlined by the fact that during the last 200 million years, more than half of Earth's surface has been created by volcanic activity along divergent plate boundaries. The mid-Atlantic ridge is a typical divergent plate boundary.

Convergent Plate Boundaries

Convergent plate boundaries, where the plates collide and one moves down into the mantle, are areas of complicated geologic processes, including igneous activity, earthquakes, metamorphism, crustal deformation, and mountain building.

The specific processes that are active along a convergent plate boundary depend on the types of crust involved in the collision of the converging plates (**Figure 17.16**). If both plates at a convergent boundary contain oceanic crust, one is thrust under the margin of the other in a process called **subduction**. A **subduction zone** is usually marked by a deep-sea trench, and the movement of the descending plate generates an inclined zone of seismic activity. The subducting plate descends into the asthenosphere, where it is heated and ultimately absorbed into the mantle. Layers of sediment may be scraped off the downgoing plate and accreted onto the continent. The island arcs of the western Pacific, including the islands of Tonga and the Marianas, formed at ocean-ocean convergent margins.

If one plate contains a continent, the lighter continental crust always resists subduction and overrides the oceanic plate. Compression may deform the continental margin into a folded mountain belt (**Figures** 17.16 and **17.17**), and the deep roots of the

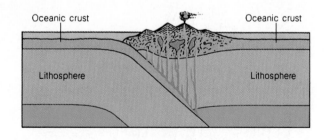

(A) The Philippine Islands represent the convergence of two oceanic plates.

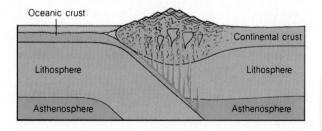

(B) South America represents the convergence of an oceanic plate and a continental plate.

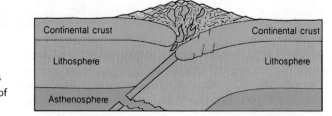

(C) The Himalaya Mountains represent the convergence of two continental plates.

Figure 17.16 Examples of the main types of convergent plate boundaries can be found today in various parts of the world. The major geologic processes at convergent plate boundaries include the deformation of continental margins into folded mountain belts, metamorphism due to high temperatures and high pressures in the mountain roots, and partial melting of the mantle over the descending plate, which produces andesitic volcanism on the overriding plate.

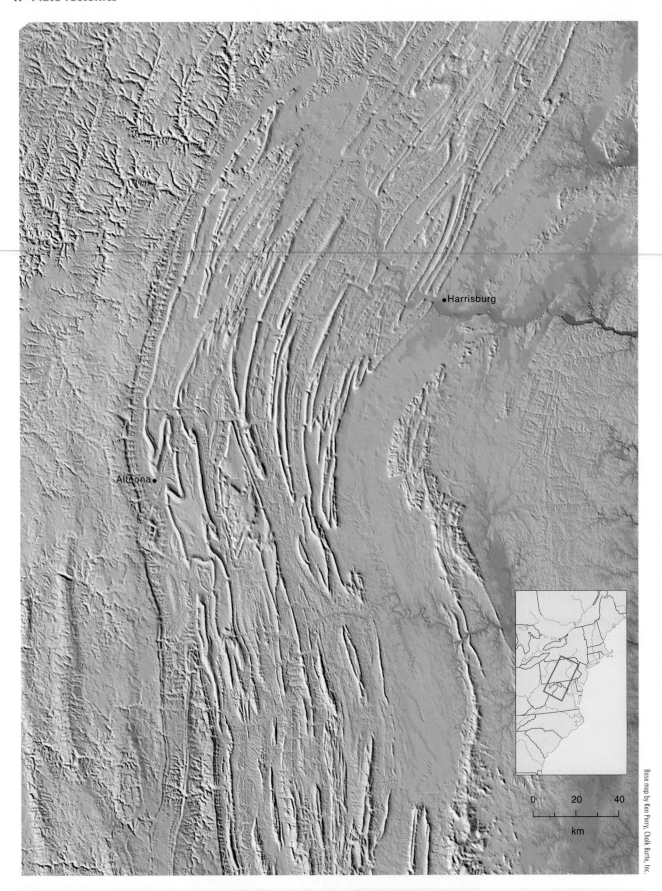

Base map by Ken Perry, Chalk Butte, Inc.

Figure 17.17 **The spectacular fold belt of the Appalachian Mountains** of the eastern United States is one of the great surface expressions of a convergent plate boundary. The deformed strata form ridges that zigzag across the terrain, showing the style of deformation produced as the African plate drifted westward and impinged against the North America plate. Deformation dies out rapidly to the northeast. Small granitic plutons intruded the mountain belt, but are not obvious on this topographic map.

mountains are intruded by magma and metamorphosed. The Cascadia subduction zone offshore of the northwestern United States is an ocean continent convergent margin.

If both converging plates contain continental crust, neither can subside into the mantle, although one can override the other for a short distance. Both continental masses are instead compressed, and the continents are ultimately "fused" into a single continental block, with a high mountain range marking the line of the suture. As a result of the collision and underthrusting, the thickness of the crust is greatly increased. The great Himalayan mountain chain formed when India and Asia collided, both of which are continental plates.

Magma is also produced at convergent plate margins. Magma extracted from the mantle above a subduction zone can differentiate to form relatively silicic magmas (like andesite and rhyolite) which have low densities. Ultimately much of this low density material is added to the continental crust as batholiths and volcanic extrusions at the surface In fact, most continental crust formed at ancient convergent plate boundaries. Steep-sided stratovolcanoes and large collapse calderas are the typical volcanoes found at convergent margins.

Transform Fault Boundaries

Transform fault boundaries are zones of shearing where plates slide past each other without diverging or converging, and without creating or destroying lithosphere (**Figure 17.18**). These boundaries occur along a special type of fault, a **transform fault**, which is simply a **strike-slip fault** between plates (that is, movement along the fault is horizontal and parallel to the fault). The term *transform* is used because the kind of motion between plates is changed—transformed—at the ends of the active part of the fault. For example, the divergent motion between plates at an oceanic ridge can be transformed along the fault to the convergent motion between plates at a subduction zone.

Transform faults can join ridges to ridges, ridges to trenches, and trenches to trenches. In all cases, transform faults are parallel to the direction of relative plate motion. The plates simply slide past each other, but as the plates move, the crust is fractured and broken. This fracturing produces the shallow earthquakes that are characteristic of transform plate boundaries. However, volcanic activity usually is not abundant at transform faults.

The San Andreas fault system of western North America is a typical continental transform fault at the boundary between the North America plate and the Pacific plate (Figure 17.13). Southwestern California is actually on the Pacific plate. The fault connects a series of short spreading ridges in the Gulf of California, with the Cascadia subduction zone that starts in northern California and extends to Canada. Shallow earthquakes are common, and often devastating, along the boundary, but there are no active volcanoes along the fault system.

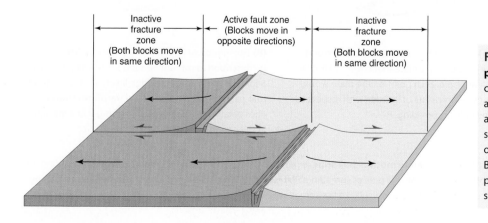

Figure 17.18 The relative movement of plates at a ridge-ridge transform fault changes along the fracture zone. The plates are moving away from the ridge, but an active fault only lies between the two ridge segments. Here plates on opposite sides of the fault move in opposite directions. Beyond the spreading ridge, however, the plates move in the same direction on both sides of the fault.

Plate Motion

> The motion of a series of rigid plates on a sphere can be complex. Each plate moves as an independent unit, in different directions and at different velocities than any other plates.

The geometry of a curved plate moving on a sphere was worked out more than 200 years ago by Swiss mathematician Leonhard Euler (1707–1783) and now provides the basis for analyzing plate motion. The basic analysis of this type of motion is illustrated in **Figure 17.19A**. In the figure, the motion of Plate 1, with respect to Plate 2, is a rotation around the axis AR (the **axis of plate rotation**), one pole of which is the point P (the **pole of rotation**). Note that the pole of the plate rotation is completely independent of Earth's spin axis and has no relation to the magnetic poles.

How do you describe the motion of plates on a sphere?

Several important facts about plate motion are immediately apparent from Figure 17.19A. First, different parts of a plate move with different velocities. Maximum velocity occurs at the equator of rotation and minimum velocity at the poles of rotation. This fact may best be understood by considering a plate so large that it covers an entire hemisphere. All motion occurs around the axis of plate rotation. The pole of rotation has zero velocity because it is a fixed point around which the hemispheric shell moves. Points Q, R, and S have progressively higher velocities, with a maximum velocity at point T, which lies on the equator of rotation.

Note also that transform faults lie on lines of latitude relative to the pole of rotation (Figure 17.19B). This condition holds for most transform faults in nature, as can be seen on a physiographic map of the Atlantic. (See the inside covers of this book.) We can thus use the orientation of transform faults to locate the pole of rotation for each plate.

What geologic features indicate direction and rates of plate motion?

Spreading ridges are linear and are usually perpendicular to plate motion. They are commonly oriented along lines of longitude relative to the plate's pole of rotation. It is

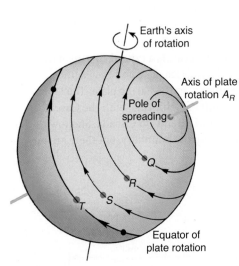

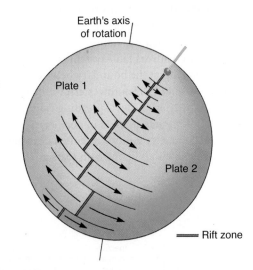

(A) Plate motion can be easily understood by considering a plate that covers an entire hemisphere. Each point on the plate would move along a line of latitude with respect to the pole of spreading, *P*.

(B) The motion of Plate 1 with respect to Plate 2 can be described as rotation around some imaginary axis. Segments of the oceanic ridge lie on lines of longitude that pass through the pole, and transform faults lie on lines of latitude. The rate of spreading is at a maximum at the equatorial line and zero at the pole.

Figure 17.19 Plate motion on a sphere requires that the plates rotate around an axis of spreading, the pole of which is called a pole of spreading. Plates always move parallel to the transform faults and along circles of latitude perpendicular to the spreading axis.

important to understand that the poles of rotation do not necessarily lie on the plate in question (Figure 17.19).

The direction of movement of the major plates, in relation to their neighbors, can be determined in several ways. As we have seen, the trends of the oceanic ridge and the associated transform faults are related to the location of the pole of rotation. Indications of movement are also drawn from seismic data, from the relative ages of different regions of the seafloor, and from the ages of chains of volcanic islands and seamounts. From these data, geologists have determined the motion of the present tectonic plates. This motion is summarized in **Figure 17.20**.

The Pacific plate is moving in a general northwesterly direction, from the East Pacific rise toward the system of trenches in the western Pacific. It is bordered by several small plates along the subduction zone, so the relative motion at each trench differs from the general trend. The America plates are moving westward from the mid-Atlantic ridge, converging with the Pacific, Cocos, and Nazca plates. The Australia plate is moving northward.

Africa and Antarctica, however, present a different situation. Both are nearly surrounded by ridges, and they have no associated subduction zones to accommodate the new lithosphere generated along the ridges. Africa moves north, toward the convergent boundary in the Mediterranean area, but this movement does not accommodate the east-west spreading from the Atlantic and Indian ridges. The Africa and Antarctica plates are apparently being enlarged as new lithosphere is generated at their margins. Without subduction zones, the ridges surrounding these plates must be moving outward from Africa and Antarctica. The Africa and Antarctica plates illustrate a very important point: *Plate margins are not fixed but can move as much as the plates themselves.* If

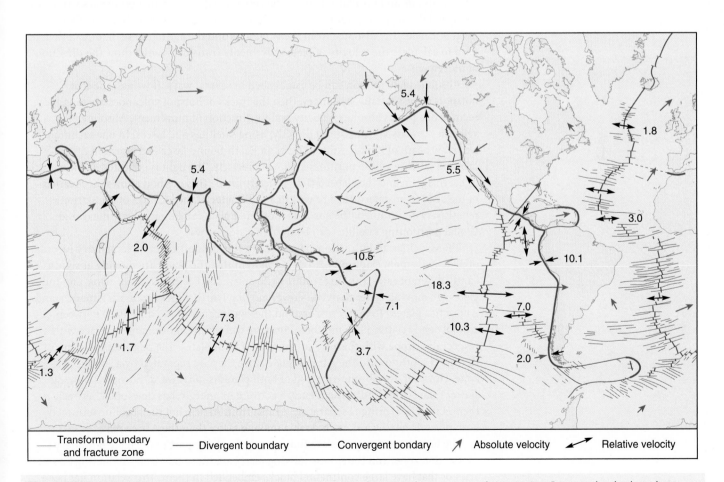

Figure 17.20 Velocities and directions of plate movement show how the major plates are currently interacting. Compare the absolute plate motions (red arrows) with the relative plate movements. The lengths of the arrows are proportional to the velocity of plate movement; the numbers represent velocity in centimeters per year.

two divergent plate margins are not separated by a subduction zone, new lithosphere is formed at each spreading axis, but none is destroyed between them. The plate between the ridges is continually enlarged, so the ridges themselves must move apart.

Another important change is in the lengths of plate margins. An oceanic ridge is essentially a fracture in the lithosphere that can grow longer. A good example is the ridge in the Atlantic Ocean. It has grown and lengthened considerably since spreading began to separate South America from Africa.

Rates of Plate Motion

Magnetic reversals on the ocean floor provide a timing mechanism to measure the relative velocity of plate motion. Absolute plate velocities can be measured compared to a fixed reference frame. The results show that plates move at different rates, ranging from 1 to 18 cm/yr.

Plate motions can be determined in two fundamentally different ways. The **relative velocity** compares the movement of one plate with respect to another plate. The **absolute velocity** compares plate movement to a fixed reference frame. The difference between the two measurements can be understood with a familiar example. Imagine you are standing on an overpass. Beneath you, two cars are traveling in opposite directions. Their speedometers say they are moving 50 km/hr. Compared to one another, their *relative velocity* is 100 km/hr, but compared to your fixed position, both cars have an *absolute velocity* of 50 km/hr.

To determine the relative velocity of a certain section of seafloor, all you need to know is its age and how far it is from the ridge. We described in a previous section how the age of the seafloor (Figure 17.11) can be estimated using the oscillation of Earth's magnetic field. The distance from the ridge axis and the age of the seafloor can then be used to calculate the velocity of plate movement. Transform faults show the direction of movement.

Absolute plate motion can be established in several ways. If we assume that hotspots are essentially stationary, then the tracks of hotspot volcanoes are tangible records of a plate's absolute velocity and its direction of movement. Absolute plate velocities can also be measured directly, using satellites and lasers. In one technique, a narrow beam of light is emitted from an Earth-bound laser and bounced off an orbiting satellite whose position is known precisely. The light is collected at the surface of Earth again, and the elapsed time is determined. This method allows the location of the laser to be determined to within a millimeter. If the location of the station is repetitively determined, the absolute motion of plate can be accurately measured. Global positioning satellites can be used in a similar way.

The velocities and directions measured in this fashion are complementary records of plate movement (Figure 17.20). It is apparent that the plates are moving at significantly different rates, ranging from 1 to 20 cm/y. The Pacific, Nazca, Cocos, and India plates are moving faster than the slower moving North America, South America, and Antarctica plates. To better understand the significance of the difference between absolute and relative plate motions, compare the relative movement of Africa with respect to Europe with the absolute motion of both plates. The relative movement of Europe is south toward the African plate since they are separated by a subduction zone. However, the absolute motion of both plates is northward. Europe is moving slower than Africa and consequently a convergent margin has developed between them. Likewise, rifting is separating Arabia from Africa, but the absolute motion of both plates is northward, with Arabia moving faster than the Africa plate.

The fastest-moving plates are those in which a large part of the plate boundary is a subduction zone, and the slower-moving plates are those that lack subducting boundaries or that have large continental blocks embedded in them. This relation has been interpreted by some geologists as evidence that the tectonic plates are part of Earth's convection system and that plate motion is largely a result of cold, dense plates sinking into the mantle.

STATE OF THE ART   The Magnetic Fabric of the Seafloor

The magnetic character of rocks on the seafloor was a major factor in deciphering the reality of plate tectonics. But given the remoteness of the seafloor, how are these paleomagnetic data collected?

Specially designed magnetometers (instruments designed to measure the strength or orientation of the present-day magnetic field) are towed behind research ships. (These magnetometers were originally designed and used to detect submarines travelling below the surface during World War II.) The position of the magnetometer and the strength of the magnetic field are simultaneously recorded. The result is a long strip map showing where the strength of the magnetic field is higher or lower than normal, as shown in the illustration.

Note that the pattern of variation in the seafloor magnetism is not as regular as a simple sine wave. The highs and lows are not separated by equal distances. In spite of the irregular widths of the bands, you can see that the patterns are symmetrical on either side of the ridge axis.

One way to interpret the map was to claim that each band had fewer (the lows) or more (the highs) magnetic minerals. But this did little to explain the overall symmetry of the patterns or the correlation of the middle high with a mid-oceanic ridge. It was soon realized that a better interpretation was that the polarity of Earth's field changed. In addition, each band must have formed anciently at a ridge and was then split apart. According to this interpretation, the magnetic highs lie over regions where the volcanic rocks on the seafloor erupted when the orientation of the magnetic field was the same as it is

today. Thus, the modern magnetic field and the paleomagnetism stored in the rocks add to one another. The lows are areas where the paleomagnetic orientation is opposite that of the present field and cancels out part of the current magnetic field. To make a map like this one, the ship must traverse back and forth across the area many times. The strips are then laid side by side and interpreted as representing stripes of rocks with different polarities and therefore different ages.

Magnetic maps can also be constructed for areas above sea level and they reveal much about the structure of the crust. The magnetic field variations shown on these maps of the continents are strongly affected by the amount of magnetite in the rocks and to a lesser degree by the polarity of the magnetic field at the time the rocks formed.

Courtesy of Lamont-Doherty Earth Observatory/Columbia University

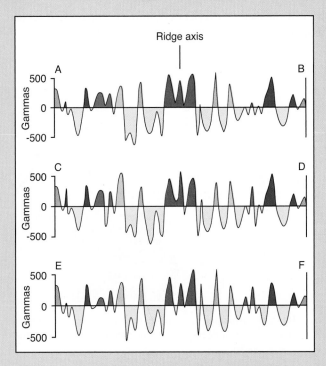

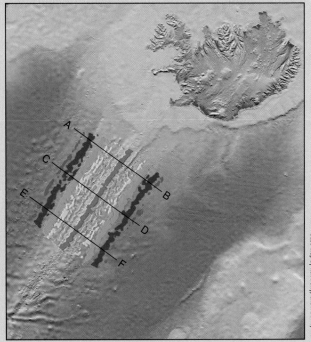

Base map by Ken Perry, Chalk Butte, Inc.

The Driving Mechanisms for Plate Tectonics

Forces that influence the motion of a plate include (1) slab-pull, (2) ridge-push, (3) basal drag, (4) friction along transform faults, and (5) friction between the converging slabs of the lithosphere in a subduction zone. Slab-pull and ridge-push probably drive plate movement.

It should be clear by now that the tectonic plates move. But why do they move? Ultimately, the energy that drives plate tectonics is heat transported out of the hot core and mantle to Earth's surface. Plate tectonics is a type of convection and is the result of Earth's effort to cool and reach thermal equilibrium with cold space.

Is convection of the mantle the only force responsible for plate movement?

One of the first models to explain the driving mechanism of plate tectonics suggested that **convection cells** within the mantle carried the plates, and that the plates played little or no active part in the convection (**Figure 17.21A**). The rising limbs of the convecting cells in the mantle would therefore determine the positions of the oceanic ridges. The convecting mantle would cause the lithosphere to split, and the moving mantle would carry the lithosphere laterally toward the subduction zone. The descending cell would mark the location of the trench and would drag the lithosphere down into the mantle. Movements in the asthenosphere were thought to be coupled strongly to the lithosphere. In other words, convection cells in the mantle supposedly *caused* ridges, trenches, and the movement of plates. The distance between plate boundaries was thought to be caused by the size of the convection cell.

A more successful model of convection theory considers the plates themselves to be active participants in the convection process (Figure 17.21B), not passive passengers on a churning mantle. In this model, the lithosphere is the cold upper layer of the convection cell. Because of its greater density, the lithosphere eventually sinks. Subduction occurs *not* because the plate is pulled down by the descending mantle, but simply because the plate becomes denser than the underlying asthenosphere. In addition, the upward flow of the asthenosphere beneath a spreading ridge is *not* the cause of seafloor spreading, but a consequence of the plates moving apart. Thus, the plate may be moved by forces that are largely independent of the convection of the mantle beneath the plate.

To better understand why the plates move, let us examine the forces that act on them. The most important forces are shown diagrammatically in **Figure 17.22**.

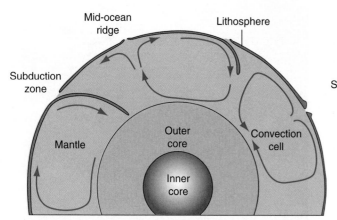

(A) Convection in the mantle drives the movement of the plates. Many characteristics of plate motion are inconsistent with this hypothesis.

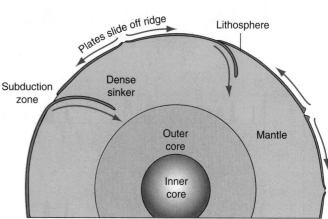

(B) Forces generated by the plates themselves cause the plates to sink into the mantle because of their density and to slide off the mid-ocean ridges.

Figure 17.21 Two suggested models of plate tectonics show how flow in the mantle might be related to plate movement.

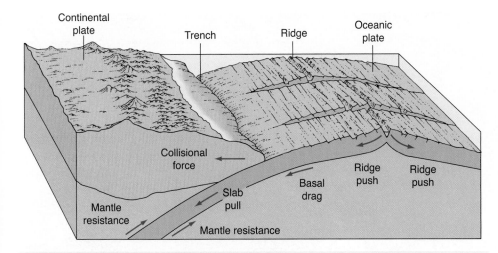

Figure 17.22 Forces active on the plates are shown with arrows on the front of this block diagram. They include slab-pull, ridge-push, basal drag, and friction along transform faults and in the subduction zone.

The forces that influence motion of the plate include:

1. **Slab-pull**: A pull exerted on the plate as the dense oceanic slab descends under its own weight into the asthenosphere in a subduction zone. In essence, the slab sinks because it is denser than the asthenosphere, and it pulls the rest of the lithosphere along with it.
2. **Ridge-push**: Gravity makes the lithosphere slip off the elevated ridge.
3. **Basal drag**: Resistance to flow exerted on the bottom of the plate by the underlying asthenosphere; shear at the base of the plate. Depending on the direction of flow in the asthenosphere, this could aid or hinder plate movement.
4. **Mantle resistance**: Frictional resistance to the movement of the subducting plate through the asthenosphere and mesosphere.
5. **Friction**: Resistance along transform faults and between the converging slabs of lithosphere in a subduction zone; shear between two plates.

Inasmuch as each plate has a nearly constant velocity—is neither accelerating nor decelerating—most researchers believe that the forces that drive the plates are approximately balanced by forces that resist their movement. Thus, the driving forces provided by slab-pull and ridge-push are nearly balanced by resisting forces such as basal drag, mantle resistance, and friction at plate boundaries.

As noted in the last section, the absolute velocity of a plate is strongly related to the proportion of its margin that is subducting. For example, plates such as the Pacific and Cocos plates, which have about 40% of their margins represented by subduction zones, have high plate velocities (greater than 5 cm/yr). Plates such as the North American, which have smaller proportions of subducting margins, move more slowly (1 to 3 cm/yr). Hence, many researchers have concluded that slab-pull is a major driving force. The correlation between the rate of plate motion and the proportion of subducting edges is so strong that other forces may be only minor. Slab-pull is aided by the metamorphic phase transitions that occur in subducting plates. As the plate moves deeper into the mantle, low-density phases convert into higher-density phases. These changes effectively create a sinker that pulls the plate deeper into the mantle.

Several examples illustrate the role of slab-pull in plate convection. If a large cauldron of molten metal is allowed to cool, a skin of solid metal forms on the surface and, because it is colder and denser than the liquid, the solid skin eventually founders and sinks into the molten liquid, thereby stirring the melt. The same process has been observed on a larger scale in lava lakes formed in the pit craters of Hawaiian volcanoes **(Figure 17.23)**. As molten lava cools, a solid layer forms over the lake, but the rigid layer splits into slabs that move about. Eventually a slab sinks, because it is cooler and

What is the difference between ridge-push and slab-pull?

Figure 17.23 Convection in a lava lake in Hawaii simulates convection and plate motion. As fresh, molten lava rises by convection, slightly older chilled lava (darker) is shoved aside to sink at some other zone in the crater (out of view). Note the transform fault near the middle of the view and the differential rate of spreading revealed by the different widths of recently chilled lava on either side of the spreading line. The front edge of the view is about 50 m wide.

denser than the underlying liquid, pulling the solid crust apart. Large slabs of the crust break up and split, causing molten lava to rise from below and create a zone of new cooling crust. Many features of plate tectonics are exhibited in Figure 17.23. Spreading ridges, transform faults, and subduction zones are all observed.

Ridge-push is probably the next most important motivator of tectonic plates. Ridge-push is related to the elevation of the ridge, which in turn is caused by the hot buoyant mantle that rises beneath it. The asthenosphere acts as a slippery layer beneath the slab, and the lithosphere simply slides downhill.

Observations of plate movement and theoretical calculations both suggest that the forces supplied by slab-pull and ridge-push are so large that the question may become: what slows the plates down? The major retarding force is probably resistance in the mantle to the sinking of a subducting slab. Friction at plate boundaries must also slow the movement of the plates. Basal drag, once thought to be the prime motivator of plate movement, is now relegated to a minor role. Basal drag may actually slow the movement of a plate.

From the perspective supplied by considering the forces acting on plates, you can think of the plates and the underlying mantle as forming a single, though complex system, with each portion of the system affecting the others. In fact, convective flow in the mantle may have a radically different aspect than the motion of lithospheric plates seen at the surface of Earth.

Key Terms

absolute velocity (p. 528)
axis of plate rotation (p. 526)
basal drag (p. 531)
continental drift (p. 508)
convection cell (p. 530)
convergent boundary (p. 523)
divergent boundary (p. 522)

friction (p. 531)
magnetic anomaly (p. 516)
magnetic reversal (p. 516)
mantle resistance (p. 531)
normal polarity (p. 515)
paleomagnetism (p. 514)
Pangaea (p. 509)

polarity chron (p. 515)
pole of rotation (p. 526)
relative velocity (p. 528)
reversed polarity (p. 515)
ridge-push (p. 531)
rift valley (p. 514)
seafloor spreading (p. 514)

slab-pull (p. 531)
strike-slip fault (p. 525)
subduction (p. 523)
subduction zone (p. 523)
tectonic plate (p. 521)
transform fault (p. 525)

Review Questions

1. Briefly explain the theory of plate tectonics.
2. Distinguish between continental drift and plate tectonics.
3. List three major evidences for continental drift.
4. Why did it take so long for the scientific community to accept the movement of the continents?
5. Why are there paleomagnetic stripes on the ocean floor?
6. Where do you expect to find the youngest oceanic lithosphere?
7. In the Pacific Ocean basin, where is the oldest oceanic lithosphere? How is its location explained by plate tectonics?
8. Describe the types of plate boundaries, and give an example of each.
9. Sketch a simple map of a part of an oceanic ridge, and draw arrows to show the relative motion along ridge-to-ridge transform faults.
10. Describe the geometry of lithospheric plate motion over the planet.

11. How is the pattern of a series of transform faults along a plate boundary related to the pole of rotation?
12. Explain how plate margins, as well as the plate itself, can migrate.
13. Explain the origin of the following features in the context of plate tectonics: (a) the Ural Mountains, (b) the Alps, (c) the Mid-Atlantic Ridge, (d) Hawaii, (e) the San Andreas Fault, (f) the Andes Mountains, and (g) volcanoes in Italy.
14. How fast are the plates moving? How do we determine rates of plate motion?
15. Draw a cross section showing a tectonic plate with a divergent and a convergent boundary, and label the major forces acting on the plate.
16. Explain the difference between (a) the convection model of plate motion, in which the mantle carries the plates, and (b) a model in which the plates themselves drive plate tectonics.

18 Seismicity and Earth's Interior

In defiance of nature, and the rule of plate tectonics, the king of Antiochus boasted that his mausoleum and these monuments would be "unravaged by the outrages of time." A few decades later, the great statues of Nemrud Dagh in eastern Turkey were toppled by an earthquake. Little did this king know, but Turkey's northern borderlands are sliced through by a great strike-slip fault and sliding underneath its southern shores is a great plate of oceanic lithosphere. In fact, Turkey projects westward toward Europe exactly because it is being squeezed out of Asia by the ongoing collision of Africa and Arabia with the Eurasian plate. Earthquakes occur with tragic regularity in Turkey; its last great earthquake shook Istanbul and the rest of northern Turkey in 1999, leaving cities in rubble and killing 20,000 people. Landslides, tsunamis, groundshaking, and liquefaction all took their toll.

Earthquakes, perhaps more than any other phenomenon, demonstrate that Earth continues to be a dynamic planet, changing each day by internal, tectonic forces. Most earthquakes occur along plate boundaries. As the plates move, these boundaries—ocean ridges, continental rifts, subduction zones, and transform faults—are the sites of the most intense earthquake activity on Earth. Earthquakes occur during sudden movements along faults.

Every year, more than a million earthquakes are recorded by the worldwide network of seismic stations and are analyzed with the aid of computers such as those at the Earthquake Information Center in Golden, Colorado. With this network, the exact location, depth, and magnitude of all detectable earthquakes are plotted on regional maps. As a result, we can monitor the details of present plate motion. But that is not all. Seismic waves also provide our most effective probe of Earth's interior, and they constitute the main method of collecting data upon which we base our present concepts of Earth's internal structure.

Indeed, earthquakes are human disasters, as the power released by a single event is staggering. When the energy stored up in deforming rocks is suddenly released, the consequences may be devastating. Many large cities lie along major faults. Thus, it is imperative that we learn as much as we can about earthquakes so that their damage can be lessened.

Major Concepts

1. Seismic waves are vibrations in Earth caused by the rupture and sudden movement of rock.
2. Three types of seismic waves are produced by an earthquake shock: (a) P waves, (b) S waves, and (c) surface waves.
3. The primary effect of an earthquake is ground motion. Secondary effects include (a) landslides, (b) tsunamis, and (c) regional or local uplift or subsidence.
4. The exact location and timing of an earthquake cannot be predicted. However, seismic risk can be evaluated and, in areas with high risk, preparations for future earthquakes made.
5. Most earthquakes occur along plate boundaries. Divergent plate boundaries and transform fault boundaries produce shallow-focus earthquakes. Convergent plate boundaries produce an inclined zone of shallow-focus, intermediate-focus, and deep-focus earthquakes.
6. The velocities at which P waves and S waves travel through Earth indicate that Earth has a layered internal structure based on composition—crust, mantle, and core. It also has a solid inner core, a liquid outer core, a weak asthenosphere, and a rigid lithosphere.
7. Plate tectonics and upwelling and downwelling plumes are the most important manifestations of Earth's internal convection. The magnetic field is probably caused by convection of the molten iron core.

Characteristics of Earthquakes

> Earthquakes are vibrations of Earth, caused by the rupture and sudden movement of rocks that have been strained beyond their elastic limits. Three types of seismic waves are generated by an earthquake shock: (1) primary waves, (2) secondary waves, and (3) surface waves.

Elastic-Rebound Theory

What causes earthquakes?

The origin of an **earthquake** can be illustrated by a simple experiment. Bend a stick until it snaps. Energy is stored in the elastic bending and is released if rupture occurs, causing the fractured ends to vibrate and send out sound waves. Detailed studies of active faults show that this model, known as the **elastic-rebound theory**, applies to all major earthquakes (**Figure 18.1**). Precision surveys across the San Andreas Fault in California show that railroads, fence lines, and streets are slowly deformed at first, as strain builds up, and are offset when movement occurs along the fault, releasing the elastic strain. The San Andreas fault is the boundary between the Pacific and North American plates. Its movement is horizontal, with the Pacific plate moving toward the northwest. On a long-term basis, the plates move quite steadily, at a rate of roughly 3 cm/yr. On a short-term however, much of the movement occurs in a series of jerks. Sections of the fault can be "locked" together until enough strain accumulates to exceed the rock's **elastic limit** and cause displacement.

The point within Earth where the initial slippage generates earthquake energy is the **focus**. The point on Earth's surface directly above the focus is the **epicenter** (**Figure 18.2**).

Types of Seismic Waves

Several different types of **seismic waves** are generated by an earthquake shock (**Figure 18.3**). Each type travels at a different speed, and each therefore arrives at a **seismograph** that might be hundreds of kilometers away at a different time. The first

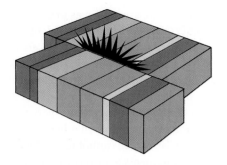

(A) Strain builds up in rocks in seismically active areas until the rocks rupture or move along preexisting fractures.

(B) Energy is released when the rocks rupture, and seismic waves move out from the point of rupture.

Figure 18.1 Earthquakes originate where rocks are strained beyond their elastic limits and rupture.

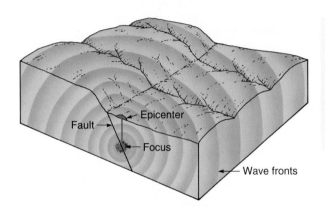

Figure 18.2 The relationship between an earthquake's focus, its epicenter, and seismic wave fronts is depicted in this diagram. The focus is the point of initial movement on the fault. Seismic waves radiate from the focus. The epicenter is the point on Earth's surface directly above the focus.

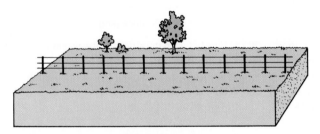

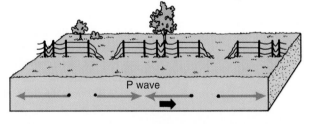

(A) Before seismic disturbance. A regular grid or a straight fence line provide good reference markers for future movement.

(B) Motion produced by a P wave. Particles are compressed and then are expanded in the line of wave progression. P waves can travel through any Earth material.

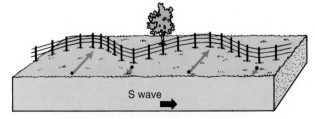

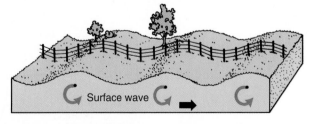

(C) Motion produced by an S wave. Particles move back and forth at right angles to the line of wave progression and are commonly called shear waves. S waves travel only through solids.

(D) Motion produced by one type of surface wave. Particles move in nearly circular paths at the surface. The motion diminishes with depth, like that produced by surface waves in the ocean.

Figure 18.3 Motion produced by various types of seismic waves can be illustrated by the distortions they produce in a regular grid. For comparison, the motion of one type of surface wave is shown.

waves to arrive are known as **primary waves** (**P waves**; **Figure 18.4**). These are a kind of **compressional wave**, identical in character to sound waves passing through a liquid or gas. The wave transmits energy by compressing and dilating the material through which it moves. Thus, the particles involved in these waves move short distances forward and backward in the direction of wave travel. P waves commonly have smaller amplitudes than the later waves. The next waves to arrive are **secondary waves** (**S waves**). In these, particles oscillate back and forth at right angles to the direction of wave travel. In other words, the particles shear or slide past one another. These **shear waves** cannot move through liquids. S waves cause a second burst of strong movements to be recorded on a seismograph (Figure 18.4A). The last waves to arrive are **surface waves**, which travel relatively slowly over Earth's surface. Particles involved in one type of surface wave move in orbits, similar to particles in water waves. They may have amplitudes up to 0.5 m and wavelengths of about 8 m.

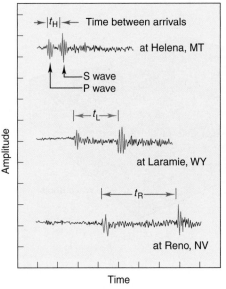

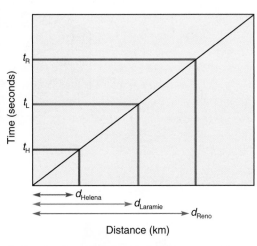

(A) The greater the distance between the seismic event and a recording seismograph station, the more time it takes for the first wave to arrive. Also, the greater the distance, the longer the interval between the arrival of P waves and S waves.

(B) The time between the arrivals of P waves and S waves is correlated with the distance between the seismic event and the recording station. For example, time at Helena, t_H, yields distance from Helena, d_{Helena}.

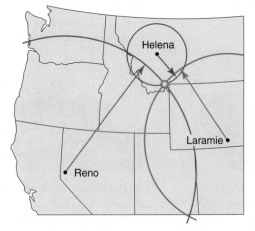

(C) The direction of the event from any single station is not known, but simply plotting the intersection of three arcs that have radii the respective distances from the three stations identifies a common point. That point lies at the epicenter of the seismic event.

Figure 18.4 Locating the epicenter of an earthquake is accomplished by comparing the arrival times of P waves and S waves at three seismic stations.

Earthquake Locations

The location of an earthquake's focus is important in the study of plate tectonics because it indicates the depth at which rupture and movement occur. Although the movement of material within Earth occurs throughout the mantle and core, earthquakes are concentrated in its upper 700 km.

Within the 700-km range, earthquakes can be grouped according to depth of focus. **Shallow-focus earthquakes** occur from the surface to a depth of 70 km. They occur in all seismic belts and produce the largest percentage of earthquakes. **Intermediate-focus earthquakes** occur between 70 and 300 km below the surface, and **deep-focus earthquakes** between 300 and 700 km. Both intermediate-focus and deep-focus earthquakes are limited in number and distribution. In general, they are confined to convergent plate margins. The maximum energy released by an earthquake tends to become progressively smaller as the depth of focus increases. Also, seismic energy from a source deeper than 70 km is largely dissipated by the time it reaches the surface. Most large earthquakes therefore have a shallow focus, originating in the crust. The depth of an earthquake's focus is calculated from the time that elapses between the arrivals of the three major types of seismic waves.

The method of locating an earthquake's epicenter is relatively simple and can be understood easily by referring to Figure 18.4. The P wave, traveling faster than the S wave, is the first to be recorded at the seismic station. The time interval between the arrival of the P wave and the arrival of the S wave is a function of the station's distance from the epicenter. By tabulating the travel times of P and S waves from earthquakes of known sources, seismologists have constructed time-distance graphs, which can be used to determine the distance to the epicenter of a new quake. The seismic records show the distance, but not the direction, to the epicenter. Records from at least three stations are therefore necessary to determine the epicenter's precise location.

Intensity

The **intensity**, or destructive power, of an earthquake is an evaluation of the severity of ground motion at a given location. It is measured in relation to the effects of the earthquake on humans. In general, destruction is described in subjective terms for the damage caused to buildings, dams, bridges, and other structures, as reported by witnesses.

The intensity of an earthquake at a specific location depends on several factors. Foremost among these are (1) the total amount of energy released, (2) the distance from the epicenter, and (3) the type of rock and degree of consolidation. In general, wave amplitude and destruction are greater in soft, unconsolidated material than in dense, crystalline rock (**Figure 18.5**).

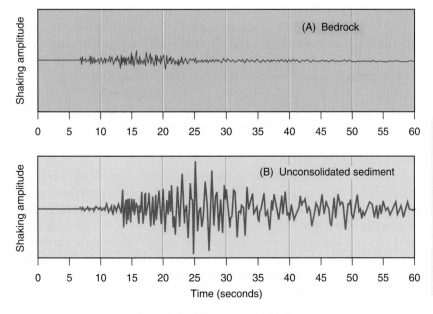

Figure 18.5 The intensity of an earthquake can be magnified in locales with unconsolidated sediment or landfill. The seismograms shown here are for the same aftershock of the 1992 Landers earthquake in California. Seismographs located on unconsolidated sediment measured much greater wave amplitudes and durations than seismometers on solid bedrock. Amplification can be by a factor of 10 or more.

Table 18.1 Earthquake Magnitude Scale

Magnitude	Approximate Number per Year
1	700,000
2	300,000
3	300,000
4	50,000
5	6000
6	800
7	120
8	20
>8	1 every few years

What is the difference between the magnitude and intensity of an earthquake?

Magnitude

The **magnitude** of an earthquake is an objective measure of the amount of energy released. It is a much more precise measure than intensity. Earthquake magnitudes are based on direct measurements of the size (amplitude) of seismic waves, made with recording instruments, rather than on subjective observations of destruction. The total energy released by an earthquake can be calculated from the amplitude of the waves and the distance from the epicenter. The first attempt by seismologists to express magnitudes of earthquakes resulted in the Richter scale (**Table 18.1**), which assigns a single number to an earthquake. Each step on the scale represents an increase in wave amplitude by a factor of 10. The vibrations of an earthquake with a magnitude of 2 are therefore 10 times as great in amplitude as those of an earthquake with a magnitude of 1, and the vibrations of an earthquake with a magnitude of 8 are 1 million times as great in amplitude as those of an earthquake with a magnitude of 2. The increase in total energy released, however, is about 30 times for each step on the scale. The largest earthquake ever recorded had a magnitude of approximately 8.8 on the Richter scale. Significantly larger earthquakes are not likely to occur because rocks are not strong enough to accumulate more energy.

Recent refinements of the earthquake magnitude scale attempt to better distinguish the differences in large earthquakes. One such modification, called the **moment magnitude scale**, is the most widely used measure of earthquake magnitude today. The moment magnitude scale is designed to reflect the amount of energy released by an earthquake. The magnitudes reported by the media are commonly from this scale.

An added advantage of the moment magnitude scale is that the magnitude of an ancient earthquake can be calculated from measurements of the amount of slip along the fault scarp. Like the standard Richter scale, moment magnitudes are logarithmic and range to about 10, but the absolute values are slightly different on the two scales, especially at the high end of the scale. Thus, the 1906 San Francisco earthquake had a Richter magnitude of 7.9 and a moment magnitude of 8.2. The largest known earthquake on the moment magnitude scale had a magnitude of 9.5. Moment magnitudes are used in the rest of this chapter.

Earthquake Hazards

In addition to ground shaking and surface faulting, earthquake hazards include submergence, liquefaction, tsunamis, and landslides. Fires and floods caused by breakage of water lines or dam failures are also important.

What kinds of secondary damage are caused by earthquakes?

Earthquakes pose a significant threat to much of the world's population. On average, several tens of thousands of people die each year because of large earthquakes. **Figure 18.6** shows the expected probability of earthquakes in the conterminous United States. The principal areas exposed to earthquake risk are in the western United States—California, Nevada, Utah, and Montana—but there are significant risks elsewhere in the country as well. The primary effect of earthquakes is the violent ground motion caused by movement along a fault. This motion can shear and collapse buildings, dams, tunnels, and other rigid structures (**Figure 18.7**). Secondary effects include soil liquefaction, landslides, tsunamis, and submergence of the land. Following are a few examples of well-documented earthquakes.

San Francisco, 1906

The most destructive earthquake in the history of the United States was the San Francisco earthquake of 1906. This shallow earthquake was located on a transform plate boundary. It lasted only a minute but had a magnitude of 8.2. The fire that followed it caused most of the destruction (an estimated $400 million in damage and a reported loss of 700 lives). From a scientific point of view, the earthquake was important because of the visible effects it produced along the San Andreas fault zone. Horizontal displacement occurred over a distance of about 400 km and offset roads, fences, and buildings by as much as 7 m.

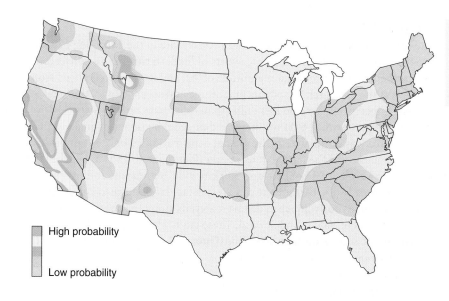

Figure 18.6 A seismic risk map shows the likelihood that an earthquake of a certain magnitude will occur. These predictions are based on the distributions and sizes of past earthquakes.

High probability

Low probability

(A) Earthquake of June 16, 1964, Niigata, Japan. Apartment houses tilted by liquefaction. About one-third of the city subsided as much as 2 m as a result of sand compaction.

(B) Earthquake of January 1994, Northridge, California. Extensive damage to freeway overpasses occurred.

(C) Earthquake of September 19, 1985, Mexico City, Mexico. A 15-story reinforced concrete building collapsed.

(D) Earthquake of February 9, 1971, San Fernando, California. Southern Pacific railroad tracks near Los Angeles were laterally displaced.

Figure 18.7 The effects of historic earthquakes are dramatically displayed in the types of damage rendered to buildings and other structures.

Alaska, 1964

The earthquake that devastated southern Alaska late in the afternoon of March 27, 1964, was one of the largest tectonic events of modern times. This convergent boundary earthquake had a moment magnitude of 9.2, and its duration ranged from 3 to 4 minutes at the epicenter. Despite its magnitude and severe effects, this quake caused far less property damage and loss of life than other national disasters (114 lives were lost and property worth $311 million was damaged) because, fortunately, much of the affected area was uninhabited. The crustal deformation associated with the Alaskan earthquake was the most extensive ever documented. The level of the land was changed in a zone 1000 km long and 500 km wide. Submarine and terrestrial landslides triggered by the earthquake caused spectacular damage to communities, and the shaking spontaneously liquefied deltaic materials along the coast, causing slumping of the waterfronts of Valdez and Seward. These landslides triggered destructive tsunamis that swept the state's southern shoreline (**Figure 18.8**). Fires started when oil tanks at Seward harbor ruptured.

Tangshan, China, 1976

The great Tangshan quake, which shook China in the early morning hours of July 28, 1976, was probably the second most devastating earthquake in recorded history. (The most destructive also occurred in China, in 1556.) In a matter of seconds, a large industrial city was reduced to rubble. The enormous shock registered 7.8 on the magnitude scale; an aftershock with a magnitude of 7.1 struck late in the afternoon, destroying structures that had withstood the main quake. The total amount of energy released by the earthquake and its aftershocks is almost unbelievable: the equivalent of 400 atomic bombs of the size dropped on Japan at the end of World War II. When the quake struck, many people were catapulted into the air, some as high as 2 m, by what were described as violent, hammer-like blows. The Tangshan quake killed 240,000 people, and many more were seriously injured. Most people were killed when their houses collapsed on them. All of the city's lifelines were destroyed, including bridges, railroads, telephone, electricity, water, and sewers. Fortunately, there were no natural gas lines in the city. Of the 680,000 residential buildings near the quake, 650,000 suffered serious damage. Eighty percent of the water reservoirs were damaged. Before the earthquake, little attention had been paid to earthquake resistance in the local building codes, and the buildings were highly vulnerable to seismic damage. Moreover,

Figure 18.8 Tsunamis from the 1964 Alaskan earthquake devastated the bay at Seward. An overturned ship, demolished truck, and torn-up dock strewn with logs and scrap metal attest to the power of the wave. A section of the waterfront slid into the bay. Waves spread in all directions, destroying railroad docks, washing out railroad and highway bridges. Flaming petroleum spread over the water, igniting homes and an electrical generation plant.

much of the city was built on young, unconsolidated sediments, which, as noted earlier, tend to amplify earthquake intensity (Figure 18.5).

The earthquake's focus was 11 km deep and caused movement in a fault zone 120 km long and 20 km wide. Ground fissures opened up, some with a lateral slip of 1.5 m. Liquefaction of wet sediments drove sand gushers from holes as wide as 1 m across. There were no foreshocks; in fact, no earthquakes of any size were detected in more than 2 months before the devastating blow came in July. In contrast, aftershocks continued for 4 years, and some 24,000 separate events were identified.

Northridge, California, 1994

Early in the morning of January 17, 1994, an earthquake of magnitude 6.6 struck Northridge, California, just north of Los Angeles. It was followed by thousands of smaller aftershocks, including at least two with magnitudes greater than 5. This earthquake, although modest compared with the much larger magnitude-8 earthquakes that can occur along the San Andreas Fault, was one of the most destructive in the region. It caused the deaths of more than 50 people and $30 billion in property damage. The earthquake occurred on a previously unknown reverse fault that does not reach the surface but dips beneath the heavily populated San Fernando Valley. The focus of the quake was at a depth of about 14 km. Shaking caused considerable damage, including the collapse of several apartment buildings (where most of the people died), a parking garage, and several freeway overpasses (Figure 18.7B). The ground surface was ruptured along a 15-km stretch, with offsets as much as 20 cm. Overall, the ground was uplifted by about 1 m, disrupting sidewalks, streets, and buildings.

The earthquake was also responsible for much secondary damage. Landslides were common on steep mountain slopes, stripped bare of their vegetation by earlier brush fires (**Figure 18.9**). Consequently, the Pacific Coast Highway and many local streets were closed, hampering rescue efforts. Much of the secondary damage was the result

What is the difference between the tectonic setting of the earthquakes in Japan and those in southern California?

Courtesy of P. Morton

Figure 18.9 The Northridge, California, earthquake of January 1994 caused several types of damage. An aftershock caused landslides and associated dust plumes on Santa Paula Ridge.

of destruction to built structures. Electrical power was cut off for more than 3 million people. An oil main and more than 250 natural gas lines were ruptured. Many homes and some businesses and university buildings burned, as sparks ignited gas leaking from the broken lines. Several homes were caught in a fireball ignited when someone tried to start a truck moments after the earthquake. Others were flooded when water lines were ruptured. Striking views of towering flames and billowing smoke above flooded streets were shown on television and in newspapers, vividly recording the significance of these secondary effects.

Kobe, Japan, 1995

The worst earthquake to hit Japan since 1923, this tremor destroyed much of the port city of Kobe on January 17, 1995. In this city of 1.4 million, nearly 5500 died, largely because of building collapse. The homes of 300,000 people were rendered unsafe. More than 600 fires started from broken gas lines and burned out of control because of ruptured water lines. The Kobe earthquake had a moment magnitude of about 7.2, and its focus was about 20 km deep.

In monetary terms, this was the largest earthquake disaster in history, causing direct losses of $140 billion. If an earthquake of similar magnitude had shaken the more densely populated Tokyo area, it could have caused direct losses of more than $1 trillion, to say nothing of casualties.

The cause of the Kobe earthquake was the subduction of the oceanic Philippine plate beneath southern Japan (**Figure 18.10**). Note in the figure that the directions of plate movement are not head-on but converge at an oblique angle. Because of this oblique collision, part of the displacement is taken up by movement along a long strike-slip fault zone. The main shock occurred along a fault in this zone.

The actual rupture of rock caused by the earthquake started 20 km from Kobe and moved rapidly toward the city. It broke the surface along a northeast-trending zone at least 9 km long. Thousands of aftershocks continued on this same trend in a zone 60 km long for several days after the quake. Both vertical and horizontal movements were about 1 or 2 m. An earthquake of this size probably occurs every 1000 to 1500 years along this strand of the main Kobe fault zone.

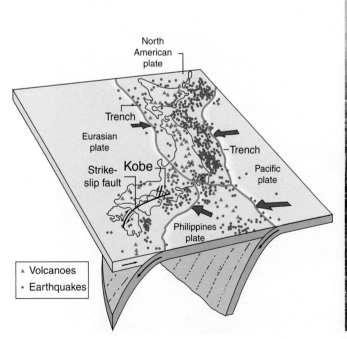

Figure 18.10   The 1995 Kobe, Japan, earthquake occurred on a strike-slip fault that accommodates some of the deformation above a northward-dipping subduction zone.

Shaking and liquefaction were the main causes of damage. **Liquefaction** occurs when unconsolidated, water-saturated regolith, soil, or landfill loses its strength and behaves like a fluid when shaken by an earthquake. Liquefied soils are unable to support buildings and other structures. In the Kobe earthquake, extensive liquefaction destroyed many buildings, bridges, highways, and utilities (sewer, gas, and water lines). Much of the city is built on human-made islands and landfill placed atop a granite basement and thus was susceptible to failure. Liquefaction most strongly affected areas with shallow water tables. As a result, geysers of wet sand erupted from fissures and covered many islands in the city. Subsidence, also a result of liquefaction, ranged from 0.5 to 3 m. Many buildings on landfills tilted because of ground settlement. However, buildings supported on deep piles sustained little or no damage.

This earthquake was about the same size as the Northridge, California, earthquake. However, many more people were killed in Kobe, and more damage was done. The likely reason was not any difference between the two earthquakes but a great difference in the human aspect: Kobe has an extremely dense population, older buildings, and construction on liquefiable landfill materials.

Why does Japan have so many earthquakes?

Izmit, Turkey, 1999

A long strike-slip fault slices across northern Turkey, connecting two convergent plate boundaries (**Figure 18.11**). Its size and deadly potential are comparable to that of California's San Andreas fault. During the early morning hours of August 17, 1999, one segment of the fault broke, producing a magnitude 7.4 earthquake. The epicenter was near the city of Izmit, 80 km south of Istanbul. A single gigantic heave created displacements of 3 to 4 m along a rupture 160 km long. The earthquake destroyed hundreds of buildings, damaged industrial and port facilities, a military base, pipelines, and roads, and was responsible for the collapse of bridges. As a result, nearly 20,000 people died and 600,000 were homeless. In some cities near the epicenter, 70% of the buildings collapsed or were uninhabitable. Surface faulting, ground shaking, subsidence, liquefaction, and even a small tsunami were responsible for most of the damage. Thousands of aftershocks plagued the area for several months after the earthquake. The large number of deaths in this heavily populated region brought the practices of contractors and building inspectors under intense scrutiny.

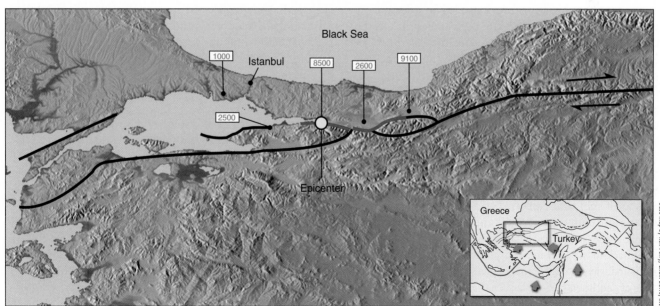

Figure 18.11 The 1999 Izmit, Turkey, earthquake occurred on a strike-slip fault system that connects two convergent plate boundaries. The red line shows the length of the fault that actually ruptured. The trace of the fault is marked by lakes, an inlet of the sea, and a line separating different types of landforms. It was responsible for the deaths of 20,000 people. The number of deaths at each location is given in the white boxes.

STATE OF THE ART   Radar Interferometry Reveals Earthquake Deformation

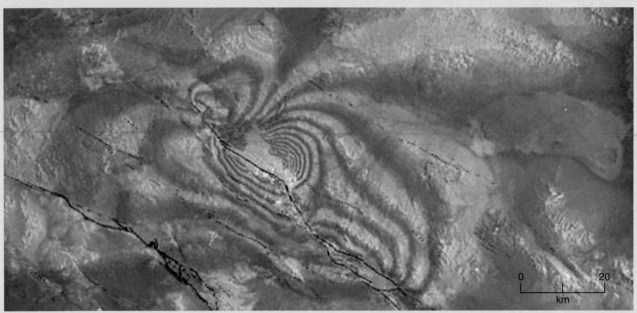

Courtesy of JPL/Caltech/NASA

On October 16, 1999, a huge earthquake was triggered at a depth of about 6 km along the infamous San Andreas fault system of southern California. With a moment magnitude of 7.1, it set off a series of waves that rumbled through the crust of the sparsely populated region. Moments later, the tear reached the surface and rapidly spread along the fault system. The maximum strike-slip displacement along the fault was 5 m, and the rocks on one side of the strike-slip fault heaved upward, forming a scarp 3 m high. Breakage extended for a distance of 50 km and to a depth of about 15 km. Beyond these limits, the earthquake energy was insufficient to actually break the rocks.

High above this scene an orbiting satellite (European Remote Sensing satellite ERS-2) snapped a picture of the Mojave desert with its radar acting as a flashlight to illuminate the scene below. By comparing this image with one taken a month earlier, a map of the extent of disruption was made—without ever setting foot in the desert.

This technique is called *satellite interferometry* and is based on the simple notion of a before-and-after picture—with an ingenious twist. Optical sensors record only the brightness (or amplitude) of light waves reflected off a surface. Radar instruments measure both the amplitude and the exact point in the oscillation cycle where the radar wave hits the surface and bounces back to the satellite. This point is called the *phase* of the returned wave. Because radar waves have wavelengths of a few tens of centimeters, a surface change of only a few millimeters causes a significant phase change. A deviation of a centimeter results in a phase change of 40%, an amount that can be easily measured, even by a satellite orbiting hundreds of kilometers above. If two satellite images are taken from exactly the same position, there should be no phase difference for any spot on the two images. But if the ground changes ever

so slightly between the two radar scans, then the phase of the wave returned from some spots in the second image are different. A map of these phase changes, called an *interferogram*, is displayed as a series of colored "fringes" that are like contour lines on a topographic map. Each fringe marks 10 cm of vertical ground motion.

The map shows a nearly continuous picture of the magnitude and distribution of the deformation caused by the Hector Mine earthquake. By starting at edges and counting the number of color bands and multiplying by 10 cm, you can see that deformation was strongest right along the surface rupture (black lines) and amounted to as much as 5 m. Clearly, displacement was not limited to a small area along the fault scarp. However, deformation declines with distance away from the fault.

This technique works best in regions where vegetation and other changes are minor. Thus, polar and arid regions are ideal. It provides many advantages over slow and expensive ground surveying, especially in remote areas. Even thick cloud cover does not obscure the surface from a probing radar system—water droplets and ice crystals do not impede the radio signals. Images can be taken in the dead of night because radar systems have their own illumination source.

Satellite radar interferometry can be used to rapidly map deformation along faults that have ruptured in earthquakes, to follow swelling volcanoes as magma accumulates beneath them, or to map their sagging as magma drains away. Other geologists use interferometry to quietly monitor landslides in remote areas or to watch the direction and amount of ice movement in glaciers or in sea ice. Ocean currents can also be measured by interferometry.

Ancash, Peru, 1970

The earthquake that shook Peru on May 31, 1970, had a magnitude of 7.8. Claiming the lives of 50,000 people, it was the deadliest earthquake in Latin American history. Much of the damage to towns and villages was caused by the collapse of adobe buildings, which are easily destroyed by ground motion. Eighty percent of the adobe houses in an area of 65,000 km^2 were destroyed. Vibrations caused most of the destruction to buildings, but the massive landslides triggered by the Andes quake were a second major cause of fatalities. In a matter of a few minutes, a huge debris avalanche fell from Peru's highest mountain; it buried 90% of the resort town of Yungay, with a population of 20,000. In 1797 similar earthquake-triggered debris avalanches killed 41,000 in Ecuador and Peru and, in 1939, killed 40,000 people in Chile.

Valdivia, Chile, 1960

Another type of earthquake hazard is exemplified by the powerful magnitude 9.5 earthquake that occurred in the Andes of Chile. It caused extensive damage from a tsunami, in addition to ground motion, landslides, and flooding. The tsunami devastated Chilean seaports with a series of waves 25 m high. This tsunami crossed the Pacific at approximately 1000 km/hr and built up to 11 m in height at Hilo, Hawaii. Nearly a day after the quake, the tsunami reached Japan, causing damage to property estimated at $70 million.

Reiteration of Earthquake Hazards

This rather extensive list of some of Earth's most destructive earthquakes is intended to reveal the most important seismic hazards. Ground shaking, surface faulting, liquefaction, tsunamis, and landslides are all major causes of destruction. Secondary damage is related to the collapse of buildings, failures of the electrical grid, broken water and gas lines, fires, and failed highways and dams.

Earthquake Prediction

Effective short-term earthquake prediction, which could save many lives and millions of dollars in property damage, is proving to be elusive. Preparation is a more achievable goal.

Considerable effort has been made to find methods of predicting earthquakes. More than a million earthquakes occur around the world each year; about 50 are large enough to cause property damage and loss of lives. Some people have tried to relate earthquakes to sunspots, tides, changes in the weather, the alignment of planetary bodies, and other phenomena but have ignored many facts about Earth's seismicity. As a result, their predictions fail. Scientists have struggled with the problem, and significant strides have been made. Yet, predicting the location and magnitude of an earthquake is still an elusive goal.

Chinese scientists claim to have been successful in predicting about 15 earthquakes in recent years. In spite of the fact that there is no scientific evidence that animals can sense the advent of an earthquake, Chinese predictions rely heavily on the centuries-old idea that animals sense various underground changes before an earthquake and hence behave abnormally. Cattle, sheep, and horses refuse to enter their corrals; rats leave their hideouts and march fearlessly through houses; shrimp crawl on dry land; ants pick up their eggs and migrate en masse; fish jump above the surface of the water; and rabbits hop about aimlessly. Chinese scientists, using a variety of precursory activities, successfully predicted the large (magnitude 7.3) Haicheng earthquake that occurred on February 4, 1975. In late 1974 the water table in this region periodically rose and fell. Well water became turbid and the ground tilted in some places. The most bizarre events occurred in mid-December, when snakes came out of hibernation and froze to death, and groups of rats appeared and scurried about. These events were followed by a swarm of small earthquakes that led to a prediction of a magnitude-6

Why is earthquake prediction so difficult?

or larger earthquake by January or February. Continued abnormal animal behavior and another swarm of earthquakes on February 1, were accompanied by variations in electrical conductivity of the ground. Some water wells spouted into the air; others stopped producing water. Finally, at 10 a.m. on February 4, seismologists convinced the provincial government to issue a warning by telephone to begin the evacuation of Haicheng. Schools, factories, and businesses were closed. Medical rescue teams were mobilized. The earthquake struck at 7:36 a.m. the same day. In some areas, more than 90% of the houses collapsed. Thousands of lives were spared, and many people were convinced that earthquakes could be predicted. This hope was dashed 2 years later by the unpredicted Tangshan earthquake already described and a decade later by the unpredicted 2008 Sichuan earthquake that took the lives of 68,000 people.

An ambitious attempt to predict earthquakes has also been made along an active strand of the San Andreas Fault system near the small town of Parkfield, California. This area was the site of a string of regularly spaced earthquakes beginning in 1857 (**Figure 18.12**). Six large earthquakes, with magnitudes of about 6, occurred at Parkfield between 1857 and 1966. An earthquake occurred on average every 21 or 22 years. Recognizing this regular spacing, geophysicists predicted that a major earthquake should occur in about 1987, and that certainly one would occur before the end of 1993. As a result, the strike-slip fault near Parkfield has become a natural laboratory, with a host of instruments arrayed along the fault to measure strain across it as well as small changes in elevation, release of gas from the soil, magnetic-field variations, electrical properties, and small earthquakes that might be foreshocks. In theory, a fault can begin to move weeks or months before the sudden rupture that marks a major earthquake.

The 1993 "deadline" came and went and no major earthquake struck Parkfield. Warnings were issued several times when small earthquakes, all less than magnitude 4.7, were triggered on the fault. The warnings were based on the hypothesis that the small quakes could be foreshocks. Aside from those events, however, no sign of slip or other changes that might precede a major earthquake were detected. Finally, a large earthquake occurred in September 2004, more than a decade after the close of the predicted window. The magnitude 6.0 earthquake triggered hundreds of aftershocks along a 35 km long stretch of the strike-slip fault.

The hypothesis that a fault is a simple system that gradually stores stress and then releases it at a specific threshold may need to be reconsidered. Some computer models of stress accumulation and relief through faulting suggest that fault behavior is likely to be so chaotic that the long-term prediction of earthquakes may be impossible. Was the monitoring and instrumentation at Parkfield a waste of money? Certainly not.

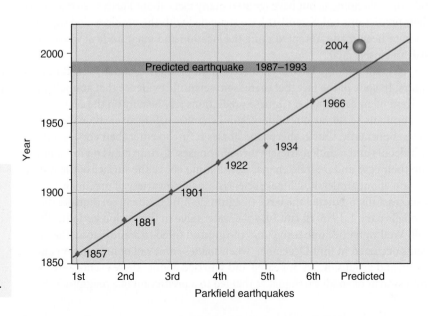

Figure 18.12 Six Parkfield earthquakes between 1857 and 1966 happened almost as regularly as the ticks of a clock. This regularity led to a prediction that a major earthquake would occur between 1988 and 1993. However, the next earthquake along the fault did not occur until 2004, showing just how difficult it is to predict earthquakes.

The information gleaned, even if it is negative with regard to earthquake prediction, has greatly expanded our understanding of the complex response of Earth's crust to tectonic stress.

Instead of attempting to predict the time, place, and magnitude of an expected earthquake, geologists are now concentrating on the more modest goal of forecasting which areas may be most susceptible to significant quakes. One approach is to calculate the probability that an earthquake of a certain magnitude will occur in a certain period ranging from decades to centuries.

Another contribution to forecasting has been the compilation of maps showing the seismic potential of major plate boundaries. **Figure 18.13** essentially shows locations along the plate boundaries, where major quakes are most likely to occur in the near future. Along the plate margins are several gaps in seismic activity, where stress may be building to a critical level. The most susceptible areas are those where major tremors have occurred in the past, but not in the last 100 years. At various times, these gaps included such heavily populated areas as southern California, central Japan, central Chile, Taiwan, and the west coast of Sumatra. Some of the gaps have now been filled with recent earthquakes. For example, the 1989 Loma Prieta earthquake in northern California occurred in a seismic gap identified by the U.S. Geological Survey. The forecast called for a 30% probability of a magnitude-6.5 earthquake within a 30-year window. Other large earthquakes occurred in seismic gaps along the western coast of Mexico in 1979 and another off the coast of Nicaragua in 1992 (magnitude 7.6). The rest of these areas appear likely to experience a major earthquake (magnitude of 7 or greater) in the next few decades.

Earthquake Preparation

Because earthquakes are still difficult, if not impossible, to predict, societies around the world need to prepare for them in an attempt to minimize destruction. Preparation involves several important steps. The first is to acquire a thorough understanding of where past earthquakes occurred, under the assumption that they will strike there in the future. Such a **seismic risk map** is shown for the United States in Figure 18.6. Maps are also made for much smaller areas that show a specific type of earthquake hazard (surface rupture, landslides, and soil liquefaction, for example). Such maps can

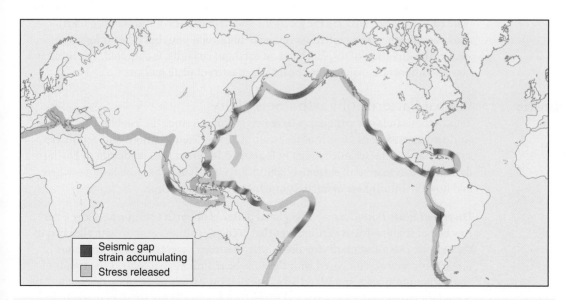

Figure 18.13 Seismic gaps are important in earthquake forecasting. Areas along plate margins that are not seismically active are believed to be building up stress and may be sites of significant seismic activity in the future. In the gray areas, earthquakes have relieved strain within the last 40 years, but in the red-shaded areas, no large quakes have occurred and strain is still building.

then be used as the basis for community zoning laws. In high-risk areas, stiffer building codes need to be in place. Critical facilities (hospitals, fire stations, and schools) should be placed where seismic hazards are lowest. Zoning laws should prohibit development in some areas. In other areas, even in the same city, where seismic hazards are smaller, zoning requirements and building codes can be more lenient, reflecting the smaller risk.

What is the best way to prepare for an earthquake?

To see how important and effective this approach is, compare the results of four earthquakes. The 1989 Loma Prieta (San Francisco) earthquake occurred in an area with strict building codes. The earthquake had a magnitude of 6.9, and about 65 people died. The 1988 earthquake in Armenia also had a magnitude of 6.9, but 28,000 people died when their poorly constructed homes and apartment buildings collapsed on them. In Latur, India, a magnitude-6.4 earthquake killed more than 11,000 people in 1993, in an area where little thought had been given to earthquake preparation. The walls in homes there were as much as 1.5 m thick, but the walls were made by stacking boulders together and filling the gaps between them with mud and pebbles. The walls crumbled as the earth shook. In the 1999 earthquake in Turkey where 20,000 people died, the importance of building with rigorous standards was revealed when whole neighborhoods built in violation of existing codes collapsed and adjacent ones built to the required standard survived. One expert concluded, "Almost all of the casualties could be attributed to buildings that collapsed because they were not built to code. Building codes save lives." A similar story of corruption and building code violations came out after 7,000 classrooms collapsed in the 2008 Sichuan earthquake.

Earthquakes and Plate Tectonics

The distribution of earthquakes delineates plate boundaries. Shallow-focus earthquakes coincide with the crest of the oceanic ridge and with transform faults between ridge segments. Earthquakes at convergent plate margins occur in a zone inclined downward beneath the adjacent continent or island arc.

A worldwide network of 125 sensitive seismic stations was established in 1961 by the U.S. Coast and Geodetic Survey. Since then, the network has been expanded greatly, and the data received are processed by computers at many seismic data centers such as the one in Golden, Colorado. From the worldwide network of seismic stations, seismologists have compiled an amazing amount of data concerning earthquakes and plate tectonics. Not only are the locations and magnitudes of thousands of earthquakes established and plotted on regional maps each year, but other information, such as the direction of displacement on earthquake faults, is collected. The result is a new and important insight into the details of current plate motion.

How are Earthquakes related to plate boundaries?

Global Patterns of Earth's Seismicity

Tens of thousands of earthquakes have been recorded since the establishment of the worldwide network of seismic observation stations. Their locations and depths are summarized in the seismicity map in **Figure 18.14**. From the standpoint of Earth's dynamics, this map is an extremely significant compilation because it shows where and how the lithosphere of Earth is moving at the present time.

Divergent Plate Boundaries   The global patterns of Earth's seismicity show a narrow belt of shallow-focus earthquakes that coincides almost exactly with the crest of the oceanic ridge and marks the boundaries between divergent plates. This zone is remarkably narrow compared with the zone of seismicity that follows the trends of young mountain belts and island arcs. The shallow earthquakes along divergent plate boundaries are usually less than 15 km deep and typically are small in magnitude. Earthquakes associated with the crest of the oceanic ridge occur within, or near, the rift valley. They appear to be associated with normal faulting and intrusions of basaltic magmas. Detailed studies indicate that earthquakes associated with the ridge crest are produced by normal faulting.

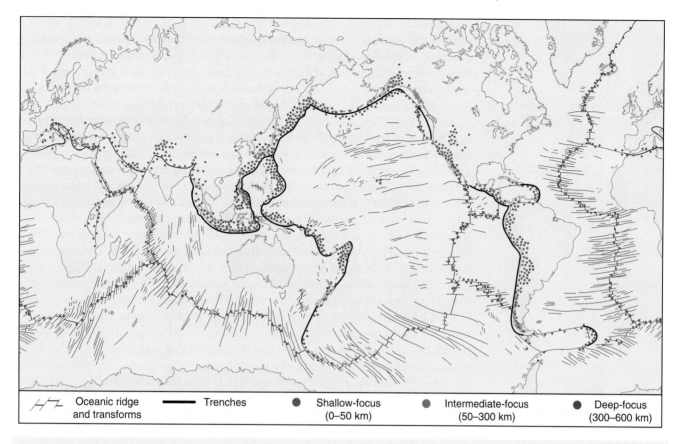

| ⌐⌐⌐ Oceanic ridge and transforms | ▬▬ Trenches | ● Shallow-focus (0–50 km) | ● Intermediate-focus (50–300 km) | ● Deep-focus (300–600 km) |

Figure 18.14 **Earth's seismicity** is clearly related to plate margins. This map shows the locations of many earthquakes that occurred during a 5-year period. Shallow-focus earthquakes occur at both divergent and convergent plate margins, whereas intermediate-focus and deep-focus earthquakes are restricted to the subduction zones of converging plates.

In June of 2011, a series of shallow (less than 10 km) extensional earthquakes, the largest a magnitude 7, occurred along the northern end of the East African rift system in Eritrea. These earthquakes were the precursors of eruptions of basaltic lava at Nabro volcano and illustrate the intimate connection between earthquakes, extension, and magmatism that occurs at divergent plate boundaries.

Transform Plate Boundaries   Shallow-focus earthquakes also follow the transform faults that connect offset segments of the ridge (Figure 18.14). Studies of fault motion indicate horizontal (strike-slip) displacement perpendicular to the ridge crest. Moreover, as is predicted by plate tectonic theory, earthquakes are restricted to the active transform fault zone—the area between ridge axes—and do not occur in inactive fracture zones.

Not all transform faults connect ocean ridge crests, however. Many long transform faults connect convergent boundaries and others slice through continents. In fact, the most damaging transform fault earthquakes are those found on land—including earthquakes on the San Andreas fault of California, the Anatolian fault of northern Turkey, and the Alpine Fault of New Zealand. The large Haitian earthquake of 2010 occurred on a complex transform fault system that separates the North American plate from the Caribbean Plate and connects a convergent boundary with a short spreading ridge in the Caribbean. The Caribbean plate is moving about 2 cm/yr to the east relative to the North American plate.

Subduction Zones   On Earth, the most widespread and intense earthquake activity occurs along subduction zones at convergent plate boundaries where slabs of oceanic lithosphere dive down into the mantle. This belt of seismic activity is immediately apparent from the world seismicity map (Figure 18.14), which shows a strong

Why aren't there as many damaging earthquakes in Iceland, which sits astride a plate boundary, as in Japan?

concentration of shallow, intermediate, and deep earthquakes coinciding with the subduction zones of the Pacific Ocean. The three-dimensional distributions of earthquakes in these belts define seismic zones that are inclined, at moderate to steep angles from the deep-sea trenches, and extend down under adjacent island arcs or continents. Along convergent plate boundaries, earthquakes occur at depths as great as 660 km. No other tectonic setting produces earthquakes as deep as this region.

The great earthquakes of the last decade—Tohoku, Japan and Sumatra, Indonesia—were both **megathrust earthquakes**, where a plate of oceanic lithosphere is thrust beneath another at a subduction zone. In the case of the 2004 Sumatran earthquake, a shallow magnitude 9.1 temblor occurred on the convergent plate boundary that separates the underthrust Indian-Australian plate from the overriding Eurasian plate (**Figure 18.15**). The seafloor ruptured at the base of the accretionary wedge and drove a devastating tsunami westward across the Indian Ocean and eastward across the Andaman Sea to the shorelines of Thailand. More than 280,000 people died from the earthquake and tsunami. The Indian-Australian plate is moving about 6 cm/yr northward, nearly parallel to the plate boundary; thus, much of the motion is oblique and both thrust and strike-slip earthquakes occur in the region. The earthquake relieved some of the stress that had built up over several decades in a matter of 3 or 4 minutes. The break between the two plates started slowly for the first minute. Then the rupture grew rapidly at a speed of about 2.5 kilometers per second to the north along the slab boundary. The final break was almost 1600 kilometers long and is marked by thousands of aftershocks. Movement along the fault at depth was about 20 m and the displacement on the seafloor itself was as much as 15 m near Banda Aceh (Figure 18.15).

How are earthquakes at convergent plate boundaries different from those at divergent boundaries?

Collision Zones   The Himalayas and the Tibetan Plateau define a wide belt of shallow earthquakes. In this area, two continents collided—India and Asia. This convergence produced the wide zone of exceptionally high topography in the Himalayas and the Tibetan Plateau, but no deep earthquakes because there is no longer a subduction zone.

An example of an earthquake in a continental collision zone is 2005 Kashmir earthquake of northern Pakistan. The earthquake occurred at a depth of 26 km on a large thrust fault in this continental collision zone. A rupture 75 km long formed in 25 seconds, growing by 1.5 km every second in both directions. In the end, the hanging wall of the thrust moved 4 m southward. A multitude of aftershocks followed the initial rupture, including 28 that were even larger than the earthquake that triggered them. Several thousand landslides, debris avalanches, and rock falls on the steep mountain slopes accompanied the earthquake. One debris avalanche buried four villages and dammed rivers to make two lakes. As a result of this tumultuous event, entire towns were destroyed when many poorly constructed buildings, including schools full of children, collapsed. More than 80,000 people died in three countries and over 3.5 million people were left homeless.

Intraplate Seismicity   Although most of the world's seismicity occurs along plate boundaries, the continental platforms also experience infrequent and scattered shallow-focus earthquakes. The zones of seismicity in East Africa and the western United States are most striking. They are probably associated with incomplete rifting. The minor shallow earthquakes, in the eastern United States (including New Madrid, Missouri, and South Carolina) and Australia, are more difficult to explain. Apparently, lateral motion of a plate across the asthenosphere involves slight vertical movement. Built-up stress can exceed the strength of the rocks within the lithospheric plate, causing infrequent faulting and seismicity along old lines of weakness such as ancient rifts. Although these continental intraplate earthquakes may be large, they are infrequent. In terms of the total energy released by seismicity each year, they account for only 0.5%.

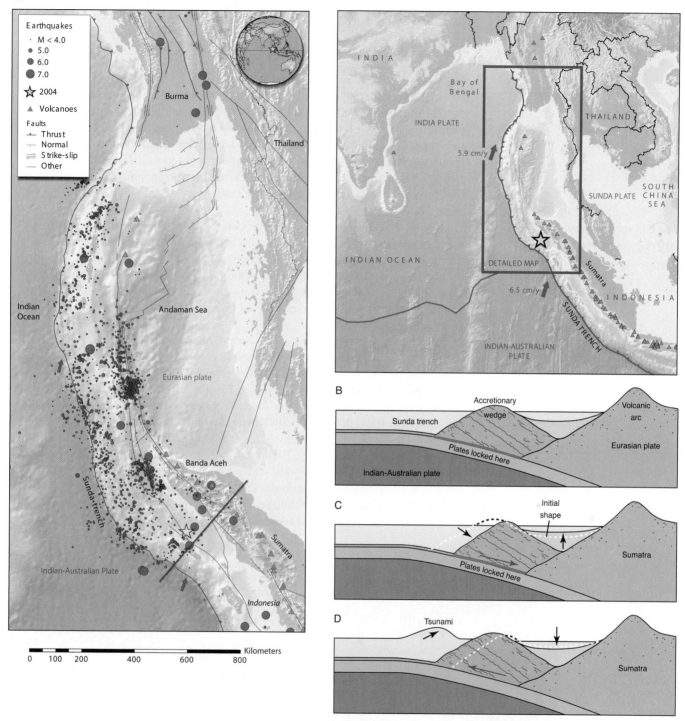

Figure 18.15 The great Sumatran Earthquake of 2004 was related to subduction of the Indian-Australian plate beneath the Eurasian plate. A wide accretionary wedge has developed above the downgoing plate and a volcanic arc lies to the east.

(A) The main earthquake (star) is on the southern end of the zone that ruptured during the earthquake—here marked by thousands of aftershocks. The Indian-Australian plate is moving northeast, almost parallel to the trench.

(B) In the past, the overriding plate may have had a shape like this with the accretionary wedge on the overriding plate extending to the left.

(C) Progressive underthrusting drug the overriding plate to the right during the decades that preceded the earthquake.

(D) Suddenly this deformation or strain was released along the low-angle fault plane (the boundary between the plates) during the 2004 earthquake. The movement of the seafloor to the left and up caused a tsunami and sent powerful seismic waves through the crust that triggered thousands of aftershocks.

Seismic Waves as Probes of Earth's Interior

> Seismic waves passing through Earth are refracted in ways that show distinct discontinuities within Earth's interior and provide the basis for the belief that Earth has a distinctive core.

Speculations about the interior of Earth have stimulated the imagination of humans for centuries, but only after we learned how to use seismic waves to obtain an "X-ray" picture of Earth were we able to probe the deep interior and formulate models of its structure and composition. Seismic waves—both P waves and S waves—travel faster through rigid material than through soft or plastic material. The velocities of these waves traveling through a specific part of Earth thus give an indication of the type of rock there. Abrupt changes in seismic wave velocities indicate significant changes in Earth's interior.

How can seismic waves "X-ray" Earth's internal structure?

Seismic waves are similar in many respects to light waves, and their paths are governed by laws similar to those of optics. Both seismic rays and light rays have velocities that depend on the kind of material through which they are transmitted. Both move in straight lines through homogeneous bodies. If the waves encounter a boundary between different substances, however, they are either reflected or refracted (bent). Familiar examples are light waves reflected from a mirror or refracted as they pass from air to water.

If Earth were a homogeneous solid, seismic waves would travel through it at a constant speed in all directions. A **seismic ray** (a line perpendicular to the wave front) would then be a straight line, like the ones shown in **Figure 18.16**. Early investigations, however, found that seismic waves arrive progressively sooner than was expected at stations progressively farther from an earthquake's source. The rays arriving at a distant station travel deeper through Earth than those reaching stations closer to the epicenter. Obviously, then, if the travel times of long-distance waves are progressively shortened as they go deeper into Earth, they must travel more rapidly at depth than they do near the surface. The significant conclusion drawn from these studies is that Earth is not a homogeneous, uniform mass, but has physical properties that change with depth. As a result of these differences with depth, seismic rays are believed to bend and follow curved paths through Earth (**Figure 18.17**).

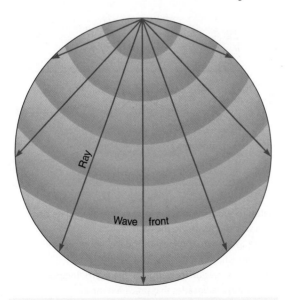

Figure 18.16 Seismic waves in a homogeneous planet would be neither reflected nor refracted. Lines drawn perpendicular to the wave fronts (rays) would follow linear paths.

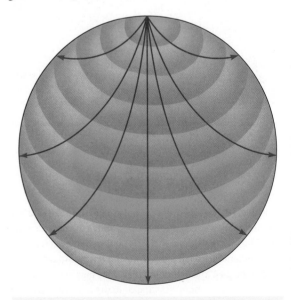

Figure 18.17 Seismic waves in a differentiated planet would pass through material that gradually increases in rigidity with depth. As a result, wave velocities would increase steadily with depth, and rays would follow a curved path.

In 1906 scientists first recognized that whenever an earthquake occurs, there is a large region on the opposite side of the planet where the seismic waves are not detectable. To better understand the nature and significance of this **shadow zone**, refer to **Figure 18.18**. For an earthquake at a particular spot (labeled 0°), a shadow zone for S waves invariably exists beyond 103° from the earthquake's focus. This huge S wave shadow zone extends almost halfway around Earth, opposite the earthquake's focus (Figure 18.18). Evidently, something stops the waves so that they do not reach the other side of Earth. This was the first evidence that Earth had a core made of something distinctly different from the rest of the planet. S waves simply do not pass through this core. One of the important properties of shear waves (like S waves) is their inability to move through liquids. S waves are transmitted only through solids that have enough elastic strength to return to their former shapes after being distorted by the wave motion. The fact that S waves will not travel through the core, therefore, is generally taken as evidence that the outer core is liquid. This, combined with Earth's magnetic field and high density, implied that the core was made of molten iron.

The effect of the core on P waves is also informative but more complex (**Figure 18.19**). The shadow zone for P waves forms a belt around the planet between 103° and 143° away from the earthquake's focus (**Figure 18.20**). Evidently, the P waves are deflected but not completely stopped by Earth's core. Consequently, they are not detected in the shadow zone. Seismic rays traveling through the mantle follow curved paths from the earthquake's focus and emerge at the surface between 0° and 103° from the focus (slightly more than a quarter of the distance around Earth). In Figure 18.19, ray 1 just misses the core and is received by a station located 103° from the focus. Ray 2, however, being steeper than ray 1, encounters the core's boundary, where it is refracted. It travels more slowly through the core, is refracted again at the core's boundary, and is finally received at a station on the opposite side of Earth. Ray 3 is similarly refracted and emerges on the opposite side, 143° from the focus. Other rays that are steeper than ray 1 are also refracted through the core and emerge between 143° and 180° from the focus. Thus, refraction at the boundary between the core and the mantle causes the P wave shadow zone.

What is the principal evidence that Earth has a core?

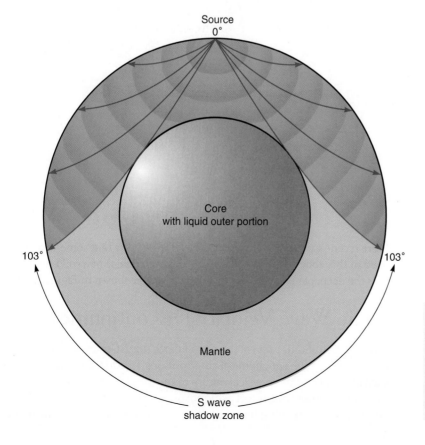

Figure 18.18 The shadow zone of S waves extends almost halfway around the globe from the earthquake's focus. This phenomenon can be explained if the outer core of Earth is liquid. Because S waves cannot travel through liquid, they do not pass through the core.

Figure 18.19 A P wave shadow zone occurs in the area between 103° and 143° from an earthquake's focus. The best way to explain the P wave shadow zone is to postulate that Earth has a central core through which P waves travel relatively slowly. Ray 1 just misses the core and is received at a station located 103° from the earthquake's focus. A steeper ray, such as ray 2, encounters the boundary of the core and is refracted. It travels through the core, is refracted again at the core's boundary, and is received at a station less than 180° from the focus. Similarly, ray 3 is refracted and emerges at the surface 143° from the focus. Other rays that are steeper than ray 1 are severely bent by the core, so that no P waves are directly received in the shadow zone. From shadow zones, seismologists calculate that the boundary of the core is 2900 km below the surface.

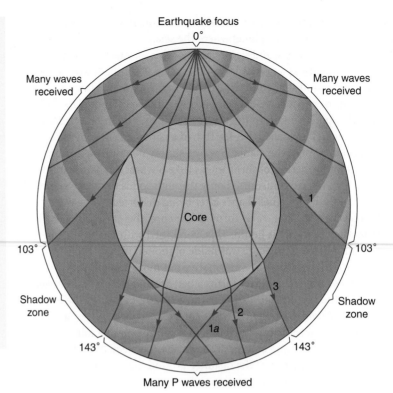

Figure 18.20 P waves are deflected by the inner core and are received in the shadow zone as weak, indirect signals. This deflection suggests that the inner core is solid.

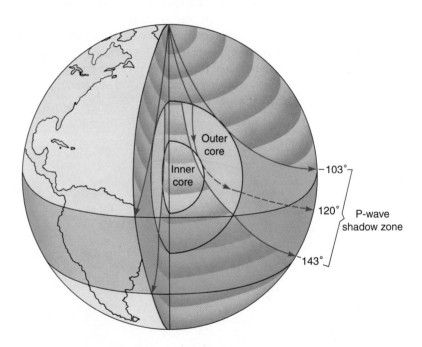

More recent studies of the P wave shadow zone show that some weak P waves are received in this zone. This is the evidence for the presence of a solid inner core, which deflects the deep, penetrating P waves in the manner shown in Figure 18.20.

Seismic Wave Velocity Discontinuities

Seismic discontinuities reveal the size of Earth's crust, mantle, and core and show that they have different chemical compositions. In addition, seismic studies reveal much about the physical nature of the interior, revealing a solid inner core, a liquid outer core, a soft asthenosphere, and a rigid lithosphere. Seismic tomography is beginning to reveal the pattern of convection in the mantle.

With the present worldwide network of recording stations, even minor variations in seismic velocities with depth, known as **seismic discontinuities**, can be determined with considerable accuracy. Seismic wave velocity versus depth curves, like the one in **Figure 18.21**, provide a huge amount of significant information about Earth's interior. The first seismic discontinuity occurs between 5 and 70 km below the surface. This is known as the **Mohorovičić discontinuity**, or simply **Moho**, after Andrija Mohorovičić, the Croatian seismologist who first recognized it. The discontinuity is considered to represent the base of the crust and heralds an important compositional change from the feldspar-rich crust to the olivine-rich mantle (**Figure 18.22**). Seismic wave velocity studies also show that the continental crust is much thicker (25 to 70 km) than oceanic crust (about 8 km).

Perhaps the most significant discontinuity, however, is the low-velocity zone from 100 to 250 km below the surface (Figure 18.22). Beno Gutenberg, a German seismologist, recognized this zone in the 1920s. The normal trend is for seismic wave velocities to increase with depth in the mantle. In the low-velocity zone, however, the trend is reversed, and seismic waves travel about 6% slower than they do in adjacent regions. The generally accepted explanation for the low seismic wave velocities is that the mantle is very near its melting point or even partially molten, with perhaps 1% to 5% liquid. A thin film of liquid around the mineral grains may slow both S and P waves. Moreover, rocks near their melting points are very weak and ductile. The low-velocity zone is embedded within the asthenosphere. The asthenosphere plays a key role in the motion of tectonic plates at Earth's surface. If Earth lacked a weak, ductile asthenosphere, the upper part of the mantle would be directly tied—frozen, if you will—to the lower part of the mantle, and plate motions would be prohibited. Apparently, the asthenosphere effectively decouples the moving lithosphere from the lower part of the mantle.

What are seismic discontinuities? What do they reveal about Earth's interior?

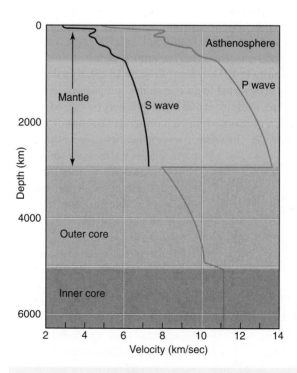

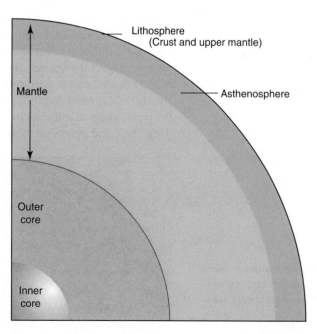

Figure 18.21 **The internal structure of Earth** is deduced from variations in the velocity of seismic waves at depth. The velocity of both P waves and S waves increases until they reach a depth of approximately 100 km. There the waves are slow until they have traveled to a depth of about 250 km. This low-velocity layer lies within the asthenosphere. Below this, the velocity of P waves and S waves increases until a depth of about 2900 km, where both velocities change abruptly. S waves do not travel through the central part of Earth, and the velocity of the P waves decreases drastically. This variation is the most striking discontinuity and indicates the boundary between the liquid outer core and the mantle. Another discontinuity in P wave velocity, at a depth of 5000 km, indicates the surface of the solid inner core.

Figure 18.22 The Moho and the low velocity zone are two important seismic discontinuities in the upper part of the planet. The Moho marks the base of the crust. The low velocity zone is revealed by a drop in the velocities of both P waves (shown here) and S waves. This marks a zone of low strength in the upper mantle between about 100 and 250 km deep. The low-velocity zone is contained in the weak asthenosphere and marks part of the mantle that is very near its melting point and may be a zone of partial melting.

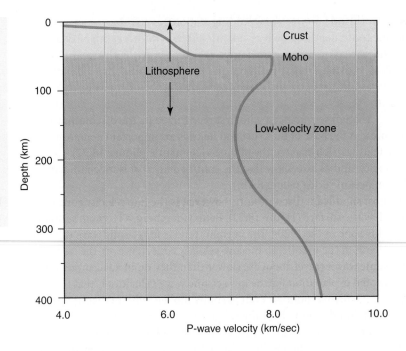

Other changes in seismic velocities occur in the mantle at depths of about 400 and 660 km to create a transition zone between the upper and lower mantle (**Figure 18.23**). These discontinuities are probably caused by phase changes: a metamorphic transformation of the minerals in the mantle (unlike the compositional change that marks the Moho). Seismic velocities suggest olivine and pyroxene are the dominant minerals in the upper mantle. With increasing depth and pressure, denser phases, with more-compact atomic structures, become stable. For example, garnet gradually replaces pyroxene between 300 and 500 km depth. Olivine is unstable at depths greater than about 400 km, where a denser polymorph of olivine (a mineral with the same composition but a different internal structure) replaces it. This metamorphic transformation probably causes the seismic discontinuity at 410 km depth (Figure 18.23). By a depth of 500 km, the olivine-polymorph has converted to an even more compact mineral structure (magnesium spinel). The discontinuity at 660 km depth may be caused by the replacement of spinel by two different magnesium-rich minerals (magnesiowustite [$(Mg,Fe)O$] and perovskite [$(Mg,Fe)SiO_3$]). The seismic wave velocities suggest that perovskite is much more abundant (Figure 18.23). At the very base of the mantle, magnesium perovskite gives way to yet another dense mineral that dominates the core-mantle transition.

The most striking variation in seismic wave velocities occurs at the core-mantle boundary, at a depth of 2900 km (Figure 18.21). There, S waves stop, and the velocities of P waves are drastically reduced. The seismic wave velocities and density of the outer core can be explained if there is a striking change in composition and physical state. Laboratory studies of seismic wave velocities and comparisons with meteorites show convincingly that the core

Figure 18.23 Discontinuities in seismic wave velocities may correspond to phase changes. The blue line shows how P wave velocities change with depth. The other lines show velocities for various minerals. The uppermost mantle is dominated by olivine. Below about 410 km a velocity increase implies olivine is replaced by denser minerals. At greater depths, magnesium-spinel is probably replaced by magnesium-perovskite. Each change increases the density and seismic wave velocity of the mantle.

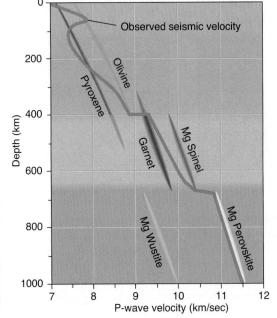

must be made mostly of iron. The outer core is most likely made of molten iron mixed with nickel, together with some lower density elements such as silicon or sulfur. The light component is needed to explain the density of the core which is about 10% less than pure iron. The deepest discontinuity is a strong increase in seismic velocity. The increase shows that the inner core is less compressible and more rigid than the outer core. Apparently, the inner core is solid (Figure 18.21). Even though the temperature must be higher, the extremely high pressures found inside the deep Earth make the more compact solid phase of iron more stable.

Convection Inside Earth

Convection of the core and mantle is the most important mechanism of heat transfer in Earth. Convection in the iron core probably creates the magnetic field, and convection in the mantle creates mantle plumes and plate tectonics.

Observations such as the three-dimensional seismic tomographs are driving a revolution in our understanding about convection in Earth's interior. Our newly acquired ability to construct three-dimensional tomographic images of Earth's deep interior opens up the breathtaking prospect of tracing tectonic plates as they plunge below the surface in a subduction zone and descend into the deep mantle. In addition, calculations of how the mantle and the core flow are becoming increasingly more realistic as more complete data about the interior are used.

Convection in the Core
Seismic velocity studies clearly show Earth's core is made mostly of iron and divided into a liquid outer core and a solid inner core. With the help of three-dimensional computer models, we are beginning to chart the flow of the molten iron alloy that forms the outer core and speculate about the origin of Earth's magnetic field. At 5000°C, the temperature of Earth's core approaches the temperature of the surface of the Sun, but it is slowly cooling as the inner core crystallizes. Consequently, a system of convective currents is established in the molten outer core. Moving metals can produce an electric current if the metal passes through a magnetic field. This principle is used to make electrical generators (dynamos) wherein conductive metallic wires are spun inside a magnetic field. In turn, an electrical current can generate a magnetic field. The magnetic field may be caused by Earth's rotation combined with convection of the molten metal in a shell surrounding the inner core (**Figure 18.24**). This creates a self-sustaining dynamo. Changes in these convective flow patterns may cause periodic polarity reversals.

Seismic maps of the core-mantle boundary show that the surface of the core is not smooth but is marked by broad swells and depressions with a difference in height of up to 20 km. A rough boundary could disturb the flow of the liquid iron in the outer core, much as a mountain influences the flow pattern of winds. Some seismic studies also suggest that Earth's solid inner core spins a bit faster than the rest of the planet.

Convection in the Mantle
Another way Earth shows its dynamic nature is by the large-scale convection of its mantle. Indeed, Earth is a large heat engine constantly churning by internal convection. The changes in mineral assemblage and density of the mantle outlined above may help control the way the interior of Earth convects. The result of combining seismic data about the mantle's layered structure with computer simulations of temperature distribution and internal convection are shown in **Figure 18.25**.

One possibility is that the upper mantle convects separately from the lower mantle because of the differences identified by seismic discontinuities (Figure 18.25). This model suggests that the mantle may convect in two more or less distinct layers that are usually separated by the 660 km discontinuity. The lower mantle may convect by generating narrow cylindrical plumes, shown in the model as yellowish mushroom-like

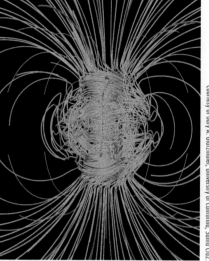

Figure 18.24 Earth's magnetic field probably forms by convection of the outer core, which is made of molten iron. A computer model of convection shows magnetic field lines as a smooth dipole with blue lines directed inward and gold field lines directed outward. The field inside the core is much more complex.

What does seismic tomography tell us about Earth's internal structure?

Figure 18.25 Earth's thermal structure and convection can be modeled using computers to complement the observations of seismic tomography. In one model, subducted slabs pass without pausing through the phase boundary at 660 km. In another model, the phase boundary is a temporary barrier that is broken down when enough subducted material accumulates and then flushes rapidly through the lower mantle. The lower mantle may convect by generating thin plumes that rise off of the core-mantle boundary. Some of the plumes may be triggered by the sinking of the dense overlying mantle.

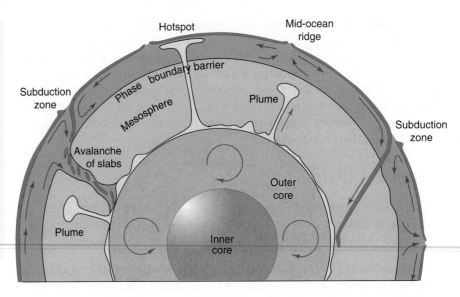

Why do we believe the mantle is moving in a slow convection system?

fingers projecting from the core-mantle boundary. In the computer models, some of these plumes penetrate the upper mantle and reach all the way to the surface, supporting the notion that volcanic hotspots are related to deep upwellings in the mantle. Convection of the upper mantle may be driven by the sinking of cold, dense slabs of subducted lithosphere. Perhaps strong rock in the lower mantle prohibits the slabs from penetrating the phase boundary at 660 km. However, the cold slabs slowly accumulating above the boundary could eventually have enough weight to break through the barrier and "flush" into the lower mantle. To understand this process, think of what happens if you place a small piece of iron on a coffee table. Iron is denser than wood, but it would not immediately sink through the table because of the strength of the wood. If you continue stacking more and more iron on the table, it will eventually break and the iron will fall to the next barrier to its movement, presumably the floor. It may take several hundred million years of subduction to amass a cold sinker that could break through the boundary. Three-dimensional numerical models show that these avalanches take the shape of broad cylindrical pipes with greatly enlarged bases, created when the cold rock hits the dense, impenetrable core. Each flushing event may trigger an upward counterflow as material from the lowest part of the mantle moves upward to balance the downward flow of the cold avalanche (Figure 18.25). Periods of enhanced volcanism on the real Earth could be a plume response to a flushing event.

In a competing model, the whole mantle convects as a single unit. Subducting slabs of oceanic lithosphere may be dense enough to pass unobstructed through the boundary between the upper and lower mantle (Figure 18.25). Recently constructed tomographic sections give some support to the suggestion. They show that inclined sheets of anomalously cold rock extend through the 660 km discontinuity and into the lowermost mantle. Ultimately, deeply subducted oceanic lithosphere must stall at the core-mantle boundary because it is less dense than the metallic core. And indeed, a large concentration of anomalously "cold" rock has been found at the core-mantle boundary. Why would cold material be present in a place where we would expect to find hot mantle created by the flow of heat from the convecting iron core? These deep-mantle cold spots lie below subduction zones. Could the cold rock correspond to ancient oceanic lithosphere, still distinctive because of its composition and lower temperature? Has dense oceanic lithosphere accumulated there over billions of years, to form a "slab graveyard"?

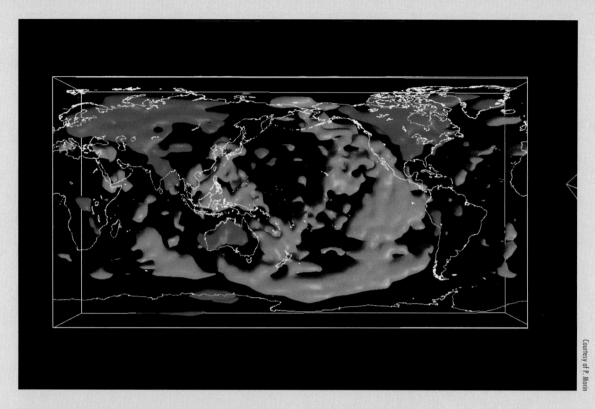

Courtesy of P. Morin

During the last decade, scientists have utilized a new analytic technique, known as **seismic tomography**, which promises to greatly enhance our knowledge of the deep internal structure of Earth, including the pattern of flow in the mantle. Seismic tomography is like its medical analog, the CAT scan (computer-assisted tomography; tomograph is based on a Greek word, *tomos*, meaning "section"). In a medical CAT scan, X rays that penetrate the body from all directions are used to construct an image of a slice (cross section) through the body. Bones, organs, and tumors are identified because they have different densities and absorb X rays differently. With the aid of a computer, these images are stacked side by side to produce a three-dimensional view.

In seismic tomography of Earth's interior, natural seismic waves from earthquakes are used in a similar way as X rays. Using the vast network of seismographs around the world, seismologists analyze the velocities of hundreds of thousands of seismic waves as they pass through Earth in different directions.

Geophysicists use computers to produce three-dimensional images of Earth's interior from the data. The results show regions where seismic waves travel faster or slower than normal. Geophysicists know from laboratory studies and from observations near volcanoes that seismic waves travel slowly through the relatively weak, hot rock and quickly through stronger, cooler rock. The tomographs can thus be interpreted as temperature maps. In addition, hot parts of the mantle, being less dense than their surroundings, will rise, whereas cool mantle rock will sink. Thus, the tomograph can be used to outline the patterns of convective flow in the mantle.

The three-dimensional view of the mantle obtained from seismic tomography provides an unparalleled view of the effects of plate tectonics on Earth's interior. At a depth of 150 km, slow seismic zones occur under most of the volcanic regions, including the mid-ocean ridges. This is evidence that the mantle directly beneath the mid-ocean ridges is hotter and probably less dense than normal. As a result, it is probably rising. At still greater depths of 350 km, there is still hot rock (shown in red on the image) concentrated beneath the ocean ridge system. Beneath the mid-Atlantic ridge this zone of hot rock is not continuous, but is broken up into isolated segments. At depths even greater than those shown here, the relationship between mantle and surface features is even weaker. This indicates that the ocean ridge system is not simply the surface expression of vertical upwelling currents from the deepest mantle. Instead, mid-ocean ridges must be fed by the movement of hot material in the *upper* mantle.

In contrast to the mantle beneath the ocean basins, the continental shields of Canada, Brazil, Siberia, Africa, and Australia are all underlain by mantle that has higher than normal velocities (blue on this tomograph) and must be some of the coldest mantle. This cold upper mantle forms a deep root beneath the continents and may not convect with the rest of the mantle, but is part of the lithosphere and moves with the continents.

GeoLogic   Inclined Seismic Zones

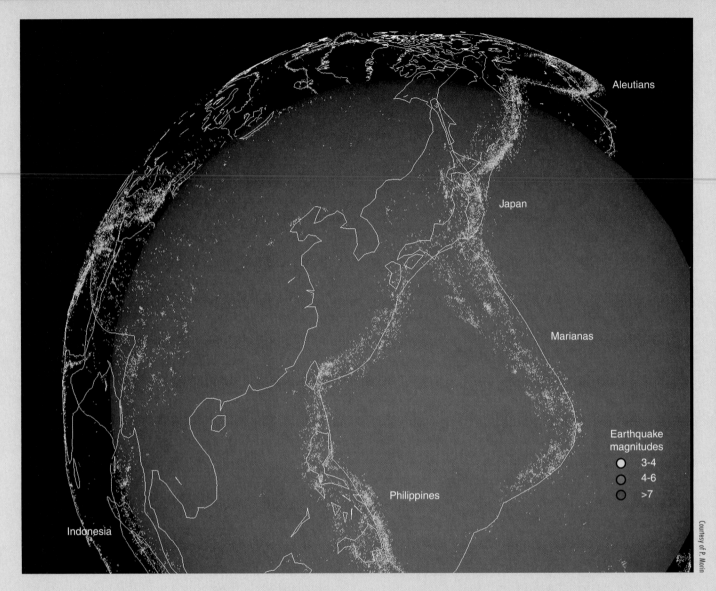

Aleutians

Japan

Marianas

Earthquake
magnitudes
○   3-4
○   4-6
○   >7

Philippines

Indonesia

Courtesy of P. Morin

Earthquakes are like a two-edged sword. On the one hand, they are devastating and kill thousands of people and cause billions of dollars of damage each year. On the other hand, they have helped us understand many of the fine points about Earth's internal structure. For example, consider the following facts about earthquake focal depths in the northwestern Pacific region shown on this globe with a transparent crust.

Observations

1. Earthquakes occur only in brittle, cool rocks.
2. Earthquakes form narrow zones that are inclined below volcanic arcs with andesitic composite volcanoes.
3. Here, and around the Pacific Ocean, these inclined earthquake zones all dip beneath the adjacent continent.
4. The earthquake zone reaches the surface at a deep oceanic trench.

5. The deepest earthquakes on Earth occur in these inclined seismic zones.
6. Earthquakes do not occur below about 600 km depth.

Interpretations

Initially, this small set of observations led geologists and geophysicists to conclude that great slabs of oceanic crust were being thrust beneath the continents. Eventually, we realized that it was not just the crust that was involved, but an entire plate of lithosphere is descending into the interior. The oceanic lithosphere is denser than the continental lithosphere; in fact, it can become denser than the mantle beneath it and subduct back into Earth's interior. Earthquakes occur at great depths only in subduction zones, because the oceanic plate is cold and brittle as it descends into the warmer mantle. Once the plate descends to depths greater than 600 km, it is too warm to rupture and form an earthquake.

Key Terms

compressional wave (p. 538)

deep-focus earthquake (p. 539)

earthquake (p. 536)

elastic limit (p. 536)

elastic-rebound theory (p. 536)

epicenter (p. 536)

focus (p. 536)

intensity (p. 539)

intermediate-focus earthquake (p. 539)

liquefaction (p. 545)

magnitude (p. 540)

megathrust earthquake (p. 552)

Moho (Mohorovičić discontinuity) (p. 557)

moment magnitude scale (p. 540)

P wave (primary wave) (p. 538)

S wave (secondary wave) (p. 538)

seismic discontinuity (p. 557)

seismic ray (p. 554)

seismic risk map (p. 549)

seismic tomography (p. 561)

seismic wave (p. 536)

seismograph (p. 536)

shadow zone (p. 555)

shallow-focus earthquake (p. 539)

shear wave (p. 538)

surface wave (p. 538)

Review Questions

1. Explain the elastic-rebound theory of the origin of earthquakes.

2. Describe the motion and velocity of the three major types of seismic waves.

3. Explain how the location of an earthquake's epicenter is determined.

4. What secondary effects commonly accompany earthquakes?

5. Describe the difficulties geologists have encountered in trying to predict earthquakes.

6. How can a seismic gap be used to predict an earthquake?

7. Describe the global pattern of earthquakes.

8. Compare the relative earthquake hazard along a divergent, transform, and convergent plate boundary.

9. How does the depth of earthquakes indicate (a) convergent plate margins and (b) divergent plate margins?

10. Draw a diagram showing the paths that would be followed by seismic rays through Earth if the core were only half the diameter shown in Figure 18.18.

11. What do seismic velocity-depth diagrams (see Figures 18.19 and 18.21) tell us about Earth's internal structure?

12. What changes could explain the transition zone between the upper and lower mantle between 400 and 660 km?

13. How does the cause of the Mohorovičić discontinuity differ from the cause of discontinuities at 400 and 660 km?

14. Where is the low-velocity zone and why is it important for plate tectonics?

15. Where and how is Earth's magnetic field thought to be generated?

16. How does the presence of anomalously hot (low seismic wave velocity) material below the oceanic ridges of the Pacific support the plate tectonic model?

17. Why are the continents underlain by cold mantle?

18. Compare convection styles in the upper and lower mantle.

19. Trace the path of a hypothetical slab of oceanic lithosphere from its production at a mid-ocean ridge to its ultimate demise.

19 Divergent Plate Boundaries

Although we have studied the ocean and used its resources for thousands of years, until now the details of the ocean floor have remained mysterious. Using sophisticated radar instruments carried by satellites, we have constructed for the first time accurate maps of the ocean floor that reveal the dominant role played by divergent plate boundaries among Earth's dynamic systems.

The ocean ridge is part of the global system of divergent plate boundaries that encircles the planet. High mountainous ridges coincide with these boundaries and extend through all of the ocean basins in a nearly continuous seam that is tens of thousands of kilometers long. Oceanic ridges are the sites of the most active volcanism on Earth. Occasionally, the rift valleys become boiling cauldrons when basalt, fresh from the mantle, erupts from long fissures onto the ocean floor. The volcanic rocks of the high ridge are so hot and porous that seawater easily circulates through the crust. Where the water exits, dense plumes of hot water belch from fragile chimneys that rise as high as skyscrapers. Here the crust is so active that its movement away from the ridge can be measured in centimeters per year.

On the northern coast of Ireland, the Giant's Causeway testifies to the floods of basaltic lava that accompanied the rifting of North America from Europe. Before 65 million years ago, these two great continents were part of Laurasia. Then the continents rifted and basalt erupted from fissures. The symmetrical columns of the Causeway formed as the mafic lavas cooled and fractured. The basaltic flows are the same age as lavas and intrusions to the north in Scotland and all are part of a province that includes a vast submarine plateau covered by similar lava flows. Eventually, a deep rift valley cut through the basalt, the crust stretched thinner and thinner and was eventually replaced by the basalts and gabbros of the Atlantic Ocean crust.

Throughout Earth's long history, the geologic processes operating at divergent plate boundaries have been among the most fundamental forces that shaped our world. It is at divergent plate boundaries that continents split and move apart, creating new continental margins, many of which contain valuable resources such as petroleum and natural gas. In this chapter, we examine these divergent plate boundaries as parts of Earth's plate tectonic system. We first study the role they play in the formation of oceanic crust and then explore how they form when continents rift and then drift apart to form new ocean basins.

Major Concepts

1. Divergent plate boundaries are zones where lithospheric plates move apart from one another. They are characterized by tensional stresses that typically produce long rift zones, normal faults, and basaltic volcanism.
2. An oceanic ridge marks divergent plate boundaries in the ocean basins. It is a broad, fractured swell with a total length of about 70,000 km. Basaltic volcanism and shallow earthquakes are concentrated along the rift zone at the ridge crest.
3. The ridge's characteristics depend upon the spreading rate. As oceanic lithosphere moves away from the ridge, it cools, becomes thicker and denser, and subsides.
4. Oceanic crust is generated at divergent plate boundaries and is composed of four major layers: (a) deep marine sediment, (b) pillow basalts, (c) sheeted dikes, and (d) gabbro. Below the crust lies a zone of sheared peridotite in the upper mantle.
5. At divergent plate boundaries, basaltic magmatism results from decompression melting of the mantle. The magma then collects into elongate chambers beneath the ridge, and some is intruded as dikes or extruded along the rift zone.
6. Seawater is heated as it circulates through the hot crust and causes extensive metamorphism. Locally, the hydrothermal fluids produce hot springs on the seafloor.
7. Continental rifting occurs where divergent plate margins develop within continents. The East African Rift, the Red Sea, and the Atlantic Ocean illustrate the progression from continental rifting to seafloor formation.
8. Continental rifting creates new continental margins marked by normal faults and volcanic rocks interlayered with thick sequences of continental sedimentary rocks. As the continental margin subsides, it is gradually buried beneath a thick layer of shallow-marine sediments.

Mid-oceanic Ridges

> A mid-oceanic ridge marks a divergent plate boundary in an ocean basin. It is a broad, fractured swell, marked by basaltic volcanism and earthquakes. As the oceanic lithosphere moves away from the ridge, it cools, becomes thicker and denser, and subsides.

The discovery of divergent plate margins only a few decades ago changed forever our understanding of Earth's dynamics. It is along divergent plate margins that the longest and most important mountain chain of our planet is located (**Figure 19.1**). Beneath divergent plate margins chambers are filled with hot magma that is episodically extruded to form new oceanic crust.

The creation of new oceanic crust is the fundamental process that occurs at a **mid-ocean ridge**. Indeed, more rock is generated at this type of divergent plate boundary than by all other processes combined. Since the early periods of Earth's history, igneous activity along divergent plate margins has generated enough basalt to cover the entire Earth with a layer about 120 km thick. As new crust forms, however, it continually spreads away from the ridge crest at rates of several centimeters per year. The newly formed basaltic crust is cooled by circulating sea-water at a rate sufficient to filter the entire ocean's water within a few million years. These circulating waters alter the hot oceanic crust, thereby creating large volumes of metamorphosed basalt.

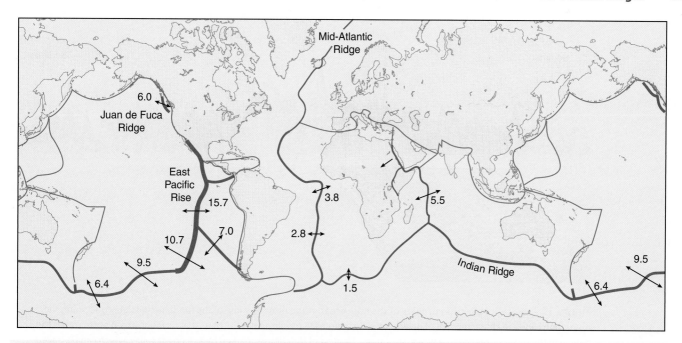

Figure 19.1 The mid-ocean ridge (red) extends as a major structural feature around the entire globe and marks divergent plate boundaries. The thickness of the line is proportional to the rate of spreading. The numbers give spreading rates in cm/y. Note the rate of spreading along the Mid-Atlantic Ridge is slow, so the ridge is high and rugged and is cut by a deep rift valley. The ridge in the eastern Indian Ocean spreads at an intermediate rate. The East Pacific Rise typically spreads at 15 cm/yr, about six times faster than the Mid-Atlantic Ridge.

Methods of Study

Our new knowledge of the ocean floor comes from a variety of direct and indirect observations. One is the use of sophisticated sonar equipment, which provide images of the ocean floor similar to relief maps of the continents made from aerial photography. In addition, the sound waves penetrate the upper layers of the oceanic crust, revealing its internal structure (**Figure 19.2**). Numerous dredge and drill-core samples have been obtained from specially designed oceanographic ships. Close-up examination of the seafloor from divers in small submarines also has revealed much about this concealed part of our planet. Only small parts of the oceanic ridge are exposed above sea level, allowing direct examination of the processes that create oceanic crust even as it forms. However, a few large segments of ancient oceanic crust have been thrust up and over continental rocks where geologists can study oceanic crust on dry land.

These direct studies are supplemented by various indirect studies using geophysical measurements such as paleomagnetism, seismicity, gravity, and heat flow. These geophysical measurements give us the ability to "see" not only the surface of the ocean floor, but also the internal structure of the oceanic crust. In the last few years, declassified sea surface measurements collected by military satellites have been converted to topographic maps covering nearly all of Earth's ocean floor (see the inside cover). All of these observations combine to provide a firm basis for constructing models of how divergent plate margins operate.

How do we "see" the topography and landforms of the ocean floor?

Topography of Mid-ocean Ridges

The oceanic ridges are the most pronounced tectonic features on Earth. If they were not covered with water, the ridges would be visible from the Moon. A mid-oceanic ridge is essentially a broad, fractured swell, a huge feature generally more than 1500 km wide (the width of Texas), with peaks rising as much as 3 km above the surrounding ocean floor. Its local relief is thus greater than some mountain ranges on land.

Remarkably, the mid-oceanic ridge is a nearly continuous feature around the entire globe, like the seam of a baseball (Figure 19.1). The ridge extends from the Arctic Basin, down through the center of the Atlantic, into the Indian Ocean, and across

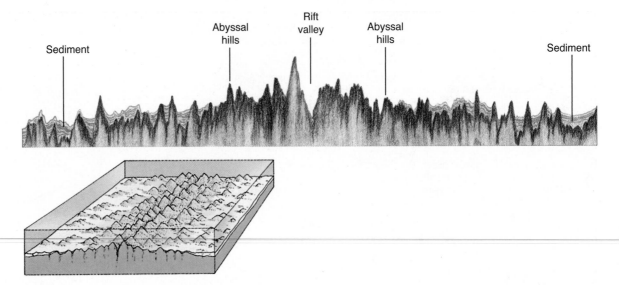

Figure 19.2 Seismic reflection profiles of the mid-ocean ridge provide some of the fundamental data concerning the topography and the characteristics of divergent plate margins. A profile across the Mid-Atlantic Ridge at 44° north shows the outline of the surface plus the subsurface structure. The crest of the ridge is marked by a deep rift valley, which can be traced along most of the Mid-Atlantic Ridge. Note the layers of marine sediment along the flanks of the ridge and their absence at the ridge crest. The vertical scale is greatly exaggerated for this 500-km-long transect.

the South Pacific, ending in the Gulf of California, a total length of about 70,000 km. Without question, it is the greatest "mountain" system on Earth. The internal structures of the "mountains" of the oceanic ridge, however, are nothing like the mountains of the continents, which largely consist of folded and metamorphosed sedimentary rocks. By contrast, the ridge is composed entirely of basalt and is not deformed by folding. Its main structures are due to extensional forces, resulting in normal faulting and basaltic igneous activity. The mid-ocean ridge is not one continuous fracture. It is broken into segments defined by offsets (**Figure 19.3B**). Many involve **transform faults** and their extensions called **fracture zones**, which are some of the most prominent features on the ocean floor. Small overlaps and minor bends also create irregularities in the trend of the ridge (**Figure 19.4**).

Many detailed characteristics of the ridge are apparent in **Figure 19.5**, which shows the Juan de Fuca Ridge. You can see that the broad ridge is arched up and broken by many faults, which form linear hills and valleys. The highest and most rugged topography is along the axis, and a prominent **rift valley** marks the crest of some ridges throughout much of their lengths. Locally, volcanoes have completely filled the rift valley and rise above its crest. The East Pacific Rise displays much less rugged topography and does not possess a prominent central rift valley (Figure 19.3A). Oceanic sediments are thickest on the flanks of the ridge, but they thin rapidly toward the crest (Figure 19.2).

The characteristics of a ridge depend on its **spreading rate**. Where the rate of spreading is slow (less than 5 cm/yr), as in the North Atlantic (Figure 19.1), the oceanic ridge is steeper and more rugged and mountainous, and a prominent rift valley develops along the axis (Figure 19.3B). At intermediate spreading rates (5 to 9 cm/yr), as in the southern part of the Indian Ocean Ridge, the topography is more subdued and the rift valley is only 50 to 200 m deep. At faster spreading rates (greater than 9 cm/yr), as in the eastern part of the East Pacific Rise, the ridge topography is relatively smooth and no rift valley develops (Figure 19.3A).

Regardless of spreading rates, however, all ridge crests are marked by a zone of volcanic activity and fissures (Figure 19.5). Away from the ridge crest, the topography is controlled by active normal faults. Some 10 to 15 km from the axis, there is little volcanic activity or faulting. Thus, the actual plate boundary, the place of intense geologic activity, is a narrow zone only 20 to 30 km wide.

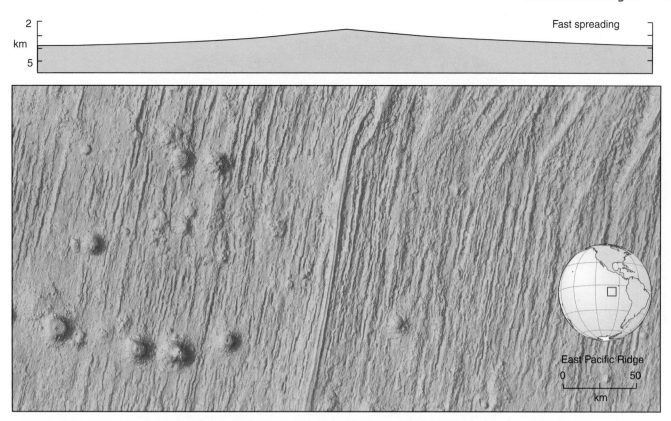

(A) Fast-spreading ridges, such as the East Pacific Ridge, usually have gentle slopes and lack a prominent rift valley at the ridge crest.

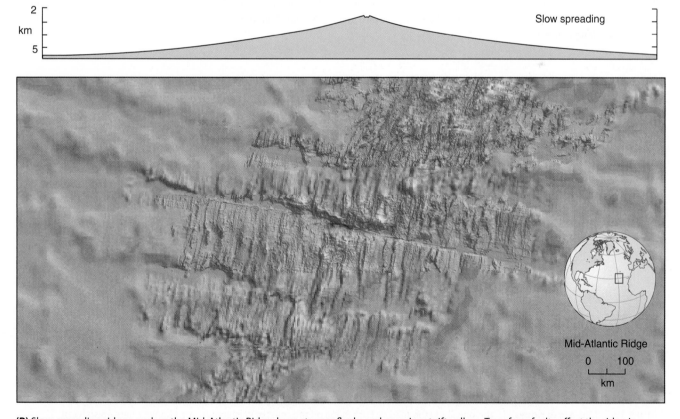

(B) Slow-spreading ridges, such as the Mid-Atlantic Ridge, have steeper flanks and prominent rift valleys. Transform faults offset the ridge in numerous places.

Figure 19.3 Spreading rate helps control many features of an oceanic ridge.

Courtesy of Lamont-Doherty Earth Observatory/Columbia University

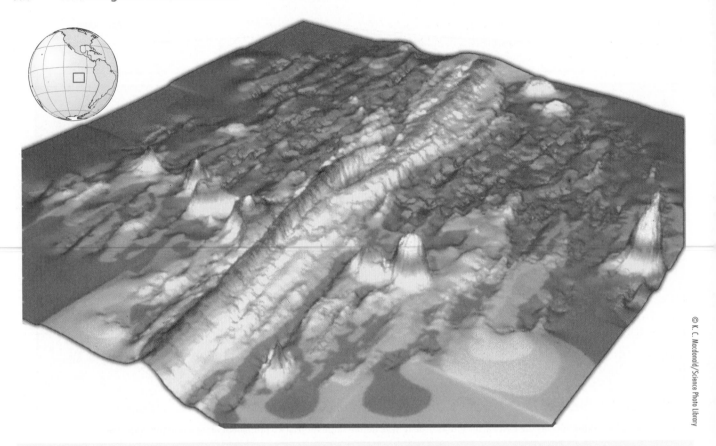

Figure 19.4 **Overlapping ridge segments along the East Pacific Rise** are shown in this color-coded topographic map. Deep-ocean floors are shown in blue, changing upward to green, yellow, red, and white. The important features in this image are the overlapping ridge segments that end in hook-shaped curves and merge with the adjacent ridge. Volcanic peaks flank the ridge. Considerable vertical exaggeration of relief is employed along this 100-km-long transect.

Figure 19.5 **Mid-ocean ridges**, like this one along the Juan de Fuca Ridge off the coast of Washington, are regions where episodes of volcanism and tectonism alternate. The high part of this ridge is a shield volcano with a shallow summit crater. The lower parts of the ridge toward the top of the image are where volcanism is less active and normal faulting, extension, and thinning are the dominant processes, as indicated by the long linear fault blocks. Volcanic peaks or seamounts flank the rise on both plates. Near the bottom of the image, a large volcano was ripped in half by rifting. The ridge terminates near the top of the image in a long transform fault marked by a deep trough, flanked by parallel ridges.

Cooling and Subsidence of Oceanic Crust   The elevation of the ocean floor is strongly influenced by its temperature. This simple observation is the result of an equally simple principle that we have repeatedly employed in our investigations of Earth systems: Hot rocks are expanded and less dense than cooler rocks of the same composition. Thus, hot rocks rise to a higher level. Because oceanic crust forms by magmatic processes at the mid-ocean ridge, the crest of the ridge is hot. As the oceanic crust moves away from the ridge, it cools. In fact, the temperature of the oceanic crust bears a simple relationship to the distance from the spreading ridge: The farther away from the ridge, the cooler the oceanic crust.

As a result of cooling, the oceanic crust becomes denser and subsides as it moves away from the ridge crest. Consequently, the depth to the ocean floor depends on the seafloor's age (**Figure 19.6**). Most studies show that the **subsidence** of oceanic crust is proportional to the square root of its age. A theoretical curve based on this heat loss model can be calculated that matches the observed depths. Water depth increases from about 2.5 km at the ridge crest to 3 km where the crust is 2 million years old, 4 km where it is 20 million years old, and 5 km where the crust is 50 million years old. In other words, the approximate age of the ocean crust can be estimated from its depth below the sea surface. Because cooling and subsidence depend on age, a fast-spreading ridge has a broader, gentler profile than a slow-spreading ridge (Figure 19.3).

An important corollary of this principle holds that there is a direct relationship between global spreading rates and sea level. If spreading rates increase, the mid-ocean ridge inflates and rises higher toward the sea surface. This expansion reduces the volume of the ocean basin, causing sea level to rise. As a result, shallow seas spill onto the stable platforms of the continents. Conversely, if seafloor spreading rates drop, sea level also drops, and the shallow seas withdraw from the continents. For example, a decrease in the spreading rate from 6 cm/yr to 2 cm/yr along a ridge that is 10,000 km long will drop global sea level by about 100 m. Such sea level changes are not rapid; they develop over tens of millions of years.

These changes in sea level may have profoundly affected the evolution of world geography throughout geologic time. Indeed, one of the great central themes of geologic history is the expansion and contraction of shallow seas over the continents, as witnessed by the numerous formations of shallow-marine sedimentary rocks that cover the stable platforms of all continents. With expanding and contracting seas, the climate and other aspects of the physical environment are significantly changed, including living space for shallow-marine and terrestrial organisms, and thus the evolution and extinction of many species are also affected.

Close-Up View of the Rift Zone   Deep-sea submersibles, both piloted and robotic, have provided spectacular close-up pictures of the rift zones at divergent plate boundaries. Thousands of photographs show that the ridge surface is covered with fresh lava flows, including thin, smooth sheet flows as well as **pillow basalt**. It is important to note that almost no sediment covers even the finest details of the lava flow surfaces (**Figure 19.7**). This lack of cover is strong evidence that the basalt is young and fresh; otherwise, oceanic sediment would cover and mask the pillow structures. Numerous open **fissures** in the crust also exist wherever the ridge has been observed at close range (**Figure 19.8**). In one small area of only 6 km², 400 open fissures were mapped, some as wide as 3 m. These are considered conclusive evidence that the oceanic crust is being pulled apart by extension.

The eruption of lava from these fractures, which parallel the rift valley, creates long, narrow ridges, sheet flows, or small mounds of pillow lavas within the rift valley. Besides the pillow basalts and fractures, many small shield volcanoes and small fissure-fed flows dot the rift valley. In places, elongate volcanic troughs or collapsed lava lakes have been discovered by detailed mapping of the rift valley. In the rift valley, springs of hot, mineral-laden waters spew from chimney-like vents (**Figure 19.9**).

A unique and previously unknown community of living organisms thrives in the darkness of the rift valley. These ecosystems are unusual in that, unlike the rest of the known biosphere, they draw their nutrients and energy from the hot-water solutions

How does the topography of the ridge in the North Atlantic differ from that of the East Pacific Rise? What causes this difference?

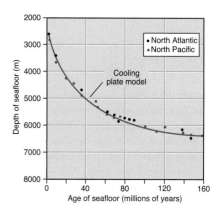

Figure 19.6 Subsidence of the ocean floor occurs as the oceanic crust cools and moves off from the ridge crest. Individual points represent the actual depth of the seafloor, and the solid line shows the calculated depth based on heat loss and contraction of oceanic lithosphere.

Figure 19.7 Pillow basalt along the Mid-Atlantic Ridge was photographed up close by scientists in the deep-diving submarine *Alvin*. Little or no sediment covers the basalt because this part of the seafloor is very young. The large elliptical pillow is approximately 1 m across.

Courtesy of Galapagos Rift 2005 Exploration/NOAA

Figure 19.8 Open fissures along the Mid-Atlantic Ridge were photographed from the *Alvin*. Hundreds of such fissures were mapped. Such fissures may be as much as one meter wide and hundreds of kilometers long. They clearly show extension along the rift zone as the oceanic crust is pulling apart.

Courtesy of Woods Hole Oceanographic Institution

Why is there little or no sediment covering the pillow basalts and sediment in the rift zone?

rising from the vents rather than from sunlight. These vent communities contain 10,000 to 100,000 times more living matter than adjacent deep-sea communities and are like isolated islands of deep-marine life. Crabs, worms, shrimp, and abundant bacteria cluster around the vents. The most spectacular organisms are giant tube worms nearly 3 m long and giant clams up to 25 cm across (Figure 19.9).

We should reemphasize the important fact that the active rift zone is extremely narrow, with volcanism and seafloor spreading occurring episodically in the same zone. During a rifting episode, the volcanoes and their flows are ripped asunder, roughly half going in each direction away from the center of the ridge. This activity imparts a striking symmetry to the topography and structure of the ridge—a symmetry also revealed by the paleomagnetic stripes.

Courtesy of Br. Robert McDermott, S.J.

Figure 19.9 Vents for hot, mineral-laden waters circulating through the hot rocks in the rift valley of an oceanic ridge cause black smokers. This warm water provides a unique habitat for exotic life that can subsist without sunlight. A black smoker spews hot water from a narrow chimney. The "smoke" is really dark minerals precipitating from the hot solution as it mixes with the cold ocean water. Giant tube worms and giant clams do not possess guts; they are nourished by bacteria that live in their tissues. Other life forms include white crabs, which swarm over the pillow basalt.

As common as seafloor eruptions must be, very few have ever been observed. However, several active eruptions on the mid-ocean ridge have been located and examined firsthand by use of underwater listening devices, formerly used only by the military to track the movement of submarines. In 1993, 1996, and 1998 scientists were able to detect new eruptions of deep-sea volcanoes on the Juan de Fuca Ridge off the Oregon coast. In 1993 a huge plume of warm water, rising above the eruption site, was discovered by ships at the surface. A remotely piloted robot submarine was lowered 2500 m to the seafloor to photograph the area and to collect samples of fresh, glassy pillow lavas. With this new monitoring system now in place, significant advances in our understanding of mid-ocean ridge volcanic systems are likely to come. We will finally know *exactly where, when, and how often volcanic eruptions occur on the ridge.*

Seismicity

A narrow belt of **shallow-focus earthquakes** coincides almost exactly with the crest of the oceanic ridge and marks the boundary between divergent plates (**Figure 19.10**). This zone is remarkably narrow compared with the zone of seismicity that follows the trends of young mountain belts and island arcs. Another very important difference between earthquakes at convergent and divergent boundaries is their depth and size. Earthquakes along divergent plate boundaries are almost always less than 10 km deep and typically are small in magnitude.

Although the zone of seismicity along the mid-oceanic ridge looks like a nearly continuous line on regional maps, two types of boundaries, based on fault motion as determined from earthquakes, can be distinguished: spreading ridges and transform faults (Figure 19.10). Earthquakes at the ridge occur within, or near, the rift valley. They are associated with intrusions of basaltic magma and normal faulting. Locally, shallow earthquakes in the rift occur in swarms related to the movement of magma in dikes. Why are there no deep earthquakes at the ridges? Earthquakes do not occur at great depth beneath the ridges, even though deformation is active there. Instead, the hotter, deeper mantle deforms ductilely and does not fracture as do the cooler and more brittle materials in the upper crust.

Shallow earthquakes also follow the transform faults that connect offset segments of the ridge (Figure 19.10), but they generally are not associated with volcanic activity. Studies of fault motion in the transform zone show strike-slip displacement in a direction away from the ridge crest, in contrast to the vertical motion on normal faults in the ridge crest.

How does seismicity on the ocean floor differ from that along a subduction zone?

Magnetic Anomalies

Magnetic surveys of the seafloor are easily accomplished, and measurements have been carried out since the mid-1950s. Magnetometers towed behind a vessel measure the magnetic field intensity. The surveys revealed a pattern of magnetic stripes

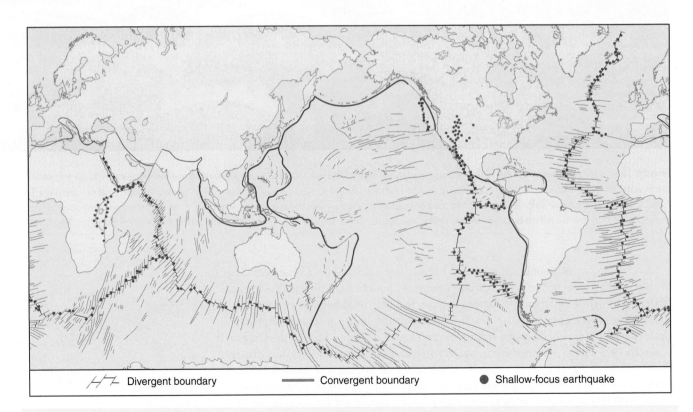

╱┼╱ Divergent boundary   ══════ Convergent boundary   ● Shallow-focus earthquake

Figure 19.10 Seismicity along divergent plate margins is concentrated along the ridge crest and along transform faults. This map shows the locations of thousands of earthquakes that occurred during a 5-year period. Shallow-focus earthquakes (generally less than 10 km deep) are associated with normal faults along the ridge crest and strike-slip movement along transform faults.

of alternating high and low **magnetic anomalies** (**Figure 19.11A**). These bands are remarkably persistent; many can be traced for hundreds of kilometers. Furthermore, the bands are parallel to the mid-ocean ridges and are offset at fracture zones just as the ridge crest is. As explained elsewhere in the text, these anomaly bands are caused by periodic reversals in the polarity of Earth's magnetic field during seafloor spreading.

Reversals of the magnetic field thus produce spectacular markers of the expansion of the ocean floors. They act as a bar code, imparting a distinctive signature of the age and spreading rate of the ocean floor. These discoveries clearly demonstrate seafloor spreading.

Heat Flow and Gravity

As you might expect for a volcanically active region, large amounts of heat are released from the mid-oceanic ridge. **Heat flow** is a measure of the amount of heat escaping per second from a given area and is usually measured in watts/m^2. Measurements show that heat flow is 10 times greater near the ridge crests than for average oceanic crust (Figure 19.11C). In addition, the numerous hot springs at active spreading boundaries show that significant heat is also carried out of the crust by convecting pore water. These data imply that a large heat source—basaltic magma—lies beneath the ridge axis. Heat flow diminishes rapidly as one moves away from the spreading center.

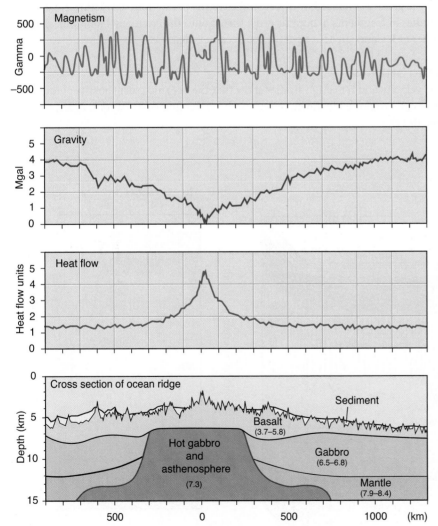

(A) Magnetic anomalies form a symmetrical pattern of highs and lows centered on the ridge crest. The pattern is caused by seafloor spreading and continuing magnetic field reversals.

(B) The gravity values measured over an oceanic ridge are lower than in adjacent areas, indicating that the ridge is underlain by lower-density rocks than the ridge flanks.

(C) Heat flow in the ocean basins peaks at the mid-ocean ridge. Over only 300 km, heat decreases to one-fifth its value at the ridge.

(D) Gravity and heat flow imply that the ridge is high because of the low density of the hot rocks and magma there. The crust (with seismic velocities of 3.7 to 6.8 km/sec) and lithosphere are very thin at the ridge. The layer with velocities of about 6.5 to 6.8 km/sec is interpreted to consist of gabbro. Seismic wave velocities are abnormally low (7.3 km/sec) beneath the ridge, probably because of partial melting. The higher mantle velocities (7.9 to 8.4 km/sec) mark lithospheric mantle.

Figure 19.11 Magnetism, gravity, heat flow, and seismic wave velocities at a mid-ocean ridge reveal much about the internal structure and origins of oceanic crust.

Gravity measurements across an oceanic ridge show low values at the crest and higher values on the adjacent flanks. This **gravity anomaly** indicates that materials below the ridge are less dense than materials of the adjacent crust and mantle (Figure 19.11B). Such measurements also suggest that the ridge is nearly in isostatic equilibrium. Consequently, differences in elevation must also reflect differences in density or thickness of the underlying lithosphere. The best explanation for the gravity anomaly is that the lithosphere is very thin just below a ridge and that hot asthenosphere with a lower density extends nearly to the surface. Putting together the gravity anomaly and heat flow data, we can construct accurate models of magma chambers and their surrounding environment beneath a mid-ocean ridge (Figure 19.11D).

Structure and Composition of the Oceanic Lithosphere

Oceanic crust is composed of four major layers: (1) deep-marine sediment, (2) pillow basalts, (3) sheeted dikes, and (4) gabbro. Below the crust lies a zone of sheared mantle peridotite.

Although oceanic crust is much more difficult to study than continental crust because it lies deep below the ocean, seismic investigations of the ocean floor enable geologists to understand the internal structure and composition of the oceanic crust. This understanding is greatly enhanced by direct studies of the ocean floor at fracture zones, by field studies of fragments of oceanic crust thrust onto the continents (ophiolites), and by studies of Iceland, an active part of the oceanic ridge.

Seismic Studies

Seismic velocity and reflection studies show that the oceanic crust consists of four major layers. From the top down, these have been designated as layers 1, 2, 3A, and 3B (**Figure 19.12**).

Figure 19.12 The major rock units in an ophiolite sequence are shown in this idealized diagram. The uppermost layer consists of deep-marine sediments. Most of the rest of the crust is made of igneous rocks. Pillow basalts and sheeted dikes form thin layers. Massive gabbro underlain by layered gabbro forms the rest of the crust. Peridotites, tectonites deformed in the mantle, are the lowest rocks found in some ophiolites. Ophiolites are thought to be fragments of the ocean floor thrust onto the continents. Correlations with seismically determined layers of the oceanic crust are shown on the right.

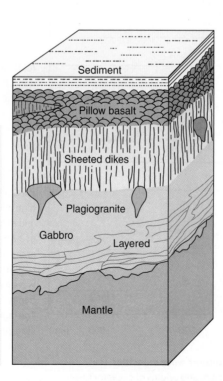

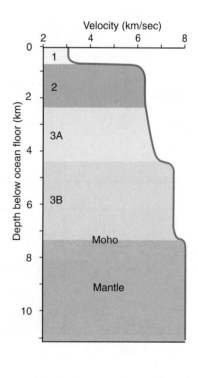

Layer 1 averages 0.4 km in thickness and has been extensively sampled by dredging and drilling. Samples show that this layer consists of fine mud that settled through the deep ocean waters. Layer 1 is thinnest near the ridge and thickens on the flanks (Figure 19.2).

Layer 2 ranges from 1 to 2.5 km thick and has P wave velocities that increase from 3.5 km/sec to 6.2 km/sec with increasing depth.

Layer 3 is the main layer of the oceanic crust and is about 5 km thick. This layer is usually subdivided into two units. *Layer 3A* has a seismic velocity of about 6.8 km/sec. The underlying *Layer 3B* has a velocity of about 7.3 km/sec.

Beneath Layer 3 lies a zone where seismic velocities are abruptly higher—about 8 km/sec. This zone is interpreted to be the upper mantle.

It is important to emphasize that the seismic structure represents units defined on variations in only one physical property—seismic wave velocity. Consequently, the boundaries between these seismic layers do not necessarily correspond exactly with the contacts between specific rock types as seen in drill holes or in ophiolites. Nonetheless, the seismic information reveals that the oceanic crust is layered and that the layers are relatively undeformed, unlike continental crust. Moreover, the seismic data provide a framework into which data from other sources can be integrated to form an accurate interpretation of the structure of oceanic crust.

How do we study the internal structure of the oceanic lithosphere?

Studies of Ophiolites

Fortunately, numerous fragments of ancient oceanic crust, with its four layers, have been thrust up on the continents. Here, geologists can study the crust's structure and rock types, gaining information needed to interpret the nature and origin of the seismic layers described above. These fragments of ancient oceanic crust are known as **ophiolites** (literally "snake rock") because of some rock's similarity to the color and texture of snakeskin. There are excellent descriptions of more than 100 ophiolites in mountain belts around the world. Most ophiolites were probably accreted onto a continent at a convergent plate margin and now exist as deformed, isolated fragments in folded mountain belts.

One of the most complete ophiolite sequences in the world is in Oman, on the Arabian peninsula off the coast of the Indian Ocean (**Figure 19.13**). This sequence of rock, as much as 15 km thick, extends about 500 km along the Arabian shore and has a width of 50 to 100 km. The absence of soil and vegetation in the desert climate provides excellent exposures. The Oman ophiolite has remained largely undeformed, unlike others that have been compressed, folded, and faulted. The sequence is only gently folded into broad anticlines. Stream erosion has cut through the slab to expose a complete sequence of oceanic lithosphere from the mantle up through the uppermost oceanic sediments. In studying the Oman ophiolite, we are able to walk across dry ocean floor and climb down into canyons to observe one layer after another in a section through the entire oceanic crust and down into the mantle below.

What is an ophiolite?

Careful field studies of the Oman ophiolite, as well as others, lead to the conclusion that the four seismic layers seen in typical oceanic crust can be identified and their origin interpreted. The correspondence between the main units in ophiolite sequences and the seismically determined layers of oceanic crust are shown in Figure 19.12 and described in the following sections.

Sediments The uppermost layer of the Oman ophiolite sequence consists of a relatively thin layer of deep-marine sediments (**Figure 19.14**). The thin layers are made of clay and calcareous and siliceous mud derived from shells of microscopic organisms (such as foraminifera, diatoms, and radiolarians). Thin beds of chert are common. Some thin graded beds of sand and mud were deposited by turbidity currents when this piece of oceanic crust was near the continental margin. In most ophiolites, the sediment layer is several hundred meters thick. This layer has been correlated with the sediments found on the ocean floor (Figure 19.12) and with the seismically defined Layer 1.

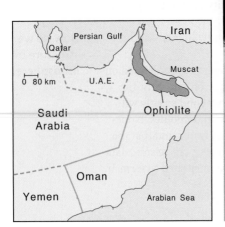

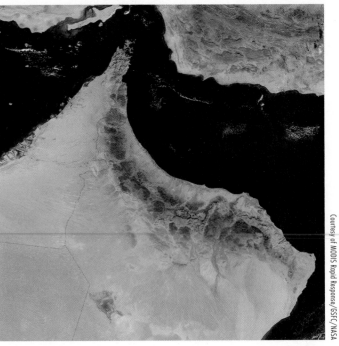

Courtesy of MODIS Rapid Response/GSFC/NASA

Figure 19.13 **The Oman ophiolite** is a large sheet of oceanic crust that was thrust onto the Arabian peninsula during its collision with Eurasia. The ophiolite is well exposed and reveals much about the internal structure of oceanic crust.

Figure 19.14 **Deep-marine sediments** at the top of the Oman ophiolite are thin beds of clay and chert. They were originally deposited horizontally but were tilted when the ophiolite sequence was thrust onto continental crust.

Basaltic Lavas Below the sediments, the Oman ophiolite has a thick layer of basalt lava flows with abundant pillows. This extensive mass of pillow basalt is unlike anything formed on the continents (**Figure 19.15**) and is evidence that ophiolites originally formed on the ocean floor. More massive sheet flows have columnar joints that developed as the lava cooled. Also common are lava breccias consisting of fragmental debris, formed as the hot lava hit cold seawater. This entire layer ranges from 1.0 to 2.5 km thick. In many places, basalt dikes and sills intrude into the lower part of the lava sequence. This dominantly volcanic layer is most like seismic Layer 2 (Figure 19.12).

Sheeted Dike Complex Below the pillow basalts, a typical ophiolite has a layer that consists almost entirely of dikes and is known as a **sheeted dike complex**. This distinctive body of rock appears to form much of seismic Layer 3A (Figure 19.12). The

Figure 19.15 **Pillow basalts** exposed in Wadi Jizzi, Oman, illustrate the characteristics of Layer 2 of the oceanic crust. This rock unit is more than 500 m thick and extends over a vast area in the Oman mountains. The pillows, or sausage-like structures, form as lava is extruded onto the seafloor and chills rapidly in the cold water under high pressure. The outer surface of the pillows is smooth and glassy, the result of rapid quenching of the lava as it contacted cold seawater.

individual dikes are vertical tabular sheets a meter or so wide. From a distance, this mass of vertical dikes looks like a sequence of sedimentary strata tilted at a high angle (**Figure 19.16**). The dikes are igneous intrusions formed by injection of magma into fissures and fractures. The total vertical thickness of this mass of dikes is about 2 km. This is strange geology; for one who has seen only continental sedimentary rocks, the sight of a huge rock mass consisting of nothing but igneous dikes seems unreal.

Another important feature of the sheeted dike complex is the age relationship of the dikes. Each vertical dike is progressively older than the adjacent one. Age does not increase from the top of the body toward its base. At the top of the sheeted dike complex, dikes intrude into the pillow basalt, but the central part of some ophiolites is 100% dikes. The structure of the sheeted dikes has been compared to a deck of cards standing on edge.

Close examination shows how the sheeted dike complex forms. Individual dikes in these swarms are so numerous that they intrude into one another. A dike intrudes along a zone of weakness, which in an area with multiple magma-filled dikes will be either along the margin or in the center of a preexisting dike. This process occurs because an earlier dike may still be hot and therefore weak in the interior, but cold and strong along its margins. Thus, a new fracture develops down the center of a dike, where the rock is weakest. As a result, a new dike splits the older dike in half (**Figure 19.17**). The final result is the development of dike-like bodies that are in essence **half-dikes** with chilled margins on only one side.

How can you recognize a half dike?

Figure 19.16 **Sheeted dikes** exposed in Wadi Hawasina, Oman, are nearly vertical, almost exactly the way they were intruded into the rift zone. Many dikes intrude into earlier dikes, forming half-dikes with chilled borders on only one margin, as shown.

Figure 19.17 **Half-dikes** form when a normal dike (A) is split down its hot center by a younger dike (B). The center may be weak because it is still hot and molten. A half-dike (C) has only one chilled margin and is thinner than a normal dike. Repeated intrusion (D) forms a sheeted dike complex above an ocean ridge magma chamber. A representative cross section (E) of the sheeted dike complex in the Troodos ophiolite complex, Cyprus, shows these relationships.

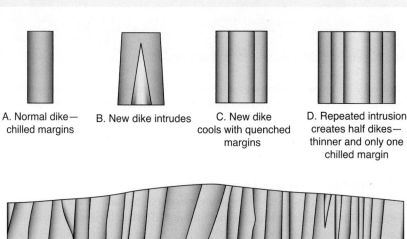

A. Normal dike— chilled margins

B. New dike intrudes

C. New dike cools with quenched margins

D. Repeated intrusion creates half dikes— thinner and only one chilled margin

E. Troodos sheeted dike complex—normal and half-dikes

Gabbro   In ophiolites, the sheeted dikes grade downward into a zone of **massive gabbro** (Figure 19.12). The gabbro may be coarse-grained. Below the massive gabbros, the Oman ophiolite is composed mostly of **layered gabbro** and lesser amounts of peridotite.

Gabbro represents the main mass of the ophiolite sequence and may be as much as 4.5 km thick. This part of an ophiolite has been correlated with the seismic Layer 3B. The seismic studies show gabbro is just as abundant in oceanic crust as it is in the ophiolite.

The layering and composition suggest that these rocks crystallized from basaltic magma contained in a magma chamber that cooled slowly at a shallow depth beneath a mid-ocean ridge. The layers probably formed as crystals accumulated on the walls and floors of the chamber to form distinct layers (**Figure 19.18**). Layers of chromite (chromium oxide) also form along with the layered gabbros and are among the richest chromium deposits in the world. The layering in these rocks is more regular than in the underlying mantle rocks.

Figure 19.18 **Layered gabbro** in the Oman ophiolite appears to have a structure like sedimentary rocks. Early-formed crystals rich in iron and magnesium settle to the base of the magma chamber and accumulate in distinct layers. Some layers are graded, and others show cross-bedding.

Tectonites (Upper Mantle) The lower part of many ophiolite sequences is made of rocks thought to have formed in the upper mantle (**Figures** 19.12 and **19.19**). These peridotites display a distinctive texture, suggesting ductile deformation at high temperature during seafloor spreading: hence the term **tectonite**. A pronounced lineation or foliation has been produced by the preferred orientation of mineral grains and by concentration of minerals into distinctive layers during deformation. The layers are typically stretched, folded, and refolded, showing that significant flowage occurred during their formation (Figure 19.19B). In many ophiolite sequences, the tectonites are between 5 and 7 km thick, but in some they are as much as 12 km thick. This material was once in the upper mantle, which is rarely exposed at the surface except in ophiolites. The change to high seismic velocities seen at the base of oceanic crust appears to occur in a layer much like the Oman tectonites.

In general, we can make strong correlations between the seismic layers of the oceanic crust and the real compositional layers seen in ophiolites (Figure 19.12). The seismic studies reveal just how far the layered character of oceanic crust extends. Moreover, both ways of studying the oceanic crust show it is very thin (about 6 to 8 km thick) when compared with crust formed at oceanic subduction zones (as much as 20 km thick) or when compared with continental crust (25 to 75 km thick).

The Oceanic Crust Observed by Deep Submersibles

Although ophiolite sequences and seismic studies of the ocean floor reveal much about the composition and structure of oceanic crust, natural cross sections have also been studied directly on the ocean floor in a very few places using submarines. One such area is the Vema Fracture Zone in the central Atlantic, a huge incision in the oceanic crust expressed as a cliff about 3 km high. In 1988 geologists in the French submersible *Nautile* observed and sampled a virtually complete section through the oceanic crust in the Vema Fracture Zone (**Figure 19.20**). Their views complement the observations made on land in ophiolites.

At the bottom of the cliff, the *Nautile* crew discovered peridotite from the upper mantle altered to serpentinite during prolonged contact with seawater. (**Serpentinite** is a metamorphic rock made of the hydrated mineral serpentine.) Continuing up the slopes and cliffs of the fracture zone, the *Nautile* traversed a section of gabbro at least 1000 m thick. Above the gabbro, the divers discovered a dike complex 1000 m high,

(A) Perhaps the best exposures of mantle material in the world form the mountains around Muscat, Oman. Because of the high iron content in these mafic rocks, weathering forms a thin layer of iron oxides ("rust") instead of a typical soil profile.

(B) Close-up view of the mantle peridotite in Oman shows a lineation and foliation resulting from plastic flow as the convecting mantle moves under the diverging plates. At the high temperatures found in the mantle, the minerals are stretched and are flattened as they slide past each other. Mineral layers are contorted and twisted as they flow. Geologists map these structures in the field to determine flow patterns in the mantle beneath the ridge.

Figure 19.19   Part of the upper mantle is included in some ophiolite complexes.

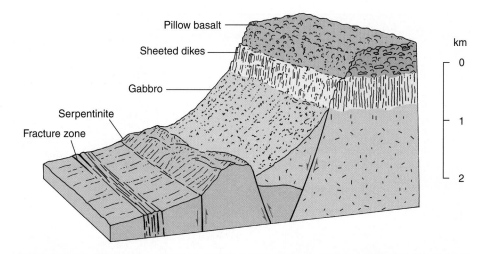

Pillow basalt
Sheeted dikes
Gabbro
Serpentinite
Fracture zone

km
0
1
2

Figure 19.20 **Geologic section along the Vema Fracture Zone** as recorded by the crew of the deep submersible *Nautile* shows a large section of the oceanic crust. Peridotite (altered to serpentinite) makes up the floor of the fracture zone. These are overlain by gabbros, sheeted dikes, and pillow basalts. This is the same sequence of rocks found in ophiolites.

made exclusively of vertical basaltic dikes intruding into each other. The last 800 m to the top of the fracture zone consisted of basaltic lava flows and many pillows.

The sequence of rocks in this seafloor exposure is nearly identical to that found in ophiolites and provides the best visual correlation between ophiolites and modern oceanic crust.

Geologic Studies of Iceland

Iceland is the best example of an oceanic ridge above sea level. Here geologists can examine in detail the surface expression of an active divergent boundary (although it is complicated by an underlying mantle plume). The island is a plateau of basalt, with a fissure-laced rift extending through the center (**Figure 19.21**). The Mid-Atlantic Ridge rises to the surface in the southwest and is offset toward the east across a diffuse volcanic and earthquake zone. Although some cinder cones and rhyolite calderas have developed, sheets of basalt extruded quietly from fissures are responsible for the largest volume of the island. Small basaltic shield volcanoes are also aligned along the rifts, showing that vertical fractures brought magma to the surface. The youngest rocks are found along the rift, with progressively older basalts occurring toward the east and west coasts (**Figure 19.22**). Of special importance are innumerable vertical basalt dikes exposed by erosion. The aggregate width of these vertical dikes is about 400 km. Because of the abundant magma in shallow chambers, groundwater is also heated to form geothermal systems and even geysers. Almost all of the faults that cut the island are normal faults. Shallow earthquakes are concentrated in the rift zone. The two halves of Iceland are separating by about 2 cm/yr.

To understand the nature of volcanism along divergent plate boundaries, let us examine the largest eruption ever recorded by humans, the Laki fissure eruption in 1783 (Figure 19.22B). Although the eruption occurred on land, it gives us important information about the nature and volume of eruptions at mid-ocean ridges.

The eruption was heralded by a series of earthquakes and the opening of several fissures 25 km long. A shallow graben formed between two of the fissures. The eruption began along part of the fissure zone, and after a day of purely explosive activity, lava reached the surface and poured out to form a huge flow that flooded a river valley. In a single day, the front of the lava flow advanced 15 km. The vent was marked by a row of lava fountains throwing red-hot molten basalt tens of meters into the air. Four more cycles of explosive activity switching to lava eruption occurred before the region

Figure 19.21 A geologic map of Iceland shows that the oldest volcanic rocks are along the eastern and western margins and the youngest rocks are near the center of the island. The youngest rocks are basaltic lava flows and dikes that lie in a zone of fissures and active volcanoes. This pattern shows that, as in the rest of the mid-oceanic ridge, new crust is created here as oceanic lithosphere spreads away from the center of the island.

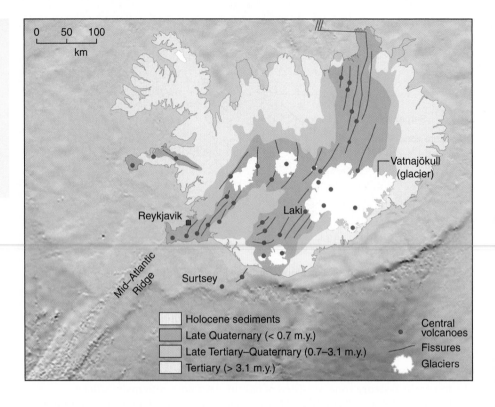

Holocene sediments
Late Quaternary (< 0.7 m.y.)
Late Tertiary–Quaternary (0.7–3.1 m.y.)
Tertiary (> 3.1 m.y.)

Central volcanoes
Fissures
Glaciers

(A) Old plateau basalts that are highly eroded by glaciers and streams are exposed in east and west Iceland. Here, on the flanks of the rift in eastern Iceland, the basaltic lavas are about 3 million years old.

(B) A fissure eruption at Laki, in Iceland's central rift zone, produced a flood of basaltic lava in 1783. This row of spatter cones marks the fissure vent for this, the largest historic eruption known.

Figure 19.22 Basaltic lava flows form the bedrock of Iceland. Old lavas are exposed on the flanks of the rift zone and young basaltic flows and other volcanic features are concentrated in the central rift zone.

became quiet again five months later. When the eruption was finished, 560 km² of land was buried under as much as 100 m of new lava, more than 100 new spatter and cinder cones had formed, and 12 km³ of magma had spilled onto Earth's surface.

The Laki eruption of 1783 was also a human catastrophe. A blanket of ash and vapors with traces of poisonous sulfur and fluorine were released from the magma. Consequently, about half the island's cattle and three-fourths of its sheep died of starvation or from fluorine poisoning. Because of these secondary effects, more than 20% of the island's human population died in a volcano-induced famine. The influence of the eruption was felt far from Iceland. The winter of 1783–1784 was colder in Europe than usual. Benjamin Franklin, then in Europe, accurately concluded that the fine ash and gases from the eruption had partially blocked the Sun's radiation and caused the cool weather.

Why is Iceland so important in understanding the mid-ocean ridges?

Visualizing the Oceanic Crust

The world beneath the sea, where oceanic crust forms, is an alien world completely unlike the surface on dry land. It may be difficult to visualize the geologic processes that occur at oceanic ridges and the nature of the rocks formed there. A significant difference in continental and oceanic environments is the direction in which rocks accumulate. Instead of widespread units of horizontal sedimentary strata stacked layer upon layer from vertical accretion, dikes intrude into dikes as they are injected from the chamber below. Thus, crustal accretion in the oceanic realm is lateral and the layers accumulate side by side. The contrast between the sheeted dikes and horizontal sedimentary rocks could not be greater. In a sequence of sedimentary rocks, time lines are nearly horizontal and crustal growth is vertical; in a body of sheeted dikes, time lines are vertical and growth is horizontal (**Figure 19.23**).

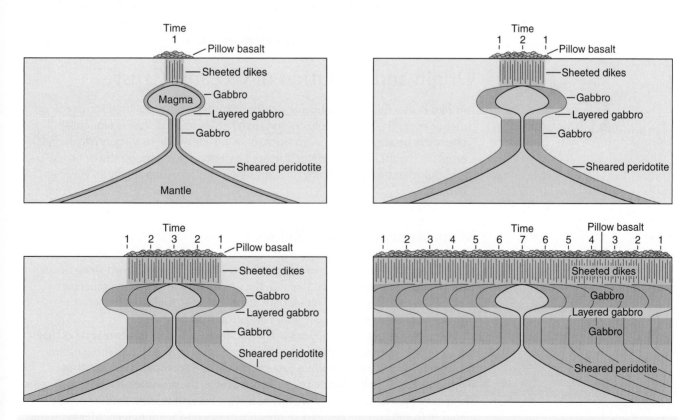

Figure 19.23 Time lines are vertical in oceanic crust, whereas time lines in continental sedimentary strata, such as those in the Grand Canyon, are horizontal. (Time lines are imaginary lines showing rocks of a given age.) These cross sections show the step-by-step construction of oceanic crust. The horizontal layering of the crust is not the result of simple superposition, but results from spreading during lateral growth of pillow basalts, sheeted dikes, and gabbro.

Figure 19.24 The oceanic crust compared with a profile of the Grand Canyon shows the magnitude of this sequence of rocks. If the oceanic crust were exposed in the Grand Canyon, the river would flow in the upper part of the sheeted dike complex, which would extend another 2 km below river level. A thick section of gabbros would underlie the entire column.

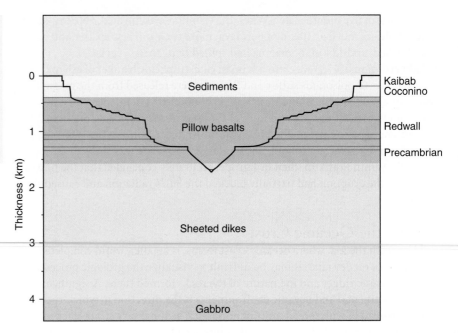

To visualize what the oceanic crust looks like, imagine a section of it compared with the Grand Canyon (**Figure 19.24**). A sequence of progressively younger horizontal sedimentary layers deposited upon a basement complex of metamorphic rocks is exposed in the real Grand Canyon. If oceanic crust were exposed, we would see thin layers of sediments extending from the canyon rim down to the base of the Coconino Formation. Lavas including pillow basalts would extend to the top of the Precambrian basement. The sheeted dikes would extend another 2000 m below the river. The thick unit of gabbro would extend another 4000 m below this.

Origin and Evolution of Oceanic Crust

At mid-ocean ridges, basaltic magma forms by decompression melting of rising mantle rock. The magma collects and then begins to crystallize in elongate chambers beneath the ridge. Some magma intrudes upward through dikes and erupts in the rift zone. Seawater is heated as it circulates through the hot crust and causes extensive hydrothermal alteration, metamorphosing large volumes of basalt.

Magmatism at Oceanic Ridges

Igneous activity is without question one of the most significant processes operating along divergent plate boundaries. Indeed, more igneous rock is formed along ocean ridges than in any other environment in the world. But why is magma generated at divergent plate boundaries rather than at other places? And what processes are involved in creating such large volumes of igneous rock?

The answers lie in what happens to the hot silicate minerals in the mantle as they rise by convection beneath the ridge. The general process is simple. Melting occurs because of a *decrease in pressure*, a process known as **decompression melting** (**Figure 19.25**). As solid mantle rises beneath a ridge, the pressure gets lower and lower. The mantle peridotite may cool slightly as it rises, but as it reaches shallow depths (about 30 to 100 km) the decrease in pressure causes it to begin to melt. Magma is thus generated beneath the ridges, and not in other places, largely because this is where mantle can rise to a zone of low pressure.

As the peridotite continues to rise, the amount of molten material increases and it becomes like slushy snow—a mixture of solid crystals and newly formed liquid. Eventually, the molten portion accumulates, first into small droplets, then merging

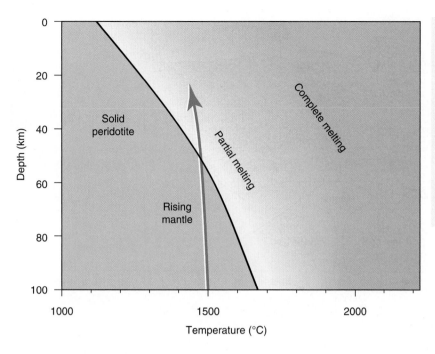

Figure 19.25 Magma forms by decompression melting under ocean ridges. The black line is the beginning-of-melting curve (the *solidus*) for mantle peridotite. The blue arrow shows the temperature-pressure path followed by mantle that rises directly below the oceanic ridge. When conditions in the upwelling mantle cross the beginning-of-melting curve, basaltic magma is produced. Melting probably occurs between 100 and 30 km deep. The melt can rise upward to form the basaltic crust of the ocean basins.

into larger and larger teardrop-shaped bodies of magma. The melt, being of lower density, then rises independently of the solid. The melt leaves behind a residue of olivine and pyroxene that becomes sheared to form tectonites (Figure 19.19) attached to the moving lithosphere. Beneath the fast-spreading East Pacific ridge, the zone of partially molten rock may be about 100 km wide. Beneath slow-spreading ridges, the zone of partial melting is much narrower or non-existent.

Much of the rising magma accumulates in a long linear chamber of molten basalt (**Figure 19.26**) directly below the ridge crest. Seismic studies suggest that the chambers are narrow, only 1 to 5 km across, but chambers filled with partially crystallized magma may be as wide as 10 km beneath fast-spreading ridges. The completely molten part is probably only several hundred meters to a kilometer thick. An active magma chamber may not be present at all times along the entire length of the ridge, especially along those that spread slowly. Periods of magma chamber development and volcanic eruption are interspersed with periods of stretching and faulting.

Because the roof of the axial magma chamber is stretched by plate divergence, vertical dikes grow upward to the floor of the ocean, removing magma from the chamber and forming a sheeted dike complex (Figures 19.16 and 19.26). Magma also erupts to form small shield volcanoes and sheets of fissure-fed lava that build a cover over the sheeted dikes and thicken the crust. Some basalt extruded onto the seafloor is quenched and forms bulbous piles of pillow lava. Most lava flows cool so rapidly in contact with seawater that they move less than 2 km before completely solidifying.

Oceanic crust develops as the magma in the axial magma chamber cools and crystallizes to form intrusive rock. The first minerals to crystallize from the basaltic magma are the dense minerals olivine and chromite, which sink to the base of the chamber and form layers (Figure 19.18)—very different from the deformed peridotite of the underlying mantle (Figure 19.19). With further cooling, crystals of pyroxene and plagioclase join olivine and chromite to form layered gabbro. The removal of these minerals from the magma (fractional crystallization) causes the residual melt to change composition. Meanwhile, new batches of magma intrude into the chamber and mix. The upper part of this magma body solidifies to form massive gabbro.

A remarkable characteristic of divergent plate margins is that all of this activity goes on in an extremely small area centered directly on the ridge. The zone where new oceanic crust is formed is only about 10 km wide and 10 km deep. Nevertheless, it extends thousands of kilometers along the mid-oceanic ridges and is responsible for generation of the entire oceanic crust in little more than 200 million years.

How is magma generated at a mid-oceanic ridge?

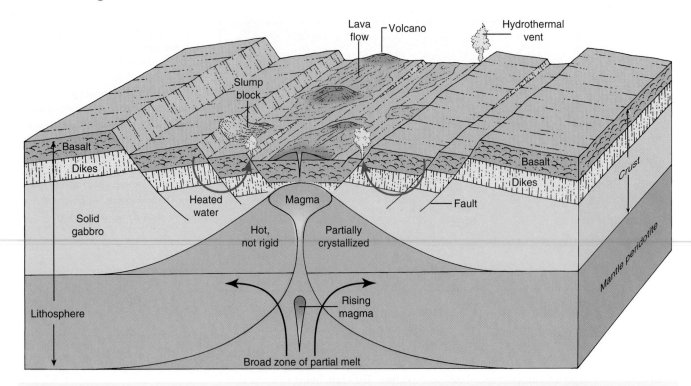

Figure 19.26 **An idealized cross section of a mid-ocean ridge** shows that hot mantle rises and then moves laterally. As it rises, pressure decreases and partial melting occurs. The less-dense magma rises buoyantly and collects in a chamber. Heat is lost by conduction and by convection of cold seawater through the hot, permeable crust. The magma consequently cools and crystallizes along the floor and walls of the chamber to form gabbro. As the roof is stretched by plate divergence, sheeted dikes propagate to the surface. The magma erupts to form pillow basalts that add to the roof of the chamber and thicken the crust. Small shields and fissure-fed flows cap the volcanic system. The hydrothermal fluids flow through small vents along the fissure systems to form submarine hot springs that cool as they mix with the surrounding seawater. Sulfides and other minerals dissolved in the fluid crystallize as the fluid cools and changes composition, and mounds of these minerals form.

Seafloor Metamorphism

The great expanses of gneiss, schist, and slate exposed in Earth's continental shields dramatically show the effects of metamorphism. Most of these were formed at convergent plate margins in the roots of ancient mountains, where dramatic changes in temperature and pressure were the agents of change. Metamorphism on the continents is typically accompanied by strong horizontal compression and structural deformation; the metamorphic rocks are folded and foliated.

But there is another type of metamorphism operating at divergent plate boundaries, the results of which we only rarely see on the continents—**seafloor metamorphism**. It is metamorphism in which the *chemical action of fluids* is the major agent of change. Metamorphism along the crest of the mid-oceanic ridges is caused by the circulation of seawater heated by igneous rocks. At the oceanic ridges, the essential tectonic process is extension that opens narrow fissures. As seawater circulates through the cracks in the hot volcanic rocks, it is heated to 400°C to 450°C. This hydrothermal water attempts to equilibrate with the crust and reacts with unstable olivine, pyroxene, and plagioclase to form the new minerals that are stable under these conditions—chlorite, epidote, sodium-rich plagioclase, talc, and serpentine—typical of greenschist facies metamorphism. The metamorphic rock is called **metabasalt** and it may be the most abundant kind of metamorphic rock exposed near Earth's surface.

The process that alters the igneous rock along ridge axes, even as they form, is a type of **hydrothermal alteration**. The circulation of hot water through the entire ridge system is a fundamental phenomenon. Lava flows are extremely permeable to the flow of water. Moreover, tectonic stretching and faulting occur at the ridge. Fissures are opened through which seawater can percolate downward, penetrating the crust to depths of 2 to 3 km. Seawater may reach the very base of the dike complex. Deeper penetration of circulating fluids is more difficult owing to the massive nature of the underlying gabbro. Locally, water penetrates deep enough to react with mantle peridotite and hydrate it to form serpentine. This metamorphic rock is neither dense nor coarsely crystalline; it is a light, weak rock capable of rising isostatically and penetrating the overlying materials like an igneous intrusion.

During the circulation, the hot water not only alters the rock, but it also dissolves minerals from the oceanic crust. Once the water is hot enough, it becomes buoyant, ascends, and is channeled toward central vents on the seafloor. Some of the dissolved material is precipitated in veins just below the seafloor. The hot water jets through springs and vents on the seafloor to form variously colored mineral-laden plumes several meters high, called **black** or **white smokers.** Copper, zinc, and lead sulfides, as well as other minerals, precipitate from the hot water as it meets the cold oxygen-rich seawater. Chimney like mounds up to 10 m high form as these minerals accumulate (Figure 19.9 and **Figure 19.27**). In fact, many of Earth's important ore deposits formed when seawater reacted with hot volcanic rocks on the ocean floor.

Hydrothermal circulation through the global ridge system is not trivial. The total amount of water circulating through the oceanic crust each year is equivalent to 2% of the annual discharge of all rivers on the planet. Hydrothermal circulation is thus a major element of hydrologic circulation. The system is large enough to recycle the entire volume of the oceans through the oceanic crust every 5 to 10 million years. This circulation has significant effects on the composition of seawater and indirectly on the composition of the atmosphere, which is in equilibrium with the circulating ocean.

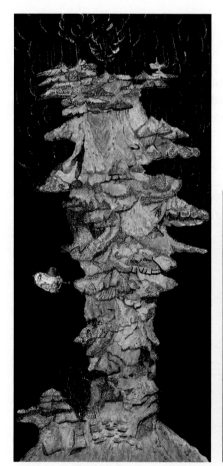

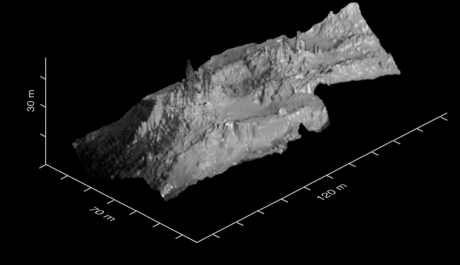

Figure 19.27 **Hot springs** form on the seafloor at mid-ocean ridges. Large mounds are created when hot fluids vent from the seafloor and react with cold seawater. This image (above) is of the Endeavour hydrothermal vent field on the Juan de Fuca ridge offshore from the state of Washington. Sulfide minerals precipitate and build up the irregular mounds and complex chimneys. Some chimneys, like this one called Godzilla (left), are as tall as skyscrapers. Note the submarine *Alvin*, for scale.

Courtesy of J. Delaney, V. Robigou, and D. Stakes, 1993. The high–rise hydrothermal vent field, Endeavor Segment, Juan de Fuca Ridge, *Geophysical Research Letters,* vol. 20, No. 17, pp. 1887–1890.

Structural Deformation and the Origin of Abyssal Hills

Abyssal hills such as those in Figure 19.5 may seem unspectacular when compared with mountain ranges, island arcs, and deep-sea trenches, but they are singularly important because they form the most abundant landforms on Earth, covering more than 30% of the ocean floor.

New high-resolution topographic data, combined with photography from deep submersibles of abyssal hills in the Pacific, shed new light on their origin. Abyssal hills are long, low ridges typically 10 to 20 km long, 2 to 5 km wide, and 50 to 300 m high (**Figure 19.28**). They are parallel to the ridge crest. Many hills are asymmetric, with steep slopes facing the ridge axis. The orientation and shape strongly suggest that the basic structures of the hills are normal faults.

Abyssal hills begin to develop when normal faulting is initiated near the crest of the ridge as the plates move away from the axis (Figure 19.28). These faults create short ridges and valleys that are really tilted fault blocks or horsts and grabens. With continued displacement, the faults grow more numerous and longer. Thus, short grabens deepen, widen, and merge with one another even while the plate moves slowly away from the ridge. The abyssal hills are the ridges between the troughs.

Apparently, abyssal hills form near the ridge crest and a short distance down the ridge flank where the crust is extending. Beyond the ridge flank, active faulting and volcanism cease, and the hills become progressively covered with sediment. Near the continents, sediment completely covers the abyssal hills to form the extensive flat **abyssal plains**.

Concluding Note

In many ways, studies of mid-ocean ridges represent a remarkable piece of scientific detective work. They give us a clear picture of a part of Earth that is extremely important but obscured and nearly as inaccessible as another planet. It is important to understand the contrasts between oceanic and continental crust thus revealed. First, much of the oceanic crust and its topographic features are related in some way to igneous activity. Moreover, the rocks of the ocean crust have been deformed by extension, not horizontal compression, so their structure contrasts markedly with the complex folds in mountains and shields of the continents. It cannot be overemphasized that the rocks of the ocean crust are all young compared with most continental crust. All oceanic rocks appear to be less than 200 million years old, whereas the great bulk of continental rocks—the ancient rocks of the shields—are more than 700 million years old.

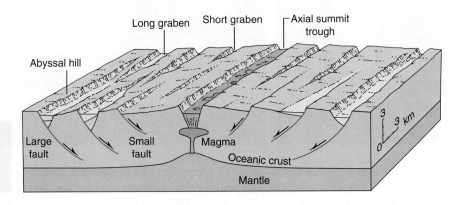

Figure 19.28 Abyssal hills form at the oceanic ridge by a combination of faulting and volcanic processes. These hills are the dominant landforms on the ocean floor.

Continental Rifts

Continental rifting occurs when divergent plate margins develop in continents. Continental rifts are typified by thin crust normal faults, shallow earthquakes, and basalt and rhyolite magmatism. Continued rifting creates new continental margins, marked by normal faults and volcanic rocks interlayered with thick sequences of continental sedimentary rocks. As the margin cools and subsides, it is overlain by a thick layer of shallow-marine sediment.

Oceanic and continental rifts are quite similar, but they also display significant differences. Continental crust is thicker and less dense than oceanic crust and is structurally much more complex. Continents have a different composition than oceanic crust, being much richer in silica. In addition, rifting of a continent is initiated on dry land, so stream erosion and sedimentation play important roles in the evolution of continental rift zones. Thus, compared to seafloor spreading, continental rifting involves different materials and different events, and it produces its own distinctive structures, landforms, and sequences of rocks.

A **continental rift** is a major elongate depression bounded by normal faults, where the entire lithosphere is deformed. It is a region where the crust has been arched upward, extended, and pulled apart and may develop into a divergent plate margin. The dominant structure is a system of parallel normal faults with large vertical displacements. Faulting produces large elongate down-dropped blocks (grabens or rift valleys) and associated uplifted blocks (horsts). Like their oceanic counterparts, continental rifts are commonly associated with volcanic rocks. The magmas are mostly basaltic, with lesser rhyolite that distinguishes them from oceanic rifts. Magmas of intermediate composition (andesite) are less common than either basalt or rhyolite at most continental rifts.

Continental rifting is not just an academic interest. The **passive continental margins** produced along mature rifts are the largest storehouses of sediment on Earth. As a result, they contain about two-thirds of the world's giant oil fields and hold more than half the world's oil reserves.

The Basin and Range Province, the East African Rift valleys, the Red Sea Rift, and the margins of the Atlantic Ocean are exceptional examples of continental rifting in various stages of development. From these examples, we can gain important insight into the characteristics of continental rifts and the processes involved in their evolution.

The Basin and Range Province

The Basin and Range Province is a large area of western North America where the crust has been uplifted and pulled apart, forming a complex rift system that extends from northern Mexico into Canada (**Figure 19.29**). Normal faulting, resulting from extension, has produced alternating mountain ranges and intervening fault-bounded basins.

There are more than 150 separate mountain ranges in this province. Some are simple tilted fault blocks with the steeper side of the range marking the side along which faulting occurred. Others are horsts with normal faults on both sides. Estimates of vertical displacement on individual faults are as much as 8 to 10 km. The upthrown blocks may be as much as 3000 m above the valley floors and are considerably dissected by erosion. Earthquakes are common along the eastern and western boundary of the Basin and Range province. The total horizontal stretching across the Basin and Range may be more than 300 km.

Structural evidence shows that the Basin and Range is an area where the continental crust has been uplifted and stretched to as much as twice its original width.

Where are the major divergent plate boundaries on the continents?

What is the dominant structure of the Basin and Range province?

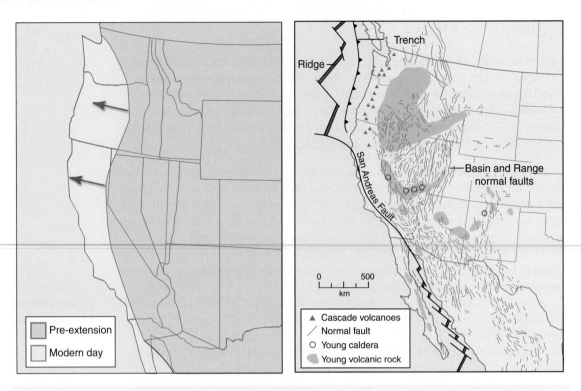

Figure 19.29 **The Basin and Range Province of western North America** extends from Mexico northward into Canada. Partial rifting of the continent has greatly extended the region since about 20 million years ago at a rate of about 1 to 5 cm/yr. Extension created normal faults. The region also has thin crust, high heat flow, and eruptions of basalt and rhyolite.

Geophysical evidence supports this conclusion. Most earthquakes are concentrated along the boundary faults in Nevada and Utah. Heat flow is as much as three times greater than normal. Seismic refraction reveals that the crust is only 25 km thick under much of the region, whereas only 25 million years ago, it may have been as much as 50 km thick. Moreover, the crust overlies a mantle with a low seismic wave velocity. There seems to be no evidence for a high-velocity lithospheric mantle directly below the continental crust. In other words, in this region, the hot asthenosphere may be in contact with the base of the continental crust. Hot springs are a natural manifestation of the high heat flow and thin crust.

Thus, the topography, structure, and geophysical evidence all show that the Basin and Range is undergoing extension that began about 20 million years ago. The overall extension rate is 1 to 5 cm/yr, compared with 2 to 20 cm/yr for mid-ocean ridges.

A distinctive sequence of sedimentary rocks forms in continental rift systems because of faulting of the crust. The sharp local relief created by normal faulting generates vigorous stream erosion. In a dry climate, such as the western United States, thick deposits of sand and gravel accumulate at the base of the mountain ranges as alluvial fans. Large rock avalanche deposits are also common next to the ranges. The sediments of the alluvial fans grade into finer sand and silt and interfinger with lake sediments that accumulate in the lower parts of the basins. In the arid Basin and Range province, the lakes may evaporate to form dry lake beds or playa mud. Thick deposits of gypsum, salt, and other evaporite minerals are commonly formed in the lakes.

Volcanic rocks, mostly basalt and rhyolite, occur both along the Basin and Range margins and near the center (Figure 19.29). Small flows of basalt are common in these localities. Locally, large calderas formed, caused by repeated eruption of rhyolite ash-flow tuff. Prominent examples include the calderas of southern Nevada and the still-active Long Valley caldera near Mammoth Mountain in eastern California.

The East African Rift

The East African Rift extends from Ethiopia to Mozambique, a distance of nearly 3000 km (**Figure 19.30**). At the north end of the rift, activity began about 30 million years ago with the eruption of flood basalts in a brief 1-million-year-long episode that formed the huge volcanic plateau. This was followed by development of several large shield volcanoes. Uplift of the area accompanied the earlier volcanism. Later the plateau was broken by extension and normal faulting and volcanism became nearly restricted to the rift itself. Today the region is a large uparched segment of the crust in which long linear blocks have faulted down near the crest. Shallow earthquakes are common along the rift and its branches to the south (Figure 19.10). Many rift valleys are closed depressions and have partly filled with water to form the large freshwater lakes of East Africa. Where evaporation is great, smaller lakes have become saline, forming significant salt deposits.

This part of Africa is underlain by thin crust and thin lithosphere (**Figure 19.31**). Recent volcanic activity is also associated with the rifting and includes such well-known volcanoes as Mount Kenya, Mount Kilimanjaro, and Nyiragongo (**Figure 19.32**). Structural and geophysical data show that the region is extending at a rate of about 0.5 cm/yr, with the total extension being slightly less than 50 km. The

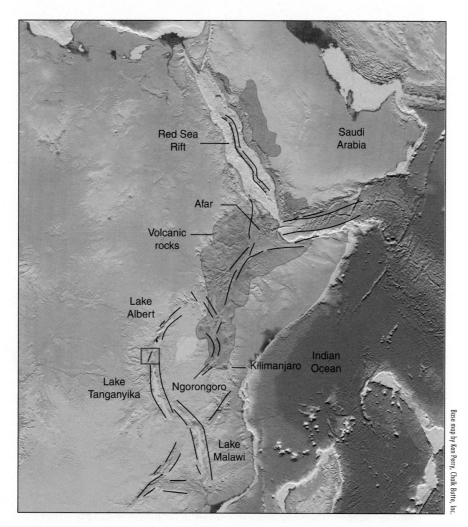

Base map by Ken Perry, Chalk Butte, Inc.

Figure 19.30 The East African Rift valleys are where the continent is being uparched and pulled apart. The black lines are the traces of normal faults. If the spreading continues, the rift system may evolve into an elongate sea like the Red Sea to the north. Volcanism started about 30 million years ago with the eruption of a 2 km thick series of flood basalts. Smaller volumes of basalt and rhyolite volcanism accompanied subsequent rifting. The blue box shows the location of Figure 19.32.

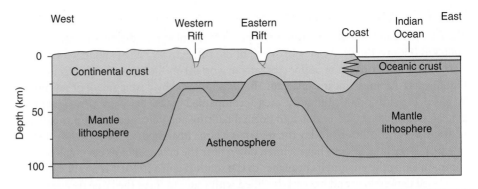

Figure 19.31 The thinning of the continental crust beneath the African Rift valleys is indicated by gravity measurements as shown in this cross section. Beneath the valleys, the top of the asthenosphere is near the base of the crust, only 25 km below the surface. The East African Rift valleys represent the first stage of continental rifting.

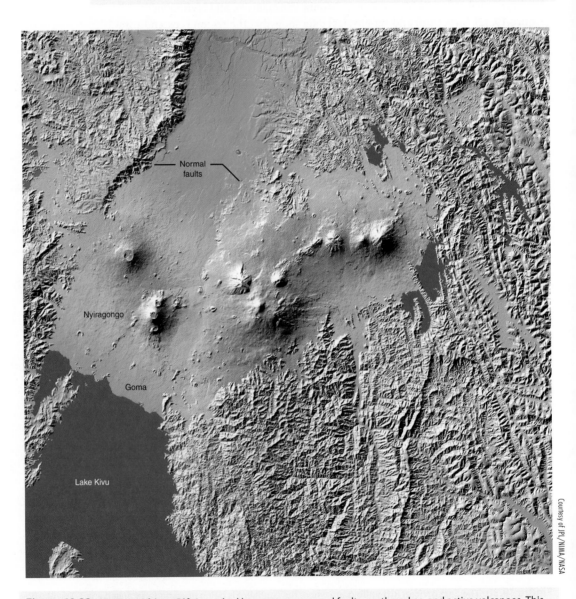

Figure 19.32 The East African Rift is marked by numerous normal faults, earthquakes, and active volcanoes. This map shows part of the western branch of the rift at the junction of Congo, Rwanda, and Uganda. The main graben is partially filled by Lake Kivu and sediment and lava covered plains. The flanks of the rift are eroded by stream valleys. Large basaltic volcanoes and small cinder cones lie on the floor of the rift and on the eastern flank. Some of the volcanoes have large calderas formed by collapse of their summits. In 2002, Nyiragongo erupted fluid lava flows that flooded the town of Goma and displaced 500,000 people.

northern end of the rift, in the Afar region, has a very thin crust (about 8 km). Here, rifting of the continental crust is nearly complete, and basalt volcanoes are abundant.

The sequence of rocks generated in the East African Rift valley is similar to that in the Basin and Range. Thick conglomerates and associated alluvial fan deposits occur near the margins of the valleys and grade into stream and lake sediments and evaporites near the center of the basins. Volcanic rocks, including lava flows and ash falls, occur both along the flanks and within the rift valleys. This characteristic rock assemblage stands out in sharp contrast to the sequence of rocks generated at the oceanic ridges, despite both being formed at divergent plate margins.

The Red Sea Rift

The Red Sea, which separates Africa from Arabia, is an important part of Earth's rift system (**Figure 19.33**). It is an extension of the divergent plate boundary in the Indian Ocean (Figure 19.1). At the northern end of the Red Sea, the rift continues to form the Gulf of Suez, and a major transform fault extends up the Gulf of Aqaba into the Dead Sea and Jordan Valley. The Red Sea Rift is about 3000 km long and 100 to 300 km wide (Figure 19.33). The margins of the Red Sea are steep normal fault scarps, as much as 3 km high, with steplike fault blocks descending abruptly down to the coast.

Much of the Red Sea is floored by continental crust that is thinner than that underlying the flanking land masses. The crust has been thinned by normal faulting and extension (**Figure 19.34**). Only the south-central part of the Red Sea reaches the abyssal depth of a deep ocean basin and is underlain by oceanic-type crust. Evaporite deposits, mostly salt interbedded with clastic sediments, cover the block-faulted continental shelf with a layer nearly 1 km thick.

Igneous rocks associated with the Red Sea Rift are mostly basalts—dikes, sills, and lava flows associated with fissures parallel to the rift (**Figure 19.35**). Individual eruptions form small shield volcanoes and cinder cones with associated lava fields. Volcanism and faulting commenced about 25 million years ago.

The small amount of oceanic crust in the deep central part of the Red Sea Rift has a pattern of symmetrical magnetic anomalies typical of the mid-ocean ridge. The magnetic patterns indicate that oceanic crust has been generated for the past 5 million years. The Red Sea is therefore an outstanding example of the initial stages of the formation of an ocean basin: the logical next step from the Basin and Range and East African Rift valleys.

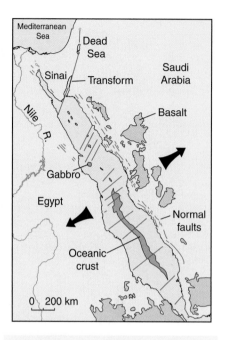

Figure 19.33 The Red Sea is a narrow ocean basin separating Arabia from Africa. Its margins are steep fault scarps, but much of the Red Sea is floored by thin continental crust. However, a narrow zone of oceanic crust (purple) extends along the Red Sea axis through most of its length. The Red Sea represents the second stage of continental rifting, in which an embryonic ocean develops.

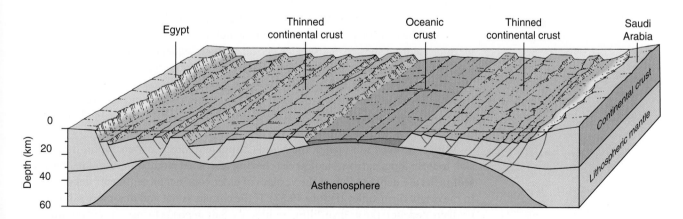

Figure 19.34 A cross section of the Red Sea illustrates the major structural elements of this stage of rifting. Continental crust is thinned by movement along a series of curved normal faults. The thinned continental crust is overlain by a salt layer up to 1 km thick. New oceanic crust occupies the central part of the rift.

Figure 19.35 Swarms of basaltic dikes along the Arabian Peninsula parallel the shore of the Red Sea. They were injected into the continental crust during the early stages of rifting.

Evolution of Continental Rifts to Passive Margins

On the basis of the examples described above, we can understand the various stages in a progression from a continental rift to a new passive margin (**Figure 19.36**). The initiating event in continental rifting is believed to be an upwarp or dome in a continent. As the lithosphere expands, it arches and thins the crust, fracturing the brittle upper part.

Extension and thinning of the continental crust ultimately create a fault-bounded rift valley. The complexly faulted edges of the rift zone gradually become a new passive continental margin. In an arid climate, the fault blocks erode to form alluvial fans, interlayered with lake deposits. Thick evaporite deposits may accumulate in playa lakes. As the crust thins, the mantle rises and decompression melting may occur. Basaltic eruptions follow. Rhyolitic magma may be produced by partial melting of the granitic crust by heat from the basalt or by differentiation of basaltic magma. These volcanic rocks are interlayered with the sediments within the rift basins.

When the continent moves away from the hot, uparched spreading ridge, the rifted margins begin to subside. Subsidence permits a thick sequence of sediment to accumulate in shallow-oceans at the new passive continental margin. Gradual subsidence of the continental margins occurs for two reasons: (1) As the lithosphere moves off from the hot rising mantle, it cools, contracts, becomes denser, and subsides isostatically; and (2) the weight of the newly deposited sediment on the continental margin causes the crust to be depressed.

With continued subsidence, the oceans ultimately enter the topographic depression to form long, narrow, shallow seas having restricted circulation. Shallow-marine sediments are deposited upon the vestiges of older continental sediments deposited in the original rift valley (**Figure 19.37**). If the climate is hot and dry, more salt may be deposited atop the graben-filling sediments. Salt deposits formed by evaporating

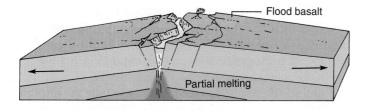

(A) Continental rifting begins when the crust is uparched and stretched, so that normal faults (red) develop. Continental sediment (yellow) accumulates in the depressions of the downfaulted blocks, and basaltic magma is injected into the rift system. Flood basalt (gray) can be extruded over large areas of the rift zone during this phase.

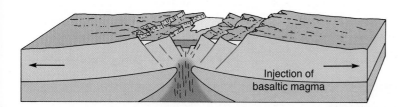

(B) Rifting continues, and the continents separate enough for a narrow arm of the ocean to invade the rift zone. The injection of basaltic magma continues and begins to develop new oceanic crust (green).

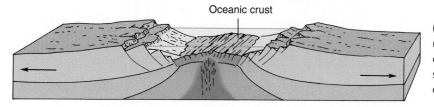

(C) As the continents separate, new oceanic crust and new lithosphere are formed in the rift zone, and the ocean basin becomes wider. Remnants of continental sediment can be preserved in the down-dropped blocks of the new continental margins.

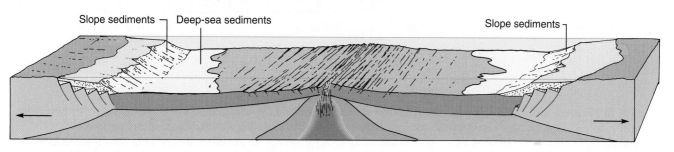

(D) Eventually a wide ocean basin opens. As the rifted margins cool and subside below sea level, a thick wedge of sediment (yellow) accumulates.

Figure 19.36 Stages of continental rifting are shown in this series of diagrams. The major geologic processes at divergent plate boundaries are tensional stress, block faulting, and basaltic volcanism.

seawater in the embryonic rifts may accumulate to more than 1 km thickness, such as those formed recently in the Red Sea. (The salt may subsequently be mobilized under isostatic pressure and rise as salt domes through the overlying strata.)

Ultimately, the rift widens enough to permit open circulation of marine water. Such narrow seas are commonly fertile ground for marine life. In tropical climates, organic reefs may flourish on the edges of the fault blocks, with associated shallow lagoons and beaches (Figure 19.37). Organic matter accumulates in shallow-marine sediments and may lead to petroleum deposits. As the margin subsides even more, large river systems are refocused to flow toward the shores of the new ocean. They bring sediment in large volumes to bury the reefs and their associated sediment under thick layers of clean, well-sorted sand, silt, mud, and limestone.

In the deeper water off the continental margins, the sequence of sediment is distinctly different. It consists of poorly sorted sandstone and shale deposited by turbidity currents (turbidites), submarine slump blocks, and rock debris from submarine landslides. As shown in Figure 19.37, turbidites grade seaward into deep-marine organic oozes that cover the basalts of the oceanic crust.

Eventually, the thinned continental crust drifts even farther away from the rift, and new oceanic crust forms in the rift zone and continues to evolve following the pattern of typical oceanic ridges.

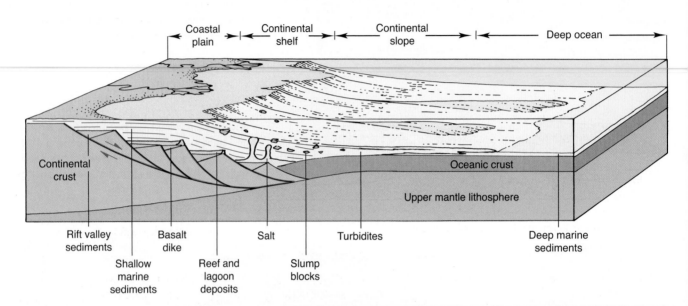

Figure 19.37 **A passive continental margin** shows features formed during rifting. Tilted fault blocks that formed during initial rifting define the margins of continental crust. (Faults in blue.) Continental sedimentary deposits consisting of alluvial fan conglomerate and playa lake evaporites may be preserved in narrow grabens. As the continent subsides, reefs and associated beach and lagoon sediments are deposited, and eventually the entire margin is covered by a thick accumulation of shallow-marine sediment that grades into deep-marine sediment. Poorly sorted dirty sandstone and shale are deposited by turbidity currents in the deep water.

STATE OF THE ART   Gravity Variations Reveal Hidden Continental Rift

Buried deep beneath the marine limestone layers and glacial till of Iowa, geologists have found evidence of a giant continental rift that is just over 1 billion years old. Highlighted in red on the map, the rift marks an aborted effort to rip North America apart. The tear is about 100 km wide and 2200 km long. It stretches from northern Kansas across parts of Nebraska, Iowa, Minnesota, Wisconsin and Michigan.

And yet, the *Midcontinent rift* is absolutely invisible through most of its length. For example, a topographic map of central Iowa reveals no elongate trough nor high mountainous flanks, only flat fields of corn and gently rolling hills carved by streams. Even careful geologic mapping shows no normal faults bounding a graben, no basaltic lava flows or intrusive gabbro, no rhyolite ash-flow tuff, no conglomerates or other sediment filling a depression. All these features were long since buried by layers of sedimentary rock that are now several kilometers thick. How then is the map showing this vast rift constructed?

The answer is to use a sensitive instrument called a *gravimeter*. Gravimeters are used to measure small variations in the strength of Earth's gravity. Most gravimeters work by measuring how much a small spring is deformed—the longer the spring stretches, the stronger the gravitational acceleration is at that spot. Like many other geophysical surveys, most gravity surveys are conducted on the ground by making repeated measurements at many different locations. The gravity at each spot on the surface varies by a minute amount from each adjacent spot because of differences in the density of the column of underlying rock. A gravity survey is an indirect way to map density (g/cm³) variations in the crust. Dense rocks near the surface create a strong gravity field; low-density rocks create a weaker gravity field.

The map of the Midcontinent shown here does not show elevation or the distribution of rock types. The different colors represent the strength of the gravity field measured (in milligals) across the region. A buried continental rift is revealed by a high "ridge" in the gravity field. A long, narrow strip of dense rocks must lie underneath the low-density sedimentary rocks at the surface. Models of the gravity field suggest a fault-bounded trough was filled with dense igneous rocks, such as basalt and gabbro, as much as 7 km thick. In places, these rift rocks are buried beneath a kilometer of sedimentary rock. However, the rift and its dense fill are actually exposed at the surface in Michigan's Keeweenawan peninsula.

Gravity maps like these are very useful for showing the structure of Earth's crust at depth and, as you can see, help us see below deposits of soil, glacial drift, sedimentary rock, or thick vegetation. They can also be used to find faults that otherwise might be undetected—until they rupture in massive earthquakes. Gravity maps are also used in the exploration for ore deposits and petroleum. In fact, following the discovery of the Midcontinent gravity high, the exploration efforts of a few oil companies proved the rift exists beneath central Iowa. Cores and chips brought to the surface from these deep holes were Precambrian rocks—dense basaltic lava flows and gabbro intrusions—formed in the ancient rift. Indeed, what your eyes see at the surface is not always the whole geologic story.

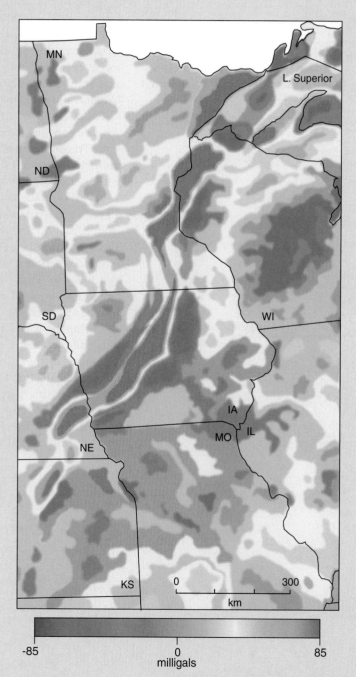

Plate Movement During the Last 200 Million Years

The considerable amount of data on plate motion enables us to trace the development of divergent plate margins during the last 200 million years of Earth history. A large continental mass (Wegener's Pangaea) rifted apart and large ocean basins formed. Subduction accommodated the growth of new ocean basins.

The tectonic system probably has operated during much of Earth's history, and it is responsible for the growth and destruction of ocean basins. Ocean basins come and go because the ancient oceanic crust is consumed at subduction zones and replaced by newer oceanic crust, created at ridges. Continents have rifted apart, drifted with tectonic plates, and rejoined a number of times, but details of the patterns of ancient plate movements are scanty.

However, the considerable amount of data on plate motion during the last 200 million years enables us to reconstruct the position of continents and to trace plate movement with some certainty (**Figure 19.38**). They indicate that a large continental mass (Wegener's Pangaea) began to break up and drift apart about 200 million years ago.

The earliest event in the splitting of Pangaea was the extrusion of large volumes of basalt along the initial continental rift zones. Remnants of these basalts are found in the Triassic basins of the eastern United States and the flood basalts of southwestern Africa, western India, and eastern Brazil. A northern rift split Pangaea along an east-west line, slightly north of the equator, and separated the northern continents (Laurasia) from the southern continents (Gondwanaland). A southern rift split South America and Africa away from the rest of Gondwanaland. Soon afterward, India was severed from Antarctica and moved rapidly northward. The plate containing Africa converged toward Eurasia, forming an east-west subduction zone. By the end of the Cretaceous period, 65 million years ago, the South Atlantic Ocean had widened to at least 3000 km. All of the major continents were blocked out by this time, except for the connection between Greenland and Europe and between Australia and Antarctica. A new rift separated Madagascar from Africa, and India continued moving northward.

During the last 65 million years, the Mid-Atlantic Ridge extended into the Arctic and finally detached Greenland from Europe. During that time, the two Americas were joined by the Isthmus of Panama, which was created by tectonism and volcanism along the subduction zone. The Indian landmass completed its northward movement and collided with Asia, creating the Himalayas. A new divergent boundary developed as Australia rifted away from Antarctica. Finally, a branch of the Indian rift system split Arabia away from Africa, creating the Gulf of Aden and the Red Sea. Another arm of the rift created the East African rift valleys.

What will be the most significant changes in the next 50 million years if the present pattern of tectonism continues?

Figure 19.38 The history of plate movement during the last 200 million years has been reconstructed from all available geologic and geophysical data. These maps show the general direction of drift from the time Pangaea began to break up until the continents moved to their present positions.

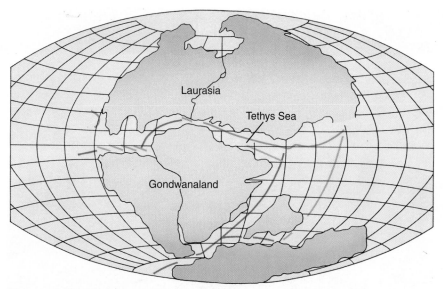

(A) Pangaea, 200 million years ago. Note the Tethys Sea between Laurasia (to the north) and Gondwanaland (to the south).

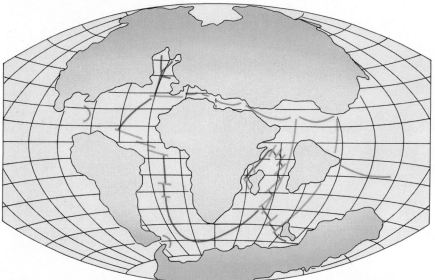

(B) Plate movement, 100 to 50 million years ago. The Atlantic Ocean is formed as North and South America drift westward. The Tethys Sea is nearly closed.

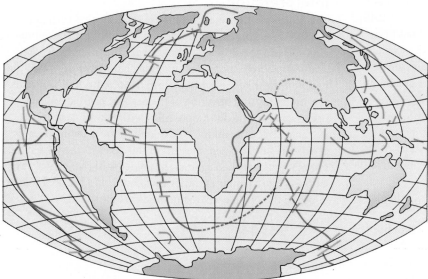

(C) Plate movement, 50 million years ago to the present.

GeoLogic   Sheeted Dikes of the Oman Ophiolite

Geologists trained on the stable platform of a continent are used to seeing vast sheets of nearly horizontal layers of sedimentary rocks stacked one upon another in what has been called layer-cake geology. Imagine the geologic consternation that might possess these same geologists when confronted by this stark desert outcrop of "layers" near Wade Hawasina in Oman.

Observations

1. At first glance, the sheeted dikes resemble a series of deformed, almost vertical, sedimentary beds. But here the "strata" are not layers of sandstone and shale.
2. Instead, they are vertical sheets made of fine-grained plagioclase, pyroxene, and olivine.
3. Most sheets are about a meter thick and have thin borders of very fine-grained rock.
4. More extensive field investigation shows that the sheets grade upward into pillow lavas and downward into gabbro plutons.

Interpretations

The geologic interpretation of these sheets is that each is a solidified dike of basalt that formed below an ancient submarine volcano composed of pillow lavas. The underlying coarse-grained gabbro is an ancient magma chamber from which the basalt in the dikes and flows was extracted. Each dike reveals the stretching and breaking of the oceanic crust at an ancient oceanic ridge. The dikes were filled with new magma freshly extracted from the mantle and added to the growing oceanic crust. The fine-grained margins of the dikes formed when the hot magma quenched against the colder wall rocks. In a way, a dike is like a planar chimney that connects a hot chamber at depth to the surface where hot fluids are extruded.

Key Terms

abyssal hill (p. 590)
abyssal plain (p. 590)
black smoker (p. 589)
continental rift (p. 591)
decompression melting
    (p. 586)
fissures (p. 571)
fracture zone (p. 568)
gravity anomaly (p. 576)

half-dike (p. 579)
heat flow (p. 575)
hydrothermal alteration
    (p. 588)
layered gabbro (p. 580)
magnetic anomaly (p. 575)
massive gabbro (p. 580)
metabasalt (p. 588)
mid-ocean ridge (p. 566)

ophiolite (p. 577)
passive continental margin
    (p. 591)
pillow basalt (p. 571)
rift valley (p. 568)
seafloor metamorphism (p. 588)
serpentinite (p. 581)
shallow-focus earthquake
    (p. 574)

sheeted dike complex (p. 578)
spreading rate (p. 568)
subsidence (p. 571)
tectonite (p. 581)
transform fault (p. 568)
white smoker (p. 589)

Review Questions

1. What major processes occur at all divergent plate boundaries?
2. Is extension or horizontal compression more common at divergent plate boundaries?
3. Why are divergent plate boundaries in the ocean basins marked by a broad rise?
4. Explain why divergent plate boundaries commonly have shallow earthquakes but lack deep earthquakes.
5. What kinds of volcanic activity are common at divergent plate boundaries?
6. Draw profiles showing the magnetic, gravity, and heat flow anomalies across a typical oceanic ridge. Explain the underlying causes of these anomaly patterns.
7. Compare and contrast a fast- and a slow-spreading ridge, and give examples of each type.
8. What is the significance of the fissure eruptions in Iceland? What structures underlie these fissure-fed lava flows?
9. Describe the typical internal structure of oceanic crust.
10. What is the role of metamorphism in the development of oceanic crust? Describe a typical ocean-floor hydrothermal system.

11. Explain the origin of abyssal hills.
12. How is basaltic magma generated at divergent plate boundaries?
13. Where is rhyolite more common, at an oceanic or a continental rift?
14. Draw a series of cross sections that outline the stages in the development of a continental rift that evolves into an ocean basin.
15. Describe a vertical sequence of sediments and rocks that might be encountered upon drilling into a continental rift basin such as that in East Africa.
16. Why do thick sequences of sedimentary rock accumulate along rifted continental margins?
17. What causes the margins of continental rifts that were once high to eventually subside below sea level?
18. Why are continental rifts developed on high bulges in the crust?
19. Which formed earlier, the North or the South Atlantic Ocean?

20 Transform Plate Boundaries

The remarkably straight valley of the San Andreas fault system of northern California, shown in the panorama above, is one of the great surface expressions of plate tectonics. The long linear valley marks faults that are part of a plate boundary that cuts through the entire lithosphere to a depth of about 100 km. The valley and spit protected Bolinas lagoon are not merely surface features. Over the last 30 million years, horizontal movements along the strike-slip faults of the San Andreas system amount to several hundred kilometers. Rocks on either side of the fault are on fundamentally different plates. The deformation is produced as the North American plate (in the upper part of the image) slips south past a narrow slice of continental crust trapped on the Pacific plate (lower). Carefully examine this panorama, because from it you can begin to develop a more accurate concept of what Earth's tectonic system does.

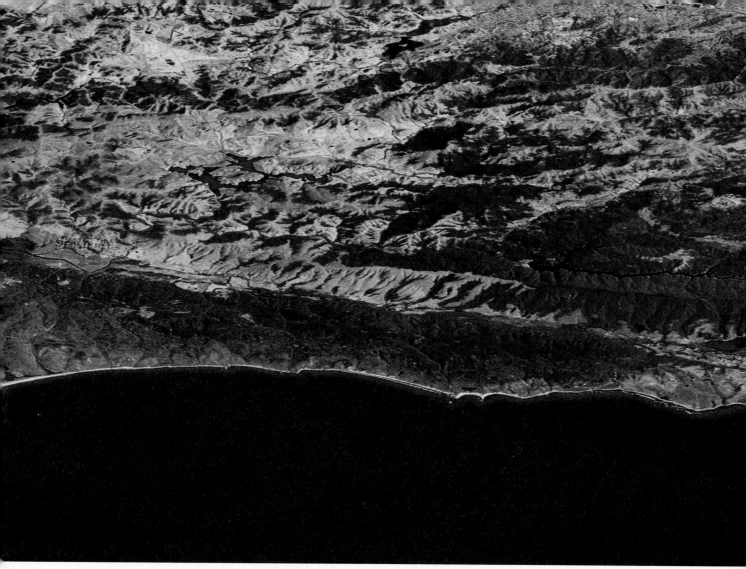

Transform faults and fracture systems are even more spectacular on the ocean floor (see the inside back cover of this book). They slice through the mid-ocean ridges in hundreds of places and sweep across the ocean floor for thousands of kilometers, imparting a remarkable fabric to the solid surface of our planet. They form the most prominent and widespread fracture system on Earth.

Although huge oceanic fracture zones are intimately related to transform fault movements, only short segments are active transform plate boundaries. The closely spaced fracture zones of the southern Atlantic Ocean grade southward into a coarser fabric in the southern Indian Ocean. There, the fracture systems have their own distinctive characteristics. In the northern Indian Ocean, fractures are short and closely spaced, whereas those in the southwestern arm of the ridge near Australia are four times as long. Some of the longest and boldest fractures are in the southern Pacific Ocean, but northward, the East-Pacific Rise is more subdued and the fracture systems have weaker topographic expressions.

Transform faults connect various kinds of plate boundaries and are essential to accommodate the different types of movement found at these boundaries. Transform plate boundaries are ubiquitous and play an important role in global tectonics. Our focus in this chapter is on the processes that occur at these plate boundaries and the features they produce.

Major Concepts

1. Transform plate boundaries are unique in that the plates move horizontally past each other on strike-slip faults. Lithosphere is neither created nor destroyed.
2. The three major types of transform boundaries are: (a) ridge-ridge transforms, (b) ridge-trench transforms, and (c) trench-trench transforms.
3. Transform plate boundaries are shear zones where plates move past each other without diverging or converging. During shearing, secondary features are created, including parallel ridges and valleys, pull-apart basins, and belts of folds. Compression and extension develop in only small areas.
4. Oceanic fracture zones are prominent linear features that trend perpendicular to the oceanic ridge. They may be several kilometers wide and thousands of kilometers long. The structure and topography of oceanic fracture zones depend largely on the temperature (or age) difference across the fracture and on the spreading rate of the oceanic ridge.
5. Continental transform fault zones are similar to oceanic transforms, but they lack fracture zone extensions.
6. Shallow earthquakes are common along transform plate boundaries; they are especially destructive on the continents.
7. Volcanism is rare along transform plate boundaries, but small amounts of basalt erupt locally from leaky transform faults.
8. Metamorphism in transform fault zones creates rocks with strongly sheared fabrics, as well as hydrated crustal and even mantle rocks.

Characteristics of Transform Plate Boundaries

> Transform plate boundaries are zones of shearing, where two plates slide horizontally past each other. Rocks in the shear zone are strongly deformed, but no new lithosphere is created and none is consumed. Transform boundaries in ocean basins and on the continents are expressed by steep, linear ridges and valleys. The major types of transform plate boundaries are ridge-ridge transforms, ridge-trench transforms, and trench-trench transforms.

Transform boundaries are strike-slip faults along which two separate tectonic plates grind horizontally past each other without forming or consuming lithosphere (**Figure 20.1**). **Transform faults** are generally vertical and parallel to the direction of movement. They are produced by **shearing**. Most transform faults are intimately related to divergent plate boundaries on the ocean floor and cut across the mid-ocean ridges. Other types of transform faults connect convergent and divergent plate boundaries, and some cut across the continents themselves. Volcanism and deformation are less common at transform boundaries than along other types, but transform faults remain complex structural and topographic features.

Oceanic transform plate boundaries are part of even longer features called **fracture zones** (Figure 20.1). Fracture zones are enormous structures that range up to 10,000 km long and may have a vertical relief of 6 km. They are remarkably narrow—only a few zones are as much as 100 km wide—but all consist of a series of parallel fractures. On regional maps, an oceanic fracture zone may appear to be a fault that offsets an oceanic ridge. However, a fracture zone (gray in Figure 20.1) extends in both directions beyond the active transform fault boundary (red). Compared with the fracture zone, the actively deforming transform plate boundary is much shorter, typically less than a few hundred kilometers. This active zone is marked by steeply dipping faults and complexly deformed oceanic crust.

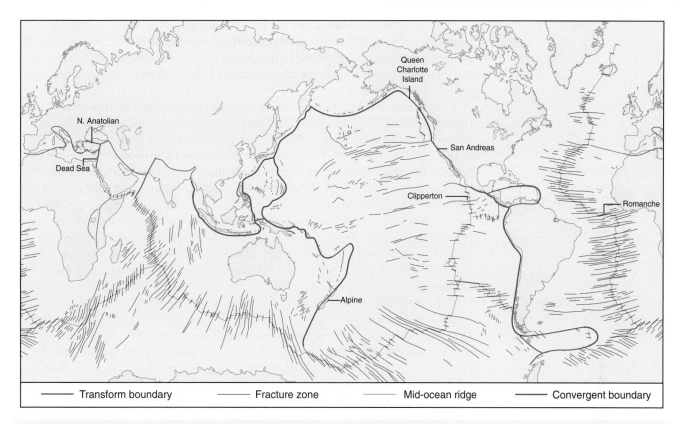

Figure 20.1 **A map of the major transform plate boundaries** and associated oceanic fracture zones shows that most are related to spreading at mid-ocean ridges. Also, most curve because they are parallel to small circles around the poles of plate rotation. Other transform boundaries are related to convergent margins in regions of complex plate movement.

Casual examination of maps of oceanic fracture zones, like Figure 20.1, may fail to convey an accurate notion of their size and length. Many are long enough that they would extend across the entire United States as systems of parallel ridges and troughs. (For example, compare the length of several Pacific fracture zones with the United States on the map.) Most of the valley walls are much higher and steeper than the ridges of the Appalachian Mountains or the Front Range of the Rockies. The relief on some fracture zones is comparable to that of the great eastern scarp of California's fault-bounded Sierra Nevada. Without erosion by river systems, the transform fault scarps on the ocean floor are quite unlike any on land. They are neither dissected with stream valleys nor greatly subdued by weathering and erosion. Most erosional modification results from mass movement along the slopes of the escarpment.

On the continents, transform plate boundaries are major strike-slip faults with displacement of hundreds of kilometers, placing rocks of widely different origins and ages side by side. This movement produces the linear cliffs, ridges, and troughs that are common surface expressions of major strike-slip faults. Like oceanic transforms, a continental transform fault is not one plane, but consists of a zone of faults that can be several kilometers wide. Stream erosion of transform faults on continents tends to subdue their topographic expression, but movement on the great Alpine Fault in New Zealand has produced mountains more than 3000 m high. Continental transform faults also pose great hazards because of the abundant earthquakes they produce.

How does a transform fault differ from the associated fracture zone?

Types of Transform Plate Boundaries

By carefully considering the map in Figure 20.1, you can see that a transform plate boundary is a special kind of fault that *transforms* one type of plate motion into another type. For example, the *diverging* motion between plates at an oceanic ridge can be changed along a transform fault to the *converging* motion between plates at a subduction zone. Transform faults can connect convergent and divergent plate boundaries in three different ways: (1) ridge-ridge transforms connect two segments of a divergent plate boundary; (2) ridge-trench transforms connect a ridge and a trench; and (3) trench-trench transforms couple trenches at two different convergent plate boundaries (**Figure 20.2**).

Ridge-ridge transform faults are by far the most abundant. It is therefore important to analyze the movements of these transforms in detail. Keep in mind that active displacement on these faults occurs only *between* the ridge segments, as shown in Figure 20.2A. The figure shows that plate movements are in opposite directions between the ridge crests. No movement occurs along the rest of the fracture zone.

Thus, the transform fault sustains formidable shearing movement. In the upper part of the plate, brittle shearing creates **mylonite** and tectonic breccias. Numerous earthquakes are produced by repeated rupturing of the brittle rocks along the upper part of the fault zone. Brittle fracturing grades into ductile deformation in the weak underlying asthenosphere; consequently, seismicity is characteristically shallow along a transform, occurring in the brittle zone. The fault cuts through the entire lithosphere.

Ridge-trench transforms are much less common (Figure 20.2B). But they form an important connection between spreading and converging plates. The longest transform faults are all of this kind. One type of ridge-trench transform connects a ridge with the overriding side of a convergent boundary, like the one east of the southern tip of South America (Figure 20.1). Here, two long transform faults connect a short ridge with the trench along the Scotia arc. Together, these plate boundaries form a huge hairpin curve. Another type of ridge-trench transform connects a ridge with the subducting side of a trench. An example is the Queen Charlotte Island Fault off the western coast of Canada (Figure 20.1).

Trench-trench transforms (Figure 20.2C) are also rare. Figure 20.1 shows two good examples. One is the Alpine Fault in New Zealand, which appears to connect

Why are ridge-ridge transform faults more abundant than other types?

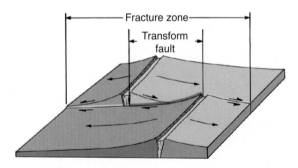

(A) Ridge-ridge transform fault.

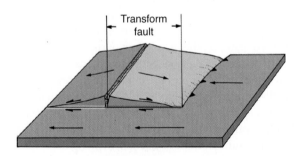

(B) Ridge-trench transform fault.

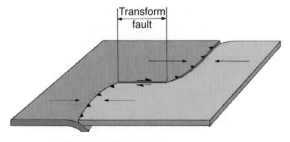

(C) Trench-trench transform fault.

Figure 20.2 **Transform faults** can connect convergent and divergent plate boundaries in various combinations. Note that relative motion occurs only along the boundary between the plates, shown in red. In all cases, the trend of a transform fault is parallel to the direction of relative motion between plates. This characteristic is helpful in determining the direction of plate motion.

two trenches in the south Pacific. The direction of subduction changes across the transform fault. Another trench-trench transform fault connects the western Aleutian trench and the Kamchatka trench on the eastern shore of Russia.

Oceanic Transform Plate Boundaries and Fracture Zones

Transform plate boundaries in the ocean basins are prominent linear features that are perpendicular to the mid-ocean ridges. They are the short, active parts of fracture zones that may be several kilometers wide and thousands of kilometers long. The characteristics of oceanic fracture zones depend on the age difference between the lithosphere on either side of the fault zone.

The long oceanic fracture zones that slice across the seafloor are not what they first seem to be. The sharp linear surface expression of fracture zones and the apparent offset of an oceanic ridge may suggest that they are simple strike-slip faults, with displacement occurring along their entire length. However, nothing could be farther from reality. To understand oceanic fracture zones, you must keep in mind the relative motion of the plates at a spreading ridge.

Active displacement on the fracture zone occurs only along the line that connects the offset ridge segments (Figure 20.2). This zone is the transform fault, and it is the only place where the fracture zone is a boundary between plates. It is along this zone that earthquakes occur. Beyond this transform zone, the plates on either side of the fracture are moving in the same direction and at the same rate; they are linked together on the same plate. There is thus an active segment of the fracture (a transform fault, which is a plate boundary) and an inactive segment (which is not a plate boundary).

As the plates slide past each other in the transform zone, their boundaries are fractured and broken (**Figure 20.3**). This fracturing produces parallel ridges and troughs in the fault zone (**Figure 20.4**). If there is no displacement along faults beyond the transform zone, then how do these valleys and ridges form? It is probably easiest

How long and wide are oceanic fracture zones?

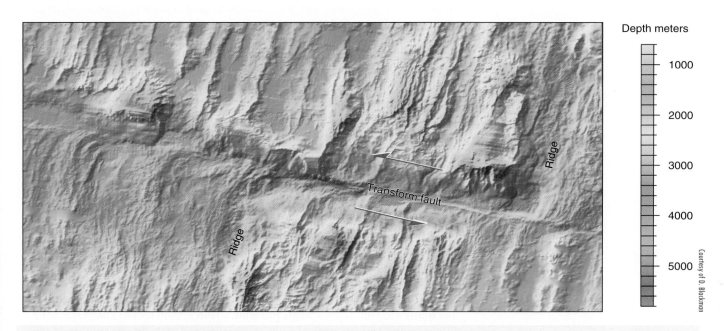

Figure 20.3 Intense shearing occurs at transform plate boundaries. This map of the flanks of the mid-Atlantic ridge shows all of the hallmarks of a transform fault. Linear valleys, depressions, and ridges are all aligned along the fault. In addition, the abyssal hills bend to make J-shaped curves as they reach the transform.

Transform fault
zone

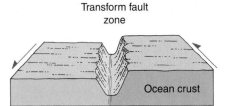
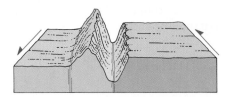
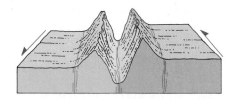

Ocean crust

Figure 20.4 Various topographic expressions of transform boundaries result from juxtaposition of rock bodies having different temperatures, ages, and internal structure. The simple-to-complex valleys and scarps shown here are common.

to envision the inactive portion of the fracture zone as a long "thermal scar" in the oceanic crust. Where a ridge segment terminates against a transform fault, the newly formed hot ocean floor lies next to older, colder ocean floor (Figure 20.2A). A steep cliff or scarp is produced if there is a large difference in the age of rocks on either side of the fault zone, as shown in the figure. The younger lithosphere near the oceanic ridge is hotter and expanded, so it stands higher than the colder and contracted lithosphere on the other side of the fracture. The higher side of the transform fault thus may change from one side to the other, as you can see in the figure.

As spreading continues, this elevation difference persists, and the scar grows longer and longer. Thus, a fracture zone is a historical recording of the relative plate motions on either side of a spreading ridge. For example, the fracture zones in the North Atlantic trace the movement of North America away from Africa.

Another distinctive characteristic of transform plate boundaries is that no significant amount of lithosphere is formed or consumed. Tremendous shearing, grinding, metamorphism, and deformation occur in these narrow zones. In fact, between 1% and 10% of the oceanic lithosphere has been deformed by processes at a transform fault. There is generally little or no volcanic or intrusive activity, although some volcanism may occur along **"leaky" transform** systems. Transform plate boundaries play an important structural role in the tectonics of our planet, and considerable movement and adjustment occur along these zones as the plates shift and move about.

Examples of Oceanic Fracture Zones
To become better acquainted with the characteristics of oceanic fracture zones, let us look more closely at three different transform systems.

Romanche Fracture Zone   The most spectacular fracture in the central Atlantic Ocean is the Romanche fracture system, which lies almost on the equator (**Figure 20.5**). The fracture zone cuts the slow-spreading Mid-Atlantic ridge and includes a ridge-ridge transform fault between the African plate on the north and the South American plate on the south. The fracture zone stretches across the entire Atlantic Ocean; even the active transform is 600 km long. Consequently, the difference in age of the oceanic crust on either side of the fracture is huge, nearly 50 million years (compare with the map on the inside front cover).

The Romanche fracture zone is a series of deep valleys separated by ridges. One narrow ridge rises above sea level to make a cluster of small islands—St. Peter and St. Paul's Rocks, one of the few oceanic island groups that did not form by volcanism. The floor of this zone is fractured and grooved and contains the deepest part of the central Atlantic Ocean (7960 m below sea level). This deep gap in the Mid-Atlantic Ridge is significant in circulation of oceanic water in that it permits cold bottom water from the ocean near Antarctica to flow from the western side of the ridge in the southern Atlantic into the deep basins of the northern Atlantic on the east side of the ridge.

The transform fault is not a single plane but a fault system tens of kilometers wide containing many separate vertical faults that branch and crosscut one another. Many slumps from the steep walls of the escarpment cover large parts of the valley floor. These deep cuts expose complete sections through the oceanic crust. Rocks dredged

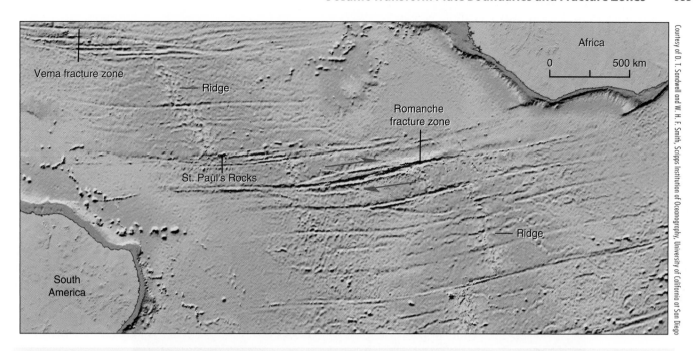

Figure 20.5 **The Romanche fracture zone** extends across most of the Atlantic Ocean, forming a huge ridge and trough system some 5000 km long and almost 100 km wide. The active transform boundary lies between the offset ridge axis.

from the floor of the fracture zone include igneous rocks such as basalt, gabbro, and mantle peridotite as well as their deformed or metamorphosed equivalents—mylonite, metabasalt, and serpentinite (hydrated peridotite). Without question, the transform is a zone of intense shearing, manifested by breccia and highly deformed serpentinite. Extensive slumping and landsliding off the steep scarps also have produced coarse talus breccias.

Clipperton Fracture Zone The Clipperton fracture zone stretches from the coast of Central America to the middle of the Pacific Ocean (Figure 20.1). The active transform boundary, however, is only 85 km long and connects two segments of the East Pacific Rise (**Figure 20.6**). The lithosphere differs in age across the fracture zone, but only by about 1.5 million years. The spreading rate on this part of the ridge is high, about 10 cm/yr. The transform fault zone is a deep cleft, bounded by high ridges on either side.

The inactive fracture zone beyond offsets magnetic stripes in the oceanic crust, indicating that the entire fracture zone originated as a ridge-ridge transform fault, but the direction of spreading changed slightly about 3 million years ago. On the north side, seismic investigations show no axial magma chamber within 70 km of the ridge-transform intersection. On the other hand, a magma chamber is present on the south side of the Clipperton fracture zone.

Transform Fault Zones in Ophiolites Some ophiolites have long, nearly vertical shear zones that slice through the igneous rocks. These zones are probably transform faults formed on the seafloor and then preserved during the thrusting of the ophiolite onto land.

For example, the Troodos ophiolite complex on the Mediterranean island of Cyprus is cut by such a vertical shear zone (**Figure 20.7**). The shear zone is about 10 km wide and cuts through the rocks of the ophiolite sequence. Dikes in a sheeted dike complex curve toward the major strike-slip fault, just as ridge tips do a transform fault zone. Outcrops of gabbro and serpentinite lie on the south side of the fault. The large mass of serpentine is interpreted to have formed when peridotite in the mantle combined with seawater flowing through the highly permeable fracture zone. Because serpentine is both buoyant and weak, it may have intruded into the fracture zone as a

Why is Cyprus important for the study of transform faults?

diapir. Talus breccias formed as steep slopes on the seafloor failed. Apparently, high ridges and deep valleys ran parallel to the shear zone. The breccias are interlayered with sediments and basaltic lava flows, which include pillow basalts. Small volumes of lava must have erupted along the shear zone. All of these features reveal that this area of the ophiolite was strongly sheared along a strike-slip fault while it was still at the bottom of the ocean and long before it was thrust onto dry land.

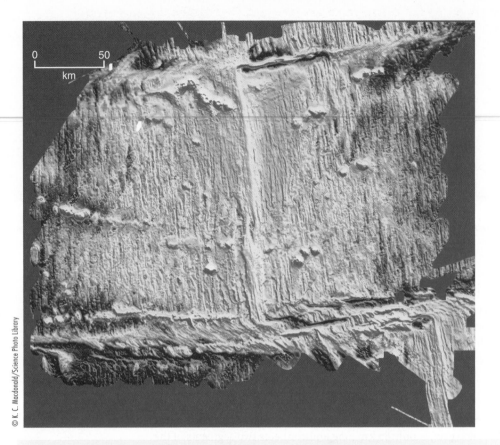

© K. C. Macdonald/Science Photo Library

Figure 20.6 The Clipperton transform fault cuts the East Pacific Rise, in the northern part of this map. It is just west of the Central American coast. The transform fault forms a series of ridges and troughs connecting two segments of the oceanic ridge. The offset is about 85 km. Note the height of the plate immediately north of the East Pacific Rise and the rugged relief in the fault zone. The fracture system where no shear occurs extends beyond the active transform fault. Blue is low elevation and light pink is highest elevation on seafloor.

Figure 20.7 The structure of a transform shear as interpreted from exposures in the Troodos ophiolite complex on Cyprus. The fault zone consists of innumerable vertical faults and is marked by complex breccias made of fragments from basaltic dikes. The large mass of serpentine intruded the fracture zone. Talus breccias are interlayered with sediments and basaltic lava flows, which include pillow basalts that erupted at the transform boundary.

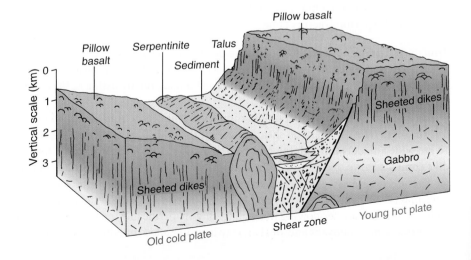

Processes at Transform Plate Boundaries

Transform boundaries strongly influence local tectonic and igneous processes and are responsible for much of the structural and topographic variation of the ocean floor. The examples described above show that most oceanic fracture zones have pronounced topographic relief, with high linear ridges and deep linear troughs, steeply inclined fault scarps, and sheared rocks. Like many strike-slip faults on land, a transform fault zone is not a single plane but a fault zone that can be as much as 100 km wide. The transforms thus contain numerous separate fault planes that branch, merge, and cross one another to make a braided network of strike-slip faults. Volcanism also diminishes near the transform intersection with an oceanic ridge.

To understand why these characteristics develop, let us explore some of the fundamental processes that take place at transform plate boundaries. Their distinctive fabric largely results from the complex interplay of strike-slip faulting, age and temperature contrasts across the transform fault, and shearing rates.

What processes occur along transform faults?

Contraction and Extension in Transform Fault Systems At many transform fault boundaries, strike-slip displacement occurs along several separate fault planes that curve slightly, merge, and separate again. Horizontal movement within this braided system of curved faults produces local zones of horizontal compression or extension (**Figure 20.8**). Horizontal compression occurs where the bends are oriented so that blocks on either side of the fault are squeezed together (Figure 20.8A). This squeezing creates uplifted regions with small folds and thrust faults that trend perpendicular to the major strike-slip faults. Because this process involves both *trans*form and com*pression* motion, it is known as **transpression**. This kind of contraction commonly creates long, low ridges, but locally, uplift may be extreme.

Areas of extension can also develop along bends in the strike-slip faults. **Transtension** (*trans*form plus ex*tension*) produces small fault-bounded troughs known as **pull-apart basins** (Figure 20.8B). This extension results in normal faulting and subsidence of a small block between the two plates. These more or less rectangular basins grow and accumulate sedimentary deposits as strike-slip movement continues. As the troughs grow, the floor may be stretched and thinned so that volcanoes may develop in the central part of the basin.

Thermal Structure and the "Cold Wall" By their very nature, ridge-ridge transform plate boundaries juxtapose a cold wall of lithosphere against a hot ridge axis. You can see in **Figure 20.9** that a relatively thick, cold, and contracted segment of older lithosphere on the flank of a ridge is adjacent to a hot ridge axis where new lithosphere is forming. This may seem at first to be a minor and insignificant characteristic, but it is a major factor in the structural and topographic evolution of the oceanic crust.

In a sense, normal processes that occur along the ridge, such as rifting and the generation of new oceanic crust, come up against this cold wall and stop. Thus, the transform faults at either end of a ridge segment form distinct boundaries to the crust-forming processes, much as the walls of a magma chamber form boundaries to igneous processes in a pluton. The cold wall cools the asthenosphere rising beneath

What effect does the "cold wall" have on the adjacent oceanic ridge?

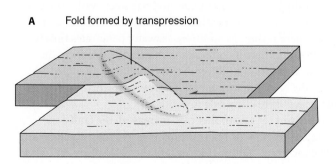

A Fold formed by transpression

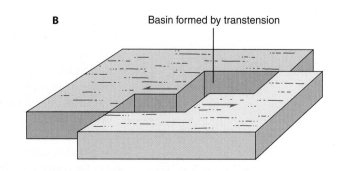

B Basin formed by transtension

Figure 20.8 Secondary compressional and extensional structures are produced by bends or offsets in the transform fault system. Small fold belts mark zones of transpression (A) and pull-apart basins mark transtensional bends (B).

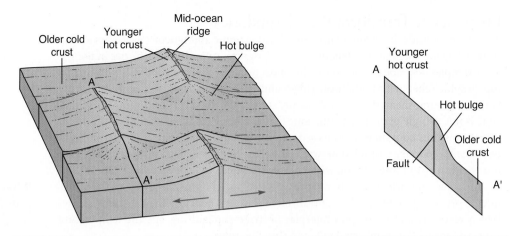

Figure 20.9 The thermal structure of a transform boundary is related to differences in age and temperature of the lithosphere across the fault. The cross section along A–A′ is parallel to a ridge segment and shows the thermal structure of a ridge-transform boundary. The older, cooler lithosphere creates a "cold wall" that inhibits magmatic processes and concentrates deformation into a narrow zone. The younger, hotter lithosphere stands higher than the older cooler lithosphere. Thus, the scarp alternates from one side of the fracture zone to the other. In addition, a hot bulge forms on the older lithosphere that is adjacent to the hot ridge.

the ridge. This cooling restricts the amount of partial melting and reduces the amount of magma that can separate from the asthenosphere. Consequently, this massive cold wall may effectively eliminate development of normal oceanic crust at the adjacent ridge crest. Near transform boundaries the oceanic crust may be very thin.

The spreading ridge also affects the cold wall. Intrusion of hot dikes or conductive heating from the hot ridge causes a bulge to develop on the cold side of the fault (Figure 20.9). This bulge is a significant topographic element of the transform. Note the bulge across from the spreading ridge along the Clipperton fracture zone (Figure 20.6).

Ridge Offset and Spreading Rate   The structure and topography of ridge-ridge transform plate boundaries vary systematically, depending on the *temperature* of the plates across the transform fault. The amount of *offset of the ridge* and the *spreading rate* control the contrast in temperature. Large ridge offsets juxtapose an old, cold section of lithosphere against a young, hot ridge. Small ridge offsets juxtapose two sections of lithosphere that have almost the same temperature, age, thickness, and strength. (For a given offset distance, ridges that spread quickly will juxtapose plates that are very similar to one another as compared to transform faults on ridges that spread slowly.)

What is the difference in the structure and topography of long offset versus small offset transform faults?

Transforms with *long ridge offsets* (or with slow-shearing rates) have a prominent cold wall (**Figure 20.10**). On one side of the fault zone, lithosphere is exceptionally thin and weak. Directly across the fault zone, the lithosphere may be 30 million years old and 50 km thick. The striking difference in the thickness of lithosphere keeps the fault zone narrow and prevents the fracture from splitting or migrating. Consequently, the fault zone is well defined, generally less than 1 km wide, and has steep walls. The cold wall also lowers the temperature of magma beneath the ridge and slows volcanism. As a result, the crust that develops here is thin. In addition, seawater penetrates through the thin crust down into the mantle. There it combines with peridotite to form the metamorphic rock serpentinite. Because of its lower density and lower strength, the serpentinite may rise through the crust and make long, narrow **serpentinite ridges** characteristic of transforms at slow-spreading ridges.

As the amount of offset decreases (or rate of shearing increases), the fault zone becomes wider and more complex (**Figure 20.11**). If the offset is less than about 50 km, little variation exists in the temperature, age, or thickness of the lithosphere on either side of the transform. Thus, the boundary is not well constrained by a wall of strong, cold lithosphere. A broad zone of deformation up to 100 km wide develops. The transform contains multiple shear zones and small ridge segments that are oblique to the ridge. In addition, volcanism is more common in the wider shear zones.

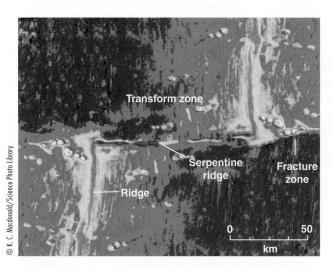

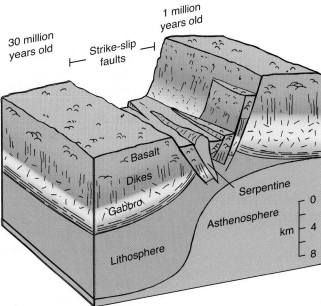

(A) The transform fault is marked by a deep linear valley. Long, narrow, linear ridges commonly parallel the faults. Also, note how the spreading ridge bends into the fault zone. Red is high elevation and deep blue is low elevation.

(B) The schematic cross section shows the pronounced contrast in thickness of the lithosphere from the accreting ridge where the lithosphere consists only of hot, new oceanic crust to the much older, colder, and thicker lithosphere on the opposite side.

Figure 20.10 **A large-offset on a transform fault** (or one that has a slow-shearing rate) has a narrow zone of deformation.

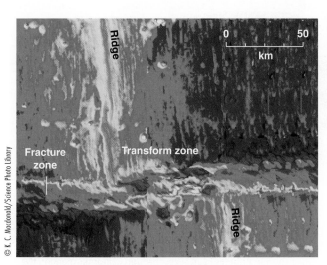

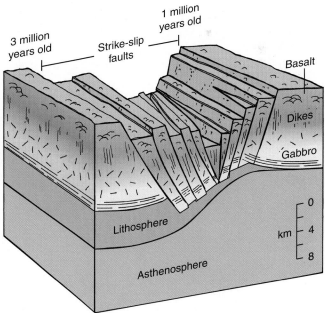

(A) The transform fault zone may be tens of kilometers wide. Several shear zones within the transform system form elongate ridges and valleys. They are linked together by extensional pull-apart basins or segments of spreading centers that trend obliquely across the shear zone. Volcanism occurs along these short ridge segments. Red is high elevation and deep blue is low elevation.

(B) This cross section shows that there is little contrast in lithospheric thickness across the transform zone. This relative uniformity permits a wide belt of deformation.

Figure 20.11 **A small offset transform fault** (or one that has a high-shearing rate) has a wide zone of deformation.

Continental Transform Faults

Continental transform faults are similar to oceanic transform faults. They are seismically active and have distinctive topographic features—fault scarps, linear ridges and troughs, and displaced stream channels formed by strike-slip faulting. Pull-apart basins and fold belts develop along bends in the faults.

Transform faults that cut continental crust are not nearly as common as oceanic transforms because most transform faults develop at oceanic ridges. Some inactive transforms are preserved as suture zones in shields or as boundaries to accreted terrains. Today's active continental transform faults are important because they help us understand global tectonics and because they are seismic hazards in populated areas.

How are transform faults expressed on continents?

Continental transform faults are similar to oceanic transforms in that their motion is essentially strike-slip and associated earthquakes are shallow. Like oceanic transforms, continental transform faults penetrate the entire lithosphere and their movement defines a plate boundary. (Most other types of faults involve only the upper part of the crust and die out at relatively shallow depths.) Continental transform faults typically have distinctive linear topographic features (**Figure 20.12**). These include relatively straight fault scarps, linear ridges and troughs, and streams and valleys that have been beheaded and displaced horizontally. The individual faults in a continental transform system therefore branch, join, bend, and sidestep each other and even establish zones of local contraction or extension.

Some differences between oceanic and continental lithosphere may be caused by the greater thickness and structural complexity of continental lithosphere. Continental transform faults may follow previous zones of weakness, such as older faults or boundaries between rock types that have contrasting strengths.

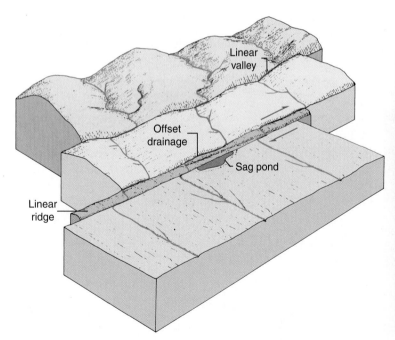

(A) Transform fault zones include strike-slip faults, fault scarps, linear ridges and valleys, offset drainages, and local elongate lakes and ponds.

(B) The San Andreas Fault slices through California, marking the transform boundary between moving tectonic plates. The great scar along the fault line is marked by linear valleys, sharp contrast in landforms, and displaced drainage.

Figure 20.12 Continental transform faults produce very distinctive landforms.

Examples of Continental Transform Faults

To appreciate these features and structures, let us consider some details of three well-known continental transform systems.

San Andreas–Gulf of California Transform System   The San Andreas–Gulf of California transform system is 3000 km long, extending from the Mendocino fracture zone off the northern coast of California southward to the tip of Baja California (**Figure 20.13**). It is a ridge-ridge transform consisting of two main parts: the San Andreas strike-slip zone and the obliquely rifting Gulf of California.

The San Andreas Fault is expressed at the surface by a great scar running through most of California, but it has many subsidiary faults. Thus, the displacement involves a zone more than 100 km wide. These faults are part of the same system that accommodates the motion of the Pacific plate past the North American plate. Undoubtedly, it is the best-known plate boundary in the world. At its southern end, the San Andreas Fault splays out into a series of subparallel strike-slip faults that extend to the head of the Gulf of California. Throughout California, the San Andreas Fault zone is marked by sharp, linear landforms, including straight and narrow valleys, linear ridges, and offset drainage patterns (Figure 20.12B). Local zones of transpression and transtension have developed deep pull-apart basins and zones of intense compression. The rocks in one folded zone have been uplifted about 15 km.

The San Andreas Fault itself is more than 1000 km long, an active boundary between the Pacific plate to the west and the North American plate to the east. The Pacific plate is moving northwestward at about 6 cm/yr relative to the North American plate. As stress builds between the plates, sudden releases cause the earthquakes for which California is noted. Earthquakes occur along its entire length and as deep as 15 km, but no deeper. The famous San Francisco earthquake of 1906 resulted from movement that produced an offset of as much as 6.4 m.

The San Andreas Fault system began to develop about 30 million years ago (in Oligocene time) and its location on the continent may have been controlled by preexisting fractures in the basement rocks. Horizontal movement along the San Andreas Fault has totaled about 300 km.

The Gulf of California segment of the transform system is a series of long transform faults that connect very short spreading ridges (Figure 20.13). The features of the Gulf of California segment are similar in many ways to those of small-offset oceanic transform systems. Most of the Gulf of California is thinned continental crust, and the floor of the ocean is very shallow. Along the spreading ridges, however, oceanic crust has formed and the bottom of the gulf attains abyssal depths. Opening of the gulf began about 5 million years ago with a spreading rate of 5 cm/yr.

The net result of movement along the San Andreas–Gulf of California transform system has been the opening of the Gulf of California and the displacement of the western block of the San Andreas northward approximately 300 km (Figure 20.13). With continued movement, Baja California and a narrow slice of western California may become an elongate continental fragment surrounded by oceanic crust. The San Andreas–Gulf of California transform system thus provides insight into how some microcontinents are formed, especially where the transforms cut the ridge obliquely, such as in the western Indian Ocean. This mechanism might explain the origin of Madagascar and the Seychelles Islands, which are continental fragments formerly attached to Africa and India.

Dead Sea Transform System   The Dead Sea transform system extends from the spreading ridge of the Red Sea northward to a zone of continent-to-continent collision in the Alpine orogenic belt in southern Turkey (**Figure 20.14**). The structure, topography, and history of the entire region are magnificent expressions of continental plates moving along a transform system. The transform zone is about 1000 km long and

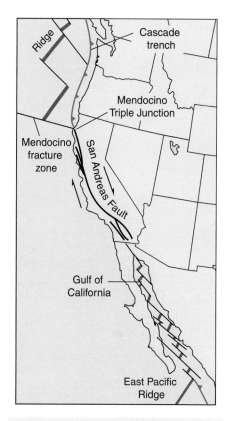

Figure 20.13 The San Andreas–Gulf of California transform system extends from northern California to just beyond the end of Baja California. It connects the Mendocino fracture zone, the Cascade trench, and the East Pacific Rise. Where the transform involves continental crust (throughout California), it forms a series of strike-slip faults with intervening pull-apart basins and compressional ridges. In the Gulf of California, where the transform system involves oceanic crust, the fault zone consists of a series of long transform faults connecting short ridge segments.

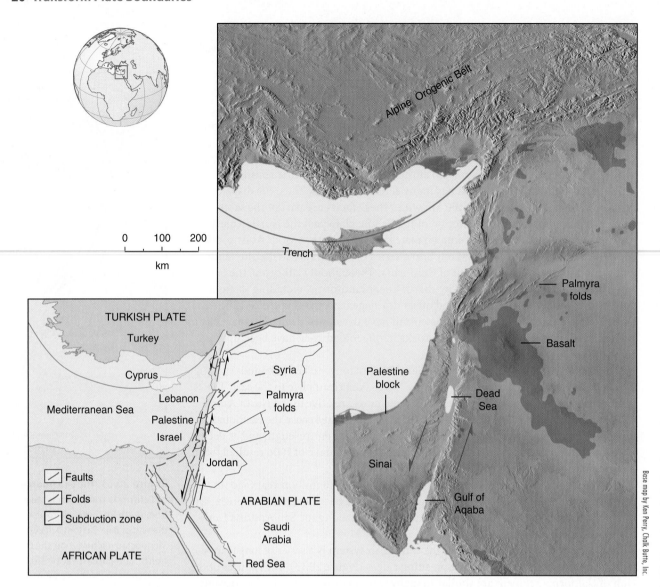

Figure 20.14 **The Dead Sea transform system** connects the Red Sea spreading ridge with the Alpine convergent belt. The movement along the transform zone has produced the long, deep, narrow pull-apart basins of the Gulf of Aqaba and the Dead Sea as well as the contractional folds of the northern Sinai and the Palmyra Mountains of Lebanon and Syria. Small eruptions of basalt occurred near the pull-apart basins.

marks the boundary between the western edge of the Arabian plate and the northern part of the African plate. It is a remarkable structure that controls the development of many features in Israel, Lebanon, Syria, and the Sinai Peninsula, as described below.

The main structure is the strike-slip fault zone, which involves several separate fault planes that slice through the entire lithosphere. These faults are not straight but have several angular bends. As a result, large, deep pull-apart basins have formed along the strike-slip fault zone (**Figure 20.15**). The Gulf of Aqaba (Elat) segment is the widest and deepest. The floor of one of these basins is more than 2000 m below sea level. Farther north, the Dead Sea trough is another pull-apart basin; it is 400 m below sea level, with water depths exceeding 300 m in places. Sediment filling the Dead Sea trough is derived from erosion of the adjacent mountains. It is several kilometers thick and continues to pour into the graben, forming alluvial fans. In a more humid climate, the Dead Sea trough would be a freshwater lake or an extension of the Gulf of Aqaba.

The structural features north and south of the Dead Sea are no less impressive. Note the direction of plate movement illustrated in Figure 20.14. The Arabian plate is

What structures north and south of the Dead Sea are produced by movement along a transform fault?

Courtesy of NASA

Figure 20.15 **Pull-apart basins** in the Gulf of Aqaba and the Dead Sea dominate this photograph taken by astronauts aboard the Space Shuttle. Such basins are caused by movement on strike-slip faults that have sharp bends and offsets of the major faults (inset). These form three deep basins along the floor of the Gulf of Aqaba and the Dead Sea basin that lies below sea level.

moving northward and the African plate southward. Two major bends in the strike-slip fault system occur, one to the north in Lebanon and Syria and the other south of the Dead Sea. As the plates move near these bends, slippage along the fault is inhibited and broad zones of transpression result. This formed compressional folds that branch off the strike-slip fault zone in the Palmyra Mountains to the north (Figure 20.14)

The Dead Sea transform system began about 25 million years ago when Arabia was still part of the African continent. As rifting began to open the Red Sea, the Arabian plate split from Africa and began to move northward. The Dead Sea transform was initiated by this movement, and vigorous tectonic activity has continued ever since. Total offset along the southern extent is at least 100 km, but to the north displacement is less, suggesting that part of the plate movement has been taken up by folding in the Palmyra fold belt. Intermittent volcanism along the transform system has occurred since Mesozoic time, and Pleistocene basaltic lava flows and cinder cones are especially obvious (Figure 20.14).

Alpine Transform System of New Zealand The Alpine Fault is a trench-trench transform system connecting the west-dipping Tonga-Kermadec Trench in the north with the east-dipping Macquarie Trench in the south. As seen in **Figure 20.16**, the Pacific plate is being consumed in the Tonga-Kermadec Trench east of the north island, and the Australian plate is being subducted south of the south island. Thus, the Alpine transform is a trench-trench transform that traverses a large segment of a continental fragment, shearing New Zealand down the middle. Consequently, this type of transform can be studied from direct observation on land.

What is the offset along the Alpine Fault?

Like the San Andreas Fault, the Alpine transform was well known as a strike-slip fault long before plate tectonic theory developed. Correlation of distinctive rock types near the ends of the fault zone indicates a huge displacement of about 480 km (note the displacement of the late Paleozoic metamorphic rocks on the map). The north and south islands of New Zealand are currently being drawn apart along the Alpine transform system. Throughout much of its length, the Alpine Fault is defined by the abrupt truncation of mountain spurs against the adjacent plains. Two large earthquakes near Christchurch were caused by motion on the strike-slip fault system. A large magnitude 7.1 earthquake in 2010 was followed by a smaller (6.7) but more damaging earthquake in 2011. Both were shallow earthquakes (less than 10 km); in contrast, much deeper earthquakes form along the subduction zone beneath northern New Zealand (Figure 20.16).

Movement along the Alpine Fault has been largely horizontal throughout most of its 40-million-year history (at a rate of up to 1 cm/yr). However, significant vertical movements associated with transpression and thrusting along the fault zone have

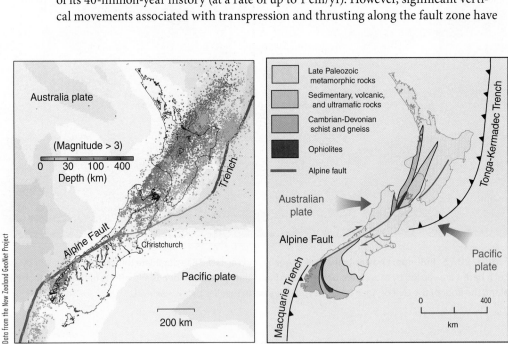

Figure 20.16 The transform system of the Alpine Fault, New Zealand, connects the west-dipping Tonga-Kermadec subduction zone to the east-dipping Macquarie subduction zone. Note the tremendous displacement of major rock units on the Alpine Fault. The northern subduction zone dips east and the southern subduction zone dips west, as shown by the pattern of progressively deeper earthquakes.

Figure 20.17 **The Alpine Fault, New Zealand,** is a strike-slip fault that connects two plate boundaries. The valley in this view was created by differential erosion along the long linear fault. Deformation along the fault created the high Alps of the southern islands.

also occurred. For example, a change in plate motions about 6 to 7 million years ago initiated horizontal compression across the fault zone. This compression caused the southern Alps to rise to their present height of over 3500 m (**Figure 20.17**).

Earthquakes at Transform Boundaries

Earthquakes at transform plate boundaries are especially abundant. The seismicity is shallow and shows strike-slip characteristics.

Earthquakes are especially common on transform plate boundaries (**Figure 20.18**). In fact, most of the earthquakes along the oceanic ridge system are actually on transform offsets to the ridge, rather than on the ridge itself. The energy released by these earthquakes is about 100 times greater than that released from earthquakes along the ridge crest. The abundance of earthquakes is related to the lithosphere temperature along the cooler transform as compared with the hotter ridge. In the relatively cold lithosphere along a transform fault, brittle fracture is common. Deformation by ductile flow is more common in the hotter rocks right at the ridge, and earthquakes are consequently less common.

Earthquakes along oceanic transform faults are shallow—most are less than 10 km deep—and small compared with those occurring at convergent plate margins (as deep as 650 km) and along continental transform faults (as deep as 20 km). Moreover, shallow earthquakes along the transforms are rarely related to magma intrusion or volcanism as are earthquakes at the ridge crest.

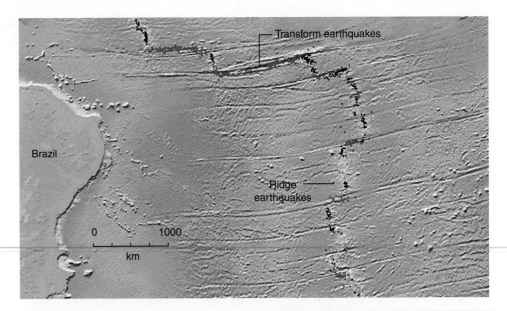

Figure 20.18 Shallow earthquakes on a mid-ocean ridge are more frequent on the transform faults, where the crust is cooler, thicker, and more brittle, than on the ridge crest itself. The region directly beneath the ridge is too hot and ductile to produce many earthquakes. The map shows strike-slip earthquakes (red) on transform faults and extensional earthquakes on normal faults (blue) on the rift valley along a portion of the Mid-Atlantic Ridge.

Courtesy of D. T. Sandwell and W. H. F. Smith, Scripps Institution of Oceanography, University of California at San Diego

In the 1960s, identification of earthquake patterns along oceanic fracture zones was fundamental to establishing our present understanding of transform fault motion. Early workers found that most of the earthquakes along transform faults are related to *strike-slip* movement on faults perpendicular to the ridge crest. In contrast, earthquakes along the ridge axis occur on *normal* faults parallel to the ridge. As predicted by the plate tectonic theory, earthquakes only occur on the active transform fault zone—typically the area between two ridge axes—and few occur along the inactive part of the fracture zone (Figures 20.2 and 20.18).

Earthquakes are also abundant along continental transform faults. The tremors that plague much of western California occur along the San Andreas transform system (**Figure 20.19**). One of the most famous earthquakes in history was the 1906 San Francisco event. This shallow earthquake lasted only a minute but had a magnitude of 8.2. The shaking ruptured gas lines, and the fire that followed caused most of the destruction (an estimated $400 million in damage—a great value at that time—and a reported loss of 700 lives). Horizontal displacement of up to 7 m occurred over a distance of about 400 km, offsetting roads, fences, and buildings. The devastating 2010 earthquake in Haiti, in which 300,000 people died, occurred on a complex transform fault system. Two parallel strike-slip faults form the boundary between the North American plate and the Caribbean Plate. They both pass through Haiti, connecting a subduction zone to the east with a short spreading ridge in the Caribbean (Figure 20.1). The destruction caused by this magnitude 7.0 earthquake was much larger than the comparable 2010 and 2011 earthquakes in New Zealand. Although the population of the Port Au Prince area is much larger than of Christchurch, damage and loss of life during the New Zealand earthquakes were limited because strict building codes are in place. Fewer than 200 people died as a result of building collapses in the New Zealand quakes.

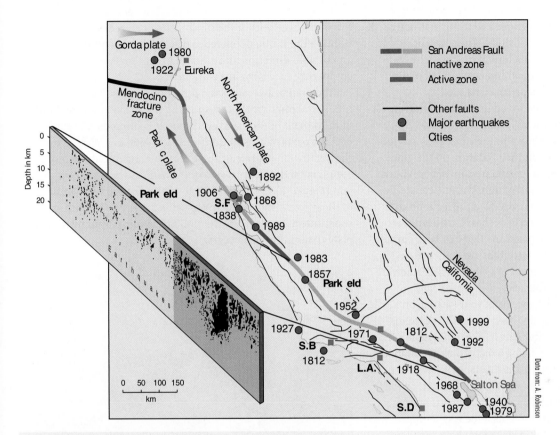

Figure 20.19 Earthquakes on the San Andreas continental transform system are concentrated along the fault and its branching subsidiaries. Almost all of the earthquakes occur at depths less than 15 km. The southern part of the fault has had many historic earthquakes while the northern segment has not. Perhaps strain is building toward a large earthquake on the northern strand.

STATE OF THE ART   Global Positioning Systems

Knowing your location accurately is vital in almost all geologic studies. It is important for research ships at sea, for oil drilling rigs on land, and even for geologists collecting an individual specimen from a rock outcrop. Traditionally, geologists have determined their locations in the field by consulting a topographic map or aerial photograph and using distinctive landmarks to approximate their position. These locations were usually accurate to within a few meters. If more precise locations were needed, laborious measurements with surveying equipment, a transit, or a steel tape could be used.

Ancient people used the stars to navigate. Stars and constellations could be found in predictable places in the sky. By knowing their own position relative to the stars, early navigators could calculate a good fix on their location on the globe. Today, the *Global Positioning System* (commonly abbreviated as GPS) uses radio signals (instead of light) from a constellation of 24 satellites (instead of stars) to find the location of a spot on the surface. In effect, GPS uses these "artificial stars" as reference points to calculate positions that can be accurate to within a few millimeters. The satellites are launched and maintained by the U.S. Department of Defense. GPS receivers are small enough to carry in the field.

A GPS unit measures the distance from your location on Earth to four or more satellites in orbit and then it calculates each distance by accurately measuring how long it takes radio signals from the satellite to reach the receiver. Using a method similar to that used to find earthquake epicenters, four "distance spheres" are formed and their mutual intersection gives your location on Earth.

Obviously, the accuracy of the location depends on the accuracy of the distance measurements. One way to improve the accuracy is to stay in the same place for a long time and allow many readings to be averaged. For precise measurements, like those described below, a GPS receiver may need to remain in the same place for an entire day.

Once a location is known, say, for a sample collection site, a geologist may be satisfied and never return to this location. But others interested in plate motion may come back to the same site again and again to measure the changing position of that spot. Remember, most plates move a few centimeters each year, a distance well within the capabilities of a precise GPS survey. Because the frame of reference (the satellite) is off the planet, an absolute direction and amount of motion can be found by comparing successive measurements.

The map below shows the power of repeated GPS surveying. The arrows are vectors showing the speed and direction of movement of each surveyed spot. It is immediately apparent that parts of coastal California are "quickly" sliding to the northwest at a rate of about 5 cm/yr. The rapidly moving region is on the western side of the San Andreas fault system (Figure 20.13). These velocity estimates are very similar to those found by measuring offset features on the fault and constitute a powerful affirmation of the role played by moving plates over millions of years.

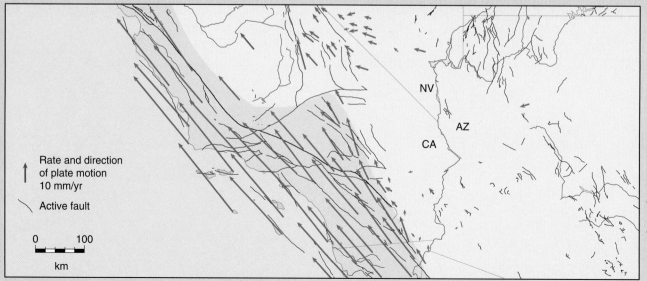

Rate and direction
of plate motion
10 mm/yr

Active fault

0   100
km

NV

AZ

CA

Magmatism and Metamorphism at Transform Plate Boundaries

Volcanoes rarely develop on transform faults, but small volumes of basalt may erupt in pull-apart basins. Metamorphism along transform fault zones creates deformation fabrics, seafloor metamorphism, and serpentinite.

At transforms that juxtapose old crust against an active spreading ridge, the generation and emplacement of basaltic magma is modified because of the cooler temperatures induced by the adjacent thick, cold lithosphere. The temperature becomes lower along the ridge axis as it approaches a transform fault (Figure 20.9). The cooler temperatures diminish volcanic activity because it disrupts the migration of the hot upwelling asthenosphere and lessens the volume of basaltic melt generated. As a result, the amount of magma supplied to the shallow magma chamber decreases as the transform boundary is approached. Consequently, the crust becomes thinner near the transform.

Some continental transform faults are locally associated with basaltic volcanism. Such "leaky" transforms usually have pull-apart basins, where small volumes of basalt erupt as lava flows and cinder cones. The lavas along the Dead Sea transform (Figure 20.14) and those near southern California's Salton Sea are good examples of this type of volcanism. Partial melting in the mantle occurs when it rises to fill the small voids created by the pull-apart basins.

The most characteristic type of metamorphism that occurs at transform plate boundaries is caused by the horizontal shearing motion. As the two plates grind against one another, fault breccias and fine-grained mylonite are created; at higher temperatures deeper in the crust, ductile deformation textures develop. Where an oceanic transform fault meets an active ridge crest, metamorphic recrystallization of basaltic lava flows and dikes is also aided by the influx of seawater into hot crust exposed by faulting in the fracture zone.

Why are earthquakes limited to shallow depths along transform faults?

GeoLogic   Transform(ation) of California

Offset pluton
160 km

North American
plate

Pacific
plate

Area of
image

Los Angeles

Courtesy of JPL/NIMA/NASA

This dramatic view of southern California is a hybrid between a map, a photograph, and a three-dimensional sculpture based on remote sensing and innumerable measurements of elevation. Clearly displayed are the features that led geologists to conclude that a transform fault system has shaped the terrain.

Observations

1. Note first of all, the sharp, nearly straight lines that separate the mountains from the basins.
2. A pronounced linear fabric in the mountains parallels these lines.
3. Geologists working on the ground find horizontal displacement and breccias along these zones showing that each one is a strike-slip fault.
4. Thus, the San Andreas is not a single fault, but a system of many faults (yellow lines).
5. A pluton is sliced and offset by one of the vertical faults in this system. Each part of the Triassic pluton is now separated by 160 km.
6. Earthquakes make this motion tragically obvious; each year 2 or 3 large earthquakes rock the San Andreas fault system causing horizontal displacements of several meters at a time.

7. Earthquake epicenters (red dots for selected earthquakes with magnitudes greater than about 4) clearly delineate the patches of broken ground found by field geologists and revealed here by satellite photography.

Interpretations

These observations clearly show that the lineaments seen on the satellite map were created by movement on a series of strike-slip faults that cut across California. As one block slides past another, strain builds up until the rocks break and release the energy during an earthquake. Repeated earthquakes along the faults, break, crush, and pulverize rocks along the fault plane. These brecciated zones are more susceptible to weathering and erosion. The linear fabric of the mountains was carved by weathering and erosion along these zones of weakness.

Long term movement on the San Andreas fault is shown by the offsets of many geological features which imply that the rate of displacement along this strand of the fault is 2 to 3 cm/y. These interpretations led to the conclusion that the San Andreas fault system is a transform plate boundary that separates the North American plate from the Pacific plate.

Key Terms

fracture zone (p. 606)
leaky transform (p. 610)
mylonite (p. 608)
pull-apart basin (p. 613)

ridge-ridge transform (p. 608)
ridge-trench transform
(p. 608)
serpentinite ridges (p. 614)

shearing (p. 606)
transform boundary (p. 606)
transform fault (p. 606)
transpression (p. 613)

transtension (p. 613)
trench-trench transform
(p. 608)

Review Questions

1. Contrast the movement of plates at a transform boundary with that at divergent and convergent boundaries.
2. How does a transform fault differ from a simple strike-slip fault?
3. What are the three main types of transform plate boundaries?
4. Draw a simple line map of the Romanche fracture zone (Figure 20.4). Show mid-ocean ridges, inactive fracture zones, the active fault zone, and the older colder side of the transform fault zone.
5. List two examples of each of the three main types of transform faults.
6. Why is there a steep cliff along one side of an oceanic fracture zone?
7. Compare the features of the Clipperton (small offset) with the Romanche (large offset) transform zones.
8. What are the effects of the cold wall of lithosphere where an oceanic spreading ridge meets a transform fault?
9. What geologic features mark the surface expression of a continental transform fault?
10. Outline the development of a pull-apart basin along a continental transform fault.
11. How can a compressional fold belt form along a transform plate boundary?
12. Where do you think oil might form in connection with a transform fault system such as the San Andreas Fault?
13. Why is seismicity common on a transform fault between two mid-ocean ridge segments but rare on the long fracture zone that extends beyond the ridge?
14. Why is volcanism rare along transform plate boundaries?
15. What kinds of metamorphic rocks would you expect to find along an oceanic transform fault zone?

21 Convergent Plate Boundaries

The convergence of two tectonic plates develops some of the most remarkable structural and topographic features on our planet. At convergent plate margins, great slabs of oceanic lithosphere slide ponderously into Earth's internal abyss—the deep mantle. As they slowly disappear from the surface, spectacularly deep trenches form graceful arcs on the seafloor. The subducted plates strive to reach mechanical and chemical equilibrium with the mantle, and, in the process, many of Earth's most dramatic landscapes and structures are created. Earthquakes, volcanic arcs, deep-sea trenches, and the continents themselves are the result of converging plates. But perhaps the most fascinating phenomena resulting from plate collision are the great mountain ranges of the world: the Alps, Andes, Rockies, and Himalayas. The internal structure of mountains shows intense folding, thrust faulting, and other features of intense horizontal compression resulting from plate collision. Young orogenic belts are elevated to great heights and subjected to vigorous erosion by streams and glaciers.

An example is the great Alpine fold and thrust belt of southern Europe. This ridge, Tête à l'Ane, in southern France, exposes thick layers of Mesozoic marine limestone and marl thrust high above sea level by the collision of two tectonic plates. It is underlain by high grade metamorphic gneisses intruded by granitic rocks—rocks that once were deep in the middle crust. Thick sequences of sedimentary strata, originally deposited on the seafloor, covered the basement. During of the collision of a series of small continental blocks with Europe, these rocks were cut by thrust faults and deformed by huge complex folds. Rocks that were once 10 km below the surface are now 5 km above sea level. Erosion has removed a large volume of rock so that what we see in the high peaks is only a small fraction of the total volume of rock.

Convergent plate margins are where continental crust is born, just as divergent plate margins are the birthplaces of oceanic crust. This is perhaps the most important fact to remember as you study these important plate boundaries. This new granitic crust is so buoyant that it can never sink into the denser mantle below. Consequently, the rocks of the continents are much older than those in the ocean basins. They preserve a record of much of Earth's ancient history—a record in the form of faults, folds, mountain belts, batholiths, and sediments.

In this chapter, we examine the basic types of convergent boundaries, how they develop, and the rocks they form. We review their intimate relationship to Earth's most destructive earthquakes and volcanoes and to deformation at all scales. In addition, we will see that the deformed rocks, high-grade metamorphic rocks, and igneous intrusions developed at convergent plate margins are the building blocks of the continents.

Major Concepts

1. Convergent plate boundaries are zones where lithospheric plates collide. The three major types of convergent plate interactions are (a) convergence of two oceanic plates, (b) convergence of an oceanic and a continental plate, and (c) collision of two continental plates. The first two involve subduction of oceanic lithosphere into the mantle.

2. Plate temperatures, convergence rates, and convergence directions play important roles in determining the final character of a convergent plate boundary.

3. Most subduction zones have an outer swell, a trench and forearc, a magmatic arc, and a back-arc basin. In contrast, continental collision produces a wide belt of folded and faulted mountains in the middle of a new continent.

4. Subduction of oceanic lithosphere produces a narrow, inclined zone of earthquakes that extends to more than 600 km depth, but broad belts of shallow earthquakes form where two continents collide.

5. Crustal deformation at subduction zones produces mélange in the forearc and extension or compression in the volcanic arc and back-arc areas. Continental collision is always marked by strong horizontal compression that causes folding and thrust faulting.

6. Magma is generated at subduction zones because dehydration of oceanic crust causes partial melting of the overlying mantle. Andesite and other silicic magmas that commonly erupt explosively are distinctive products of convergent plate boundaries. At depth, plutons form, composed of rock ranging from diorite to granite. In continental collision zones, magma is less voluminous, dominantly granitic, and probably derived by melting of preexisting continental crust.

7. Metamorphism at subduction zones produces low-temperature–high-pressure facies near the trench and higher-temperature facies near the magmatic arc. Broad belts of highly deformed metamorphic rocks mark the sites of past continental collision.

8. Continents grow larger as low-density silica-rich rock is added to the crust at convergent plate boundaries and by terrane accretion.

Types of Convergent Plate Boundaries

> Three distinctive types of convergence are recognized: (1) the convergence of two oceanic plates, (2) the convergence of a continental plate and an oceanic plate, and (3) the convergence of two continental plates.

We have learned a great deal about convergent boundaries from geophysical studies that include gravity surveys, measurements of heat flow, and seismic-reflection profiles. In addition, geochemical studies of the igneous rocks erupted at these boundaries tell the tale of their generation, rise, and differentiation. These results, combined with field studies of ancient and modern mountain belts and arcs, give us an integrated picture of the geologic features of convergent plate margins and the processes that have shaped them.

If you study **Figure 21.1**, you will observe that **convergent boundaries** involve either the convergence of two oceanic plates (numerous trenches in the Pacific), the convergence of an oceanic and a continental plate (the western margin of South America), or two continental plates (India and Asia).

Convergence of Two Oceanic Plates

The simplest type of convergent plate margin—**ocean-ocean convergence**—consists of two oceanic plates. As the plates collide, one is thrust under the other, forming a **subduction zone.** The subducting plate descends into the mantle, where it is heated,

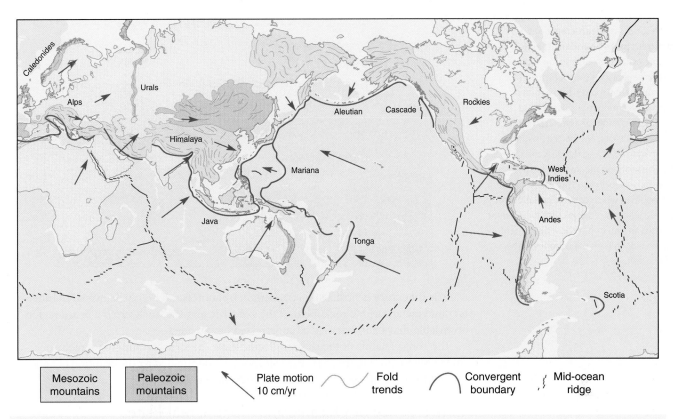

Figure 21.1 Convergent plate margins are marked in two ways: either by deep trenches, where plates of oceanic lithosphere converge and one descends to be recycled into the mantle, or by high folded mountain belts. In both cases, earthquakes and magma are generated. Absolute plate motions are shown with arrows. Trenches and mountain belts are labeled.

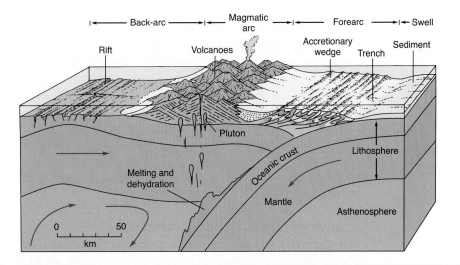

Figure 21.2 Ocean-ocean convergence is dominated by volcanic activity and construction of an island arc. Features developed include an outer swell, a forearc, a volcanic arc, and a back-arc basin. The forearc is underlain by an accretionary wedge—sediment scraped off the downgoing slab. Widespread metamorphism and large granitic intrusions are rare or absent.

triggering the generation of magma. The magma, being less dense than the surrounding rock, rises and erupts on the seafloor, ultimately building an arc of volcanic islands (**Figure 21.2**). Andesite is the volcanic rock that characteristically forms at such sites.

Several important structural and topographic features form at many subduction zones (Figure 21.2). A broad rise or bulge in the downgoing plate, known as an **outer swell**, commonly develops where the plate bends to dive down into the mantle. Closer to the island arc, a deep **trench** and a **forearc ridge** form. The forearc commonly

What are the typical features produced by ocean-ocean convergence?

Figure 21.3 At ocean-continent convergent plate boundaries, major geologic processes include formation of an accretionary wedge, deformation of the continental margin into a folded mountain belt, metamorphism due to high pressures and high temperatures in the mountain roots, and partial melting of the mantle overlying the descending plate. The resulting magmas commonly differentiate to form andesite and even more silicic magmas, which cool to form plutons. Explosive volcanism is also common. Granitic batholiths and metamorphosed sedimentary rocks develop in the deeper zones of the orogenic belt.

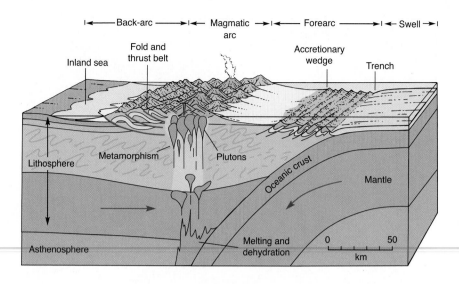

traps sedimentary deposits and is underlain by faulted and highly deformed sedimentary and metamorphic rock. Behind the **volcanic arc**, the **back-arc** is a broad region of variable character that may be compressed or extended.

The Tonga Islands in the western Pacific show the structure and topography of a simple island arc (Figure 21.1). The volcanoes are dominated by the eruption of andesite, and the back-arc region is extending to form a basin.

Convergence of Oceanic and Continental Plates

A subduction zone also develops where oceanic and continental plates converge (**Figure 21.3**). The less-dense continental crust always resists subduction into the dense mantle and overrides the oceanic plate. Consequently, the volcanic arc forms on the continent, and compression may deform the continental margin into a **folded mountain belt**. Moreover, the deep mountain roots are intruded by granitic plutons and metamorphosed. In this setting, a trench, a deformed forearc, and a back-arc region of deformation are all important.

Ocean-continent convergence has created the Andes Mountains of western South America. The Cascade Range of western North America is another example, forming above an east-dipping subduction zone. An older example is the Rocky Mountain chain of western North America, which was deformed during late Mesozoic and early Tertiary time (about 150 to 50 million years ago). The Appalachian Mountains in the eastern United States were deformed several times in the Paleozoic Era (about 500 to 300 million years ago). During some part of their histories, all of these mountain chains experienced subduction of oceanic lithosphere beneath a continental margin.

Convergence of Two Continental Plates

If both converging plates contain continental crust, neither is subducted into the mantle, because continental crust is too buoyant (**Figure 21.4**). In **continent-continent convergence,** one plate overrides the other for a short distance. In sharp contrast to the other two types of convergence, there is no outer swell, deep subduction zone, trench, or forearc wedge. Instead, both continental masses become compressed, and the continents ultimately are "fused" into a single block, with a folded mountain belt marking the line of the suture. **Orogenic metamorphism** and granite generation mark this kind of convergence.

What structures and rock sequences are produced by continent-continent convergence?

Collisions between two continents are occurring in several places. The most dramatic is the one that produced the Himalaya Mountains and Tibetan Plateau of southern Asia. The Himalayas, Earth's highest mountain belt, is a wide and highly deformed zone of mountains that rose as India collided with the Eurasian continent. Russia's Ural Mountains also formed during late Paleozoic time when the Siberian continental

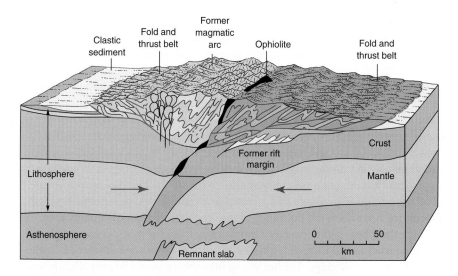

Clastic sediment · Fold and thrust belt · Former magmatic arc · Ophiolite · Fold and thrust belt · Former rift margin · Crust · Mantle · Lithosphere · Asthenosphere · Remnant slab

0   50
km

Figure 21.4 Continent-continent collision is marked by complete subduction of the oceanic crust. A high mountain belt forms by folding, thrust-faulting, and doubling of the crustal layers as one continent is thrust beneath the other. Ophiolites are thrust into the suture zone. Granite magma and high-grade metamorphic rocks form deep in the mountain belt.

(A) The young Himalaya mountain chain formed as a result of the ongoing collision of India and Eurasia. It is Earth's highest range, with some peaks more than 7000 m above sea level, but deep valleys have been cut by river and glacier erosion.

(B) The Ural Mountains formed in the late Paleozoic (about 350 million years ago) when Europe collided with Asia. The mountains have been deeply eroded so that no peaks are higher than 2000 m, but the internal structure reveals its origin.

Figure 21.5 High folded mountain belts are formed during the collision of two continents but are gradually eroded away, as seen in this comparison of the Himalayas and Urals.

mass collided with Europe. The Urals are not tectonically active and are deeply eroded, so they are much lower than the Himalayas (**Figure 21.5**).

Factors Influencing the Nature of Convergent Plate Boundaries

> Plate buoyancy, convergence rates and directions, and the thermal structure of a subduction zone are all important to the development of convergent plate boundaries.

Plate Buoyancy

Many geologic processes at convergent plate boundaries are influenced by density differences that make one plate more buoyant than the other. The most obvious expression of this fact is that subduction occurs because oceanic plates cool and become denser than the underlying mantle. Another important example is the sharp contrast in density between oceanic plates and continental plates. Oceanic crust is composed mostly of basalt (about 3.0 g/cm³) and is much denser than continental crust (about 2.8 g/cm³). Thus, at ocean-continent plate boundaries, the oceanic plate descends beneath the continental plate.

Less obvious, but nevertheless important, are density differences caused by differences in the thickness of the crust. For example, seamounts and thick plateaus of basaltic lava that erupted on the seafloor can make the lithosphere slightly more buoyant than sections of oceanic lithosphere that lack this thick crust and are made mostly of dense mantle instead. Consequently, oceanic lithosphere with thick crust resists subduction and may bend into the mantle at a relatively low angle. Moreover, as a seamount chain approaches a subduction zone, it may clog up the subduction process or become scraped off the oceanic lithosphere and accreted to an island arc or continental margin.

Temperature also affects the buoyancy of the lithosphere. We have already seen how oceanic lithosphere gets denser as it moves away from the mid-oceanic ridge and cools. Imagine what would happen if a continental subduction zone developed immediately next to an oceanic ridge. The hot young lithosphere would be only slightly less buoyant than the old cold continental lithosphere. Consequently, it would not subduct readily and the hot slab might dip into the mantle at a low angle.

The Thermal Structure of Subduction Zones

The physical and chemical behaviors of most materials are profoundly affected by temperature. There are many familiar examples. Warm honey flows much more readily than refrigerated honey. Snow a few degrees below freezing is crisp and brittle, but near its melting point, it is slushy and flows. A similar situation occurs in lithospheric plates. The temperature variations at convergent plate margins exert strong controls on rock dynamics.

The Cold Slab   The most obvious feature of the thermal structure of a subduction zone (**Figure 21.6**) is the deep penetration of the cold subducting plate into the hot asthenosphere. Rocks are very good insulators, and heat diffuses very slowly through them. Consequently, subducted lithosphere heats very slowly as it moves down through the hot mantle. As a result, temperatures as low as 400°C may be found in the plate at a depth of 150 km. This is a strikingly anomalous situation; in an area with a normal temperature gradient, the temperature would be as high as 1200°C at this

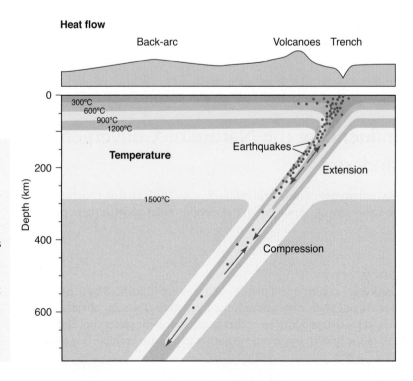

Figure 21.6 The thermal structure of a subduction zone is dominated by the underthrusting of a thick, cold slab of oceanic lithosphere into the hot mantle. The descending lithosphere remains cold compared with the surrounding rock to considerable depth. Gradually, the slab heats as it dives deeper and deeper into the hot mantle. The heat flow from the volcanic arc is higher than adjacent regions because heat is carried upward by magma and perhaps by convection in the wedge-shaped area above the slab. Note that earthquakes (dots) occur only in the cold, brittle parts of the slab.

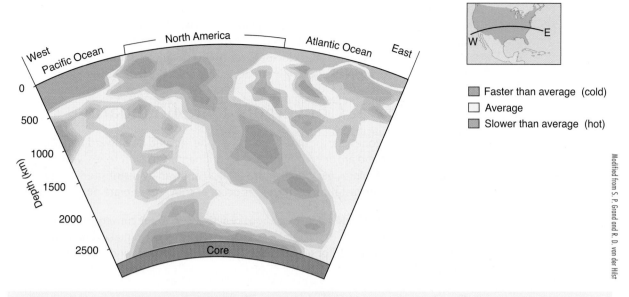

Figure 21.7 **The trace of a cold subducted slab** appears to extend from the continental margin all the way to the core-mantle boundary in this seismic tomograph of Earth's interior beneath North America. Blue represents the cold parts of the mantle, which have high seismic wave velocities; red represents the warmest parts of the mantle, which have lower seismic velocities.

depth. Even at a depth of more than 600 km, the central zone of the subducted plate is still as much as 600°C cooler than the surrounding mantle. Because rocks are such poor conductors of heat, it would take an estimated 12 million years for the plate to reach even this relatively low temperature.

The fact that the subducting lithosphere is so much colder than the hot asthenosphere through which it moves explains a great deal about the slab's behavior. The cold slab is much more brittle, stronger, and resistant to ductile flow. It resists mixing with the rest of the mantle and continues to move downward as a discrete plate for hundreds of kilometers. Recent seismic investigations show that the cool temperature of a subducting slab persists all the way to the bottom of the mantle—a distance of about 2700 km and a time of more than 100 million years—before it completely warms up (**Figure 21.7**).

The Hot Arc   A second important feature of the thermal structure of a convergent plate boundary is the elevation of the heat flow in the volcanic arc (Figure 21.6). The high heat flow in this region reflects the large amount of heat transported from the mantle by magmas generated in the subduction zone.

In addition to this magmatic heat transport, flow in the asthenosphere above the subducting plate may enhance heat flow. Because the subducting plate drags the overlying asthenosphere downward, hotter asthenosphere from a greater depth must flow upward to replace it, as shown by the arrows. This convective movement delivers extra heat to the region immediately above the descending plate.

Why are a cold slab and a hot arc produced at convergent margins?

Plate Motions: Directions and Velocities

As you might expect, the velocity and direction of plate motion play important roles in the dynamics of convergent plate margins (Figure 21.1). Plates moving directly toward each other (such as the Pacific and Eurasian plates) converge with high energy and develop long, continuous subduction zones, intense compressive deformation, and vigorous igneous activity.

Plates converging at oblique angles (such as the North American and the Pacific plates) tend to slide past each other and have a strong shearing component. They develop short discontinuous convergent boundaries interspersed with long transform faults. In fact, many convergence zones are like this and have components of extension or transform movement along the plate boundary.

Modified from S. P. Grand and R. D. van der Hilst

Extremely rapid plate motion develops extensive geologic activity, such as the "Ring of Fire" around the fast-moving Pacific plates. Likewise, the Australian-Indian plate is moving at a high rate northward into the nearly fixed Eurasian continent, explaining the high Himalaya range. Plate velocities are also important for the angle of subduction. Rapidly moving plates generally subduct at lower angles than slower plates.

Seismicity at Convergent Plate Boundaries

In subduction zones, earthquakes occur in a zone inclined downward beneath the adjacent island arc or continent. But in continental collision zones, where subduction is minimal, earthquakes are shallow and widely distributed. Many of Earth's most devastating earthquakes occur at convergent plate boundaries.

The most widespread and intense earthquake activity occurs at convergent plate boundaries. Almost 95% of the total energy released by all earthquakes comes from these margins. These belts of seismic activity are obvious on the world seismicity map (**Figure 21.8**).

Earthquakes and Subduction Zones

A strong concentration of shallow, intermediate, and deep earthquakes coincides with the descending plate in a subduction zone. Earthquakes beneath the Tonga arc in the South Pacific illustrate this point nicely (**Figure 21.9**). The **inclined seismic zone**

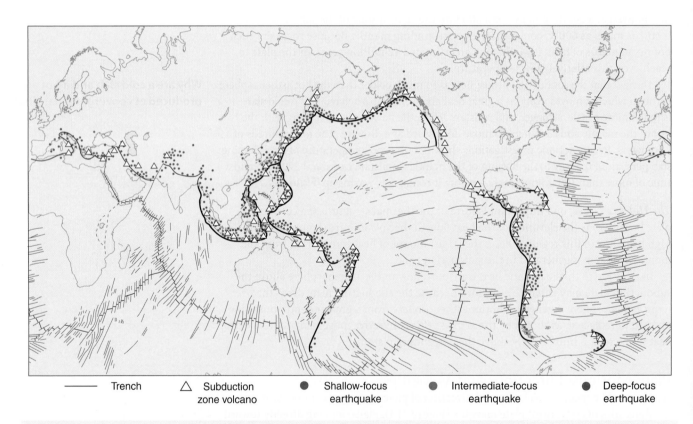

| —— Trench | △ Subduction zone volcano | ● Shallow-focus earthquake | ● Intermediate-focus earthquake | ● Deep-focus earthquake |

Figure 21.8 Earthquakes and volcanoes at convergent plate boundaries are common. Earthquakes occurring here are the most devastating. This map shows the locations of some of the tens of thousands of earthquakes that occurred during a 5-year period. Shallow-focus, intermediate-focus, and deep-focus earthquakes form the inclined zones of earthquakes characteristic of subduction zones. Volcanoes at subduction zones are also the most destructive kind. Subduction zone volcanoes form the "Ring of Fire" around the Pacific Ocean and the arcs of the Mediterranean and Indonesia. Shorter volcanic arcs are found in the Caribbean and South Atlantic. Zones of continental collision, such as the Himalayan region, are quite different from subduction zones. They have abundant, but shallow, earthquakes, and they lack prominent active volcanoes.

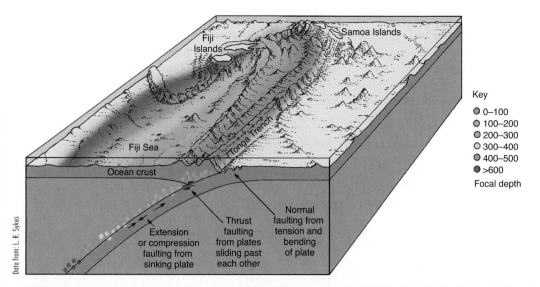

Data from: L. R. Sykes

Figure 21.9 **Earthquake foci in the Tongan region** in the South Pacific occur in a zone inclined from the Tonga Trench toward the Fiji Islands. The top of the diagram shows the distribution of earthquake epicenters, with focal depths represented by different-colored bands. The cross section on the front of the diagram shows how the seismic zone is inclined from the trench. The colored dots represent different focal depths. This seismic zone accurately marks the boundary of the descending plate in the subduction zone.

plunges into the mantle to a depth of more than 600 km. The angle of the seismic zone is usually between 40° and 60°.

Near the top of the subducting slab, a zone of shallow earthquakes forms where the downgoing slab shears against the overriding plate. Deeper in the subduction zone, earthquakes originate within the slab and not in the surrounding asthenosphere (Figure 21.9). These earthquake zones must correspond to the regions of most intense shearing in the cold, brittle part of the plate.

Why do earthquakes at subduction zones occur so deeply within the mantle? In no other tectonic setting do we detect earthquakes much deeper than 25 km. The answer seems to lie in the thermal structure of a subduction zone (Figure 21.6). Cold slabs of lithosphere plunge down into the mantle in these zones. These "cold" rocks break by brittle fracture when stress exceeds their elastic limits, generating earthquakes. The surrounding warmer mantle also deforms, but it does so by slow ductile flow. In addition, some deep earthquakes may be caused by abrupt metamorphic mineral changes in the subducting slab.

Not all earthquakes in subduction zones are generated by simple compression. Studies of seismic waves indicate that the type of faulting varies with depth. For example, near the walls of the trench, normal faulting is typical, resulting from tensional stresses generated by the bending of the plate as it enters the subduction zone (Figure 21.9). In the zone of shallow earthquakes, thrust faulting dominates as the descending lithosphere slides beneath the upper plate. At intermediate depths, extension and normal faulting result when a descending plate that is denser than the surrounding mantle sinks under its own weight. Compression results when the mantle resists the downward motion of the descending plate. In many subduction zones, the deepest earthquakes result from compression, indicating that the mantle material at that depth resists the movement of the descending plate.

How does an earthquake zone at a convergent margin differ from that of a divergent margin?

Earthquakes and Continental Collision

Earthquakes in continental collision zones are spread out over much broader areas than those involving subduction of oceanic crust and do not form an inclined zone of seismicity. For example, the Himalaya and adjacent Tibetan Plateau are marked by a wide belt of shallow earthquakes that reveals the northward movement of the Indian plate, currently about 0.5 cm/yr. This east-west zone of earthquakes is about 2500 km long and between 1200 and 2000 km wide (Figure 21.8). Moreover, because there is no longer a cold, brittle slab subducting beneath this region, almost all of the earthquakes are relatively shallow. Similarly, the convergent boundary between the Africa-Arabia and the Eurasian plate coincides with a wide seismic belt in the Mediterranean region and across Turkey, Iraq, and Iran (Figure 21.8).

Deformation at Convergent Boundaries

Intense deformation occurs along convergent plate margins. At subduction zones, mélange is produced in the forearc accretionary wedge. In some arc and back-arc regions, compression creates folds and thrust faults, but in others extension causes rifting. Collision of two continents is marked by strong horizontal compression that causes folding and thrust faulting.

The single most distinctive feature of the rocks at a convergent plate margin is their structural deformation by folding and faulting. The scale of deformation ranges from small wrinkles in mineral grains or fossils to huge folds and faults tens of kilometers wide that combine to form mountain belts hundreds of kilometers wide and thousands of kilometers long. Understanding this deformation is key to understanding the structure of most continental crust, because convergent plate boundaries are where most continental crust forms.

Compression at Subduction Zones

Strong contraction caused by horizontal compression occurs at many convergent plate boundaries, including ocean-continent and continent-continent boundaries. Especially striking is the deformation in the forearc region of a subduction zone. Unconsolidated sediment is scraped off the descending plate and piles up into a long wedge—called the **accretionary wedge**—in front of the overriding plate (**Figure 21.10**). The structures in an accretionary wedge are similar to those in the deformed snow that accumulates in front of a moving snow plow. The weak layer of snow is sliced off the solid road surface and stacks up in a thick, internally deformed pile immediately in front of the rigid blade. In this analogy, the overriding plate is the snowplow.

How is an accretionary wedge formed?

Folds of all sizes form in the accretionary wedge (Figure 21.10). As expected from the orientation of the applied stresses, the hinge planes of these folds are parallel to the trench and dip in the same direction as the subduction zone. Temperature and pressure changes during compression cause metamorphism, and slaty cleavage develops during folding. Thrust faults also cut through the soft sediment (Figure 21.10). These generally dip in the same direction as the subduction zone. As subduction continues and more sediment is scraped off, the deformed mass grows toward the trench, simultaneously shortening and thickening. Consequently, the front of the accretionary wedge grows steeper until it becomes unstable. It then collapses along faults that allow extension and thinning of the deformed mass. This cycle of tectonic shortening followed by gravitational collapse occurs repeatedly. The net result is uplift of deeper rocks to the surface.

This complicated deformation produces an accretionary wedge that is a chaotic mixture of rock types known as **mélange**, one of the most structurally complex rock bodies in the crust. The deformed rocks are mostly sediments, but in some subduction zones, volcanic seamounts or other fragments of igneous oceanic crust also are scraped off the downgoing plate and incorporated into the wedge.

Not all of the sediment on the oceanic crust is scraped off to form an accretionary wedge. Apparently some is subducted deep into the mantle. As much as 20% to 60% of

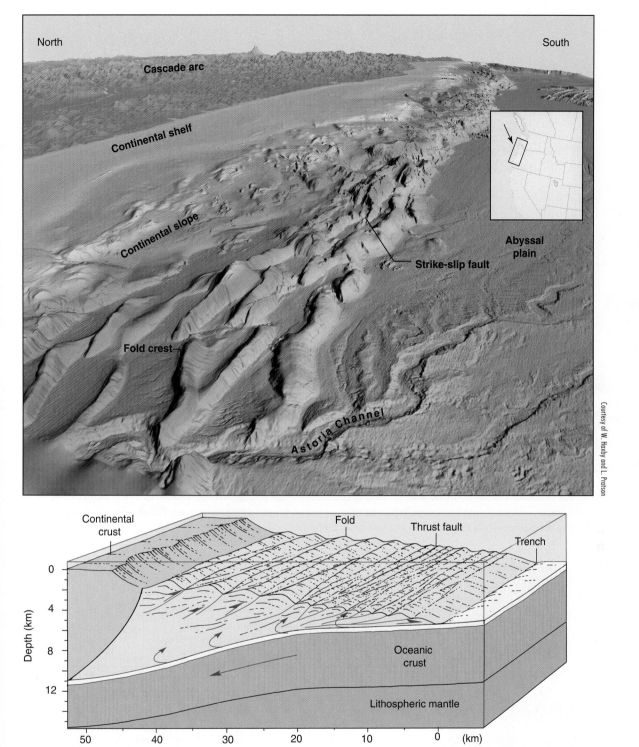

Courtesy of W. Haxby and L. Pratson

Figure 21.10 Accretionary wedges form at convergent plate margins as sediment and some igneous rock are scraped off the downgoing slab of oceanic rock. A southward-looking sonar image of the accretionary wedge along the coast of Oregon reveals that its surface is like a folded carpet. The ridges mark anticlines and areas where thrust sheets are stacked on top of one another. Transverse faults are strike-slip faults. Here the Juan de Fuca plate is being subducted beneath the North American plate. The block diagram shows the internal structure of an accretionary wedge derived from a seismic reflection profile. Faults and folds deform the rocks in the wedge. As sediment is removed from the downgoing plate, it is added to the base of the accretionary wedge. Stacks of thrust faults form above the downgoing oceanic crust. Folds of all sizes form between the thrust faults.

© Hubert Stadler/CORBIS

Figure 21.11   The Andes Mountains of South America are forming by subduction of oceanic lithosphere beneath continental crust. Here, in the Atacama Desert of northern Chile, you can see a row of andesitic stratovolcanoes towering over an intensely deformed series of layered sedimentary rocks. Deep in the mountain belt, metamorphic rocks are probably forming today.

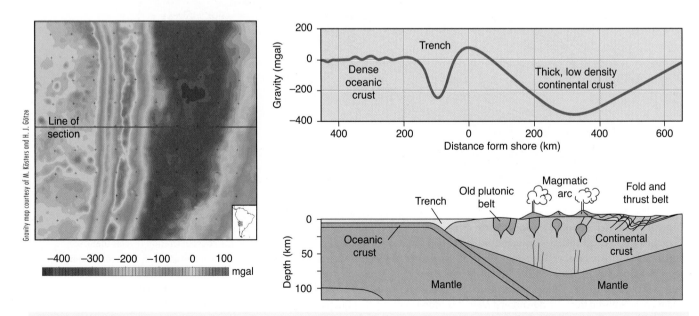

Gravity map courtesy of M. Kösters and H. J. Götze

Figure 21.12   The thick crust beneath the Andes is revealed by the gravity field. The gravity profile and geologic cross section show that the outer bulge on the downgoing slab is marked by a positive gravity anomaly (red tones). Gravity is lower over the deep trench (light blue), which is filled with low-density water, and over the accretionary wedge, which is made of low-density sediment. Gravity is highest in the part of forearc that is underlain by the cold, dense subducting slab. The volcanic arc and folded mountain belt are marked by the lowest gravity anomalies (deep purples) because of the great thickness (as much as 70 km) of low-density crust beneath the Andes.

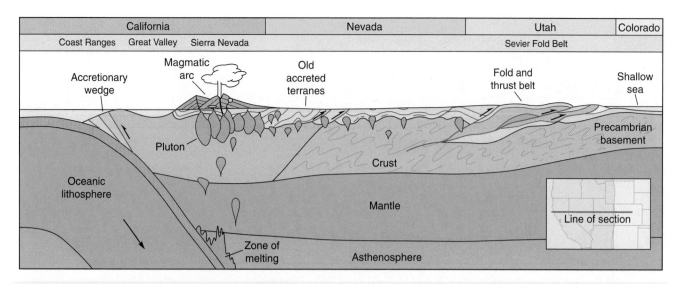

Figure 21.13 **Much of western North America developed at a convergent plate margin** 150 to 60 million years ago. The former locations of the accretionary wedge, magmatic arc, and folded mountain belt are shown. A sedimentary basin formed in the back-arc region because of tremendous weight added by the mountain belt. The eroding mountains supplied sediment for the thick deposits that have accumulated along the margins of the mountains. With increasing distance from their sources, the sedimentary environments include alluvial fans, floodplains, deltas, and shallow-marine settings where shale interfingered with carbonates.

the sediment on the oceanic crust is subducted. Some sediment may be carried deep into the mantle, possibly to the core-mantle boundary.

If a subduction zone lies along a continental margin, compression produces a long **orogenic belt** (folded mountain belt) parallel to the continental margin. The Andes, parts of the Rockies, and the European Alps are classic examples.

In the central Andes, horizontal compression and crustal elevation are active today in the arc and back-arc (**Figure 21.11**). Here, compression is causing pronounced folding, thrust faulting, and thickening of the crust. The fold hinges and thrust faults trend parallel to the arc and the linear belts of granitic batholiths. Intrusion of granitic plutons has added to the deformation. Not surprisingly, much of the lower and middle crust has been metamorphosed from the tremendous heat and pressure during compression and intrusion. The crust beneath the Andes is 75 km thick, probably thickened by intrusion of mantle-derived magmas as well as by folding and contraction (**Figure 21.12**). The deformed continental margin and its volcanic crest rise up to 6 km above sea level. Rapid erosion of such mountain belts occurs simultaneously with volcanism, intrusion, and deformation.

Another good example of compression at an ocean-continent boundary is provided by the Mesozoic history of western North America (**Figure 21.13**). During the Early Mesozoic, as oceanic lithosphere was subducted beneath the continent, a long magmatic arc developed along the western margin of North America. Behind the arc, thrusting and folding created a mountain belt. This extra load depressed the crust in much the same way that thick glacial ice sheets make the crust subside. Because of this subsidence, a shallow sea expanded onto the continent behind the thrust belt. Clastic sediments were shed from the thrust belt, forming coarse alluvial fans, floodplains, deltas, and shallow-marine shales that interfinger with carbonates deposited in deeper water. Thin beds of volcanic ash, erupted from volcanoes in the arc, also fell into the shallow sea. Today, these sedimentary basins are rich in natural resources, including coal, oil, and natural gas.

The European Alps have a long and complicated history that started with convergence of the oceanic part of the African plate beneath Eurasia (see the photograph that opens this chapter). Convergence has not yet completely consumed the oceanic crust

What is distinctive about compression along a subduction zone at a continental margin?

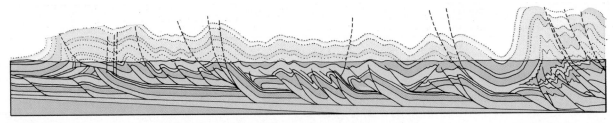

(A) The Canadian Rockies contain both folds and thrust faults.

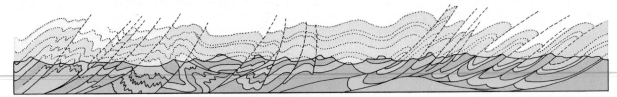

(B) The Appalachian Mountains consist of tight folds and thrust faults. These geologically older mountains have been eroded to within 1000 to 3000 m of sea level. Resistant sandstones form the remaining mountain ridges.

(C) The Alps are a young range that consist of complex folds, many of which are overturned.

Figure 21.14 **The structure of folded mountain belts** reflects intense compression at convergent plate boundaries. Yet, each range can have its own structural style, as shown in these cross sections.

of the Mediterranean Sea. Remnant volcanic arcs and scraps of unsubducted oceanic crust mark places where suturing of the two continents is still incomplete. In the Alps, great overturned folds called **nappes** (French, "tablecloths") show enormous amounts of crustal shortening (**Figure 21.14C**). The rocks are so strongly deformed that spherical pebbles were stretched into rods as much as 30 times longer than the original pebble diameters! Deformation is most intense near the continental margin and dies out northward toward the continental interior.

Compression in Continental Collision Zones

When two lithospheric plates that carry continents converge, the orogenic belt has very different characteristics than those formed during ocean-continent convergence. An excellent example of deformation related to continental collision is the Himalaya chain of southern Asia. The course of this collision is shown in **Figures 21.15** and **21.16.**

The vast Himalaya orogenic belt formed during the past 100 million years, as oceanic lithosphere that was carrying India moved northward and was subducted beneath Asia. As a result, an accretionary wedge developed on the southern edge of Asia. The sediments along the continental margin were also folded and faulted (Figure 21.15A). Simultaneously a magmatic arc developed as oceanic lithosphere was progressively consumed at the subduction zone. When the two continental masses began to collide about 50 million years ago, subduction-zone volcanism ceased (Figure 21.15B). (The floors of the Black Sea and the Caspian Sea are remnants of oceanic lithosphere

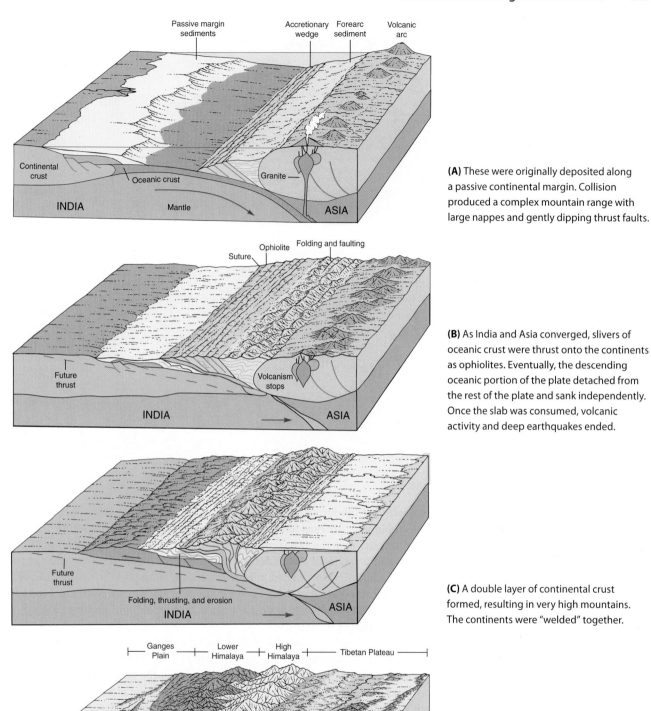

Passive margin sediments

Accretionary wedge

Forearc sediment

Volcanic arc

Continental crust

Oceanic crust

Granite

INDIA

Mantle

ASIA

(A) These were originally deposited along a passive continental margin. Collision produced a complex mountain range with large nappes and gently dipping thrust faults.

Suture

Ophiolite

Folding and faulting

Future thrust

Volcanism stops

INDIA

ASIA

(B) As India and Asia converged, slivers of oceanic crust were thrust onto the continents as ophiolites. Eventually, the descending oceanic portion of the plate detached from the rest of the plate and sank independently. Once the slab was consumed, volcanic activity and deep earthquakes ended.

Future thrust

Folding, thrusting, and erosion

INDIA

ASIA

(C) A double layer of continental crust formed, resulting in very high mountains. The continents were "welded" together.

Ganges Plain

Lower Himalaya

High Himalaya

Tibetan Plateau

Crust melts to form granite

INDIA

ASIA

(D) During high-grade metamorphism in the roots of the mountain range, the continental crust itself may partially melt to form granite with distinctive compositions; these are found in no other tectonic setting.

Figure 21.15 Continental collision formed the Himalaya Mountains and involved the deformation of oceanic and shallow marine sedimentary rocks.

Figure 21.16 The Himalaya mountain belt formed by the collision of the Indian and Eurasian plates. One hundred thirty million years ago, India rifted away from Antarctica and Africa and moved northward until it collided with Asia. The collision started about 50 million years ago and built the high Himalaya range and the Tibetan Plateau to the north.

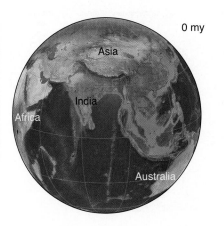

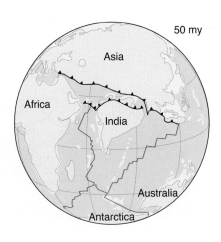

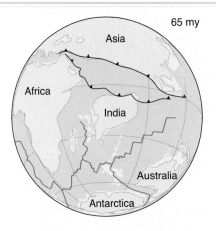

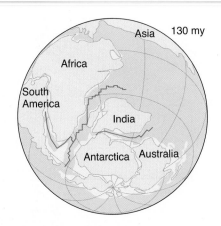

that were not subducted.) Slices of oceanic crust were thrust onto the continent. These ophiolites are now found north of the Himalayas along the suture marking the boundary between the two continents (Figure 21.15C). As the two continents collided, India was thrust under Asia, effectively doubling the thickness of the crust to about 70 km. Its buoyancy prevented it from descending deep into the mantle. Thrust faults and folds formed a belt of deformed mountains, mainly in the overriding Asian plate. Much of the deformation was ductile and accompanied high-grade metamorphism in the deeper part of the crust.

Because of compression and crustal thickening, the Himalayas and the extensive highlands of the Tibetan Plateau rose (Figure 21.15D). Mount Everest, Earth's highest peak, lies in the High Himalaya that developed over the greatly thickened part of the crust. The continental masses were welded together to make a single large continent with an internal mountain range. At some point, the slab of descending oceanic lithosphere must have become detached and then sank, independent of the Indian continent. When it sank, volcanic activity and deep earthquakes ceased.

Deformation associated with the collision drove Southeast Asia and parts of China eastward along strike-slip faults that fan away from India (**Figure 21.17**). Of course, erosion of the mountain belt continued throughout this long process.

It is interesting to contrast compression in the Himalayas with compression in the Appalachian Mountains of the eastern United States. Compression in the Appalachians was produced by several episodes of subduction of oceanic lithosphere, followed by collision between North America and Europe or Africa. These collisions occurred during the Paleozoic Era, making the Appalachian Mountains more than 300 million years old. A cross section of the Appalachians shows a different style and extent of deformation than in the Himalayas (Figure 21.14B). The major structural features are tight folds and thrust faults. Orogenic (or regional) metamorphism accompanied

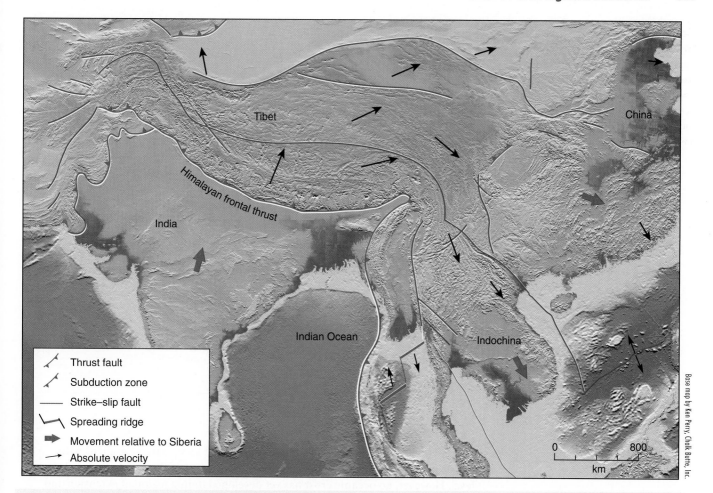

Figure 21.17 Complex folds, mountains, and plateaus mark the collision zone between India and Eurasia, as shown on this digital shaded relief map. The collision also drove parts of Southeast Asia and China eastward along lengthy strike-slip faults or shear zones.

Legend:
- Thrust fault
- Subduction zone
- Strike–slip fault
- Spreading ridge
- Movement relative to Siberia
- Absolute velocity

Labels: Tibet, China, India, Himalayan frontal thrust, Indian Ocean, Indochina

the collisions, but most of the plutonic rocks are subduction-related and preceded the collisions.

Extension at Convergent Boundaries

It may seem paradoxical that extension occurs in regions where two plates of lithosphere converge. However, mild extension—not compression—is the dominant deformation occurring at most oceanic volcanic arcs and at some continent arcs as well. Before we consider why, let us describe the evidence for extension and its effects.

Extension Above Subduction Zones In island arcs and some continental arcs, grabens and normal faults typically are centered on the active volcanic region. For example, the modern-day andesitic composite volcanoes of Ecuador lie in a graben hundreds of kilometers long. The trends of these faults are parallel to the trench, indicating that the direction of extension is perpendicular to the trench. Extension like this is very common in island arcs. Compression, marked by thrust faults and folding, is found in few modern-day island arcs.

The back-arc basins found behind most oceanic island arcs also reveal the effects of extension. These shallow oceanic basins are traversed by normal faults and have high heat flow and active seafloor volcanism. Extension may ultimately lead to **back-arc spreading**. The Mariana and Tonga-Kermadec arcs of the western Pacific have such basins. The back-arc regions of many continental arcs are also marked by extension and subsidence like that behind oceanic island arcs. For example, back-arc extension is active in the Aegean Sea behind a volcanic arc rooted in continental

Figure 21.18 The Aegean back-arc basin developed in continental crust above a subduction zone in the eastern Mediterranean Sea. Like back-arc basins formed within ocean basins, the Aegean basin has subsided, and normal faulting has caused extension. However, no oceanic crust has yet developed.

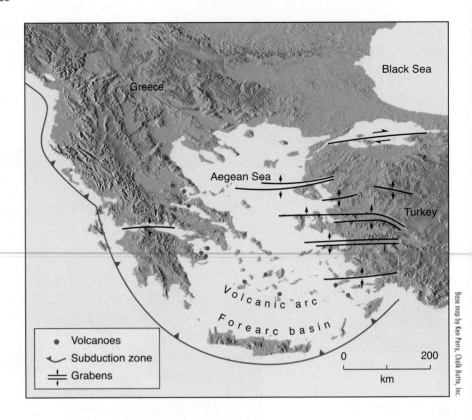

Volcanoes
Subduction zone
Grabens

0   200
km

Base map by Ken Perry, Chalk Butte, Inc.

crust (**Figure 21.18**). Normal faulting has produced many narrow grabens with trends that are parallel to the arc and has allowed most of the area to drop below sea level. However, no oceanic crust has yet developed in this region of extension and subsidence.

The islands of Japan, with their striking composite volcanoes and active subduction zones, were once part of the Asian mainland before back-arc extension opened the sea of Japan. This sea marks more extensive development of rifting above a subduction zone and is underlain by oceanic crust.

Extension and Continental Collision   As a secondary effect of the collision of two continents, extensional tectonics may also develop. The Rhine Graben of central Europe was created by the collision that raised the Alps. Lake Baikal in southern Siberia lies in one of Earth's deepest continental rifts, but the stresses responsible for it are related to the collision that formed the Himalayas.

Compression Versus Extension   What controls whether a particular convergent margin experiences horizontal compression or extension? No one yet knows for sure, but several possible causes have been identified. One important variable might be the angle of the subducting plate. For example, the Mesozoic fold and thrust belt of western North America formed during a period of rapid convergence. The subducting plate was probably inclined at a low angle, perhaps even dragging along the base of the overriding lithosphere. Such a close coupling between subducting plate and overriding plate may cause contraction in the back-arc.

In addition, the subduction of young, hot, and buoyant oceanic lithosphere that drags along the base of the overriding slab may be important for the development of some fold and thrust belts. Another way to cause a compressional fold and thrust belt is by collision of a continent with minor arcs or continental fragments. On the other hand, back-arc extension may be dominant where these factors are missing.

Why does extension occur in back-arc basins?

One cause of extension at convergent plate boundaries may be the convective flow of the mantle beneath the arc (Figure 21.2). As described above, the downgoing slab may drag the viscous mantle with it, causing hot asthenosphere from deeper in the mantle to flow upward and take its place. A convection pattern is thus formed in the asthenosphere above the subducting plate. It thins the lithosphere and may cause spreading behind the volcanic arc.

The absolute motion of the overriding plate (rather than the more obvious relative movement) also may play a role in developing extensional structures (Figure 21.1). For example, extension is common in arcs where the overriding plate is moving away from the trench.

Magmatism at Convergent Boundaries

Magma in a subduction zone is probably generated when water in the descending oceanic crust is driven out and rises into the overlying mantle. The addition of water lowers the melting point of the mantle rock and causes partial melting. Differentiation of this magma produces andesite and rhyolite, which rise and intrude as plutons or extrude to make long-lived composite volcanoes or calderas.

Most volcanoes erupting above sea level are clearly associated with subduction zones at convergent plate boundaries (Figure 21.8). The geographic setting for volcanic activity along such zones depends on the type of plate interaction. Where two oceanic plates converge, an island arc forms, as you have seen. Where a continent is on the overriding plate, similar volcanic activity develops in a folded mountain belt. In both cases, the close association of volcanism and convergent plate boundaries is clear. In contrast, where two continents collide, volcanism is rare, although magmas form granite plutons at depth.

What produces a magmatic arc at convergent margins?

Island Arc Magmatism

On a map, the most obvious manifestation of convergence of two oceanic plates is an arcuate chain of volcanoes that rise from the seafloor (Figure 21.8). The volcanic arc forms on the overriding plate and is parallel to the curving trench. Typical island arcs are the Tonga Islands, the Aleutian Islands, and the West Indies (Antilles) Islands (Figure 21.1).

These volcanoes lie about 100 km from the trench and 100 km above the inclined seismic zone that shows the location of the subducted slab. They mark a zone of voluminous magma production and high heat flow (Figure 21.6). The volcanoes are built on igneous intrusions and deformed metamorphic rocks, including remnants of oceanic crust formed at a mid-oceanic ridge. Most of these volcanoes are large composite volcanoes that erupt large volumes of andesite and lesser amounts of basalt and rhyolite.

The major volcanoes rise 1 to 2 km above their surroundings and are quite regularly spaced, about every 50 to 75 km. Smaller extrusions build a nearly continuous ridge connecting the major cones. Most island arcs are several hundred kilometers wide and extend discontinuously along the length of the trench.

Continental Arc Magmatism

A continental volcanic arc is a chain of many composite volcanoes on the margin of the continent above a subduction zone. The active volcanoes and underlying plutons are about 100 to 200 km landward from the trench. Subsidiary vents, lava domes, cinder cones, and fissure vents dot the landscape between the major volcanoes. The deep part of the arc consists largely of plutonic rocks that are the roots of volcanic systems. Multiple plutons intrude one another and form long, linear batholiths. The plutons are typically diorite to granite in composition. These plutons are commonly larger and more silicic than those found in island arcs. Moreover, they intrude into preexisting continental crust that is made of folded and thrust-faulted sedimentary rocks overlying a basement of older igneous and metamorphic rocks.

Generation of Magma in Subduction Zones

Magmatism along subduction zones is quite different from the basaltic fissure eruptions of divergent plate boundaries. The magmas generated at subduction zones are characteristically andesite or even rhyolite. They are richer in silica than basaltic magma and thus are more viscous. Consequently, water dissolved in the magma cannot escape easily to form gas bubbles. Moreover, subduction zone magmas contain more water and other dissolved volatiles than do those formed at divergent plate boundaries. These two characteristics result in violent, explosive eruptions from central vents. They commonly produce ash flows, composite volcanoes, and collapse calderas, as well as viscous lava flows and domes. These important differences are probably caused by dramatically different mechanisms of magma generation.

How do andesitic and silicic magmas form in subduction zones?

What triggers the generation of magma at subduction zones? How does insertion of a *cold* slab into the mantle create *hot* molten magma? The probable answer to this paradox lies not in the temperature of the plate, but in its high water content. The water in the descending plate was incorporated into the oceanic crust by metamorphism at a mid-oceanic ridge. This water is eventually released from the subducting oceanic lithosphere and rises into the overlying wedge of mantle. There the water lowers the melting point of the mantle rock.

Figure 21.19 illustrates how the release of water from the subducting slab is involved in the generation of magma. Water moves along with the descending plate into a subduction zone as a pore fluid in sediments and as water included in the structures of minerals formed by ocean ridge metamorphism. The descending slab is subjected to progressively higher temperatures and pressures (red line in **Figure 21.20**). Water trapped in the sediments is probably squeezed out at shallow depths beneath the accretionary wedge, but some of the water tied up in the chemical structures

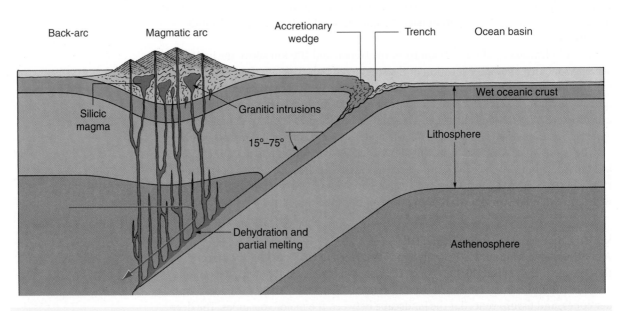

Figure 21.19 Magma at convergent plate boundaries is generated at depths of about 100 to 150 km. Subduction of oceanic crust carries sediment and basalt into the hot asthenosphere. (This sediment and basalt were altered earlier by ocean ridge metamorphism at a divergent plate boundary.) The descending slab is slowly heated; eventually, the hydrous minerals in the crust decompose and release water. At this critical depth, the water rises into the overlying mantle, causing it to melt partially. This basaltic magma rises buoyantly into the crust, where it may differentiate to form andesite and rhyolite. The magmas may crystallize to make plutons or erupt at the surface.

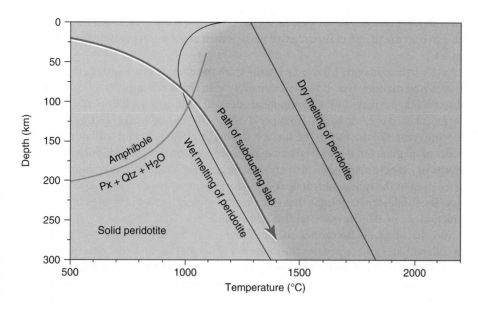

Figure 21.20 The generation of magma in a subduction zone is primarily due to the role played by water. As a descending plate slides into the mantle, it follows a path of increasing pressure and temperature (red arrow). Where the path crosses the breakdown curve for amphibole (blue line), an important mineral in metamorphosed oceanic crust, water is released. The buoyant fluid rises into the overlying mantle and there induces partial melting. Wet peridotite begins to melt at a temperature nearly 500°C lower than dry peridotite. This new mafic magma is wetter and more oxidized than magma produced at mid-ocean ridges and may differentiate to make silicic magma such as andesite or rhyolite.

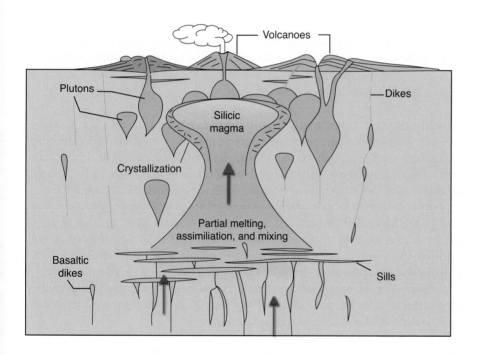

Figure 21.21 Intrusions at convergent plate margins are one of the major ways that continental crust is produced. Hot basaltic magma rising from a subduction zone may assimilate crustal rocks or mix with other magma simultaneous with fractional crystallization. Through these processes, basaltic magma differentiates to make andesite and rhyolite. Magma may rise in teardrop-shaped diapirs or through fractures or dikes that merge into larger and larger masses.

of minerals may remain in the oceanic crust until depths greater than 100 km are reached.

Figure 21.20 shows a typical temperature-pressure path encountered by the upper part of the cold slab as it descends into the hot mantle. Somewhere between 100 and 200 km deep, the slab has warmed enough that the hydrated minerals are no longer stable. Consequently, they break down to form new minerals that lack water, together with a separate water-rich fluid. Figure 21.20 shows that the mineral amphibole in wet oceanic crust may **dehydrate** to form pyroxene, quartz, and water.

The water released rises buoyantly into the hot mantle peridotite overlying the cold slab, causing the peridotite to partially melt (Figure 21.19). Careful examination of Figure 21.20 reveals why melting happens: Water dramatically lowers the temperature at which peridotite begins to melt. You can see this relation by comparing the melting temperature at 100 km on the dry-melting curve with the melting temperature on the wet-melting curve. Consequently, the wet mantle above the subducting slab may

Why is magma generated at convergent margins?

melt without increasing its temperature. (Here, water acts as a flux, much like fluorite added to iron ore in a steel furnace will make it melt at a lower temperature than pure iron ore.)

As this hybrid magma (derived partially from the oceanic crust and partially from the overlying mantle), rises it reacts extensively with the overlying crust (**Figure 21.21**). Thus, this magma may contain components derived from oceanic sediments, from metamorphosed oceanic basalt, from peridotite in the mantle wedge, and from the overlying crust. On the way to the surface, the magma also may mix with other batches of magma. Or, it may cool and experience fractional crystallization to form andesite or rhyolite magmas, which are even richer in silica. Eventually, magma that started deep in the subduction zone cools to form plutons or extrudes as lava or ash flows. The fundamental point is that silica-rich continental crust is formed by the extraction of low-density material from the mantle.

Note that at subduction zones, magma is generated by *partial melting*—a process that differentiates and segregates the materials of Earth. Here, magma rich in silica is produced. It is concentrated in island arcs or in granite plutons in mountain belts of the continents. Unlike basalt, which may become dense enough to be subducted, this silica-rich material cannot sink into the mantle. It becomes concentrated to form additional continental crust.

Subduction zone magmas are distinct from those in most other tectonic settings. We have already emphasized that the typical subduction zone magma is andesitic in composition, but the full spectrum of igneous rock compositions occurs. Moreover, it is very important to remember that subduction zone magmas are characteristically enriched in the water, as well as other volatile components such as chlorine, sulfur, and oxygen. These elements were probably extracted from the subducted oceanic crust that had been altered by ocean ridge metamorphism.

Magmatism in Continental Collision Zones

Where two continents collide, hot mantle-derived magmas do not form after the subduction zone disappears (Figure 21.15). Nonetheless, small volumes of silicic magma are produced. A distinctive granite is the most important igneous rock. In this setting, continental rocks, including metamorphosed shale and other clastic sedimentary rocks, can partially melt to form granitic magmas that are rich in silica and aluminum (**Figure 21.22**). Many of these granites contain minerals rarely found in other types of granite (such as muscovite, garnet, tourmaline, and cordierite). The magmas do not rise far from their sources before they crystallize and only rarely do they erupt to form lava or ash flows. The heat for melting probably comes from deep burial by tectonic underthrusting of the continental crust. In the Himalayas, for example, partial melting of continental crust has produced sill-like sheets of muscovite granite that intrude into previously folded and metamorphosed rocks.

Volcanic Eruptions at Convergent Boundaries

Volcanoes above subduction zones commonly erupt violently to form viscous lava flows, lava domes, or ash flows or ash falls. Tsunamis, lahars, and debris avalanches are also common. Although the volcanoes erupt infrequently, some eruptions can be predicted.

There have been many volcanic eruptions at convergent plate boundaries over the past 1000 years. Many have been fatal. In the last 100 years, about 100,000 people have died as a result of volcanic eruptions. Historical accounts of a few of these eruptions help us understand the nature of volcanic activity associated with converging plates. The following sections recount four of the most spectacular, devastating eruptions in recorded history: Mount Vesuvius, Krakatau, Mont Pelée, and Mount St. Helens.

Courtesy of Michael J. Dorais

(A) Sills and dikes of younger light-colored granitic rock cut across darker metasedimentary rocks metamorphosed during the collision between India and south Asia. The granitic magma formed by melting metasedimentary rocks and has not moved far from its source. The thickest sill is about 10 m thick.

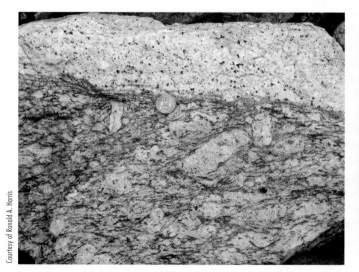

Courtesy of Ronald A. Harris

(B) The dark gneiss is cross-cut by younger and lighter-colored granite. The granite contains the distinctive dark mineral tourmaline.

Courtesy of Ronald A. Harris

(C) Garnet (red) and muscovite mica (small shiny grains) are evidence that the magma was Al-rich, a key characteristic of granites formed during continental collision. Foliation was produced by ductile shear.

Figure 21.22 Magmas generated during continental collision are different than those formed at other types of plate boundaries.

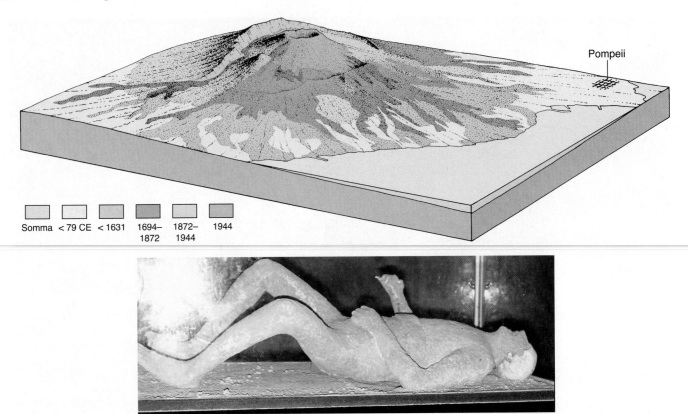

Somma < 79 CE < 1631 1694– 1872– 1944
                        1872   1944

Figure 21.23 **Vesuvius** erupted and buried Pompeii, Italy, with ash in 79 CE. It is one of several composite volcanoes that lie above a westward-dipping subduction zone beneath Italy. People asphyxiated by poisonous gas during the eruption were buried in the ash. Eventually, the bodies decomposed, leaving cavities in the ash. By filling these cavities with plaster, archeologists have made detailed casts. Excavations provide important insights into the hazards posed by volcanic activity at convergent plate margins.

79 CE—Mount Vesuvius, Italy

The spectacular cone of Mount Vesuvius looms above the skyline of Naples, Italy. Over the ages, it has repeatedly erupted magma generated in a subduction zone beneath the Italian peninsula (**Figure 21.23**). In 79 CE, Mount Vesuvius erupted catastrophically. An extraordinarily vivid eyewitness account of the eruption was recorded by Pliny the Younger, then a 17-year-old boy. He related details of how his famous uncle, Pliny the Elder, died while observing the volcano. Along with him, many perished in the destruction of the two cities of Pompeii and Herculaneum.

Beginning in 63 CE and continuing for 16 years, earthquakes shook the western coast of Italy. Then, on the morning of August 24, 79 CE, Mount Vesuvius exploded with a devastating eruption of white-hot ash and gas. Within two days, ash falling from the cloud buried Pompeii, which was directly downwind and near the volcano. Many people suffocated by sulfurous fumes or were burned by the searing heat of brief pyroclastic surges that swept through the town; others died in their homes when roofs collapsed from the weight of the tephra. The entire town and most of its 20,000 inhabitants were buried by ash and forgotten for more than 1000 years, until Pompeii was excavated in 1748 (Figure 21.23). By contrast, the town of Herculaneum was buried nearly instantaneously by rapidly moving ash flows that accumulated to a depth of 20 m.

The ash fall and flow that buried Pompeii and Herculaneum are first-class examples of the type of violent eruption that is common in volcanoes along convergent margins. The once-smooth and symmetrical cone of Mount Vesuvius was shattered by the explosion, which created a large caldera where a peak once existed. Subsequent eruptions have built a new cone inside the older caldera.

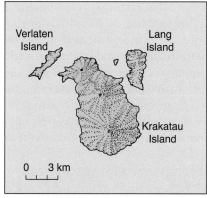

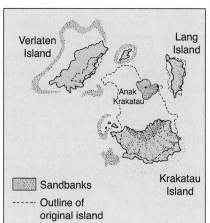

Figure 21.24 Maps of Krakatau before (top) and after (bottom) its 1883 eruption show the force of violent volcanic eruptions at convergent plate boundaries. Krakatau is a composite volcano along the Indonesian arc. All that remains of the volcano are several small islands like the one in the background. A small volcanic cone (Anak Krakatau) has been rebuilt over the center of the old volcano.

1883—Krakatau, Indonesia

Krakatau is a small volcanic island west of Java, part of an island arc along the subduction zone associated with the Java Trench (Figure 21.1). After remaining dormant for two centuries, Krakatau began to erupt on May 20, 1883. The eruption culminated in a series of four great explosions on August 26 and 27. One of them was heard in Australia, 4800 km away. The explosions are considered the greatest in recorded history. The whole northern part of the island, which stood about 600 m high, was blown off, forming a huge caldera, 300 m below sea level (**Figure 21.24**). Tremendous quantities of ash were thrown high into the atmosphere, and some circled the globe for 2 years. Krakatau was uninhabited, but more than 36,000 people were killed in Java and Sumatra by the huge tsunami produced by the eruption. Today, a new volcano is developing in the middle of the submarine caldera.

1902—Mont Pelée, West Indies

Ash flows are an important part of volcanism along convergent plate margins, and the great eruption of Mont Pelée, on the island of Martinique in the West Indies, helped initiate an understanding of this type of eruption. Pelée's eruption was preceded by nearly a month of warnings, in the form of extrusions of steam and fine ash from the volcanic vent, accompanied by numerous small earthquakes.

Then, on May 8, 1902, a gigantic explosion blew ash and steam thousands of meters into the air. The dense, hot ash fell back out of the atmosphere and moved as an ash flow, sweeping down the mountain's slopes like a high-speed avalanche. In fewer than 2 minutes, the hot, incandescent ash flow moved 10 km from the side vent on Pelée and swept over the city of St. Pierre. It annihilated the entire population of more than 30,000 people, except for one man, a prisoner who was being held underground in the city jail. Every flammable object was instantly set aflame. The rushing cloud of ash moved over the waterfront and across the sea surface, capsizing all of the ships.

What characteristics do volcanoes at convergent plate margins have in common?

The ash flow was a mixture of hot glass shards and pumice that flowed at the base of a billowing cloud of gas. The fundamental force that causes ash to flow rapidly is simply the pull of gravity, just like an avalanche of snow. But the key to the speed of the cloud is that such a mixture of hot gas and fragments of ash is highly mobile.

Intermittent ash-flow eruptions continued on Mont Pelée for several months. By October, a bulbous dome of lava, too viscous to flow very far, had formed in the crater. A spire of solidified lava was then slowly pushed up from a vent in the dome, like toothpaste from a tube. The spire repeatedly crumbled and grew again from the lava dome. Each collapse produced a hot avalanche that rushed down the slopes of the volcano.

1980—Mount St. Helens, Washington State

The best-documented eruption of a composite volcano related to a subduction zone is Mount St. Helens in Washington State. On May 18, 1980, it erupted with a force estimated to be 500 times greater than the atomic bomb that destroyed Hiroshima, Japan, at the end of World War II (**Figure 21.25**).

Mount St. Helens is part of the Cascade Range, which extends about 1500 km from British Columbia to northern California. Mount St. Helens is the youngest of the 15 major volcanoes in the Cascade Range. It consists of coalesced dacite domes, lava, and interlayered ash deposits.

Mount St. Helens had been dormant for 123 years, but on March 20, 1980, it began to stir with a series of small earthquakes. After a week of increasing local seismicity, it began to eject steam and ash. A series of moderate eruptions continued intermittently for the next six weeks. Within a few days after these first eruptions started, the U.S. Geological Survey issued warnings. During the weeks to come, the U.S. Forest Service and Washington State officials closed all areas near the mountain, undoubtedly saving thousands of lives.

By the second week of activity, more than 30 geologists had gathered to conduct a wide variety of studies. In particular, they monitored the development of a large bulge on the north flank of the mountain. By the end of April, the bulge was 2 km long and 1 km wide and was expanding horizontally at a steady rate of 1.5 m/day. Clearly, the mountain was being inflated by magmatic intrusion.

By monitoring the bulge, seismicity, and the gas emissions, geologists believed they would detect some significant change to warn of an imminent large eruption. However, no anomalous activity occurred. In fact, seismic activity decreased. Thirty-nine earthquakes were recorded on May 15 and only 18 on May 17.

On Sunday morning, May 18, the mountain was silent. Only minor plumes of steam rose from two vents. David Johnston, a 30-year-old geologist, was monitoring gas emissions and observing 8 km northwest of the volcano's crater. Abruptly, he cried over a two-way radio: "Vancouver! Vancouver! This is it!" Moments later, Johnston vanished in the blast of hot ash and gas as more than 4 km³ of material was thrown from the blast on the north side of the mountain.

The best way to understand the nature of this eruption is to study the sequence in **Figure 21.26**. At 8:32 a.m., the mountain was shaken by an earthquake with a magnitude of approximately 5. The bulge on the north slope destabilized and then moved downslope as a great landslide. This uncapped the bottled-up magma and gas bubbles formed and explosively expanded in the low atmospheric pressure. Consequently, the eruption blasted horizontally across the collapsing slope. This lateral blast of rock, ash, and gas caused most of the destruction and loss of life. The blast wave leveled the forest in an area 35 km wide and 23 km outward on the mountain's north flank (Figure 21.26).

The eruption caused three separate, but interrelated, processes: (1) lahars, (2) ash flows, and (3) ash falls. The lahars originated largely from water-saturated ash on the mountain's upper slopes. Most of the lahars were hot. At the height of the flow, the lower Toutle River was heated to 90°C. The debris flows swept up 123 homes, as well as cars, logging trucks, and timber, and carried them downstream, destroying bridges and other constructions. So much sediment from the Toutle was carried into the

Photograph courtesy of U.S. Geological Survey

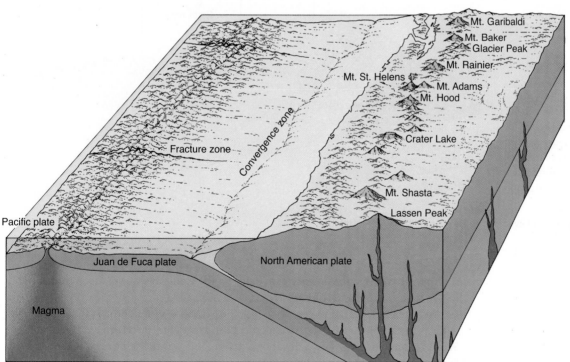

Figure 21.25 The 1980 eruption of Mount St. Helens in Washington State was one of the largest and most scientifically important to occur in the United States. The eruption and devastation by explosive blasts and ash are typical of composite volcanoes built above subduction zones. The Cascade Range contains 15 large composite volcanoes, extending in a line from British Columbia to northern California. These volcanoes are formed by subduction of the Juan de Fuca plate beneath the North American plate.

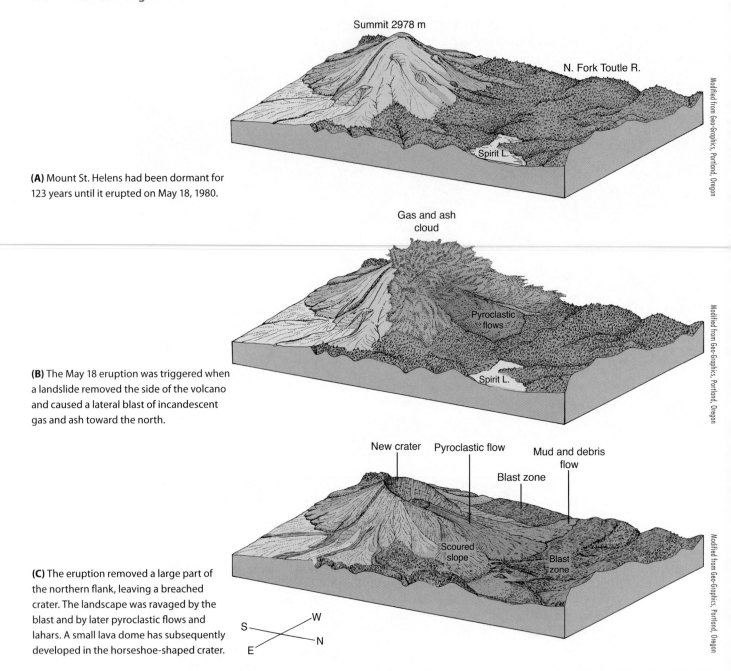

Summit 2978 m

N. Fork Toutle R.

Spirit L.

(A) Mount St. Helens had been dormant for 123 years until it erupted on May 18, 1980.

Gas and ash cloud

Pyroclastic flows

Spirit L.

(B) The May 18 eruption was triggered when a landslide removed the side of the volcano and caused a lateral blast of incandescent gas and ash toward the north.

New crater   Pyroclastic flow   Mud and debris flow

Blast zone

Scoured slope

Blast zone

(C) The eruption removed a large part of the northern flank, leaving a breached crater. The landscape was ravaged by the blast and by later pyroclastic flows and lahars. A small lava dome has subsequently developed in the horseshoe-shaped crater.

W
S
E   N

Modified from Geo-Graphics, Portland, Oregon

Figure 21.26   The sequence of events in the eruption of Mount St. Helens.

Columbia River that the downstream depth of the Columbia was reduced from 12 m to 4 m within a day, and ships upstream were trapped.

An important part of the eruption was the extrusion of numerous ash flows. Traveling as fast as 130 km/hr, the incandescent ash and debris (at a temperature of about 500°C) extended northward a distance of 9 km. This was the same kind of devastating ash flow that came from Mont Pelée. Some ash flows reached Spirit Lake where, together with debris flows that had been deposited earlier, they blocked the lake's outlet. Consequently, the level of water in the lake rose 60 m.

Immediately after the lateral blast, a vertical ash cloud rose to 18 km altitude. The ash cloud then fanned out downwind (eastward), and ash began to settle like a soft, gray snow. The cloud then moved in a broad arc across the United States and had completely circled Earth by June 5.

Other smaller ash-flow eruptions and lahars continued throughout the summer of 1980. The closing of the eruption episode was marked by the slow rise of magma through the central conduit to form a new lava dome in the summit crater. After a pause in volcanism, a new dome grew on the flank of the older dome. It remained active from 2004 to 2008.

A Summary of Volcanic Eruptions

The violent eruptions that characterize volcanoes at convergent plate boundaries result from the magma's high silica content, which makes it strong and viscous. Moreover, subduction zone magmas are rich in dissolved water. The dissolved gases cannot escape easily from the viscous melt. As a result, tremendous pressure builds up in the magma, and when eruptions occur, they are highly explosive. The eruptions produce huge quantities of ash, often in hot ash flows. Eruption of thick, viscous lava commonly precedes or follows the explosive activity. Tsunamis, lahars, and debris avalanches (caused by collapse of the sides of the volcanoes) are other significant hazards associated with steep stratovolcanoes like those formed at convergent plate margins.

Metamorphism at Convergent Margins

> In the forearc of a subduction zone, metamorphism occurs at high-pressure–low-temperature conditions. In a magmatic arc, or in a zone of continental collision, metamorphism occurs at higher temperatures and lower pressures. Most metamorphic rocks in the continental crust were formed at convergent plate boundaries.

Why are metamorphic processes associated with plate convergence? Recall that metamorphism is driven by *changes* in a rock's environment, mainly changes in temperature and pressure. Because systems always seek equilibrium, these changes cause new minerals to form that are in equilibrium with the new conditions. At convergent plate boundaries, the unique tectonic and magmatic processes create dramatic temperature and pressure changes. Hence, metamorphism is a major process at convergent boundaries. Most metamorphic rocks in the continents were created at convergent plate boundaries.

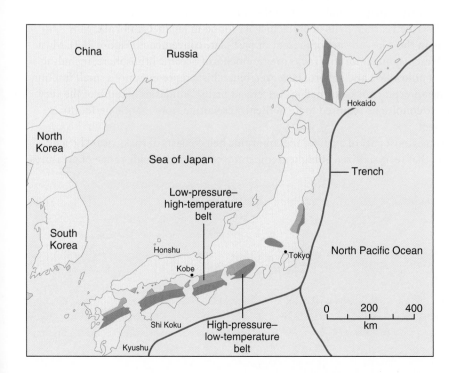

Figure 21.27 Metamorphism at convergent plate margins is an important process. Paired metamorphic belts formed in Japan during Mesozoic subduction. A high-pressure–low-temperature belt formed near the trench in the accretionary wedge and a belt of low- to intermediate-pressure metamorphism formed in the region of the magmatic arc.

Figure 21.28   Blueschist belts form by high-pressure–low-temperature metamorphism in accretionary wedges near subducting plates of oceanic lithosphere. The irregular blocks of blueschist shown here is part of a mélange formed in a Paleozoic subduction zone along the east coast of New Zealand. The blocks are metamorphosed fragments of oceanic crust.

At both modern and ancient subduction zones, two distinctive types of metamorphic rocks are juxtaposed to create **paired metamorphic belts** (**Figure 21.27**). The pair consists of an outer and an inner belt. The outermost metamorphic belt forms in the accretionary wedge. Here fine-grained schists and slates contain the distinctive blue amphibole called glaucophane and other minerals indicative of the **blueschist facies** (**Figures 21.28** and 6.16). These minerals form here because they are stable under the conditions unique to the forearc region: high pressure and relatively low temperature (less than 300°C). These unique high-pressure–low-temperature conditions are explained by the thermal structure of a subduction zone (Figure 21.6). Low temperatures result directly from the cold slab of subducting oceanic lithosphere. High pressure is attained because the slab drags cold oceanic rocks as deep as 30 to 50 km in the mantle, where the pressure is 10 to 15 kilobars. Once formed, blueschist facies rocks are brought rapidly back to the surface of the accretionary wedge by faulting, as described earlier. Belts of blueschist are found in Japan, California, New Zealand, and in the Alps, all sites of present or past plate convergence (Figure 21.1). Most of the blueschist metamorphic rocks are fragments of oceanic lithosphere, including chunks of pillow lava and gabbro. However, blueschists represent only a small fraction of the metamorphic rocks formed at convergent plate boundaries because of the very distinctive conditions required to form them. Serpentine from mantle peridotite is also present.

The innermost part of a paired metamorphic belt consists of rocks near the magmatic arc that recrystallized at higher temperatures and over a wide range of pressures or depths. In this zone of high heat flow (Figure 21.6), metamorphism is driven by heat released from cooling magmas that intruded to form plutonic belts. Where plutons intrude into sedimentary or other rocks near the surface, narrow contact metamorphic aureoles form, and hornfels dominates. Wider zones of orogenic metamorphism develop at moderate depths. In these belts, thrusting and folding also bury rocks to great depths where they become hotter. The metamorphic rocks are typified by mineral assemblages of the greenschist and amphibolite facies. These are the most common kind of metamorphic rocks found on the continents.

Because of the higher temperatures, intensive plastic deformation accompanies orogenic metamorphism at convergent plate margins. The original sedimentary and volcanic rocks become strongly foliated schists and gneisses. The horizontal stresses generated by the converging plates cause foliation that is perpendicular to the direction of stress. Hence, slaty cleavage, schistosity, and gneissic layering in the deeper parts of a mountain range are characteristically vertical or dip at high angles. In a few collision zones, sedimentary rocks containing carbon were thrust so deeply (more than 100 km) that diamond formed from the carbon. Diamonds like this have been found in China, the Alps, Kazakhstan, and Norway. However, it is the depth of the rock, not the horizontal pressure from collision, that produces the pressure that drives recrystallization.

In the deeper parts of a mountain belt, metamorphism can become intense enough to produce migmatite—a complex mixture of thin layers of once-molten granitic material sandwiches between sheets of schist or gneiss. It develops largely from the partial melting of preexisting rocks. In these zones, high temperatures and pressures soften the entire rock body, which then behaves like a highly viscous liquid if it is subjected to stress. As a result, metamorphic rocks in deeper parts of orogenic belts exhibit complex flow structures. Migmatites are probably the sources of some granitic magma.

Formation of Continental Crust

Continents grow by accretion at convergent plate boundaries. New continental crust is created when silicic magma is added to deformed and metamorphosed rock in a mountain belt.

We have already seen how orogenic belts form at convergent boundaries. Strong deformation of rocks on a continental margin occurs during subduction and collision. These rocks are metamorphosed and intruded by silicic magmas that have low densities. Portions of this magma are extracted from the mantle or from subducted oceanic crust and form new additions to the continental crust. By virtue of their low densities, these rocks cannot be subducted and must remain in the continental crust. The gradual growth of the continents by addition of magma and deformation of preexisting crust is known as **continental accretion**. Consequently, both the origin and the evolution of Earth's continental crust are intimately tied to the processes that occur at convergent plate margins.

STATE OF THE ART   Magnetic Maps Show Accreted Terranes

Geologists are always trying to find ways to see below the obscuring skin of soil and vegetation and discover the nature of the rocks hidden below. Geophysical techniques provide tools that can sense important characteristics of rocks beneath such a cover and for some depth into the interior. The strength of Earth's magnetic field is not uniform. Its strength varies because of the changing pattern of flow inside Earth's core, because of changes in the solar wind, and, of importance here, due to differences in the magnetic properties of rocks in the crust.

Magnetic minerals (such as magnetite) in rocks cause distortions in the magnetic field. Magnetic minerals induce local magnetic fields that either add or subtract from Earth's field. *Magnetic susceptibility* is a measure of how much magnetism can be induced in a rock. In general, sedimentary and most metamorphic rocks have relatively low magnetic susceptibilities. Igneous rocks, on the other hand, tend to have more magnetite and to be more magnetic.

By measuring the strength of the magnetic field at a multitude of individual spots, a map of the distribution of magnetic and nonmagnetic materials in the crust forms a powerful interpretive tool for geologists. For land-based magnetic surveys, the most commonly used *magnetometer* is the proton precession magnetometer that measures only the strength of Earth's magnetic field, not its polarity. This kind of magnetometer contains a cylinder filled with water that is surrounded by a coil of conductive wire. The hydrogen nuclei (protons) in the water behave like tiny spinning dipole magnets. Because Earth's magnetic field applies a force to these protons, they begin to precess or wobble as they spin. This precession induces a small but measurable current in the coil. The frequency of this current correlates with the strength of the local magnetic field. Airborne magnetometers are usually towed behind aircraft or mounted on wing tips. By repeatedly flying back and forth across a region, a magnetic map is constructed. Line spacings may be as little as 200 m, but more typically they are a kilometer or so apart.

An aeromagnetic map of Alaska shows the power of this technique. Much of Alaska's landscape is difficult to traverse. High mountains, glaciers, surging rivers, short field seasons, and hordes of mosquitoes make normal field geologic investigations difficult. A series of airborne magnetic surveys, however, can be conducted quickly and stitched together with a computer to see the magnetic fabric of the state and nearby offshore areas. The magnetic variations reveal the distribution of various rock types, ages, and the folded structures of the mountain belts. They define the boundaries of arcuate accreted terranes that were added to Alaska during millions of years of plate convergence. The aeromagnetic map also shows the striped fabric of the seafloor before it subducts down the oceanic trench.

Courtesy of K. Nyman, Geological Survey of Finland

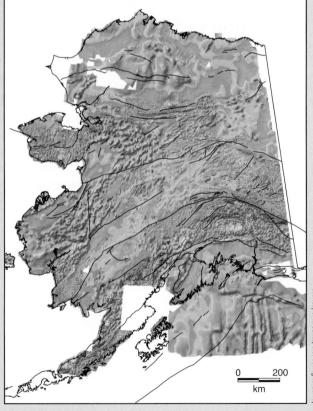

Courtesy of R. Saltus, U.S. Geological Survey

Accreted Terranes

Studies worldwide reveal that many continental margins consist of a multitude of separate crustal blocks, each with its own distinctive origin and history. These blocks have been juxtaposed against one another by major faults (**Figure 21.29**). Each block is a distinctive terrane, a term that refers to a region or group of rocks sharing a common age, structure, stratigraphy, and origin. These exotic segments of the orogenic belt are called **accreted terranes**. The terranes vary in size, and their rocks, fossils, histories, and magnetic properties contrast sharply. Fossils indicate that each terrane formed at different times and in very different environments than any other; paleomagnetic data show that the various terranes originated at different latitudes thousands of kilometers away.

Accreted terranes in the orogenic belt of western North America are a prime example (Figure 21.29). In this region, many independent terranes are squeezed together, each with its own internal structure, rock types, and fossils. Each terrane contrasts sharply with adjacent segments. One contains remnants of ancient seamounts related to mantle plumes; another, segments of shallow-marine limestones like those forming in the Bahamas; and still others are pieces of old volcanic arcs formed by subduction. Some are crustal fragments of metamorphosed basement complexes. Each terrane is separated from the next by long major fault systems, most of which show evidence of strike-slip movement. For example, paleomagnetic properties and fossils suggest that many of the terranes in British Columbia originated far to the south and traveled hundreds of kilometers northward, along the continental margin during the Mesozoic Era.

The Appalachian Mountains, which stretch from Newfoundland to Alabama, are another example of continental accretion. They contain slices of ancient Europe, Africa, and oceanic islands accreted to the continent during the Paleozoic Era (**Figure 21.30**).

Why are accreted terranes a common feature in many mountain belts?

Accretion of North America

Evidence for even older orogenic and accretion events can be seen on the map of North America in **Figure 21.31**. It summarizes many radiometric ages for the metamorphic and granitic basement, as well as structural details compiled from years of field mapping and geophysical studies. The ages of the basement rocks of North America form a regular pattern. The oldest rocks found thus far are in northern Canada and are about 4 billion years old in the Slave Province. Other ancient scraps of continental crust are in southern Greenland, the Superior and Wyoming Provinces, where granites and metamorphic rocks have ages between 2.5 and about 3.8 billion years.

Surrounding the Superior Province to the south, west, and north is a vast area of gneiss and granite from 1.8 to 1.9 billion years old. In addition, its structural trends are oriented differently than in the older terrane. To the southeast, the rocks are younger still; the granitic intrusions and metamorphic rocks are as young as 1.0 billion years old.

Continental Growth Rates

The crudely concentric pattern of the basement age provinces in North America is strong evidence that the continent grew by the accretion of material around its margins during a series of mountain-building events. Each province probably represents a period of relatively rapid crustal growth during mountain building, related to convergent margin tectonics and magma production. The growth of North America is typical of other continents. Although the details are still sketchy, the amount of continental crust must have gradually grown during Earth's long history. Geological studies are starting to reveal the rate of growth of the continents (**Figure 21.32**). Continental crust grew slowly for the first billion years of Earth's history. Initially, much of the crust may have been swept back into the mantle. As Earth matured, its continents appear to have grown rapidly between about 3.5 and 1.5 billion years ago. Subsequently, growth of continental crust was slower. During both of these time periods, however, subduction and plate tectonics seem to be the principal cause of crustal growth.

Thus, the origin of continental crust can be related to the evolution of convergent plate margins. In an analogous manner, the origin and evolution of the oceanic crust is intimately connected to the processes at divergent plate boundaries.

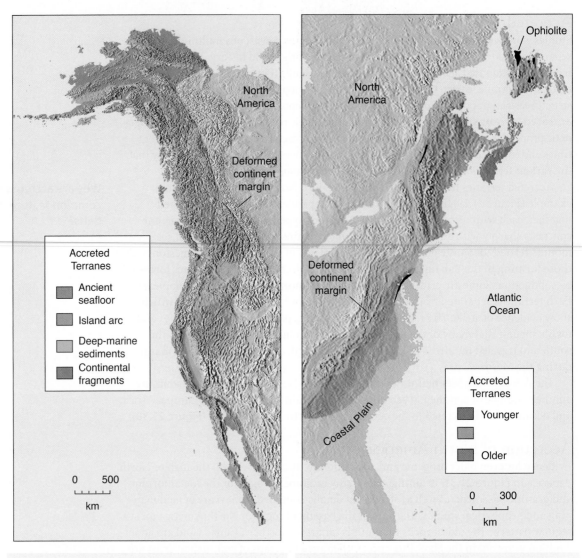

Figure 21.29 Accreted terranes along convergent plate margins are an important component of most continents. Western North America is composed of rocks that moved and became attached during episodes of convergence in the Mesozoic and Cenozoic eras. Before reaching their present positions, these rocks were island arcs, fragments of rifted continents, or oceanic plateaus. After accretion, they were shuffled along the margin by strike-slip faults.

Figure 21.30 Accreted terranes form much of eastern North America. The Appalachian Mountains, which extend from Newfoundland to Alabama, contain terranes that were once parts of ancient Europe, Africa, island arcs, and even oceanic islands. These terranes were accreted to the continent during plate convergence and continental collision in the Paleozoic Era, millions of years before the western North American accretions.

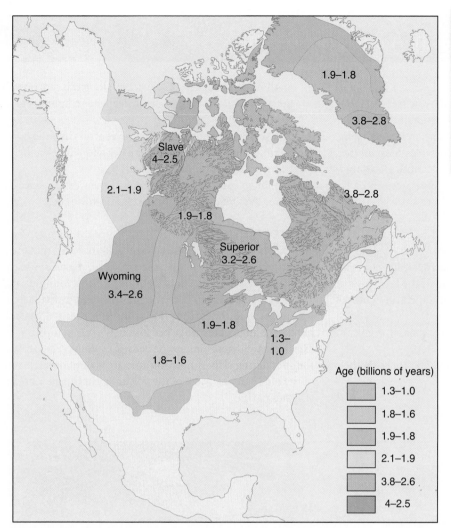

Figure 21.31 Radiometric ages of basement terranes in North America show several geologic provinces, each representing a mountain-building event. The ages of the major granitic intrusions are in billions of years, and the lines represent the trends of the folds and structural trends in the metamorphic rocks. The continent apparently grew by accretion as new mountain belts formed along its margins.

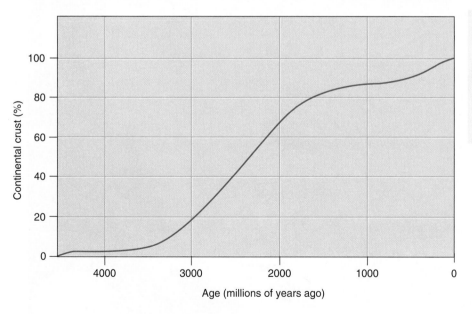

Figure 21.32 The amount of continental crust has grown over the last 4 billion years of Earth's history. The curve shows our best estimate of the rate of growth. During the last few billion years, the rate of growth was not as high as during the earlier history. Today, most continental crust forms at subduction zones.

GeoLogic   Lau Basin: A Back-Arc Basin

The Tonga-Kermadec arc of the southwest Pacific is like many other arcs related to subduction. A deep narrow trench lies to the east of the arc. Subduction carries the Pacific Ocean lithosphere beneath the arc and is marked by an inclined zone of intense earthquakes. Explosive andesitic volcanoes dot the length of the arc, but many of the active volcanoes are young and still below sea level. However, unlike some arcs, a broad submarine basin—the Lau Basin—lies west of the island arc. What could have formed this basin?

Observations

1. The broad basin is corrugated with narrow ridges and valleys that more or less parallel the trend of the arc.
2. A narrower, slightly deeper valley runs down the center of the basin and forms a rift (marked in red).
3. Rocks along the submarine rift are mostly basaltic volcanic rocks.
4. The central rift is normally magnetized and rocks on the flanks are reversely magnetized.

5. Shallow earthquakes are clustered along the central valley, but are not as numerous as the deeper earthquakes along the subducting slab.
6. Hot springs are aligned along the central rift valley.
7. Subducting oceanic lithosphere lies about 300 km below the central part of the basin.
8. A high volcanic plateau also marks the western side of the basin. It is made of inactive andesitic volcanoes.

Interpretations

Geologists think the Lau Basin is a back-arc basin that formed by rifting apart an older volcanic arc. The Tonga ridge and the Lau Ridge mark the flanks of this rift. The youngest, most active part of the basin is a rift-valley, marked by extension, volcanism, shallow earthquakes, and hydrothermal vents. It is much like an oceanic ridge because new crust is forming here. The back-arc extension may be caused by convection driven by the subducting slab. A new volcanic arc is developing just west of the old high arc. Most of the volcanoes in this new arc are still submarine.

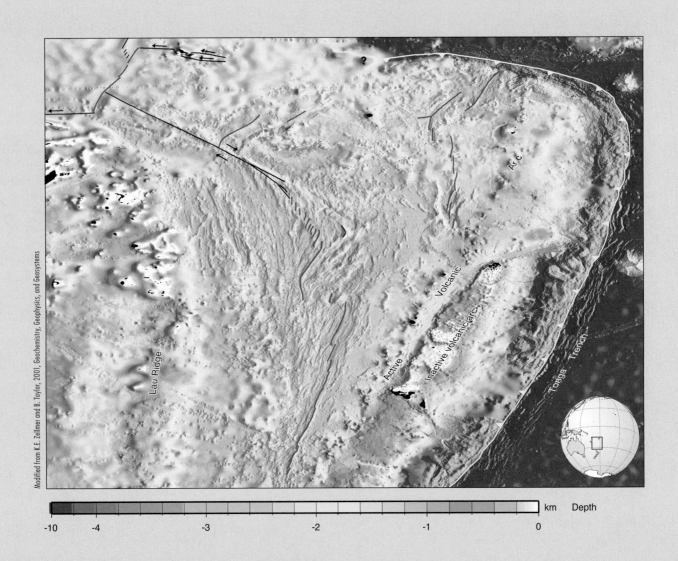

Modified from K.E. Zellmer and B. Taylor, 2001, Geochemistry, Geophysics, and Geosystems

km   Depth

-10   -4   -3   -2   -1   0

Key Terms

accreted terrane (p. 661)
accretionary wedge (p. 638)
back-arc (p. 632)
back-arc spreading (p. 645)
blueschist facies (p. 658)
continental accretion (p. 659)
continent-continent
convergence (p. 632)

convergent plate boundary
(p. 630)
dehydrate (p. 649)
folded mountain belt (p. 632)
forearc ridge (p. 631)
inclined seismic zone (p. 636)
mélange (p. 638)
nappe (p. 642)

ocean-continent conver-
gence (p. 632)
ocean-ocean convergence
(p. 630)
orogenic belt (p. 638)
orogenic metamorphism
(p. 632)
outer swell (p. 631)

paired metamorphic belt
(p. 655)
subduction zone (p. 630)
trench (p. 631)
volcanic arc (p. 632)

Review Questions

1. Draw a simple cross section across a subduction zone showing the outer swell, trench, accretionary wedge, and magmatic arc.
2. Contrast the features of a continent-continent convergent plate boundary with those of an ocean-ocean plate boundary.
3. Why are trenches so deep?
4. Where do you expect to find the highest heat flow at a subduction zone? Where would you find the lowest gravity anomalies? Explain.
5. Describe how an accretionary wedge grows. How do the highly deformed rocks in the wedge differ from those in a folded mountain belt?
6. Is extension ever found at convergent plate boundaries?
7. Describe the absolute plate motions along the convergent plate boundaries: Australian-Indonesian, Indian-Eurasian, North American-Pacific, and Nazca-South America. How do you think these differences are reflected in the nature of the plate boundary?
8. Explain how magma is generated at subduction zones and contrast that with magma generation at a mid-oceanic ridge.
9. What explains the origin of paired metamorphic belts at ocean-continent convergence zones?

10. Contrast the type of metamorphism that is found at convergent plate boundaries with metamorphism that occurs at divergent plate boundaries.
11. In which kind of convergent plate boundary would you expect to find the strongest compressional deformation, at an ocean-ocean boundary, an ocean-continent boundary, or a continent-continent boundary? Why?
12. Why are earthquakes found deep (more than 300 km) in the mantle only at convergent plate boundaries?
13. Give two reasons why volcanoes at convergent plate margins are so explosive.
14. How would you discriminate an ancient accretionary wedge from the sediments that form on a rifted plate margin?
15. What would a slice of oceanic crust found in the middle of a continent imply about the tectonic history?
16. What are the characteristic sedimentary rocks formed along convergent plate margins? How do they compare with those found in a continental rift?
17. What processes lead to the growth of continents?
18. The average chemical composition of the continental crust corresponds to that of an andesite. Can you explain this observation?

22 Hotspots and Mantle Plumes

Earth's tectonic system is dominated by geologic processes that occur at the margins of large plates of lithosphere, and our main focus has been on dynamics of these plate boundaries. But plate interiors are not entirely quiet. Several currently active volcanic systems, such as Hawaii and Yellowstone National Park, are far from plate boundaries. Moreover, many large flood basalt provinces, such as those in the Pacific Northwest or central Siberia, have no explanation in simple plate margin dynamics.

Instead, these "hotspots" are believed to be surface manifestations of mantle plumes—long, narrow columns of hot material that flow upward from deep in the mantle—and appear to be independent of plate movements. Apparently, this type of convection stirs the mantle from deep below the shallow zone that feeds the mid-ocean ridges. Thus, mantle plumes provide new information about the inaccessibly deep portions of our planet.

In Yellowstone National Park, a hotspot is manifest by numerous geysers and hot springs, like the Grand Prismatic Spring, shown above. The dramatic colors are from primitive bacteria that can survive the harsh conditions. The magnificent geothermal activity in this area lies within a huge caldera 70 km across that last erupted about 600,000 years ago and spread ash over much of the western United States. The caldera has been filled in by thick rhyolite lava flows. Outside the park, basaltic lava related to the Yellowstone plume has erupted as recently as 2,000 years ago. The hotspot is manifest by geophysical and geochemical studies, but more visibly by numerous geysers, boiling mud pots, and hot springs. More than anything else, these mantle plumes reveal themselves as tremendous thermal anomalies. Not only is water heated to the point that it can flash to steam and create a geyser, but at deeper levels in the middle crust even solid rock may reach its melting point and become partially molten. Rhyolitic magma still resides in chambers just a few kilometers below some of the world's most spectacular scenery.

In this chapter, we discuss hotspots and the mantle plumes that appear to lie below them as large and important geologic systems. We examine the major effects of mantle plumes that ascend beneath the ocean basins and then turn our attention to those that rise beneath the continents. We also explore what causes plumes to move upward and why they are intimately related to intraplate volcanoes, earthquakes, and broad swells and basins.

Major Concepts

1. Mantle plumes appear to be long columns of hot, less dense solids that ascend from deep in the mantle. Mantle plumes create hotspots with high heat flow, volcanism, and broad crustal swells.
2. A plume evolves in two stages. When a plume starts, it develops a large, bulbous head that rises through the mantle. As the head deforms against the strong lithosphere, crustal uplift and voluminous volcanism occur. The second stage is marked by the effects of a still rising but narrow tail.
3. Basaltic magma is created because of decompression of the rising hot plume. Magmas formed in mantle plumes are distinctive and show hints of being partially derived from ancient subducted slabs that descended deep into the mantle.
4. A starting plume that rises beneath the ocean floor produces a large plateau of flood basalt on the seafloor. Subsequently, a narrow chain of volcanic islands forms above the tail of the plume, revealing the direction of plate motion.
5. If a plume develops beneath a continent, it may cause regional uplift and eruption of continental flood basalts. Rhyolitic caldera systems develop when continental crust is partially melted by hot basaltic magma from the plume. Continental rifting and the development of an ocean basin may follow.
6. Plumes may affect the climate system and Earth's magnetic field.

Hotspots and Mantle Plumes

> Mantle plumes appear to be long, nearly vertical columns of hot, upwelling materials that buoyantly rise from deep in the mantle. At the surface, plumes are marked by hotspots with high heat flow, volcanic activity, and broad crustal swells.

The first ideas about hotspots and mantle plumes emerged in 1963, from geologic observations of the Hawaiian Islands. It was well known that Hawaii, the largest of the islands, had active volcanoes, few strong earthquakes, and extremely high heat flow. The linear chain of volcanoes lies on a broad rise in the middle of the Pacific Ocean floor.

Significantly, geologists noted an absence of tectonic contraction to form belts of folded strata on Hawaii. Even strong extension is missing, although narrow rift zones emanate from the summits of the **shield volcanoes** and feed lava flows. In addition, it was discovered that the volcanoes on the islands are progressively older toward the northwest (**Figure 22.1**). For example, the island of Hawaii consists of several still-active shield volcanoes. The volcanoes on Maui, an island about 80 km to the northwest, are barely extinct. Farther to the northwest, the chain of islands becomes still older, and many volcanoes are deeply eroded; some are submerged below sea level.

Other linear chains of volcanic islands and seamounts in the Pacific, Atlantic, and Indian oceans show similar trends: An active or young volcano is at one end of the chain, and the series of volcanoes becomes progressively older toward the other end (**Figure 22.2**). For these reasons, the volcanically active parts of these chains came to be called **hotspots**. We are not completely sure, but the simple idea of **mantle plumes** rising as narrow columns from the deep interior explains many such features and has become a generally accepted part of global tectonic theory. Hotspots appear to be the surface expression of mantle plumes.

Evidence for Mantle Plumes

It must be emphasized that mantle plumes have not been observed directly. But the indirect evidence for their existence is substantial:

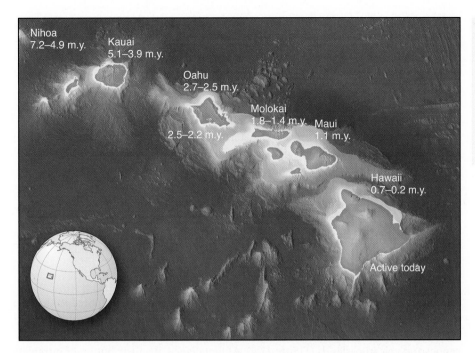

Figure 22.1 The volcanic islands of Hawaii are progressively older and more eroded to the northwest. Active volcanoes erupt periodically on the southeasternmost island, Hawaii. A linear chain of seamounts extends even farther to the northwest. Ages of volcanic rocks from the main shield-building stage are given in millions of years before present.

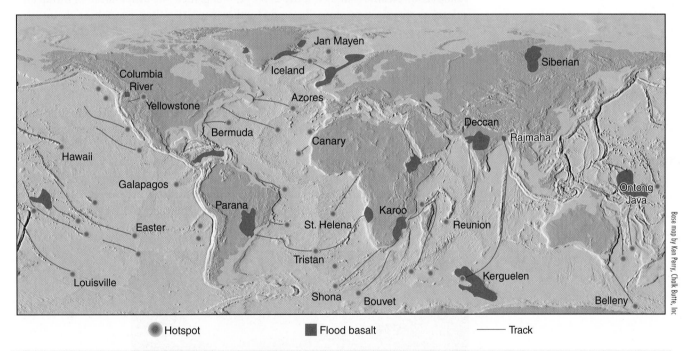

Figure 22.2 Hotspots, oceanic plateaus, and continental flood basalts related to mantle plumes are shown on this map. Basaltic volcanism in the ocean basins has formed hundreds of islands, seamounts, and plateaus. Flood basalts, shield volcanoes, and large rhyolitic calderas may form above plumes that lie beneath the continents. Ancient flood basalt provinces (gray) are connected to currently active hotspots by linear chains of seamounts (red lines). For example, the currently active volcanoes on the island of Tristan da Cunha in the South Atlantic mark the site of a hotspot whose initiation erupted flood basalts in South America and Africa approximately 125 million years ago. In the North Atlantic, two flood basalt provinces, 65 million years old, are linked to a plume beneath Iceland.

1. Locally, zones of high heat flow and associated volcanism (hotspots) occur far from plate boundaries.
2. These hotspots do not drift with the plates. Many are more-or-less stationary, suggesting that they are rooted deep in the mantle far below the moving lithosphere.
3. Geochemical studies show that the basalts erupted from hotspot volcanoes are different from those that come from the upper mantle at divergent plate

boundaries. The evidence suggests that the lavas are derived from deep in the mantle, below the asthenosphere.

4. Oceanic islands at hotspots are associated with large topographic swells. This association is evidence for an extra source of mantle-derived heat to expand the lithosphere.

5. Perhaps the most convincing evidence for mantle plumes comes from recent advances in seismic studies of Earth's interior. Tomographic images of the mantle beneath Iceland reveal that a narrow column of material with low seismic wave velocities extends to at least 400 km beneath the island (**Figure 22.3**). Other investigations suggest it extends even deeper, to at least 700 km. The plume has a diameter of about 300 km. High temperatures of the material in the plume probably cause the low seismic wave velocities. In fact, the plume may be as much as 200°C warmer than the surrounding mantle. Further refinements and higher resolution images are revealing more about the precise shapes and depths of mantle plumes.

What is the evidence for the existence of mantle plumes?

Characteristics of Hotspots and Mantle Plumes

The volume of volcanic rock produced at mantle plumes is small compared to that produced along divergent and convergent plate boundaries (**Figure 22.4**). Most intraplate volcanoes lie on the floor of the South Pacific, which is dotted by many submarine volcanoes and volcanic islands (Figure 22.2). At first glance, the distribution of intraplate volcanoes may seem random. But upon further inspection, linear trends or chains become apparent, especially in the Pacific Ocean (see the maps on the inside covers).

Volcanism over a mantle plume produces a submarine volcano, which can grow into an island. Steps in this process are shown in **Figure 22.5**. If the plume's position in the mantle does not change for a long time, the moving lithosphere carries the active volcano beyond its magma source. This volcano then becomes dormant, and a new one forms in its place over the "fixed" plume. A continuation of this process builds one volcano after another, producing a linear chain of volcanoes parallel to the direction of plate motion.

What generates a mantle plume?

From the evidence provided by studies of hotspots, it appears that mantle plumes may have a variety of shapes and sizes. They may consist of hot mantle material rising as blobs rather than in a continuous streak. For the most part, mantle plumes can be envisioned as long, slender columns of hot rock that originate deep inside Earth's

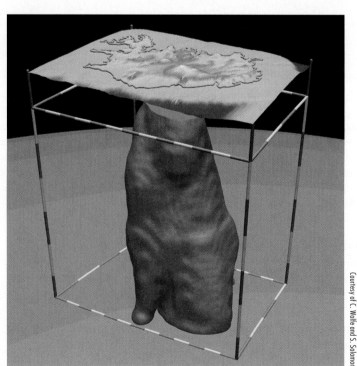

Figure 22.3 A mantle plume beneath Iceland is revealed by anomalously low seismic wave velocities in a cylindrical mass below the island. The low seismic wave velocities show that the plume has a higher temperature than the surrounding mantle. The plume extends to at least 400 km depth and has a diameter of about 300 km.

mantle (**Figure 22.6**). They rise slowly toward the surface, arching the overlying lithosphere, forming volcanoes and plutons, and causing small, shallow earthquakes. Some plumes may have diameters of as much as 1000 km, but most are only several hundred kilometers across. The material in the plume appears to rise at rates of perhaps 2 m/yr. Plumes rise under continents and oceans alike, and they occur in the center of plates and along some mid-oceanic ridges. The extra heat they bring to the lithosphere commonly produces domes up to 1000 km in diameter, with uplift ranging from 1 to 2 km at the center of the dome.

Some geologists think that mantle plumes originate at depths of at least 700 km and perhaps as deep as 2900 km at the core-mantle boundary. Their positions appear to be relatively stationary, as the lithospheric plates move over them. Plumes are thus independent of the crust's major tectonic elements, which are produced by plate movement. As a result, hotspots provide a reference frame for determining the absolute, rather than the relative, motion of tectonic plates. However, plumes are not absolutely immovable. Some appear to wave slightly in the "mantle wind" (**Figure 22.7**).

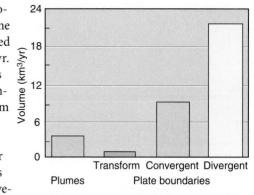

Figure 22.4 The volume of magma produced from mantle plumes is much smaller than that produced at divergent or convergent plate boundaries.

The Evolution of Mantle Plumes: Heads and Tails

Like the plate tectonic system, mantle plumes are a type of convection that slowly stirs the mantle. However, plate tectonics and mantle plumes are related to two distinct types of convection. One is the convection involved in plate motion, wherein material rises at divergent plate boundaries and descends at convergent boundaries. The other is the rise of thin columns of material in slender plumes from deep in the mantle. Although plumes clearly transport much less heat than the processes at tectonic plate boundaries, mantle plumes are also largely driven by internal heat. Plumes probably arise from a hot layer at the *base* of the mantle. Because it is so hot, the boundary layer surrounding the molten iron core must have much lower viscosity (100 to 1000 times less) and slightly lower density than the mantle above it. As heat from the fluid iron core flows into this boundary layer, parts of the mantle expand and become less dense. When a small portion becomes lighter than the cooler mantle above it (a difference of about 200°C and 0.1 g/cm³ may be sufficient), it becomes buoyant and begins to rise.

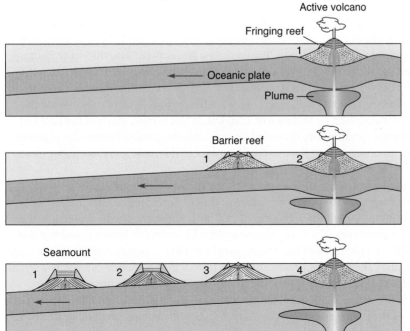

(A) A volcanic island forms above a more-or-less stationary mantle plume. As the volcano grows, its base subsides because of the added weight of the basalt. The volcano forms on top of a broad swell in the lithosphere caused by the heat carried in the plume.

(B) As the plate moves, the first volcano is carried away from the source of magma and stops growing. The island then gradually erodes to sea level. Meanwhile, over the plume, a new volcanic island forms.

(C) Continued plate movement produces a chain of islands. Reefs can grow to form an atoll. As the plate cools and subsides, the volcano may drop below sea level.

Figure 22.5 A linear chain of volcanic islands and seamounts results from plate movement above a mantle plume. The string of volcanoes produced reveals the path of the moving plate.

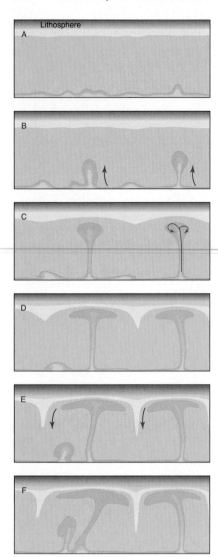

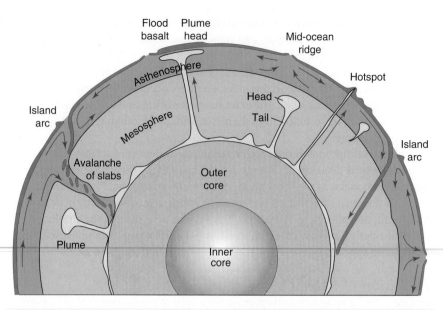

Figure 22.6 Plumes rise from the core-mantle boundary, according to current theory, and are an important type of mantle convection. Some mantle material in the hot boundary layer between the mantle and core becomes hot, and therefore more buoyant, and rises upward in a cylindrical plume. A new plume starts with a large head, behind which is a slender tail. When the plume reaches the cold, rigid lithosphere, it flattens and spreads outward. Flood basalts may erupt from the plume head. Hotspot islands may form from the narrower, long-lived tail. Oceanic crust subducted deep into the mantle at some ancient subduction zone (right) may be part of the source material of mantle plume.

Figure 22.7 The development of a mantle plume, shown in these cross sections, is related to the rise of low-density material from deep in the mantle. The colors represent temperature, with yellow hottest and brown the coolest. (A) As a plume rises, its head enlarges and a narrow tail develops (B, C). When the plume head hits the base of the lithosphere, it flattens (D). Cold, dense material also sinks (tan). Gradually, each plume cools—note the lower temperatures in the central plume—and new plumes develop (E).

How does the shape of a plume change as it rises through the mantle?

Consequently, small bumps form in the boundary layer (Figure 22.7). These may form **diapirs** that ultimately enlarge and develop into buoyant mantle plumes.

Laboratory experiments suggest that a new **starting plume** ascends through the mantle with a large bulbous **head** fed by a long narrow pipe, or **tail**, that extends to greater depth (Figure 22.7). As a new plume rises through the mantle, resistance to its flow causes the head to rise more slowly than the material in the tail. Consequently, the plume head enlarges as it is fed by material flowing up through the long narrow tail; the rising plume head inflates like a balloon. Enlargement of the head also occurs because material from the surrounding cooler mantle is swept into the rising plume.

Because the plume head grows as it moves through the mantle, a large plume head can develop only if it traverses a large distance. Using this relationship, we can estimate that plumes must rise thousands of kilometers, probably from the core-mantle boundary to the surface—a distance of 2700 km. If this is true, much of the heat lost from the molten metallic core is carried away by mantle plumes. Plumes are probably responsible for about 10% of Earth's total heat loss; plate tectonics accounts for more than 80% of the heat lost from the mantle.

When the head of a starting plume nears the surface and encounters the strong lithosphere, it appears to spread out to form a disk of hot material 1500 to 2500 km across and 100 to 200 km thick (Figure 22.7). This is about the size of most continental **flood basalt** provinces. The rising plume uplifts the surface to form a broad low dome (**Figure 22.8**). If a plume head has a temperature of 1400°C (about 200°C warmer than normal mantle), the buoyancy force and extra heat of the hot plume can create a broad swell hundreds of kilometers across and as much as 1 km high (Figure 22.8). This uplift can cause extension, normal faulting, and rifting of the overlying lithosphere. Moreover, as the plume rises to shallow depths, the reduced pressure allows it to melt partially and to produce basaltic magma. The larger the plume head, the larger the volume of basalt that can form.

Eventually, the plume head dissipates by cooling or mixing with the shallow asthenosphere (Figure 22.8). The rest of the plume's history is dominated by flow through the long tail. In contrast to a plume head, the narrow tail is thought to be only about

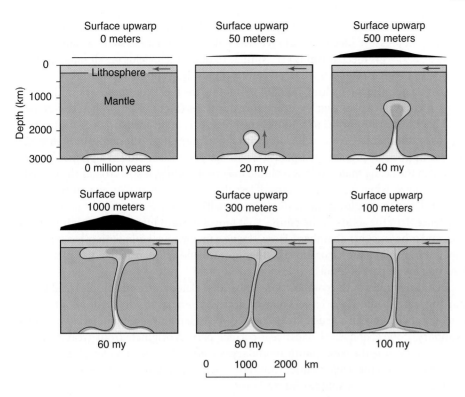

Surface upwarp
0 meters

Surface upwarp
50 meters

Surface upwarp
500 meters

Depth (km)
0
1000
2000
3000

Lithosphere
Mantle

0 million years

20 my

40 my

Surface upwarp
1000 meters

Surface upwarp
300 meters

Surface upwarp
100 meters

60 my

80 my

100 my

0   1000   2000   km

Figure 22.8 A starting plume has a large head and a narrow tail as shown in these cross sections. Each step shows the size and shape of an evolving mantle plume, along with a profile showing uplift at the surface. The plume's head is enlarged because of entrainment of material from the surrounding mantle and because of its slow movement relative to the material in the tail. Once the head of the plume hits the strong lithosphere, it deforms by flattening into a thinner and wider disk. Uplift of the surface is caused by the buoyancy and heat of the mantle plume. Here, 1 km of uplift occurred over the center of the plume during the first few tens of million years of its history. Gradually, the plate moves and the plume head dissipates, leaving a narrower tail. Eventually, this too disappears.

300 km in diameter. Uplift, extension, and basaltic magmatism will also be associated with this part of a plume's evolution, but all will be less than during the starting phase. As the lithosphere moves away from the focus of a plume, it cools, contracts, and subsides. This cooling phase may persist for hundreds of millions of years and may be accompanied by the slow subsidence of the crust and the development of a large sedimentary basin.

Ultimately, the plume itself also loses thermal energy and dies, as new plumes form elsewhere and continue to carry heat from the interior to the surface. A typical life span may be about 100 million years. In short, mantle plumes are temporary features that form and ultimately fade and die.

Plates and plumes are complementary, each involved in a different form of mantle convection. Plumes probably come from a hot boundary layer at the *base* of the mantle, whereas tectonic plates are the cool boundary layer at the *top* of the mantle. As the core loses heat, part of the overlying mantle becomes buoyant and rises in a plume. In contrast, as plates cool, they become denser than the underlying mantle and sink. Thus, in addition to the plate tectonic system, there is also a plume tectonic system. It involves mostly vertical movements of the lithosphere accompanied by volcanism. These processes are superimposed on the constantly moving tectonic plates.

Making Magma in Mantle Plumes

Basaltic magma is generated in a rising mantle plume by decompression melting. Magmas formed in mantle plumes are distinctive and show hints that they are partially derived from vestiges of ancient slabs subducted deep into the mantle. In continental settings, rhyolite and granite may form above a mantle plume by partial melting of the crust or by fractional crystallization of basalt.

The various components of a magma system can be compared to those of a river system. Each magma system has a *source* of magma, a *path* along which the magma is transported, and a final site of *emplacement* where the magma erupts as a lava flow or where the magma crystallizes to form a pluton. Understanding the magma source—where, how, and why it is generated—is key to understanding and predicting the behavior of the entire magma system.

The source of magma in a rising plume is probably related to melting caused by a drop in pressure (decompression) as the hot material rises to shallow depth (**Figure 22.9**). This mechanism is very similar to that which yields basalt magma at mid-oceanic ridges. **Decompression melting** must be a common phenomenon in the mantle because the melting point of peridotite, the most common rock in the upper mantle, decreases slightly as pressure decreases (Figure 22.9). When low-density, solid mantle moves upward in a plume, the pressure is reduced faster than the plume can lose its heat and reach equilibrium with its surroundings. The arrow in Figure 22.9 shows one possible pressure-temperature path for peridotite rising in a plume. Although the rising material in the plume may cool slightly, it still crosses the partial melting curve, when it reaches depths of less than 100 km or so. Consequently, part of the peridotite in the plume melts and forms basaltic magma. This basaltic melt is even less dense than the solids in the plume, so it moves upward into the crust, filling dikes, sills, and other magma chambers. Here, the new magma may mix with other magma, cool to form solid plutons, or move farther upward and erupt, usually as a quiet lava flow that forms part of a shield volcano.

Detailed studies of the compositions of basalts related to mantle plumes lead to a startling conclusion. These basalts may be derived from part of the mantle contaminated by ancient oceanic crust, including oceanic basalt and sediment. Here we have a quandary. If the geophysical evidence is correct, plumes originate from great depths, perhaps as deep as the core-mantle boundary. So, how did crustal materials including sediment get into the deep mantle? Could the sources of plumes be the graveyards for subducted oceanic lithosphere (Figure 22.6)?

Oceanic crust, with its basalt and its marine sediment, undergoes a significant density increase during subduction when garnet and other dense high-pressure minerals form. Perhaps the density increase allows the oceanic crust to travel all the way to the core boundary, like a rock sinking slowly in thick mud. There, it can subduct no further, because the core is much denser. So the subducted lithosphere may reside there for millions of years and mix with mantle rocks already at the boundary. If this mixture becomes expanded by heat escaping from the iron core, it may rise slowly back toward the surface and eventually melt to create basalt magma.

Such a conclusion provides the final element for the vast recycling system that is Planet Earth. Partial melts of the shallow mantle rise buoyantly and then erupt at a mid-ocean ridge. The crust created at the ridge becomes buried by deep-marine sediment before it subducts back into the mantle. On the way down, it is metamorphosed, dehydrated, and incompletely mixed with mantle rocks. Eventually, the oceanic lithosphere may reach the core-mantle boundary. There it resides until heat lost from the core makes it buoy back to the surface again, but this time it rises as a part of a long narrow plume rather than sinking as a stiff sheet (Figure 22.6).

Is there a difference between basalts erupted at mid-oceanic ridges and those erupted on oceanic islands?

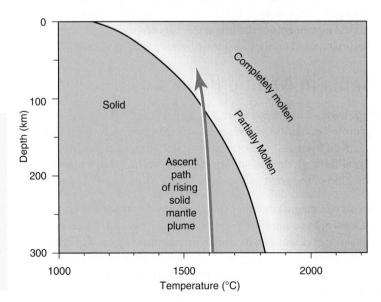

Figure 22.9 Magma in a rising mantle plume is produced as a result of decompression melting. The black line marks the temperature at which melting begins for mantle peridotite. The blue arrow shows the pressure and temperature path followed by a rising plume. Basaltic magma is produced when conditions in the plume cross the melting curve at a shallow depth. The partial melt can be extracted and rises to erupt as an ocean island or continental flood basalt.

STATE OF THE ART   X-Ray Fluorescence Spectrometry

As you read this chapter, you may wonder exactly how lavas that erupt above mantle plumes are different from those that erupt at mid-ocean ridges or at island arcs. To find out, the lava flows may be studied using *X-ray fluorescence spectrometry*—a widely used technique to find the elemental composition of rocks.

Inside an X-ray fluorescence spectrometer, powerful X rays are beamed onto a specially prepared sample of rock. In turn, "secondary" X rays are released from all the atoms in the rock as electrons move from shell to shell. These secondary X rays have wavelengths that are characteristic of each element. Thus, a silicon atom will yield X rays that are different from those emitted by iron or by rubidium. In addition, the strength of the silicon X rays is directly proportional to the amount of silicon in the specimen. The various wavelengths of X rays are separated from each other by diffraction as they pass through specially designed crystals. The intensity of each X-ray wavelength is then measured and converted to an element concentration by comparison with X-ray intensities from rocks of known composition.

Chemical analyses of basalt lavas from three different settings are shown in the table. There are only small differences in the major element concentrations, but large differences in the trace element concentrations in the three basalts.

Geochemists have devised many graphical ways to compare such data. One of the most useful is to divide the element concentrations in a rock by those estimated for Earth's primitive mantle. On this kind of graph (sometimes called a spider diagram), the differences among the basalts are obvious. Mid-ocean ridge basalts have relatively smooth patterns but are poor in the elements plotted on the left. Basalts from island arcs have higher concentrations of these elements and decided depletions in two highly charged elements—niobium (Nb) and titanium (Ti). Ocean island basalts, erupted above mantle plumes, have higher concentrations of most of the trace elements, and the curve actually peaks at Nb.

Now that we can see the differences, what do they mean? The differences show that the mantle sources of the basalts differ widely from one another. For example, the source of mid-ocean ridge basalts has been depleted in many of the trace elements (especially Ba—barium, Rb—rubidium, and Th—thorium) over the eons by the gradual extraction of continental crust from the mantle. Island arc basalts are enriched in elements (such as Ba, Rb, and K—potassium) that are soluble in water and poor in elements like Nb and Ti that are not soluble. The soluble elements are carried into magmas when a subducting slab dehydrates. Finally, ocean island basalts have sources like mid-ocean ridge basalts, but smaller amounts of melting have enriched the trace elements in the partial melts. In this way, the compositions of rocks erupted at the surface reveal much about Earth's deep interior.

Composition of Basalt from Different Settings

	Mid-ocean Ridge	Island Arc	Ocean Island
Major oxides in weight percent			
SiO_2	50.7	49.2	49.2
Al_2O_3	15.5	15.3	12.8
Fe_2O_3	10.6	9.9	12.5
Trace elements in parts per million			
Rb	1	14	25
Nb	9	1	50
La	3	10	35
Zr	85	50	220

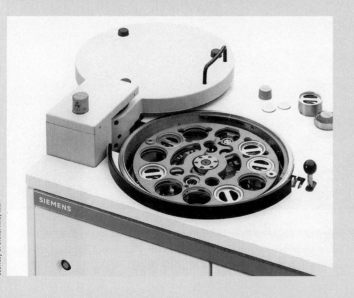

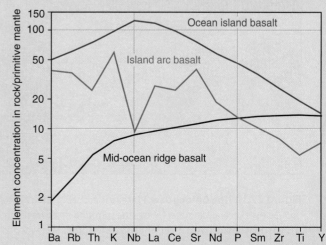

Courtesy of Bruker AXS, Inc.

Mantle Plumes Beneath the Ocean Basins

> A starting plume may yield flood basalt flows that erupt on the ocean floor and form a large oceanic plateau. As the lithospheric plate continues to move over the plume, a narrow chain of volcanic seamounts forms, with the active volcanoes lying directly over the tail of the plume. If a plume is centered on a mid-oceanic ridge, an elongate volcanic plateau forms.

Many mantle plumes rise to Earth's surface beneath the ocean basins. Each of these plumes has a discrete history, with a distinct beginning and an end. What, then, is produced when a new plume with its large head rises beneath an ocean basin? And what happens as the long-lived plume tail evolves? Some oceanic plumes are even centered on mid-ocean ridges, creating an exceptionally rich mixture of volcanic and tectonic features. In the sections that follow, we will examine each of these three types of oceanic volcanism: a starting plume, volcanism related to a tail, and a plume on a mid-ocean ridge.

How are mantle plumes expressed on the ocean floor?

Starting Plumes: Oceanic Plateaus and Flood Basalts

The first type of oceanic volcanism produces distinctive underwater landforms. Scattered across the ocean floor are several broad plateaus that rise thousands of meters above their surroundings (Figure 22.2). These **oceanic plateaus** are not easily explored, and consequently, little is known about them. However, oceanic plateaus may form by some of the most spectacular volcanic events on the planet.

The largest oceanic plateau is the Ontong-Java Plateau (**Figure 22.10**); it is two-thirds the size of Australia. A coral-capped bit of the plateau rises above sea level to form the Ontong-Java atoll, the largest atoll in the world. The surrounding region of the equatorial western Pacific is underlain by oceanic crust that is 25 to 43 km thick—as much as five times thicker than typical oceanic crust. The thick crust is apparently made of about 36 million cubic kilometers of basalt lava flows, enough to cover the entire conterminous United States with a layer 5 m thick. The lavas buried older oceanic crust, with its magnetic stripes, which originally formed at an oceanic ridge. There appear to be no large shield volcanoes or calderas on the plateau. Instead, lava must have erupted from long fissures on the ocean floor, probably as flood basalts, quite unlike the small eruptions that occur at a mid-ocean ridge. If these oceanic flood lavas are similar to those found on the continents, individual lava flows may have been hundreds of kilometers long.

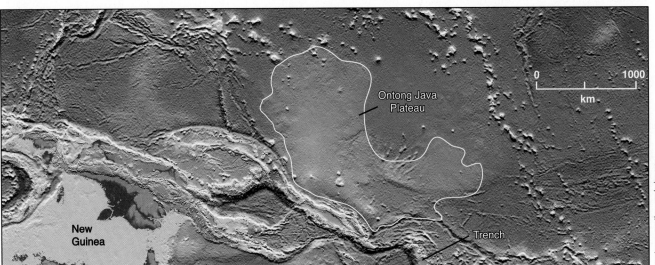

Base map by Ken Perry, Chalk Butte, Inc.

Figure 22.10 The Ontong-Java Plateau is probably a huge accumulation of submarine flood lavas erupted during the Cretaceous. The plateau rises several kilometers above the surrounding abyssal depths and is underlain by crust that may be 40 km thick, about five times thicker than normal oceanic crust. Oceanic plateaus are probably related to eruptions from the enlarged head of a new plume.

The paleomagnetic character of the lavas on the plateau and a few recently acquired radiometric ages indicate that the Ontong-Java Plateau formed in at least two episodes at about 120 million years ago and about 90 million years ago; during the Cretaceous Period. Much of the plateau was probably created in less than 3 million years. If that estimate is correct, the vents on the Ontong-Java Plateau must have erupted between 15 and 20 km^3 of lava each year. That amount is comparable to the volume of new crust formed by the entire oceanic ridge system in a year, and it dwarfs the 1980 Mount St. Helens eruption of less than 1 km^3 of volcanic rock.

From a geologic perspective, such a rapid outpouring of such a huge volume of lava is truly remarkable. The entire submarine landscape of this large area was changed in only a few million years. Most geologic processes that affect such large regions take tens of millions of years to accomplish such changes. For example, the Rocky Mountains have been rising for more than 40 million years, and the Andes have been building for at least 30 million years.

Such a vast oceanic plateau may represent a spasm of igneous activity related to the initiation of a new mantle plume. As the enlarged plume head rose beneath the ocean floor, partial melting produced huge volumes of basaltic lava that erupted over a geologically short period of time. Eventually, the heat from the plume head was lost through this volcanism, and the amount of partial melting declined. Oceanic lithosphere continued to move over the plume tail, and a **hotspot track** formed. The Louisville hotspot (Figure 22.2) is the most likely current location of the plume that fed the Ontong-Java oceanic plateau.

How does the Ontong-Java Plateau differ from an oceanic ridge?

Plume-Tail Volcanism: Hotspot Island Chains

A second type of oceanic volcanism produces hotspot island chains. Presently, most intraplate volcanism is dominated by the construction of large shield volcanoes over plume tails. Most eruptions are relatively quiet flows of basaltic lava from vents at a volcano's summit or on its flanks. Large collapse calderas are common at the summits of these shield volcanoes. In addition, heat from the plume and the weight of the volcano drive a variety of vertical tectonic processes. Hotspot volcanic systems can be explored by referring to the Hawaiian island chain.

The Hawaiian Plume   Hawaii is the best known example of volcanic activity above the still-rising tail of a mantle plume beneath oceanic lithosphere (**Figure 22.11**). Hawaii is the active area of a series of otherwise extinct volcanoes stretching across the Pacific seafloor to the Aleutian Trench (**Figure 22.12**). The lava that has erupted from the hotspot is more than enough to cover the entire state of California with a layer 1.5 km thick. Most of the volcanoes are below sea level now. The islands northwest of Hawaii are all deeply eroded extinct volcanoes.

The island of Hawaii consists of five major volcanoes, each built up by innumerable eruptions. The largest active volcano on Earth, Mauna Loa, dominates the big island of Hawaii. It rises 9000 m above the ocean floor (**Figure 22.13**). This large shield volcano was built by repeated eruptions of lava during the past million years, and it is still active. Many recent flows can be seen along Mauna Loa's flanks, extending as dark lines from a series of fissures, or **rift zones,** along the crest of a ridge (Figure 22.11). The oval caldera formed by the repeated collapse of Mauna Loa's summit when dense intrusions sank under their own weight or as magma withdrew from subterranean chambers.

Southeast of Mauna Loa is the younger Kilauea volcano, where young lava flows erupt primarily from rift zones. An eruption along the East Rift Zone has continued almost without pause since 1986. As a result, an extensive lava field and a small shield volcano have formed, and tube-fed flows extend from the vent to the sea. Kilauea is growing higher, and the island is growing larger on its margin. Farther southeast is an even younger submarine shield volcano, Loihi (Figure 22.13). It, too, is an active volcano and will one day rise to the surface as it progressively grows higher by repeated eruption and intrusion.

The volcanoes of Hawaii are shaped like huge rounded plateaus. The classic shield shape of Hawaiian volcanoes describes only that part of the volcano above sea level.

How does a mantle plume produce a chain of volcanic islands?

Figure 22.11 **The island of Hawaii** as seen from space. Two old eroded volcanoes form the northern (top) part of the island. Mauna Loa is the large volcano in the foreground; it has erupted many times in the past 150 years. The individual flows appear as thin dark lines extending from fissures that emanate from a summit caldera. Kilauea (right) is the youngest of the volcanoes that has reached sea level. Its most recent eruptions began in 1983.

They are much flatter above sea level than below. The shape of the entire volcano is complex because subaerial lavas and submarine flows behave differently. Subaerial lavas are more fluid and form gently inclined slopes, whereas submarine lavas do not flow as freely as those on land. They are quenched rapidly by the cold seawater and some also become granulated as they erupt into the cold seawater. These factors cause submarine lava to pile up and produce steeper slopes.

These steeper submarine slopes are susceptible to gravitational failure and mass movement. Vast landslides steepen the slopes of the submarine portions of the volcanoes. Most of the submarine flanks of Mauna Loa and Kilauea are actually huge landslide scars (Figure 22.13). The landslides range from slow-moving slumps bounded by curved faults to fast-moving debris avalanches. At the surface, landslides are marked by high scarps separated by flat benches. In fact, entire volcanoes are being ripped apart by their own weight (**Figure 22.14**). The active volcanoes are spreading seaward away from the molten or plastic cores of their rift zones. This extension is driven by gravity and is one cause of the rift zones. In turn, the rift zones become sites for intrusion of magma that causes further spreading of the island. These giant submarine landslides pose a hazard to people who live on other islands as well. Catastrophic failure of a giant debris avalanche could cause a tsunami to sweep across the Pacific basin.

The hot plume tail beneath Hawaii has also created a large swell in the oceanic lithosphere (Figure 22.12). This broad uplift is probably caused by the plume's buoyancy and by heating from the mantle plume. The swell is about 1500 km across, more than 4000 km long, and nearly 1 km high. It is elongated along the track of the hotspot

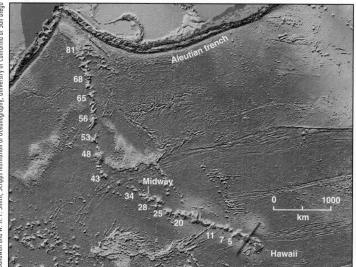

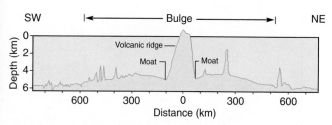

(A) The volcanic islands and seamounts form the most obvious part of the long chain. A bend in the chain marks a change in the direction of plate movement, which is presently to the northwest. An elongate rise marks the hotspot trail; it is highest near the plume beneath Hawaii and is progressively lower toward the northwest. The narrow trough or moat that lies on both flanks of the island is due to subsidence from the weight of the volcano. Numbers are ages in millions of years for volcanic rocks along the seamount chain.

(B) This cross section shows the volcanoes sitting on a broad bulge caused by heating from the underlying mantle plume. On either side, a narrow moat or trough has formed because the weight of the volcanic islands bends the Pacific plate downward.

Figure 22.12 **The topography of the Hawaiian-Emperor Island chain** shows the critical elements of the evolution of a hotspot chain.

and drops gently to the northeast where the lithosphere cooled and contracted after it passed the plume. The high volcanic ridge caps the swell and rises another 5 km or so to the ocean surface.

So great is the weight of these volcanoes that a narrow depression or moat lies at their base (Figure 22.12). Moreover, because of the load, the islands are slowly sinking by making isostatic adjustments. The island of Hawaii is sinking about 3.5 mm/yr; Maui is sinking 2.2 mm/yr. Farther up the island chain and made of older inactive volcanoes, Oahu is sinking even more slowly. Large submergence rates for the island of Hawaii make it difficult for coral reefs to grow around its shorelines, whereas on the older, more slowly sinking islands of Maui and Oahu, organic reefs flourish in the clear warm water. Consequently, white sand beaches like those of Oahu, derived from the breakup of the coral, are absent on Hawaii.

Hawaiian Earthquakes   Earthquakes in Hawaii (**Figure 22.15**) are typical of those that form above mantle plumes. Because no plate boundary is involved, earthquakes are relatively small and infrequent. The most common earthquakes are shallow and are related to the movement of magma or to slumping along the landslide-bounding faults. They typically have magnitudes of less than 4.5 and depths of less than 10 km. Another smaller family of deeper earthquakes arises in the mantle. The largest and most recent of this type was a magnitude-6.2 tremor on the island of Hawaii in 1973, occurring about 40 km deep. The earthquake caused about $5.6 million in damage and injured 11 people.

What causes these deeper earthquakes if no plate interactions are involved? Most of the deeper ones probably release strain built up by the enormous load of the volcanoes on the lithosphere. As the volcanoes grow, they add ever more weight to the lithosphere, bending it downward. Consequently, the earthquakes are most common beneath the actively growing part of the volcano chain. The older parts of the hotspot chain are seismically inactive.

Why do we believe the island of Hawaii is located above a mantle plume?

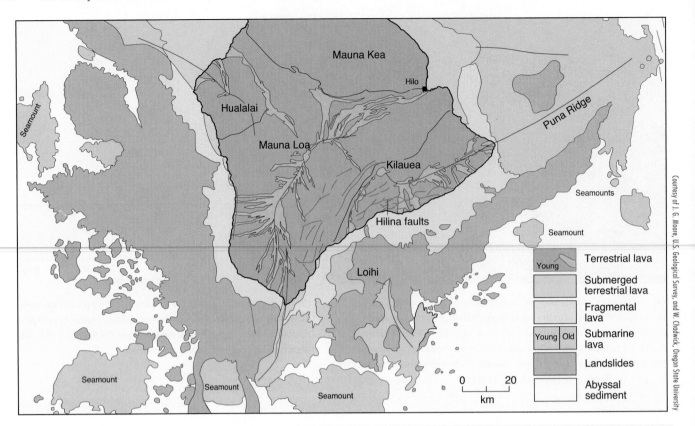

Figure 22.13 **The island of Hawaii** consists of several volcanoes that rise from the floor of the Pacific. They include Mauna Loa, the tallest volcano on Earth, and Loihi, a submarine volcano still growing toward the surface. The map shows the importance of eruptions from narrow rift zones in the building of Mauna Loa and Kilauea. The arcuate shape of Loihi is also controlled by submarine eruptions from rift zones. Note that huge landslides have slipped from the submerged flanks of the volcanoes. Normal faults on the island and some rift zones are controlled by movement of these landslides.

Hawaiian Volcanism Like other oceanic hotspots, Hawaii's volcanic rocks are largely basalt. Silica-rich rocks, such as andesite and rhyolite, are extremely rare. Magmas formed by partial melting of the mantle are overwhelmingly basaltic and poor in volatiles, such as water vapor, compared with those formed in subduction zones. The absence of silicic continental crust beneath the island chain may partly explain the lack of rhyolite: There are no granitic rocks to assimilate and enrich the magmas in silica.

Eruptions on Hawaiian volcanoes are relatively quiet and predictable but, nonetheless destructive. During the most recent eruptions of the Kilauea volcano, destruction of homes, buildings, and roads cost millions of dollars. Most eruptions start by the opening of a short, narrow fissure, only a meter or so wide, that rapidly lengthens to as much as several kilometers. Along the rift, lava erupts as a series of fountains and locally forms a nearly continuous "curtain of fire." The eruptions usually focus on one point on the fissure, where a cinder cone or low shield volcano subsequently develops. Individual flows are usually only tens of kilometers long and eruption rates are much slower than those inferred for flood basalts. Small earthquakes accompany the eruptions because the volcano inflates and deflates as magma is intruded into or flows out of the volcano. During a long eruptive episode, activity may shift up and down the rift system as lava breaks out at different places. Where the lava flows enter the sea, small yet dangerous explosions may occur. Occasionally, the summits of the volcanoes have collapsed to form small calderas several kilometers across.

Volcanic gases, consisting mostly of water and carbon dioxide but including noxious sulfurous gases as well, may mix with humid air to form a type of volcanic smog. However, Hawaiian eruptions are usually too weak to inject aerosols into the atmosphere and affect the climate.

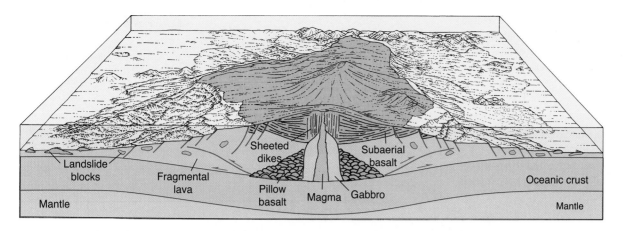

Figure 22.14 Magma system at an oceanic mantle plume, as deduced from geologic, earthquake, and geochemical studies. After forming in the mantle, basalt magma rises buoyantly through dikes into small magma chambers, where it mixes with magma still present from earlier episodes of intrusion. The magma in the chamber cools and crystallizes to form gabbro along the walls. The movement of large slump blocks rips the volcano apart. Magma fills these rift zones and may erupt to form pillow basalt or fragmental lava under water or, if the volcano is high enough, subaerial lavas above the water. The weight of the volcano causes the underlying oceanic crust to subside.

Evolution of Seamounts and Islands   The volcanic islands and **seamounts** scattered across the ocean floors are the largest volcanic edifices on the planet. The factors that control their evolution are quite unlike those for composite volcanoes at convergent plate margins or the lava fields erupted from fissures at divergent plate margins. Most of the volcanic activity that forms the islands and seamounts occurs under water, so the temperature and pressure conditions are quite different from the conditions where subaerial extrusions occur. Also, a submarine volcano is not subjected to contemporaneous stream erosion as it grows. The following summarizes our understanding of the origin of these volcanoes (**Figure 22.16**).

Magma extracted from a mantle plume moves upward through the lithosphere, eventually reaching the ocean floor. It rises through the brittle crust along fractures or dikes, extruding not only through a central summit vent, but also from fissures or rifts on the flanks. Therefore, a seamount grows upward and outward by extrusion of lava over various parts of its surface (Figure 22.16A). Magma also flows nearly horizontally through the rift zones as it moves from a summit magma chamber to the volcano's flanks. Thus, intrusive dikes make up a large fraction of the total volume of the shield volcano. Gabbro crystallized in small magma chambers also must constitute an important part of the interior. These three types of rock—submarine lavas, dikes, and gabbro—make up most of the volcano (Figure 22.14).

During the evolution of a large submarine volcano, isostatic balance requires the base of the volcano to subside while its top grows upward. The compensating root is about twice as thick as the overlying mass. Thus, a basaltic volcano with a relief of 3 km must have grown upward 9 km because its base simultaneously had to subside 6 km. Figure 22.16A–C shows the subsidence and growth of a typical seamount during the first million years of its evolution.

Other factors are important in the growth of islands and seamounts. Submarine lavas erupt in two very different forms. *Pillow lava* might be considered the subaqueous equivalent of *pahoehoe*. Cold seawater chills the lava so rapidly that a crust forms instantly. Each flow advances in a complex multitude of repeatedly budding pillows. Another type of flow leaves beds of *tuff*, fragmented glassy material formed by the explosion and granulation of hot lava when it hits cold seawater. In addition, the abundance of vesicles, and consequently the density of the lava, is directly related to water pressure. At depths of about 1000 m, vesicles form only about 5% of the rock, whereas at depths of 100 m they may form as much as 40%. Consequently, basalt extruded at oceanic depths is denser than basalt extruded on land.

How does a volcanic island change with time?

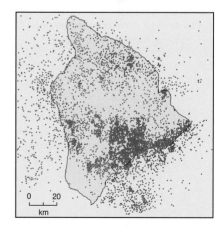

Figure 22.15 Shallow earthquakes are common in volcanic islands above mantle plumes. Earthquake epicenters for Hawaii are shown here. Most of the earthquakes are related to the movement of magma or to slippage on faults related to large landslides. A few deeper earthquakes are caused by bending of the lithosphere under the weight of the volcanoes.

(A) The first 4000 years of eruption makes a volcano 1000 m high, but it has only 0.4% of its ultimate volume. Because of subsidence, its volume is much greater than its height suggests. Pillow lava and fragmental lava dominate.

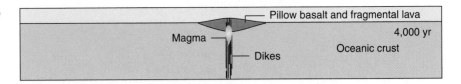

(B) After 400,000 years, the seamount is 4000 m high, and it has about 40% of its ultimate volume. The volcano also grows by diking and by crystallization of small plutons. The volcano spreads under its own weight.

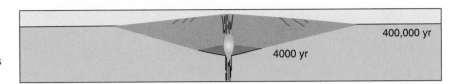

(C) After about 1 million years, the volcano reaches sea level, and erosion joins subaerial eruptions. A typical shield volcano develops, with gentle slopes, rift zones, and a summit caldera.

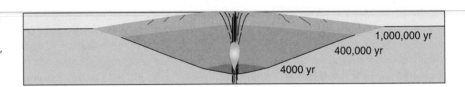

(D) Summit and flank eruptions continue to build the shield volcano above sea level, but wave action and erosion begin to overwhelm the construction phase. Wave-cut platforms and sea cliffs enlarge, slump blocks develop, and stream erosion is vigorous.

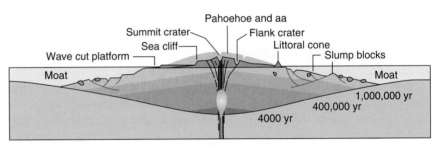

(E) Within a few million years after the volcano drifts beyond the hotspot, erosion develops a wave-cut platform. Subsidence follows, as the volcano drifts beyond the uparched area above the mantle plume. In tropical areas, coral reefs may develop a flat limestone cap on the eroded volcano.

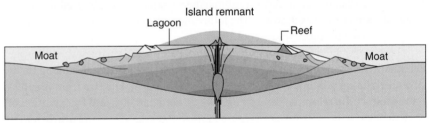

Figure 22.16 The evolution of a hypothetical submarine volcano related to an oceanic hotspot. Most volcanoes in the Pacific drift away from their source plume in fewer than a million years.

Every volcano is subject to the force of gravity, and submarine mass movement can be quite spectacular on seamounts and islands. As magma works its way upward, the volcano swells and radial cracks may develop. In addition, the very weight of the volcano itself begins to tear its fabric apart (Figure 22.16D; Figure 22.14). The fractures may become faults and a large block may slump downward. This movement leaves a gigantic scar near the shoreline. The great Hilina fault scarp on the southern slopes of Hawaii's Kilauea volcano is believed to be such a slump scar (Figure 22.13). Mapping the topography of submarine volcanoes shows that the slump blocks are vast, with scars 30 to 40 km wide and slump deposits covering areas of more than 10,000 km².

Eventually, some submarine volcanoes reach sea level, where coastal processes combine to influence the volcanic eruptions (Figure 22.16D). When lava, with a temperature of about 1100°C, erupts on land and then hits cold seawater, it explodes by creating clouds of expanding steam, and the lava shatters into fragments. This unconsolidated material, which is easily eroded and reworked by waves, is deposited along the shore to form a broad platform. Waves can erode this sea-quenched lava with remarkable speed: Wave-cut platforms more than 3 km wide have been eroded in less than 250,000 years. This is roughly equivalent to the extreme rate of erosion in the Himalaya Mountains, where, on average, a layer of rock a meter thick is removed every thousand years. Thus, for a seamount to grow and become an island, rates of extrusion must be greater than rates of erosion.

Once a volcano rises above sea level (Figure 22.16D), pahoehoe and aa flows dominate. The volcano grows rapidly, as lava is extruded from summit and flank eruptions. A steep-sided circular caldera may form near the summit.

The magma supply is gradually cut off when the volcano is carried away from the relatively fixed hotspot (Figure 22.16E). The volcano becomes deeply eroded by streams, waves, and sometimes glaciers (**Figure 22.17**). Its summit is soon eroded away and a broad, flat platform forms near sea level. As the volcano moves farther from the uparched hotspot, it subsides below sea level. Once the summit subsides to 200 m below sea level, it remains essentially unchanged by erosion. In tropical regions, however, the growth of coral reefs may add an important structure to the volcanic edifice. A reef typically begins as a fringe around a young volcano, evolving into a barrier reef and eventually into an atoll. This reef material forms a limestone platform capping the top of the eroded basaltic volcano.

Ultimately, the eroded volcano is transported to a subduction zone, where it is either consumed along with the oceanic crust into the mantle or accreted onto a continental margin. After a trip of thousands of kilometers, the seamounts of the Hawaiian-Emperor chain are currently being consumed by subduction down the Aleutian Trench north of Japan. If an oceanic plateau ever formed above the starting Hawaiian plume, it has been long since subducted.

Plumes at Mid-ocean Ridges

The third type of oceanic volcanism shows that not all plumes lie stranded in the middle of oceanic plates. For reasons discussed below, some plumes lie directly beneath divergent plate boundaries. At least six plumes lie on or near the Mid-Atlantic Ridge. From north to south, they are Jan Mayen, Iceland, Azores, Tristan da Cunha, Shona, and Bouvet (Figure 22.2). Iceland is the best known, so we discuss it to illustrate the characteristics of plumes that are centered on mid-ocean ridges.

(A) Kauai is deeply dissected by stream erosion. Layer upon layer of basaltic lava exposed in the Grand Canyon of Hawaii. Volcanism from the main shield stage ranges from about 3.9 to 5.1 million years ago.

(B) Nihoa is only a tiny remnant of the shield volcano that once existed here. The lavas here are over 7 million years old. The island, less than 2 km across, is the exposed part of an eroded platform that is about 30 km across.

Figure 22.17 The old islands of the Hawaiian chain are old and deeply eroded.

The Iceland Plume   Iceland lies at the intersection of the north-trending, seismically active Mid-Atlantic Ridge and the seismically inactive Greenland-Faeroe ridge or plateau. If you look at the seafloor map on the inside cover, you might wonder why Iceland is the only great island along a mid-oceanic ridge. The answer is that more is involved beneath Iceland than a simple divergent plate boundary. The east-trending plateau is a hotspot trail that has been torn asunder at the spreading ridge and carried away on two separate plates.

If not for the mantle plume that lies beneath Iceland, this area would be an obscure, submerged part of the global ridge system. Instead, it is a zone of compromises and contrasts. The compositionally distinctive sources of basalts erupted at mid-oceanic ridges are mixed deep in the mantle with those typical of plumes, yielding an intermediate mixture. These hybrid basalts erupt along the Mid-Atlantic Ridge axis as much as 200 km north and south of Iceland.

What evidence indicates a plume exists beneath Iceland?

Volcanism on Iceland, caused by the plume, formed basaltic crust more than 30 km thick, four times thicker than typical oceanic crust. Locally, this thick crust has partially melted near new injections of hot basalt, creating rhyolitic magma. In other volcanic centers, mantle-derived basalts have experienced fractional crystallization to create rhyolite. As you know, rhyolite is extremely rare at normal mid-ocean ridges. Because of diverse magmas and a unique tectonic setting, volcanic eruptions on Iceland have created flood basalts, shield volcanoes, fissure eruptions, composite volcanoes, rhyolite domes, and ash flow calderas—a variety of volcanic features quite unlike those at a normal mid-oceanic ridge!

In addition to these subaerial eruptions, submarine and subglacial eruptions are both common. In the fall of 1996, a small fissure eruption beneath the Vatnajôkull ice cap melted much of the overlying ice. Some eruptions broke through the ice when hot magma contacted ice-cold water, exploding and showering the ice cap with black basaltic ash (**Figure 22.18**). After about a month, enough water accumulated to float the glacier off its floor. Water catastrophically burst from the base of the glacier and inundated the outwash plain. Bridges were destroyed as the floods drained the subglacial meltwater. House-sized blocks of ice were ripped from the glacier and tumbled down the plain to the ocean.

In 2010, a small eruption of another Icelandic volcano disrupted air traffic in Europe for months. Let's review the story of this eruption. Icelandic volcanologists closely monitor all types of volcanic activity to watch for the precursors to eruptions (**Figure 22.19**). Beginning in the middle of 2009, shallow earthquakes focused on Eyjafjallajökull subglacial volcano (*jökull* means "glacier" in Icelandic). In January 2010, GPS measurements and satellite interferometry and tilt studies combined to show that the volcano was swelling—as it turns out from the intrusion of a magma into the crust beneath the volcano at about 4 to 5 km depth. The earthquakes increased in number and intensity; most of the earthquakes occurred between 8 and 12 km deep marking the path of the magma to the roots of the volcano. During this time the earthquakes became shallower and localized into a linear zone on the flank of Eyjafjallajökull. An eruption was predicted and then started on 20 March when a linear fissure opened and a short *basaltic* lava flow formed low on the flank of the volcano. The initial eruptions were small and relatively quiet. The eruption was fed by a narrow vertical dike that connected the magma chamber with the surface. After a two week pause, the eruptions moved from the flank to the glacier clad summit of the volcano. Moreover, the composition of the magma changed from basalt to a more differentiated *alkali-rich andesite*. This magma was what was left over from an earlier eruptive period and had changed composition from basalt to a more silica- and water-rich composition by fractional crystallization. Eruptions from the summit were highly explosive and drove fine ash to heights of 6 to 9 km where the jet stream picked it up and carried the ash over much of Europe. Because volcanic ash damages aircraft engines, many flights were cancelled and travelers stranded. The change to explosive eruptions had two fundamental causes: (1) the differentiated andesitic magma contained much more water than typical basalt; and (2) at the summit, the magma melted glacial ice

(B) About 1 month after the eruption started, the meltwater burst from the base of the glacier and flooded the outwash plain of the glacier, carrying large blocks of ice and destroying bridges like this one.

(A) The overlying ice collapsed into the subglacial lake. Eventually, explosions caused by the contact of hot basalt with the cold water sent low plumes of basaltic ash over the glacier.

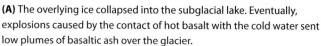

Figure 22.18 Eruption of basalt beneath a glacier on Iceland caused the base of the glacier to melt.

in the caldera and the ensuing contact of hot lava with water explosively converted it to steam. Eventually, the explosive phase subsided and by October the eruption was over. If not for a unique combination of factors, a highly differentiated water-rich magma, steam explosions, and location of the jet stream, the eruption may have gone unnoticed by most of the world.

The Iceland Ridge began to form about 60 million years ago, when Greenland was rifted away from Europe to open the North Atlantic Ocean (**Figure 22.20**). Rifting was apparently assisted by the development of a new mantle plume. Large continental flood basalt provinces in Greenland and the northern British Isles mark the position of the starting plume when the continents were still attached (Figure 22.2). Gradually, an open ocean developed between the two continents. Rising high from the seafloor between the continents, active volcanoes on proto-Iceland continued to form above the still-rising tail of the plume. The volcanoes and lava flows were aggressively eroded by streams, glaciers, and waves, keeping much of the island near sea level. Moreover, because the Iceland plume is centered on a ridge, the volcanoes eventually drifted away from the ridge axis and became inactive. As the newly formed lithosphere moved away from the ridge and cooled, it subsided to make the long plateau that links Greenland to the Faeroe Islands. Unlike a spreading ridge, this ridge lacks active volcanoes and has no earthquakes.

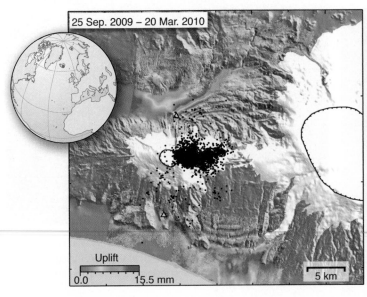

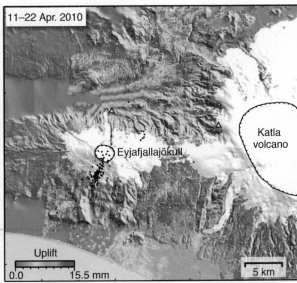

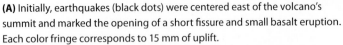

(A) Initially, earthquakes (black dots) were centered east of the volcano's summit and marked the opening of a short fissure and small basalt eruption. Each color fringe corresponds to 15 mm of uplift.

(B) After the fissure eruption stopped, earthquakes moved to a linear belt that terminated at the summit caldera. Here explosive eruptions of andesitic magma started. Interferometry shows that uplift continued and became centered on the volcano.

(C) The explosive phase of the summit eruption was dramatic and sent fine ash high into the atmosphere. It was driven by the expansion of magmatic gases and by steam explosions in the crater of the volcano. Lightning is the result of the buildup of static electricity in the eruption cloud.

Figure 22.19 **The eruption of Eyjafjallajökull** that disrupted air traffic for weeks in 2010 owes many of its distinctive features to the Icelandic hotspot.

At some ridge-centered mantle plumes, a "sudden" shift in the position of the ridge—a so-called **ridge jump**—may isolate a plume tail on one side of the ridge. Half the hotspot trail is stranded on the other side and ceases to grow. For example, the plume beneath the Azores was once on the Mid-Atlantic Ridge but is now east of the ridge (Figure 22.2). The Tristan da Cunha plume is also on the opposite side of the ridge from half of its hotspot trail. A ridge jump may explain why the Ninetyeast Ridge lies north of the Indian Ocean ridge and its parental plume is now centered beneath Kerguelen Island, south of the ridge (Figure 22.2). The ridge apparently drifted off the plume about 37 million years ago.

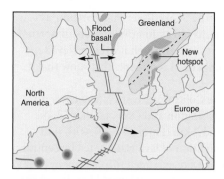

(A) 60 million years ago. A new mantle plume rose to create a large continental flood basalt province on the margins of what are now Europe and Greenland.

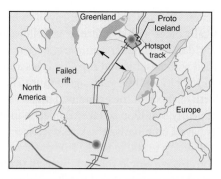

(B) 30 million years ago. Greenland rifted away from Europe and a mid-ocean ridge formed. Volcanoes formed above the plume, but drifted away to make the aseismic ridge (green) between Greenland and Europe.

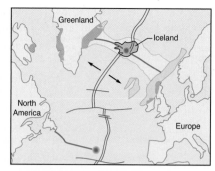

(C) Today, Greenland and Europe are far apart and volcanically active Iceland sits astride the Mid-Atlantic Ridge. It is being rifted apart, but it is still underlain by an active mantle plume.

Figure 22.20 **The history of Iceland** stretches back 60 million years and involves a mantle plume and a mid-ocean ridge.

Mantle Plumes Beneath Continents

If a plume develops beneath a continent, it may cause regional uplift and eruption of flood basalt from fissures and rhyolite from calderas. Continental crust is not strongly deformed above a mantle plume, but the lithosphere bends to form broad swells and troughs; this bending may trigger shallow earthquakes. Sometimes, continental rifting and the development of an ocean basin may follow the development of a new plume.

The vast sheets of basaltic lavas—continental flood basalts—that cover large areas of the continents have puzzled geologists for more than a century. Their origins were not explained by any aspect of plate tectonic theory, and they remained a major geologic mystery until it was suggested that they might be caused by mantle plumes. Magmatic systems above subcontinental plumes are quite different from oceanic hotspots in composition, eruption and intrusion style, and the nature of the volcanic deposits. The reasons are that (1) the continental crust is thicker and less dense than oceanic crust; (2) the continents' silica-rich rocks may become assimilated and change the magma's composition; and (3) continental crust responds to stress quite differently than oceanic crust.

The Yellowstone Plume

Yellowstone National Park, in northwestern Wyoming, is best known for its spectacular scenery—hot springs, geysers, deep canyons—and abundant wildlife. However, the rocks tell an even more dramatic story. Three huge volcanic calderas form the heart of Yellowstone. From these, more than 8500 km³ of rhyolite ash-flow tuff erupted over the past few millions of years, including some of the largest ash-flow tuffs known. The most recent major eruption occurred only 620,000 years ago, and its ash buried parts of every state west of the Mississippi River (**Figure 22.21**). It is the heat from this still-active volcanic system that drives the flow and eruption of groundwater as thermal springs and geysers.

Southwest of Yellowstone, the Snake River Plain is a depression 800 km long and 80 km wide, slicing across north-trending mountain ranges (**Figure 22.22**). The plain descends from 2500 m at the margin of Yellowstone Park to 1200 m in southwestern Idaho. The Snake River Plain is covered by small basalt lava flows and is dotted with small shield volcanoes, some as young as 2000 years old. The basaltic lava flows cap thick accumulations of rhyolite ash-flow tuff, similar to tuffs at the surface in

What is the evidence that Yellowstone overlies a mantle plume?

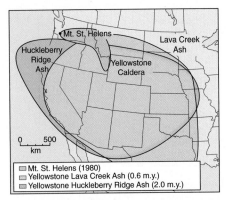

Figure 22.21 Rhyolite eruptions from Yellowstone calderas have buried the western United States with ash several times. The last large eruption from this caldera system ejected more than 3000 km³ of ash as fall and flow deposits. The volume of ash from the devastating eruptions of Mount St. Helens in 1980 is small by comparison: Less than 1 km³ of magma was erupted.

Yellowstone, but much older. The rhyolites below the Snake River Plain become systematically older with distance from Yellowstone, as old as 16 million years in northern Nevada. Even farther southwest, a narrow rift slices through northern Nevada. It is filled with basaltic and rhyolitic lavas and associated sediments that show that it formed 16 or 17 million years ago (**Figure 22.23**).

North of this rift, the great flood basalts of the Columbia River Plateau cover an area of nearly 5 million square kilometers in Washington and Oregon. These flood basalts have a cumulative thickness between 1 and 2 km (**Figure 22.24**), with individual flows up to 100 m thick. This great accumulation of lava was not fed by central eruptions from a single volcano. Instead, the lavas were extruded through many fissures. Such continental flood basalts may erode into plateaus of layered basalt and are therefore also known as **plateau basalts**. Vast **dike swarms** now mark the fissures through which the lava extruded. The largest eruptions occurred about 17 million years ago from long fissure vents along Idaho's western border; these parallel the rift in northern Nevada. Some basalt lavas flowed all the way to the Pacific Ocean, in some cases as much as 500 km.

How are these volcanic and tectonic features related? One hypothesis holds that all are related to a mantle plume that presently lies beneath Yellowstone. According to this theory, about 17 million years ago a mantle plume rose beneath what is now the common border of Idaho, Oregon, and Nevada. The enlarged head of this starting plume fed the Columbia River flood basalts. Eruption rates were very high and some lava flows were exceptionally long. Simultaneously, uplift accompanied by extension created the rift through northern Nevada (Figure 22.23).

As the North American plate moved southwestward (at about 3.5 cm/yr) over the plume's tail, a trail of huge rhyolite calderas formed sequentially atop a broad crustal swell across southern Idaho (Figure 22.23). Each rhyolite caldera was much like the currently active Yellowstone caldera. Rhyolite magma was probably formed by partial melting of a mixture of old continental crust and young basalt. The basalt had been intruded to form dikes and sills above the plume. Large granite plutons crystallized beneath the calderas. In the wake of the plume, the lithosphere cooled, contracted, and subsided, and was covered by younger basaltic lavas forming the broad depression that is the Snake River Plain (Figure 22.22). The eruption rates and volumes of basalt on the Snake River Plain were much smaller than on the Columbia River Plateau. Individual flows are all less than 75 km long, and many are much shorter.

The Snake River Plain is a depression atop a broad arch. The elevated region is about 600 to 1000 km across, comparable in size to oceanic swells related to mantle plumes. The epicenters of earthquakes (up to magnitude 6) form a crescent around the front of the crustal swell (Figure 22.22). These earthquakes are related to normal faults and crustal extension. Some shallow earthquakes in Yellowstone may also be related to magma movement below the surface.

Figure 22.22 The Snake River Plain and Yellowstone calderas form a dramatic scar across the mountainous terrain of the western United States. During the last 17 million years (late Cenozoic), silicic volcanism swept across the region as North America moved westward over a nearly stationary mantle plume. Later eruptions of basalt formed small shield volcanoes and fissure-fed flows. The huge Yellowstone caldera marks the present site of the Yellowstone plume. A string of rhyolite calderas like the Yellowstone caldera lies underneath the Snake River Plain to the west. Earthquake locations (dots) are superimposed.

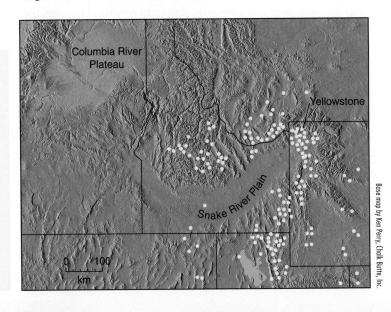

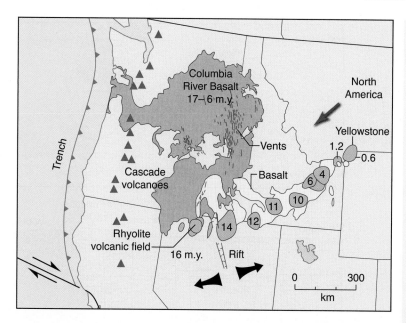

Figure 22.23 **The Cenozoic features of the northwestern United States** may be related to the development of the Yellowstone mantle plume. About 16 million years ago, rhyolite calderas formed near the common borders of Oregon, Idaho, and Nevada, where the plume was probably centered. A narrow rift developed in central Nevada. Simultaneously, the eruption of flood basalts from long fissures on the Idaho border formed the Columbia River Plateau. As the plume head dissipated and North America moved over the plume tail, rhyolite volcanic fields (green) formed in a narrow strip across the Snake River Plain of southern Idaho (the numbers are ages in millions of years ago). Later, small eruptions of basalt covered the Snake River Plain (yellow). Today, the Yellowstone hotspot lies beneath Yellowstone National Park, where large rhyolite eruptions blanketed much of the region.

Figure 22.24 **Flood basalts of the Columbia Plateau** formed between 17 and 6 million years ago. Locally, several lava flows are stacked one upon the other. Erosion that formed the Channeled Scablands of Washington has exposed this sequence of flows. Individual flows may be more than 100 m thick and as long as 500 km. They provide important information about the style of volcanic activity above mantle plumes.

Interpretations of gravity variations across the Yellowstone Plateau and the Snake River Plain reveal much about the subsurface structure (**Figure 22.25**). A large gravity low marks the Yellowstone caldera, because it is partly filled by low-density ash and lava and large bodies of hot rock and rhyolitic magma. In addition, seismic wave velocities are anomalously low beneath the caldera. Most likely the low velocities are caused by this hot (probably still molten) rock just below the caldera. In contrast, the Snake River Plain is marked by a gravity high, probably caused by dense basalts at the surface and the accumulation of basalt dikes and sills in the lower and middle crust. No still-molten magma chambers have been discovered beneath the Snake River Plain.

The fundamental cause of the huge geologic anomaly at Yellowstone is the focused flow of heat from the mantle (Figure 22.25), probably from a rising mantle plume. Heat

Figure 22.25 High heat flow and topography are associated with the Yellowstone plume. The Snake River Plain is a zone of subsidence formed in the wake of the plume. Gravity studies help geologists construct a cross section through the region. Rock densities are given in grams per cubic centimeter and are lowest in the region of the mantle affected by the passage over the plume. Crustal density beneath the Snake River Plain may be higher because of the intrusion of dense basaltic dikes and sills.

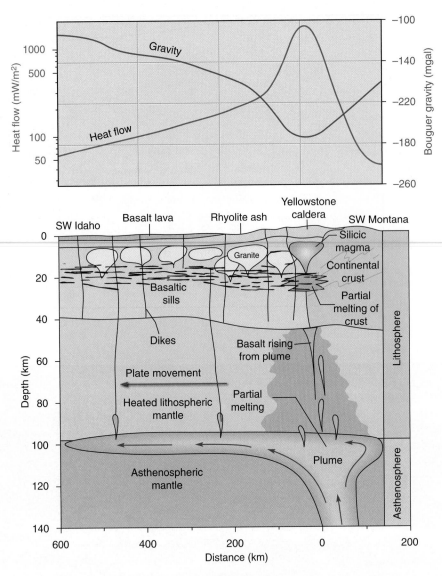

from the plume, probably transferred to the crust by basalt magmas, has (1) uplifted the entire region, (2) caused a multitude of shallow earthquakes, (3) created many separate basalt and rhyolite magma systems of different ages, and, consequently, (4) dramatically modified the structure and composition of the lithosphere. Measurements of the heat flow (2000 mW/m²) in Yellowstone show that it is about 30 times the continental average. The total thermal energy released at Yellowstone in one year is 5% of that released from the entire rest of the western United States. Heat flow across the Snake River Plain in the wake of the plume is still high, but it dwindles to normal heat flow near the Oregon border. Northeast of the plume, where its geologic effects are not yet manifest, the heat flow is much lower (Figure 22.25).

Continental Rifting, Flood Basalts, and Mantle Plumes

Many geological observations suggest genetic links among continental rifts, flood basalts, and mantle plumes. Many episodes of continental rifting have been preceded by crustal uplift and the outpouring of huge volumes of continental flood basalt. Is there any explanation for this relationship?

As discussed earlier in this chapter, a plume begins as a small bump on an internal boundary layer that rises and develops a very large mushroom-shaped plume head and a narrow tail (Figure 22.7). The plume head may be several thousand kilometers across. As the plume head encounters the overlying rigid lithosphere, its ascent slows, and it flattens against the strong, rigid lithosphere. Heat and buoyancy from the plume may cause the overlying lithosphere to dome and weaken. As a result, it may stretch and ultimately rift. While this is happening, copious continental flood basalt may erupt over a short period and spread over large areas. Unlike the slow, steady eruption of basalt at mid-oceanic ridges, the eruption of lava generated from starting plumes is rapid and episodic.

The connection between a flood-basalt province, continental rifting, and a hotspot track is exemplified by the opening of the southern Atlantic Ocean (Figure 22.2). The flood basalt provinces lie along the present-day continental margins, but they originally extruded above mantle plumes that later developed rifts and seafloor spreading. About 125 million years ago, when South America and Africa were still connected, huge eruptions formed the Etendeka flood basalt province of southern Africa and the Parana basalts of South America. In southern Brazil and adjacent Paraguay, more than 1 million cubic kilometers of basalt were extruded in about 10 million years.

This volcanic episode probably marked the arrival of the head of a new plume at the base of the lithosphere. Rifting of the continent started at about the same time, and the rift grew northward and southward from the site of the plume-related volcanism. Because of rifting, South America slowly slipped away from Africa. The oceanic lithosphere of the southern Atlantic basin formed in the wakes of these two continental fragments. A hotspot near the mid-oceanic ridge left a narrow track of seamounts on each side of the ridge. The modern volcanic island of Tristan da Cunha lies over the current location of this plume (Figure 22.2). Volcanism along the hotspot trace was much less voluminous than during the initial stage and apparently records the continued rise of mantle material in the tail of the plume.

Elsewhere, several major plumes have risen beneath continental crust during the past 250 million years, creating flood basalt provinces and continental rifts. An example is the vast basalt flood that formed India's Deccan Plateau (1 to 2 million cubic kilometers) as India rifted away from Africa. In North Africa the floods of basalt in the Ethiopian plateaus extruded as the Red Sea Rift developed. Precambrian flood basalts that were probably caused when an aborted rift developed are found in the Lake Superior area of northern Michigan. Continental flood basalts are, therefore, important records of the role played by mantle plumes in initiating divergent plate boundaries.

Plumes alone, however, probably cannot cause continents to break apart. The Siberian flood basalts (Figure 22.2) of the latest Paleozoic age are some of the greatest outpourings of lava the world has known, but the continent did not break up. Likewise, the Yellowstone plume may have aided in the extensional disruption of western North America, but it did not cause complete rifting and the creation of new seafloor. In situations where the plate motions are already suitably established, the additional stress generated by uplift above a plume, or the lowered viscosity it creates in the mantle, may be sufficient to let rifting proceed when it otherwise might not. In addition, the presence of the plume may cause an already active rift to shift over the center of the plume.

Plumes, Climate Change, and Extinctions

Mantle plumes may affect Earth's climate system and magnetic field.

From the foregoing, it should be clear that, like the theory of plate tectonics, the model of a mantle plume is a simple but powerful concept. It explains much of the geologic activity in the central parts of plates that never seemed to fit a simple interpretation of plate tectonics. Volcanic islands, rifts in continents, flood basalts, and continental calderas find explanations in the mantle plume model. Recently, mantle plumes have been used to explain another class of phenomena, including climate change, mass extinctions, and even changes in Earth's magnetic field.

Hypothetically, the effects of mantle plumes may extend far beyond the limits of the flood basalts and rhyolite ash that periodically pour from them. For example, the volcanic activity associated with a starting plume, either beneath a continent or on an ocean basin, occurs in a short, dramatic episode. These spasms of volcanic activity

What effect could a mantle plume have on climate?

and rapid extrusion of lava may change the composition and circulation of the oceans and the atmosphere. During eruptions, huge volumes of volcanic aerosols and gases, including carbon dioxide (a greenhouse gas), are released.

Because a series of plumes developed during the latter part of the Mesozoic Era (particularly in the Pacific Ocean, such as the one that developed the Ontong-Java Plateau), some scientists have speculated that enough carbon dioxide was released to raise global temperatures by several degrees. Thus, the warmth that typified the Cretaceous may have had its roots deep in the mantle. Some of the environmental adjustments may have contributed to mass extinctions, including the one in which dinosaurs vanished.

The Deccan flood basalts of India erupted at the boundary between the Cretaceous and Tertiary Periods—a time marked by extinctions that included dinosaurs and many other species. The Siberian flood basalts have also been correlated with extinctions at the very end of the Paleozoic Era. These environmental shifts may have helped promote the origin of new species. These provocative hypotheses need further investigation; perhaps within your lifetime we shall establish the cause of these great extinctions.

A secondary result of the release of carbon dioxide from mantle plumes may have been the deposition of organic carbon in marine sediments, especially as black carbon-rich shales and as beds of coal. Plants can convert carbon dioxide into organic carbon molecules and release oxygen gas into the atmosphere. If the carbon is locked in sediments, the atmosphere may eventually become enriched in oxygen. Some paleontologists suggest that oxygen contents higher than those in today's atmosphere were important for the evolution of anomalously large animals, such as dinosaurs in the Mesozoic and large insects in portions of the Paleozoic. Some scientists contend that such high oxygen contents may have been produced during times of enhanced plume development. The excessive carbon dioxide from the mantle could have released oxygen by the mechanism described above.

Another effect of mantle plumes may also have an example in the Cretaceous, a period when several large plumes formed. A significant decrease in the number of reversals in the polarity of the magnetic field marks this part of Earth's history. Magnetic field reversals are probably related to changes in the convective pattern in the metallic outer core. If large amounts of heat are drained from the core during the development of several mantle plumes during a short interval, the convection patterns in the core might be changed. These changes might have diminished the number of field reversals, perhaps by slowing convection in the core.

Key Terms

decompression melting (p. 674)

diapir (p. 672)

dike swarm (p. 688)

flood basalt (p. 672)

hotspot (p. 668)

hotspot track (p. 677)

mantle plume (p. 668)

oceanic plateau (p. 676)

plateau basalt (p. 688)

plume head (p. 672)

plume tail (p. 672)

ridge jump (p. 686)

rift zone (p. 677)

seamount (p. 681)

shield volcano (p. 668)

starting plume (p. 672)

Review Questions

1. What type of volcanic activity occurs within the central parts of tectonic plates, beyond the active plate margins?

2. Explain the origin of chains of volcanic islands and seamounts.

3. Outline the evidence that suggests that mantle plumes are real.

4. What causes a mantle plume to rise?

5. Describe the probable shape and size of a mantle plume. Does the shape of the plume change during its history?

6. How is the production of magma in a mantle plume similar to its production at a divergent plate boundary? If magma is produced in similar ways under the two conditions, why does it have different compositions? How are its sources different?

7. Compare the size, composition, and structure of a typical subduction-related volcano with those of a plume-related volcano.

8. What causes the large topographic swell that surrounds an active ocean island volcano?

9. What does the study of flood basalt provinces tell us about

volcanic systems that are related to mantle plumes?

10. Compare the style of volcanism related to a starting plume with that related to a plume tail.

11. What kinds of earthquakes are related to mantle plumes?

12. If Iceland is part of the oceanic ridge system, why is it so much higher than the rest of the Mid-Atlantic Ridge? Does this situation occur anywhere else?

13. What evidence is there that Yellowstone National Park is underlain by an active mantle plume? How does it differ from the Hawaiian mantle plume?

14. Why is rhyolite more common above continental plumes than above oceanic plumes?

15. Compare the possible contrasts between the dominant modes of convection in the upper and in the lower mantle.

16. What represents a more universal type of mantle convection, plumes or plate tectonics?

17. If a starting plume rises 2 m/yr, how long will it take to rise from the core-mantle boundary to the base of the lithosphere?

18. How could a mantle plume affect Earth's global climate?

23 Tectonics and Landscapes

In the preceding chapters, we have seen that the landscape is shaped by Earth's geologic processes and evolves systematically. The processes of landscape development, however, are complex because tectonism produces a variety of structural settings upon which erosion occurs. The geologic structures of the shields are different from those of the stable platforms, folded mountain belts, or island arcs. Consequently, the landforms that develop in each tectonic setting have many distinctive characteristics. Moreover, climate influences the types of processes that operate within a region. Nonetheless, a landscape has many distinctive characteristics that reveal its history.

This panoramic photograph of Monument Valley of the Colorado Plateau illustrates this point very nicely. At first glance, this image appears to show the intricate details of a "lost world" with mesas, buttes, sheer cliffs, gentle slopes, complex spires and towering pinnacles. This landscape has fascinated humans since before historic times and many myths and legends are associated

with it. Indeed, this landscape may appear to be incomprehensible. However, there is a system and a beauty in the evolution of the land that promises an intellectual reward for those who take time to understand and appreciate it.

The landscape of Monument Valley, like any other, can be explained in light of Earth's tectonic and hydrologic systems. The events that produced this unique landscape began hundreds of millions of years ago when the colorful sedimentary rocks were deposited on the continent and in shallow marine waters. Gradually, the sediment lithified and turned to rock as it was buried. Joints broke through the rocks. Sometime in the last few tens of millions of years the region was uplifted above sea level and streams carved intricate systems of valleys. Weathering ate away at the cliffs and slopes, enlarging fractures and isolating columns of sandstone. Eventually, the details of the landscape were sculpted out by differential erosion of once continuous layers of sedimentary rock. Resistant formations eroded to form the pinnacles, steep cliffs, and mesa walls. Easily eroded shales form the gentler slopes.

You can see from the photograph that the landscape has much to tell about its history, but to read this history, you must learn a new language—a language written in the mesas, canyons, and rocks.

Major Concepts

1. The most important factors in the evolution of continental landscapes are tectonic setting, climate, and differential erosion.
2. The surface of a continental shield evolves through erosion and isostatic adjustment of a mountain belt. Ultimately, a surface of low relief near sea level is produced and equilibrium is reached. Local relief on a shield is largely the result of differential erosion on complexly deformed metamorphic and igneous rocks and is usually less than 100 m.
3. Stable platforms result from deposition of sediment in shallow seas that transgress and regress across the shields. The rocks are nearly horizontal or are warped into broad domes and basins. Erosional features on stable platforms are circular or elliptical cuestas and strike valleys and rolling hills developed by dendritic drainage patterns.
4. Landscapes developed on folded mountain belts are typically controlled by deformational structures, such as folds and thrust faults. Ridges and valleys carved on plunging folds are the most common landform.
5. Horsts and grabens are the major structural features formed in rift systems. The landscapes developed on these structures are eroded mountain ranges and basins partly filled with sediment.
6. Flood basalts formed at hotspots and rifts bury the previous landscape and form a completely new surface that is subsequently modified by weathering and erosion. Uplift and erosion of this surface commonly produce basaltic plateaus.
7. Landforms developed in magmatic arcs are dominated by volcanic features, such as composite cones and andesitic lava flows. These disrupt the drainage and create temporary lakes. Ultimately the original volcanic features are eroded away and deeper granitic intrusions are exposed and etched out by differential erosion.

Factors Influencing Continental Landscapes

Tectonic setting, climate, and differential erosion are the most important controls on the evolution of continental landscapes on Earth.

Compared with other planets in the solar system, Earth is unique because of its consistently changing landscapes. The surfaces of the Moon, Mercury, Mars, and other planetary bodies are dominated by meteorite impact structures formed more than 4 billion years ago. But mountain building, volcanism, erosion, and sedimentation constantly change the surface of Earth. Indeed, most of Earth's continental landforms are very young and formed during the last 2 million years.

The constant resurfacing of our planet by both erosion and deposition results in a changing landscape that may seem at first to be unorganized and chaotic, but if you study Earth's surface from a variety of perspectives, you will find system and order in every feature of the landscape (**Figure 23.1**). Nothing is random. Every valley, plateau, volcano, and sand dune was produced by a geologic system, and every landscape preserves some record of its history and how it was formed.

We have emphasized in previous chapters how the hydrologic system operates and how each agent, including running water, glaciers, and wind, produces distinctive landforms. However, to read the stories told by the landscape more effectively, we must consider how the hydrologic system operates in different tectonic settings and how erosion on different rock sequences and structural features develops distinctive features. In every region, tectonism is important in determining the underlying

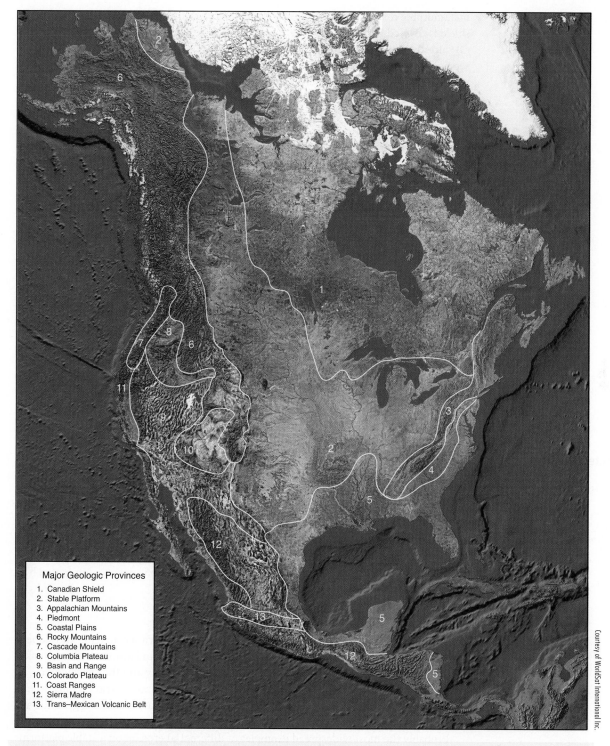

Figure 23.1 **The major geologic provinces of North America** are shown on this map. The tectonic setting or structure of the rocks just below the surface is a major control on the character of the landscape. Most other continents also have shields, stable platforms, and folded mountain belts.

Major Geologic Provinces

1. Canadian Shield
2. Stable Platform
3. Appalachian Mountains
4. Piedmont
5. Coastal Plains
6. Rocky Mountains
7. Cascade Mountains
8. Columbia Plateau
9. Basin and Range
10. Colorado Plateau
11. Coast Ranges
12. Sierra Madre
13. Trans–Mexican Volcanic Belt

foundation of Earth's surface features. The main focus in this chapter is on how tectonism influences the evolution of the landscape by providing several fundamental structural settings upon which the agents of the hydrologic system operate to produce their distinctive landforms.

Let us emphasize a few important facts about the continents. Although the various continents may appear to be unique in size, shape, and surface features, they all have three basic components: (1) a large, relatively flat area of ancient complex igneous and

What are the three major structural components of all continents?

metamorphic rocks known as a **shield**, (2) a broad, low platform or plain where the shield is covered with a veneer of sedimentary rocks known as the **stable platform**, and (3) **folded mountain belts** formed above a subduction zone or where two continental segments have been sutured together during continental collision (Figure 23.1). The geologic differences between continents are mostly in the size, shape, and proportions of these three components. In addition, some continents may be split by a rift system and some may be involved with a magmatic arc produced at a subduction zone. The important underlying theme in this chapter is that each of these major tectonic elements provides a different substructure on which the hydrologic system operates. Shields, platforms, mountain belts, and continental rifts, with their distinctive rock types and structure, will each tend to develop a distinctive type of landscape. Details of the landforms formed on each major tectonic setting will of course be determined by climatic factors that govern the type and intensity of the processes operating in the hydrologic system.

Climate

Climate is a major factor in landscape development because it controls the action of the hydrologic system and, therefore, of many types of geologic processes in a given region. Weathering, slope stability, river erosion, groundwater, glaciers, and wind are all subject to climatic conditions. A continent's climate is controlled mostly by latitude, but also by topography. For example, high mountains have low temperatures and create rain shadows downwind from them. Although the structural components of the continents are fundamentally the same, details of the landforms developed on them depend to a considerable degree upon climate. The landscape of the Canadian shield is modified by glaciation, the Brazilian shield is covered by a thick soil developed by tropical weathering, and the landforms of the shield of the Sinai Peninsula are different still because this shield is in an arid climate that imposes distinctive processes of weathering and erosion.

How does climate influence landscape?

The topography of the stable platforms and mountain ranges are likewise affected by climate. Limestone regions in humid climates develop karst topography, but in arid regions they are resistant to erosion and form ridges, ledges, and cliffs (**Figure 23.2**). In the tropics the high rainfall accelerates weathering, and thick soils (some more than 100 m thick) may develop. These soils mask structural details in the underlying rock that would be etched out in relief in arid regions. The complete disruption of river systems by glaciation results in many distinctive landforms produced both by glacial erosion and glacial deposition. Many landforms in arid regions may be dominated by eolian processes, and sand seas may cover extensive areas.

Differential Erosion

Differential erosion occurs on all scales, from a mountain range, to cliffs and slopes formed on alternating hard and soft rock bodies, down to thin laminae within a rock. Differential erosion is thus responsible for much of the beauty and the spectacular scenery of Earth. Differential erosion generally is well expressed in arid regions, where differences in rock type, jointing, and the availability of surface water and groundwater combine to produce fascinating details of the landscape.

Probably the most widespread examples of differential erosion on a stable platform are the alternating cliffs and slopes that develop on sequences of alternating hard and soft sedimentary rocks. Soft shales typically form slopes, and the more-resistant sandstones and limestones produce cliffs. The height of a cliff and the width of a slope are largely functions of the thickness of the layers involved. Similarly, if the series of resistant and nonresistant rocks are tilted, the nonresistant rocks are quickly eroded to form lowlands or valleys, leaving the resistant rocks as hills or ridges (**Figure 23.3**).

How is differential erosion expressed on a sequence of tilted strata?

The main point here is that erosion is a selective process. It rapidly removes weak rock to form valleys or depressions in the landscape and leaves resistant rock bodies standing in relief as mountains, hills, and ridges. In this way, landscapes commonly reflect the structure of the rocks exposed at the surface.

(A) In humid environments, such as southeast China, limestone is nonresistant to erosion and weathers to form karst terranes.

(B) In an arid environment, such as the western United States, limestone is resistant to weathering and erosion and forms cliffs or ledges.

Figure 23.2 Climate plays a major role in the evolution of all landscapes. The same rocks in different climates will erode and weather very differently.

Figure 23.3 **Differential erosion of a sequence of tilted strata** produces ridges called hogbacks on resistant formations and long, narrow valleys in nonresistant rocks.

Evolution of Shields

A continental shield results from the formation of mountain belts at convergent plate margins and from the subsequent erosion and isostatic adjustment, which reduce the mountain to a broad, flat surface near sea level. Local relief is usually less than 100 m and depends upon differential erosion on the igneous and metamorphic rocks.

Continental shields are distinctive in that their surface features are formed on ancient and complex metamorphic rocks—gneisses, schist, slates, quartzites, and marbles—all of which have been intruded by granitic batholiths, stocks, and dikes. In most regions, the shields have been eroded down to near sea level and differential erosion has etched out the nonresistant structural features of the various rock units. Thus, shields have a very distinctive landscape, although climatic influences (glaciation, desert, tropical forests, and so on) may leave their particular imprint upon the surface.

How does a mountain belt evolve into a segment of a shield?

Shields and their associated stable platforms are the fundamental tectonic components of continents, so an understanding of how they developed is essential in understanding the origin of the surface features of our planet. A general model showing how the basement rock evolves from mountain building is shown in **Figure 23.4**. Two major factors control this process: (1) erosion of the mountain belt by running water and (2) contemporaneous isostatic adjustment of the mountain belt as a result of the removal of material by erosion. Both erosion and isostatic adjustment continue until equilibrium is reached—a condition in which the topographic relief is eroded down to sea level and the mountain root has rebounded to a state of gravitational equilibrium. Under these conditions, large-scale erosion cannot occur because the surface is at sea level and uplift does not occur because of isostatic equilibrium.

In Figure 23.4A a new mountain belt has been formed by plate convergence. It is important to note that there are significant changes in the dominant structural features of the mountain belt, from the surface down to the deep roots. Andesitic volcanism occurs at the surface. At shallow depths, where the confining pressure is low, the rocks are relatively brittle, and the compression that formed the mountain belt has developed thrust faults (green). At greater depth, the rocks are under greater confining pressure, and they yield to plastic flow, which produces tight folds (tan). At still greater depths, complex folds are formed. Silicic magma, generated in the lower crust and involving magmas from the subduction zone, rises because it is less dense than the surrounding rock and reaches a level where it spreads out to form large plutons. In the

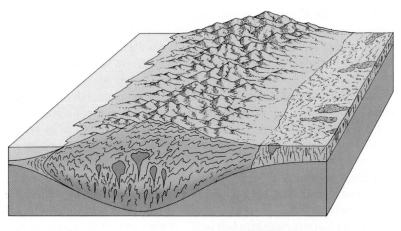

(A) In the early stages of development, there is maximum relief. Some andesitic volcanoes may still remain at the surface. Thrust faults and folds are the dominant structures exposed at the surface, and headward erosion of tributary streams begins to adjust the stream pattern to the major structural trends. The Andes of South America are in this stage of their development.

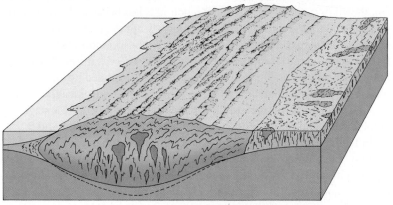

(B) As erosion removes material from the mountain belt, isostatic adjustment causes the mountain root to rebound. Tight folds formed in the deeper part of the mountain system are exposed at the surface, and headward erosion adjusts the stream pattern to the folded rocks. Many tributaries flow parallel to the structural grain of the mountain belt. The central Appalachian Mountains of eastern North America are similar to this.

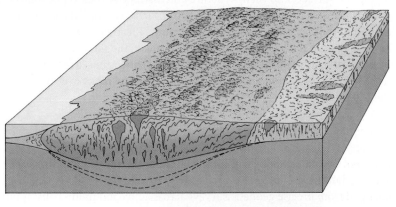

(C) Continued erosion and isostatic adjustment reduce the topographic relief and the size of the root of the mountain belt below. Complex folds and granitic igneous intrusions, originally formed deep in the mountain belt, are now exposed at the surface. The stream patterns adjust to the new structure and rock types. Local relief and rates of erosion are greatly reduced. The eroded mountain belts of Scotland and Norway have structures like these.

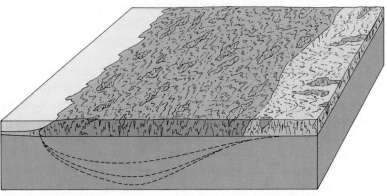

(D) Isostatic equilibrium is ultimately reestablished. The rocks formed deep in the mountain root are exposed to the surface, and local relief is only a few meters. Stream patterns adjust to the structural trends in the metamorphic terrain. At this stage, the mountain belt constitutes a new segment of the shield. The Canadian shield has developed by the processes outlined here.

Figure 23.4 A continental shield develops from a folded mountain belt. Erosion occurs during crustal deformation, so by the time mountain building terminates, the mountain range is already carved into a rugged terrain. After deformation, a mountain root extends down into the mantle to compensate for the high topography. Note that the style of structure in a mountain belt changes with depth. Andesitic volcanic features may dominate at the surface. Thrust faults and folds occur at shallow depths. Tight folds and granitic intrusions occur at intermediate depths. In the deeper roots, metamorphic rocks intruded by small granitic bodies dominate.

How does isostasy influence the development of a shield?

deeper roots of the mountain belt, metamorphic rocks, including granulites and migmatites, dominate and are intruded by smaller bodies of granite (reddish brown). The topography in Figure 23.4A is young. It is controlled by andesitic volcanism, thrust faulting, and folds. The topographic relief is high, and headward erosion is beginning to adjust the drainage pattern to the structural trends of the mountain belt.

In the second stage (Figure 23.4B) the upper segment of the mountain belt has been removed by erosion, but isostatic rebound causes the mountain belt to rise continually. The dominant structure, now exposed at the surface, is a series of tight folds (tan) etched into relief. Erosion on the folds produces a series of ridges on resistant rocks that zigzag across the landscape. Long strike valleys erode on less resistant strata such as shales. The result is a "valley-and-ridge" type of topography.

At a later stage (Figure 23.4C) erosion has removed the zone of folded strata, but isostatic adjustment continues to elevate the mountain belt. Note the position of the root, or base, of the mountain belt in (C), compared to its former position in (A) and (B), shown in dashed lines. Isostatic rebound is less because the mountain root is not so deep, and topographic relief is consequently not as great as before. Complex folds and granitic intrusions, which formed deep in the mountain belt, are now exposed at the surface, and the landforms are controlled by these features. Resistant granitic rock bodies typically form elliptical low mountains surrounded by lowlands of metamorphic rock. Small-scale structural features in the metamorphic terrain may be eroded into relief, forming a complex of low ridges and lowlands. At this stage, the topographic relief and rate of erosion are much less than in stages A and B.

In Figure 23.4D erosion and isostatic adjustments have reached a state of equilibrium. There is no mountain root extending down as a bulge below the mountain topography, and as a result there is no isostatic uplift. Metamorphic rocks and igneous intrusions (reddish brown), which were formed in the deep mountain root, are now at the surface, and their structure and distinctive rock types control the topographic features to be developed. The entire surface is eroded close to sea level. The mountain root is now in isostatic equilibrium. The area is tectonically stable, and a new segment of the basement complex is formed. The most significant event in the subsequent history of the region is that slight changes in sea level may cause the sea to spread across the region and deposit shallow-marine rocks over the shield to form a stable platform.

Rates of Uplift and Erosion

The rate of uplift during mountain building is variable and difficult to measure. However, estimates can be made on the basis of the age of strata that were originally deposited in the sea but are now found high in mountain ranges, together with data obtained from precise geodetic surveys across active mountain belts. A number of such observations give a general rate of uplift of 6 mm/yr. If erosion did not occur contemporaneously with uplift, a mountain summit could rise 6 km in 1 million years. If these measurements are correct, a full-size mountain belt could be created in a relatively short time (5 to 10 million years).

Current estimates of rates of erosion are based on extensive measurements of the volume of sediment carried by the major rivers of the world. In mountainous areas, rates of erosion range from 1 to 1.5 m/1000 yr. From these data we can draw a generalized graph showing rates of uplift and the relation of rates of erosion to elevation (**Figure 23.5**). The rate of uplift is roughly 5 to 10 times faster than the maximum rate of erosion. The relatively rapid rate of deformation and uplift is shown by the steep line on the graph and can occur in a time span of 5 million years. Erosion would occur during uplift, so by the time deformation and uplift terminate, the mountain range would already be carved into a rugged terrain, and perhaps as much as 1 km of rock would have been removed. The main idea that this graph emphasizes is not the absolute rate of erosion but the rapid decrease in the rate with a decrease in elevation. From the regional viewpoint, the rate of erosion depends on the height of the landmass above sea level.

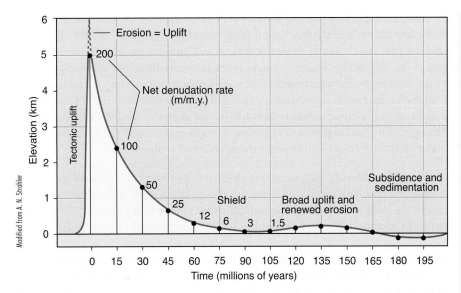

Modified from A. N. Strahler

Figure 23.5 Rates of erosion of a mountain belt decrease exponentially with time. The period of mountain building is shown by the steep line at the beginning of the graph. The tectonic deformation is shown to last 5 million years, but most of the uplift may occur within 2 million years. Erosion proceeds contemporaneously with uplift, increasing in intensity with increasing elevation. By the time deformation ends, uplift of 6 km has occurred, but the surface is already eroded down to 5 km. The initial rate of erosion is 1 m/100 yr, but isostatic adjustment occurs at a ratio of 4:5. The initial rate of new lowering of the surface is thus 200 m per 1 million years (200 m/m.y.). In 15 million years, one-half of the mass is removed, and the net rate at which the surface is lowered is reduced to 100 m/m.y. After 30 million years, only one-quarter of the mass remains, and the average altitude is 1.25 km. In 75 million years, the mountain belt may be reduced to a new segment of the shield.

As erosion removes material from the mountain belt, the mountain root rebounds in an attempt to reestablish the balance. In the early stages of erosion, the removal of 500 m of rock is generally compensated for by an isostatic uplift of about 400 m, so there is a net lowering of only 100 m of the mountain surface. If we assume that isostatic adjustment occurs constantly at a ratio of 4:5, the initial rate of net lowering of the surface will be 0.2 m/1000 yr or 200 m/m.y., as shown at the top of the curve in Figure 23.5. In contrast, at the end of 15 million years, the net rate of lowering of the surface has been reduced to approximately 100 m/m.y.

Erosion and isostatic adjustment continue to reduce the topographic relief. By the end of 30 million years, the elevation and rate of erosion are again halved, to one-quarter of the initial value. Approximately three-quarters of the original landmass has now been removed, and the structures of the deep mountain roots are exposed. The regional surface is a broad, nearly flat plain. Local relief of a few tens of meters is produced by differential erosion of belts of different metamorphic and igneous rock types. Erosion of the surface and the associated isostatic adjustment have declined at a rapid rate, probably exponential, so that a near balance is reached. Once the original mountains are eroded almost to sea level, the exposed roots of the mountain belt are a new segment of the continental shield.

The subsequent evolutionary history of the landscape is intimately related to broad uplift and subsidence and to changes in sea level. Sea level is important in our model of erosion because it is the ultimate level to which stream erosion can effectively lower the continental surface. Both erosion and isostatic adjustment combine to produce a flat slab of continental crust, the upper surface of which is eroded to near sea level. There are, however, several reasons why the continental lithospheric plate can be expected to move both up and down with respect to sea level even though it is in a state of near isostatic equilibrium. For example, the top of the asthenosphere is not a perfectly smooth surface but undulates in swells and depressions. Bulges in the upper surface of the asthenosphere may also result from hotspots in the mantle. As the plate moves over these highs and lows, the continental crust may be upwarped or depressed with respect to sea level. This change would cause the sea to expand or contract across the stable platform.

Changes in sea level can also result from changes in the rate of spreading at mid-ocean ridges. If spreading is rapid, the oceanic ridge swells and is arched upward, reducing the volume of the ocean basins and causing expansion of the sea over the flat continental surface. Slow spreading deflates the oceanic ridge, causing the sea to withdraw from the continents. Transgressions and regressions of the sea may thus be related to plate tectonics.

Another important point is that any change in sea level, regardless of the cause, may affect the erosional processes on the continent. If sea level is lowered (or the continent is upwarped), the processes of erosion will be rejuvenated. An increase in elevation of the land brings a rapid increase in the rate of erosion. An uplift of 1 km will be followed by accelerated erosion lasting probably 5 to 10 million years. If sea level rises (or a continent is depressed), the sea will advance over the land. Erosion terminates, and shallow-marine sediments are deposited on the eroded surface of the shield. Transgressions and regressions of the sea over the continental platform have been the major recurring events in this area through most of geologic time.

The Canadian Shield   The Canadian shield covers about one-fourth of the North American continent, more than 3 million square kilometers. Although it has been glaciated, the basic structural and topographic features are well expressed. On a regional basis, the surface of the shield resembles a vast saucer with the center in Hudson Bay. The shield extends northward to Baffin Island, eastward to Labrador and Newfoundland, southward to the Great Lakes, and westward to the interior plains of Canada. The landscape of this vast area is truly remarkable. As can be seen in **Figure 23.6**, the most striking characteristic is the vast expanse of the low flat surface, and throughout thousands of square kilometers the shield is barely above sea level. The only surface features that stand out in relief are the resistant rock formations that rise 30 to 100 m

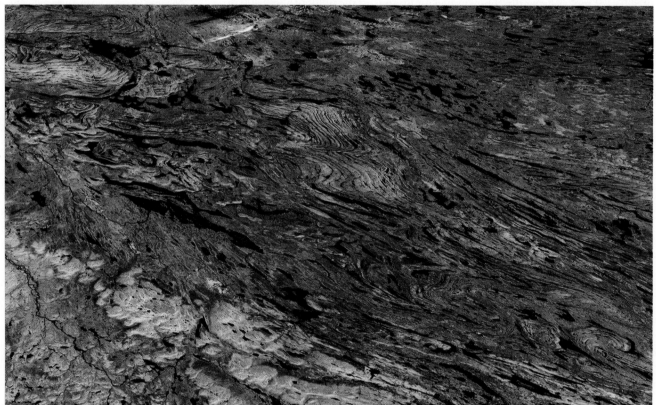

Courtesy of U.S. Geological Survey and EROS Data Center

Figure 23.6 The Canadian shield is a broad, flat surface carved on a complex of igneous and metamorphic rocks that were originally formed deep in the roots of a mountain belt. This view shows that many structural features are eroded into relief. Erosion by a continental ice sheet during the last glacial advance is responsible for most of the small landscape features. Linear faults and zones of nonresistant rock are occupied by lakes.

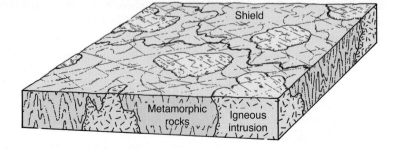

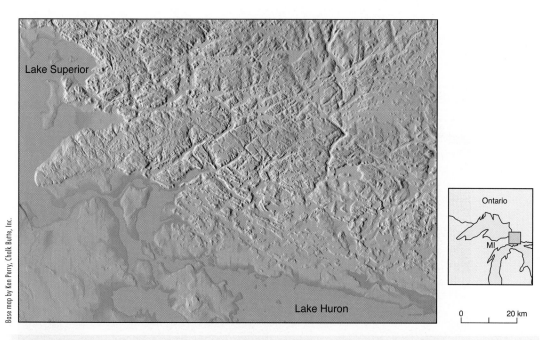

Lake Superior

Lake Huron

Ontario

MI

0   20 km

Base map by Ken Perry, Chalk Butte, Inc.

Figure 23.7 The complex structures of the shield are etched out by differential erosion of the igneous and metamorphic rocks, as shown in this topographic map of the Canadian shield in southern Ontario. The major fracture systems are eroded into narrow valleys, although the regional topography is eroded to within a few hundred meters of sea level.

above the adjacent surface. The structural complexity of the shield (**Figure 23.7**) is shown by the complex patterns of erosion, alignment of lakes, and differences in tone and texture of the landscape. The major structural elements of the Canadian shield have been mapped, and several distinct geologic provinces were discovered. Each represents a different mountain system of a different age, and each has its own characteristics of rock type and structure.

The basic elements of the shields of other continents are quite similar to those of the Canadian shield. Conditions of the local climate, however, impart distinctive characteristics to the landscape. In the shields of desert regions, such as North Africa, Arabia, and Australia, the structural trends and local landforms are partly covered by desert features such as wind-blown sand, alluvial fans, and lag gravels. Shields in the tropics are commonly covered with thick soil and tropical vegetation. Those in the polar regions, such as Scandinavia's Baltic shield, carry the strong imprint of glaciation. Yet the fundamental topographies of all shields are remarkably similar.

Stable Platforms

Landforms developed on stable platforms result from differential erosion on horizontal to gently dipping sedimentary rocks. Dendritic drainage is common on horizontal strata and produces rolling hills. Cuestas and strike valleys typically form on the low dipping strata on the flanks of structural domes and basins.

The stable platform is simply that part of the basement complex covered with sedimentary strata. As we have seen, erosion and isostatic adjustment combine to produce a flat slab of continental crust that is very near sea level. Thus, any change in sea level causes the sea to expand or recede far across the surface of the low, flat continent. The history of the stable platform over hundreds of millions of years has been that of repeated transgression and regression of shallow seas in which were deposited cycles of sandstone, shale, and limestone. Each of these sedimentary sequences is bounded by an erosional unconformity. The total thickness of sedimentary strata on a stable platform rarely exceeds 2000 to 3000 m, so the sedimentary rocks form only a thin veneer

Figure 23.8 The stable platform is an area in which the basement complex is covered with a veneer of sedimentary rocks that are gently inclined or essentially horizontal. Stable platforms constitute much of the world's flat lowlands and are known locally as plains, low plateaus, and steppes.

covering the underlying igneous and metamorphic basement. Earth movements on the stable platform are mostly broad regional undulations that produce gentle domes (upwarps) and basins (downwarps) in the otherwise nearly horizontal strata. Thus, the landscapes on all stable platforms throughout the world consist of landforms developed on horizontal or gently inclined sedimentary rocks (**Figure 23.8**). These form the vast plains of the interior of North America, the great steppes of the Ukraine, the central lowlands of China, and the vast flat interior of Australia.

Major Landforms of the Stable Platform

One of the most fundamental characteristics of a stable platform is the distinctive stream drainage pattern that develops on horizontal strata. A given layer of sedimentary rock forms a large area that is essentially homogeneous in all directions. Such a surface presents no appreciable structural control over the development of the drainage systems. Tributaries are free to develop and grow with equal ease in all directions so that **dendritic drainages** typically develop on horizontal rocks (**Figure 23.9**). Seen from the ground, the landscape seems to be a broad expanse of low "rolling hills."

Another important factor in the development of the major landscapes of the stable platform is differential erosion on the horizontal layers of sedimentary rock. Sedimentary strata deposited by the expanding and contracting seas consist of sequences of sandstone, shale, and limestone formations. Of these, shale is by far the most abundant and the least resistant. Layers of sandstone and limestone are thinner but are much more resistant in most climates. Differential erosion on these rocks is responsible for much of the landscape of the stable platform.

How does the landscape of a stable platform differ from that of a shield?

Resistant rock layers such as sandstone and limestone commonly form a resistant cap rock where interbedded with shale. The cap rock forms a low **plateau**. Erosion and slope retreat along the edges of the plateau create alternating cliffs (on the resistant unit) and slopes (on the nonresistant layers). As stream erosion eats headward, a large portion of the plateau can be detached from the main plateau to form a small **mesa** (**Figure 23.10**).

Where a sequence of alternating resistant and nonresistant strata is tilted, the nonresistant units are eroded into long **strike valleys** or lowlands trending parallel to the

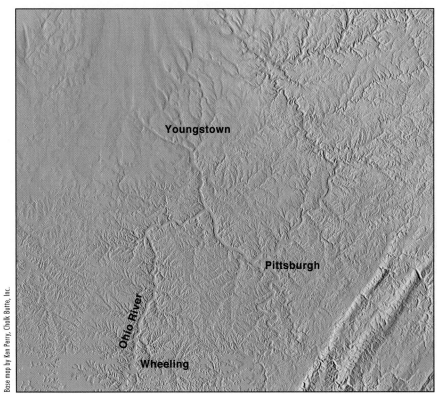

Youngstown

Pittsburgh

Ohio River

Wheeling

PA

OH

WV

0   20   40
km

Figure 23.9 The stable platform of the east-central United States, shown on this shaded relief map, is underlain by flat-lying sedimentary rock layers dissected by dendritic drainage systems. Seen from the ground, this area appears as low rolling hills. The westernmost part of the Appalachian Mountains lie in the southeastern corner of the map.

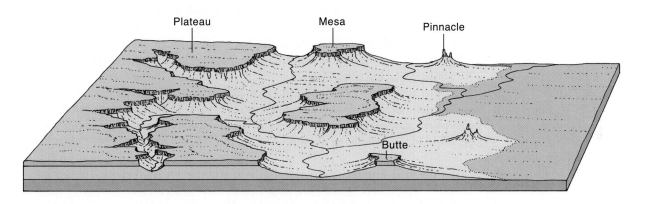

Plateau   Mesa   Pinnacle

Butte

Figure 23.10 Differential erosion of horizontal strata is characteristic of a continent's stable platform and forms plateaus, mesas, buttes, and pinnacles. Resistant beds of jointed sandstone commonly cap plateaus and control the erosion patterns.

strike, or trend, of the rock layers. The resistant layers are left standing as long, asymmetrical ridges. Ridges formed on gently inclined strata are known as **cuestas**. Sharp ridges formed on steeply inclined layers are **hogbacks** (Figure 23.3).

The major structural features of the stable platform are broad upwarps and swells that form structural **domes**, and downward movements that create **basins**. These warps in the crust are large, ranging from a few hundred to thousands of kilometers across. They may form while shallow seas cover the area so that sedimentary rocks may be thicker in the basins and thinner across the crest of the domes.

On the flanks of these structures, the strata may dip at angles of 20° or 30° or more. As stream erosion proceeds, strata are removed from the top of a dome, which is eroded outward to form a concentric series of sharp-crested cuestas or hogbacks with intervening strike valleys (**Figure 23.11**). The drainage pattern developed by erosion along strike valleys is commonly circular. If the older rocks in the center of the dome

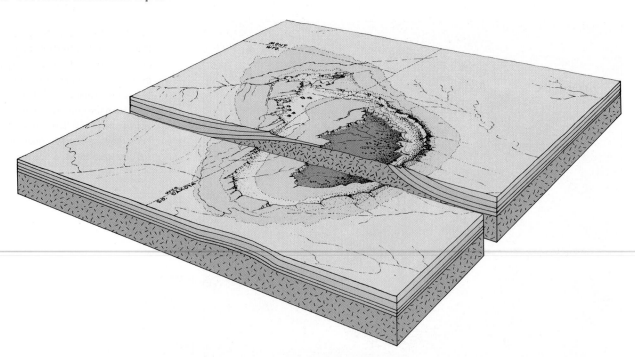

Figure 23.11 Hogbacks and intervening strike valleys form in domal structures in the stable platform. Differential erosion of a structural dome in the Black Hills of South Dakota is shown in this diagram.

are nonresistant, the center of the uplift may be eroded into a topographic lowland bordered by inward-facing cliffs formed on the younger, resistant units. If the older rocks are more resistant, the center of the dome remains high, forming a dome-shaped hill or ridge. Large structural domes have inward-facing cliffs, whereas large basins have outward-facing cliffs.

Domes and basins in the stable platform may form in a variety of ways. They may result from mild compression associated with a mountain-building event at a distant convergent plate margin. Also, domes and basins may be the result of changing temperatures in the underlying mantle.

Differential Erosion on the Stable Platform in Arid Climates

Differential erosion of nearly horizontal strata also produces many fascinating small landforms. These landforms are best displayed in arid regions, where details of the topography are not obscured by vegetation. **Buttes**, **pinnacles**, and **pillars** are some of the most spectacular of these small features (Figure 23.10). They are simply the result of differential erosion on receding cliffs composed of stratified sedimentary rocks. If a massive cliff has well-developed joints, vertical **columns** develop. Where stratification produces alternating layers of hard and soft rock, additional detail can be etched out by differential weathering and erosion.

Why didn't a "Grand Canyon" develop in Kansas?

Jointing commonly plays an important role in the evolution of these landforms, for it permits weathering to attack a rock body from many sides at once. The famous columns and pillars in Bryce Canyon National Park, Utah, for example, result from differential weathering along a set of intersecting vertical joints and horizontal bedding planes (**Figure 23.12**). The joints divide the rock into columns. Nonresistant shales separate the more resistant sandstone and limestone, and differential erosion forms deep recesses in the columns, producing the fascinating slopes and landforms. This configuration is in striking contrast to the thick massive homogeneous sandstone that forms nearby Zion National Park, Utah, where joints are the only significant zones of weakness. As a result, erosion along the joints separates the rock into large blocks with steep cliffs.

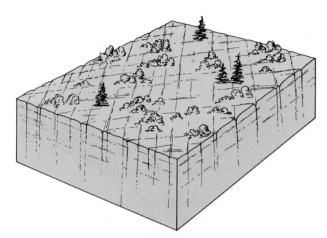

(A) Initial stage. Intersecting joints separate the rocks into columns.

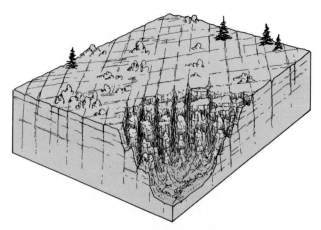

(B) Intermediate stage. Weathering and erosion along the joints accentuate the columns, which erode into various forms as a result of alternating hard and soft layers.

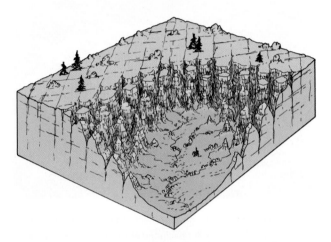

(C) Final stage. As weathering and erosion proceed, the cliff retreats. Old columns are completely destroyed, but new ones are continually created.

(D) Eroded columns provide the spectacular scenery in Bryce Canyon National Park, Utah. The columns were eroded into colorful Tertiary sedimentary rocks.

Figure 23.12 The evolution of columns by differential erosion along a receding cliff commonly is controlled by intersecting joint sets. Rapid erosion along joint systems separates the columns from the main cliff. Differential erosion, accentuating the difference between rock layers, produces the fluted columns.

Natural arches are also products of differential erosion in arid climates. The best examples are found in the massive sandstones of the Colorado Plateau, in the western United States. The diagrams in **Figure 23.13** illustrate how natural arches may be formed. This arid region receives little precipitation, and much of the surface water seeps into the thick, porous sandstone. Groundwater is most abundant beneath dry stream channels, and its movement follows the general surface drainage lines. Groundwater, emerging as a seep in a cliff beneath a dry waterfall, dissolves the cement in that area. Loose sand grains are washed or blown away, so that an **alcove** soon develops at the base of the normally dry waterfall. If the sandstone is cut by joints, a large block can be separated from the cliff as the joints are enlarged by weathering processes. The alcove and joint surface continue to enlarge, and an isolated arch is eventually produced (Figure 23.13). Weathering then proceeds inward from all surfaces until the arch is destroyed, leaving only columns standing.

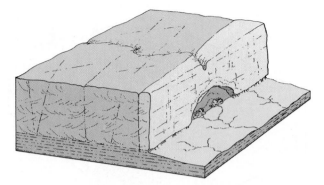

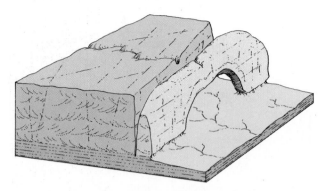

(A) Initial stage. In arid regions, much of the surface water seeps into the ground below a stream channel. This water may move laterally above an impermeable layer and eventually emerge as a spring at the base of a cliff. Cement holding sand grains together is soon dissolved in this area of greatest moisture; the sand grains fall away, so that a recess, or alcove, forms beneath the dry falls from the intermittent stream above.

(B) Intermediate stage. If a joint system in the sandstone is roughly parallel to the cliff face, the joints can be enlarged by weathering, which separates a slab from the main cliff.

(C) Final stage. An arch is produced as the alcove enlarges. Weathering then proceeds inward from all surfaces until the arch collapses.

(D) The initial stage in the development of a natural arch in Zion National Park, Utah. A well-developed alcove formed beneath a dry waterfall. Weathering along a large joint will soon separate the alcove from the cliff to make a natural arch.

Figure 23.13 **Natural arches** develop in massive sandstone formations by selective solution activity in nearly horizontal rocks of the semiarid or arid parts of a stable platform.

The Stable Platform of North America

The stable platform of North America consists of three distinctive parts: (1) the area underlain by Paleozoic strata in the east, (2) the Great Plains region underlain by Mesozoic and Cenozoic strata to the west, and (3) the Atlantic and Gulf coastal plains. The region north of the Ohio and Missouri rivers has been glaciated and is now largely covered with glacial moraine. However, to the south, the classic landforms of a stable platform are well developed—erosional landscapes on horizontal and gently inclined strata.

The major domal structures are the Cincinnati arch in Ohio, Kentucky, and adjacent areas; the Wisconsin dome; and the Ozark dome in Missouri. Large basins formed between the upwarps in Illinois, Michigan, and Ohio. The extension of the Cincinnati arch southward forms the Nashville dome in which weaker rocks in the center have been eroded away to form a topographic basin. Here limestones are abundant and details of the topography developed on the dipping strata are greatly influenced by groundwater erosion with resulting typical karst landforms.

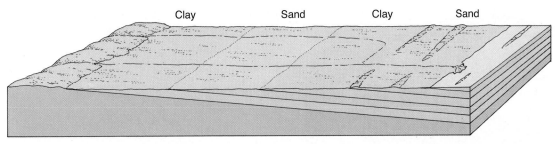

Clay   Sand   Clay   Sand

(A) As the coastal plains emerge above sea level, the drainage system is simply extended downslope directly toward the new shoreline. These are called consequent streams.

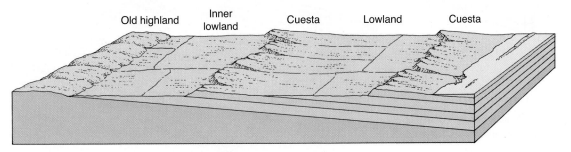

Old highland   Inner lowland   Cuesta   Lowland   Cuesta

(B) Headward erosion of tributary streams along the nonresistant shale units produces linear lowlands (strike valleys). The more-resistant sandstone units remain as linear ridges called cuestas.

Figure 23.14 The Atlantic and Gulf coastal plains consist of Tertiary and Cretaceous rocks that are inclined toward the sea. The evolution of landforms developed on the inclined strata is shown in idealized block diagrams.

Throughout the Great Plains, the sedimentary strata dip gently westward. Erosion on the tilted strata has formed cuestas and intervening lowlands in Kansas, Oklahoma, and adjacent areas. A notable exception is the Black Hills dome in western South Dakota, which is a classic domal structure with surrounding elliptical hogbacks and strike valleys (Figure 23.11).

In the coastal plains of the Atlantic and Gulf Coast states, alternating layers of sandstone and shale deposited during the Mesozoic and Cenozoic Eras dip gently seaward. The topography of the coastal plains therefore consists of a series of low cuestas on the sandstone layers and broad, low strike valleys in the soft shales (**Figure 23.14**). The major streams developed as the seas receded, following the direction of the initial slope. These streams are called **consequent streams**. A younger set of streams eroded headward along the weak shale formations and excavated a belt of lowlands. These are called **subsequent streams**. The resulting drainage system in the coastal plains has a **trellis pattern**.

Folded Mountain Belts

Differential erosion of folded mountain belts produces a series of ridges and valleys conforming to the structural trends of the folds. The landscape generally reflects the specific structural style of the mountain belt.

Folded mountains have complex structures that commonly include tight folds, thrust faults, accreted terranes, igneous intrusions, and andesitic volcanic rocks. Landforms are therefore quite variable and the features of mountainous topography will vary according to the age of the mountain and stage of development and the unique structure and rock assemblages that may exist. There are, however, some basic trends in the development of mountain landscapes. The fundamental factor for landscape

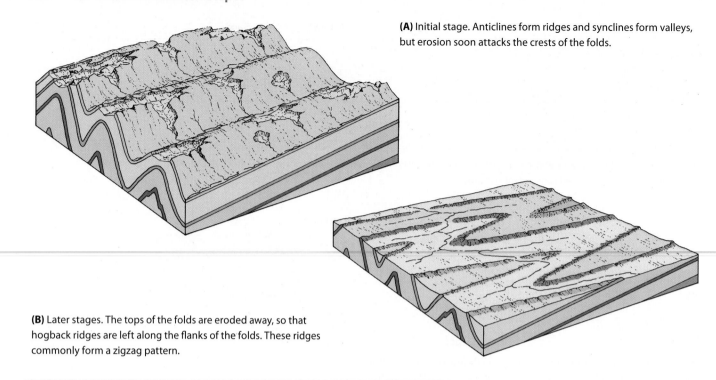

(A) Initial stage. Anticlines form ridges and synclines form valleys, but erosion soon attacks the crests of the folds.

(B) Later stages. The tops of the folds are eroded away, so that hogback ridges are left along the flanks of the folds. These ridges commonly form a zigzag pattern.

Figure 23.15 **Differential erosion of folded rocks** typically forms alternating ridges and valleys.

development is for differential erosion to etch out the structural fabric of the mountain belt and form linear valleys along weak rocks and fault zones and linear ridges along the more resistant rock units. Because folds are the dominant structures in mountain belts, the nature and style of folds greatly influence the style of landforms to be developed.

Landforms that typically develop on folded strata are shown in **Figure 23.15**. In the initial stages of erosion, the anticlines may form ridges, and the synclines may form long valleys. Some major streams may be superposed across the anticlinal ridges. As erosion proceeds, the crests of the anticlines are cut by the narrow valleys that grow along the flanks of the ridge. As the crest of the ridge is breached, "anticlinal valleys" are enlarged and deepened so that the crest of the anticline becomes open along its length.

As erosion proceeds rapidly headward along the nonresistant formations, the crests of the anticlines are eroded away, and the surface topography bears little resemblance to the underlying folds. Differential erosion effectively removes the weak rock layers to form long strike valleys. Resistant rock bodies stand up as narrow hogback ridges. The ridges thus mark the limbs of the folds. The pattern of topography is typically one of alternating valleys and ridges that zigzag across the landscape, and the drainage system generally forms a trellis pattern.

It is important to note that folded and eroded formations of sedimentary rock imply the erosion and removal of huge volumes of rock from the landscape. Excellent examples are the hogbacks along the Front Range of Colorado (**Figure 23.16**). Hogbacks and strike valleys formed on Paleozoic and Mesozoic strata along the front of the Rocky Mountains throughout Colorado and into Wyoming. Sedimentary rocks once covered the older Precambrian granites and metamorphic rocks now exposed in the present mountain range. They have since been eroded, leaving the tilted and eroded edges of the strata as hogbacks and strike valleys.

In young folded mountain belts, especially those that are still active, andesitic volcanoes may dominate the landscape to produce features characteristic of a magmatic arc. Also, thrust faults may greatly complicate the structure so that the resulting landscape is more complex.

Figure 23.16 Hogbacks along the Front Range of Colorado are the remnants of a thick sequence of Paleozoic and Mesozoic strata that were deformed into a large dome. The sequence at the end of the diagram shows the amount of material removed by erosion.

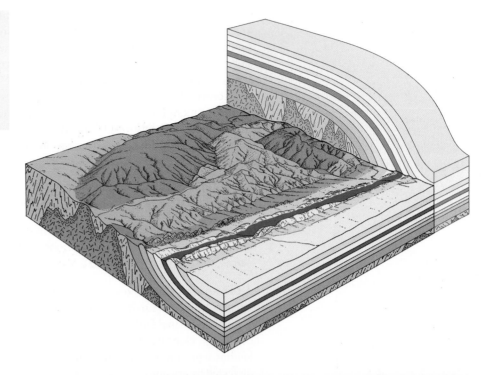

Figure 23.17 The Appalachian Mountains in eastern North America are a classic example of the topography formed by erosion of a folded mountain belt. The linear ridges are formed on resistant quartzite formations along the flanks of long, narrow folds. The ridges are about 300 m high.

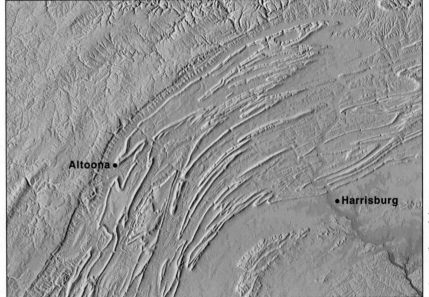

Altoona •

• Harrisburg

0   20   40

km

The Appalachian Ridge and Valley Province

The Ridge and Valley Province of the eastern United States is a classic example of a landscape formed on folded and thrust-faulted strata (**Figure 23.17**). It extends from Pennsylvania to central Alabama, a distance of more than 2000 km. In Alabama, the folded structures plunge beneath the younger sediments of the Coastal Plain. Similar topography is exposed again in the Ouachita Mountains in Arkansas and Oklahoma and in the Marathon Mountains in west Texas. A large segment of the folded Appalachian mountain belt is, therefore, still buried beneath younger sediments.

The deformation that produced the folds occurred during the late Paleozoic Era (more than 200 million years ago). The mountain belt was then deeply eroded, and most of the region was covered by Cretaceous and possibly Tertiary sediments. Regional uplift ensued, and renewed erosion began to remove the sedimentary cover.

What landforms typically develop on folded mountain belts?

The major east-flowing streams were superposed on the northeast-trending structures, and differential erosion began to cut the extensive valley system between the resistant sandstone ridges (Figure 23.15). Extension of the drainage system by more rapid erosion of the nonresistant shale and limestone formations resulted in stream capture and the development of a trellis drainage pattern. (Note that in humid regions limestones are eroded rapidly, whereas in drier regions they are resistant.)

Folded mountain belts similar to the Appalachian are found in many areas of the world, such as in southwest China, the Zagros Mountains of Iran, the Urals of Russia, and parts of the Andes, as well as the foothills of the Himalayas.

Continental Rifts

> Continental rifts produce horsts and grabens that are rapidly modified by erosion and sedimentation. Erosion dissects the uplifted blocks, and the sediment is deposited in the grabens as alluvial fans and lake deposits. With time, the rift may evolve into a continental margin.

A **continental rift** is a region where the crust has been arched upward, thinned, stretched, and fractured. The dominant structures in this tectonic setting are parallel systems of normal faults, with large vertical displacements. Typically, the faults produce large, elongate, down-dropped grabens and associated uplifted horsts. The major landform produced by normal faulting is a steep cliff, or **fault scarp**, which is soon dissected by erosion.

The evolution of erosional landforms developed on a normal fault block is shown in **Figure 23.18**. As soon as uplift occurs, stream erosion begins to dissect the cliff

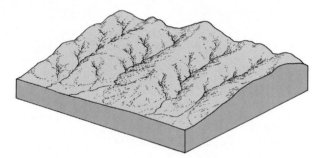

(A) The original dissected upland, before faulting, consists mostly of valley slopes.

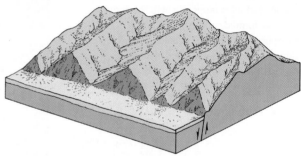

(B) The first major period of faulting is accompanied by accelerated stream erosion. Valleys are cut through the scarp produced by faulting to form triangular faceted spurs.

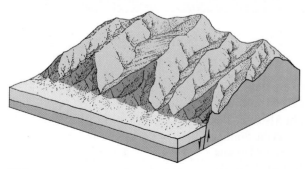

(C) Recurrent movement along the fault can produce a series of fresh scarps, which are subsequently dissected by stream erosion. Older faceted spurs recede and are worn down.

(D) Faceted spurs on the Wasatch Mountains, central Utah. A normal fault lies at the base of the range. Stream erosion during uplift created valleys that bound triangular faceted spurs.

Figure 23.18 Erosion of fault-block mountains follows a series of stages in which faceted spurs are a prominent landform. This landscape is typical of the Basin and Range province of the western United States and illustrates the evolution of landforms in continental rifts.

and produces a series of triangular faces known as **faceted spurs**. Erosion also forms gullies along the blunt face of the faceted spurs, so the cliff produced by faulting is considerably modified. Recurrent movement along the fault can produce a fresh scarp at the base of the older faceted spurs, but it is also rapidly modified by gullying to form a series of compound-faceted spurs. When movement on the fault ceases, the cliff continues to erode down and back from the fault line.

The idealized stages in landscape development by block faulting in an arid region are illustrated in **Figure 23.19**. In the initial stage, maximum relief is produced by the uplift of fault blocks. Relief diminishes throughout subsequent stages, unless major uplift recurs and interrupts the evolutionary trend by producing greater relief during later stages. In arid regions, depressions between mountain ranges generally do not fill completely with water. Large lakes do not form because of the low rainfall and great evaporation. Instead, shallow, temporary lakes, known as **playa lakes**, form in the central parts of basins and fluctuate in size during wet and dry periods. They may be completely dry for many years and then expand to cover a large part of the valley floor during years of high rainfall. When rainfall is high enough to maintain a permanent body of water, the lake is commonly saline, owing to the lack of an outlet. Great Salt Lake, in Utah, is an example. Most basins in the Basin and Range Province have dry playas during most of the year; however, during the cooler, wetter climatic periods that

Why is the topography of the Appalachian Mountains so different from that of the Rocky Mountains?

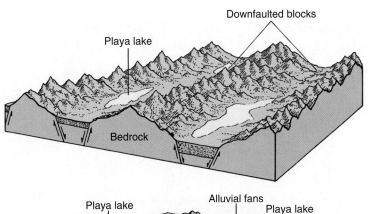

(A) Initial stage. Faulting produces maximum relief. Initially, some areas in the mountains are undissected. Playa lakes may develop in the central parts of the basins.

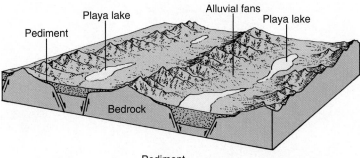

(B) Intermediate stage. The mountain range is completely dissected, and the mountain front retreats as a pediment develops. Alluvial fans spread out into the valley.

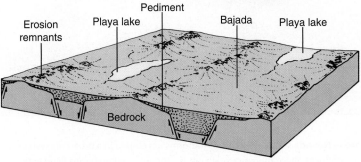

(C) Late stage. The basins become filled with sediment. Erosion wears down the mountain ranges to small, isolated remnants. The pediments expand and are buried by the alluvial fans, which merge to form bajadas. Most of the surface is an alluvial slope.

Figure 23.19 A model of landscape development in the Basin and Range Province of the United States. Continental rift systems such as this evolve through a series of stages until the mountains are consumed.

accompanied the Pleistocene glaciers, lakes with surface areas of thousands of square kilometers developed in many of these valleys.

Weathering in the uplifted mountain mass produces more sediment than can be carried away by the intermittent streams, which may flow only during spring runoff. The overloaded streams commonly deposit much of their load where they emerge from the mountain front. This sediment accumulates in broad alluvial fans. Throughout their history, the mountain ranges are eroded and the debris is deposited in the adjacent basin. As erosion continues (Figure 23.19B), the mountain mass is dissected into an intricate network of canyons and is worn down. At the same time, the mountain shrinks farther as the front recedes through the process of slope retreat.

In the final stage (Figure 23.19C) of this model, the fans along the mountain front grow and merge to form a large alluvial slope, a **bajada**. As the mountain front retreats, an erosion surface, known as a **pediment**, develops on the underlying bedrock. It expands as the mountain shrinks. Pediments are generally covered by a thin, discontinuous veneer of alluvium. As shown in Figure 23.19C, pediments may become completely buried.

What landforms typically develop in continental rifts?

Basin and Range Province

This tectonic and erosional model effectively explains much of the landscape in the Basin and Range Province of the western United States (**Figure 23.20**). The Basin

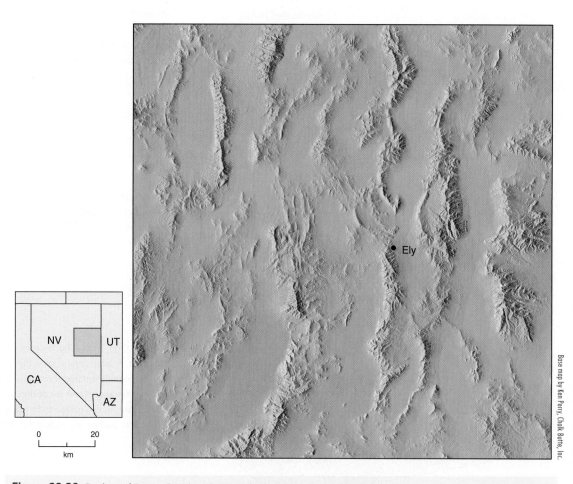

Base map by Ken Perry, Chalk Butte, Inc.

Figure 23.20 Basin and Range Province in Nevada shows the typical landforms developed in a broad rift system in an arid region. Erosion of the fault blocks produces alternating mountain ranges and intervening basins filled with sediment.

and Range is a large area where the crust has been uplifted and extended, forming a complex rift system that reaches from northern Mexico to southern Idaho and Oregon. The block faulting, resulting from extension, has produced alternating mountain ranges and intervening basins. There are more than 150 separate ranges in this province. Some are simple tilted fault blocks that are asymmetrical in cross section, the steeper side marking the side along which faulting occurred. Others are faulted on both sides. The internal structure of the fault blocks is complex, with folds, faults, and igneous intrusions recording an earlier history of crustal deformation.

To the north, throughout much of western Utah and all of Nevada, the area is in the initial stage of development. The basins occupy about half the total area, and the pediments are small. The relief of the mountain ranges is high, with most alluvial fans just beginning to coalesce into broad alluvial slopes. Recurrent movement along many fault systems is indicated by complex faceted spurs, similar to those illustrated in Figure 23.18D. Continued movement in recent times is clear from faulted alluvial fans and recurring earthquakes in the area.

In Arizona and Mexico, erosion in the Basin and Range has proceeded much further, and the area is in the late stage of development. The ranges are eroded down to small remnants of their original size. Extensive bajadas, spreading over approximately four-fifths of the area, cover wide pediments, through which isolated remnants of bedrock protrude.

Flood Basalts: Plains and Plateaus

The extrusion of large volumes of basalt may flood large areas of the landscape to produce a basaltic plain that commonly develops a series of distinctive landforms and ultimately evolves into a dissected basaltic plateau.

The tectonic setting for extrusion of continental flood basalts is commonly a rift system or a hot spot. The unique feature of flood basalts is that, as the name implies, huge volumes of fluid basaltic lava are extruded and flood the landscape, covering extensive areas of preexisting landforms. The floods of basalt thus provide a new surface that is modified by erosion through a series of stages resulting in a distinctive landscape quite unlike that of a shield, stable platform, or folded mountain belt.

Evolution of Basaltic Plains

The general trends in the evolution of **basaltic plains** might best be understood by considering what develops in areas of local volcanism because the local areas provide a small-scale model of the major trends in landform evolution of regional basaltic floods. The sequence of landforms resulting from erosion in areas where minor volcanic activity occurs is shown in **Figure 23.21**. Upon extrusion, lava follows established river channels and partly filled stream valleys. The lava flows disrupt drainage in two principal ways: (1) Lakes are impounded upstream, and (2) the river is displaced and forced to flow along the margins of the lava flow. Subsequent stream erosion is then concentrated along the lava margins.

As erosion is initiated in the displaced drainage system (Figure 23.21B), new valleys are cut along the flow margins and become gradually deeper and wider with time. Cinder cones, if formed during the volcanic activity, are soon obliterated because the unconsolidated ash is easily eroded. Only the conduit through which the lava was extruded remains as a resistant **volcanic neck**. With time, the area along the margins of the lava flow is eroded, so the old stream now flooded with lava is left standing higher than the surrounding area as an **inverted valley**. In the final stage of erosion (Figure 23.21C), the inverted valley is reduced in size and ultimately becomes dissected into isolated mesas and buttes.

(A) Initial stage. Lavas extruded from volcanic vents flow down existing rivers and streams and block the normal drainage system. Lakes commonly form upstream, and new stream channels develop along the margins of the lava flow. Volcanic cones are fresh and relatively untouched by erosion.

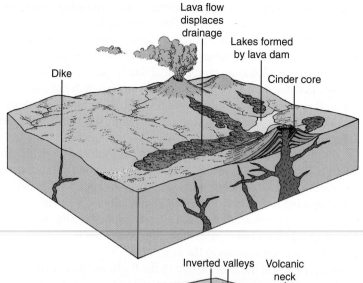

(B) Intermediate stage. The cinder cones are worn down until only volcanic necks are left standing. Erosion along the margins of the lava flow removes the surrounding rock, so the flow forms a sinuous ridge, or inverted valley.

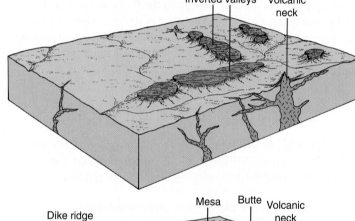

(C) Late stage. Inverted valleys are eroded to mesas and buttes. Volcanic necks and dikes commonly form peaks and isolated ridges.

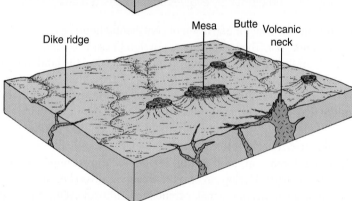

Figure 23.21 **A model of landscape development in an area of local basaltic volcanism** shows how inverted topography may develop.

In regions of extensive volcanism, such as those related to mantle plumes and flood basalts, large areas may be completely buried by lava flows. In the initial stage, basaltic lava covers the lowlands, forming a lava plain. Drainage usually is displaced to the margins of the lava plain, but some rivers may migrate across the plain. When extrusion ceases, stream erosion begins to dissect the lava plain and eventually cuts it into isolated plateaus and mesas. These, in turn, are ultimately eroded away.

Foremost among the older basaltic plains that are now uplifted and eroded into dissected plateaus is the Ethiopian Plateau in northeast Africa. Here, floods of basalt associated with the East African Rift valleys are eroded into plateaus more than 4000 m above sea level. The Deccan basalts of India, the Parana basalts of South

America, and the basaltic lava flows of northern Siberia have developed landscapes typical of plateau basalts.

Basaltic Plains of North America

An example of a landscape developed on basaltic plains is the Columbia Plateau and Snake River Plain of Washington, Oregon, and Idaho. This region of more than 400,000 km² represents one of the world's major accumulations of lava on a continent (**Figure 23.22**). Approximately 100,000 km³ of basaltic lava was extruded in this area through fissure eruptions and small shield volcanoes. The lavas were very fluid and completely covered preexisting mountains and valleys with a local relief of more than 750 m. The age of the basalts ranges from late Cenozoic (about 17 million years) to Recent, with the youngest flows in Craters of the Moon National Monument being less than 2000 years old. These lavas are probably related to the passage of North America over a mantle plume. Individual lava flows vary from 2 m to as much as 50 m in thickness. In places, the total sequence of basalt flows is more than 4 km thick.

The extrusion of the lavas produced a new surface that is currently being eroded and modified by the Columbia and Snake river systems. In southern Idaho, where the lavas are young, the region is in the initial stage of development, with large areas of basalt flows essentially untouched by stream erosion. In the eastern part of the region, the Snake River has been forced to flow along the southern margins of the plain because new young flows cover the central part of the area (Figure 23.22). In western Idaho, the Snake River has cut a deep canyon in the basalts. Hell's Canyon of the Snake, along the Idaho–Oregon border, is deeper than the Grand Canyon of the Colorado.

The Columbia Plateau in Oregon and Washington is made of much older basaltic lavas and is more deeply eroded. Erosion of the Columbia Plateau has been complicated by catastrophic flooding during the last ice age. But here again, the major drainage is displaced to the margins of the plains, where it is actively cutting deep canyons.

What are inverted valleys? How do they originate?

Figure 23.22 The eastern Snake River Plain in Idaho is underlain by extensive floods of basalt that form a broad, smooth plain. The young lava flows of the Snake River Plain are not eroded. Note how the Snake River has been displaced to the southern margin of the plain by repeated volcanism.

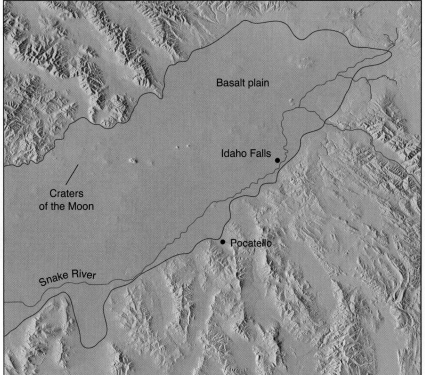

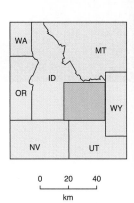

Magmatic Arcs

The landscapes of magmatic arcs formed above subduction zones are dominated by volcanic landforms that evolve by uplift and erosion until most of the volcanic features are erased. Ultimately, the volcanoes are eroded away to expose the deeper igneous intrusions.

Island arcs are the surface manifestation of magmatic activity produced at convergent plate boundaries. The most characteristic landform is a group of large composite andesitic volcanoes that rise thousands of meters above their surroundings. The volcanoes extend in a linear zone parallel to the convergent plate margin. The general trend in landform evolution in a magmatic arc is for streams and glaciers to systematically erode the volcanoes even as eruptions continue. Thus, the high volcanic peaks and collapse calderas are soon eroded down to circular or elliptical remnants and rounded hills. Circular landforms dominate the landscape (**Figure 23.23**). Repeated extrusion occurs in magmatic arcs, however, so volcanic features representing a variety of ages occur within a given area. Drainage systems are constantly being displaced by new volcanoes, and large segments of a river system may be completely obliterated as extrusion continues. New drainage patterns develop in its place. A well-integrated drainage system is therefore difficult to establish in an active volcanic arc and many small lakes commonly develop. Glaciers also form on the high volcanic peaks, even in low latitudes such as the central Andes.

When igneous activity terminates, erosion and associated isostatic uplift occur so that, ultimately, the volcanic material is removed and granitic intrusions are exposed at the surface. Thus, the landscape evolves from a volcanic terrain to one formed on igneous intrusions.

Figure 23.23 Two of the hundreds of volcanoes in the Kuril Island arc of the western Pacific show the typical landforms developed on an active magmatic arc. Large calderas occur on the opposite ends of the island, each with a younger cone in the middle. Older volcanoes can be recognized by their circular forms.

Courtesy of JSC/NASA

The Cascade Volcanic Chain

The Cascade Mountains of northwestern United States is a magmatic arc built on continental crust (**Figure 23.24**). Here, famous composite volcanoes such as Mount Shasta, Crater Lake, Mount St. Helens, and Mount Rainier rise above the surrounding landscape, and hundreds of smaller volcanoes form a belt roughly 80 km wide and more than 500 km long. Between the volcanoes are beds of lava and tuff that mask all the other rocks and surface features. In many areas the average distance between these small volcanoes is little more than 5 km. Volcanic eruptions and contemporaneous erosion have been the dominant geologic processes for most of the last several million years.

California's Mount Shasta exceeds 4000 m in elevation and Mount Lassen, which last erupted in 1914 and 1915, rises to more than 3000 m. Much of the stream drainage has been disrupted by the flows and volcanoes. Lakes are numerous, but many are dry much of the time because of the porous nature of the lava and tephra. During the last ice age, glaciers formed on all of the volcanic peaks higher than about 3000 m, and many glaciers still exist today.

Cycles of eruption and erosion have persisted since 15 million years ago. In the middle Cascades, an imposing row of huge Quaternary (less than 1 million years old) volcanoes dominate the skyline. Volcanoes and volcanic deposits represent all ages from middle Cenozoic to the present. For example, the dome at the summit of Mount St. Helens is only a few years old. Crater Lake, in southern Oregon, was once a similar huge volcano, but 7000 years ago a cataclysmic eruption fractured the upper part of the cone as great quantities of ash erupted. The summit collapsed to form a caldera containing a lake about 600 m deep. The western Cascades are so completely eroded that no trace of their original volcanic landscape remains. North of Washington's Mount Rainier, the Cascades have been uplifted and deeply eroded to expose early Cenozoic granitic batholiths that intrude even older metamorphic and sedimentary rock.

We thus see in the Cascades examples of practically every phase of landscape developed in magmatic arcs, from active volcanoes to exposed granitic intrusions. This is America's segment of the great Ring of Fire that essentially surrounds the Pacific Ocean, and similar landforms can be found from the southern end of the Andes to the Aleutian Islands, Japan, the Philippines, and Indonesia.

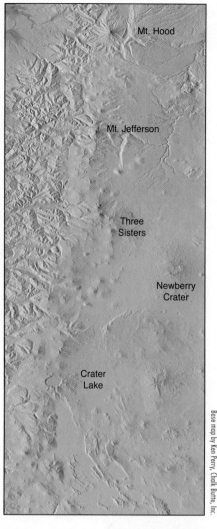

Base map by Ken Perry, Chalk Butte, Inc.

Figure 23.24 Erosion of a magmatic arc produces a landscape marked by circular forms that are the remnants of volcanoes and collapse calderas. Radial drainages commonly develop on the volcanoes. This shaded relief map shows the Cascade Mountains in Oregon. The young volcanoes are being eroded but their circular form is still preserved. An older volcanic terrain to the west is dissected into deep valleys and canyons.

GeoLogic   Landscape of Snow Canyon, Utah

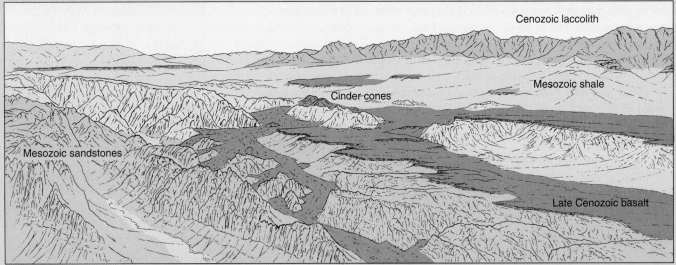

This area is a land of colorful mesas, cliffs, and plateaus but there is much more to this area than spectacular scenery. The sketch will help you identify the major geologic features.

Observations

1. The red and white cliffs are made of Mesozoic sandstones that are resistant to erosion.
2. The tree-covered slope behind them is underlain by late Mesozoic shales that are less resistant to erosion.
3. The mountains in the background expose a middle Cenozoic granitic laccolith.
4. Late Cenozoic basalt flows bury part of the countryside. A road winds its way across one of them and another basalt flow forms a low inverted valley on top of the red sandstone in the central part of the scene. A third basalt flow forms a high inverted valley on top of the white cliffs and black cones of talus spill from it.

Interpretations

Initially, the layers of sedimentary rock were deposited by the wind, in rivers, or in shallow seas. At the time, this area was east of an active convergent plate margin. Eventually, subduction-related magmatism created granitic magma, which intruded into the sequence of sedimentary rock. But convergence ceased and was replaced by extension and rifting on a grand scale. As a result, the sequence of sedimentary rocks and the laccolith were raised several kilometers above sea level. Erosion by running water then stripped the sedimentary cover off the intrusion, and carved the rugged red and white cliffs and deep canyons. The nonresistant rock layers were easily stripped away to create the tree-covered slope. Continental rifting was accompanied by several episodes of basaltic volcanism. Lava flows cascaded down the white cliffs and flowed into the red canyons below. Continued uplift and erosion during the volcanism created the inverted valleys from the resistant lava flows.

Key Terms

alcove (p. 709)	cuesta (p. 707)	inverted valley (p. 717)	shield (p. 698)
bajada (p. 716)	dendritic drainage (p. 706)	mesa (p. 706)	stable platform (p. 698)
basaltic plain (p. 717)	differential erosion (p. 698)	natural arch (p. 709)	strike valley (p. 706)
basin (p. 707)	dome (p. 707)	pediment (p. 716)	subsequent stream (p. 711)
butte (p. 708)	faceted spur (p. 715)	pillar (p. 708)	trellis pattern (p. 711)
column (p. 708)	fault scarp (p. 715)	pinnacle (p. 708)	volcanic neck (p. 717)
consequent stream (p. 711)	folded mountain belt (p. 698)	plateau (p. 706)	
continental rift (p. 714)	hogback (p. 707)	playa lake (p. 715)	

Review Questions

1. What are the major factors that influence the evolution of continental landscapes?
2. Contrast the landscapes formed on a shield found in a tropical climate with one found in a polar region.
3. How does differential erosion produce alternating cliffs and slopes?
4. Describe the model of evolution of a mountain belt into a new segment of the shield. At what stage are erosion rates highest?
5. Explain the origin of columns and pillars, such as those in Bryce Canyon National Park.
6. How are natural arches formed?
7. Describe and illustrate, with sketches, the landforms that typically develop on a stable platform.
8. Why is the Coastal Plain of the Atlantic so low and smooth?
9. What changes would be necessary for deep canyons to form in Kansas?
10. Describe and illustrate, by means of a cross section, the landforms that typically develop by erosion of folded sedimentary rocks found along a convergent plate margin.
11. What processes are responsible for the adjustment of a drainage pattern to flow along structural trends such as those found in folded mountain belts?
12. Describe and illustrate, by means of a cross section, the origin of landforms in a rift system.
13. How do the landforms of a magmatic arc differ from those of a folded mountain belt formed by continental collision? What landscape features might they have in common?
14. Why is the western slope of the Cascade Mountains (Figure 23.24) more intricately eroded than the eastern side?
15. How do rivers adjust in areas where eruptions of basaltic lavas are important?
16. On a map of North America, outline the extent of the shield, stable platform, folded mountain belts, and basaltic plains.
17. List several ways in which tectonics influence or control the nature of landscapes.

Part IV

Epilogue

It is appropriate to look back at Earth and it's dynamic geologic systems now that you have a firmer grasp on how it works. One way of doing this is by examining the relationship between people and the planet. Nowhere is this relationship more clear than in the realm of our natural resources. Just like the planet itself, human societies rely on energy and a host of materials, all of which are derived from Earth. Another fruitful way to look back at Earth is from the vantage point of space. Earth is but one planet among many that orbit our Sun, having formed from a swirling mass of gas and dust 4.55 billion years ago. By comparing and contrasting Earth with these other bodies, you will see how truly unique Earth is. You will see how a delicate balance between planet size, location, and composition has allowed Earth to be a haven for life. With only a few minor changes in any of these factors, Earth would have been a radically different place, inhospitable for life as we know it.

Data courtesy of NASA/Goddard Space Flight Center

24 Earth's Resources

Our modern society is completely dependent on a wide variety of mineral resources. Often we take for granted that these resources will always be available. The sobering fact is that most of our natural resources are finite and nonrenewable. History records the fall of cultures and civilizations in which a depleted resource base played a major role. An example is the decline of Mesopotamia 1500 years ago. As the people of southern Asia began to irrigate their arid lands, salt became concentrated in the soil when irrigation waters evaporated. Agricultural production gradually diminished. The destruction of one of their most valued resources led to the decline of their entire civilization.

Our problem is not just one of finding more and more resources, but of balancing those resources with a burgeoning human population. In the early 1900s, Earth's population was about 1.7 billion people. Now the population is more than 7 billion. Every 3 years the equivalent of the population of the United States is added to the planet. In 100 years, our numbers have increased more than threefold. The increase in population has not come without its costs.

During the next decade, we will use more oil, gas, iron, and other mineral resources than were consumed throughout previous human history. These facts lead us to ask a series of

important questions. Is there really a finite supply of mineral resources? Will we completely consume the metallic resources discovered only in the last century? Will we be able to find and extract even deeper deposits? What about oil resources—will they be gone in another 30 years? Do we have enough agricultural lands to support ourselves? Will there be ample fresh water for billions of additional people? In short, what will be the fate of our resource-dependent civilization?

One of those critical materials is potash—the commercial name for water soluble potassium. Its most important use is in fertilizer for agriculture. The photograph above shows a potash mining operation in southern Utah. Potash is extracted by injecting brine into buried layers of potassium-rich salt about 800 m deep. The layers originally formed as evaporites in Paleozoic times. The solution is then pumped back to the surface and into the evaporation ponds where the potassium crystallizes as a solid. A blue dye is used to enhance the absorption of solar energy in the ponds. Without such extraction of resources, it would be impossible for us to sustain our current standard of living.

Ore deposits and other types of earth resources are not equally distributed around the world but are rare and only concentrated by very specific geologic conditions. All mineral resources develop slowly and eons pass before a new deposit forms. In this chapter we review the origin of these resources and examine some of the implications of our increasing rate of consumption of our natural resources.

Major Concepts

1. Mineral resources are concentrated by geologic processes operating in the hydrologic and tectonic systems. Many require long periods to form; these resources are finite and nonrenewable.
2. Ore deposits are formed by igneous, sedimentary, metamorphic, and weathering processes. Many metallic ores involve transport and deposition of metals in a hydrothermal fluid.
3. Earth's principal nonrenewable energy resources include coal, oil, natural gas, and nuclear power. Renewable energy resources include solar energy, wind power, hydroelectric power, tidal power, and geothermal energy. At present these renewable forms of energy provide only a small fraction of our energy needs.
4. The location and richness of most of Earth's natural resources are directly or indirectly controlled by plate tectonics.
5. There are limits to population growth on Earth imposed by the finite nature of many of our natural resources.

Mineral Resources

> The present store of mineral resources has been concentrated very slowly by a variety of geologic processes related to the plate tectonic and hydrologic systems. Most mineral resources are therefore finite and nonrenewable.

Mineral resources range from the soils that support agriculture to metals such as silicon, which is used in high-technology applications such as computers. Though technically not minerals, oil, natural gas, coal, and some other sources of energy are also included as mineral resources because they are extracted from Earth. Mining worldwide produces about $500 billion worth of metallic ore each year; another $700 billion of energy minerals are produced.

Why are mineral resources distributed so unevenly throughout the world?

The world's valuable deposits of minerals and energy fuels were formed slowly by the major geologic systems during various periods in the geologic past (**Table 24.1**). Their formation required very long intervals of time and occurred under specific geologic conditions. Some metallic mineral deposits were formed in such restricted geologic settings that they approach uniqueness. For example, 40% of the world's reserves of molybdenum are in one igneous intrusion in Colorado; 77% of tungsten reserves are in China; more than 50% of tin reserves are in Southeast Asia; and 75% of chromium reserves are in South Africa. If mineral resources are depleted, we cannot just go out and find more. More of many deposits simply do not exist. Fortunately, most metals, unlike fossil fuels, can be **recycled**.

In contrast to some energy resources and most biological resources (such as agricultural crops and forest products), very few mineral resources are **renewable**, meaning they are replenished in a short period of time. The processes that form mineral resources operate so slowly (by human standards) that their rates of replenishment are infinitesimally small in comparison to rates of human consumption. For example, the generation of oil from sedimentary rocks may take more than 10 million years. Consequently, mineral deposits are finite and therefore are exhaustible or **nonrenewable**. Most of our mineral resources are like a checking account that will never receive another deposit. The faster we withdraw, or the larger the checks we write, the sooner the account will be depleted. Moreover, today, few areas remain completely unexplored for mineral deposits. Most of the continents have been mapped and studied extensively, so the inventory of natural resources is nearly complete. We can accurately estimate the extent of many of our mineral resources and their rates of consumption. With these estimates, projecting how long they will last is not especially difficult.

Table 24.1 The Major Geologic Processes That Form Mineral Resources

Process	Deposits Formed	Mineral Resource
	Igneous processes	
	Magmatic segregation	Chromium, vanadium, nickel, copper, cobalt, platinum
	Pegmatites	Beryllium, lithium, tantalum
	Hydrothermal deposits	Copper, lead, zinc, molybdenum, tin, gold, silver
	Sedimentary processes	
Clastic rocks	Stream deposits	Sand, gravel
	Placer deposits	Gold, platinum, diamonds, tin, ilmenite, rutile, zircon
	Dune deposits	Sand
	Loess deposits	Soil
Chemical precipitates	Evaporite deposits	Halite, sylvite, borax, gypsum, trona
	Marine sediment	Banded iron formation, phosphate, limestone
Organic precipitates	Hydrocarbon deposits	Oil, natural gas, coal
	Marine deposits	Limestone
	Metamorphic processes	
	Contact metamorphism	Tungsten, copper, tin, lead, zinc, gold, silver
	Regional metamorphism	Gold, tungsten, copper, talc, asbestos
	Weathering and groundwater	
	Soil	Agriculture
	Residual soils	Clay
	Residual weathering deposits	Nickel, iron, cobalt, aluminum, gold
	Groundwater deposits	Travertine, uranium, sulfur
	Brines in basins	Lead, zinc, copper
	Geothermal wells	Hot water, electricity
	Water	Drinking water, irrigation

Data from: Kesler, S. E. *Mineral Resources, Economics, and the Environment.* New York: Macmillan, 1994.

Processes That Form Mineral Deposits

The origin of most ore deposits is related to fundamental geologic processes. We recognize four major groups of mineral deposits formed by (1) igneous processes, (2) metamorphic processes, (3) sedimentary processes, and (4) weathering and groundwater processes.

The important minerals on which modern civilization depends are an extremely small part of Earth's crust. For instance, copper, tin, gold, and other metallic minerals occur in quantities measured in parts per million (and, in most cases, a very few parts per million). To form a mineral deposit, some metals must be concentrated thousands of times beyond their "normal" concentrations in rocks. In contrast, the rock-forming minerals (such as feldspar, quartz, calcite, and clay) are abundant and widely distributed. The important question, then, is how these very small quantities of important minerals are concentrated into deposits large enough to be used.

Below, we will consider some principles governing the concentration of rare minerals in ore deposits and the origin of some nonmetallic resources. It may surprise you to learn that essentially every geologic process—including igneous activity, metamorphism, sedimentation, weathering, and deformation of the crust—plays a part in the genesis of some valuable mineral deposits (Table 24.1). The occurrence or absence of most mineral deposits, therefore, is controlled by a region's specific geologic conditions and plate tectonic setting (**Figure 24.1**).

Igneous Processes

Many mineral resources are formed by magmatic processes. Magmas have higher concentrations of some elements than most other rocks, and some minerals can reach even higher concentrations in specific areas of an igneous rock. Prime examples are the exotic ultramafic volcanic rocks that host diamonds. Diamond crystals were probably ripped from diamond-bearing wall rocks by magma rising through the deep mantle (**Figure 24.2**). Laboratory experiments show that diamond is stable at depths of at least 150 to 200 km. At low pressure, the stable form of carbon is the soft mineral graphite, but the reaction of diamond to form graphite proceeds very slowly at the low temperatures found at Earth's surface. Besides its use as a gem, diamonds have found industrial uses as abrasives and as strong coatings. Diamond deposits are limited to regions underlain by Precambrian crust (Figure 24.1). The richest deposits are found in South Africa and Australia, but diamonds have been discovered recently

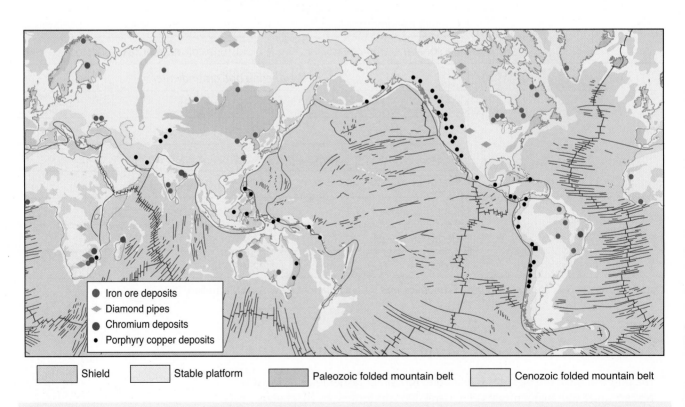

Shield   Stable platform   Paleozoic folded mountain belt   Cenozoic folded mountain belt

- Iron ore deposits
- Diamond pipes
- Chromium deposits
- Porphyry copper deposits

Figure 24.1 Major ore deposits are related to specific tectonic settings. Most important iron ores are in sedimentary rocks of Precambrian age exposed in the shields. Chromium and diamond deposits are also concentrated on the shields and formed in ancient basaltic intrusions related to mantle plumes or rifts. In contrast, many copper (and lead, zinc, and silver) deposits form around intrusions in young mountain belts at convergent plate boundaries.

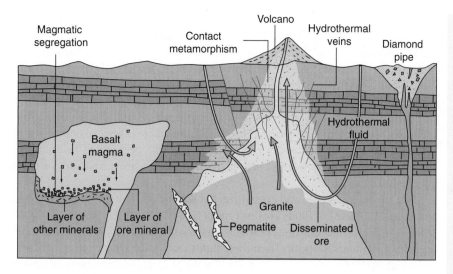

Figure 24.2 **Igneous processes** produce many metallic ore deposits as illustrated in this cross section. Magmatic segregation (left) occurs as heavy crystals sink through fluid magma and accumulate in layers near the base of a magma chamber. Ore in pegmatites consists of beryllium and lithium, which are concentrated in residual melts during fractional crystallization. Other elements, such as gold, silver, copper, lead, and zinc, are carried by hydrothermal fluids (blue and purple arrows) through fractures in the surrounding rock. Eventually, the metallic elements are precipitated as hydrothermal vein deposits. The concentration of ore deposits by contact metamorphism occurs as fluids from the cooling magma replace parts of the surrounding rock. Iron and tungsten deposits may be formed by this process. Diamonds, brought from deep in the mantle, are mined from volcanic pipes in which the upper part is a complex breccia (right).

Why are there no diamond mines in Iowa?

in the Northwest Territories. Diamond prospecting in the United States has identified diamond-bearing igneous pipes on the Colorado-Wyoming state line and in Arkansas, but none of these is currently economic to mine. Industrial diamonds can also form in large impact craters when the mineral graphite is converted to diamond during high pressure metamorphism caused by meteorite impact. Popigai crater in Arctic Russia may have minable concentrations.

Concentrations of other ores result when minerals forming in a magma have different temperatures of crystallization and density. One concentration process is **magmatic segregation**, in which dense mineral grains accumulate in layers near the base of an igneous body (Figure 24.2). To form ore, the layer must form when only one kind of mineral is crystallizing. This requirement calls for unusual conditions because most magmas have several different kinds of minerals crystallizing simultaneously. Once they crystallize, the dense minerals can sink to the bottom of the chamber to form layers. Deposits of chrome, nickel, vanadium, and platinum form in some mafic intrusions by this process (**Figure 24.3**). The largest of these are in the Bushveld Complex of South Africa. The rich nickel and copper sulfide ores of Sudbury, Ontario, also sank to the base of a very unusual, but large body of mafic magma. Some geologists have concluded that the magma formed because of a meteorite impact about 1.7 billion years ago.

Late-stage segregation and crystallization is another process by which rare minerals are concentrated (Figure 24.2). In an original magma, many elements occur in amounts of only a few parts per million but do not fit readily into the crystalline structure of the rock-forming minerals. As a silica-rich magma crystallizes, these rare

Courtesy of Kevin Walsh

Figure 24.3 **Layers of dense minerals such as chromite** may form by crystallization and settling to the floor of large mafic igneous intrusions. The layers shown here are from the Bushveld Complex, South Africa. The black layers are chromite and the tan layers are made of olivine.

elements are concentrated in the last remaining melt. Examples of such rare element segregations include beryllium, lithium, uranium, and tantalum. These late-stage, water-rich magmas can then intrude into fractures in a pluton or in the surrounding rock. As they cool, the rare elements crystallize as minerals in a coarse-grained rock called **pegmatite**. Individual quartz crystals several meters long have formed in pegmatites. These rare elements have important uses in electronics, lightweight metal fabrication, and nuclear reactors. Several gemstones are found in granitic pegmatites, including topaz, beryl, and tourmaline. In addition, large masses of common silicate minerals, such as feldspar and mica, can be mined from pegmatites. Feldspars are used in ceramics, and muscovite mica is used in makeup, insulators, and various construction materials.

As some magmas cool and crystallize, hot water-rich solutions are concentrated and released. These hot fluids are one type of **hydrothermal fluid** and can become laden with high concentrations of soluble materials, including many economically important elements. Deposits from these fluids are part of a large category of ores known as **hydrothermal ore deposits** (Figure 24.2). Hydrothermal fluids can be injected into fractures in the rock surrounding an intrusion to form **veins** with minerals containing gold, silver, copper, lead, zinc, molybdenum, and other elements. The fluids can also permeate small fractures and grain boundaries in the already-solid part of the intrusion to form metallic minerals distributed in the intrusion. This process produces large deposits of low-grade ore (0.2% to 2% copper); much of the pluton may be mineralized. If the intrusion is exposed near the surface, these low-grade deposits can be removed profitably by open-pit mining. An important example of this type of mineral deposit is the copper found in some porphyritic igneous intrusions, often of dioritic or granitic composition. **Porphyry copper** deposits currently account for more than 50% of the world's copper production. Utah's Bingham copper mine is one the largest of this type in the world **(Figure 24.4)**.

Other types of hydrothermal fluids are formed by a variety of processes and tectonic settings; not all hydrothermal fluids are released directly from magmas. Intrusions may heat groundwater, causing it to convect and carry ore metals that may become concentrated in veins. The penetration of water along faults in rift zones into the deep and hot parts of the crust can also create hydrothermal solutions that carry gold and other metals (Figure 24.2).

A variety of igneous materials are also used as **industrial minerals**—generally nonmetallic minerals or rocks that are used in industry. The cost per ton for these

Figure 24.4 Porphyry copper deposits form around shallow igneous intrusions where hydrothermal fluids deposit copper and alter the surrounding rocks. Most are mined as large open pits with huge tailing (waste rock) piles like these near Bingham, Utah.

© Lee Prince/ShutterStock, Inc.

materials is generally low. Building stone, for example, is commonly quarried from homogeneous unfractured plutons exposed at the surface. Road aggregate with very specific properties is sometimes formed by crushing intrusive igneous rocks of known strength, especially in areas where gravel is not common.

Metamorphic Processes

Metamorphism changes the texture and mineralogy of rocks and in the process can form important new mineral resources.

Regional Metamorphism   Important deposits of industrial minerals, such as asbestos, talc, and graphite, have formed during the metamorphism of large regions in the roots of mountain belts. Marble and serpentine are commonly used to face buildings and in other construction projects. Corundum (Al_2O_3) and garnet are common metamorphic minerals used as industrial abrasives. On the other end of the value scale, these same two minerals also form gems—ruby and sapphire are colored varieties of corundum, and garnet is a semiprecious gem. Colombian emeralds (gem form of beryl) appear to have formed during metamorphism of sedimentary rocks.

Metallic ore deposits are also created by regional metamorphic processes. Gold, copper, and tungsten bearing hydrothermal fluids are also expelled from rocks during regional metamorphism of the continental crust. Fault zones, formed deep in the crust and now exposed in the continental shields, have been mineralized when such fluids flowed along them.

Contact Metamorphism   Some ore deposits form by contact metamorphism—metamorphism along the contact between an igneous intrusion and the surrounding rocks. In this process, heat and the flow of chemically active fluids from a cooling magma alter the adjacent rock by adding or removing elements. Limestone surrounding a granite pluton is particularly susceptible to alteration by hot, acidic hydrothermal solutions related to the intrusion. For example, large volumes of calcium can be replaced by iron in a hydrothermal fluid to form valuable ore deposits that contain tungsten or tin.

Seafloor Metamorphism   Hot hydrothermal fluids circulating through the oceanic crust cause seafloor metamorphism. These fluids leach metals (such as manganese, iron, copper, zinc, lead) and sulfur from the crust and transport these elements to hot spring vents on the ocean floor. Minerals precipitate when the hydrothermal fluids mix with seawater and cool. Mounds of sulfide ores collect on the seafloor where the hot waters are released.

Sedimentary Processes

A significant result of the erosion, transportation, and deposition of sediments is the segregation and concentration of mineral grains according to size and density (**Figure 24.5**). As we have emphasized elsewhere, soluble minerals are transported in solution; silt and clay-sized particles are transported in suspension, and sand and gravel are moved mostly as bed load by strong currents. Each of these forms of sediment transport may create mineral deposits.

Clastic Sediments   Sand and gravel are concentrated in river bars, beaches, and alluvial fans. These deposits, both modern and ancient, are valuable resources for the construction and glass-making industries. In the United States alone, more than $1 billion worth of sand and gravel is mined each year, making it the largest mineral industry not associated with fuel production in the country.

Clastic sedimentary processes also concentrate gold, diamonds, and tin oxide. Originally formed in veins, volcanic pipes, and intrusions, these minerals are eroded and transported by streams. Because they are much denser than most silicate minerals, they are deposited and concentrated where current action is weak, such as on the insides of meander bends or on protected beaches and bars (Figure 24.5). Such layers and lenses of valuable minerals are known as **placer** deposits, and they are mined

What important ore deposits are formed by metamorphic processes?

How can diamonds be concentrated in both igneous and sedimentary processes?

Figure 24.5 The concentration of heavy minerals in placer deposits occurs by stream action. Gold, diamonds, and tin, for example, if eroded from their original deposits, will accumulate in areas where stream currents are weak, such as at the base of a waterfall or inside a meander bend.

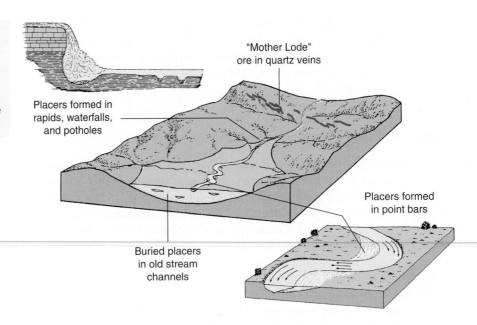

"Mother Lode" ore in quartz veins

Placers formed in rapids, waterfalls, and potholes

Placers formed in point bars

Buried placers in old stream channels

from both modern and ancient rivers and beaches. Some placers in modern rivers have been traced upstream to the source of the ore in vein deposits higher in the drainage basin. Placer ore deposits include gold and diamonds; some ancient placers also contain uranium minerals.

Chemical Precipitates   Chemically precipitated sediments are also important sources of minerals. Chemical precipitates from seawater form the great bulk of iron ore mined today. **Banded iron formations** of Precambrian age are especially significant because of their abundance (**Figure 24.6**). These deposits consist of alternating layers of iron oxide and chert, formed during a unique period of Earth's history, from 1.8 to 2.2 billion years ago, when the oxygen in Earth's atmosphere was building up to modern levels. Consequently, much iron was removed from solution in seawater. They cannot form under present environmental conditions. Many of these sedimentary sequences were subsequently metamorphosed.

Another way in which sedimentary processes concentrate valuable minerals is in the evaporation of saline waters in restricted embayments of the ocean or in large lakes where little clastic sediment is deposited. Valuable evaporites are formed in many playa lakes. As evaporation proceeds, dissolved minerals are concentrated and eventually precipitated as solid crystals. These **brines** and **evaporite** deposits include elements of commercial value, such as potassium, sodium, and magnesium salts, sulfates, borates, and nitrates (**Figure 24.7**). Gypsum is a common sulfate mineral formed by evaporation and is an especially important industrial mineral used in home construction as wallboard and as the base for plaster of Paris.

Figure 24.6 Banded iron formations formed as iron minerals interlayered with chert precipitated from shallow Precambrian oceans. Today, they form our major source of iron. These banded iron formations are in Australia.

© Blue Gum Pictures/Alamy

Figure 24.7 Brines form by strong evaporation. These ponds on the shores of Great Salt Lake are sources of magnesium as well as salt.

Less exotic chemical precipitates are also important mineral resources. Limestone, deposited by organic and inorganic processes, is a key component of agricultural lime, concrete, and other building materials. Limestone is crushed to make aggregate for roads and as a flux in steel smelting. The phosphate mineral apatite precipitates from upwelling seawater and is also mined to make fertilizers. Carbon-rich organic deposits are discussed below as energy minerals.

Sedimentary Fluids   Hydrothermal fluids may also form in sedimentary basins as the strata subside to deeper and deeper levels in the crust. If meteoric water penetrates these deep basins, it may be heated to as high as 300°C. At this temperature, soluble elements may be extracted from the rocks and then concentrated by crystallizing out of the fluid at shallower and cooler levels of the crust. There, ore minerals may precipitate in cavities and veins. Important deposits of lead and zinc in Paleozoic limestone of the upper Mississippi Valley of Missouri and Wisconsin formed in this way.

Weathering and Groundwater Processes

The simple process of chemical weathering also concentrates valuable minerals (Table 24.1). For example, water removes soluble material such as sodium, potassium, calcium, and magnesium from rocks, leaving insoluble compounds as a residue. Weathering, therefore, can enrich ore deposits that were originally formed by other processes (**Figure 24.8**). For example, gold may occur as small inclusions in the sulfide mineral pyrite. Weathering may destroy the pyrite, concentrating the gold in a surface layer.

Chemical weathering can also concentrate, in the regolith, an element that was originally dispersed throughout a rock body. For example, extensive weathering of granite in tropical and semitropical zones commonly concentrates relatively insoluble metallic oxides and hydroxides in the thick regolith as it removes the more-soluble material. Because aluminum is relatively insoluble in groundwater, deposits of aluminum called *bauxites* may form in this way. Aluminum ores form where other elements

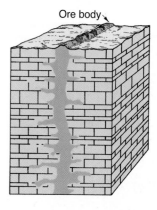

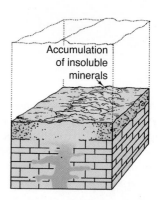

Figure 24.8 The concentration of ore deposits by weathering processes occurs as soluble rock, such as limestone, is removed in solution, leaving insoluble minerals concentrated as a residue.

are efficiently flushed out by a combination of high rainfall and good drainage due to high relief. Ophiolites exposed to deep tropical weathering may concentrate residual deposits of nickel, cobalt, and iron.

Perhaps the most important resource formed by weathering is agricultural soil. Physical and chemical weathering processes, as well as biological activity, are important in soil formation. Physical weathering breaks rocks up into small fragments; chemical weathering attacks minerals in these fragments and produces new clay minerals. The voids around the fragments allow nutrients and water to seep into the ground, and the clays absorb water and nutrients so that plants can gradually use them. A mixture of clay and silt with decayed organic materials creates productive agricultural soil. In many places, however, soil is being stripped away by erosion much faster than it is being formed. Even in a humid climate, it may take thousands of years to produce a few centimeters of good soil.

Even the clay formed by chemical weathering is a valuable resource and may be used in ceramics and to make tile and bricks (Table 24.1).

Another important resource controlled by surface processes is the groundwater itself. Groundwater is an important source of drinking and irrigation water in many regions of the world. In many of the world's aquifers, groundwater is a nonrenewable resource. In these aquifers, recharge rates are so slow that groundwater is pumped out much faster than it can be replenished. We are therefore essentially mining these groundwater resources, and once depleted, they cannot be replenished in a human time frame. Elsewhere, aquifers behave more like reservoirs used to temporarily store water for later use. Recharge occurs over the course of the year, and the water table remains essentially constant.

Energy Resources

Renewable energy resources include solar energy, hydropower, tidal energy, wind energy, and geothermal energy. Together they cannot be expected to fulfill our long-term energy needs. Fossil fuels form slowly from sedimentary rocks and are being rapidly depleted. Nuclear energy may become a major energy source in the future.

Modern society's technological progress and standard of living are intimately related to energy consumption. Until recently, energy resources and the capacity for growth seemed unlimited. Now, we are beginning to appreciate that energy sources are limited. Understanding the sources of energy and how they can be used most effectively will be among the most pressing problems of the twenty-first century.

Our sources of energy are found in both renewable and nonrenewable forms. A renewable energy source is either one that is available in unlimited amounts, for all practical purposes, or one that will not be appreciably diminished in the foreseeable future. Solar energy, tidal energy, and geothermal energy are the most important examples. In contrast, nonrenewable energy sources, such as mineral resources, are finite and exhaustible. They cannot be replaced once they have been consumed. Coal and petroleum, the fossil fuels on which modern culture relies so much, are nonrenewable. These energy sources have been concentrated by geologic processes that operated over vast periods of geologic time. Although the same processes function today, they operate too slowly to replenish these fuels.

An important aspect of today's energy picture is that more than 90% of the energy we use is produced from nonrenewable fossil fuels (**Figure 24.9**). The exponential growth in consumption of the world's fossil fuels has brought on the present energy crisis. Few technical analysts—whether economists, geologists, or engineers—doubt that the problem exists. They see the trends and events that indicate, in the near future, difficulties ranging from an awkward situation to disaster and economic peril. By contrast, most of the general public—if we are to accept the results of public opinion polls—do not believe that a problem exists.

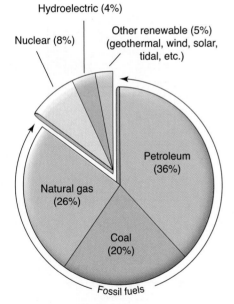

Figure 24.9 The energy consumed in the United States is provided from a variety of resources, but is predominantly from petroleum. Over the last two decades, however, our reliance on oil has diminished and the use of coal and natural gas has increased.

Are oil production and consumption increasing in the United States?

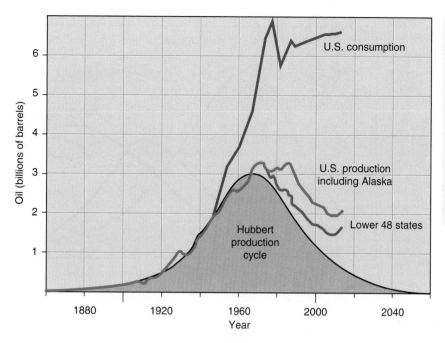

Figure 24.10 **Projected rates of**
petroleum consumption in the United
States were calculated in 1956 by M. King
Hubbert. His analysis of reserves, production,
and consumption rates predicted a decrease
in production beginning in 1970, with
supplies declining to near exhaustion by the
year 2070. The major oil crisis that occurred
in 1974 substantiates his projections. U.S.
production (green and red curves) decreased
soon after 1970 but the amount of oil
produced in the United States has increased
slightly since 2008.

The energy crisis, however, was accurately predicted almost 60 years ago. In 1956 M. King Hubbert, an eminent research geologist for Shell Oil Company, analyzed the reserves, production, and rate of consumption of petroleum in the United States. From these data, he predicted a continuous decline in U.S. production beginning in 1971 (**Figure 24.10**). Take a moment and study these curves and observe that they show the decline actually started about 1975. If our actual reserves exceeded this estimate by a third, the huge increase would postpone the day of reckoning by only 5 years. More recent forecasts of world oil production also generally paint a gloomy picture. Most geologic forecasts predict that world oil production will peak sometime between 2015 and 2030. From then on, there will be less available in each year than in the previous year. According to these estimates, petroleum production could decline to near exhaustion by 2070. Obviously, if these predictions are correct, oil prices will rise. Analysis of the data by the U.S. Department of Energy and the International Energy Agency, on the other hand, predict no drop in world oil production until sometime after the end of the period they scrutinized (ending in 2035). Obviously, predicting any future events is difficult.

Visualizing the volume of a natural resource buried beneath the ground in limited parts of the world can be difficult. To get some idea of the finite nature of the world's petroleum, imagine the Great Lakes drained of water. Then imagine pouring into those basins all the oil we know of or anticipate ever finding. A small puddle in the Lake Superior basin—less than 5% of the lake's volume—would represent the world's oil for all time. In the future, coal might be expected to replace petroleum as the main hydrocarbon resource, but natural gas, solar energy, and hydrogen fusion will certainly be needed (**Figure 24.11**). It is quite clear, therefore, that we must greatly curtail our reliance on fossil fuels and replace them with other energy sources.

Renewable Energy Sources

Solar Energy   Solar radiation is the most important renewable, or sustained-yield, energy source. It has the added benefits of being clean. If our present understanding of the evolution of stars is correct, the Sun will continue to shine for the next several billion years. The major problem, of course, is that **solar energy** is distributed over a broad area. To be used effectively in our urban societies, it must first be concentrated in a small control center, where it can be converted to electricity and then distributed.

Solar energy can be converted to electricity in several ways. One technology uses a very large array of mirrors and lenses to focus the Sun's energy to heat water, turn it

Figure 24.11   A world without oil. The history of energy sources for humans is marked by changes from wood, to coal, to oil. Future sources may be natural gas, hydrogen, nuclear, solar, and other renewable sources.

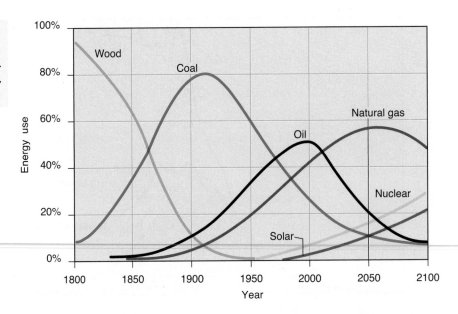

into steam, and use the steam to turn turbines that generate electricity. A more familiar, but technically more advanced, method involves the use of **photovoltaic cells** to convert sunlight directly into electricity (**Figure 24.12**). A variety of familiar electronic devices, especially calculators and outdoor lights, are powered by small arrays of these cells. The same technique is used to generate electricity for space satellites. Larger panels with many individual cells for collecting solar radiation have been mounted on homes to provide some of the electrical needs of a household, but collecting systems for large-scale solar energy use, although technically feasible, are not yet economical. Although the technology is improving, photovoltaic cells are relatively inefficient, and large collectors are still necessary. For example, to satisfy the present needs for electrical energy in the United States, a collecting system covering 7000 km² (about twice the size of the state of Rhode Island) would be required. Experimental systems with many solar panels have been built in California. Moreover, experimentation with new materials may lead to more efficient photovoltaic cells. Currently, conventional solar cells can extract only about 10% to 20% of the energy that falls on them, but specialized research versions extract nearly 45%.

More passive uses of solar power are becoming increasingly important for heating homes and businesses and producing hot water. Solar collectors can be mounted on south-facing roofs. These solar collectors consist of black panels filled with narrow tubes through which fluids circulate to collect heat from the Sun. Water can be heated and piped into the home, or air can be warmed by heat from the fluid in the collector.

Limitations on the usefulness of solar energy include the number of sunny, cloud-free days in an area. More critical is the inefficiency of the current solar collectors. Large-scale solar energy use, therefore, is a long way away, although local use in individual homes and buildings reduces the need for other forms of energy.

Biomass—materials of biological origin—provides another way to collect, store, and use solar energy. Crops can be grown specifically for the production of fuel. For example, corn can be converted into liquid alcohol for use in automobiles. Alternatively, woody plants, grains, and even municipal garbage can be combusted to drive electrical generating plants. This technology has the added benefit of reducing the amount of landfill needed for waste, but on the other hand, combustion and fuel production facilities create pollutants, including carbon dioxide, that enter the biosphere.

Hydroelectric Power   **Hydroelectric power** is another sustained source of energy. In a sense, this is also energy extracted from the Sun, because the gravitational potential energy acquired by water is provided by the Sun as it heats the ocean surface to produce buoyant air laden with water vapor. Subsequently, the kinetic energy of water flowing downhill is used to turn turbines that generate electricity. The power plants

What are the principal sources of renewable energy?

Figure 24.12 **Solar energy can be converted directly to electricity** in photovoltaic cells, but the process is still quite inefficient. Large arrays of cells must be used to produce electricity for modern urban societies. This experimental array of solar cells in California is being used to test the technology for large-scale production of electricity.

are usually built in dams or at waterfalls on rivers (**Figure 24.13**). Hydroelectric power is relatively inexpensive and clean. As a side benefit, the dams are important flood-control devices and the reservoirs behind them serve as sources of irrigation water. Hydroelectric power constitutes about 4% of U.S. energy consumption, but it has been developed in the United States to approximately 25% of its maximum capacity. With full development, hydroelectric power would still provide only about 15% of the energy needed in the United States. To reach this level of production, however, large areas of wilderness and farmland would have to be flooded. A problem with hydroelectric generators is that, when a river system is modified by a dam, many unforeseen side effects may occur, including the destruction of natural habitats upstream and downstream from the dam. Moreover, the reservoir behind the dam is only a temporary feature, destined to become filled with sediment. The useful life expectancy for most large reservoirs is only 20 to 200 years, so this source of energy is, in reality, limited.

Wind Energy   Another indirect form of solar energy can be extracted from the wind. Windmills have been used for centuries to drive grain mills and pump water from shallow aquifers. Only recently, however, has the wind been used to generate electricity on a significant scale. For example, Denmark has embarked on a vigorous plan to exploit **wind energy** and has a goal of producing 50% of its electricity in this way. The country currently derives 30% of their electrical energy from the wind. Individual generators are scattered across the countryside, but there are also large offshore wind farms. Elsewhere, large wind farms, with hundreds of windmills, have been constructed in remote areas as in eastern California (**Figure 24.14**). Passes between mountains that funnel winds to high velocities are particularly favorable locations. By 2020, California intends to generate one-third of its energy from renewable resources, including wind energy.

Figure 24.13 Hydroelectric power is produced by channeling dammed rivers through turbines inside dams such as this one. Hoover Dam impounds the Colorado River to form Lake Mead in Nevada and Arizona.

Figure 24.14 Wind energy can be extracted to produce electricity using modern propeller-driven turbines but has been used for centuries to drive windmills to pump water and grind wheat into flour. This field of turbines is in California.

Like many other solar-based energy sources, wind power has many advantages. It is pollution-free, releases no carbon dioxide or other greenhouse gases, requires no mining or processing of fuel, and has no radiation dangers. Moreover, virtually every country has plentiful wind resources that are free. However, individual windmills are moderately expensive, so wind energy is not an immediate solution to energy problems in developing countries. Moreover, much of our energy consumption is as portable fuels for automobiles and other transportation vehicles. Like all forms of solar energy, no portable fuel is a direct product of wind power, although electricity could be used to hydrolyze water and create gaseous hydrogen fuel. Wind power could also be used for electric vehicles if they become accepted and technical advances allow them to compete with gasoline-powered vehicles. In more-developed countries, the unseemly appearance, noise pollution, and large tracts of land set aside for power generation are of concern. If people come to accept these as trade-offs for a clean source of energy, wind power may become an increasingly important part of a diverse set of energy sources.

What limits the potential for wind energy to become an important energy source?

Geothermal Energy   Earth has its own internal source of heat, which is expressed on the surface by hot springs, geysers, and active volcanoes (**Figure 24.15**). In general, temperature increases systematically with depth, at a rate of approximately 3°C/100 m at shallow levels. Temperatures at the base of the continental crust can range from 400°C to 1000°C, and at the center of Earth they are perhaps as high as 4500°C.

Figure 24.15 Geothermal energy stored in heated groundwater is extracted in facilities such as this one in New Zealand to generate electricity.

Unfortunately, most of Earth's heat is far too deep to be artificially tapped, and the heat we can reach by drilling is typically too diffuse to be of economic value. Like ore deposits, however, **geothermal energy** can be concentrated locally and has been used for years in Iceland and in areas of Italy, New Zealand, and the United States. In most cases, this thermal energy is focused by hot groundwater or steam circulating at shallow levels in permeable rocks (**Figure 24.16**). Geothermal energy is concentrated where magma is relatively close to the surface. In addition, deep circulation of groundwater along normal faults in rifts produces concentrations of hot springs used to produce electricity. Geothermal systems are similar to hydrothermal fluids that produce ore deposits. Hot springs and geysers are common manifestations of buried geothermal systems. These geothermal areas are most abundant above subduction zones, along continental rifts, and above mantle plumes.

Geothermal systems require a combination of special conditions. In addition to a large source of concentrated heat, such as a shallow magma chamber, a commercial geothermal system typically has to have a large reservoir of permeable rock with ample fractures and other pore spaces filled with water and steam. The heated water convects through the pore spaces, becoming hot as it nears the heat source, then rising and eventually cooling. In the most favorable settings, the permeable zone is capped by rocks of low permeability. This seal helps to contain the fluid, so that all of the energy is not expelled by convection of the water to the surface. Where geothermal water temperatures are above 150°C, electricity can be produced by pumping the fluids to the

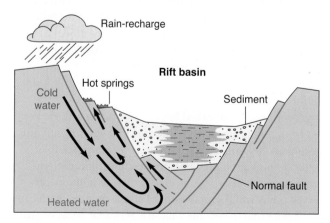

(A) In a fault-bounded rift basin, cold, near-surface water flows to great depth, where it is heated, and eventually returns to the surface along faults.

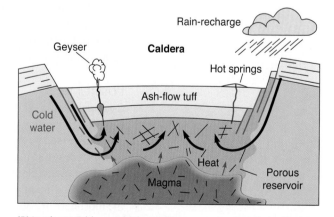

(B) In a large caldera, cold descending groundwater is heated when it gets near a hot magmatic intrusion. Hot springs and geysers form where the water returns to the surface.

Figure 24.16 Geothermal energy can be extracted from subsurface hot water tapped by drilling. The hot water forms by the deep circulation of water along normal faults in rifts or by heating above shallow, still-molten magma bodies.

Can geothermal energy provide all of our future energy needs?

surface, where they flash to steam at low pressures. This energy can be used to drive a turbine and produce electricity. Cooler geothermal waters can be used for space heating. The homes in Reykjavik, Iceland, for example, are heated by geothermal water piped throughout the city. Elsewhere, geothermal water is used to heat greenhouses for growing vegetables and flowers year-long.

The United States, Philippines, Mexico, and Italy lead the world in the production of electricity from geothermal power plants. A few large geothermal systems, such as the Geysers of northern California or the Lardarello field of Italy, produce enough energy to power large cities, but geothermal energy currently accounts for about 0.2% of that used worldwide. Estimates indicate that, at its maximum worldwide development, geothermal energy would yield only a small fraction of the world's total energy requirements.

Geothermal energy has several advantages over more traditional sources of energy. First, electricity produced by geothermal systems is relatively nonpolluting. Geothermal plants do not produce air pollution or carbon dioxide as do plants that burn oil or gas. Second, geothermal energy is renewable in the sense that the heat source is typically long-lived. A large magma chamber may take thousands of years to cool below the temperature needed to sustain an active geothermal system. A more immediate cause of the death of a geothermal system may be artificial or industrial extraction of the groundwater or steam faster than it can be resupplied. A diminishing supply of steam related to overproduction in the Geysers geothermal system of northern California limits the energy output of the power plants there. On the negative side, hot geothermal fluids carry high concentrations of dissolved salts and metals and are very corrosive to the pipes and other equipment that contains them. The discharge of the hot geothermal fluid can pollute neighboring streams, lakes, or aquifers. Moreover, as in any case where groundwater is removed, the withdrawal of geothermal fluids can cause subsidence of the surface over the circulating system.

Tidal Energy   Another sustained energy source is the ocean tides. **Tidal energy** can be harnessed by a dam built across the mouth of a bay where the tidal range is high. At the narrowed entrance to the bay, the rise and fall of the tides produce a strong tidal current that can be channeled through the dam and used to turn turbines that generate electricity. Tidal power cannot be generated along most continental margins, however. It is practical only where the tidal range is large, greater than 8 m or so. Large tidal ranges are enhanced by narrow, nearly enclosed bays. Tidal power plants have been built along the Bay of Fundy in Nova Scotia, Ireland, Russia, and China; the largest facility is in France at the mouth of the Rance River. No tidal power plants exist in the United States, but test facilities exist in Oregon, Maine, and New York. With maximum development, tidal power could supply 15% of the energy needed in the United States. Tidal power is renewable, as long as the Moon continues its gravitational tug-of-war with Earth's oceans; consequently, it is also more consistent than wind and solar plants. Moreover, it can be extracted without air or water pollution.

Fossil Fuels

Coal, petroleum, and natural gas commonly are called **fossil fuels** because they contain solar energy preserved from past geologic ages. The idea that we currently use energy released by the Sun more than 200 million years ago may seem remarkable. Energy from the Sun is converted by biological processes into combustible, carbon-rich substances (plant and animal tissues). This organic matter may be subsequently buried by sediment and preserved. Only small proportions of all the organic matter in the biosphere are buried with the potential to become a natural resource. Most organic materials decay by combining with the oxygen in the atmosphere to produce carbon dioxide and water. In a manner of speaking, the storage of this ancient solar energy is a type of savings account we inherited from the distant past.

Coal   Extensive **coal** deposits originate from plant material that flourished in ancient temperate swamps, typically found in low-lying floodplains, deltas, and coastal barrier islands. Modern examples include the coastal swamps of Sumatra and the Great Dismal Swamp along the coast of Virginia and North Carolina. In this area, the lush growth of vegetation has produced a layer of **peat** more than 2 m thick, covering an area of more than 5000 km². In this environment, the layer of peat will eventually be covered with sand and mud from an adjacent lagoon and beach, as sea level slowly rises (**Figure 24.17**). Because of increased temperature and pressure from the overlying sediment, water and organic gases are cooked and squeezed out of the plant debris, causing the percentage of carbon to increase. By this process, peat is compressed and is eventually transformed into coal in a series of steps (**Figure 24.18**). The sequence lignite, subbituminous, bituminous, and anthracite marks this increase in rank, or metamorphic grade, of coal. Experimental studies show that this process proceeds at about 200°C to 300°C; some anthracite coals have experienced even higher temperatures during tectonic folding and low-grade metamorphism. The coal itself is a complex mixture of graphitic carbon and more complicated hydrocarbons—hydrogen-carbon compounds made of molecular chains and rings. Some coals contain abundant plant fossils, including bark, leaves, and wood. If sea level rises and falls repeatedly, a series of coal beds can develop, interbedded with beach sand and near-shore mud.

Coal deposits are restricted to the latter part of the geologic record, when plant life became plentiful. The most important coal-forming periods in Earth's history were

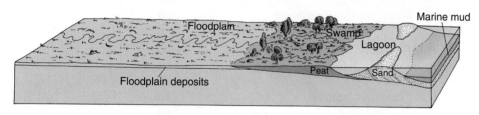

(A) The sequence of sedimentary environments along a coast grades seaward from floodplain to swamp and lagoon, to barrier bar, to offshore mud.

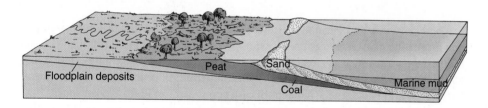

(B) As the sea expands inland, each sedimentary environment shifts landward. Swamp vegetation is deposited over previous floodplain deposits, the sand of the barrier island is deposited over the previous swamp (peat) muck, and marine mud is deposited over the coastal sand. The heat and pressure of the overlying sediment change the peat to coal.

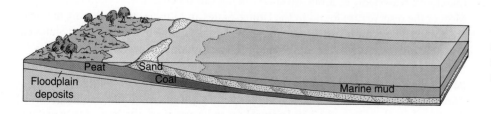

(C) Continued expansion of the sea superposes swamp deposits over floodplain sediments, beach sand over swamp material, and mud over beach sand.

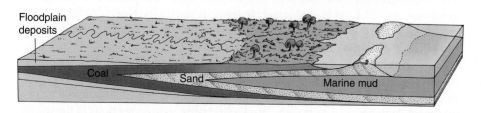

(D) As the sea recedes, the sequence is reversed: The sand of the barrier island is deposited over offshore mud, coal is deposited over the sand, and floodplain sediments are deposited over coal. By expansion and contraction of the sea, layers (lenses) of sediments and coal are thus deposited in an orderly sequence.

Figure 24.17 Coal deposits are commonly formed by the expansion and contraction of a shoreline and involve the associated movement of a swamp and barrier bar.

Figure 24.18 **The origin of coal** involves burial, compaction, and induration of plant material. The process begins in extensive swamps. Plant material produced in the swamp decomposes to form peat (about 50% carbon). Subsidence causes the peat to be buried with sediment, and the resulting increase in temperature and pressure compacts the peat, expelling water and gases and thus forming lignite and brown coals (about 72% carbon). With continued subsidence and deeper burial, the lignite is compressed into bituminous coal (about 85% carbon). Further compression (commonly induced by tectonism) drives out most of the remaining hydrogen, nitrogen, and oxygen, producing anthracite coal, which is about 93% carbon.

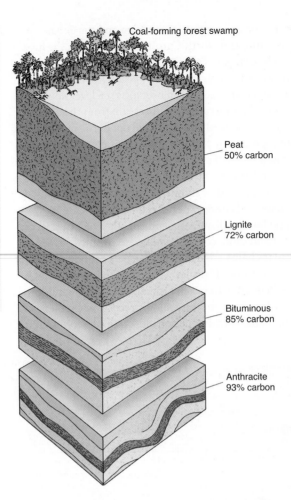

Coal-forming forest swamp

Peat
50% carbon

Lignite
72% carbon

Bituminous
85% carbon

Anthracite
93% carbon

What problems result from relying more and more on coal as an energy source?

the Pennsylvanian and Permian Periods about 300 million years ago. Great swamps and forests then covered large parts of most of the continental platforms. (The Western industrial nations, such as Great Britain, Germany, and the United States, developed with energy from these coals.) Other important periods of coal formation were the Cretaceous and Paleogene. Coal deposits are nonexistent in Precambrian rocks.

Coal is an important energy resource because there are large reserves of it. With the completion of at least reconnaissance geologic mapping of most of the continents, all of the major coalfields are believed to have been discovered (**Figure 24.19**), and therefore a reasonably accurate inventory of the world's coal reserves can be made. Most of the reserves are found in the United States and Russia, so developing nations are not likely to grow by using vast amounts of coal. Coalfields in the United States can probably sustain our current rate of energy consumption for several hundred years.

Just over 20% of U.S. energy needs are met by using coal (Figure 24.9). In the early part of this century, coal played a much more important role as an energy source, providing more than three-quarters of the country's needs. With the technological developments that allowed oil and natural gas to be exploited, the importance of coal diminished. However, with the depletion of oil and gas, we are again witnessing a shift to reliance on coal. Coal is used largely to generate electricity. Undoubtedly, increased use of coal will bring with it serious environmental problems related to **strip mining** and acid rain produced by burning sulfur-rich coals.

Coal is probably the dirtiest of all fossil fuels. Considerable quantities of sulfur compounds are emitted by burning. Left unchecked, this sulfur is released into the atmosphere and there combines with water to make sulfuric acid. This acid is an important component of **acid rain**. Two highly industrialized portions of the world, eastern North America and northern Europe, have rain with a pH of less than 4.2. Normal precipitation has a pH of about 5.7. Where the acid precipitation is not neutralized by reaction with carbonate rocks, the pH of streams and lakes also drops.

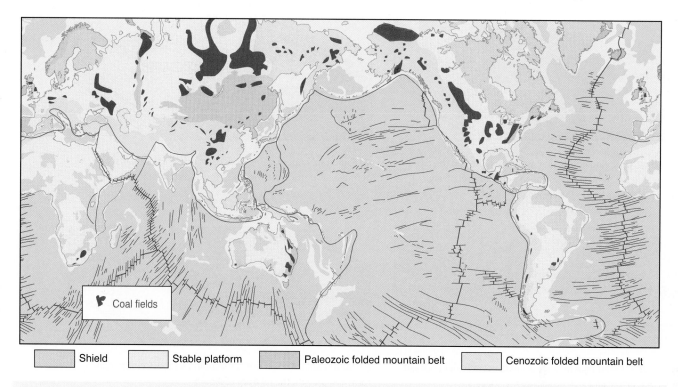

Coal fields

| | Shield | | Stable platform | | Paleozoic folded mountain belt | | Cenozoic folded mountain belt |

Figure 24.19 **The principal coal deposits** are concentrated in fluvial and deltaic sedimentary rocks on the continental platforms.

Acid buildup is blamed for the near extinction of fish populations in southern Scandinavia. Fortunately, new technologies allow sulfur to either be removed from coal before it is burned or scrubbed from smoke before it is released to the atmosphere. Environmental acids are also created by the mining of coal. Pyrite is a common mineral in many coal layers. When it is exposed to oxygen-rich groundwater in open mines, the pyrite dissolves to create sulfuric acid that enters stream drainages (**Figure 24.20**). This is one form of **acid-mine drainage**. Present mining regulations in the United States and Canada have largely alleviated this problem. Another problem with the use of coal as a source of energy is that large volumes of ash are left after burning. Ash may be as much as 30% of the coal. It consists of clastic material such as clay and quartz that were included in the ancient swamps. Large quantities of gypsum sludge are also created in removing sulfur from the smokestacks. Other problems with coal mining are related to the destruction of the natural landscape in strip mines and subsidence over some underground mines.

Petroleum and Natural Gas   Petroleum and natural gas are hydrocarbons. Natural gas consists largely of the simple organic molecule known as methane (CH_4). Crude oil is made of larger and more complex molecules composed only of hydrogen and carbon linked together in chains and rings. In contrast to coal, the hydrocarbons forming oil and gas deposits originate largely from microscopic algae and other plants and animals that once lived in the oceans and in large lakes. The remains of these organisms accumulated with mud on the seafloor. Because of their rapid burial, they escaped complete decomposition.

Deposits of oil and gas form if several basic conditions are met. *First*, the source beds must have sufficient organic material in fine-grained sediments. The environment of deposition must be poor in oxygen to prevent destruction of the organic materials before burial. *Second*, the beds must be buried deep enough (usually at least 2 km) for heat and pressure to compress rock and cause the chemical transformations that break down organic debris in the **source rocks** to form hydrocarbons. Oil generation usually begins when temperatures reach 100°C to 150°C. Between 150°C and 200°C, methane (and the mineral graphite) is produced as complex organic molecules break

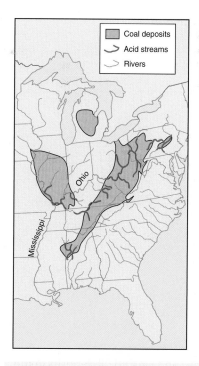

Figure 24.20 **Streams in the eastern United States became acidic** when pyrite was oxidized during the weathering of coal mines. Modern environmental regulations have returned these streams to more normal conditions.

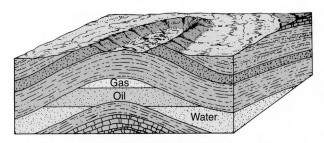

(A) Anticline. Oil, being lighter than water, migrates up the dip of permeable beds and can be trapped beneath a relatively impermeable shale bed in the crest of an anticline.

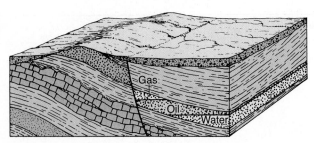

(B) Fault trap. Impermeable beds can be displaced against a permeable stratum and then trap the oil as it migrates up dip.

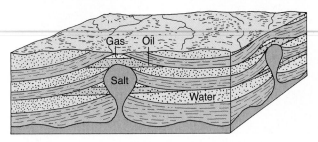

(C) Salt dome. Oil and gas may accumulate near the flanks of salt diapirs that pierce and arch up sedimentary layers.

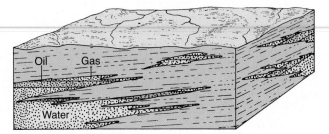

(D) Stratigraphic trap. Shale surrounding a sandstone lens can form and prevent the oil from escaping.

Figure 24.21   The accumulation of oil and gas requires (1) a reservoir rock (a permeable formation, such as a porous sandstone) into which the petroleum can migrate and (2) a barrier (an impermeable cap rock) to trap the fluids. Some of the geologic structures that trap oil and gas are shown here.

down to form simpler ones. *Third*, the materials must migrate from the scattered pores in the source beds to become concentrated. Once formed, the hydrocarbon fluids are squeezed by compaction out of the shales. Because of their low densities, these fluids generally migrate upward from source beds into more porous and permeable rock (usually sandstone or porous limestone or dolomite) called **reservoir rocks**. *Fourth*, as the oil and gas migrate through the reservoir beds, a physical barrier, or **hydrocarbon trap**, must cause the oil or gas to accumulate. If the reservoir beds provide an unobstructed path to the surface, the oil and gas seep out at the surface and are lost. This is one reason why most oil and gas deposits are found in relatively young rocks. In older rocks, there has been more time for erosion and Earth movements to provide a means for the oil and gas to escape. Traps in the path of oil's upward migration can result from a variety of geologic conditions, such as those shown in **Figure 24.21**. Most traps involve some sort of permeability barrier. For example, shales are much less permeable than sandstones and provide effective cap rocks for many oil and gas deposits. Exploration for oil and gas, therefore, is based on finding sequences of sedimentary rocks that provide good source and reservoir beds and then finding an effective trap. Gas, oil, and water all migrate together, but once trapped, they separate from one another on the basis of their densities. Natural gas is the lightest and fills the pore spaces in the uppermost part of the reservoir; oil occupies an intermediate position and floats on water (Figure 24.21).

Why can't we find oil deposits in the shields?

If any component of this sequence is missing, a hydrocarbon deposit will not form. For example, only about 250 of the 800 sedimentary basins produce oil or gas (**Figure 24.22**). Some basins may not have organic-rich source rocks. Others have adequate organic materials, but the sediments have not yet been buried deep enough for oil and gas to form. In many others, oil and gas formed but have subsequently leaked from the rocks because traps were inadequate. Groundwater may permeate a basin and strip the reservoirs of oil. The age of source rocks is not as critical as for coal deposits,

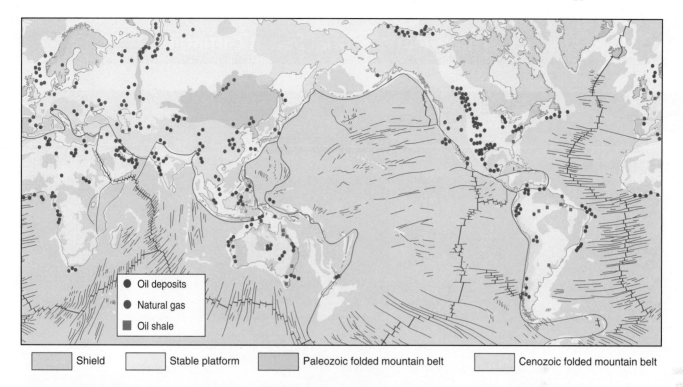

Shield | Stable platform | Paleozoic folded mountain belt | Cenozoic folded mountain belt

Figure 24.22 **Major oil and gas fields and deposits of oil shale** are found on all of the continents except Antarctica. Most form in thick sedimentary deposits on continental crust. The rifted margins of Africa, South America, and Europe are important oil-producing environments. Other important fields lie in the sedimentary basins of the platform and the flanks of folded mountain belts.

Figure 24.23 **Oil wells** such as this in the North Sea are important symbols of our modern society. Trillions of dollars of oil and gas are produced each year. Oil is the most expensive basic commodity of our modern society.

but because of progressive leakage, young sedimentary rocks are more productive than older ones. Algae, the principal source of oil, have been on Earth for a very long time and oil of Precambrian age is known.

Oil and gas are commonly extracted from Earth through holes drilled through the trap rock and into the permeable reservoir. This simple fact makes oil and gas much more economical to produce than many types of solid mineral resources. The fluids may be under sufficient pressure to flow to the surface on their own, or they may be pumped out (**Figure 24.23**). In some fields, water, natural gas, or even steam is pumped into the subsurface to flush more oil or gas from a reservoir.

STATE OF THE ART   Three-Dimensional Seismic Imaging

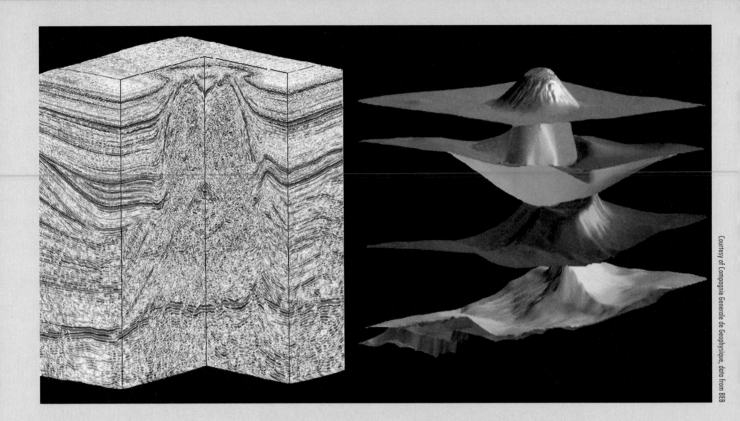

Courtesy of Compagnie Generale de Geophysique, data from BEB

Geologists are trained to look at a two-dimensional surface, such as a landscape or road cut, and construct a mental image of the third dimension. For example, a team of geologists may make a geologic map showing the distribution of various geologic units on the surface—a two-dimensional view (as described elsewhere in the text). They may then, by a process of interpretation, draw a vertical cross section showing how the outermost layers, faults, and folds extend below the surface where no one can see. Every geologist longs for a type of "X-ray vision" that would reveal what the structure of the interior is really like. Seismic investigations are one way that this can be done.

Seismic geophysicists specialize in these techniques and their skills are widely employed in oil exploration. A marine survey is conducted on a ship with an air gun that sends a series of acoustic (sound) pulses toward the bottom of the sea. This energy travels as compressional waves to the seafloor, where some of it enters the sediment and rock as seismic waves. Portions of this seismic energy are reflected back to the surface by the layers of sediment. Sensors towed behind the ship detect the reflected energy and time its arrival back at the surface. In this way, an image of the discontinuities (usually faults and sedimentary layers) in the crust can be constructed. Such information is usually shown as a vertical cross section—a *seismic section*—much like one that might be constructed from interpretation of a map.

In the last 10 years, the capabilities of computer processing have increased so rapidly that a single vertical section is no longer the goal of a seismic survey. Instead, the surveyors make a series of traverses across a region to create an array of sections that can be combined in a computer to construct a three-dimensional image that reveals the subsurface geology.

Using powerful computers, these blocks can be sliced anywhere and at any angle to allow a geologist to see the structure from any perspective. Multiple "maps" can be viewed; vertical sections can be stacked together and played back like a movie while a geologist traces faults or distinctive beds through the three-dimensional image. In fact, some oil companies have developed special rooms with multiple projectors that allow teams of geologists with special headsets to "walk through" a virtual-reality model of the crust. They explore details of the structures that might help them find deposits of oil that can later be drilled and exploited. The diagram above is a three-dimensional seismic image sliced to show the subsurface structure of a salt dome in Germany. The layers of sedimentary rock show up as red, blue, and white stripes, cut by the irregular dome. The colored planes to the right show the configuration of several deformed layers as revealed by the seismic image. The "surface" of this block is actually a horizon at depth. These tools have dramatically changed the way an oil company looks for new deposits and increased their success rate.

A recent boom in natural gas production has been driven by the exploitation of **shale gas**—natural gas trapped within fine-grained shale (**Figure 24.24**). This gas can be extracted by **hydraulic fracturing** (*fracking* for short). In this process, water is injected deep into layers of organic-rich shale to create small fractures. The injected fluid is laced with sand grains and chemical agents that prop open the cracks and allow the natural gas in the shale to accumulate in the pores and then be extracted. Without the fracking process, the gas would remain in the tight, nearly impermeable shale unable to flow into a structural or stratigraphic trap. Shales with the right amount of organic material and at the right depths are relatively common, including beds that underlie Texas, North Dakota, Pennsylvania, and New York, as well as many other countries.

In some geologic environments, hydrocarbons remain as solids in the shale in which the organic debris originally accumulated. These deposits are known as **oil shales**. They are reservoirs of oil that may become important in the future. There are huge reserves of oil shale (Figure 24.22). The United States has more than 10 times more oil that could be recovered from shale than it does from conventional wells. Most of the oil shale in North America lies in the Green River Formation of Colorado, Wyoming, and Utah. The shale was deposited in a series of shallow middle Cenozoic lakes. The problem with all oil shale is that it must be mined and heated to extract the oil. This process requires considerable energy and water resources and is not yet economically feasible. Reclamation of strip mines and the safe disposal of processing wastes are also costly. Another similar resource that is currently being exploited is **tar sand**.

What are the limitations of using oil shale as an energy source?

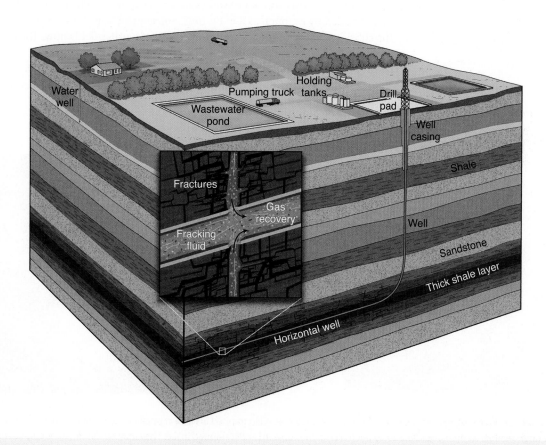

Figure 24.24 **Shale gas** can be extracted through the process of hydraulic fracturing (or "fracking"). First, a vertical hole is drilled to reach a shale layer rich in natural gas. The drill hole is then turned and continued horizontally through the shale for a kilometer or more. Water, sand and various additives are then pumped under high pressure into the drill hole. This fractures the shale and releases the natural gas that is then pumped out of the hole. Waste water is held in ponds at the surface or trucked offsite for disposal.

In Alberta, Canada, for example, hydrocarbons are found in loose, unconsolidated sand deposits. The sand is mined in large open pits or hot steam is used to extract the oil from the sand. For every barrel (42 gallons or 160 liters) of crude oil produced, one metric ton of sand is mined. The oil from these sands is a large fraction (7%) of the total United States use and exceeds the amount of oil imported from Saudi Arabia, for example.

Oil and gas are convenient forms of energy because they are easy to handle and transport. Currently, about 36% of the U.S. total energy needs comes from petroleum, and another 25% comes from natural gas (Figure 24.9). Unfortunately, at the present rate of consumption, the known reserves of oil will be depleted in about 50 years. In the United States, proven oil reserves are only about three times our annual consumption. In part because of the extraction of shale gas by hydraulic fracturing, world gas reserves are projected to last much longer.

The real lifetime of our oil and gas reserves depends upon the further development of technology, energy conservation efforts, world politics, and of course, the price we are willing to pay. If the current trends continue, we may need to begin large-scale gasification and liquefaction of coal and oil shale deposits and to rely more on coal- and gas-fired electrical plants. Since natural gas burns cleaner than oil or coal and produces less greenhouse gas, its use may be a bridge into a future that relies on nuclear, wind, and solar energy. Clearly, we should use renewable sources of energy whenever possible.

Environmental problems associated with oil and gas production are not as severe as for coal mining. However, oil extraction can lead to subsidence, just like that caused by the extraction of groundwater. For example, parts of Long Beach, California, subsided by as much as 8 m by 1967. Subsequently, water injection has stopped the subsidence. Injection is a common practice in other fields now. The extraction of shale gas is not without problems and controversies. Some people are concerned that the injected fluids could contaminate water supplies found in aquifers overlying the shales. In a few cases, some of the liberated natural gas escaped extraction and flowed

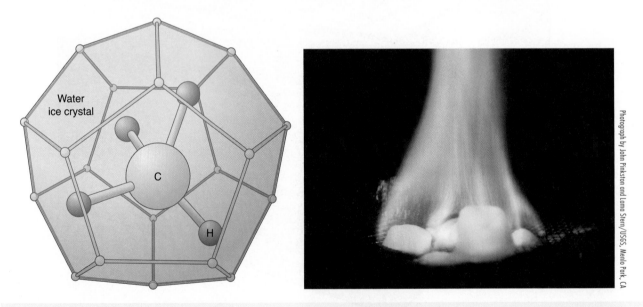

Photograph by John Pinkston and Laura Stern/USGS, Menlo Park, CA

Figure 24.25 Methane hydrates consist of a methane molecule (CH_4) trapped inside a crystalline "cage" of frozen water (left). They form on the deep seafloor as methane released by bacteria bubbles upward and is trapped in ice. They may be abundant enough to form an important source of natural gas and will burn if ignited with a match (right). However, they will be difficult to mine and pose environmental hazards because methane is a greenhouse gas.

upward, contaminating water supplies. Finally, oil spills are another environmental problem; they can happen during production and during the transportation of oil in pipe lines and large sea-going tankers. The magnitude of this threat is exemplified by the Deepwater Horizon oil spill in the Gulf of Mexico. A fire and explosion on a drilling platform in the summer of 2010 killed eleven men and released a literal gusher of oil into the ocean. Oil streamed out of a broken pipe near the seafloor for three months before it was finally capped. Five million barrels of oil were released. This is the equivalent of 25% of the United States daily use. The oil contaminated deep and shallow waters and washed onto the shorelines for months. In the end, legal penalties to the oil companies involved were about $4 billion; another $4 billion was paid out in settlements for damages.

Methane Hydrates   This icy substance forms where rising bubbles of methane gas, given off as bacteria digest organic matter in mud, react with cold seawater (**Figure 24.25**). The ices are concentrated in a layer about 100 m thick where pore spaces between sediment particles are filled with the hydrate. Gaseous methane is trapped beneath this impermeable cloak.

Methane hydrates were first discovered on the seafloor in the 1970s. They are now known to cover vast areas of the seafloor where the water is deep and cold enough. These ices may contain twice as much carbon as all other petroleum, coal, and natural gas sources put together. However, the highly volatile hydrates will be very difficult to mine. The gas cannot simply be drilled into and then pumped on shore. Even if a way can be devised to economically extract these valuable materials from the seafloor, there are concerns about the effect of releasing methane into the atmosphere, either during mining or if seawater warms by continued global warming.

Methane is a greenhouse gas and decomposition of hydrates may contribute to global warming. In turn, release of methane gas from the icy hydrates could be triggered by warming of the seafloor related to climate change. Several large submarine landslides may have been triggered when hydrates decomposed as the ocean warmed at the end of the last ice age. One landslide on the continental slope off Norway slid 800 km and probably induced a large tsunami that inundated Scotland. Elsewhere, craters as large as 700 m across pockmark the floor of the polar Barents Sea and may mark places where methane gas vented explosively to the surface as climate changed. It is conceivable that a large gas release could warm the planet further and help to create other landslides and release more gas from the seafloor, forming a complicated feedback cycle. Some scientists have speculated that a sharp rise in sea temperatures, extinctions of tiny marine animals, and fossil composition changes were all related to a massive release of seafloor methane 55 million years ago.

Nuclear Energy

The ever-increasing demands for energy and the decreasing supply of fossil fuels put the spotlight on nuclear power as an answer to our energy requirements. The technology of nuclear energy production is well developed. Nuclear energy is commonly generated by the controlled fission of uranium-235 in what is known as a nuclear reactor. The fission is produced by bombarding the uranium in a fuel rod with neutrons. As a result of the splitting of the uranium nucleus into lighter isotopes of other elements, a small mass is converted to heat energy. This heat is used to drive steam turbines that generate electricity.

The key element in the development of nuclear energy is uranium, because it can be induced to split by fusion. The average uranium content in the rocks of Earth's crust is only 2 parts per million. At current prices, uranium concentrations must be as high as 5000 parts per million before the ore can be mined profitably. Uranium is concentrated by a variety of igneous, metamorphic, and sedimentary processes. For

What are the advantages of nuclear energy?

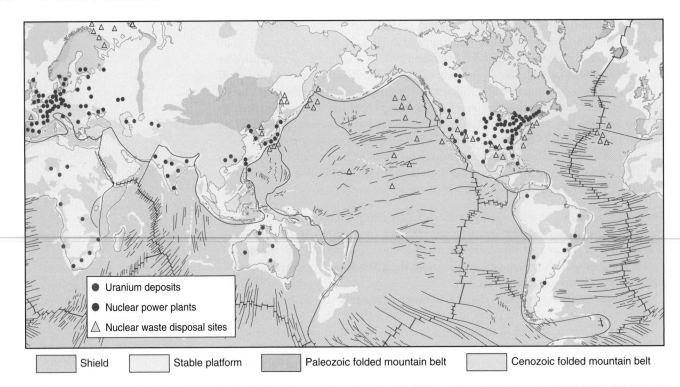

Figure 24.26 Uranium deposits and nuclear reactors may become a key part of the future production of energy, but only if the public becomes convinced that nuclear reactors are safe and that their waste products can be safely stored in repositories for long periods of time.

example, uranium is concentrated in rhyolitic magma by fractional crystallization. Explosive eruptions may shower the rhyolite as ash over large areas, where it weathers by reactions with the atmosphere and water. As the glassy ash reacts with oxygen, the uranium becomes oxidized and water-soluble. Uranium ions are then leached out of the rock and transported by surface water and groundwater. It can later be deposited and highly concentrated if a barrier to its transport is encountered. In this case, the barriers are usually chemical rather than physical. In permeable sedimentary rocks, uranium may be absorbed by clay minerals and reduced to its insoluble state by carbonaceous organic matter. The rich uranium deposits in the Colorado Plateau of the western United States probably formed in this way and are concentrated in ancient stream channels, especially where fossil wood is found. Other important uranium deposits form as vein deposits in granite and as placers of heavy uranium oxides in ancient (older than about 2.2 billion years) stream channels. Canada, Russia, the United States, Australia, Namibia, and France are all important producers of uranium ores (**Figure 24.26**).

Contemporary society, however, has hesitated to move toward large-scale production of nuclear energy because of the possibility of environmental problems and reactor accidents. Radiation hazards during mining and energy production, problems of waste disposal, and thermal pollution of fresh and marine waters, as well as potential terrorist activities, are among the greatest concerns. For example, if the rate of fission is not carefully controlled, the reaction may become a rapid and destructive chain reaction. Enough heat may be released to melt the floor of the reactor, dropping hot radioactive materials to the water table. If the water table is shallow, steam explosions, like those generated by magma-water interactions, could shower radioactive debris over large areas. Even venting steam from small accidents can release harmful doses of radioactive materials.

A small accidental leak of radioactive materials at the Three Mile Island reactor in Pennsylvania in 1979 turned public opinion against the use of nuclear energy in the United States. The failure of the Chernobyl reactor, in what is now Ukraine, showed even more people about the potential dangers associated with nuclear reactors. A partial meltdown at this reactor in 1986 released a cloud laden with radioactive by-products across much of central Europe. An estimated 50 people died of radiation exposure. Nuclear energy was regaining a level of acceptance, when another reactor accident turned public sentiment again. The megathrust earthquake that struck Japan in 2011 damaged the Fukushima nuclear plant when the tsunami overtopped barriers intended to protect it. The sea walls were about 6 m high, but the tsunami was about 9 m high. Seawater flooded emergency generators and cut off power to pumps that moved coolant through the hot reactor core. As a result, continued radioactive decay caused the fuel rods to melt in 3 of the 6 reactors. Small steam explosions dispersed radioactive material into the atmosphere; some of the material fell onto the surrounding land and sea. Water with dissolved radioactive elements was also dumped into the ocean. Fortunately, no one died as a result of exposure to radiation. Although such calculations are extremely difficult to make, it has been predicted that the release of radiation may lead to "extra" cancer deaths in the region—the numbers range from 0 to 100 in an overall population of several million. Significantly, a second set of reactors at Fukushima II escaped meltdown because their emergency generators were in a watertight reactor building and pumps for cooling had extra protection from flooding and were quickly repaired.

Only when these hazards and disposal problems are solved will nuclear energy become a more important source of energy. France is the one country that has moved toward large-scale production of electricity from nuclear energy. About 75% of its electricity is produced in nuclear reactors. Belgium (54%), Ukraine (47%), Switzerland (41%), Sweden (40%), South Korea (35%), Bulgaria (33%), and Finland (32%) all produce major proportions of their electricity in nuclear reactors. In contrast, the United States produces only about 20% of its electricity in this way and Canada only about 15%. A few countries, most notably Italy, have complete bans on the production of nuclear energy. After the accident in Japan, Germany also plans to phase out nuclear energy.

Plate Tectonics and Mineral Resources

> Earth's system of plate tectonics creates many different environments favorable for the creation of mineral deposits and accumulations of fossil fuels. Understanding these associations has led to the discovery of new deposits that sustain a high standard of living for many people.

Plate tectonics is important to understanding the genesis of and exploration for new mineral resources. Plate margins are where many of Earth's dynamic processes prevail, including magma intrusion, faulting, and folding. Moreover, the tectonic setting determines the type of sedimentary rocks that accumulate in a region. Even weathering and stream processes are controlled by a continent's climate belt, a product of plate movement. Following, we briefly review some of the most important types of mineral resources found in each of the four types of tectonic settings discussed earlier.

Divergent Plate Margins

Basaltic volcanism, normal faulting, and sedimentation in closed basins are the products of continental rifts. A distinctive suite of ore deposits forms because of these geologic processes (**Figure 24.27**).

For example, the great intrusions of mafic magmas developed in rifts have produced magmatic segregation deposits of sulfide ore rich in platinum and copper. Basaltic lavas erupted in continental rifts are also rich in copper that can be extracted by groundwater circulating on deep faults and later deposited in enriched zones as native copper. Many native copper deposits of northern Michigan formed in a rift that

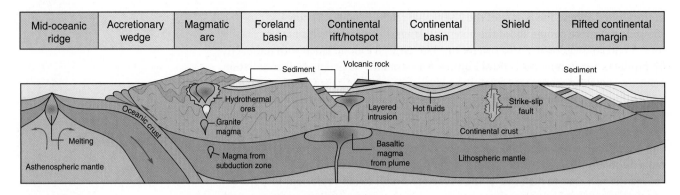

Mid-oceanic ridge	Accretionary wedge	Magmatic arc	Foreland basin	Continental rift/hotspot	Continental basin	Shield	Rifted continental margin

Magmatic segregation	**Oceanic ridge deposits in mélange**	**Hydrothermal vein deposits**	**Sedimentary**	**Magmatic segregation**	**Sedimentary**	**Metamorphic shear zone**	**Sedimentary**
chromium platinum	copper zinc nickel chromium	gold silver	coal oil oil shale gas	platinum copper, nickel	oil, gas, coal, salt	gold tungsten	oil, gas, coal, evaporites, beach placers
Hydrothermal sulfides		**Contact metamorphic**		**Hydrothermal**			
copper zinc		copper, lead, zinc, gold, silver, iron, tin, tungsten, molybdenum		copper, lead, zinc, silver	**Hydrothermal**		**Precambrian**
				Sedimentary	lead, zinc fluorite		Banded iron formations
				evaporites, brines sand, gravel			

Figure 24.27 Mineral resources are intimately related to the plate tectonic system. Some of the major ore-forming environments are shown on this schematic cross section. Plate tectonics controls igneous, sedimentary, and metamorphic processes and even the climate; it therefore exerts a major control on the types of ore deposits formed at any location and time.

is more than 1 billion years old. Proterozoic rifting in what is now the Yukon and the Northwest Territories of Canada also produced native copper deposits. In this narrow rift, more than 3 km of basaltic flows are overlain by 4 km of clastic sediments including conglomerates. The entire sequence is capped by evaporites to form a classic rift sequence of rocks.

In other continental rifts, the evaporites are the principal mineral resource. Rifts may contain evaporites related to playa lakes. Deposits of salt, soda, potash, and gypsum, as well as mineral-laden brines, are extracted from rifts in East Africa and the Basin and Range Province of the western United States.

Divergent boundaries on the ocean floor also have a characteristic suite of mineral resources that owe their origin to basaltic magmatism, high heat flow, deep fractures, and abundant seawater. For example, pods or layers of chromium and platinum minerals form in or below oceanic-ridge magma chambers. A slice of mineralized oceanic crust that is thrust onto a continent as an ophiolite may become an ore deposit.

In addition, seawater becomes a hot brine when it circulates through the oceanic crust and then vents onto the seafloor. Ancient hydrothermal mounds are important sulfide deposits sought for their copper and zinc. Only those brought to the surface in ophiolite complexes can be mined today. The copper deposits of the ophiolite on the island of Cyprus are a characteristic example of this seafloor mineralization.

Rifted continental margins gradually evolve into broad submarine shelves where shallow-marine sediment collects. The cooling and subsidence of the once-hot elevated margins are the important geologic controls here. Many of Earth's most important resources are formed in this divergent margin environment. Large deposits of oil, gas, and coal form in sediments deposited in tropical or subtropical environments such as this. During the Proterozoic, banded iron formations formed as sedimentary precipitates along continental shelves. At other times, ocean upwellings formed phosphorus deposits on the shelves that are now mined as fertilizer on the coastal plains of the eastern United States.

Many important beach placer deposits also form on rifted continental margins. Erosion of a continental shield and concentration by longshore drift have formed ore deposits of dense minerals with titanium, niobium, zirconium, and rare earth elements.

Convergent Plate Margins

The role of convergent margin processes in producing many deposits is exemplified by the distribution of porphyry copper deposits (Figure 24.1). Many hydrothermal deposits of copper and other metallic ores (molybdenum, lead, zinc, silver, and gold) are associated with diorite to granite intrusions formed above subduction zones. Porphyry copper deposits are found in island arcs and in continental volcanic arcs (Figure 24.27). In some cases, the volcanoes themselves host the ore deposits. The copper ores of western North and South America, Iran, Pakistan, Philippines, and New Guinea are examples of relatively young porphyry copper deposits.

The collision of two continents produces fewer ore deposits, but distinctive granites do form by partial melting of the continental crust. Hydrothermal and placer deposits of tin form in and near these granites. The tin deposits of Malaysia, Indonesia, and southeastern China may have formed in this tectonic setting.

Oil and gas deposits are commonly related to convergent plate boundaries. As noted above, organic-rich sedimentary rocks are deposited on rifted continental margins. In time, many of these rift-margin sequences become involved in plate collisions of various sorts. Convergent margin thrust systems may bury organic-rich sedimentary rocks to sufficient depths to cause oil or gas formation in sediments that otherwise would remain cool and unproductive. Deformation also creates pathways for the oil and gas to migrate and, ultimately, become trapped. The most obvious of these structures are found in fold belts, where anticlines may be natural petroleum traps.

Transform Plate Boundaries

Mineral resources along transform boundaries are represented by the oil and gas fields of southern California. Small pull-apart basins between offsets in the strike-slip faults of the San Andreas transform fault system are ideal for accumulation of clastic sedimentary rocks rich in organic materials. Continued faulting, subsidence, and sedimentation bury these strata to depths where oil and gas can form and migrate to a variety of stratigraphic and structural traps.

A few deep shear zones in the continental shields may be ancient transform faults (Figure 24.27). Some have gold deposited by hydrothermal fluids that passed along the permeable fault zones. Otherwise, metallic ore deposits are rare along transform plate boundaries, mainly because igneous rocks are also rare.

Intraplate Settings

Mineral deposits found in plate interiors are also distinctive. Two different types are worth considering here. The first type is found in stable platforms where sedimentary rocks accumulate in broad, slowly subsiding basins. Hydrothermal fluids formed here may concentrate a variety of elements, including lead, zinc, and fluorine. Important deposits of this type are found in Paleozoic sedimentary rocks of Missouri, Wisconsin, and Tennessee and in the Pine Point region of the Northwest Territories of Canada. Other sedimentary basins, such as the Witwatersrand of South Africa, contain ancient placer deposits of gold and, if old enough, uranium minerals. Intracontinental basins also contain many of the world's important evaporite deposits, including those in the Michigan Basin, central Europe, and the Paris Basin. These same sedimentary basins, if they were also filled with sufficient organic materials, may lead to coal, oil, and gas deposits.

The second type of resource is associated with hotspots or mantle plumes—in either oceanic or continental settings. The vast magma bodies formed from plume heads have been important sources of some metallic resources. For example, the large Noril'sk ore deposit of Siberia formed at the same time that the flood basalts of the Siberian traps erupted. The ores formed by magmatic segregation of nickel, copper, and platinum.

Mantle plumes beneath continents may also produce granites that have hydrothermal deposits of tin. Such granites are important sources of tin minerals that were further concentrated in placer deposits in Nigeria, for example.

What mineral resources would you expect to find in the area where you live?

Limits to Growth and Consumption

Rapid population growth and the associated industrial expansion cause consumption of natural resources to increase at an exponential rate. We are finding that there are limits to growth. These limits will probably be reached through the depletion of natural resources.

The consumption of natural resources is proceeding at a phenomenal rate. The rapid population growth that has prevailed during the last few hundred years is unprecedented (**Figure 24.28**). There are currently about 7 billion people on the planet. A population of 1 billion was not reached until 1830 CE. The population doubled in the next 100 years and doubled again in the next 45 years.

Just how many people can Earth hold? Estimates range widely, but most hover near 10 to 15 billion people, surprisingly close to the estimate of Leeuwenhoek, the Dutch naturalist famous for his use of early microscopes. In 1679 he suggested that Earth could sustain no more than 13.4 billion people. He based his estimate on the population and area of the Netherlands and extrapolated that population density over two-thirds of Earth's land area. Obviously, this must be an upper limit, because the Netherlands sustains such a high population density by importing many of its resources. However, if the growth rate for the last century persists, the population will reach his estimated limit in the next century.

What are the limits to growth?

The challenge is basically one of changing from a period of growth to a period of nongrowth. This change will require a fundamental revision of current popular economic and social thinking, which is based on the assumption that growth must be permanent and the even more basic assumption that growth must occur for society to prosper. It is important, therefore, to consider how resources are used and the rates at which they are being depleted. Perhaps the most critical point to make is that the rate of resource consumption increases exponentially. The exponential rate results both from population growth and from the growth of the average annual consumption per person, which increases yearly. In other words, growth is exponential because of an increasing population and a rising standard of living.

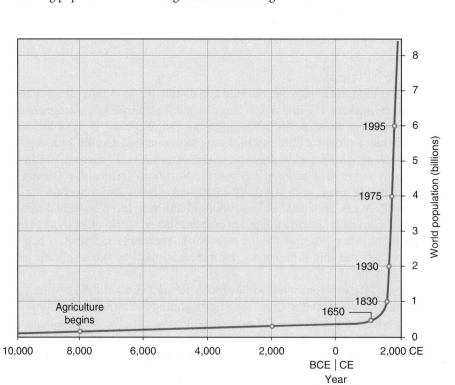

Figure 24.28 Earth's human population has grown exponentially. This growth has resulted in a similar increase in the rate of consumption of our natural resources.

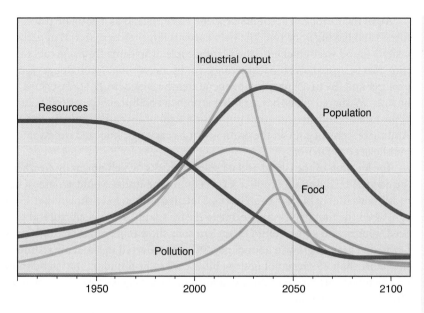

Figure 24.29 A computer model of resource consumption and its influence on other variables assumes no major changes in the physical, economic, and social relationships that historically have governed the development of the world system. All variables plotted here follow historical values from 1900 to 1970. Food, industrial output, and population grow exponentially until the rapidly diminishing resource base forces a slowdown in industrial growth. Because of natural delays in the system, both population and pollution continue to increase for some time after the peak of industrialization. Population growth is finally halted by a rise in the death rate due to decreased food and medical supplies.

Data from: D. H. Meadows and others, *Limits to Growth*, and Graham Turner

Studies show that the limits of population growth probably will not be imposed by pollution. The limits will be established by the depletion of natural resources. The projected interaction of some major variables, as the population grows to its ultimate limit, is shown in **Figure 24.29**. Food, industrial output, and population will continue to grow exponentially until the rapid depletion of resources forces a sharp decline in industrial growth.

On our finite Earth, unlimited population growth is impossible. In fact, the transition to a stable or declining phase has already begun. This does not pose insurmountable technological, biological, or social problems. It does, however, require some fundamental adjustments in our present growth culture. If we can achieve appropriate cultural adjustments, a steady-state population could be one of humanity's greatest advances. The alternative could be catastrophic.

Easter Island, Earth Island

The history of Easter Island shows the effects of overuse of natural resources and disregard for the limitations of a natural environment.

Three thousand kilometers from the western shores of South America lies Easter Island (Rapa Nui), a small volcanic island that is part of a long chain of hotspot islands and seamounts in the South Pacific (**Figure 24.30**). Easter Island formed like many other hotspot islands. Submarine basaltic eruptions gradually built a large volcano that eventually rose above the crashing waves. The most recent eruptions formed several small cinder cones that are less than 1 million years old.

The subtropical climate of the island ensures that it is warm and wet. The mean temperature is 20°C, and the island receives about 120 cm of rain annually. Weathering of the basaltic lavas created a rich volcanic soil. Eventually, the island was colonized by plants and animals floating or flying to this lonely island paradise. By the time the first Polynesians landed on the island between 400 and 700 CE, Easter Island was richly forested. Several species of trees, notably the date palm, bushes, and other plants blanketed the landscape. However, 1000 years later, when the first European visitors arrived, the landscape and environment of Easter Island were quite different. The climate had not changed. The basic geologic processes that shaped the island, produced soil, and allowed abundant rain to fall had not changed. But, apparently, its human population had heedlessly consumed its resources.

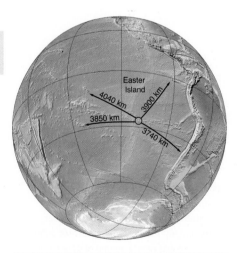

Figure 24.30 Easter Island provides lessons about the potentially catastrophic impact of humans on their environment. Easter Island is a small hotspot island that lies in the subtropical part of the South Pacific, about 3000 km from South America.

All of the European accounts describe Easter Island as a desolate place, lacking forests and having poor soil. The only animals larger than insects that early European visitors found were chickens. The island people, who came from a great seafaring culture, met early explorers in leaky canoes or by swimming. Jacob Roggeveen, a Dutch explorer and the first European to encounter the island, in 1722, described what he saw: "Withered grass, hay, or other scorched and burnt vegetation . . . its wasted appearance could give no other impression than of a singular poverty and barrenness." And there were, of course, the ominous stone carvings, knocked over, some left in the middle of construction.

The location of the island and its volcanic rocks, which normally weather to produce fertile soils, should produce a rich environment that would provide a healthy and prosperous life for human inhabitants. Detailed studies have shown that 1500 years ago, when the first Polynesians came to Easter Island, it was a subtropical paradise. Seed, spores, and pollen found in ancient soils show that palms, tree daisies, ferns, herbs, and grasses grew in abundance. Bones uncovered in the same sediments show that more than 25 species of nesting birds used to be found on the island (more than on any other Polynesian island). In addition, the fossil evidence shows that many sea birds bred on the island.

What happened to this island and its people? Ancient pollen found in the layers of soil give a detailed account. Recorded in the soil is the destruction of an entire ecosystem and the demise of a culture. The first Polynesians reached the island between 400 and 700 CE. The early colonizers enjoyed an abundance of food and lumber; in their abundance, the population increased. But the pollen records show dramatic changes in the plant life on the island as early as 900 CE. (**Figure 24.31**). In only a few centuries, the islanders had begun the process of overharvesting the forests, one of their most important natural resources. The fossil pollen shows that by the 1400s the Easter Island palm had disappeared, probably consumed by humans and unable to regenerate itself because introduced rats devoured the palm seeds. Other trees that were needed by the islanders to make ropes and fibers show a similar decline, although they took longer to disappear.

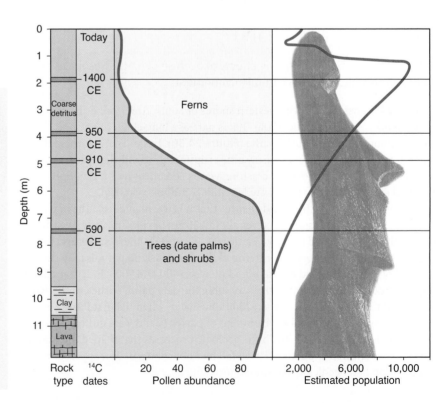

Figure 24.31 Pollen fossils collected from Easter Island paint a grim picture of the demise of the natural resource base for the people of Easter Island. Pollen was collected from cores taken from sediments that accumulated in a small volcanic crater. The record suggests that when people colonized the island about 400 CE, date palms were abundant, as indicated by the amount of palm pollen preserved in sediment. By 1000 CE, pollen records show, the island was nearly devoid of date palms. By the time Europeans visited the island in the 1700s, there were none. As a result of the overuse of their primary resources such as the date palm, the population of the island probably collapsed from a high of about 10,000 to as few as 1000 in the course of a few generations.

There were several reasons the inhabitants deforested the island and many unalterable consequences. People on the island used wood both for construction materials to make canoes and houses and for fuel. In addition, they cleared land so that they could plant crops to feed the growing population. Statue construction, which required logs and ropes made from tree fibers to transport the huge basalt images up to 10 km, was at a peak during the years 1200 to 1500. Deforestation led to destruction of plant and animal habitat. Land birds, which provided food for the islanders, disappeared with the forests. Soil erosion increased with the destruction of the forests, and, in turn, growing crops became difficult and ultimately impossible.

With dwindling forest resources, islanders appear to have turned to exploiting the bounties of the sea. Archeological remains suggest a shift in the islander's diets—from porpoise (which could be harpooned from good seafaring canoes), to sea birds and shellfish, to small sea snails, chickens, and rats. As food supplies declined and warfare broke out, the people of Easter Island fell to cannibalism—a fact recorded in the archeological remains and oral traditions of the people.

Hand in hand with the environmental degradation came the demise of the island's society. Excessive consumption within the society depleted the small island's natural resources and intensified deforestation. Resources decreased and food became scarcer; canoes could no longer be constructed, so people could not escape to other islands. A complex centralized society was apparently replaced with feuding and chaos. Warrior societies formed and, between 1770 and 1864, the huge stone statues were toppled by rival gangs. It is estimated that the human population of the island fell to between one-quarter and one-tenth of its prior size.

Some have asked, "Why didn't the Easter Islanders understand what they were doing and stop before it was too late? What were *they* thinking when they cut down the last palm tree?" But a better question might be, "What can *we* learn from the history of Easter Island?" In myriad ways, the history of Easter Island is a microcosm of our own planet. Viewing Earth as our island in the vastness of space, we can see a potential repeat of the disasters that happened on this lonely Pacific island. Today, we see a human society that largely disregards the long-term effects of its actions and increasingly consumes many of its natural resources. The people of Easter Island had nowhere to go once they exhausted most of their resources. And where do we, the inhabitants of "Earth Island," have to go if we exhaust our resources and change our global environment?

GeoLogic   The Origin of Coal

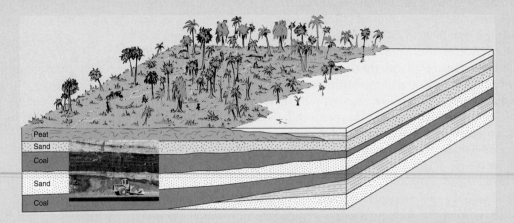

Thick layers of black coal are some of our most valuable resources. Many also preserve abundant evidence of their origin. Careful observations of the character of the coal and its enclosing rocks allow us to understand the sequence of events involved in its formation.

Observations

1. Coal is made of the carbon mineral graphite mixed with a variety of organic molecules, many of which have counterparts in still living materials.
2. Hosts of plant fossils are still clearly visible in most coal layers—tiny grains of pollen, leaves, bark, and wood from large trees related to those found today in tropical and temperate rain forests.
3. Coal beds are universally contained within layers of sedimentary rock, clearly indicating their formation and accumulation at the surface.
4. The thicknesses of the overlying strata show that coal beds were buried to depths of several kilometers.

Interpretations

With these facts in mind, what do geologists see when they look at a coal seam like that mined here? Many envision a vast nearly impenetrable swamp with closely spaced trees with their roots standing in water. Rotting plant material forms a black organic-rich mud in the ponds. Fallen trees clog the forest and thick layers of leafy litter are everywhere. The ocean is not far away and the quiet lapping of the waves can be heard. Nearby, a slow-moving river wends its way to the ocean and some of its banks are lined with white sand.

The fossils and organic molecules show that coal is plant derived. The sands and shales that enclose coal deposits were formed in beach, delta, and shallow marine environments. They typically show the shoreline was moving back and forth across a subsiding continental margin when the coal layer was deposited. Eventually, as burial depths increased, the organic material was converted by metamorphism to coal of various grades. Tectonic uplift and erosion brought the coal beds back to the surface where they could be mined to support our energy-hungry modern lifestyle.

Key Terms

acid-mine drainage (p. 745)
acid rain (p. 744)
banded iron formation
     (p. 734)
biomass (p. 738)
brine (p. 734)
coal (p. 742)
evaporite (p. 734)
fossil fuel (p. 742)
geothermal energy (p. 741)

hydraulic fracturing (p. 749)
hydrocarbon trap (p. 746)
hydroelectric power (p. 738)
hydrothermal fluid (p. 732)
hydrothermal ore deposit
     (p. 732)
industrial mineral (p. 732)
magmatic segregation (p. 731)
methane hydrates (p. 751)
mineral resource (p. 728)

nonrenewable (p. 728)
nuclear energy (p. 751)
oil shale (p. 749)
peat (p. 743)
pegmatite (p. 732)
photovoltaic cell (p. 738)
placer (p. 733)
porphyry copper (p. 732)
recycle (p. 728)
renewable (p. 728)

reservoir rock (p. 746)
shale gas (p. 749)
solar energy (p. 737)
source rock (p. 745)
strip mining (p. 744)
tar sand (p. 749)
tidal energy (p. 742)
vein (p. 732)
wind energy (p. 739)

Review Questions

1. List the ways in which mineral resources are concentrated.
2. Why are most mineral resources considered nonrenewable if their formation processes are continuing?
3. Explain how magmatic segregation concentrates ores such as chromium.
4. How are minerals concentrated by streams?
5. Explain how some mineral resources are concentrated by weathering processes.
6. List five different ways in which hydrothermal fluids can be generated.
7. Which energy sources are renewable? Which are nonrenewable?
8. Contrast the advantages and disadvantages of solar power, hydroelectric power, and wind energy.
9. What are fossil fuels?
10. Explain how coal originates.
11. Describe various kinds of petroleum traps.
12. What kinds of strata serve as (a) source beds and (b) reservoir beds for petroleum?
13. Are there similarities between the reservoir rocks for oil and for geothermal fluids?
14. Why is oil rarely found in the center of a syncline?
15. Compare the maps in Figures 24.22 and 24.19. Why are coal and oil commonly found in the same regions?
16. What are the major problems with nuclear energy?
17. What hydrocarbons may become more important fuel sources in the future?
18. Give an example of an important mineral resource whose origin is strongly controlled by plate tectonics.
19. Hydrothermal deposits formed on oceanic crust are rich in copper and nickel, whereas similar deposits formed in continental crust are poor in nickel but richer in lead and zinc. Speculate about the causes of this association between metal deposits and kinds of crust.
20. List the mineral resources that you have used today. Which were renewable and which were not?
21. Discuss the factors that limit the growth of population and industrialization.

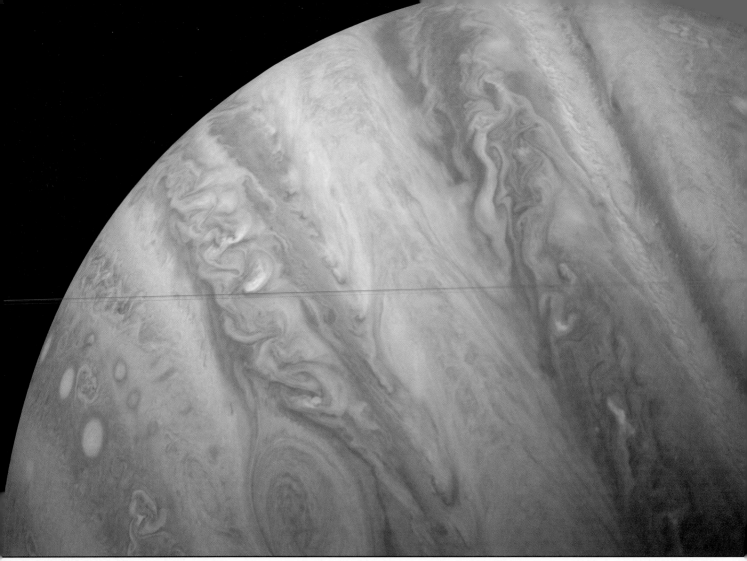

25 Other Planets

The exploration of the solar system is one of the most exciting adventures ever experienced by humankind; for the first time, we are exploring whole new worlds and can compare them with Earth. The study of planetary geology, however, does much more than merely satisfy scientific curiosity. By comparing in detail the geologic nature and evolution of different planets, we can better recognize those principles and processes that are fundamental to the geology of Earth and those that are of secondary importance.

Take for example, the two planetary bodies shown above. On the left is Jupiter and on the right is Mars. Jupiter is truly gigantic. If all of the other planets were taken together to form a single planet, it would still not be as massive as Jupiter. But size is not the only difference between Jupiter and Mars. The colorful bands and delicate swirls revealed in this image are fleeting, temporary features of its flowing atmosphere. The rapid spin on its axis and the uneven distribution of solar heating have created a huge convective system in Jupiter's atmosphere, somewhat like Earth's but more organized into belts of winds blowing in different directions and shearing past one another. The details of the convection are brought out by colorful particles of sulfurous aerosols entrained in the swirling gas. But even these gases aren't like those in Earth's atmosphere. Instead of nitrogen and oxygen, Jupiter's air consists of two gases extremely rare on Earth—hydrogen and helium.

Courtesy of JPL/USGS/NASA

On the right is Mars, a planet that is much smaller than Jupiter. It would fit neatly into the orange core of the giant turbulent storm in the lower left of the image. Mars is even smaller than Earth; its diameter is only about half of Earth's. Its density is less than Earth's and it orbits the Sun at a distance that is greater than Earth but less than Jupiter. Look carefully at this true color image of Mars. Obviously, Mars lacks the colorful atmosphere that is so characteristic of Jupiter. The martian atmosphere is thin and made mostly of carbon dioxide. Mars has a solid surface pockmarked by a multitude of circular impact craters. The large number is inherited from its early days. Mars has an old surface. But some of the smooth, less cratered plains are volcanic lava flows, and just out of view on the other side of the planet are some of the Solar System's largest shield volcanoes with diameters of hundreds of kilometers. If you look carefully you will see that volatiles have also had a great effect. Long thin streaks decorate the dark plains. These streaks are created by wind erosion and deposition. Look even more carefully and you will find that the impact crater rims are eroded and not just by the wind. Locally, dry river beds dissect the landscape. Once Mars had a hydrologic system somewhat like Earth's, with rivers, lakes, polar ice caps, and ground water. But over the eons that has changed.

While you study this chapter, try to find out why these planets are so different. Keep in mind that each planet, moon, or asteroid was shaped by geologic systems that were created by the unique conditions found on each body. Understanding the conditions, such as surface temperature, composition of the rocks and atmosphere, extent of internal differentiation, impact history, size, and especially the amount of internal heat, is critical for developing a true appreciation of the geologic systems of other planets. You will find that this comprehension will reinforce the fundamental concepts you have already learned about Earth's dynamic geologic systems.

Major Concepts

1. Impact cratering was the dominant geologic process in the early history of all planetary bodies in the solar system.
2. Earth, the Moon, Mercury, Venus, and Mars form a family of related planets, known as the inner planets, which experienced similar sequences of events in their early histories.
3. Both the Moon and Mercury are primitive bodies, and their surfaces have not been modified by hydrologic and tectonic systems. Much of their surfaces are ancient and heavily cratered.
4. Mars has had an eventful geologic history involving crustal uplift, volcanism, stream erosion, and eolian activity. Huge tracts of cratered terrain remain, but they are intensely eroded. Liquid water may have existed on its surface, and there is controversial new evidence that life may have evolved there.
5. The surface of Venus is dominated by relatively young volcanic landscapes and such tectonic features as faults and folded mountain belts. The crust of Venus does not appear to be broken into tectonic plates, however, and much of its evolution is related to the development of mantle plumes.
6. Cratering on the icy moons of Jupiter, Saturn, Uranus, and Neptune suggests that a period of intense bombardment affected the entire solar system more than 4 billion years ago.
7. Most of the icy moons of Uranus and Neptune show evidence of geologic activities, such as volcanic extrusions of slushy ice and rifting.
8. Asteroids and comets are the smallest members of the solar system. They appear to be remnants of the bodies that accreted to form the larger planets.
9. The planets formed in a thermal gradient around the Sun. The inner planets are thus rich in silicates and iron, which are stable at high temperature, and the outer planetary bodies have large amounts of ice, which is stable at low temperature.
10. The geologic evolution of a planet depends on its source of heat energy, its size, and its composition.

The Solar System

Three types of planets formed in our solar system. The inner planets are small and made mostly of silicates and iron metal. The outer planets are large and made largely of gaseous hydrogen and helium. The icy planets also lie in the outer solar system but are small and have surfaces dominated by water ice.

Review in your mind the intellectual journey we have taken in space and in time as we studied Earth. Now contrast our home planet with the other worlds of the solar system. The planets, moons, and other objects that orbit the Sun have an almost infinite variety if we dwell on the details of their surfaces, interiors, sizes, and densities. However, if we step back, we recognize only three fundamentally different types of planets. The differences between these groups can be appreciated by looking at the simplified model planets in **Figure 25.1** and **Table 25.1**.

The **inner planets** are small rocky bodies composed mostly of silicates and iron metal—materials that solidify at high temperatures. Their interiors are probably differentiated by density into metallic cores, mantles, and crusts. Several of these planets

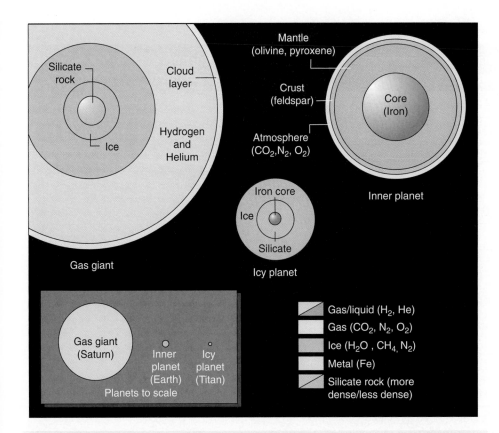

Figure 25.1 **Three different types of planets** are found in the solar system as shown in these cross sections. The inner planets are small and made mostly of silicates and iron metal. The outer planets are large and made largely of gaseous hydrogen and helium. The icy planets also lie in the outer solar system but are small and have surfaces dominated by water ice. An example of each is drawn to scale in the inset.

also have atmospheres composed of volatile gases; the most common are carbon dioxide, nitrogen, and oxygen. Io, a satellite of Jupiter, is the only sizable object in the outer solar system that falls into this group.

The large **outer planets** are **gas giants**—large balls of hydrogen and helium—and lack solid surfaces entirely. Obviously, they are extremely rich in volatile materials—compounds that refuse to solidify except at very low temperatures. These planets are also layered on the basis of density. Buried deep inside their interiors, smaller Earth-sized masses of silicate, metal, or even water ice constitute their cores.

Transitional between these two extremes, at least in terms of density and abundance of volatile elements, are the **icy planetary bodies**. Here we include, along with Pluto and comets, almost all of the moons of the outer planets. Water ice blankets these relatively small bodies, but ices of ammonia, methane, and nitrogen may also be present. Denser silicate minerals and perhaps metals are mixed with the ices or form discrete cores, as illustrated in Figure 25.1.

Thus, a simple pattern emerges out of this complexity. Only a few materials are really important: silicate rocks, iron metal, ice, and gas. The proportions of these materials inside a planet are reflected by its density. More important, the types of planets show a systematic relationship with distance from the Sun. Keep these characteristics in mind as we review the nature of the planets. We first discuss the inner planets in order of their sizes (Table 25.1), because size exerts such a strong control on their histories.

What are the major differences between the inner and outer planets?

Table 25.1 Physical Characteristics of the Planets and Selected Moons

Planetary Body	Density (g/cm³)	Diameter (km)	Surface Composition	Atmosphere Composition	Known Moons
Mercury	5.43	4,880	Silicate		0
Venus	5.24	12,104	Silicate	CO_2	0
Earth	5.52	12,756	Silicate and water	N_2 and O_2	1
Moon	3.34	3,476	Silicate		
Mars	3.93	6,787	Silicate	CO_2	2
Asteroids					
Eros	2.7	33	Silicate		
Vesta	3.5	549	Silicate		
Ida	2.6	56	Silicate and iron		1
Ceres	2.1	1,020	Silicate and carbon		
Jupiter	1.33	143,800		H_2 and He	67
Io	3.53	3,640	Silicates and sulfur	SO_2 (thin)	
Europa	3.01	3,120	Water ice		
Ganymede	1.94	5,260	Water ice		
Callisto	1.83	4,820	Water ice		
Saturn	0.69	120,660		H_2 and He	62
Mimas	1.14	400	Water ice		
Enceladus	1.61	500	Water ice		
Tethys	0.98	1,060	Water ice		
Dione	1.48	1,120	Water ice		
Rhea	1.24	1,530	Water ice		
Titan	1.88	5,150	Water ice	N_2	
Uranus	1.32	51,120		H_2 and He	27
Miranda	1.20	470	Water ice		
Ariel	1.67	1,150	Water ice		
Umbriel	1.40	1,170	Water ice		
Titania	1.71	1,580	Water ice		
Oberon	1.63	1,520	Water ice		
Neptune	1.64	49,530		H_2 and He	13
Triton	2.06	2,700	Nitrogen, Water ice	N_2 (thin)	
Pluto	2.03	2,310	Nitrogen and methane ice	N_2 and CH_4 (thin)	5
Charon	1.65	1,210	Water ice		
Eris	2.52	2,320	Nitrogen and methane ice		1

The Inner Solar System

The inner planets are composed of rocky material that condensed near the Sun. The Moon and Mercury were too small to generate enough heat to sustain a tectonic system and ceased to be active after their first major thermal event. Mars, being larger, developed a more prolonged period of tectonism. Venus, nearly as large as Earth, has continent-like highlands; a young, dominantly volcanic surface; and folded mountain belts.

The Moon

With the development of the space program, the Moon has become one of the best-understood planetary bodies in the solar system. As curious as it may seem, only a few decades after putting astronauts on the Moon, we probably understand the Moon's earliest history better than Earth's, for Earth lacks a rock record of the first 800 million years of its history. The Moon is a comparatively small planetary body (Table 25.1), with a diameter only about one-fourth of Earth's. Moreover, it is less dense than Earth, with a density (3.3 g/cm³) that suggests it is made almost entirely of silicate rocks, which have densities of about 3.0 g/cm³. If there is an iron core, it must be very small.

Impact Processes   One of the most important results of the exploration of the Moon is the discovery that cratering, from the impact of meteorites and comets, is a fundamental and universal process in planetary development. The Moon is pockmarked with billions of **impact craters**, which range in size from microscopic pits on the surface of rock specimens to huge, circular basins hundreds of kilometers in diameter. How are impact craters formed, and what is their geologic significance?

Conceptually, the process is relatively simple, as is illustrated in **Figure 25.2**. As a **meteorite** strikes the surface, its kinetic energy is almost instantaneously transferred to the ground as a shock wave that moves downward and outward from the point of impact. This initial compression wave is followed by a relaxation wave as rocks decompress back to low pressure, causing material to be ejected from the surface along ballistic trajectories. This fragmented material accumulates around the crater, forming an **ejecta blanket** and a system of splashlike **rays**. A central peak on the crater floor can result from the rebound, and rocks in the crater rim can be overturned. Its steep walls rapidly slump and move downslope by mass movement. This process results in partial filling of the excavation and forms concentric terraces (slump blocks) inside the crater rim. Many large craters (more than 300 km in diameter) contain a series of concentric ridges and depressions and hence are known as **multiring basins**. Meteorite impacts create new landforms (craters) and new rock bodies (ejecta blankets), so records of the events are preserved.

What role does a shock wave play in the origin of an impact crater?

After a crater is formed on the Moon, it is subject to modification. Isostatic adjustment can arch up the crater floor to compensate for the removal of material in the crater's formation. The rays and eventually the entire crater may gradually become obliterated by subsequent bombardment. In addition, the crater and ejecta can be buried by lava flows.

How is an impact structure modified with time?

The Surface of the Moon   Study the surface of the Moon in **Figure 25.3** and you will see two contrasting types of landforms; these reflect two major periods in its history. The bright, densely cratered highland resulted from an intense bombardment of meteorites, most of which impacted more than 4 billion years ago. The dark, smooth areas that mostly occupy low regions, such as the circular interiors of impact basins, are maria (singular **mare**). We know from rock samples brought back from the *Apollo* missions (1969–1974) that the maria resulted from great floods of basaltic lava that filled many large craters and spread out over the surrounding area. Most of this volcanic activity, therefore, occurred after the formation of the densely cratered terrain.

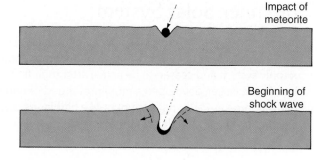

(A) The impact of a meteorite causes the rock to be instantly fractured, fused, and partly metamorphosed.

Impact of meteorite

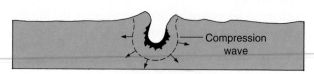

(B) A shock wave is propagated downward and outward from the point of impact.

Beginning of shock wave

(C) The shock wave expands.

Compression wave

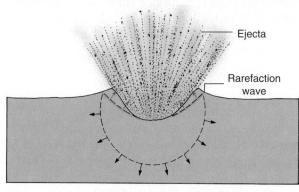

(D) The shock wave is reflected back toward the surface. The crater begins to form, and material is fragmented. The result is similar to that produced by an explosion.

Ejecta

Rarefaction wave

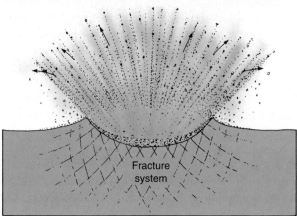

(E) The fragmented material is thrown upward and outward. Solid bedrock is fractured and forced upward to form a crater rim.

Fracture system

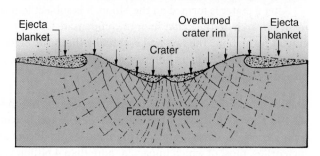

(F) The crater rim may be overturned and a peak may develop in the center of the crater floor. Ejected particles fall back to the surface to form a blanket of debris and a system of rays.

Ejecta blanket

Overturned crater rim

Ejecta blanket

Crater

Fracture system

Figure 25.2 Hypothetical stages in the formation of a meteorite impact crater. The kinetic energy of the meteorite is almost instantly transferred to the ground as a shock wave that moves out, compressing the rock. At the point of impact, the rock is intensely fractured, fused, and partly vaporized by shock metamorphism. The shock wave is reflected back as a rarefaction wave that throws out large amounts of fragmental debris, and the solid bedrock is forced upward to form the crater rim. A large amount of fragmental material falls back into the crater.

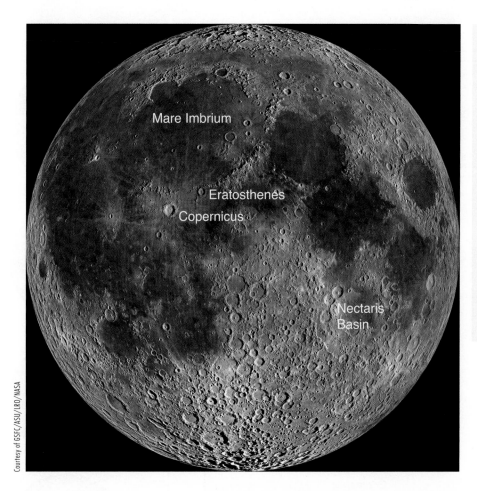

Courtesy of GSFC/ASU/LRO/NASA

Figure 25.3 The surface of the Moon shows two contrasting types of landforms: densely cratered highlands, called terrae, and dark, smooth areas of lava plains, called maria. We know from rock samples brought back by the *Apollo* missions that the maria resulted from great floods of basaltic lava filling many large craters and spreading over the surrounding area. The volcanic activity thus occurred after the formation of the densely cratered terrain. These relationships between surface features imply that the Moon's history involved three major events: (1) a period of intense bombardment by meteorites, (2) a period of volcanic activity, and (3) a subsequent period of relatively light meteorite bombardment (resulting in young, bright-rayed craters). The lunar surface has a very low level of erosion and has not been modified by wind, water, or glaciers.

Radiometric dates on samples brought back from the Moon indicate that most of the lavas are between 4 and 3 billion years old. Almost no global tectonic activity has occurred on the Moon during the last 3 billion years; in fact, very little has occurred in its entire history. We find no evidence of intense folding or thrust faulting and no indication of major rifts. The main lunar features that can be attributed to structural deformation are narrow grabens, formed by minor extension, and wrinkle ridges, formed by minor compression. Nor has the lunar surface been modified by wind, water, or glaciers. Without an atmosphere, it has no hydrologic system, and its surface is strikingly different from Earth's.

Lunar History One of the most significant results of geologic exploration of the Moon has been the construction of a lunar geologic time scale. This was done by using the same principles of superposition and crosscutting relations devised in the early nineteenth century by geologists studying Earth. In Figure 25.3 you can read much of the record of lunar history yourself.

A period of intense bombardment is recorded in the densely cratered terrain. The heavy cratering gives us an important insight into the origin of all planetary bodies. Apparently, they formed by **accretion** when one body after another collided and merged into a growing planet. So much heat was released by this process that much of the outer layers of the Moon melted during a process that led to the internal differentiation of the Moon into layers of different density.

Accretion and differentiation were followed, or accompanied, by the impact of large asteroid-sized bodies that produced the huge multiring basins such as Imbrium Basin. A subsequent event is evident in the floods of lava that filled the lowlands and spread over parts of the densely cratered terrain. A period of light bombardment by

What principles did scientists use to develop a geologic time scale for the Moon?

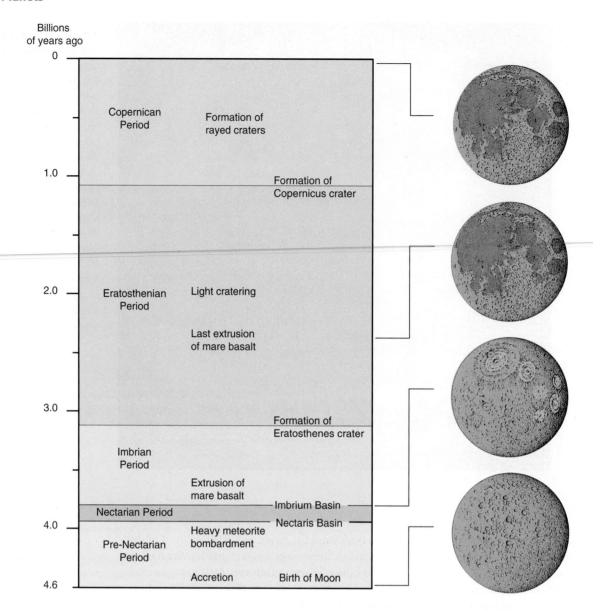

Figure 25.4 The lunar time scale was constructed on the principle of superposition. Five major periods have been recognized. They are (1) the origin of the Moon and the Pre-Nectarian Period, (2) the Nectarian Period (intense bombardment), (3) the Imbrian Period (extrusion of basalt), (4) the Eratosthenian Period (light bombardment), and (5) the Copernican Period (light bombardment).

meteorites followed that formed craters upon the lava flows of the maria. The light cratering episode reveals that most of the debris that formed the planets had been swept up by this time. Radiometric dates of lunar rock samples provide benchmarks of absolute time, and the major events in lunar history have been outlined as shown in **Figure 25.4**.

Perhaps the most important aspect of the Moon's geologic evolution is that the most dynamic events occurred during the early history of the solar system, even before the oldest rock preserved on Earth was formed. The Moon thus provides important insights into planetary evolution that are unobtainable from studies of Earth.

Mercury

Mercury is nearly 1.5 times larger in diameter than the Moon and has a much greater density (Table 25.1), suggesting that much of its interior is made of dense iron metal.

What evidence suggests that the Moon and Mercury had similar histories?

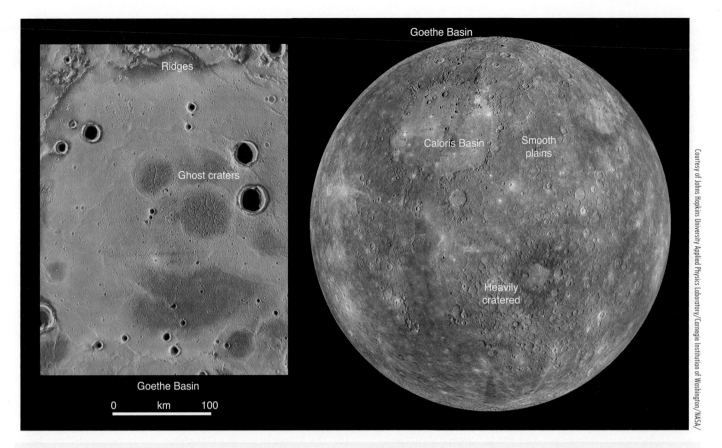

Figure 25.5 **Mercury** and the Moon are strikingly similar as you can see in this global mosaic acquired by the *Messenger* spacecraft. Each has a densely cratered terrain with huge multiring impact basins like Caloris, younger smooth volcanic plains, and even younger rayed craters. The topographic map on the right shows Goethe Basin contains buried "ghost" impact craters with radiating fractures and extensive wrinkle ridges. The grabens are probably the result of cooling and fracturing of thick flood lavas that filled the craters. On the other hand, the elongate ridges probably reflect the global cooling of Mercury which caused it to shrink slightly and contract in a compressional state. The biggest fractured crater is 60 km across. Purple is low and white is the highest elevation; the total range of heights on this map is about 1 km.

Thus, Mercury's interior is quite different from the Moon's. However, Mercury's surface features are strikingly similar to the Moon's, as is evident from the images in **Figure 25.5**. Most of Mercury's surface is nearly saturated with craters, whose range of sizes is similar to that of the Moon's craters. Some are obviously old because they are battered by the impact of other meteorites; others are younger and have bright rays extending out from the point of impact.

The largest impact structure seen on Mercury is the Caloris Basin. It is a multiring basin similar in size and form to the Imbrium Basin on the Moon. Smooth plains cover the floor of the Caloris Basin as well as much of the lowlands beyond. The similarity to the lunar maria suggests that the plains also formed by the extrusion of fluid basaltic lava, but we have no samples from Mercury to prove that they did.

Geologists have established a preliminary geologic time scale for Mercury and have developed a hypothesis for its geologic evolution. Mercury's surface features indicate a sequence of four major events, broadly similar to the sequence of events recorded on the Moon: (1) accretion, planetary differentiation, and intense meteorite bombardment; (2) formation of multiring basins; (3) flooding of basins by the extrusion of basaltic lava; (4) global contraction as it cooled; and (5) light meteorite bombardment. Because there is no atmosphere or water on Mercury, its surface has not been modified by a hydrologic system.

STATE OF THE ART   Geologists on Other Worlds: The Apollo Program

Sometimes it seems hard to believe, but 40 years ago astronauts from Earth stepped from a spindly legged spacecraft out onto a truly alien world, the Moon, and declared that this was, "One small step for man, one giant leap for mankind." Indeed, many people think that this event was the most important in the century or perhaps of the entire millennium. What event could match this, the first time humans left their home planet?

Why did we go to the Moon? Although there were many political justifications for pursuing a lofty goal as part of a Cold War strategy, certainly one goal of the Apollo Project was to better understand the Moon, and thereby come to a better understanding of our home planet.

To accomplish this latter task, all of the tools of the geologist were brought to bear. Aerial photographs, rock hammers, drills, magnetometers, seismographs, and many other instruments were used on the Moon's surface. Astronauts, specially trained in lunar geology, worked in nine different field areas on the near side of the Moon (*Apollo 11, 12, 14, 15, 16,* and *17*). They returned with more than 380 kg of rock and soil. Back in the laboratory these special rocks were carefully analyzed with the most sophisticated instruments to determine their physical characteristics, ages, and precise chemical compositions, all in an attempt to decipher the Moon's history.

Top Ten Discoveries of Lunar Exploration

1. **The Moon Is Old.** Radiometric ages show that the Moon formed 4.6 billion years ago, at the same time that Earth formed.
2. **Accretion.** The Moon formed when a multitude of smaller objects, in orbit around the ancient Earth, collided with one another and by mutual gravitational attraction collected into a large body.

3. **The Moon's Outer Layers Melted.** Geochemical studies show that the Moon probably melted to a depth of several hundred kilometers during its early history.
4. **Crust Formation.** The bright lunar highlands are made of anorthosite that formed when plagioclase feldspar floated to the top of this huge global magma ocean.
5. **Meteorite Impact.** Meteorite impact, not volcanism, is the principal cause of lunar craters. Heavy meteorite bombardment lasted until about 3.8 billion years ago.
6. **Basalt Is Common.** The vast dark lunar maria formed from a series of flood basalt eruptions—most of them 3 to 4 billion years ago.
7. **The Moon Is Dead.** No lava flows younger than about 3 billion years old have been found. The small Moon quickly cooled after its hot start.
8. **The Moon Is Dry.** No water was found in any lunar rocks. Even less volatile elements, such as potassium, are found in low abundances on the Moon. However, the Moon's poles may harbor ice in permanently shadowed areas.
9. **Small Lunar Core.** Seismic and geochemical studies show that if the Moon has an iron core, it is very small.
10. **The Moon Is Earth's Daughter.** The facts revealed by study of the lunar rocks, suggest that the Moon formed when a Mars-sized object crashed into Earth, ejecting part of its mantle (see Figure 25.29).

One day astronauts from Earth will return again to the Moon and continue the geologic exploration they began so long ago. One day we will have permanent bases on the Moon that will serve as outposts for furthering our understanding of our neighbor in space.

Courtesy of Eugene Cernan/NASA

Mars

A fascinating red planet, Mars orbits the Sun beyond Earth (**Figure 25.6**). For years it was a planet of mystery and intrigue, and there was much speculation that Mars might host life. Telescopic observations revealed polar ice caps and shifting markings that often darkened during the martian spring. Before the space program, some observers thought that life on Mars had evolved to a civilized state. Streaks were believed to be canals or vegetated land alongside canals. As it turned out, all this speculation was fanciful.

Courtesy of NASA/JPL/Caltech

Figure 25.6 The planet Mars, as photographed from the *Viking* spacecraft, has a surface dramatically different than the Moon's. Although it has some heavily cratered regions, Mars has many features indicating that its surface has been modified by atmospheric processes, recent volcanic activity (circular volcanoes on left), and crustal deformation (the Valles Marineris rift extends across the bottom of the image).

Still, the Mars we have just explored by our space probes and landers is a wondrous place, more like Earth than any other planet (**Figure 25.7**). Many of its Earthlike surface features are not only large, but gigantic (**Figure 25.8**). There are huge rift valleys, volcanoes, global dust storms, polar ice caps, and dry river beds. Mars has been eroded by enormous floods of water that once flowed across its surface (**Figure 25.9**), but presently, temperatures and atmospheric pressures are such that water can exist only as vapor or as ice. Thus, nearly all of the water on Mars is frozen as ice caps or is locked up as ground ice. Now, wind alone is the major process altering the landscape of Mars. Some dust storms grow to such proportions that at times they blanket the entire planet.

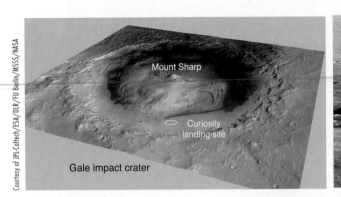

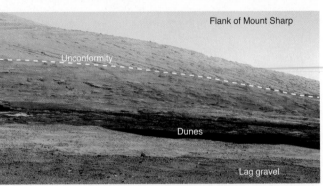

Figure 25.7 **The martian surface** as photographed by the *Curiosity Rover* (right) which landed in Gale crater (left). Gale was filled by a thick succession of sedimentary deposits and then erosion exposed the layering. Within the sedimentary rocks on the flanks of Mount Sharp, a pronounced unconformity separates lower horizontal beds from an upper series of inclined beds. The upper strata look like foreset beds in a delta or perhaps large scale cross strata like those found in eolian sandstones. The floor of the crater is covered by dark dunes and a lag gravel.

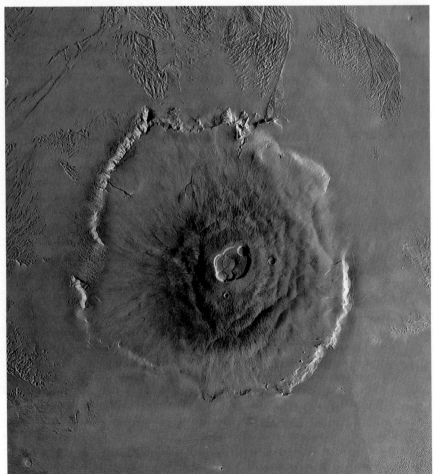

Figure 25.8 **Volcanoes in the Tharsis region of Mars** include huge structures, much larger than any found on Earth. Olympus Mons, the largest volcano on Mars, is shown here. It is 700 km across at the base and 23 km high. The complex caldera at the summit is 65 km in diameter.

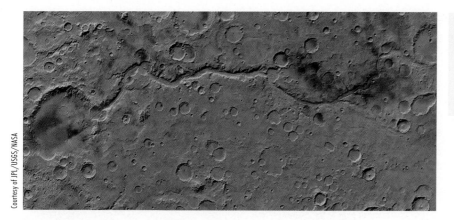

Figure 25.9 Dry stream channels on Mars are similar in many respects to dry riverbeds in arid regions on Earth. Some have typical braided patterns; others meander. This photomosaic shows a drainage system more than 500 km long.

The huge dry river channels on Mars (Figure 25.9) also create an intriguing mystery. Under what ancient conditions did liquid water once flow in great floods across the surface of Mars? Why did things change on such a global scale? Some answers to these questions may be suggested when we consider the size and thermal history of Mars.

Mars has a diameter about half that of Earth or Venus but almost twice that of the Moon (Table 25.1). Thus, Mars generated more internal heat, leading to more geologic activity than experienced by Mercury and the Moon. Like all other planetary bodies in the solar system, Mars experienced an early period of intense bombardment, followed by volcanic activity in which floods of lava were extruded. Two major provinces create a global dichotomy on Mars. A heavily cratered southern hemisphere contrasts with a relatively young, smooth northern hemisphere where the impact craters were buried or destroyed by younger events.

How do river channels on Earth differ from those on Mars?

Apparently, Mars cooled more slowly than the smaller Moon, producing younger giant shield volcanoes and younger deformation of its surface. Huge domes in the lithosphere are also cut by deep rifts, such as Valles Marineris (**Figure 25.10**).

Mars started much as Earth did, developing a core, mantle, crust, and even a relatively dense atmosphere early in its history. Mars may once have had a moderate climate and abundant liquid water. Small seas may have formed. During this time, rainfall and flooding created stream channels and other features somewhat similar to those on Earth. But Mars is much smaller than Earth, has less internal heat, cooled more quickly, and was geologically active over a much shorter period. Its internal convection appears to have been stirred by mantle plumes rather than plate tectonics. Moreover, Mars is farther from the Sun than Earth. Billions of years ago, Mars grew cold and dry. All of its water became locked into ice caps or frozen in the pore spaces of rocks and soil, and stream erosion ceased. As it cooled inside, its volcanoes ceased erupting several hundred million years ago. Although its present carbon dioxide atmosphere is very thin (exerting only 6/1000 of Earth's pressure), great dust storms now rage on Mars as the major process altering its surface.

How does Mars differ from the Moon and Mercury?

Life on Mars?   In 1996, dramatic headlines and television stories flashed the news around our planet that Earth might not be the only planet inhabited by living things. The putative Martians were not ominous invaders from space, but tiny microscopic blebs seen with a powerful electron microscope (**Figure 25.11**). Evidence for life on Mars comes from the intensive study of a few of the 14 meteorites that are believed to have come from Mars. The martian origin of these meteorites is based on very young radiometric ages (less than 1 billion years old), mineral compositions, and especially on the composition of the inert gases trapped inside the meteorites. The ratios of the gases extracted from them match ratios measured by spacecraft for the atmosphere of Mars, but not Earth, Venus, or any other meteorite types.

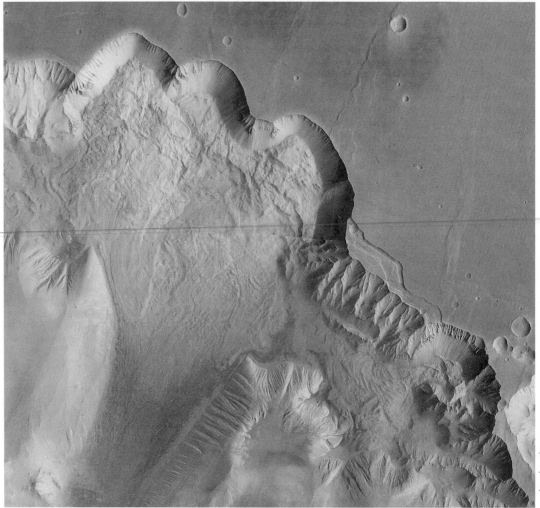

Courtesy of JPL/USGS/NASA

Figure 25.10   Landslides in Valles Marineris. The landslide on the far wall has two components: an upper blocky portion, which is probably disrupted cap rock, and a finely striated lobate extension, which is probably debris derived from the old cratered terrain exposed in the lower canyon walls. Similar lineations are found on terrestrial landslides and show direction of movement. This part of Valles Marineris is about 5 km deep. This image is approximately 200 km wide.

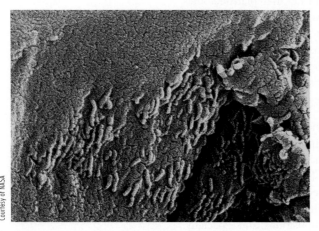

Courtesy of NASA

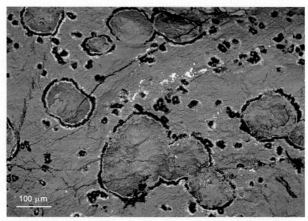

Courtesy of NASA

(A) Scanning electron microscope view of shapes like bacteria (colored) found on the fractured surface of the meteorite.

(B) This false-color microscope image shows carbonate globules (brown) that have rims of iron oxides and sulfides (black) like those formed by some terrestrial bacteria. The carbonate blebs are enclosed in a crystal of pyroxene (green).

Figure 25.11   Evidence for life has been found in a meteorite thought to have come from Mars.

At least one of these meteorites has several bits of evidence of past life on Mars. First, complex organic molecules (long chains of hydrocarbons) are found in these meteorites. Second, tiny globules found on broken surfaces of the meteorite and made of carbonate minerals have wormlike shapes similar to fossils of terrestrial bacteria (Figure 25.11). Moreover, the carbonate blebs have rims that are a mixture of magnetite and iron sulfide minerals; these minerals are also formed by some terrestrial bacteria. Third, the isotopic composition of carbon in the meteorites is like that which is created by living things on Earth.

The hypothesis that these tiny features are actually fossilized bacteria was naturally intensely scrutinized. Evidence against past life in this meteorite includes the very small size of the "bacteria." No fossil bacteria found on Earth are this small. The objects in Figure 25.11 are 10 to 100 times smaller than terrestrial bacteria and are about the same size as viruses. Moreover, carbonate globules with iron minerals can form by inorganic processes in hot springs on Earth. Finally, hydrocarbons of the sort identified in these Martian meteorites are not particularly rare. Many different kinds of meteorites contain the same kinds of molecule but clearly did not host living organisms.

All of the evidence collected by studying the surface of Mars and by carefully probing these meteorites makes it clear that Mars is rich in water compared with the Moon, Mercury, or Venus. We think that liquid water is critical for the evolution of life on any planet, and the presence of water has led to much speculation about life on Mars. Moreover, there is abundant evidence that Mars once had a more temperate climate and perhaps even higher atmospheric pressures. Nonetheless, only ambitious new missions to Mars in the coming decades will help us resolve the supremely important questions about life there. Ultimately, we need samples actually collected on Mars to understand the details of its early evolution.

Is there any evidence for ancient life on Mars?

Venus

Of all the planets, Venus is most like Earth in both size and density (Table 25.1). Thus understanding the similarities and differences between these two "twins" yields important clues about what controls the major characteristics of a planet. Unfortunately, the surface of Venus is totally obscured by clouds of sulfuric acid in a thick atmosphere made mostly of carbon dioxide. Nonetheless, Soviet and U.S. spacecraft used radar to reveal features as small as a football field. From these data an outline of the planet's history is emerging, though many important facets of its history are still mysterious.

Volcanic plains, mountain belts, volcanoes, and high "continents" that rise several kilometers above vast rolling lowlands show that Venus has a surface similar, in some ways, to the surface of Earth with its continents and ocean basins (**Figure 25.12**). However, Venus has almost no water, and because of an enhanced greenhouse effect in its carbon dioxide–rich atmosphere, surface temperatures (almost 500°C) are higher than on Mercury. Its atmosphere exerts a pressure almost 90 times that of Earth's.

Venus does not have a heavily cratered terrain dominated by ancient impact structures like those of the Moon, Mercury, and Mars. This is the single most important fact we know about the surface of Venus. It clearly shows that the surface of Venus is young, perhaps only 500 million years old, and has been repeatedly modified by tectonism and extensive volcanism. However, Venus has about 1000 relatively young impact craters.

How do the craters on Venus differ from those on the Moon and Mars?

Although erosional and depositional features dominate the surface of Earth and are common on Mars, they appear to be relatively insignificant on Venus. Eolian and mass movement processes are the only effective sedimentary processes. Wind streaks and dune fields have been identified, but they are not widely distributed.

Volcanic features dominate the landscape of Venus. Smooth volcanic plains make up more than 80% of the planet. The plains are built by thin, fluid lava flows. In addition, thousands of small shield volcanoes, generally 2 to 8 km in diameter, with summit craters, are scattered across the plains and concentrated in local clusters.

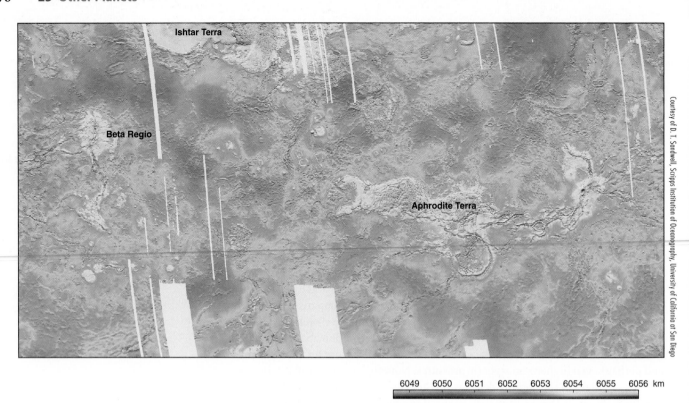

6049   6050   6051   6052   6053   6054   6055   6056 km

Figure 25.12 Topographic map of Venus shows that it consists mostly of low plains and several continent-sized highlands, including Ishtar Terra and Aphrodite Terra. Colors show elevation changes, with purple lowest and red highest.

What are the major products of volcanic activity on Venus?

These features are similar to seamounts on Earth. In some areas, eruptions built large volcanoes, as much as 225 km in diameter. Some steep-sided volcanic domes suggest the presence of lavas with higher viscosities and possibly more silicic compositions (**Figure 25.13B**). Another distinctive volcanic landform is called a corona (Figure 25.13D). It consists of a large (500 km across) raised wreath of faults and fractures surrounding a central volcanic plain that may be dotted by smaller volcanoes and scarred by a large collapse caldera. Many of these and other volcanoes lie on broad structural domes that are cut by long rift valleys. These structures may indicate the presence of broad mantle upwellings or plumes like those on Earth.

Most of the volcanism on Venus appears to be quite young. None of the plains are as heavily cratered as the lunar maria. Lacking confirming radiometric dates from samples collected on the surface, planetary scientists think that the average age of the surface may be only about 500 million years. Compare this with the Moon, where nearly the entire surface is more than 3 billion years old.

The surface of Venus has tectonic features that are spectacularly displayed because they are essentially unmodified by erosion. Some uplifted domes are completely laced with polygonal patterns of faults and fractures (Figure 25.13). In other areas, deformation of the plains occurs in linear belts of narrow ridges, similar to the wrinkle ridges on the Moon, but wider and longer. Folds and thrust sheets form parallel ridges and troughs that clearly represent compressional deformation.

It appears that Venus has had a long and especially eventful geologic history. Like its sister planet Earth, Venus lacks heavily cratered terrain formed early in the history of the solar system. But billions of years of volcanism and tectonism have erased any vestige of its battered crust. Nonetheless, Venus has not developed a system of plate tectonics to recycle its lithosphere and rid its interior of heat. Instead, Venus, like Mars, seems to be losing heat via hotspot development.

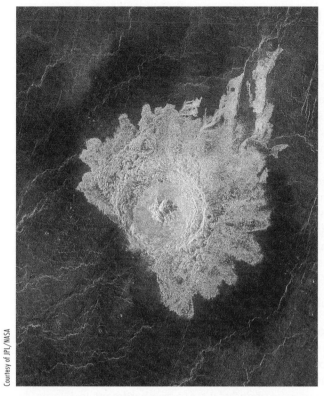

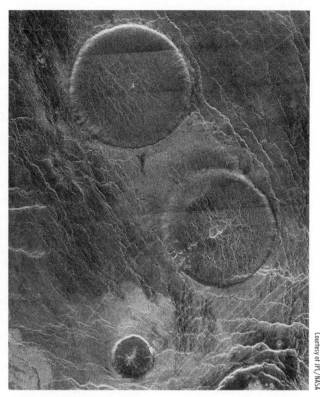

(A) Impact craters on Venus are unique because of the high surface temperature and high atmospheric pressure. This crater is about 30 km in diameter. The asymmetrical, radial-lobate ejecta pattern suggests that the ejected material was fluid, like a mudflow. Note the bright flows extending to the upper right.

(B) Steep-sided, flat-topped volcanic domes are similar to silicic domes on Earth. Their structure and morphology suggest that they were formed by viscous magma.

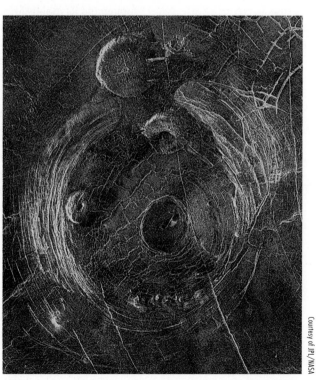

(C) This large structural dome is cut by a complex fault system and flanked by a series of folds. The width of the image is about 125 km.

(D) Coronas are important volcano-tectonic features found on Venus. They may be underlain by mantle plumes.

Figure 25.13 The surface of Venus as revealed by Magellan radar images.

How does the thermal history of a planet influence its tectonic system?

Thermal Histories of the Inner Planets

A major, yet simple, lesson has been learned in the few decades of space exploration—the thermal history and resulting geologic activity of a planet are largely dependent on the planet's size. Small bodies cool rapidly because they have large surface areas compared with their masses. Larger bodies retain heat longer and, as a result, have prolonged periods of internal geologic activity, such as volcanism and crustal deformation. Thus, we see a natural sequence arising by examining the ages of the rocks and surfaces of the inner planets. Meteorites are derived from the smallest bodies in the solar system—asteroids—and have crystallization ages clustering between 4.5 and 4.6 billion years old. The Moon preserves many areas of ancient heavily cratered terrains that are more than 3.9 billion years old, but some parts are flooded with younger basaltic lavas, perhaps as young as 3 billion years old. Mercury, Mars, Venus, and finally Earth have progressively younger surfaces, with very young (less than 0.5 billion years old) volcanic rocks forming much of the surfaces of both Venus and Earth. Most of this difference is simply related to the size and slower cooling rates, and hence prolonged tectonic and volcanic activity, on the larger planets.

The Outer Solar System

Planetary bodies in the outer solar system were formed mostly of the lighter elements: hydrogen, helium, and oxygen. Therefore, the satellites of the giant planets are composed mostly of ice, not of rock, as are the inner planets. They are relatively small and did not generate enough internal heat to sustain geologic activity much beyond the period of intense bombardment. Io, Europa, Ganymede, Enceladus, Miranda, and Triton are notable exceptions, in that each has a distinctive tectonic style resulting from unique energy systems.

Jupiter and Its Satellites

Jupiter and its moons form a planetary system of incredible beauty and intrigue. The giant planet has a volume 1300 times greater than Earth's (Table 25.1). The surface we see with telescopes is a layer of colorful clouds swirling in complex patterns (**Figure 25.14**). The exploration of Jupiter proved to be one of the most significant results of the *Voyager* mission (1979–1989) and was the main goal of *Galileo* (1995–2003). Jupiter has no solid surface and hence no record of a geologic history. The density of Jupiter is only 1.3 g/cm^3, slightly more than that of liquid water, and there is conclusive evidence that silicate rock is not its most important constituent. Rather, Jupiter and the rest of the large outer planets are composed dominantly of hydrogen and helium that cloak Earth-sized masses of silicates and metals (Figure 25.1).

Jupiter's moons, however, are solid planetary bodies containing geologic wonders that reveal fundamental ideas about the origin of planets. Jupiter's four large moons (Io, Europa, Ganymede, and Callisto) are called the Galilean satellites because they were discovered by Galileo in 1610. Each of Jupiter's moons shows a diverse landscape resulting from impact, volcanism, and fracturing (**Figure 25.15**). Three of these moons have surfaces composed mostly of water ice, and one is probably made mostly of silicates and iron.

Io   Io is Jupiter's innermost major satellite. In contrast to every other large satellite in the outer solar system, it lacks an icy outer layer and is composed mostly of silicates like the inner planets. It is only slightly larger and denser than Earth's Moon (Table 25.1). However, unlike the Moon, Io is volcanically active (Figure 25.15A). More than a dozen erupting volcanoes were spotted by the *Voyager* cameras as they plummeted past Jupiter (**Figure 25.16**). More were discovered by *Galileo*, including an erupting fissure vent. The volcanoes are so active that no impact craters have yet been discovered. There is no ancient cratered terrain; all of it was engulfed long ago by volcanic materials.

Courtesy of NASA/JPL/Caltech

Figure 25.14 Jupiter and its four major moons. Not visible are Jupiter's faint rings, seen for the first time on the *Voyager 1* mission.

Io is the most volcanically active body in the solar system and probably has been throughout much of geologic time. The question is why? The Moon, which is about the same size as Io, cooled quickly because of its small size and has been volcanically dead for almost 3 billion years. Theoretically, Io should be dead, too. However, energy from tidal forces may give it a continual energy boost. Europa and Ganymede, as well as Jupiter, exert a strong gravitational pull on Io, forcing it into an eccentric orbit. Io is therefore constantly massaged by tidal forces as it moves closer to and then farther from Jupiter. These conditions cause the satellite to be heated by internal friction. Indeed, **tidal heating** may make Io molten at a depth of only 20 km, so that volcanic activity would naturally be a dominant surface process.

Why does Io have so many active volcanoes?

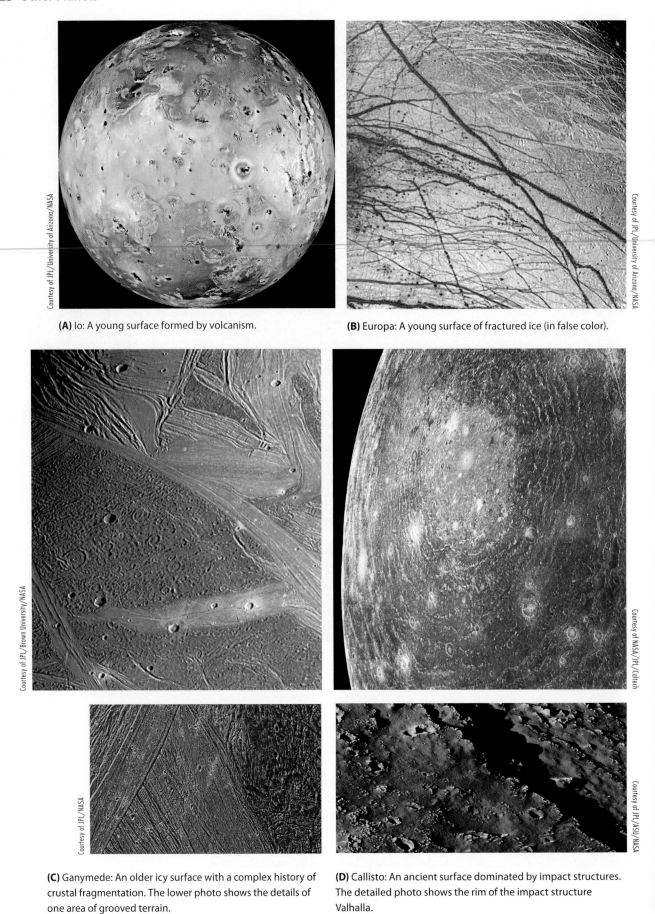

Courtesy of JPL/University of Arizona/NASA

Courtesy of JPL/University of Arizona/NASA

(A) Io: A young surface formed by volcanism.

(B) Europa: A young surface of fractured ice (in false color).

Courtesy of JPL/Brown University/NASA

Courtesy of NASA/JPL/Caltech

Courtesy of JPL/NASA

Courtesy of JPL/ASU/NASA

(C) Ganymede: An older icy surface with a complex history of crustal fragmentation. The lower photo shows the details of one area of grooved terrain.

(D) Callisto: An ancient surface dominated by impact structures. The detailed photo shows the rim of the impact structure Valhalla.

Figure 25.15 The Galilean moon of Jupiter.

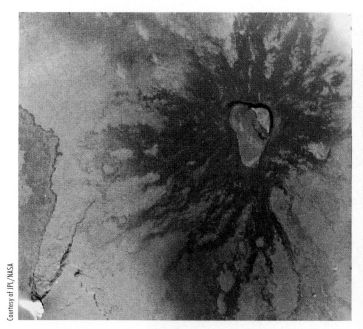

Courtesy of JPL/NASA

Courtesy of NASA

(A) Maasaw Patera is a large central volcano capped by a triangular caldera. Lava flows erupted from near the summit flowed down the flanks of the volcano.

(B) The volcano Pele is shown in eruption by the *Voyager* cameras. The eruption produced a spray similar in shape to that from a lawn sprinkler. From this vertical view, the margins of the ash cloud form a dark elliptical pattern.

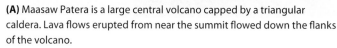

Figure 25.16 Volcanoes are Io's most spectacular landforms. Many are active today.

Europa   Europa has a density of about 3 g/cm³ and must therefore be composed mostly of dense silicate rock (Table 25.1). However, spectroscopic measurements show that Europa is surrounded by a frozen ocean of ice. Europa represents the class of icy planets (Figure 25.1). Its surface is distinctive in that it is essentially free from large impact craters. Watery lava erupted through cracks and fissures in the crust and repeatedly coated the surface with fresh ice. The most obvious features on Europa are sets of tan streaks or bands that are probably fractures formed by its constant gravitational tug-of-war with Jupiter (Figure 25.15B).

The near-absence of impact craters on Europa shows that the surface is very young, formed after the early periods of heavy meteorite bombardment. Resurfacing by the eruption of lavas must have continued until very recently. This scenario is very similar to what we just described for Io. Yet, Io and Europa have distinctly different surfaces and would not be confused by anyone. What makes the difference? Europa has a surface of solid water ice that probably overlies a deep ocean (**Figure 25.17**). Parts of Europa's surface are similar to fractured ice packs in Earth's polar regions. An ocean of liquid water must lie at a shallow depth to explain the shapes and sizes of the ice floes. Could Europa, then, host life that evolved in such an ocean? Could internal heat, released from tidal heating, drive communities of organisms like those found clustered around hydrothermal vents in Earth's deep seas?

How is the composition of Europa different from the composition of the inner planets?

Ganymede   Jupiter's largest satellite is Ganymede, whose diameter is approximately 1.5 times that of Earth's Moon (Table 25.1). It is even larger than Mercury. It has a bulk density of only 1.9 g/cm³ and may consist of about 50% water ice surrounding a rocky core. Ganymede has a cratered surface with two contrasting terrain types (Figure 25.15C). The older terrain is nearly saturated with impact craters. It is dark and is believed to be composed of "dirty" ice, containing fragments of dust and particles from outer space. This older crust appears to have fractured and split apart, and many of the fragments have shifted. The younger terrain is brighter and is crossed

Courtesy of JPL/University of Arizona/NASA

Figure 25.17 Europa's surface is made of ice. These blocks of fractured ice floated on a liquid ocean, which must lie at a shallow depth, before they refroze. Note the absence of abundant craters on this young surface.

by closely spaced, nearly parallel grooves. There are fewer craters on the bright grooved terrain than on the darker surface, so the grooved terrain is believed to be much younger. Apparently, the old, cratered terrain fractured at some time late in the period of intense bombardment, and cleaner ice from below was extruded into the fractures to form the grooved terrain. The breaking and movement of the crustal fragments is a type of plate tectonics on a frozen world with a lithosphere of ice.

Callisto   Callisto is the outermost Galilean satellite. It is only slightly smaller than Ganymede (Table 25.1) and is also believed to consist of a rocky core surrounded by a thick mantle of ice. Callisto, in contrast to the other Galilean satellites, is saturated with craters (Figure 25.15D). The surface of Callisto is very old, recording events during the early history of the solar system. Callisto is covered with ancient dark, dirty ice, similar to the old terrain on Ganymede. Many craters have bright interiors and ejecta blankets. The bright material is probably clean, melted ice from below the dirty crust, ejected onto the surface during impact. Aside from the densely cratered terrain, the most striking feature on Callisto is a large multiring basin. This feature is reminiscent of the multiring basins on the planets of the inner solar system. Important differences exist, however, in the numbers of rings, their spacing, and their elevation. Apparently, ice responds quite differently to the shock of impact than rock.

What geologic features are unique on each of the Galilean satellites of Jupiter?

Saturn and Its Satellites

Saturn is similar to Jupiter in many ways (Table 25.1). It is a gigantic ball of gas, mostly hydrogen and helium, and is the center of a miniature planetary system with an elaborate family of at least 18 satellites. Its atmosphere is not as colorful as Jupiter's but is marked by dark bands alternating with lighter zones. Saturn's rings, of course, have long been considered its most dramatic feature; they have intrigued astronomers for more than 300 years (**Figure 25.18**). They extend over a distance of 40,000 km and yet are only a few kilometers thick. The rings are probably made up of billions of particles of ice and ice-covered rock, ranging from a few micrometers to a meter or more in diameter. Each particle moves in its independent orbit around Saturn, producing an extraordinarily complex ring structure.

Except for Titan, the seven largest moons of Saturn are small, icy bodies. They range from 390 to 1530 km in diameter. In other words, they are only one-half to one-tenth the diameter of the Moon. The surfaces of most of Saturn's icy satellites are saturated with impact craters (Figure 25.18). Many of the moons have large fracture systems and other strange surface markings, probably resulting from an exotic type of icy volcanism.

Two of Saturn's moons warrant separate discussion here. Tiny Enceladus has a rifted terrain and young, smooth plains that may have been produced by "lavas" of slushy water that erupted from fissures (**Figure 25.19**). The ridged terrain is probably similar in its origin to the grooved areas of Ganymede. Why would such a small body have smooth young plains? The recent heating of Enceladus is probably related to the same type of tidal heating that warms Io.

Are the satellites of Saturn more like those of Jupiter or Earth's Moon?

Courtesy of NASA/JPL/Caltech

Figure 25.18 Saturn and its major moons assembled as a composite picture. Saturn's icy moons are a geologically diverse group of mostly small satellites. They show an amazing variety of young and old surfaces, impact craters, evidence of icy volcanism, and global fracture systems.

Courtesy of JPL/Space Science Institute/NASA

Courtesy of JPL/Space Science Institute/NASA

Figure 25.19 Enceladus is one of the most interesting of the satellites in the solar system. Although it is tiny (less than 500 km across), it shows a remarkable record of geologic activity. In this false color view, its smooth, uncratered surface is geologically young, which indicates that it has experienced a relatively recent thermal event and an exotic form of volcanic activity in which floods of water and icy slush were extruded. The smooth plains are in turn cut by large extensional fractures. Near the south pole, geyserlike jets of water erupt from some of the fractures (bottom inset) and create one of Saturn's icy rings.

Titan, larger than the planet Mercury, is the only moon in the solar system that has a substantial atmosphere. Surprisingly, the atmosphere of Titan is more like Earth than any other object in the solar system; it is composed mainly of nitrogen (N_2). There is no oxygen, however, but there are traces of methane (CH_4). The orange color of Titan comes from a haze or smog of hydrocarbon particles in the atmosphere. Solar heating drives convection in the atmosphere and radar images collected by *Cassini* show that dark sand dunes stretch across much of the surface, but especially near the equator. The dunes are linear dunes like those found in some of Earth's deserts, but instead of being made of quartz sand they appear to be made of dark grains of frozen hydrocarbons. Moreover, Titan has a young surface with only a few impact craters. Apparently,

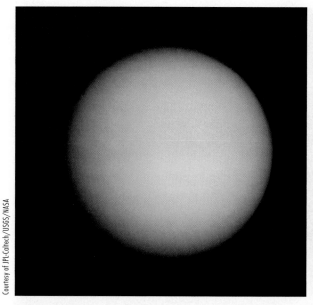

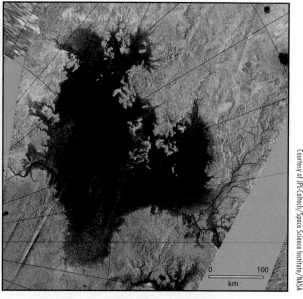

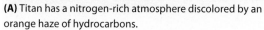

(A) Titan has a nitrogen-rich atmosphere discolored by an orange haze of hydrocarbons.

(B) Radar has been used to penetrate the atmosphere, revealing that Titan has a young surface and is surprisingly Earthlike with dunes, mountain belts, and even dendritic rivers and lakes filled with hydrocarbon liquids.

Figure 25.20 Titan is one of the largest moons in our Solar System—just a bit larger than Mercury.

older craters have been destroyed, not just by eolian erosion or burial, but also by some sort of a tectonic system. Long ridges parallel to the equator were formed by compression and there is speculation that there are icy volcanoes on its surface, too. But the most remarkable of all is the discovery that Titan has rivers and large lakes through which liquid hydrocarbons flow (**Figure 25.20**). Thus, Titan has enough internal energy to continually reshape its surface through tectonism and volcanism and it has an envelope of volatiles—both liquids and gases—that flow to shape its surface.

Uranus and Its Satellites

Like the other giant planets, Uranus has no solid surface but is enveloped by a thick atmosphere of hydrogen and helium (**Figure 25.21**). It is much smaller than Saturn and only about four times as large in diameter as Earth (Table 25.1). Its cloud layer is bland and bluish, because of the presence of methane. It also has a thin dark system of rings that are almost invisible. It is unique among all other planets in the solar system, because its spin axis is tipped on its side; that is, its axis of rotation lies nearly in the plane of its orbit. Thus, it rolls, like a ball, as it moves on its orbital path around the Sun, whereas other planets spin like tops.

Uranus has five major moons (Table 25.1). Each moon occupies a nearly circular orbit, lying in the plane of Uranus's equator. Their orbits share the unusual axial inclination of the planet itself. Oberon, Titania, Ariel, and Umbriel are quite similar in size (1100 to 1600 km in diameter) and are approximately the size of the intermediate moons of Saturn (Tethys, Dione, and Rhea). Their surfaces are nearly saturated with craters (Figure 25.21) and are composed mostly of water ice.

Neptune and Its Satellites

Neptune is only slightly smaller than Uranus and is similar to its neighbor in composition (Table 25.1). Both planets, called the "twins of the outer solar system," are thought to have large cores of water ice and rock, surrounded by thick atmospheres of hydrogen, helium, and minor methane. However, Neptune's cloud layers are banded

Figure 25.21 Uranus and its five major satellites. Uranus is smaller than Saturn, but still has a thick atmosphere of hydrogen and helium. The blue color is caused by methane. The moons are relatively small and all have icy surfaces.

How could volcanism occur on the frigid bodies of the outer solar system?

in various shades of blue (**Figure 25.22**). Oval storm systems spin in the atmosphere. Clouds of bright methane ice tower above the storm systems. Like the other gas giants, Neptune has a system of rings made of ice particles in orbit around the planet.

Triton is the most interesting and largest moon of Neptune. It is only slightly smaller than Earth's own Moon (Table 25.1). Triton has an extremely tenuous atmosphere of nitrogen and methane, and its exotic landscape is formed from ices of those gases. Triton has a surprisingly large variety of geologic features, including ice caps, fractured terrain, "lava" lakes, and volcanic or geyser eruptions (**Figure 25.23**). It is not

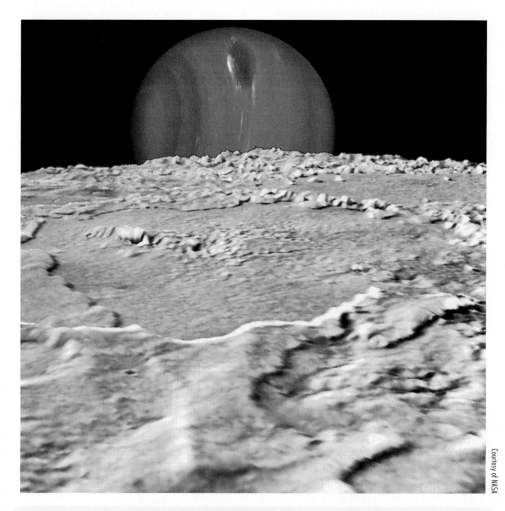

Courtesy of NASA

Figure 25.22 Neptune and its largest satellite, Triton. Neptune is about the size of Uranus but has a banded blue atmosphere decorated with brilliant white clouds of methane ice. Triton is so cold that nitrogen ice forms on its surface.

Courtesy of JPL/USGS/NASA

Figure 25.23 Triton is so far from the Sun and its surface temperature is so low (40°K), that nitrogen is frozen solid to form a large ice cap shown on the bottom of this photo. A fractured terrain with many crisscrossed linear features appears to be the result of rifting. Floods of "lava" (probably mixtures of water, nitrogen, and methane) formed smooth plains and lava lakes. Dark streaks are formed from geyserlike volcanic eruptions from beneath the bright ice cap.

Figure 25.24 Pluto, no longer considered a planet by many astronomers, is nonetheless representative of a large class of icy objects that orbit the Sun beyond Neptune. Our best views of Pluto have come from the Hubble Space Telescope which show that it is has its own set of satellites— 3 of the 5 known are shown here.

Courtesy of ESA, H. Weaver (JHU/APL), A. Stern (SwRI), the Hubble Space Telescope Team, and the Pluto Companion Search Team/NASA

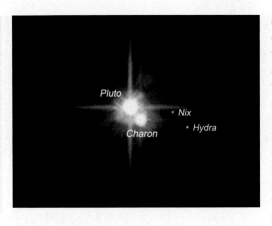

a simple cratered body. Triton is so cold (only 40° above absolute zero) that nitrogen is frozen solid on its surface. However, there is enough warmth from the Sun to cause the nitrogen to vaporize seasonally. Consequently, nitrogen ice caps shift from pole to pole. Its constantly changing ice caps of nitrogen and methane ice mark it as one of the most distinctive planetary bodies in the solar system. Its odd retrograde orbit and young surface combine to suggest that it formed elsewhere in the outer solar system but was then captured and tidally melted as it went into orbit around Neptune.

Pluto

Discovered in 1930, Pluto is tiny, about one-half the diameter of the Moon, and so far from the Sun that it takes nearly 250 years to complete one orbit. Pluto is also distinct from the large outer planets in that it lacks a thick, hydrogen-rich atmosphere. Instead, it is made principally of rock and ices of water, methane, and nitrogen (Table 25.1). Its only atmosphere is created by the sublimation of these ices. (At this distance from the Sun, Earth's nitrogen-rich atmosphere would freeze solid to form a thin layer of ice.) Indeed, Pluto is much more similar to the moons of Neptune that to any of the major planets. However, Pluto has several of its own moons (**Figure 25.24**). Because Pluto is such an oddity—an icy body among the gas-rich outer planets—there has been much conjecture about its origin. Perhaps, Pluto and Triton both accreted in the same frigid part of the outer solar system as Sun-orbiting planetesimals. Triton was then captured by Neptune. Pluto remained in Sun orbit, but Charon, its largest moon, may have formed when Pluto collided with another object and fragmented.

In 2006, the International Astronomical Union demoted Pluto from its status as a planet to that of **dwarf planet**. This is a class of small, nearly spherical bodies that also includes the largest asteroid Ceres. Pluto is not considered a planet because it has not cleared its orbit by accretion. Indeed, there are a few planetary bodies even larger than Pluto in the outer solar system. The discovery of one of them, Eris, precipitated the decision to reclassify Pluto (Table 25.1).

Small Bodies of the Solar System: Asteroids and Comets

Although among the smallest members of the solar system, asteroids and comets hold answers to some of the biggest questions regarding the origin of the solar system.

Asteroids

Besides the eight major planets, thousands of smaller planetoids are also part of the solar system. These minor planets are called **asteroids** (**Figure 25.25**). There are more than 10,000 known asteroids, but many others are far too small to be seen even through the best telescopes. Most are found between the orbits of Jupiter and Mars, where the gravitational force of Jupiter prevented them from accreting to form a single larger planet. The largest asteroid is only about 1000 km across.

Much of our best information about asteroid surfaces came from the photographs taken by the *NEAR* spacecraft which orbited and then landed on the small asteroid named Eros. Craters of every size are visible on its surface. Incomplete crater walls define its irregular shape. The abundance of craters suggests that the surface formed

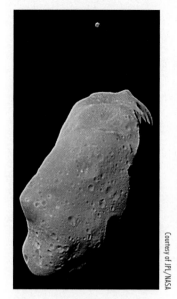

(A) Ida is only 56 km long and 24 km across. Ida, like many other asteroids, is not large enough to be spherical. It was shaped by impact with other asteroids, but it does have its own tiny moon Dactyl.

Courtesy of JPL/NASA

(B) Eros, photographed by the *NEAR* spacecraft, is only about 20 km long and heavily cratered.

Courtesy of NASA/JPL/Caltech and the Advanced Physics Laboratory, Johns Hopkins University

(C) Close-up photograph of the surface of Eros. The regolith and boulders were created by multiple impacts. This photograph shows an area only 12 meters across and was taken from 250 m above the surface of Eros.

Courtesy of JPL/JHUAPL/NASA

Figure 25.25 **Asteroids** are among the smallest members of the inner Solar System. Most are irregularly shaped and orbit between Mars and Jupiter.

billions of years ago, perhaps because of massive fragmentation of a once-larger body. Eros, like most of the other asteroids, is not big enough to sustain active geologic systems driven by internal heat. Perhaps it never was. No lava flows or tectonic features have formed on this small body. Eros lacks the gravitational energy to pull itself into a sphere.

Galileo photographed two asteroids. Ida (Figure 25.25) has a long axis that only measures 56 km, but it has its own tiny moon. Ida is also irregularly shaped and heavily cratered. Impact cratering is the main process that shapes asteroids today.

The *Dawn* spacecraft has given us a detailed view of Vesta—one of the largest asteroids with a diameter of about 525 km. *Dawn* orbited the asteroid for over a year and collected information on its surface, its composition, and its internal structure. Unlike Eros and Ida, Vesta was big enough to become warm and internally differentiate. Long ago, blobs of dense molten iron sank to form a metallic core about 100 km across and basaltic lavas erupted on its surface to form its crust. Thus, Vesta is like other terrestrial planets in many ways. Nonetheless, these events happened so long ago that Vesta's surface is still dominated by impact craters. It must have cooled quickly because of its small size. Huge chunks have been knocked off this miniature planet and it has a highly irregular shape (**Figure 25.26**). Its size and gravity were never large enough to pull it back into a sphere. Bits and pieces of igneous rocks have been knocked off Vesta and fell to Earth as distinctive types of stony meteorites.

In fact, most meteorites that fall to Earth as shooting stars or meteors are probably fragments of asteroids. By carefully studying the composition of meteorites, we have learned that many came from asteroids that had differentiated anciently, formed iron cores and silicate mantles, and Vesta developed a crust made of basaltic lava flows. These meteorites reveal that the parent asteroids are much like the inner planets in their compositions. Radiometric dates of meteorites have also established that the solar system formed during a short interval between 4.6 and 4.5 billion years ago.

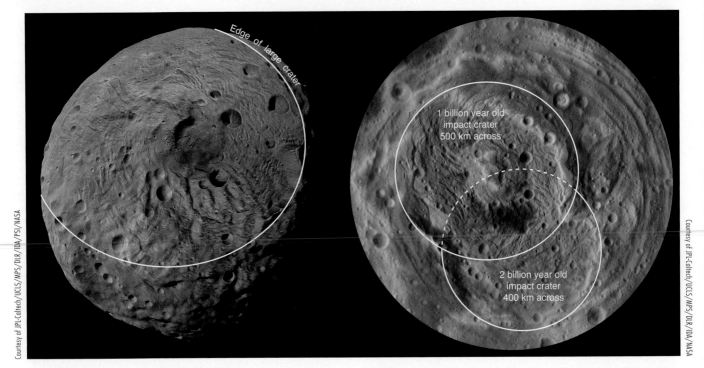

(A) The battered crust of Vesta is composed at least partly of basaltic lavas that erupted 4.5 billion years ago. A white line marks the rim of the large crater shown in the colored map (B).

(B) The rims of the large impact craters are revealed on this colored shaded relief map. The largest is almost 500 km across, has a large central peak, and may be only 1 billion years old. Colors show the range of elevations with blue (low) and red (high).

Figure 25.26 Vesta is one of the largest asteroids, but it still has an ancient surface marked by multiple impact craters and fractures.

Figure 25.27 Comets are actually small ice bodies that formed in the outer solar system. Some have elliptical orbits that take them near the Sun. When they enter the warmth of the inner solar system, the ice sublimes to vapor and forms a long tail that streams behind it. Hale-Bopp (1997) was probably the most spectacular comet seen in the latter half of the twentieth century.

It is commonly concluded from this information that asteroids and meteorites are remnants of a swarm of small bodies from which the inner planets formed. Thus, by carefully studying the meteorites, we can understand more about the conditions of formation and differentiation of planetary bodies.

Comets

Comets are the most distant members of the solar system. Some have orbits that take them so far from the Sun that it takes tens of thousands of years to complete a single revolution. Although we have not collected samples of any comet, telescopic and space probe studies of these small bodies show that they are composed basically of ice and dust and have a kinship with the icy bodies of the outer solar system (Figure 25.1). The ices of water, carbon dioxide, carbon monoxide, methane, and ammonia have been identified. These are mixed with various silicate minerals and metal particles. The mixture leads to the common notion that comets are "dirty snowballs."

Because comets have strongly elliptical (elongated) orbits with the Sun at one focus, they occasionally enter the inner solar system, where it is much warmer than in the outer solar system. The icy nucleus of a comet partially vaporizes when it comes close to the Sun, forming a large diffuse coma (the sphere of gas and dust around the nucleus) and spectacularly long tails of gas and dust. During the early months of 1997, Comet Hale-Bopp moved through the inner solar system (**Figure 25.27**). It treated stargazers to a spectacular view. As it swept through the inner solar system, its head enlarged and its tail became longer and longer as heat from the Sun vaporized ice inside the comet. (Comets do not glow from internal energy. They simply reflect sunlight off the molecules of gas and dust.) Even though the icy nucleus is only a few tens of kilometers across, the bright coma of this comet was as large as Jupiter and its tail extended millions of kilometers behind it. As Hale-Bopp moved back out of the inner solar system, the ice recondensed and the tail disappeared.

The origin and history of comets are very enigmatic. Comets may have originally formed near Uranus and Neptune, but subsequent gravitational perturbations from Jupiter must have ejected them to distant orbits that presently envelop the solar system. Periodically, some comets are gravitationally forced, perhaps by a passing star, into shorter elliptical orbits that take them into the inner solar system. Comets must be remnants of the planetesimals that accreted to form the outer planets and their satellites.

Using the Hubble Space Telescope, several of these remnants have been identified in the outer solar system. Ranging in size to as much as 200 km across, these icy bodies orbit beyond Neptune. More than 1200 have been discovered, and some estimates suggest that as many as 200 million may exist. In fact, Pluto and Triton may be large members of this group of outer solar system planetesimals.

Origin of the Solar System

The solar system probably formed by gravitational collapse of a huge cloud of gas and dust. The inner planets formed from dense silicates and metals that crystallized at high temperatures near the forming Sun, while the outer planets additionally included elements that form solids at low temperatures. Dense atmospheres became attached to the large icy cores of the outer planets.

Most scientists believe that the universe began about 15 billion years ago, in what has become known as the Big Bang. This gigantic explosion caused matter to expand outward from one point to form the billions of swirling galaxies and, in time, the stars and their planets. It is generally thought that our solar system was spawned in a cold, diffuse cloud of gas and dust, or a **nebula**, deep within a spiral arm of the Milky Way galaxy. The huge cloud was made up largely of the two lightest elements, hydrogen and helium, along with lesser oxygen and even smaller quantities of heavy elements, such as silicon and iron. The nebula rotated slowly about a central concentration of mass and contained a system of complicated eddies. Under the force of gravity, the giant

(A) A slowly rotating portion of a large nebula becomes a distinct globule as a mostly gaseous cloud collapses by gravitational attraction.

(B) Rotation of the cloud prevents collapse of the equatorial disk while a dense central mass forms.

(C) A protostar "ignites" and warms the inner part of the nebula, possibly vaporizing preexisting dust. As the nebula cools, condensation produces solid grains that settle to the central plane of the nebula.

(D) The dusty nebula clears by dust aggregation into planetesimals or by ejection during a T-Tauri stage of the star's evolution. A star and a system of cold bodies remain. Gravitational accretion of these small bodies leads to the development of a small number of major planets.

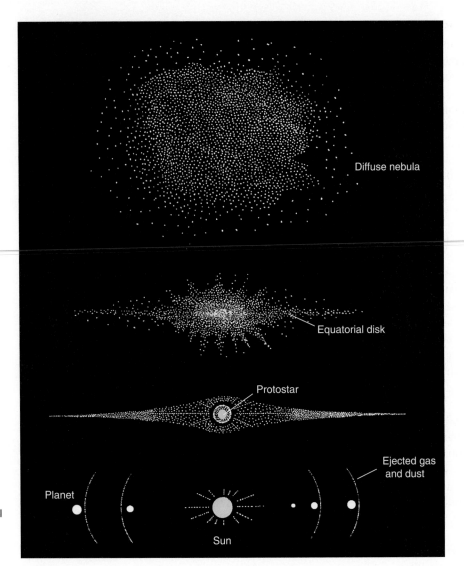

Figure 25.28 The evolution of a dusty nebula to a star with a surrounding system of orbiting planets.

How did silicates become concentrated in the inner planets and water ice in the planetary bodies of the outer solar system?

cloud began to collapse and assume the shape of a rotating disk, with an increasingly hot and dense mass at the center (**Figure 25.28**).

During the collapse, much of the cloud's matter swirled toward the dense central core to form the Sun. The outer part of the cloud was naturally the coldest, so substances there—such as water, ammonia, and methane—solidified as low-density ices. Nearer the Sun, those materials remained as vapor, but silicon, iron, aluminum, and similar materials could combine with oxygen and crystallize at high temperatures into solids, to form dense rocky material. However, these elements were not as abundant as the ice-forming materials. Thus, early in the history of the solar system, there was a separation and differentiation of material. Silicate minerals stable at high temperature were concentrated in the central region, whereas icy solids dominated near the fringes of the cloud.

Over a relatively short period (possibly as short as 100,000 years), the small particles in the embryonic solar system accreted into larger and larger particles, until asteroid-sized bodies of rock and ice called **planetesimals** formed. As the planetesimals orbited the infant Sun, the larger bodies grew by accretion as smaller objects repeatedly slammed into them. These planetesimals became the principal planets.

A planet's size and composition were therefore determined to a considerable degree by its distance from the Sun. In the high-temperature regions near the Sun, only materials such as the scarce metals and silicates crystallized into solids and accreted to form planets. Proceeding outward toward cooler and cooler temperatures, materials with lower crystallization temperatures, such as water and then methane and finally nitrogen, also became solid (ices) and accumulated to form planets. Because these volatile elements are much more abundant than the silicates, large icy bodies formed in the outer solar system. Huge amounts of gaseous hydrogen and helium became gravitationally anchored to these giant planets.

Most of the material of the nebula swirled inward toward the very center of the solar system. The intense pressure raised the temperature to a point where it became a vast nuclear furnace—a new star, the Sun. By this time, the principal planets and their satellites already orbited the Sun, and swept up most of the remaining debris in their orbital paths. This final stage of planetary accretion is clearly recorded as densely cratered terrain on the surfaces of the Moon, Mercury, Mars, and most other planetary bodies.

All planetary bodies were heated to some degree because of the impact of the numerous planetesimals that formed them. If heated sufficiently, much of the planet melted and the constituent materials became differentiated—that is, denser materials were separated and concentrated in the core and lighter materials were concentrated near the surface. This process is known as planetary **differentiation** and led to the layered internal structure of the solid inner planets and icy satellites of the outer planets (Figure 25.1).

What is the importance of planetary differentiation?

The Role of Impact Processes in the Origin of the Planets

With all these images of the planets before you, you can probably come to a simple but dramatic conclusion about the fundamental geologic processes in the solar system. Impact cratering may be the most important process in the origin and subsequent evolution of the planets. To understand further and emphasize the role of impact cratering in our solar system, let us consider several examples.

Impact Origin of the Moon   A dramatic hypothesis for the origin of the Moon involves a glancing collision of Earth with a Mars-sized object that vaporized and ejected material from the already differentiated Earth (**Figure 25.29**). The refractory

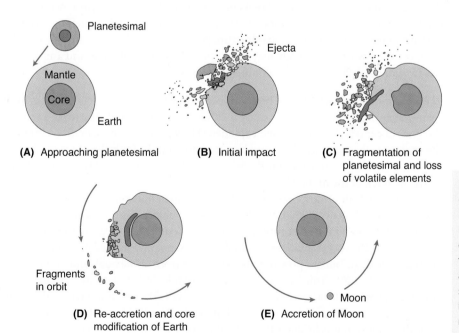

(A) Approaching planetesimal

(B) Initial impact

(C) Fragmentation of planetesimal and loss of volatile elements

(D) Re-accretion and core modification of Earth

(E) Accretion of Moon

Figure 25.29 A giant collision of the early Earth with a body the size of Mars may have ejected material into orbit, where it accreted to form the Moon. The iron core of the impacting body would have plunged through Earth's mantle and merged with the already formed core. Earth may have been stripped of its primordial atmosphere and been left with a globe-encircling ocean of magma.

(A) The track the meteorite was composed of particles and vapor shed from the meteor during its frictional heating.

(B) Fragments of the meteorite fell over a large area. At least one fell through the ice on this lake.

Figure 25.30 A large fireball exploded in the air over central Russia in 2013.

silicate portion of this material could have become solid again and accreted while in orbit around Earth to create a small water- and iron-poor natural satellite—the Moon.

Other Large Impacts   The role of similar large-body impacts in the evolution of other planets is a topic of increasing speculation. For example, Mercury's relatively high density may be explained partly as the result of a giant impact that stripped away the outer silicate layers of the already-differentiated planet, leaving it enriched in the dense metallic iron that formed its core. A late, large impact on Venus may have slowed its spin and reversed its rotational direction, as compared with that of all other planets. The global dichotomy between Mars's heavily cratered northern hemisphere and its relatively young, smooth northern plains may be traced back to a giant impact basin in the northern hemisphere. The small icy satellites of Saturn and Uranus, some scientists conjecture, were fragmented several times, only to re-accrete later. Moreover, the rings that encircle the outer planets may be created again and again by the collisional fragments of small icy moons. At the very least, several icy satellites sustained massive impacts that created global fracture systems, as well as large craters. Finally, a giant collision with a large body may have tipped Uranus on its side, and another may have fragmented Pluto to form a double-planet system.

What role has impact played in the evolution of the planets and moons of the solar system?

The Russian Fireball of 2013   The continuing importance of impact processes in the solar system was dramatically revealed in 2013 when a large fireball streaked across the skies of central Russia and then exploded (**Figure 25.30**). The shockwave from the explosion shattered windows and knocked down walls for miles. Amazingly, no one was critically injured, but hundreds were hurt by the broken glass in several cities along the track of the fireball.

As it entered the atmosphere, the meteor is estimated to have been about 20 m across and traveling at 65,000 km/h. The bus-sized object was probably the largest meteor to hit Earth in over a century. Friction caused the temperature to reach almost 20,000°K and the rocky chunk left behind a trail of hot pressurized gas and swirls of meteoritic dust. Intense pressure built up at the front of the body and eventually exceeded its strength. This caused the meteor to explosively burst into a multitude of fragments when it was between 15 and 20 km high. The explosion released 20 times more energy than the atomic bombs used in World War II. Momentarily, the light from the explosion was brighter than the Sun. A powerful shock wave and sonic boom closely followed the fireball. The trail of dust and vapor disappeared in a matter of minutes.

Most of the meteor vaporized as it went through the atmosphere, but small pieces showered the region. The fragments are the most common type of meteorite, so common that its official name is "ordinary" chondrite. This class of stony meteorite is the most common to fall to Earth and consists of a mixture of silicates and flecks of native iron.

In an amazing cosmic coincidence, on this same day scientists from around the world were tracking an asteroid that came closer to Earth than any other in recent history. It passed within 27,000 km of Earth, closer than some satellites, but it never posed a threat of actually striking the planet. The asteroid was measured by radar and is about 30 m in diameter. If it had hit Earth, it would have excavated a crater roughly 1 km across. The orbits of the small asteroid and the Russian meteor show that they are not related. The orbit of the asteroid takes it between Venus and Earth, but the calculated orbit for the meteor was much more eccentric and stretched beyond Mars into the asteroid belt

Impact is a fundamental process in planetary formation and evolution. It is not the dominant process it was 4.5 billion years ago, but meteorite impact still occurs today.

Conclusions

We live in an extraordinary period of geologic exploration. In a single lifetime, we have explored all of the planets except Pluto, with spacecraft flybys or orbital missions. By studying other planetary bodies, we gain a greater understanding of our own Earth. Its size and composition are just right for the development of a tectonic system that recycles the lithosphere, creates continents and ocean basins, and concentrates ores and minerals. Earth's gravitational field is strong enough to hold an atmosphere. Earth is just the right distance from the Sun so that water can exist as solid, liquid, and vapor and can move in a hydrologic cycle. If the planet were a little closer to the Sun, our oceans would evaporate; if farther from it, the oceans would freeze solid. Studying other planets has taught us that Earth is a small place, an oasis in space, a home we are still trying to understand.

GeoLogic   Chicxulub: Smoking Gun?

A major extinction of many forms of life, ranging from dinosaurs to microscopic marine plankton, marks the boundary between the Cretaceous and Paleogene time periods. What could have caused the extinction?

Observations

1. Iridium (Ir) concentrations are high in meteoritic material, but low in rocks found at Earth's surface.
2. High concentrations of the element iridium occur in sedimentary layers formed at the transition between the Cretaceous and Paleogene time periods (65 million years ago).
3. Fine fragments of minerals in the iridium layer have planar microfeatures that form only at intense but short-lived pressures.
4. The iridium layer is thickest in the areas near the Yucatan peninsula of southern Mexico.
5. Gravity surveys show that a large (nearly 200 km diameter) circular basin is buried below several kilometers of sediment in the Yucatan peninsula.

Interpretations

Though they took decades to accumulate, once in place these facts rapidly led to the Earth-shaking interpretation that our planet was struck by a large asteroid 65 million years ago. According to this widely accepted theory, its impact scattered Ir-rich ejecta across much of the world. The shock wave created microscopic shock features in mineral grains and excavated a large depression that rapidly collapsed as a central peak and surrounding moat formed. Many scientists are also convinced that the fine dust blasted into the atmosphere blocked the Sun and helped cause the **mass extinction** that included the dinosaurs.

Are there problems with this interpretation? Of course there are. For example, why was the extinction selective, taking some of the tiniest marine plankton and the biggest animals on the continents, but leaving others unscathed? Nonetheless, the evidence is overwhelming that an impact occurred. The possible relation of the impact to extinction will drive continued research in attempts to strengthen or ultimately reject the impact-extinction connection.

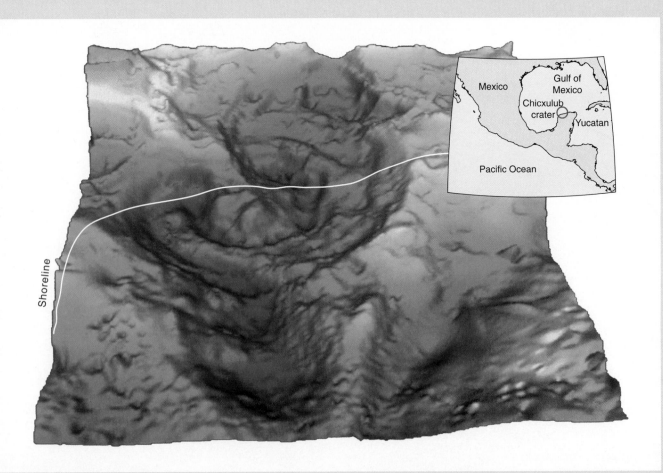

Shoreline

Mexico

Gulf of Mexico

Chicxulub crater

Yucatan

Pacific Ocean

Key Terms

accretion (p. 769)
asteroid (p. 790)
comet (p. 793)
differentiation (p. 795)
dwarf planet (p. 790)

ejecta blanket (p. 767)
gas giant (p. 765)
icy planetary body (p. 765)
impact crater (p. 767)
inner planet (p. 764)

mare (p. 767)
mass extinction (p. 798)
meteorite (p. 767)
multiring basin (p. 767)
nebula (p. 793)

outer planet (p. 765)
planetesimal (p. 794)
ray (p. 767)
tidal heating (p. 781)

Review Questions

1. Explain the meaning of the color code in the color bars in Table 25.1.
2. What geologic process has been most significant in modifying the surfaces of the Moon, Mercury, and Mars?
3. Outline the stages in the production of a crater by the impact of a meteorite. What geologic features are produced by impact?
4. How are craters modified with time?
5. Explain how a geologic time scale was developed for events in the Moon's history.
6. Outline the major events in lunar history.
7. Compare and contrast the geology of Mercury with that of the Moon.
8. Why is the Moon so poor in water and iron if it formed close to Earth, where both materials are abundant?
9. Describe the volcanoes on Mars.
10. What tectonic features are found on Mars?
11. Describe the fluvial features on the surface of Mars. How do they compare with fluvial features on Earth?

12. Describe the surface features generated by wind on Mars.
13. Compare and contrast the surface features of Venus with those on Earth and Mars.
14. Compare and contrast the sizes, densities, compositions, and surface features of the four large moons of Jupiter.
15. Why are the surfaces of the Galilean satellites of Jupiter so different in age? In composition and density?
16. Explain why Io is still volcanically active, whereas the Moon, which has a similar size and density, is not.
17. What is the significance of the major surface features on the Saturnian moon Enceladus?
18. How is Earth geologically unique among the planetary bodies of the solar system?
19. Contrast the compositions of asteroids and comets.
20. Why is there such a great composition and size difference between the inner and the outer planets?
21. What do you think is the most common rock type on the surfaces of the inner planets? On the moons of the outer planets?

Glossary

aa flow A lava flow with a surface typified by angular, jagged blocks. Contrast with pahoehoe flow.

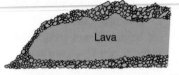

Lava

aa flow

A horizon The uppermost horizon or topsoil layer of Earth's surface, a thin, humus-rich dark layer down to a light, bleached layer. Compare with B and C horizons.

ablation Reduction of a glacier by melting, evaporation, iceberg calving, or deflation.

abrasion The mechanical wearing away of a rock by friction, rubbing, scraping, or grinding.

absolute age Geologic time measured in a specific duration of years (in contrast to relative time, which involves only the chronologic order of events). Also called numerical age.

absolute velocity Tectonic plate movement to a fixed reference frame.

abyssal Pertaining to the great depths of the oceans, generally 1000 fathoms (2000 m) or more below sea level.

abyssal floor Broad, relatively smooth deep-ocean basins between oceanic ridges. It includes the abyssal hills and plains.

abyssal hills The part of the ocean floor consisting of hills rising as much as 1000 m above the surrounding floor. They are found seaward of most abyssal plains and occur in profusion in basins

isolated from continents by trenches, ridges, or rises.

abyssal plains Flat areas of the ocean floor, having a slope of less than 1:1000. Most abyssal plains lie at the base of a continental rise and are simply areas where abyssal hills are completely covered with sediment.

accreted terrane A region or group of rocks sharing a common age, structure, stratigraphy, and origin.

accretion A process by which material is added to a tectonic plate or landmass or, in planetary geology, the growth of a planetary body by meteor impact.

accretionary heat The heat deposited from the impact of meteorites during the formation of planetary bodies.

accretionary wedge A wedge-shaped body of faulted and folded material scraped off subducting oceanic crust and added to an island arc or continental margin at a subduction zone.

acid mine drainage Water flow from mines that has a low pH, commonly as a result of the weathering of sulfide to sulfate minerals followed by their partial dissolution.

acid rain Rain with a low pH, commonly as a result of combining with sulfur dioxide pollutants in the atmosphere.

aftershock An earthquake that follows a larger earthquake. Generally, many aftershocks occur over a period of days or even months after a major earthquake.

agate A variety of cryptocrystalline quartz in which colors occur in bands. It is commonly deposited in cavities in rocks.

alcove A very shallow cave at the base of the normally dry waterfall.

alluvial fan A fan-shaped deposit of sediment built by a stream where it emerges from an upland or a mountain range into a broad valley or plain. Alluvial fans are common in arid and semiarid climates but are not restricted to them.

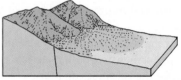

Alluvial fan

alluvium A general term for any sedimentary accumulations deposited by comparatively recent action of rivers. It thus includes sediment laid down in river beds, floodplains, and alluvial fans.

amorphous solid A solid in which atoms or ions are not arranged in a definite crystal structure. Examples: glass, amber, obsidian.

amphibole An important rock-forming mineral group of mafic silicates. Amphibole crystals are constructed from double chains of silicon-oxygen tetrahedra. Example: hornblende.

amphibolite A metamorphic rock consisting mostly of amphibole and plagioclase feldspar.

amphibolite facies A metamorphic facies formed with medium pressure and average to high temperature; in metamorphosed mafic rocks amphibole is a typical mineral.

andesite A fine-grained igneous rock composed mostly of plagioclase feldspar and from 25% to 40% pyroxene, amphibole, or biotite, but no quartz or K-feldspar. It is abundant in mountains bordering the Pacific Ocean, such as the Andes Mountains of South America, from which the name was derived.

angle of repose The steepest angle at which loose grains will remain stable without sliding downslope.

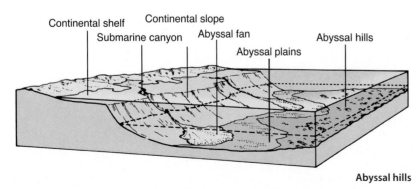

Continental shelf
Continental slope
Submarine canyon
Abyssal fan
Abyssal plains
Abyssal hills

Abyssal hills

angular unconformity An unconformity in which the older strata dip at a different angle (generally steeper) than the younger strata.

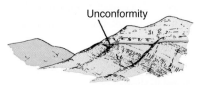

Angular unconformity

anion A negatively charged ion.

anomaly A deviation from the norm or average.

anorthosite A coarse-grained intrusive igneous rock composed primarily of calcium-rich plagioclase feldspar.

anticline A fold in which the limbs dip away from the hinge. After erosion, the oldest rocks are exposed in the central core of the fold.

Anticline

aphanitic texture A rock texture in which individual crystals are too small to be identified without the aid of a microscope. In hand specimens, aphanitic rocks appear to be dense and structureless.

aquifer A permeable stratum or zone below the Earth's surface through which groundwater moves.

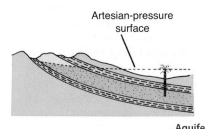

Aquifer

arch An arch-shaped landform produced by weathering and differential erosion.

Archean The eon of geologic time from 4.0 (or 3.8 on some scales) to 2.5 billion years ago.

arête A narrow, sharp ridge separating two adjacent glacial valleys.

arid A dry climate such as exists in deserts.

aridosol A type of soil that forms in arid climates where weathering is weak. These soils are dry and have little organic matter.

arkose A sandstone containing at least 25% feldspar.

artesian-pressure surface The level to which water in an artesian system would rise in a pipe high enough to stop the flow. Synonymous with potentiometric surface.

artesian water Groundwater in a confined aquifer and under pressure great enough to cause the water to rise above the top of the aquifer when it is tapped by a well.

ash Volcanic fragments the size of dust particles.

ash flow A turbulent blend of unsorted pyroclastic material (mostly fine-grained) mixed with high-temperature gases ejected explosively from a fissure or crater.

ash-flow caldera The largest silicic volcanoes on Earth; these are collapse craters surrounded by sheets of tuff, and form very low, broad shields.

ash-flow tuff A rock composed of volcanic ash and crystals, formed by deposition and consolidation of ash flows; also known as ignimbrite.

assimilation The process by which hot magma incorporates or dissolves the surrounding solid country rock.

asteroid A small, rocky planetary body orbiting the Sun. Asteroids are numbered in the tens of thousands. Most are located between the orbit of Mars and the orbit of Jupiter. Their diameters range downward from 1000 km.

asthenosphere The weak zone inside Earth directly below the lithosphere. Seismic velocities are distinctly lower in the asthenosphere than in adjacent parts of Earth's interior. The material in the asthenosphere is therefore believed to be soft and yielding to plastic flow.

asymmetric fold A fold (anticline or syncline) in which one limb dips more steeply than the other.

Asymmetric fold

atmosphere The mixture of gases surrounding a planet. The Earth's atmosphere consists chiefly of oxygen and nitrogen, with minor amounts of other gases. Synonymous with air.

atoll A ring of low coral islands surrounding a lagoon.

atom The smallest unit of an element. Atoms are composed of protons, neutrons, and electrons.

atomic mass The sum of the number of neutrons and protons.

atomic number The number of protons in the nucleus of an atom. It uniquely defines an element.

atomic weight The mass of one atom of an element, essentially the sum of the protons and neutrons in the nucleus of an atom.

avulsion The shifting of a river's course during the construction of a delta, generally occurring during a flood.

axis **1** (crystallography) An imaginary line passing through a crystal around which the parts of the crystal are symmetrically arranged. **2** (geophysics) A straight line about which a planet or moon rotates or spins.

axis of plate rotation Used to describe the motion of a plate moving on a sphere and measured as a rotation around an imaginary axis. Compare with pole of plate rotation.

B horizon The subsoil layer below the Earth's surface, composed of fine clays and colloids washed down from the topsoil. One of three primary soil profiles. Compare with A and C horizons.

back-arc basin The area behind a subduction-related volcanic arc where folds and faults form. Most oceanic back-arcs are extending.

back-arc spreading Extension behind a subduction-related volcanic arc typically related to submarine volcanism.

backswamp The wet area of a floodplain at some distance beyond and lower than the natural levees that confine the river.

backwash The return sheet flow down a beach after a wave is spent.

badlands An area nearly devoid of vegetation and dissected by stream erosion into an intricate system of closely spaced, narrow ravines.

bajada The surface of a system of coalesced alluvial fans.

Atoll

banded iron formation   Sedimentary deposits that consist of alternating layers of iron oxide and chert. They are the major sources of iron ore.

bar   An offshore, submerged, elongate ridge of sand or gravel built on the seafloor by waves and currents.

barchan dune   A crescent-shaped dune, the tips or horns of which point downwind. Barchan dunes form in desert areas where sand is scarce.

barrier island   An elongate island of sand or gravel formed parallel to a coast.

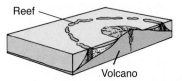

Barrier island

barrier reef   An elongate coral reef that trends parallel to the shore of an island or a continent, separated from it by a lagoon.

Barrier reef

basal drag   Resistance to flow exerted on the bottom of a tectonic plate by the underlying asthenosphere; shear at the base of the plate.

basalt   A dark-colored, fine-grained, mafic volcanic rock composed of plagioclase (over 50%) and pyroxene. Olivine may or may not be present.

basaltic plain   An area of low relief underlain by basaltic lava flows.

base level   The level below which a stream cannot effectively erode. Sea level is the ultimate base level, but lakes form temporary base levels for inland drainage systems.

basement complex   A series of igneous and metamorphic rocks lying beneath the oldest stratified rocks of a region. In shields, the basement complex is exposed over large areas.

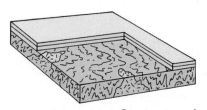

Basement complex

basin   **1** (structural geology) A circular or elliptical downwarp. After erosion, the youngest beds are exposed in the central part of the structure. **2** (topography) A depression into which the surrounding area drains.

batholith   A large body of intrusive igneous rock exposed over an area of at least 100 km².

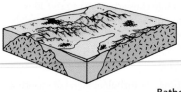

Batholith

bathymetry   The measurement of ocean depths and mapping of the topography of the ocean floor.

bauxite   A mixture of various amorphous or crystalline hydrous aluminum oxides and aluminum hydroxides, commonly formed by intense chemical weathering in tropical and subtropical regions. Bauxite is the principal ore of aluminum.

bay   A wide, curving recess or inlet between two capes or headlands.

baymouth bar   A narrow, usually submerged ridge of sand or gravel deposited across the mouth of a bay by longshore drift. Baymouth bars commonly are formed by extension of spits along embayed coasts.

beach   A deposit of wave-washed sediment along a coast between the landward limit of wave action and the outermost breakers.

beach drift   Occurs when sediment is moved laterally by oblique waves. It works together with the longshore current to create longshore drift.

bed   A layer of sediment 1 cm or more in thickness.

bedding plane   A surface separating layers of sedimentary rock.

bed load   Material transported along the bottom of a stream by rolling or sliding, in contrast to material carried in suspension or in solution.

bedrock   The continuous solid rock that underlies the regolith everywhere and is exposed locally at the surface. An exposure of bedrock is called an outcrop.

benioff zone   A zone of earthquakes that dips away from a deep-sea trench and slopes beneath the adjacent continent or island arc.

biochemical rock   A type of sedimentary rock formed during the growth of organisms such as algae, coral, or swamp vegetation.

biochemical sediment   A sediment made of material precipitated as a result of biological processes, such as shells made of calcium carbonate.

biomass   Materials of biological origin.

biosphere   The totality of life on or near Earth's surface.

biotite   "Black mica." An important mafic silicate with silicon-oxygen tetrahedra arranged in sheets.

bird-foot delta   A delta with distributaries extending seaward and in map view resembling the claws of a bird. Example: the Mississippi Delta.

black smoker   Hot water and dissolved ions from the oceanic crust are released at central vents on the seafloor to form dark-colored plumes of mineral precipitates dominated by sulfides. Compare with white smoker.

block faulting   A type of normal faulting in which segments of the crust are broken and displaced to different elevations and orientations.

blowout   A dune shaped like a parabola with the concave side toward the wind. Commonly formed along shorelines (same as a parabolic dune).

blueschist   A fine-grained schistose rock characterized by high-pressure, low-temperature mineral assemblages and typically containing the blue amphibole glaucophane.

blueschist facies   A metamorphic facies formed by recrystallization at relatively low temperature but high pressure, for example, in a subduction zone. In metamorphosed basalts, its blue color comes from the mineral glaucophane.

boulder   A rock fragment with a diameter of more than 256 mm (about the size of a volleyball). A boulder is one size larger than a cobble.

bracketed intrusion   An intrusive rock that was once exposed at the surface by erosion and was subsequently covered by younger sediment. The relative age of the intrusion thus falls between, or is bracketed by, the ages of the younger and older sedimentary deposits.

braided stream   A stream with a complex of converging and diverging channels separated by bars or islands. Braided streams form where more sediment is available than can be removed by the discharge of the stream.

breaker   A collapsing water wave.

breccia   A general term for sediment consisting of angular fragments in a matrix of finer particles. Examples:

sedimentary breccias, volcanic breccias, fault breccias, impact breccias.

brine   Saline water rich in dissolved constituents, commonly formed by evaporation of surface waters but they can also form in the subsurface.

brittle deformation   Occurs under certain strain conditions when rock bodies change shape by breaking to form continuous fractures and lose cohesion. Broken or fractured in contrast to plastic flow.

butte   A somewhat isolated hill, usually capped with a resistant layer of rock and bordered by talus. A butte is an erosional remnant of a formerly more extensive slope.

Butte

C horizon   A zone of partially disintegrated and decomposed bedrock below the B horizon of the Earth's surface. The rock fragments are often weathered, spheroidal boulders that may be completely decomposed.

calcite   A mineral composed of calcium carbonate ($CaCO_3$).

caldera   A large, more or less circular depression or basin associated with a volcanic vent. Its diameter is many times greater than that of the included vents. Calderas are believed to result from subsidence or collapse and may or may not be related to explosive eruptions.

calving   The breaking off of large blocks of ice from a glacier that terminates in a body of water.

capacity   The maximum quantity of sediment a given stream, glacier, or wind can carry under a given set of conditions.

carbon 14   A radioactive isotope of carbon (^{14}C). Its half-life is 5730 years.

carbonaceous   Containing carbon.

carbonate mineral   A mineral formed by the bonding of carbonate ions (CO_3^{2-}) with positive ions. Examples: calcite ($CaCO_3$), dolomite [$CaMg(CO_3)_2$].

carbonate rock   A rock composed mostly of carbonate minerals. Examples: limestone, dolomite.

catastrophism   The belief that geologic history consists of major catastrophic events involving processes that were far more intense than any we observe now. Contrast with uniformitarianism.

cation   A negatively charged ion.

cave   A naturally formed subterranean open area, chamber, or series of chambers, commonly produced in limestone by solution activity or in basalt flows as lava tubes.

cement   Minerals precipitated from groundwater in the pore spaces of a sedimentary rock and binding the rock's particles together.

cementation   The process of binding the clasts in a sedimentary rock together to make solid rock by precipitation of minerals from pore water.

Cenozoic   The era of geologic time from the end of the Mesozoic Era (65 million years ago) to the present.

chalcedony   A general term for fibrous cryptocrystalline quartz.

chalk   A variety of limestone composed of shells of microscopic oceanic organisms.

Channeled Scablands   An area in eastern Washington marked by a network of braided channels from 15 to 30 m deep. They are created by catastrophic flooding when dams of glacial ice broke to release impounded lake waters.

chemical precipitate   Sedimentary rocks form by the precipitation of inorganic material from lakes or shallow seas. Limestones and evaporites are two common types.

chemical weathering   Chemical reactions that act on rocks exposed to water and the atmosphere so as to change their unstable mineral components to more stable forms. Oxidation, hydrolysis, carbonation, and direct solution are the most common reactions.

chert   A sedimentary rock composed of granular cryptocrystalline silica.

cinder   A fragment of volcanic ejecta from 0.5 to 2.5 cm in diameter.

cinder cone   A cone-shaped hill composed of loose volcanic fragments erupted from a central vent.

cirque   An amphitheater-shaped depression at the head of a glacial valley, excavated mainly by ice plucking and frost wedging.

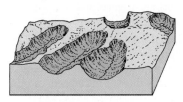

Cirque

clastic   **1** Pertaining to fragments (such as mud, sand, and gravel) produced by

the mechanical breakdown of rocks. **2** A sedimentary rock composed chiefly of consolidated clastic material.

clastic texture   The texture of sedimentary rocks consisting of fragments of minerals, rocks, and organic skeletal remains.

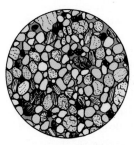

Clastic texture

clay   Sedimentary material composed of fragments with a diameter of less than 1/256 mm. Clay particles are smaller than silt particles.

clay minerals   A group of hydrous silicates formed by weathering of minerals such as feldspar, pyroxene, or amphibole. Silicate tetrahedra are arranged in sheets.

claystone   A compact, very fine-grained rock consisting of consolidated clay-sized particles.

cleavage   The tendency of a mineral to break in a preferred plane in the crystal lattice.

climate   The long-term average of precipitation, temperature, and wind direction and orientation.

climate system   The dynamic system of moving water and air that controls Earth's weather (temperature, precipitation, and wind) on a long-term basis.

closed system   In geology, a system that exchanges only energy (usually heat), but no matter with its surroundings.

coal   A common fuel mineral made mostly of carbon resulting from the metamorphic decomposition of the remains of terrestrial plants. Found in sedimentary rock.

coastal upwelling   Rise of deep ocean water along the coasts of continents. These waters are cold and rich in nutrients.

cobble   A rock fragment with a diameter between 6.4 cm (about the size of a tennis ball) and 25.67 cm (about the size of a volleyball). Cobbles are larger than pebbles but smaller than boulders.

cohesive strength   The strength caused by cohesive forces between atoms in a mineral or rock. This strength is exceeded when the rock fractures without plastic deformation.

collecting system   The network of tributaries in the headwater region that collect and funnel water and sediment to the main stream.

color   An obvious, yet not diagnostic, property of a mineral.

column   Vertical rock pillars that are vestiges of a massive cliff with well-developed joints that remain after weathering and erosion.

columnar jointing   A system of fractures that splits a rock body into long prisms, or columns. It is characteristic of lava flows and shallow intrusive igneous flows.

Columnar joint

comet   A small icy object in orbit around the Sun. The orbits of many comets are elliptical and when they near the Sun, the ice sublimes to make a fuzzy head and long tail of gas and dust.

compaction   A reduction in volume caused by the weight of overlying materials. For example, loose sediment may be compacted into a tight, coherent mass by the accumulation of more sediment on top of it.

competence   The maximum size of particles that a given stream, glacier, or wind can move at a given velocity.

composite volcano   A large volcanic cone built by extrusion of ash, lava, and shallow intrusions. Synonymous with stratovolcano.

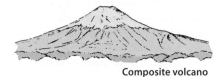

Composite volcano

compound   A substance made of two or more elements bound together.

compressing flow   Glacial ice experiences compression below the snow line because upvalley ice is continually pushing against slow moving ice at the front of the glacier. Compare with extending flow.

compression   A system of stresses that tends to reduce the volume of or shorten a substance.

compressional wave   See primary waves (P waves).

conchoidal fracture   A type of fracture that produces a smooth, curved surface. It is characteristic of quartz and obsidian.

concretion   A spherical or ellipsoidal nodule formed by accumulation of mineral matter after deposition of sediment.

condensation   The process by which a vapor becomes a liquid or a solid.

conduction   Transmission of heat energy by the impact of moving atoms. Contrast with convection.

cone of depression   A conical depression of the water table surrounding a well after heavy pumping.

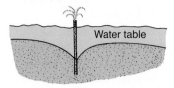

Cone of depression

confined aquifer   Water-bearing permeable rock units enclosed between impermeable strata such as shales. See aquifer.

confining pressure   The pressure or stress caused by the weight of overlying material.

conglomerate   A coarse-grained sedimentary rock composed of rounded fragments of pebbles, cobbles, or boulders.

consequent stream   On a coastal plain, the major streams developed as sea level decreases; they follow the direction of the initial slope.

contact   The surface separating two different rock bodies.

contact metamorphism   Metamorphism of a rock near its contact with a magma.

continent   A large landmass composed mostly of granitic rock. Continents rise abruptly above the deep-ocean floor and include the marginal areas submerged beneath sea level.

continental accretion   The growth of continents by incorporation of deformed sediments, arc magmas, and accreted terranes along their margins.

continental crust   The type of crust underlying the continents, including the continental shelves. The continental crust is commonly about 35 to 70 km thick. Its density is typically 2.7 g/cm³, and the velocities of primary seismic waves traveling through the crust are less than 6.2 km/sec. Contrast with oceanic crust.

continental drift   The theory that the continents move in relation to one another.

continental glacier   A thick ice sheet covering large parts of a continent. Present-day examples are found in Greenland and Antarctica.

continental margin   The zone of transition from a continent to the adjacent ocean basin. It generally includes a continental shelf, continental slope, and continental rise.

continental rift   Breakup of a continent by extension, normal faulting, and thinning; may eventually form a divergent plate.

continental rise   The gently sloping surface located at the base of a continental slope (see diagram for abyssal hills).

continental shelf   The submerged margin of a continental mass extending from the shore to the first prominent break in slope, which usually occurs at a depth of about 120 m.

continental slope   The slope that extends from a continental shelf down to the ocean deep. In some areas, such as off eastern North America, the continental slope grades into the more gently sloping continental rise.

continent-continent convergence
Occurs when continents on two different tectonic plates collide. The Himalaya mountain belt is produced by this process.

contraction   In tectonics, deformation causing a body or rocks to shorten and fold or thicken. Contrast with extension.

convection   Transmission of heat energy by the rise of buoyant hot material and sinking of cold material. Contrast with diffusion and conduction.

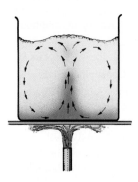

Convection

convection cell   The space occupied by a single convection current.

convergent plate boundary   A plate boundary at which plates collide. Convergent plate boundaries are sites of considerable geologic activity and are characterized by volcanism, earthquakes, and crustal deformation. See also subduction zone.

coquina   A limestone composed of an aggregate of shells and shell fragments.

Coquina

coral   A bottom-dwelling marine invertebrate organism of the class *Anthozoa*. Most build hard skeletons of calcium carbonate.

core   The central part of the Earth below a depth of 2900 km. The core is thought to be composed mostly of iron, in contrast to the overlying mantle of silicate rock.

Coriolis effect   The tendency of moving fluids on Earth's surface to be deflected to the right in the Northern Hemisphere and to the left in the Southern Hemisphere. Caused by Earth's spin.

correlation   Tieing rock units together in time by using their physical attributes; for example, fossils can show that two outcrops are the same age.

country rock   A general term for rock surrounding an igneous intrusion.

covalent bond   A chemical bond in which electrons are shared between different atoms so that none of the atoms has a net charge.

crater   An abrupt circular depression formed by extrusion of volcanic material, by collapse, or by the impact of a meteorite.

Crater

craton   The stable continental crust, including the shield and stable platform areas, most of which have not been affected by significant tectonic activity since the close of the Precambrian Era.

creep   The imperceptibly slow down-slope movement of material as a result of gravity.

crevasse   **1** (glacial geology) A deep crack in the upper surface of a glacier. **2** (natural levee) A break in a natural levee.

cross-bedding   Stratification inclined to the original horizontal surface upon which the sediment accumulated. It is produced by deposition on the slope of a dune or sand wave.

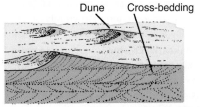

Cross-bedding

crosscutting relations, principle of   The principle that a rock body is younger than any rock across which it cuts.

crust   The outermost compositional layer, or shell, of Earth (or any other differentiated planet). The crust consists of low-density materials compared to the underlying mantle. Earth's crust is generally defined as the part of the Earth above the Mohorovičić discontinuity. It represents less than 1% of Earth's total volume. See also continental crust, oceanic crust.

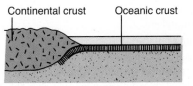

Crust

cryptocrystalline texture   The texture of rocks composed of crystals too small to be identified with an ordinary microscope.

crystal   A solid, polyhedral form bounded by naturally formed plane surfaces resulting from growth of a crystal lattice.

crystal face   A smooth plane formed by growth of the surface of a crystal.

Crystal face

Creep

crystal form   The geometric shape of a crystal. Examples: cubic, prismatic.

crystal lattice   A systematic, symmetrical network of atoms within a crystal.

crystalline texture   The rock texture resulting from simultaneous growth of crystals.

crystallization   The process of crystal growth. It occurs as a result of condensation from a gaseous state, precipitation from a solution, or cooling of a melt.

crystal structure   The orderly arrangement of atoms in a crystal.

cuesta   An elongate ridge formed on the tilted and eroded edges of gently dipping strata.

daughter isotope   An isotope produced by radioactive decay of its parent isotope. The quantity of a daughter isotope continually increases with time.

debris flow   The rapid downslope movement of debris (rock, soil, and mud).

declination, magnetic   The horizontal angle between true north and magnetic north at a given point on Earth's surface.

decompression melting   Melting of rock caused by a release of pressure. Common where the mantle rises beneath mid-ocean ridges and in mantle plumes.

deep-focus earthquake   An earthquake that occurs between 300 and 700 km below Earth's surface; generally confined to convergent plate margins.

deep-marine environment   The sedimentary environment of the abyssal plains.

deep-ocean basin   Half of Earth's surface is composed of these enormous expanses of sediment-covered basaltic crust under the sea.

deep-sea fan   A cone-shaped or fan-shaped deposit of land-derived sediment located seaward of large rivers or submarine canyons. Synonymous with abyssal cone, abyssal fan, and submarine cone.

deep-sea trench   See trench.

deep water   The thick zone of cold water that lies below the surface and transitional layers of the ocean.

deflation   Erosion of loose rock particles by wind.

deflation basin   A shallow depression formed by wind erosion where groundwater solution activity has left unconsolidated sediment exposed at the surface.

dehydration　Loss of water.

delta　A body of sediment deposited at the mouth of a river. Many are roughly triangular in shape.

Delta

dendritic drainage pattern　A branching stream pattern, resembling the branching of certain trees, such as oaks and maples.

density　Mass per unit volume, expressed in grams per cubic centimeter (g/cm^3).

density current　A current that flows as a result of differences in density. In oceans, density currents are produced by differences in temperature, salinity, and turbidity (the concentration of material held in suspension).

deposition　The process by which sediment settles out of a transporting fluid.

deranged drainage　A distinctively disordered drainage pattern formed in a recently glaciated area. It is characterized by irregular direction of stream flow, few short tributaries, swampy areas, and many lakes.

Deranged drainage

desert　Earth's regions of low precipitation. They are often dry, barren areas covered with eolian sand or lag gravels.

desert climate　A climate with generally high temperatures, high rates of evaporation, and low precipitation. Most deserts lie at about 30 degrees north or south of the equator.

desertification　The process of transforming arid land into a barren desert. Often induced by human activities or climate change.

desert pavement　A veneer of pebbles left in place where wind has removed the finer material.

detachment fault　A low angle fault into which more steeply dipping faults merge. It is generally associated with normal faults formed during extension.

detrital　**1** Pertaining to detritus. **2** A rock formed from detritus.

detritus　A general term for loose rock fragments produced by mechanical weathering.

diapir　Masses of mobile rock shaped somewhat like inverted teardrops. Some form in thick sequences of sedimentary rock, when lighter weight material buoys to the surface to form structural domes. Rising magma may also move in diapirs.

differential erosion　Variation in the rate of erosion on different rock masses. As a result of differential erosion, resistant rocks form steep cliffs, whereas nonresistant rocks form gentle slopes.

differential stress　A condition in which the stress applied to a rock body is not the same in all directions.

differential weathering　Occurs when different rock masses weather at different rates.

differentiated planet　A planetary body in which various elements and minerals are separated according to density and concentrated at different levels. Earth, for example, is differentiated, with heavy metals (iron and nickel) concentrated in the core; lighter minerals in the mantle; and still lighter materials in the crust, hydrosphere, and atmosphere.

differentiation　The process by which magma changes composition usually by fractional crystallization, magma mixing, or assimilation of wall rocks. See magmatic differentiation and planetary differentiation.

diffusion　The movement of ions through a solid, liquid, or gas, typically driven by pressure, temperature or composition differences. Contrast with convection.

dike　A tabular intrusive rock that cuts across strata or other structural features of the surrounding rock.

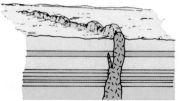

Dike

dike swarm　A group of associated dikes.

diorite　A phaneritic intrusive igneous rock consisting mostly of intermediate plagioclase feldspar and pyroxene, with some amphibole and biotite.

dip　The angle between the horizontal plane and a structural surface (such as a bedding plane, a joint, a fault, foliation, or other planar features).

disappearing stream　A stream that disappears into an underground channel and does not reappear in the same, or in an adjacent, drainage basin. In karst regions, streams commonly disappear into sinkholes and follow channels through caves.

discharge　Rate of flow; the volume of water moving through a given cross section of a stream in a given unit of time.

disconformity　An unconformity in which beds above and below are parallel.

discontinuity　A sudden or rapid change in physical properties of rocks within Earth. Discontinuities are recognized by seismic data. See also Mohorovičić discontinuity.

dispersing system　A network of distributaries at the mouth of a river, where sediment and water are released into an ocean, a lake, or a dry basin.

dissolution　The process by which materials are dissolved.

dissolved load　The part of a stream's load that is carried in solution.

distributary　Any of the numerous stream branches into which a river divides where it reaches its delta.

divergent plate boundary　A plate margin formed where the lithosphere splits into plates that drift apart from one another. Divergent plate boundaries are areas subject to tension, where new crust is generated by igneous activity. See also oceanic ridge.

divide　A ridge separating two adjacent drainage basins.

dolomite　**1** A mineral composed of $CaMg(CO_3)_2$. **2** A sedimentary rock composed primarily of the mineral dolomite.

dolostone　A sedimentary rock composed mostly of the mineral dolomite. Sometimes referred to simply as dolomite.

dome　**1** (structural geology) An uplift that is circular or elliptical in map view, with beds dipping away in all directions from a central area. **2** (topography) A general term for any dome-shaped landform.

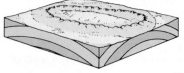

Dome

downcutting   The erosion process of abrasion of a stream channel, gully, or canyon floor by sand and gravel as they are swept downstream by the flowing water.

downwarp   A downward bend or subsidence of a part of Earth's crust.

drag   In blowing wind, caused by the impact of air molecules on grain surfaces. May cause particles to shake and then lift off, spinning into the air.

drainage basin   The total area that contributes water to a single drainage system.

drift   A general term for sediment deposited directly on land by glacial ice or deposited in lakes, oceans, or streams as a result of glaciation.

drip curtain   A thin sheet of dripstone hanging from the ceiling or wall of a cave.

dripstone   A cave deposit formed by precipitation of calcium carbonate from groundwater entering an underground cavern.

drumlin   A smooth, glacially streamlined hill that is elongate in the direction of ice movement. Drumlins are generally composed of till.

Drumlin

ductile   Easily bent. Contrast with brittle.

ductile deformation   Permanent deformation of a rock body without fracturing or losing cohesion. Contrast with brittle deformation.

dune   A low mound of fine-grained material that accumulates as a result of sediment transport in a current system. Dunes have characteristic geometric forms that are maintained as they migrate. Sand dunes are commonly classified according to shape. See also barchan dune, parabolic dune, seif dune, star dune, and transverse dune.

dwarf planet   A planetary body that directly orbits the Sun and is massive enough to be spherical or ellipsoidal in shape, but that has not cleared its orbital region of other objects. Pluto, Eris, and Ceres are examples.

dynamic system   An interacting group of objects and processes that cause change when materials and energy move and convert from one form to another.

earthquake   A series of elastic waves propagated in Earth, initiated where stress along a fault exceeds the elastic limit of the rock so that sudden movement occurs along the fault.

eclogite   A high-grade metamorphic rock made of garnet and pyroxene and lacking plagioclase. As a result, eclogite has a high density.

eclogite facies   A metamorphic facies formed at relatively high pressure and temperature at the base of the continental crust or in a slab of subducting oceanic lithosphere. In rocks with basaltic composition, its typical mineral assemblage is garnet plus clinopyroxene without plagioclase.

ecology   The study of relationships between organisms and their environments.

ecosphere   A small, sealed globe containing a closed system in which plants and animals are self-sustaining.

ejecta   Rock fragments, glass, and other material thrown out of an impact crater or a volcano.

ejecta blanket   Rock material (crushed rock, large blocks, breccia, and dust) ejected from an impact crater or explosion crater and deposited over the surrounding area.

elastic deformation   Temporary deformation of a substance, after which the material returns to its original size and shape. Example: the bending of mica flakes.

elastic limit   The maximum stress that a given substance can withstand without undergoing permanent deformation either by solid flow or by rupture.

elastic-rebound theory   The theory that earthquakes result from energy released by faulting; the sudden release of stored strain creates earthquake waves.

electron   A negatively charged subatomic particle.

elevated marine terrace   A wave-cut platform and marginal sea cliff lifted by tectonic forces above sea level.

El Niño   A warm ocean current on the west coast of South America that occasionally forms, when the normally strong trade winds weaken, allowing warmer water to approach the shore, disrupting the upwelling of cold nutrient-rich water and leading to weather changes over much larger areas.

end moraine   A ridge of till that accumulates at the margin of a glacier.

energy   A measure of the amount of work that can be done, usually measured in ergs (cgs) or joules (mks).

entrenched meander   A meander cut into the underlying rock as a result of regional uplift or lowering of the regional base level.

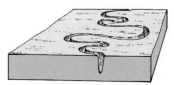

Entrenched meander

eolian   Pertaining to wind.

eolian environment   The sedimentary environment of deserts, where sediment is transported and deposited primarily by wind.

eolian system   The system of wind and the sediment it moves driven by the circulation of the atmosphere.

eon   A major subdivision of geologic time consisting of eras. Example: Phanerozoic Eon.

epicenter   The area on Earth's surface that lies directly above the focus of an earthquake.

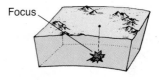

Epicenter

epoch   A division of geologic time; a subdivision of a period. Example: Pleistocene epoch.

equilibrium   A condition of the lowest possible energy, where the result of the forces acting on a system is zero.

era   A division of geologic time; a subdivision of an eon. Example: Mesozoic Era.

Eratosthenian Period   The period of lunar history when large craters, the rays of which are no longer visible, such as Eratosthenes, were formed (from 3.1 billion to 0.8 billion years ago).

erg   A large area covered with sand dunes. A sand sea such as those found in Earth's large deserts.

erosion   The processes that loosen sediment and move it from one place to another on Earth's surface. Agents of erosion include water, ice, wind, and gravity.

erratic   A large boulder carried by glacial ice to an area far removed from its point of origin.

Erratic

escarpment   A cliff or very steep slope.

esker   A long, narrow, sinuous ridge of stratified glacial drift deposited by a stream flowing beneath a glacier in a tunnel or in a subglacial stream bed.

estuary   A bay at the mouth of a river formed by subsidence of the sand or by a rise in sea level. Fresh water from the river mixes with and dilutes seawater in an estuary.

eustatic change of sea level   A worldwide rise or fall in sea level resulting from a change in the volume of water or the capacity of ocean basins.

evaporite   A rock composed of minerals derived from evaporation of mineralized water. Examples: rock salt, gypsum.

exfoliation   A weathering process by which concentric shells, slabs, sheets, or flakes are successively broken loose and stripped away from a rock mass.

Exfoliation

exposure   Bedrock not covered with soil or regolith; outcrop.

extending flow   Glacial ice experiences extension above the snow line because velocities of ice flow increase progressively from the head to the snow line. Thus, the ice is under tension and is constantly pulling away from upvalley ice. Compare with compressing flow.

extension   In tectonics, deformation causing a body of rocks to lengthen, stretch, or thin. Contrast with contraction. Commonly associated with normal faulting and divergent plate boundaries.

extrusive rock   A rock formed from a mass of magma that flowed out on the surface of Earth. Example: basalt.

faceted spur   A spur or ridge that has been beveled or truncated by faulting, erosion, or glaciation.

facies   A distinctive group of characteristics within part of a rock body (such as composition, grain size, or fossil assemblages) that differ as a group from those found elsewhere in the same rock unit. Examples: conglomerate facies, shale facies, and brachiopod facies.

fan   A fan-shaped deposit of sediment. See also alluvial fan and deep-sea fan.

fault   A surface along which a rock body has broken and been displaced.

fault block   A rock mass bounded by faults on at least two sides.

fault scarp   A cliff produced by faulting.

faunal succession, principle of   The principle that fossils in a stratigraphic sequence succeed one another in a definite, recognizable order.

feldspar   A mineral group consisting of silicates of aluminum and one or more of the metals potassium, sodium, or calcium. Examples: K-feldspar, Ca-plagioclase, and Na-plagioclase.

felsic   The minerals feldspar and quartz or an igneous or metamorphic rock made predominantly of feldspar and quartz. Contrast with mafic.

firn   Granular ice formed by recrystallization of snow. It is intermediate between snow and glacial ice. Sometimes referred to as neve.

fissure   An open fracture in a rock.

fissure eruption   Extrusion of lava along a fissure.

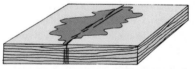

Fissure eruption

fjord   A glaciated valley flooded by the sea to form a long, narrow, usually steep walled inlet, extending inland as much as 100 km.

flint   A popular name for dark-colored chert (cryptocrystalline quartz).

flood basalt   An extensive flow of basalt erupted chiefly along fissures. Synonymous with plateau basalt.

floodplain   The flat, occasionally flooded area bordering a stream.

fluvial   Pertaining to a river or rivers.

fluvial environment   The sedimentary environment of river systems.

flux   **1** A substance that when added to another lowers its melting point. Water is a common flux inside Earth, or **2** flow.

flux melting   Melting caused by the addition of water (or other volatiles). Common in subduction zones.

focus   The area within Earth where an earthquake originates.

fold   A bend, or flexure, in a rock.

Fold

folded mountain belt   A long, linear zone of Earth's crust where rocks have been intensely deformed by horizontal stresses and generally intruded by igneous rocks. The great folded mountains of the world (such as the Appalachians, the Himalayas, the Rockies, and the Alps) are believed to have been formed at convergent plate margins.

foliation   A planar feature in metamorphic rocks, produced by the secondary growth of minerals. Three major types are recognized: slaty cleavage, schistosity, and gneissic layering.

footwall   The block beneath a dipping fault surface.

foraminifer   Single-celled organisms that secrete calcium carbonate shells. They are an important source of biochemical sediment in the oceans.

forearc   At a convergent plate margin, the region between the trench and volcanic arc. The forearc is underlain by a long sedimentary basin and accretionary prism.

forearc ridge   A topographical feature between a trench and an associated volcanic arc, underlain by the accretionary wedge of sediment scraped off the seafloor during subduction.

foreshore   The seaward part of the shore or beach lying between high tide and low tide.

formation   A distinctive body of rock that serves as a convenient unit for study and mapping.

fossil   Naturally preserved remains or evidence of past life, such as bones, shells, casts, impressions, and trails.

fossil fuel   A fuel containing solar energy that was absorbed by plants and animals in the geologic past and thus is preserved in organic compounds in their remains. Fossil fuels include petroleum, natural gas, and coal.

fractional crystallization   The separation of crystals and melt that causes the residual magma to progressively change its composition. Early crystallized mafic minerals commonly are separated by gravitational settling, so that the residual magma is left enriched in silica, sodium, and potassium.

fracture   An irregular break in a rock or a break in a crystal that is not parallel to a crystal face.

fracture zone   **1** (field geology) A zone where the bedrock is cracked and fractured. **2** (tectonics) A zone of long, linear fractures on the ocean floor, expressed topographically by ridges and troughs. Fracture zones are the topographic expression of transform faults.

friction   **1** Resistance along transform faults and between the converging slabs of lithosphere in a subduction zone. **2** Shear between two plates.

fringing reef   A reef that lies alongside the shore of a landmass.

Fringing reef

frost heaving   The lifting of unconsolidated material by the freezing of subsurface water.

frost wedging   The forcing apart of rocks by the expansion of water as it freezes in fractures and pore spaces.

gabbro   A dark-colored, coarse-grained rock composed of Ca-plagioclase, pyroxene, and possibly olivine, but no quartz.

gas   The state of matter in which a substance has neither independent shape nor independent volume. Gases can be compressed and tend to expand indefinitely.

gas giant   A large planet composed mainly of hydrogen and helium; it may also have a core of ice and rocky and metallic materials. In our solar system, Jupiter and Saturn are good examples.

gelisol   A type of soil found in cold climates that contains permafrost.

geode   A hollow nodule of rock lined with crystals; when separated from the rock body by weathering, it appears as a hollow, rounded shell partly filled with crystals.

Geode

geologic column   A diagram representing divisions of geologic time and the rock units formed during each major period.

geologic cross section   A diagram showing the structure and arrangement of rocks as they would appear in a vertical plane below Earth's surface.

geologic map   A map showing the distribution of rocks at Earth's surface.

geologic time scale   The time scale determined by the geologic column and by radiometric dating of rocks.

geology   The study of Earth and other planets.

geosphere   That part of the Earth that is solid, in contrast to the atmosphere, hydrosphere, and biosphere.

geothermal   Pertaining to the heat of the interior of Earth.

geothermal energy   Energy extracted from steam and hot water found within Earth's crust.

geothermal gradient   The rate at which temperature increases with depth inside a planet.

geyser   A thermal spring that intermittently erupts steam and boiling water.

glacial environment   The sedimentary environment of glaciers and their meltwaters.

glacial plucking   The process of glacial erosion by which large rock fragments are loosened by ice wedging, become frozen to the bottom surface of the glacier, and are torn out of the bedrock and transported by the glacier as it moves. The process involves the freezing of subglacial meltwater that seeps into fractures and bedding planes in the rock.

glacial striation   Numerous parallel, shallow scratches, several tens of centimeters long, formed by angular particles dragged across a rock surface by flowing ice.

glacier   A mass of ice formed from compacted, recrystallized snow that is thick enough to flow plastically.

glacier system   The interconnected components of the climate, hydrologic, and tectonic system involved with moving ice.

glass   **1** A state of matter in which a substance displays many properties of a solid but lacks crystal structure. **2** An amorphous igneous rock formed from a rapidly cooling magma.

glassy texture   The texture of igneous rocks in which the material is in the form of natural glass rather than crystal.

global change   A worldwide change, usually referring to a change in climate of the entire planet and not of just a local area or region.

global warming   A rise in the average temperature of the atmosphere.

glossopteris flora   An assemblage of late Paleozoic fossil plants named for the seed fern *Glossopteris*, one of the plants in the assemblage. These flora are widespread in South America, Africa, Australia, India, and Antarctica and provide important evidence for the theory of continental drift.

gneiss   A coarse-grained metamorphic rock with a characteristic type of foliation (gneissic layering), resulting from alternating layers of light-colored and dark-colored minerals.

gneissic foliation   The type of metamorphic foliation characterizing gneiss, resulting from alternating layers of silicic and mafic minerals.

Gondwanaland   The ancient continental landmass that is thought to have split apart during Mesozoic time to form the present-day continents of South America, Africa, India, Australia, and Antarctica.

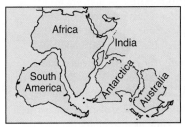

Gondwanaland

graben   An elongate fault block that has been lowered in relation to the blocks on either side.

Graben

graded bedding   A type of bedding in which each layer is characterized by a progressive decrease in grain size from the bottom of the bed to the top.

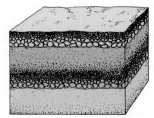

Graded bedding

graded stream   A stream that has attained a state of equilibrium, or balance, between erosion and deposition, so that the velocity of the water is just great enough to transport the sediment load supplied from the drainage basin, and neither erosion nor deposition occurs.

gradient   (stream) The slope of a stream channel measured along the course of the stream.

grain   A particle of a mineral or rock, generally lacking well-developed crystal faces.

granite   A coarse-grained igneous rock composed of K-feldspar, plagioclase, and quartz, with small amounts of mafic minerals.

granular disintegration   A process of weathering that produces loose accumulations of isolated grains. It is common in granite, where feldspars weather to clay and quartz grains. The dissolution of calcite cement in sandstone also causes granular breakdown.

granulite   A high-grade metamorphic rock that typically lacks hydrous minerals like micas and amphibole.

granulite facies   A high grade of metamorphism producing medium- to coarse-grained rocks that lack hydrous minerals like biotite and amphibole.

gravel   The coarsest (greater than 2 cm across) clasts found in clastic sedimentary rocks, includes cobbles and boulders.

gravity   The force of attraction between two bodies; the magnitude depends on the mass of the bodies and the distance between them. On Earth, the attraction is toward the center of the planet.

gravity anomaly   An area where gravitational attraction is greater or less than its normal value.

graywacke   An impure sandstone consisting of rock fragments and grains of quartz and feldspar in a matrix of clay-size particles.

greenhouse effect   The warming of a planet's atmosphere caused when certain gases (especially water vapor and carbon dioxide) absorb of solar energy reflected off the surface.

greenhouse gas   A gas that absorbs infrared radiation and helps control the temperature of a planet's atmosphere. The most important are carbon dioxide, water, and methane.

greenhouse warming   An increase in the surface temperature caused by the accumulation of greenhouse gases (e.g.,

carbon dioxide, water, and methane) in the atmosphere.

greenschist facies   Metamorphic conditions typified by low temperature and low pressure.

greenstone   A low-grade metamorphic rock that commonly has green minerals such as chlorite and talc.

groundmass   The matrix of relatively fine-grained material between the phenocrysts in a porphyritic rock.

ground moraine   Till (debris) that has been transported by a glacier and deposited at the base of the ice.

groundwater   Water below Earth's surface; generally in pore spaces of rocks and soil.

groundwater system   The water that seeps into the ground and moves slowly through the pore spaces in soil and rocks.

guyot   A seamount with a flat top.

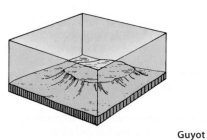

Guyot

gypsum   An evaporite mineral composed of calcium sulfate with water ($CaSO_4 \cdot 2H_2O$).

Hadean   The eon of Earth's history between its formation and the beginning of the Archean (about 4 billion years ago). Very few rocks are preserved from this period.

half-dike   A split dike of igneous rock formed when a fracture inside an older dike permits magma flow to develop a secondary dike in the interior of an older dike. They form during spreading at oceanic ridges.

half-life   The time required for half of a given sample of a radioactive isotope to decay to its daughter isotope.

halite   An evaporite mineral composed of sodium chloride (NaCl).

hanging valley   A tributary valley with the floor lying ("hanging") above the valley floor of the main stream or shore to which it flows. Hanging valleys commonly are created by deepening of the main valley by glaciation, but they can also be produced by faulting or rapid retreat of a sea cliff.

hanging wall   The surface or block of rock that lies above an inclined fault plane.

Hanging wall

hardness   **1** (mineralogy) The measure of the resistance of a mineral to scratching or abrasion. **2** (water) A property of water resulting from the presence of calcium carbonate and magnesium carbonate in solution.

headland   An extension of land seaward from the general trend of the coast; a promontory, cape, or peninsula.

headward erosion   Extension of a stream headward, up the regional slope of erosion.

heat flow   The flow of heat from the interior of Earth.

high-grade metamorphism   Metamorphism that occurs under high temperature and high pressure.

hinge line   The line where folded beds show maximum curvature. The line formed by the intersection of the hinge plane with the bedding surface.

hinge plane   The plane formed by connecting the hinge lines in a fold. It divides the fold into two separate parts. Also called an axial surface.

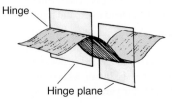

Hinge

hogback   A narrow, sharp ridge formed on steeply inclined, resistant rock.

Hanging valley

horizon   **1** (geologic) A plane of stratification assumed to have been originally horizontal. **2** (soil) A layer of soil distinguished by characteristic physical properties. Soil horizons generally are designated by letters (for example, A horizon, B horizon, C horizon).

horn   A sharp peak formed at the intersection of the headwalls of three or more cirques.

Horn

hornblende   A variety of the amphibole mineral group.

hornfels   A nonfoliated metamorphic rock of uniform grain size, formed by high-temperature metamorphism. Hornfelses typically are formed by contact metamorphism around igneous intrusions.

horst   An elongate fault block that has been uplifted in relation to the adjacent rocks.

Horst

hotspot   The expression at Earth's surface of a mantle plume, or column of hot, buoyant rock rising in the mantle beneath a lithospheric plate.

hotspot track   An aligned series of volcanic centers in which the age of volcanism becomes progressively older away from one end and thought to be related to the movement of lithosphere over a hot mantle plume.

humidity   The water vapor in the air.

hummock   A small, rounded or cone-shaped, low hill or a surface of other small, irregular shapes. A surface that is not equidimensional or ridgelike.

hydration   Chemical combination of water with other substances.

hydraulic   Pertaining to a fluid in motion.

hydraulic fracturing   The process of breaking rocks by injecting a fluid under high pressure. Commonly used to release methane and oil from tight rocks for extraction. Also called fracking.

hydraulic head   The pressure exerted by a fluid at a given depth beneath its surface. It is proportional to the height of the fluid's surface above the area where the pressure is measured.

hydrocarbon trap   A physical barrier to the flow of oil and gas that forces oil and gas to accumulate rather than disperse.

hydroelectric power   An energy resource using the kinetic energy of water flowing downhill to turn turbines that generate electricity.

hydrologic system   The system of moving water at Earth's surface.

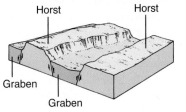

Hydrologic system

hydrolysis   A chemical reaction wherein hydrogen ions replace other ions in a mineral. Commonly results in the production of hydrous minerals such as clay or complete dissolution of calcite.

hydrosphere   The waters of Earth, as distinguished from the rocks (geosphere), the air (atmosphere), and living things (biosphere).

hydrostatic pressure   The pressure within a fluid (such as water) at rest, exerted on a given point within the body of the fluid.

hydrothermal alteration   The process of hot water and other fluids reacting with solid rock during the formation of metasomatic rocks, such as metallic ore deposits.

hydrothermal deposit   A mineral deposit formed by hot water. The high temperature commonly is associated with emplacement of a magma.

hydrothermal fluid   Hot water–rich solutions; some are heated by or released from magma.

hydrothermal ore deposits   Economically valuable accumulation of minerals formed by the movement of hot fluids.

ice sheet   A thick, extensive body of glacial ice that is not confined to valleys. Localized ice sheets are sometimes called ice caps.

ice wedging   A type of mechanical weathering in which rocks are broken by the expansion of water as it freezes in joints, pores, or bedding planes. Synonymous with frost wedging.

icy planetary body   A small planetary body in the outer solar system whose interior and surface are dominated by water ice. Pluto, Eris, and the moons of Saturn are good examples.

igneous rock   Rock formed by cooling and solidification of molten silicate minerals (magma). Igneous rocks include volcanic and plutonic rocks.

ignimbrite   A volcanic rock formed by pyroclastic flows—particles that moved laterally across the surface in a hot, rapidly moving, and gas-charged flow. Synonym: ash-flow tuff.

impact crater   A depression in a planetary body's surface formed by collision with a meteor or comet.

inclination, magnetic   The angle between the horizontal plane and a magnetic line of force.

inclined seismic zone   An inclined sheet-like zone of earthquakes that marks a descending plate in a subduction zone.

inclusion   A rock fragment incorporated into a younger igneous rock.

index mineral   A mineral that forms at a specific metamorphic grade.

industrial mineral   Generally nonmetallic minerals or rocks that are used in industry. The cost per ton for these materials is generally low. Examples include limestone for cement production, building stone, sand, and gravel.

inner core   The solid innermost part of Earth's extremely hot center, composed mostly of iron and about 2400 km in diameter.

inner planets   The rocky innermost bodies of the solar system: Mercury, Venus, Earth, the Moon, and Mars.

intensity   Evaluation of the destructive power or severity of an earthquake's ground motion at a given location.

interlocking texture   The texture typical of igneous rocks in which mineral grains fit together like pieces in a jig-saw puzzle.

intermediate-focus earthquake   An earthquake with a focus located at a depth between 70 and 300 km.

intermittent stream   A stream through which water flows only part of the time.

internal drainage   A drainage system that does not extend to the ocean.

internal heat   The heat found within a planet that comes from planetary accretion and natural radioactivity.

interstitial Pertaining to material in the pore spaces of a rock. Petroleum and groundwater are interstitial fluids. Minerals deposited by groundwater in sandstone are interstitial minerals.

intertropical convergence zone The band encircling Earth near its equator where the northeast and southeast trade winds come together. Also called the doldrums.

intrusion **1** Injection of a magma into a preexisting rock. **2** A body of rock resulting from the process of intrusion.

intrusive rock Igneous rock that, while it was fluid, penetrated into or between other rocks and solidified. It can later be exposed at Earth's surface after erosion of the overlying rock.

inverted valley A valley that has been filled with lava or other resistant material and has subsequently been eroded into an elongate ridge.

ion An atom or combination of atoms that has gained or lost one or more electrons and thus has a net electrical charge.

ionic bond A chemical bond formed by electrostatic attraction between oppositely charged ions.

ionic substitution The replacement of one kind of ion in a crystalline lattice by another kind that is of similar size and electrical charge.

island arc A chain of volcanic islands. Island arcs are generally convex toward the open ocean. Example: the Aleutian Islands.

isostasy A state of equilibrium, resembling flotation, in which segments of Earth's crust stand at levels determined by their thickness and density. Isostatic equilibrium is attained by flow of material in the mantle.

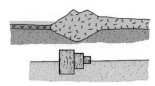

Isostasy

isotope One of the several forms of a chemical element that have the same number of protons in the nucleus but differ in the number of neutrons and thus differ in atomic weight.

jet stream A fast-moving stream of cold air that marks the boundary between the polar and temperate air masses. Its location changes and strongly affects the weather in the northern continents.

joint A fracture in a rock along which no appreciable displacement has occurred.

joint-block separation The breakdown of a rock along a system of fracture planes.

kame A body of stratified glacial sediment. A mound or an irregular ridge deposited by a subglacial stream as an alluvial fan or a delta.

karst topography A landscape characterized by sinks, solution valleys, and other features produced by groundwater activity.

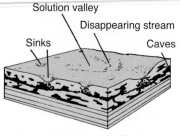

Karst topography

kettle A closed depression in a deposit of glacial drift formed where a block of ice was buried or partly buried and then melted.

komatiite A rare extrusive rock rich in magnesium and the mineral olivine. It is found mostly in ancient volcanic successions in the continental shields. The old komatiite lava flows are evidence for the high temperatures of volcanism early in Earth's history.

laccolith A concordant igneous intrusion that has arched up the strata into which it was injected, so that it forms a pod-shaped or lens-shaped body with a generally horizontal floor.

Laccolith

lag deposit A residual accumulation of coarse fragments that remains on the surface after finer material has been removed by wind.

lagoon A shallow body of seawater separated from the open ocean by a barrier island or reef.

lahar A volcanic debris flow.

lamina (pl. laminae) A layer of sediment less than 1 cm thick.

laminar flow A type of flow in which the fluid moves in parallel lines. Contrast with turbulent flow.

land degradation The processes by which land becomes unproductive. See desertification.

landform Any feature of Earth's surface having a distinct shape and origin. Landforms include major features (such as continents, ocean basins, plains, plateaus, and mountain ranges) and minor features (such as hills, valleys, slopes, drumlins, and dunes). Collectively, the landforms of Earth constitute the entire surface configuration of the planet.

landslide A general term for relatively rapid types of mass movement, such as debris flows, debris slides, rockslides, and slumps.

lateral moraine An accumulation of till deposited along the side margins of a valley glacier. It accumulates as a result of mass movement of debris on the sides of the glacier.

lateral slip Nearly horizontal movement (shear) of blocks on either side of a vertical fault.

laterite A soil that is rich in oxides of iron and aluminum formed by deep weathering in tropical and subtropical areas.

Laurasia The ancient continental landmass that is thought to have split apart to form Europe, Asia, North America, and Greenland.

lava Magma that reaches Earth's surface.

lava dome Bulbous lava flow or viscous plug of lava piled near its vent. Most are made of rhyolite.

lava tube A cave in a mafic lava flow formed when the sides and top freeze solid, and the interior remains fluid. The still molten lava may flow out, leaving a tubular cavity.

layered gabbro Layered masses of mafic igneous rock that crystallized from a fluid basaltic magma chamber. Layers probably formed as crystals accumulated sequentially on the walls and floors of the chamber.

leaky transform At transform plate boundaries, where basalt magma erupts from faults (rare).

lee slope The part of a hill, dune, or rock that is sheltered or turned away from the wind. Synonymous with slip face.

levee, natural A broad, low embankment built up along the banks of a river channel during floods.

lift Caused by wind forces flowing over grains of sand, creating a zone of low air pressure over the grain and picking them up when the grains' resistance to movement is overcome.

limb   The flank, or side, of a fold.

limestone   A sedimentary rock composed mostly of calcium carbonate ($CaCO_3$).

lineament   A topographic feature or group of features having a linear configuration. Lineaments commonly are expressed as ridges or depressions or as an alignment of features such as stream beds, volcanoes, or vegetation.

linear dune   An elongate sand dune oriented in the direction of the prevailing wind.

liquefaction   Occurs when unconsolidated, water-saturated regolith, soil, or landfill loses its strength and behaves like a fluid when shaken by an earthquake.

liquid   The state of matter in which a substance flows freely and lacks crystal structure. Unlike a gas, a liquid retains the same volume independent of the shape of its container.

lithification   The processes by which sediment is converted into sedimentary rock. These processes include cementation and compaction.

lithosphere   The relatively rigid outer zone of Earth, which includes the continental crust, the oceanic crust, and the part of the upper mantle lying above the weaker asthenosphere.

load   The total amount of sediment carried at a given time by a stream, glacier, or wind.

loess   Unconsolidated, wind-deposited silt and dust.

longitudinal profile   The profile of a stream or valley drawn along its length, from source to mouth.

longitudinal wave   A seismic body wave in which particles oscillate along lines in the direction in which the wave travels. Synonymous with P wave.

longshore current   A current in the surf zone moving parallel to the shore. Longshore currents occur where waves strike the shore at an angle.

longshore drift   The process in which sediment is moved in a zigzag pattern along a beach by the swash and backwash of waves that approach the shore obliquely.

low-grade metamorphism   Metamorphism that is accomplished under low or moderate temperature and low or moderate pressure.

luster   The appearance of the light reflected from a mineral surface, described, for example, as dull, glassy, or metallic.

mafic   A mineral or rock rich in iron and magnesium silicates such as olivine and pyroxene.

magma   Molten rock, generally a silicate melt with suspended crystals and dissolved gases.

magma mixing   The partial or complete blending of two or more different magmas.

magmatic differentiation   A general term for the processes by which magmas differentiate or change composition. It includes fractional crystallization, magma mixing, and assimilation.

magmatic segregation   Separation of crystals of certain minerals from magma as it cools. For example, some minerals (including certain valuable metals) crystallize while other components of the magma are still liquid. These early formed crystals can settle to the bottom of a magma chamber and thus become concentrated there, forming an ore deposit.

magnetic anomaly   A deviation of observed magnetic inclination or intensity (as measured by a magnetometer) from a constant normal value.

magnetic reversal   A complete 180-degree reversal of the polarity of Earth's magnetic field.

magnetism   A physical phenomenon produced by the motion of electric charge that results in attractive and repulsive forces between objects.

magnetosphere   A region of the extreme upper atmosphere that is dominated by the magnetic field and charged particles are trapped in it. It acts as a type of radiation shield.

magnitude   A measure of the size of an earthquake, usually calculated from the common logarithm of the largest ground motion observed and corrected for distance from the earthquake focus.

mantle   The zone of the Earth's interior between the base of the crust (the Moho discontinuity) and the core.

mantle plume   A buoyant mass of hot mantle material that rises to the base of the lithosphere. Mantle plumes commonly produce volcanic activity and structural deformation in the central part of lithospheric plates.

mantle resistance   Frictional resistance to the movement of a subducting plate through the asthenosphere and mesosphere.

marble   A metamorphic rock consisting mostly of metamorphosed limestone or dolomite.

mare   (pl. maria) Any of the relatively smooth, low, dark areas of the Moon. The lunar maria were formed by extrusion of lava.

mass extinction   The extinction of a large number of species in a relatively short period of time, theorized to be caused by large meteorite impact, climate change, disease, etc.

massive gabbro   A homogeneous, coarse grained mafic igneous rock. In ophiolites, sheeted dikes grade downward into massive gabbro. Compare with layered gabbro.

mass movement   The transfer of rock and soil downslope by direct action of gravity without a flowing medium (such as a river or glacial ice). Synonymous with mass wasting.

matrix   The relatively fine-grained rock material occupying the space between larger particles in a rock. See also groundmass.

meander   A broad, looping bend in a river.

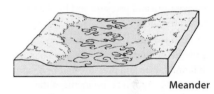

Meander

medial moraine   A ridge of till formed in the middle of a valley glacier by the junction of two lateral moraines where two valley glaciers converge.

megathrust earthquake   An earthquake the occurs along the boundary between a subducting plate and the overriding plate at a convergent plate boundary.

mélange   A mixture of diverse deformed rocks formed in the accretionary prism at a convergent plate margin.

melt   The molten liquid part of magma or, as a verb, to become liquified by heat.

Mercalli scale   A measure of earthquake intensity determined from the effects on people and buildings, ranges from I (low) to XII (nearly total destruction).

mesa   A flat-topped, steep-sided highland capped with a resistant rock formation. A mesa is smaller than a plateau but larger than a butte.

mesosphere   The part of the mantle between the asthenosphere and the core, also known as the "middle sphere."

Mesozoic   The era of geologic time from the end of the Paleozoic Era (about 250 million years ago) to the beginning of the Cenozoic Era (about 65 million years ago).

metabasalt   Metamorphosed basalt. It is the most abundant type of metamorphic rock exposed near Earth's surface, because much is formed on the seafloor.

metaconglomerate   A metamorphosed conglomerate, under differential stress, individual pebbles are stretched into distinctive linear blades.

metallic bond   A chemical bond in which shared electrons move freely among the atoms.

metamorphic facies   An environment of metamorphism described in terms of pressure and temperature and corresponding to distinctive mineral assemblages.

metamorphic grade   An estimate of the intensity of metamorphism ranging from low (low temperature and pressure of equilibration) to high (high temperature and pressure).

metamorphic rock   Any rock formed from preexisting rocks by solid state recrystallization driven by changes in temperature and pressure and by chemical action of fluids.

metamorphism   Alteration of the minerals and textures of a rock by changes in temperature and pressure and by a gain or loss of chemical components in the solid state.

metasomatism   A change in the chemical composition of a rock during metamorphism, usually caused by the transport of ions by fluids.

metastable   An object or mineral that is not in equilibrium with its environment; it exists outside of it stability field. An additional force may be needed to make it reach equilibrium. For example, a boulder sitting in a slight depression on steep slope is in a metastable position and diamonds are metastable at Earth's surface.

meteoric water   Water derived from the atmosphere, such as rainwater, snow, or hail.

meteorite   Any particle of solid matter that has fallen to Earth, the Moon, or another planet from space.

methane hydrates   An icy substance consisting of methane trapped with a framework of water ice. It forms where rising bubbles of methane gas, given off as bacteria digest organic matter in mud, react with cold seawater.

mica   A group of silicate minerals exhibiting perfect cleavage in one direction.

mid-ocean ridge   Broad fractured swell in the ocean basins. New oceanic crust is formed at this type of divergent plate boundary. Synonymous with oceanic ridge.

migmatite   A mixture of igneous and metamorphic rocks in which thin dikes and stringers of granitic material interfinger with metamorphic rocks.

Milankovitch cycle   Cyclical climatic changes caused by variations in Earth's orbital characteristics—eccentricity of the orbit and tilt (obliquity) and precession (wobble) of the spin axis.

mineral   A naturally occurring inorganic solid having a definite internal structure and a definite chemical composition that varies only within strict limits. Chemical composition and internal structure determine its physical properties, including the tendency to assume a particular geometric form (crystal form).

mineral resource   Any soil, mineral and other rock, oil, or natural gas that is extracted from Earth for commercial use.

Mohorovičić discontinuity   The first global seismic discontinuity below the surface of Earth. It lies at a depth varying from about 5 to 10 m beneath the ocean floor to about 35 km beneath the continents. Commonly referred to as the Moho.

Mohs hardness scale   A scale of mineral hardness ranging from 1 for soft minerals to 10 for very hard minerals.

mollisol   A type of soil common in semi-humid or temperate grasslands. These soils are typically deep, high organic matter and nutrient rich.

moment magnitude scale   A logarithmic measure of the amount of energy released by an earthquake. It is the most widely used measure of earthquake severity today.

monocline   A bend or fold in gently dipping horizontal strata.

Monocline

moraine   A general term for a landform composed of till.

mountain   A general term for any landmass that stands above its surroundings. In the stricter geological sense, a mountain belt is a highly deformed part of Earth's crust that has been injected with igneous intrusions and the deeper parts of which have been metamorphosed. The topography of young mountains is high, but erosion can reduce old mountains to flat lowlands.

mud crack   A crack in a deposit of mud or silt resulting from the contraction that accompanies drying. Commonly preserved in sedimentary rock, they show that the sedimentary environment was occasionally exposed to the air during deposition, and suggests that the original sediment was deposited in shallow lakes, on tidal flats, or on exposed stream banks.

mudflow   A flowing mixture of mud and water.

Mudflow

mudrock   A fine-grained sedimentary rock made of clay and silt-size particles. Shale is a finely laminated type of mudrock.

multiring basin   A large crater (on the Moon they are more than 300 km in diameter) containing a series of concentric ridges and depressions. Example: the Orientale basin on the Moon.

muscovite   [$KAl_3Si_3O_{10}(OH)_2$] One of two types of mica, it is white or colorless and found along with felsic minerals.

mylonite   A foliated metamorphic rock formed by intense shearing and deformation of preexisting grains. Formed in the transition between brittle fracture and ductile flow.

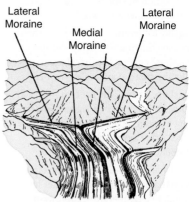

Lateral Moraine   Medial Moraine   Lateral Moraine

Moraines

nappe   A large thrust sheet with overturned folds.

natural arch   An arch-shaped landform produced by weathering and differential erosion.

natural levee   A broad, low embankment built up along the banks of a river channel during floods.

nebula   A vast, cold, diffuse cloud of gas and dust in space.

Neogene   The second geologic time period of the Cenozoic Era from about 34 million years to 2 million years ago and ending with the Quaternary Period.

neutron   A subatomic nuclear particle that has no electrical charge but a mass almost the same as a proton.

neve   Granular ice formed by recrystallization of snow. Synonymous with firn.

nonconformity   An unconformity in which stratified rocks rest on eroded granitic or metamorphic rocks.

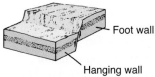

Nonconformity

nonfoliated   A metamorphic rock that lacks any preferred orientation of its mineral grains.

nonrenewable   Substances that are finite and destructible, and therefore exhaustible.

normal fault   A steeply inclined fault in which the hanging wall has moved downward in relation to the footwall.

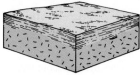

Foot wall

Hanging wall

Normal fault

normal polarity   Periods when the magnetic field was oriented as it is today with the north magnetic pole in the north.

nuclear energy   Energy released by controlled fission or fusion or atomic nuclei.

nucleus   Tightly packed protons and neutrons in an atom.

numerical age   Geologic time measured in a specific duration of years (in contrast to relative time, which involves only the chronologic order of events). Synonym: absolute age.

obsidian   A glassy igneous rock with a composition equivalent to that of granite.

ocean basin   A low part of the lithosphere lying between continental masses. The rocks of an ocean basin are mostly basalt with a veneer of oceanic sediment.

ocean-continent convergence   A subduction zone where oceanic and continental plates abut. The plate with the less-dense continental crust always resists subduction into the dense mantle and overrides the oceanic plate, often deforming into a folded mountain belt.

oceanic crust   The type of crust that underlies the ocean basins. It is generally less than 8 km thick, composed predominantly of basalt and gabbro. Its density is about 3.0 g/cm³. The velocities of compressional seismic waves traveling through it exceed 6.2 km/sec. Compare with continental crust.

oceanic plateau   A part of the ocean floor that is elevated compared to its surroundings. The crust is unusually thick beneath a plateau. They are most likely the result of submarine volcanism related to a starting plume.

oceanic ridge   The continuous ridge, or broad, fractured topographic swell, that extends through the central part of the Arctic, Atlantic, Indian, and South Pacific oceans. It is several hundred kilometers wide, and its elevation above the ocean floor is 600 m or more. The ridge marks a divergent plate boundary where new oceanic lithosphere is being formed.

ocean-ocean convergence   A subduction zone where two adjoining oceanic plates collide, with one thrust under the other.

ocean ridge metamorphism   The most characteristic type of metamorphism in the oceanic crust, formed by the circulation of hot seawater through cold oceanic crust, converting olivine and pyroxene into hydrated silicate minerals, including serpentine, chlorite, and talc.

oil   A mixture of liquid hydrocarbons extracted from the ground and used for fuel. Synonymous with petroleum.

oil reservoir   Porous rock that can contain oil.

oil shale   Shale that is rich in hydrocarbon derivatives. In the United States, the chief oil shale is the Green River Formation in the Rocky Mountain region.

oil trap   Impermeable rocks or structures that block the flow of oil and force it to accumulate into larger bodies.

olivine   An important silicate mineral with magnesium and iron [$(Mg,Fe)_2SiO_4$].

oolite   A limestone consisting largely of spherical grains of calcium carbonate in concentric spherical layers.

Oolite

ooze   (marine geology) Marine sediment consisting of more than 30% shell fragments of microscopic organisms.

open system   A system that exchanges both heat and matter with its surroundings.

ophiolite   A sequence of rocks characterized by ultramafic rocks at the base and (in ascending order) gabbro, sheeted dikes, pillow lavas, and deep-sea sediments. The typical sequence of rocks constituting the oceanic crust.

ore deposit   A mass of rock containing metal (or some other commodity like diamonds) of sufficient abundance to be extracted at a profit.

organic chemical   A carbon-hydrogen molecule.

organic sediment   A sediment deposited through biological means and rich in hydrocarbons, such as coal.

orogenic   Pertaining to deformation of a continental margin to the extent that a mountain range is formed.

orogenic belt   A mountain belt.

orogenic metamorphism   Metamorphism the occurs in folded mountain belts at convergent plate margins when surface rocks become deeply buried and intruded by magma.

orogeny   A major episode of mountain building.

outcrop   An exposure of bedrock.

outer core   The extremely hot, liquid part of the core. It is made mostly of iron located between the inner core and the mesosphere, about 2270 km thick.

outer planets   The planets farthest from the Sun: Jupiter, Saturn, Uranus, and Neptune.

outer swell   A broad rise or bulge in the downgoing tectonic plate that lies outboard of the trench.

outlet glacier   A tonguelike stream of ice, resembling a valley glacier, that forms where a continental glacier encounters a mountain system and is forced to move through a mountain pass in large streams.

outwash   Stratified sediment washed out from a glacier by meltwater streams and deposited in front of the end moraine.

outwash plain   The area beyond the margins of a glacier where meltwater deposits sand, gravel, and mud washed out from the glacier.

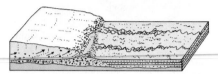

Outwash plain

overturned fold   A fold in which at least one limb has been rotated through an angle greater than 90 degrees.

Overturned fold

oxbow lake   A lake formed in the channel of an abandoned meander.

oxidation   Chemical combination of oxygen with another substance.

oxide mineral   A mineral lacking silicon, but containing oxygen bound to a metal. Examples: hematite and magnetite.

oxisol   A type of red tropical soil rich in oxidized iron and clays and poor in soluble elements.

ozone   A gas made of three oxygen atoms (O_3).

ozone hole   A zone in the stratosphere where ozone concentrations are lower than normal, which allows ultraviolet radiation to reach Earth's surface. It can be enhanced by pollutants.

ozone layer   A zone within the stratosphere where ozone (O_3) is abundant and forms a protection from some of the Sun's harmful ultraviolet radiation.

pahoehoe flow   A lava flow with a billowy or ropy surface. Contrast with aa flow.

paired metamorphic belt   At both modern and ancient subduction zones, two distinctive types of metamorphic zones are juxtaposed with a high P/T zone close to the trench and a parallel zone of moderate P/T metamorphism farther from the trench in a folded mountain belt.

paleocurrent   An ancient current, which existed in the geologic past, with a direction of flow that can be inferred from

cross-bedding, ripple marks, and other sedimentary structures.

Paleogene   The first geologic time period of the Cenozoic Era from about 65 million years to 34 million years.

paleogeography   The study of geography in the geologic past, including the patterns of Earth's surface, the distribution of land and ocean, and ancient mountains and other landforms.

paleomagnetism   The study of ancient magnetic fields, as preserved in the magnetic properties of rocks. It includes studies of changes in the position of the magnetic poles and reversals of the magnetic poles in the geologic past.

paleontology   The study of ancient life.

Paleozoic   The era of geologic time from the end of the Precambrian (about 540 million years ago) to the beginning of the Mesozoic Era (about 250 million years ago).

Pangaea   A former continent from which the present continents originated by plate movement from the Mesozoic Era to the present.

parabolic dune   A dune shaped like a parabola with the concave side toward the wind. Blowout dune.

parent isotope   In radioactive decay reactions, the unstable atom that breaks down into another isotope. Compare with daughter isotope.

partial melting   The process by which minerals with low melting points melt while other minerals in the rock are still solid. The composition of the melt can be quite different from that of the parent rock.

passive margin   (plate tectonics) A lithospheric plate margin at which crust is neither created nor destroyed. Passive plate margins generally are marked by transform faults.

patterned ground   A glacial or polar landform in which ice-bearing ground is split into crude polygons by alternating freezing and thawing.

peat   An accumulation of partly carbonized plant material containing approximately 60% carbon and 30% oxygen. It is considered an early stage, or rank, in the development of coal.

pebble   A rock fragment with a diameter between 2 mm (about the size of a match head) and 64 mm (about the size of a tennis ball).

pediment   A gently sloping erosion surface formed at the base of a receding mountain front or cliff. It cuts across

bedrock and can be covered with a veneer of sediment. Pediments characteristically form in arid and semiarid climates.

pegmatite   A very coarse grained igneous rock typically with a granitic composition.

pelagic sediment   Deep-sea sediment composed of fine-grained detritus that slowly settles from surface waters. Common constituents are clay, radiolarian ooze, and foraminiferal ooze.

peninsula   An elongate body of land extending into a body of water.

perched water table   The upper surface of a local zone of saturation that lies above the regional water table.

Perched water table

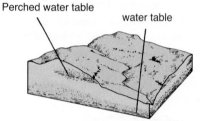

Perched water table

peridotite   A dark-colored ultramafic igneous rock of coarse-grained texture, composed of olivine, pyroxene, but with essentially no feldspar and no quartz.

period   A division of geologic time smaller than an era and larger than an epoch. Example: Cretaceous Period.

permafrost   Permanently frozen ground.

permanent stream   A stream or reach of a stream that flows continuously throughout the year. Synonymous with perennial stream.

permeability   The ability of a material to transmit fluids.

petroleum   Liquid mixture of hydrocarbon molecules extracted from the ground and used for fuel.

phaneritic texture   The texture of igneous rocks in which the interlocking crystals are large enough to be seen without magnification.

Phanerozoic   The eon following the Precambrian and extending from about 540 million years ago to the present.

Pediment

phenocryst   A crystal that is significantly larger than the crystals surrounding it. Phenocrysts form during an early phase in the cooling of a magma when the magma cools relatively slowly.

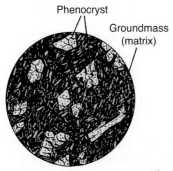

Phenocryst

Groundmass (matrix)

Phenocryst

photovoltaic cell   A device used to convert sunlight directly into electricity.

phyllite   A foliated metamorphic rock intermediate between slate and schist. Small mica crystals give broken surfaces a silky sheen.

physical weathering   The breakdown of rock into smaller fragments by physical processes such as frost wedging. Synonymous with mechanical weathering.

physiographic map   A map showing surface features of Earth.

pillar   A column of rock produced by differential weathering or erosion, intersecting joint planes.

pillow basalt   See pillow lava.

pillow lava   An ellipsoidal mass of igneous rock formed by extrusion of lava underwater.

pinnacle   A high tower or spire shaped pillar of rock alone or cresting in a summit.

placer   A mineral deposit formed by the sorting or washing action of water. Placers are usually deposits of heavy minerals, such as gold.

plagioclase   A group of feldspar minerals with a composition range from $NaAlSi_3O_8$ to $CaAl_2Si_2O_8$.

planetary differentiation   The processes by which the materials in a planetary body are separated according to density, so that the originally homogeneous body is converted into a zoned or layered (shelled) body with a dense core, a mantle, and a crust.

planetesimal   Asteroid-sized bodies of rock and ice.

plankton   Collective term for very small plants and animals that drift near the surface of water. Phytoplankton include

bacteria, algae (including diatoms), and fungi. The small animals are called zooplankton.

plastic deformation   A permanent change in a substance's shape or volume that does not involve failure by rupture.

plate   (tectonics) A broad segment of the lithosphere (including the rigid upper mantle, plus oceanic and continental crust) that floats on the underlying asthenosphere and moves independently of other plates.

plateau   An extensive upland region.

plateau basalt   Basalt extruded in extensive, nearly horizontal layers, which, after uplift, tend to erode into great plateaus. Synonymous with flood basalt.

Plateau basalt

plate tectonics   The theory of global dynamics in which the lithosphere is believed to be broken into individual plates that move in response to convection in the upper mantle. The margins of the plates are sites of considerable geologic activity.

platform reef   A type of reef that grows in isolated oval patches in warm, shallow water on the continental shelf. See reef.

playa   A depression in the center of a desert basin, the site of occasional temporary lakes.

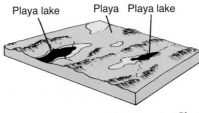

Playa lake   Playa   Playa lake

Playa

playa lake   A shallow temporary lake formed in a desert basin after rain.

Pleistocene   The epoch of geologic time from the end of the Pliocene Epoch (about 2.6 million years ago) to the beginning of the Holocene Epoch (about 10,000 years ago). The major event during the Pleistocene was the expansion of continental glaciers in the Northern Hemisphere. Synonymous with glacial epoch, ice age.

plucking   (glacial geology) The process of glacial erosion by which large rock fragments are loosened by ice wedging, become frozen to the bottom surface of the glacier, and are torn out of the

bedrock and transported by the glacier as it moves. The process involves the freezing of subglacial meltwater that seeps into fractures and bedding planes in the rock.

plume   See mantle plume.

plume head   Bulbous head of a diapiric mass of solid mantle rising through Earth's interior and fed by a long narrow pipe, the plume tail.

plume tail   The long narrow pipe through which buoyant, but still solid, mantle flows to the plume head and which persists for tens of millions of years after the plume head has dissipated.

plunge   The inclination, with respect to the horizontal plane, of any linear structural element of a rock. The plunge of a fold is the inclination of the axis of the fold.

plunging fold   A fold with its axis inclined from the horizontal.

pluton   Igneous rock formed beneath Earth's surface.

pluvial lake   A lake that was created under former climatic conditions, at a time when rainfall in the region was more abundant than it is now. Pluvial lakes were common in arid regions during the Pleistocene.

point bar   A crescent-shaped accumulation of sand and gravel deposited on the inside of a meander bend.

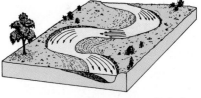

Point bar

polar climate   The climate that prevails at Earth's poles, with temperatures commonly below freezing and precipitation low.

polarity chron   The major intervals of alternating magnetic polarity (each about 1 million or so years long).

polarity epoch   A relatively long period of time during which Earth's magnetic field is oriented in either the normal direction or the reverse direction.

polarity event   A relatively brief interval of time within a polarity epoch; during a polarity event, the polarity of Earth's magnetic field is reversed with respect to the prevailing polarity of the epoch.

polar wandering   The apparent movement of the magnetic poles with respect to the continents.

pole of rotation   A pole of the imaginary axis about which a tectonic plate rotates.

polymorphism   The ability of a chemical compound to crystallize with more than one kind of crystal structure. For example, Al_2SiO_5 may crystallize as three different minerals, depending on the prevailing temperature and pressure.

pore fluid   A fluid, such as groundwater or liquid rock material resulting from partial melting, that occupies pore spaces of a rock.

pore space   The spaces within a rock body that are unoccupied by solid material. Pore spaces include spaces between grains, fractures, vesicles, and voids formed by dissolution.

porosity   The percentage of the total volume of a rock or sediment that consists of pore space.

porphyritic texture   The texture of igneous rocks in which some crystals are distinctly larger than others.

porphyry copper   Deposits of copper disseminated throughout a porphyritic granitic rock.

potentiometric surface   The level to which water in an artesian system would rise in a pipe high enough to stop the flow.

pothole   A hole formed in a stream bed by sand and gravel swirled around in one spot by eddies.

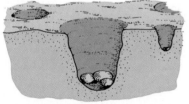

Pothole

Precambrian   The division of geologic time from the formation of Earth (about 4.5 billion years ago) to the beginning of the Cambrian Period of the Paleozoic Era (about 540 million years ago). Also, the rocks formed during that time. Precambrian time constitutes about 90% of Earth's history.

precipitation   Condensation of a liquid from a vapor as when the air becomes oversaturated with water, or when a solid crystallizes from a liquid as when quartz forms from molten magma.

pressure ridge   An elongate uplift of the congealing crust of a lava flow, resulting from the pressure of underlying and still fluid lava.

primary sedimentary structure   A structure of sedimentary rocks (such as cross-bedding, ripple marks, or mud cracks) that originates contemporaneously with the deposition of the sediment (in contrast to a secondary structure, such as a joint or fault, which originates after the rock has been formed).

primary wave   See P wave.

Proterozoic   The eon of geologic time from about 2.5 to 0.54 billion years ago. The final eon of the Precambrian.

proton   A positively charged nuclear particle.

pull-apart basin   Small fault-bounded trough produced by transtension along an irregular strike-slip fault.

pumice   A light-colored volcanic rock with abundant vesicles in natural glass.

P wave (primary seismic wave)   A type of seismic wave, propagated like a sound wave, in which the material involved in the wave motion is alternately compressed and expanded.

pyroclastic   Pertaining to fragmental rock material formed by volcanic explosions.

pyroclastic-fall tuff   A volcanic rock made of volcanic ash that fell more or less vertically out of the atmosphere.

pyroclastic texture   Volcanic texture marked by of a mixture of glassy fragments (shards) and broken crystals and produced when explosive eruptions blow bits of still-molten magma and crystals into the air as a mixture of hot fragments (ash).

Pyroclastic flow

pyroxene   A group of rock-forming silicate minerals composed of single chains of silicon-oxygen tetrahedra. Compare with amphibole, which is composed of double chains.

quartz   An important rock-forming silicate mineral composed of silicon-oxygen tetrahedra joined in a three-dimensional network. It is distinguished by its hardness, glassy luster, and conchoidal fracture.

quartzite   A sandstone recrystallized by metamorphism.

Quaternary   The period of geologic time from about 2.6 million years ago to the present. It follows the Neogene.

radioactivity   The spontaneous disintegration of an atomic nucleus with the emission of energy.

radiocarbon   A radioactive isotope of carbon, ^{14}C, which is formed in the atmosphere and is absorbed by living organisms.

radiogenic heat   Heat generated by radioactivity.

radiometric dating   Determination of the age in years of a rock or mineral by measuring the proportions of an original radioactive material and its decay product.

rain shadow   A dry area lying downwind from a high mountain chain.

ray   A long streak of ejecta that radiates from the rim of a meteorite impact crater.

ray   The path along which a seismic wave travels. Seismic rays are perpendicular to the wave crest.

rayed crater   A meteorite crater that has a system of rays extending like splash marks from the crater rim.

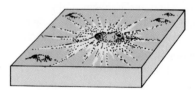

Rayed crater

recessional moraine   Hills and ridges generated in periods of stabilization during the recession of glacial ice behind a terminal moraine. See terminal moraine.

recharge   Replenishment of the groundwater reservoir by the addition of water.

recrystallization   Reorganization of elements of the original minerals in a rock resulting from changes in temperature and pressure and from the activity of pore fluids.

recycle   After use, materials that can be used again after minimal reprocessing. Example: most metals.

reef   A solid structure built of shells and other secretions of marine organisms, particularly coral.

regional metamorphism   Metamorphism of large areas of crust, usually during mountain building at convergent plate margins. Contrast with contact metamorphism.

regolith   The blanket of soil and loose rock fragments overlying the bedrock.

regression   A drop in sea level causes the shoreline to move downslope.

relative age   The age of a rock or an event as compared with some other rock or event.

relative dating   Determination of the chronologic order of a sequence of events in relation to one another without reference to their ages measured in years. Relative geologic dating is based primarily on superposition, faunal succession, and crosscutting relations.

relative time   Geologic time as determined by relative dating, that is, by placing events in chronologic order without reference to their ages measured in years.

relative velocity   The rate of movement of one tectonic plate with respect to another plate. Compare with absolute velocity.

relief   The difference in altitude between the high and the low parts of an area.

renewable   Energy resources that can be continually regenerated or are basically infinite, including biomass, solar energy, wind power, hydroelectric power, tidal power, and geothermal energy.

reservoir rock   The rock host for accumulations of oil and gas. Usually permeable sandstone, limestone, or dolomite that can store fluids in the pore spaces where outward flow is blocked by an impermeable rock.

reverse fault   A fault in which the hanging wall has moved upward in relation to the footwall; a high-angle thrust fault.

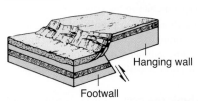

Reverse fault

reverse polarity   A period in time when polarity of Earth's magnetic field was the opposite of today's polarity. Compare with normal polarity.

rhyolite   A fine-grained volcanic rock composed of quartz, K-feldspar, and plagioclase. It is the extrusive equivalent of a granite.

Richter scale   A logarithmic scale for expressing the magnitude of an earthquake in terms of the energy dissipated in it. A modified version of this scale, the moment magnitude scale, is commonly used today.

ridge jump   A "sudden" shift in the position of the mid-ocean ridge; it may isolate a plume tail on one side of the ridge.

ridge-push   Gravitational force that makes the lithosphere slip off an elevated mid-ocean ridge.

ridge-ridge transform   A type of strike-slip fault joining two mid-ocean ridges and where active movement occurs only between the ridge segments. Beyond these ridge intersections no movement occurs along the fracture zone.

ridge-trench transform   A strike-slip fault that connects a mid-ocean ridge to a subduction zone trench.

rift system   A system of faults resulting from extension.

rift valley   1 A valley of regional extent formed by block faulting in which tensional stresses tend to pull the crust apart. Synonymous with graben. 2 The down-dropped block along divergent plate margins.

rift zone   An area with a series of parallel normal faults and fissures produced by the intrusion of magma or by tectonic extension.

rip current   A current formed on the surface of a body of water by the convergence of currents flowing in opposite directions. Rip currents are common along coasts where longshore currents move in opposite directions.

ripple marks   Small waves produced on a surface of sand or mud by the drag of wind or water moving over it.

river system   An integrated system of tributaries and a trunk stream, which collect and funnel surface water to the sea, a lake, or some other body of water. A river with all of its tributaries.

roche moutonnée   An abraded knob of bedrock formed by an overriding glacier. It typically is striated and has a gentle slope facing the upstream direction of ice movement.

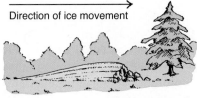

Direction of ice movement

Roche moutonnée

rock   An aggregate of minerals that forms an appreciable part of the lithosphere.

rock avalanche   A large rockfall where a whole hillside breaks off from the face of a mountain and evolves into a flowing mass of particles rather than moving as a coherent sheet.

rockfall   The most rapid type of mass movement, in which rocks ranging from large masses to small fragments are loosened from the face of a cliff.

rock flour   Fine-grained rock particles pulverized by glacial erosion.

rock glacier   A mass of poorly sorted, angular boulders cemented with interstitial ice. It moves slowly by the action of gravity.

rock salt   Made of the mineral halite (NaCl), it crystallizes when evaporation concentrates sodium and chlorine ions to the point that salt is stable in a brine. Strong evaporation creates saline lakes in closed desert basins and in restricted bays along the shore of the ocean.

rockslide   A landslide in which a newly detached segment of bedrock suddenly slides over an inclined surface of weakness (such as a joint or bedding plane).

rule of Vs   Used to interpret the structure of dipping layers of rock, where the V-shaped notches cut by erosion point in the direction the bed dips.

runoff   Water that flows over the land surface.

sag pond   A small lake that forms in a depression, or sag, where active or recent movement along a fault has impounded a stream.

salinity   The saltiness or dissolved salt content in a soil or body of water.

saltation   The transportation of particles in a current of wind or water by a series of bouncing movements.

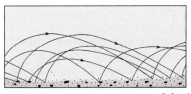

Saltation

salt dome   A dome produced in sedimentary rock by the upward movement of a body of salt.

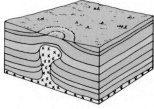

Salt dome

saltwater encroachment   Displacement of fresh groundwater by salt water in coastal areas, due to the greater density of salt water.

salt wedging   A type of physical weathering in which dissolved salt crystallizes and grows in pore spaces in rocks and pushes it apart.

sand   Sedimentary material composed of fragments ranging in diameter from 0.0625 to 2 mm. Sand particles are larger than silt particles but smaller than pebbles. Much sand is composed of quartz grains, because quartz is abundant and resists chemical and mechanical disintegration, but other materials, such as shell fragments and rock fragments, can also form sand.

sand sea   A large accumulation of wind-blown sand dunes and sheets formed in desert regions where rainfall is low. Also called ergs.

sand sheet   Deposits of gently undulating but nearly flat layers of eolian sand formed of grains that are too big to saltate and form dunes. May grade in to higher sand dunes.

sandstone   A sedimentary rock composed mostly of sand-sized particles, usually cemented by calcite, silica, or iron oxide.

sand wave   A wave-shaped deposit of sand grains moved by wind or water. They range from small ripples on stream beds less than a centimeter high to giant sand dunes several hundred meters high. They are typically asymmetrical, with the gentle slope facing the moving current.

saturated   **1** The condition wherein the pore spaces in a rock are completely filled with water. **2** The point at which a condensed phase forms by precipitation from a solution; the point of maximum concentration of dissolved ions, in which no more solute may be dissolved in the solution. For example, a brine is saturated with salt when crystals of halite form in equilibrium with the liquid.

saturated zone   The zone in the subsurface in which all pore spaces are filled with water. Contrast with the overlying zone of aeration.

scarp   A cliff produced by faulting or erosion.

schist   A medium-grained or coarse-grained metamorphic rock with strong foliation (schistosity) resulting from parallel orientation of platy minerals, such as mica, chlorite, and talc.

schistosity   The type of foliation that characterizes schist, resulting from the parallel arrangement of coarse-grained platy minerals, such as mica, chlorite, and talc.

scoria   A dark colored volcanic rock containing abundant vesicles.

sea arch   An arch cut by wave erosion through a headland.

Sea arch

sea cave   A cave formed by wave erosion.

sea cliff   A cliff produced by wave erosion.

seafloor metamorphism   Solid state recrystallization of rock along the crest of the mid-ocean ridges as a result of equilibration with seawater heated by igneous rocks.

seafloor spreading   The theory that the seafloor spreads laterally away from the oceanic ridge as new lithosphere is created along the crest of the ridge by igneous activity.

sea ice   Ice formed by freezing of seawater as opposed to icebergs that come from glaciers.

seamount   An isolated, conical mound rising more than 1000 m above the ocean floor. Seamounts are probably submerged shield volcanoes.

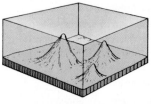

Seamount

sea stack   A small, pillar-shaped, rocky island formed by wave erosion through a headland near a sea cliff.

Sea stack

secondary wave   See S wave.

sediment   Material (such as gravel, sand, mud, and lime) that is transported and deposited by wind, water, ice, or gravity; material that is precipitated from solution; deposits of organic origin (such as coal and coral reefs).

sedimentary environment   A place where sediment is deposited and the physical,

chemical, and biological conditions that exist there. Examples: rivers, deltas, lakes, and shallow-marine shelves.

sedimentary rock   Rock formed by the accumulation and consolidation of sediment.

sedimentary structure   In sedimentary rocks, the stratification, cross-bedding, graded bedding, ripple marks, and mud cracks that provide information about how the sediments accumulated.

sedimentary system   The physical, chemical, and biological conditions necessary for the formation of sedimentary rocks at or near Earth's surface involving fluvial, alluvial-fan, eolian, glacial, delta, shoreline, organic-reef, and marine environments.

seep   A spot where groundwater or other fluids (such as oil) is discharged at Earth's surface.

seif dune   A linear dune of great height and length.

seismic   Pertaining to earthquakes or to waves produced by natural or artificial earthquakes.

seismic discontinuity   A surface within Earth at which seismic wave velocities abruptly change.

seismic gap   The part of an active fault that has experienced little or no seismic activity for a long time.

seismic ray   The path along which a seismic wave travels. Seismic rays are perpendicular to the wave crest.

seismic reflection profile   A profile of the configuration of the ocean floor and shallow sediments on the floor obtained by reflection of artificially produced seismic waves.

seismic risk map   A map showing the future potential for earthquakes commonly based on where past earthquakes occurred.

seismic tomography   Study of the internal structure of Earth based on examination of the characteristics (e.g., velocities, reflections, directions) of earthquake waves.

seismic wave   A wave or vibration produced within Earth by an earthquake or artificial explosion.

seismograph   An instrument that records seismic waves.

sequence   For sedimentary rocks, a group of sedimentary units that is bounded on top and bottom by regional unconformities.

serpentinite   A metamorphic rock made of the hydrated mineral serpentine.

serpentinite ridges   Narrow ridges characteristic of transform fault boundaries near spreading ridges. They form where the less dense and weaker metamorphic rock serpentinite rises.

settling velocity   The rate at which a grain falls through water or air.

shadow zone   (seismology) An area where there is very little or no direct reception of seismic waves from a given earthquake because of refraction of the waves in Earth's core. The shadow zone for P waves is between about 105 and 142 degrees from the epicenter.

shale   A fine-grained clastic sedimentary rock formed by consolidation of clay and mud.

shale gas   Natural gas (methane) trapped inside shale.

shallow-focus earthquake   Earthquakes that occur from the surface to a depth of 70 km.

shallow-marine environment   The sedimentary environment of the continental shelves, where the water is usually less than 200 m deep.

shear   Stress that causes two adjacent rock bodies to slide past one another.

shear wave   A type of seismic wave wherein the elastic vibrations of the particles are transverse to the direction the wave is moving. Shear waves cannot pass through liquids. Synonym: S wave.

sheeted dike complex   A body of rock consisting almost entirely of dikes in vertical tabular sheets a meter or so wide. In the oceanic crust, they are found below and thought to feed the upper layer of basaltic lavas.

sheeting   A set of joints formed essentially parallel to the surface. It allows layers of rock to break off as the weight of overlying rock is removed by erosion. It is especially well developed in granitic rock.

shield   An extensive area of a continent where igneous and metamorphic rocks are exposed and have approached equilibrium with respect to erosion and isostasy. Rocks of the shield are usually very old (that is, more than 600 million years old).

Shield

shield volcano   A volcano shaped like a flattened dome and built up almost entirely of numerous flows of fluid basaltic lava. The slopes of shield volcanoes seldom exceed 10 degrees, so that in profile they resemble a shield or broad dome.

shore   The zone between the waterline at high tide and the waterline at low tide. A narrow strip of land immediately bordering a body of water, especially a lake or an ocean.

shoreline of equilibrium   A shoreline where the energy of the waves and longshore drift is just sufficient to transport the sediment that is supplied.

shoreline system   The interconnected series of processes, materials, and energy transformations that shape the shoreline of large lakes and the ocean.

silica   Silicon dioxide; quartz is the most common mineral form.

silicate   A mineral containing silicon-oxygen tetrahedra, in which four oxygen atoms surround each silicon atom.

silicon–oxygen tetrahedron   The structure of the ion SiO_4^{2-}, in which four oxygen atoms surround a silicon atom to form a four-sided pyramid, or tetrahedron.

sill   A tabular body of intrusive rock injected between layers of the enclosing rock.

silt   Sedimentary material composed of fragments ranging in diameter from 1/265 to 1/16 mm. Silt particles are larger than clay particles but smaller than sand particles.

siltstone   A fine-grained clastic sedimentary rock composed mostly of silt-sized particles.

sinkhole   A depression formed by the collapse of a cavern roof.

Sinkhole

slab-pull   A pull exerted on a tectonic plate as the dense oceanic slab descends under its own weight into the asthenosphere in a subduction zone.

slate   A fine-grained metamorphic rock with a characteristic type of foliation (slaty cleavage), resulting from the parallel arrangement of microscopic platy minerals, such as mica and chlorite.

slaty cleavage   The type of foliation that characterizes slate, resulting from the parallel arrangement of microscopic platy minerals, such as mica and chlorite. Slaty cleavage forms distinct zones of weakness within a rock, along which it splits into slabs.

slip face   See lee slope.

slope retreat   Progressive recession of a scarp or the side of a hill or mountain by mass movement and stream erosion.

slope system   The series of processes and materials that drive the evolution of slopes. Gravity is the most important energy source.

slump   A type of mass movement in which material moves along a curved surface of rupture.

slump block   A landslide block that moves as a unit along a definite fracture or shear plane.

snowline   The line on a glacier separating the area where snow remains from year to year from the area where snow from the previous season melts.

soil   The surface material of the continents, produced by disintegration of rock; regolith that has undergone chemical and physical weathering and includes organic material.

soil profile   A vertical section of soil showing the soil horizons and parent material.

solar energy   Radiation from the Sun.

solid   The state of matter in which a substance has a definite shape and volume and some fundamental strength.

solid-state diffusion   The movement of ions through a solid, an important mechanism in metamorphism.

solifluction   A type of mass movement in which material moves slowly down-slope in areas where the soil is saturated with water. It commonly occurs in permafrost areas.

solution   Typically a fluid containing dissolved ions, but some homogeneous solid mixtures are also called solutions.

solution valley   A valley produced by solution activity, either by dissolution of surface or subsurface materials, such as limestone, gypsum, or salt.

sorting   The separation of particles according to size, shape, or weight. It occurs during transportation by running water or wind.

source rock   For magmas, the solid rock that melts to give rise to magma. It can vary in composition from the ultramafic mantle to silicic continental crust and controls the composition of the resultant partial melt.

spatter cone   A low-steep-sided volcanic cone built by accumulation of splashes and spatters of lava (usually basaltic) around a fissure or vent.

speleothem   Term for all types of deposits formed in caves. See stalagmite, stalactite.

spheroidal weathering   The process by which corners and edges of a rock body become rounded as a result of exposure to weathering on all sides, so that the rock acquires a spheroidal or ellipsoidal shape.

Spheroidal weathering

spit   A sandy bar projecting from the mainland into open water. Spits are formed by deposition of sediment moved by longshore drift.

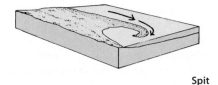

Spit

splay   A small deltaic deposit formed on a floodplain where water and sediment are diverted from the main stream through a crevasse in a levee.

spreading rate   The rate of separation of two plates at a divergent plate boundary, usually measured in cm/y.

spring   A place where groundwater flows or seeps naturally to the surface.

stability range   The spectrum of conditions over which a mineral or rock body is in equilibrium with its environment. For example, many minerals are stable over a wide range of temperatures, but above a specific temperature, the mineral melts to form a liquid.

stable   A condition in which a rock or mineral exists in equilibrium with its environment.

stable platform   The part of a continent that is covered with flat lying or gently tilted sedimentary strata and underlain by a basement complex of igneous and metamorphic rocks. The stable platform has not been extensively affected by crustal deformation.

stalactite   An icicle-shaped deposit of dripstone hanging from the roof of a cave.

stalagmite   A conical deposit of dripstone built up from a cave floor.

star dune   A mound of sand with a high central point and arms radiating in various directions.

starting plume   The initial stage of a type of convection that involves the slow rise of less dense but solid mantle through the mantle. See plume head, plume tail.

stock   A small, roughly circular intrusive body, usually less than 100 km² in surface exposure.

storm surge   The temporary rise of the sea as a result of the low atmospheric pressure and wind associated with a storm such as a hurricane. Storm surges can be several meters high and inundate the shoreline to cause great damage and geologic change.

strain   A change in shape or position when differential stress is applied to a rock body. Contrast with stress.

strata   (plural of **stratum**) Layers of rock, usually sedimentary.

stratification   The layered structure of sedimentary rock.

stratosphere   The portion of Earth's atmosphere between about 11 km to 50 km and in which temperature increases gradually to about 0°C and clouds rarely form.

stratovolcano   A steep-sided volcano built up of ash, lava flows, and shallow intrusions. Synonymous with composite volcano.

streak   The color of a powdered mineral.

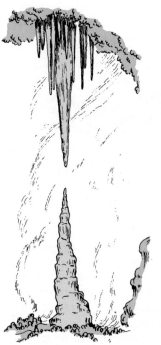

Stalactite

stream load   The total amount of sediment carried by a stream at a given time.

stream order   The hierarchical number of a stream segment. The smallest tributary has the order number of 1, and successively larger tributaries have progressively higher numbers.

stream piracy   Diversion of the headwaters of one stream into another stream. The process occurs by headward erosion of a stream having greater erosive power than the stream it captures.

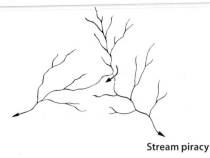

Stream piracy

stream terrace   One of a series of level surfaces in a stream valley representing the dissected remnants of an abandoned floodplain, stream bed, or valley floor produced in a previous stage of erosion or deposition.

strength   A rock's natural resistance to deformation.

stress   Force (pressure) applied to a material that tends to change its dimensions or volume; force per unit area.

striation   A scratch or groove produced on the surface of a rock by a geologic agent, such as a glacier or stream.

strike   The bearing (compass direction) of a horizontal line on a bedding plane, a fault plane, or some other planar structural feature.

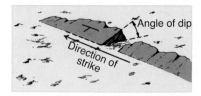

Strike

strike-slip fault   A fault in which movement has occurred parallel to the strike of the fault.

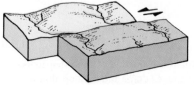

Strike-slip fault

strike valley A valley that is eroded parallel to the strike of the underlying nonresistant strata.

strip mining A method of mining in which soil and rock cover are removed to obtain the sought-after material.

subaerial Occurring beneath the atmosphere or in the open air, with reference to conditions or processes (such as erosion) that occur on the land.

subaqueous Occurring beneath water, with reference to conditions or processes that occur on the floors of rivers, lakes, and oceans.

subduction Subsidence of the leading edge of a lithospheric plate into the mantle.

subduction zone An elongate zone in which one lithospheric plate descends beneath another. A subduction zone is typically marked by an oceanic trench, lines of volcanoes, and crustal deformation associated with mountain building. See also convergent plate boundary.

subglacial volcano A volcano that forms beneath a glacier. Pillow basalt and steam explosions are important.

sublimation The conversion of a solid to a gas without melting. Example: carbon dioxide passes directly from dry ice to a vapor that is heavier than air.

submarine canyon A V-shaped trench or valley with steep sides cut into a continental shelf or continental slope.

submarine fan A cone of debris at the foot of a continental slope formed of turbidites deposits.

subsequent stream A younger set of streams eroded headward along weak formations to excavate a belt of lowlands. Contrast with consequent streams. See trellis.

subsidence A sinking or settling of a part of Earth's crust with respect to the surrounding parts.

supernova Large explosion of a short-lived massive star. The explosion recycles newly formed elements back into space.

superposed stream A drainage system that retains its original form as it downcuts during uplift.

superposition, principle of The principle that, in a series of sedimentary strata that has not been overturned, the oldest rocks are at the base and the youngest are at the top.

supervolcano A volcano with very large eruptions, exceeding 1000 km^3 in volume in a single deposit. They typically are calderas.

surface current An ocean current in the upper 100 meters driven by the wind.

surface water The thin, generally warm upper layer of the ocean.

surface wave A seismic wave that travels along Earth's surface. Contrast with P waves and S waves, which travel through the Earth and at higher velocities.

suspended load The part of a stream's load that is carried in suspension for a considerable period of time without contact with the stream bed. It consists mainly of mud, silt, and sand. Contrast with bed load and dissolved load.

suspension Elevation of small grains into a moving fluid. For example, dust is carried by wind turbulence and clays are transported in suspension in streams.

suture A belt of intensely deformed rocks that marks the site of continental collision.

swamp A seasonally flooded lowland dominated by woody plants in contrast to a marsh.

swash The rush of water up onto a beach after a wave breaks.

S wave (secondary or shear seismic wave) A seismic wave in which particles vibrate at right angles to the direction in which the wave travels. Contrast with P wave.

syncline A fold in which the limbs dip toward the axis. After erosion, the youngest beds are exposed in the central core of the fold.

system A group of interdependent materials that interact with energy to form a unified whole.

tailings Waste rock from a mine.

talus Rock fragments that accumulate in a pile at the base of a ridge or cliff.

Talus

tar sand Sand or sandstone impregnated with viscous petroleum or its breakdown products

tectonic creep Slow, apparently continuous movement along a fault (as opposed to the sudden rupture that occurs during an earthquake).

tectonic plate One of the mosaic of seven major plates and several smaller subplates comprising the lithosphere of Earth.

tectonics The branch of geology that deals with regional or global structures and deformational features of Earth.

tectonite Rocks with a distinctive texture suggesting ductile deformation at high temperature during seafloor spreading.

temperate climate A moderate climate found at mid-latitudes on Earth, with adequate precipitation for plant growth and no extreme temperatures.

tension Stress that tends to pull materials apart.

tephra A general term for pyroclastic material ejected from a volcano. It includes ash, dust, bombs, and other types of fragments.

terminal moraine A ridge of material deposited by a glacier at the line of maximum advance of the glacier. Contrast with recessional moraine.

Terminal moraine

terra (pl. terrae) Densely cratered highlands on the Moon.

terrace A nearly level surface bordering a steeper slope, such as a stream terrace or wave-cut terrace.

Terrace

terrestrial planet The inner "Earth-like" planets, Mercury, Venus, Earth, and Mars.

texture The size, shape, and arrangement of the particles that make up a rock.

thermohaline circulation The convection of the ocean caused by differences in temperature and salinity.

thermosphere The outermost shell of the atmosphere, between the mesosphere and outer space.

thin section A slice of rock mounted on a glass slide and ground to a thickness of about 0.03 mm, thin enough for light to pass through many kinds of minerals.

threshold velocity The velocity at which a grain of a certain size will be picked up by a flowing fluid such as water. The threshold velocity for large particles is higher than for small particles.

thrust fault   A low-angle fault (45 degrees or less) in which the hanging wall has moved upward in relation to the footwall. Thrust faults are caused by horizontal compression.

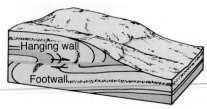

Thrust fault

tidal energy   A renewable energy resource using the rise and fall of ocean tides to power turbines and generate electricity.

tidal flat   A large, nearly horizontal area of land covered with water at high tide and exposed to the air at low tide. Tidal flats consist of fine-grained sediment (mostly mud, silt, and sand).

tidal heating   Planetary heating caused by tidal deformation of a planet and the resultant frictional heating. It is only important in unusual situations where gravitational tides raised by a large body affect a much smaller one (for example, Jupiter on its moon Io).

tide   The alternate rise and fall of the water level on coastlines that is produced by the gravitational attraction of the Moon and Sun on Earth's oceans. More generally, it is the force of gravity of one planetary body on another.

till   Unsorted and unstratified glacial deposit.

tillite   A rock formed by lithification of glacial till (unsorted, unstratified glacial sediment).

tombolo   A beach or bar connecting an island to the mainland.

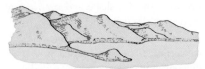

Tombolo

topography   The shape and form of Earth's surface as expressed in elevations above or below sea level.

tower karst   Topography characterized by steep, cone-shaped hills produced by groundwater dissolution of the surrounding rock.

transform fault   A special type of strike-slip fault forming the boundary between

two moving lithospheric plates, usually along an offset segment of the oceanic ridge.

transform plate boundary   See transform fault.

transgression   An advance of the sea across a continental margin caused by changing climates or tectonics.

transpiration   The process by which water vapor is released into the atmosphere by plants.

transporting system   The main trunk stream, which functions as a channel through which water and sediment flow from the collecting area toward the ocean.

transpression   Horizontal compression along an irregular strike-slip fault to create uplifted regions with small folds and thrust faults. This process involves both transform and compression motion. Contrast with transtension.

transtension   Horizontal extension along an irregular strike-slip fault. Extension occurs in zones where the bends are oriented so that blocks on either side of the fault are pulled apart, creating fault-bounded depressions or basins. This process involves both transform and extension motion. Contrast with transpression.

transverse dune   An asymmetrical dune ridge that forms at right angles to the direction of prevailing winds.

travertine terrace   A terrace formed from calcium carbonate deposited by water on a cave floor.

tree ring dating   A numeric dating technique that examines the annually formed rings in cross sections of tree trunks.

trellis drainage pattern   A drainage pattern in which tributaries are arranged in a pattern similar to that of a garden trellis.

Trellis drainage pattern

trench   A narrow, elongate depression of the deep-ocean floor oriented parallel to the trend of a continent or an island arc.

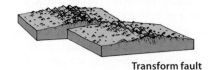

Transform fault

trench-trench transform   A transform fault that connects two subduction-related trenches.

tributary   A stream flowing into or joining a larger stream.

tropical climate   A climate that is frost-free with temperatures high enough to support year-round plant growth and abundant precipitation. This climate prevails near the equator.

troposphere   The lowermost zone of the atmosphere, where most of the weather occurs.

tsunami   A seismic sea wave; a long, low wave in the ocean caused by an earthquake, faulting, or a landslide on the sea floor. Its velocity can reach 800 km per hour. Tsunamis are commonly and incorrectly called tidal waves.

tuff   A fine-grained rock composed of volcanic ash.

turbidite   A sedimentary rock deposited by a turbidity current. Graded bedding is characteristic.

turbidity current   A current in air, water, or any other fluid caused by differences in the amount of suspended matter (such as mud, silt, or volcanic dust). Marine turbidity currents, laden with suspended sediment, move rapidly down continental slopes and spread out over the abyssal floor.

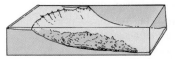

Turbidity current

turbulent flow   A type of flow in which the path of motion is very irregular, with eddies and swirls. Contrast with laminar flow.

ultramafic rock   An igneous rock composed entirely of mafic minerals.

unconfined aquifer   A subsurface zone saturated with water that is connected to the surface by open pore spaces through which it can be recharged. Also called a water table aquifer. Contrast with confined aquifer.

unconformity   A discontinuity in the succession of rocks, containing a gap in the geologic record. A buried erosion surface. See also angular unconformity, nonconformity.

uniformitarianism   The theory that geologic events are caused by natural processes, many of which are operating at the present time.

upwarp   An arched or uplifted segment of the crust.

U-shaped valley   A valley with a U-shaped profile caused by glacial erosion. Contrast with the V-shape of a typical stream valley.

valley glacier   A glacier that is confined to a stream valley. Synonymous with alpine glacier and mountain glacier.

varve   A pair of thin sedimentary layers, one relatively coarse-grained and light-colored, and the other relatively fine-grained and dark-colored, formed by deposition on a lake bottom during a period of one year. The coarse-grained layer is formed during spring runoff, and the fine-grained layer is formed during the winter when the surface of the lake is frozen.

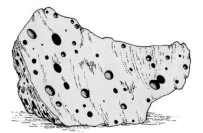

Varve

vein   A tabular rock body deposited in a fracture. Many ore minerals were deposited in veins when hot fluids flowed through fractures.

ventifact   A pebble or cobble shaped and polished by wind abrasion.

Ventifact

vesicle   A small hole formed in a volcanic rock by a gas bubble that became trapped as the lava solidified.

Vesicle

viscosity   The tendency within a body to resist flow. An increase in viscosity implies a decrease in fluidity, or ability to flow.

volatile   1 Capable of being readily vaporized. 2 A substance that can readily be vaporized, such as water or carbon dioxide.

volcanic arc   An arcuate chain of volcanos on the margin of an overriding plate at a convergent boundary with a subduction zone.

volcanic ash   Dust-sized particles ejected from a volcano.

volcanic bomb   A hard fragment of lava that was liquid or plastic at the time of ejection and acquired its form and surface markings during flight through the air. Volcanic bombs range from a few millimeters to more than a meter in diameter.

Volcanic bomb

volcanic neck   The solidified magma that originally filled the vent or neck of an ancient volcano and has subsequently been exposed by erosion.

Volcanic neck

volcanism   The processes by which magma and gases are transferred from Earth's interior to the surface.

wash   A dry stream bed.

Wash

water gap   A pass in a ridge through which a stream flows.

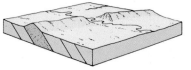

Water gap

water table   The upper surface of the zone of saturation.

Water table

wave base   The lower limit of wave transportation and erosion, equal to half the wavelength.

wave-built terrace   A terrace built up from wave-washed sediments. Wave-built terraces usually lie seaward of a wave-cut terrace.

wave crest   The highest part of a wave.

wave-cut cliff   A cliff formed along a coast by the undercutting action of waves and currents.

wave-cut platform   A terrace cut across bedrock by wave erosion. Synonymous with wave-cut terrace.

wave-cut terrace   See wave-cut platform.

wave height   The vertical distance between a wave crest and the preceding trough.

wavelength   The horizontal distance between similar points on two successive waves, measured perpendicular to the crest.

wave period   The interval of time required for a wave crest to travel a distance equal to one wavelength; the interval of time required for two successive wave crests to pass a fixed point.

wave refraction   The process by which a wave is bent or turned from its original direction. In sea waves, as a wave approaches a shore obliquely, part of it reaches the shallow water near the shore while the rest is still advancing in deeper water; the part of the wave in the shallower water moves more slowly than the part in the deeper water. In seismic waves, refraction results from the wave encountering material with a different density or composition.

wave trough   The lowest part of a wave, between successive crests.

weathering   The processes by which rocks are chemically altered or physically broken into fragments as a result of exposure to atmospheric agents and the pressures and temperatures at or near Earth's surface, with little or no transportation of the loosened or altered materials.

welded tuff   A rock formed from particles of volcanic ash that were hot enough to become fused together.

white smoker   Warm water and dissolved ions from the oceanic crust are released at central vents on the seafloor to form light-colored plumes of mineral precipitates dominated by sulfates. Compare with black smoker.

wind energy   Wind used to turn turbines and generate electricity.

wind gap   A gap in a ridge through which a stream, now abandoned as a result of stream piracy, once flowed.

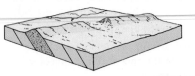

Wind gap

wind shadow   The area behind an obstacle where air movement is not capable of moving material.

X-ray diffraction   In mineralogy, the process of identifying mineral structures by exposing crystals to a beam of X rays and studying the resulting diffraction patterns.

yardang   An elongate ridge carved by wind erosion. The ridges are parallel to the prevailing winds in arid regions with soft sediment at the surface.

yazoo stream   A tributary stream that flows parallel to the main stream for a considerable distance before joining it. Such a tributary is forced to flow along the base of a natural levee formed by the main stream.

zeolite facies   A metamorphic facies formed at relatively low temperature and pressure where zeolite minerals are stable.

zone of ablation   Reduction of a glacier by melting, evaporation, iceberg calving, or deflation.

zone of accumulation   The area above the snow line where there is a net gain of glacial ice from snow.

zone of aeration   In a water table or unconfined aquifer, the area at the top where the pore space is filled partly with air and partly with water.

zone of saturation   In a water table or unconfined aquifer, the area where all of the openings in the rock are completely filled with water. The water table lies at the top of the zone of saturation.

Index

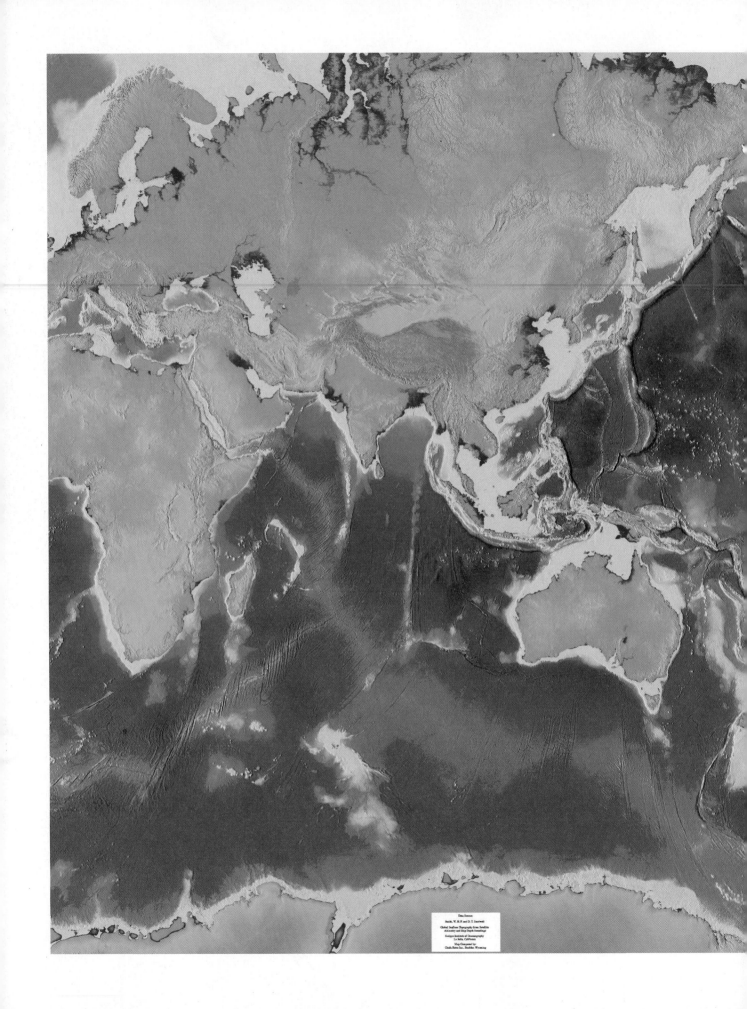

Data Source:

Smith, W. H. F. and D. T. Sandwell

Global Seafloor Topography from Satellite
Altimetry and Ship Depth Soundings

Scripps Institution of Oceanography
La Jolla, California

Map Computed by:

Chalk Butte Inc., Boulder, Wyoming